Mathematical
Surveys
and
Monographs

Volume 238

Algebraic Geometry Codes: Advanced Chapters

Michael Tsfasman
Serge Vlăduţ
Dmitry Nogin

AMERICAN MATHEMATICAL SOCIETY
Providence, Rhode Island

2010 *Mathematics Subject Classification.* Primary 14Hxx, 94Bxx, 14G15, 11R58; Secondary 11R04, 11T71, 11M38, 11R37, 11R42, 14D22, 14J20.

For additional information and updates on this book, visit
www.ams.org/bookpages/surv-238

Library of Congress Cataloging-in-Publication Data

Names: Tsfasman, M. A. (Michael A.), 1954– author. | Vlăduţ, S. G. (Serge G.), 1954– author. | Nogin, Dmitry, 1966– author.

Title: Algebraic geometry codes: Advanced chapters / Michael Tsfasman, Serge Vlăduţ, Dmitry Nogin.

Other titles: Title of previous publication: Algebraic geometric codes

Description: Providence, Rhode Island: American Mathematical Society, [2019] | Series: Mathematical surveys and monographs; volume 238 | Algebraic-geometric codes / by M.A. Tsfasman, published: Dordrecht: Kluwer Academic Publishers, 1991. Republished as: Algebraic geometric codes: Basic notions (Providence, R.I.: American Mathematical Society, 2007). Algebraic geometry codes: Advanced chapters, continues 2007 publication, beginning at chapter 5. | Includes bibliographical references and indexes.

Identifiers: LCCN 2019003782 | ISBN 9781470448653 (alk. paper)

Subjects: LCSH: Coding theory. | Number theory. | Geometry, Algebraic. | AMS: Algebraic geometry – Curves – Curves. msc | Information and communication, circuits – Theory of error-correcting codes and error-detecting codes – Theory of error-correcting codes and error-detecting codes. msc | Algebraic geometry – Arithmetic problems. Diophantine geometry – Finite ground fields. msc | Number theory – Algebraic number theory: global fields – Arithmetic theory of algebraic function fields. msc | Number theory – Algebraic number theory: global fields – Algebraic numbers; rings of algebraic integers. msc | Number theory – Finite fields and commutative rings (number-theoretic aspects) – Algebraic coding theory; cryptography. msc | Number theory – Zeta and L-functions: analytic theory – Zeta and L-functions in characteristic p. msc | Number theory – Algebraic number theory: global fields – Class field theory. msc | Number theory – Algebraic number theory: global fields – Zeta functions and L-functions of number fields. msc | Algebraic geometry – Families, fibrations – Fine and coarse moduli spaces. msc | Algebraic geometry – Surfaces and higher-dimensional varieties – Arithmetic ground fields. msc

Classification: LCC QA268 .T755 2019 | DDC 003/.54–dc23

LC record available at https://lccn.loc.gov/2019003782

Visit the AMS home page at `https://www.ams.org/`

10 9 8 7 6 5 4 3 2 1 24 23 22 21 20 19

Contents

Preface

The first volume of this book, *Algebraic Geometric Codes: Basic Notions*, was published by the AMS in 2007[1]. In that book we promised to complete it with the second one, *Algebraic Geometry Codes: Advanced Chapters*. Partly because of objective difficulties and partly due to our laziness, writing this book took us ten years.

For those who have already forgotten the content of the first volume, we have written an Appendix, where we present the basics of the theory and concisely recall main results.

We tried to make the text as clear and self-contained as possible; though this volume is addressed to a more experienced reader than the first one, we cannot claim that we have quite achieved this goal.

Somewhat to our own surprise we manage to cover almost all topics that we consider to be the main topics of the theory, and to fulfill the promises that we gave in the Preface to *Basic Notions* (with the exception of graphs without short cycles and codes over rings). Of course, there are many results which we were unable to cite in detail, as well as many co-adjacent domains.

We treat the following subjects: Curves with many $\mathbb{F}_q$-points, especially those attaining the Drinfeld–Vlăduţ bound, namely modular curves, classical and Drinfeld, and Elkies's explicit constructions. Class field theory and towers of curves therefrom. Oesterle bounds for the number of points. Examples of good curves, in particular curves of small genera, Deligne–Lusztig curves, recursive García–Stichtenoth towers, etc. Then we present our theory of infinite global fields, including the basic inequality and generalized Brauer–Siegel theorem, class field examples of towers, and so on. Decoding of algebraic geometry codes. Dense sphere packings, especially those that are good asymptotically, Elkies–Shioda lattices, class field constructions, and other links between number theory and algebraic geometry on one hand and lattices and sphere packings on the other. Then we touch on an intriguing subject of the number of $\mathbb{F}_q$-points on surfaces and multidimensional varieties, much less known than that on curves. And, of course, codes obtained from these varieties. There are some applications either of algebraic geometry codes directly or stylistically similar to them to fast multiplication, cryptography, quantum codes, and the like.

We hope that the book will be useful for the reader interested in algebraic geometry and number theory, as well as to one interested in what it gives for coding theory and other applications. The two volumes together form a mixture of a textbook for graduate and best undergraduate students, a handbook for specialists, and a reading for mathematicians specializing in other domains.

[1]See `https://bookstore.ams.org/surv-139`

* * * * *

Let us explain the choice of topics, chapter by chapter. Both volumes of this book are written for mathematicians of different specialties and basic education. We keep in mind specialists in coding theory, algebraic geometry, number theory, combinatorics, geometry, and so on.

In the first volume, Chapter 1 is an introduction to coding theory for those who do not know it, and a geometric view of it for those who do.

Chapter 2 is a compendium of algebraic geometry of curves we need.

Chapter 3 treats curves over finite fields, a subject that is in a sense closer to number theory than to classical algebraic geometry.

In Chapter 4 we skim off the cream, presenting what can be gotten for codes using algebraic curves.

In this volume we fill in the gaps and keep promises given in the first one.

Algebraic geometry codes with good parameters need curves with many points. There are several types of such constructions. Historically the first and one of the most beautiful is that of modular curves (Chapter 5).

Unfortunately, modular curves are not good over finite fields whose cardinality is not a square. And we know no analogue for number fields. The construction that works in these cases is that of class field towers (Chapter 6).

In Chapter 7 we treat other questions concerning curves with many points, in particular, the question of how to construct modular curves by explicit equations.

One of the most exciting sides of the theory and, in some sense, its *raison d'être*, is the parallelism between fields of functions of curves over finite fields and number fields. Together they are called *global fields*, and their theory is very beautiful and well developed. We try to go further. Being motivated initially by questions of the asymptotic behaviour of code parameters, and thus of the asymptotic behaviour of the number of points on curves in towers and families, in Chapter 8 we present the theory of infinite global fields and their zeta functions, which constitutes a very interesting part of the whole theory.

Chapter 9 is devoted to decoding of algebraic geometry codes, which is of primary importance for applications and also poses some interesting questions for algebraic geometers.

An old, nice problem of geometry—that of dense sphere packing (Chapter 10)—is linked to codes, as well as to number and function fields. The constructions that we present are close in spirit to those of algebraic geometry codes.

Multidimensional varieties over finite fields are much less studied than curves. In Chapter 11 we expose what is known about them and construct codes therefrom.

The last chapter (Chapter 12) presents some other applications and analogues of algebraic geometry codes, like fast multiplication in large finite fields, cryptography, quantum codes, and so on.

Since the first volume was published long ago, for the convenience of the reader we sum up its contents in the Appendix.

* * * * *

We permit ourselves to give to the reader some advice concerning different chapters of the present volume.

If you are interested in coding theory, you need constructions of good codes, their parameters, encoding, and decoding. Then you can read Chapter 9 (Decoding: Some Examples), Chapter 11 (Codes from Multidimensional Varieties, though it needs more knowledge of algebraic geometry), and, finally, Chapter 12 can also be interesting for you. As for the other chapters, they form the algebraic geometry and number theory base and expose the deep nature of the studied objects; correspondingly, the algebraic geometry and arithmetic prerequisites are more wide. If you are mostly interested in arithmetic geometry, you would like Chapters 6, 7, and especially Chapter 8. Chapter 10 gives a beautiful application to sphere packings.

For an algebraic geometer, Chapter 11 may be of utmost interest; however, it is better to read after Chapters 5 and 7.

If you are interested in number theory, read Chapters 6 and 8.

Those who like all these domains may wish to read the whole book.

* * * * *

We are especially interested in the links and interchanges of different parts of mathematics that we see in the domain. Let us list some of these meeting-points.

1. *Codes and algebraic varieties over finite fields.* This is the main topic of the first volume [**TV91**], especially in the case of curves; see the Appendix. A linear code is equivalent to a multiset of points in $\mathbb{P}^{k-1}$; see Sec. A.1. Then it is natural to consider codes corresponding to $\mathbb{F}_q$-points of algebraic varieties, curves in particular. This leads us to many questions about curves—but also about multidimensional varieties—over finite fields.

2. *Codes and packings.* A good error-correcting code is a dense packing of equal spheres in the Hamming space $\mathbb{F}_q^n$. Thus, we are bound to look at the sphere packing problem in other spaces, in particular, that in the Euclidean space $\mathbb{R}^n$. In this setting a linear code corresponds to a lattice packing, and there are many ways to construct rather dense packings from codes with good parameters; see Chapter 10.

3. *Number fields, curves, and sphere packings.* Natural constructions of lattices from number fields known for two centuries extend to constructions of lattices from curves. Both happen to lead to dense sphere packings; see Chapter 10.

4. *Algebraic curves over finite fields and number fields.* The main question we get is to find out the number of points of a curve over the ground field and over its finite extensions. They are assembled together in the notion of the zeta function of the curve. With this notion, calculus enters into the picture together with algebra and geometry. If we look at the theory of curves from an algebraic point of view, we see one of the utmost gems of modern mathematics, the parallelism between curves over finite fields and number fields, i.e., finite extensions of $\mathbb{Q}$. We see this in Chapters 6 and 8, and keep it in mind everywhere.

5. *Curves with many points and modular functions.* This same theory leads to modular curves, giving us examples of large genus curves with many points; see Chapter 5. In geometric language this is a part of another gem and the center of modern mathematics, moduli spaces. Note that the recent solution of the densest sphere packing problem in dimensions 8 and 24 also uses modular functions.

6. *Curves, varieties, and group theory.* The group theory and the algebraic group theory appear when we consider class field towers (Chapter 6) and when

we look for varieties with many points (Deligne–Lusztig varieties, Grassmannians, and so on); automorphism groups of varieties, especially those of curves, are also sometimes important for applications.

7. *Asymptotic parameters of families of codes and limit zeta functions.* To find and study codes of large length, we set asymptotic questions: what happens when the parameters tend to infinity? This corresponds to the asymptotic study of zeta functions of families of algebraic varieties and number fields as the genus of the curve (the discriminant of the field) tends to infinity. At present we have a beautiful—though highly incomplete—theory of limit zeta functions; see Chapter 8. Since a zeta function is characterized by the set of its zeroes, in the limit we get a measure on the critical line, and a study of such measures adds another analytic taste to our geometric and arithmetic objects.

* * * * *

It may be useful to point out the interdependence of these topics. Roughly speaking, the first four chapters (Chapters 5–8, the enumeration being continuous throughout both volumes of the book) are linearly ordered, i.e., each chapter uses something from the preceding ones. On the contrary, Chapters 9–12 are essentially independent of each other, as well as of Chapters 5–8.

Acknowledgments. Our research while working on this book was supported by the French National Scientific Research Center (CNRS), in particular by the Institut de Mathématique de Luminy, the Institut de Mathématiques de Marseille, the French–Russian Poncelet laboratory, and the Laboratoire de Mathématique de Versailles; by the Institute for Information Transmission Problems; and by the Independent University of Moscow. It was also supported in part by the Russian Science Foundation (project no. 14-50-00150) and by ANR project FLAIR (ANR-17-CE40-0012).

CHAPTER 5

Curves with Many Points. I: Modular Curves

Curves with many points over finite fields are of interest from both a pure mathematics perspective and coding theory viewpoint. There are many methods to construct them, which can roughly be divided into the following classes:

1. Modular curves of various types;
2. Class field theory constructions;
3. Explicit recursive curve towers;
4. Curves coming from Artin–Schreier or Kummer covers;
5. Other constructions.

Note however that there are numerous interconnections between these classes.

The present chapter is devoted to modular curves and codes arising from them. We consider two types of curves: classical modular curves (their reductions in fact) and Drinfeld modular curves together with their reductions. For a prime p, classical modular curves yield codes over $\mathbb{F}_{p^2}$, while Drinfeld curves yield codes over $\mathbb{F}_q$, where $q = p^{2a}$ is an arbitrary even power of p.

We are going to show that both types of modular curves have a lot of rational points, and therefore the corresponding codes have good asymptotic parameters. Moreover, for codes of both classes there exist polynomial construction algorithms. Unfortunately, this result, being of major theoretical importance, does not lead to any practically useful algorithm, since the complexity is rather high. For Drinfeld curves these algorithms are much more conceptual and simple. A natural question arises as to whether it is useful to consider codes on classical modular curves while those on Drinfeld modular curves are in many aspects simpler and exist for any even power of a prime. However, having the same asymptotic parameters, the two classes of codes greatly differ in many other aspects. For example, it is rather easy to obtain codes of moderate length on classical modular curves, which rarely happens for the case of Drinfeld curves.

The second subject of this chapter is the possibility of explicit constructions of modular curve towers, discovered by Elkies. These constructions concern both types of the modular curves, classical and Drinfeld, and in fact some other types of curves close to modular ones.

The plan of the chapter is as follows. In Sec. 5.1, based on the study of moduli schemes of elliptic curves, we consider codes on reductions of classical modular curves. Section 5.2 is devoted to Drinfeld curves, representing moduli varieties of elliptic modules, and the corresponding codes. In Sec. 5.3 it is shown that in many cases the curves from Secs. 5.1 and 5.2 can be constructed explicitly as curves in recursive towers, e.g., the García–Stichtenoth towers. Note that this alternative description of modular curves allows us to obtain a low-complexity algorithm for constructing codes on modular curves (see Sec. 7.4.3 below).

5.1. Classical Modular Curves

This section is devoted to classical modular curves or, more precisely, their reduction to a finite characteristic and to codes arising from these reductions. As we have shown in Chapter 4 (see *Basic Notions*), the properties of an algebraic geometry code constructed from a curve are determined by its geometry and arithmetic. We begin with some necessary information on the structure of modular curves and their reductions and then give corollaries for codes on such curves.

Section 5.1.1 contains an analytic description of modular curves, their morphisms, and their fields of rational functions. In Sec. 5.1.2 we present the theory of moduli schemes of elliptic curves; this theory implies that reductions of modular curves modulo p are good and have many $\mathbb{F}_{p^2}$-rational points. Section 5.1.3 discusses codes on these reductions of modular curves: we estimate their parameters taking into account some improvements due to Weierstrass points and also describe group properties of a subclass of such codes; we also explain why some of these codes can be constructed with polynomial complexity. We do not give exact estimates for the complexity, but we should note that in any case one *cannot* obtain along these lines a construction or algorithm of practical utility, due to unpleasant singularities of the models of modular curves used in Secs. 5.1 and 5.2: the best presently known estimate for this complexity is $O(n^{12})$, where n is the code length, and surely cannot be lowered, say to $O(n^6)$, which would be still useless in practice.

5.1.1. Modular Curves

We begin with an analytic description of modular curves as factors of the upper half-plane over discrete subgroups of its automorphisms.

Riemann surface $\Gamma\backslash H^*$. By H we denote the upper half-plane of the complex plane $\mathbb{C}$:

$$H = \{z \in \mathbb{C} \mid \operatorname{Im} z > 0\}.$$

The *special linear group* (over $\mathbb{Z}$)

$$\mathrm{SL}_2(\mathbb{Z}) = \left\{ \begin{pmatrix} a & b \\ c & d \end{pmatrix} \,\middle|\, a, b, c, d \in \mathbb{Z},\ ad - bc = 1 \right\}$$

has a natural left action on H. The action is given by

$$\gamma \colon z \mapsto \frac{az+b}{cz+d},$$

where $\gamma = \begin{pmatrix} a & b \\ c & d \end{pmatrix} \in \mathrm{SL}_2(\mathbb{Z})$. Thus, the element $\begin{pmatrix} -1 & 0 \\ 0 & -1 \end{pmatrix}$ acts trivially, so that the action of $\mathrm{SL}_2(\mathbb{Z})$ induces an action of the *modular group*

$$\Gamma(1) = \mathrm{PSL}_2(\mathbb{Z}) = \mathrm{SL}_2(\mathbb{Z})/\{\pm 1\},$$

which is the factor group of $\mathrm{SL}_2(\mathbb{Z})$ modulo its center. In what follows, we write $\gamma = \begin{pmatrix} a & b \\ c & d \end{pmatrix} \in \Gamma(1)$, meaning that γ is the image of $\begin{pmatrix} a & b \\ c & d \end{pmatrix} \in \mathrm{SL}_2(\mathbb{Z})$ in $\Gamma(1)$.

We need the following standard subgroups of $\Gamma(1)$. Fix a positive integer N and let

$$\Gamma(N)=\left\{\begin{pmatrix} a & b\\ c & d\end{pmatrix}\in\Gamma(1)\;\middle|\; a,d\equiv 1 \pmod N,\ b,c\equiv 0 \pmod N\right\},$$
$$\Gamma_0(N)=\left\{\begin{pmatrix} a & b\\ c & d\end{pmatrix}\in\Gamma(1)\;\middle|\; c\equiv 0 \pmod N\right\}.$$

Sometimes the definitions of $\Gamma(N)$ and $\Gamma_0(N)$ are written as

$$\Gamma(N)=\left\{\begin{pmatrix} a & b\\ c & d\end{pmatrix}\;\middle|\;\begin{pmatrix} a & b\\ c & d\end{pmatrix}=\begin{pmatrix} 1 & 0\\ 0 & 1\end{pmatrix} \pmod N\right\},$$
$$\Gamma_0(N)=\left\{\begin{pmatrix} a & b\\ c & d\end{pmatrix}\;\middle|\;\begin{pmatrix} a & b\\ c & d\end{pmatrix}=\begin{pmatrix} * & *\\ 0 & *\end{pmatrix} \pmod N\right\},$$

where the asterisks stand for arbitrary elements of $\mathbb{Z}$. Clearly, $\Gamma(N)\subset\Gamma_0(N)$ for $N\neq 1$. Note that for $N\geq 3$ we have

$$\begin{pmatrix} -1 & 0\\ 0 & -1\end{pmatrix}\notin\Gamma(N);$$

$\Gamma(N)$ is called the *principal congruence subgroup of level N*.

LEMMA 5.1.1. *There exist*

(a) *a natural group isomorphism*

$$\Gamma(1)/\Gamma(N)\xrightarrow{\sim}\mathrm{PSL}_2(\mathbb{Z}/N\mathbb{Z}),$$

where $\mathrm{PSL}_2(\mathbb{Z}/N\mathbb{Z})$ is the factor group of the special linear group $\mathrm{SL}_2(\mathbb{Z}/N\mathbb{Z})$ modulo its center; in particular, $\Gamma(N)$ is normal in $\Gamma(1)$;

(b) *a natural bijection of sets*

$$\Gamma(1)/\Gamma_0(N)\xrightarrow{\sim}\mathbb{P}^1(\mathbb{Z}/N\mathbb{Z}),$$

where $\mathbb{P}^1(\mathbb{Z}/N\mathbb{Z})$ is the projective line over the ring $\mathbb{Z}/N\mathbb{Z}$, i.e., the set of one-dimensional free $\mathbb{Z}/N\mathbb{Z}$-submodules of $(\mathbb{Z}/N\mathbb{Z})^2$.

SKETCH OF THE PROOF. (a) The map $\begin{pmatrix} a & b\\ c & d\end{pmatrix}\mapsto\begin{pmatrix} a & b\\ c & d\end{pmatrix} \pmod N$ is by definition a surjective group homomorphism $\Gamma(1)\to\mathrm{PSL}_2(\mathbb{Z}/N\mathbb{Z})$. Its kernel is clearly $\Gamma(N)$, which gives (a).

(b) Consider the map $\Gamma(1)\to\mathbb{P}^1(\mathbb{Z}/N\mathbb{Z})$ which associates with $\begin{pmatrix} a & b\\ c & d\end{pmatrix}\in\Gamma(1)$ the free $\mathbb{Z}/N\mathbb{Z}$-submodule of $(\mathbb{Z}/N\mathbb{Z})^2$ generated by $\begin{pmatrix} a & b\\ c & d\end{pmatrix}\begin{pmatrix} 1\\ 0\end{pmatrix}=\begin{pmatrix} a\\ c\end{pmatrix} \pmod N$. It is surjective, and under the induced action of $\Gamma(1)$ on $\mathbb{P}^1(\mathbb{Z}/N\mathbb{Z})$ the element $\begin{pmatrix} 1\\ 0\end{pmatrix}$ has stabilizer $\Gamma_0(N)$, which gives (b). □

COROLLARY 5.1.2. *Indices of the subgroups $\Gamma(N)$ and $\Gamma_0(N)$ in $\Gamma(1)$ satisfy*

(a)
$$[\Gamma(1):\Gamma(N)]=\begin{cases}\dfrac{N^3}{2}\displaystyle\prod_{p\mid N}(1-p^{-2}) & \textit{for } N\geq 3,\\ 6 & \textit{for } N=2;\end{cases}$$

(b)
$$[\Gamma(1):\Gamma_0(N)]=N\prod_{p\mid N}(1+p^{-1}).$$
□

The proof can be obtained by induction on the number of distinct divisors of N, the case $N = p$ being (almost) trivial.

LEMMA 5.1.3. *For $N \geq 3$ the group $\Gamma(N)$ contains no element of finite order.* □

A *congruence subgroup* Γ of $\Gamma(1)$ is a subgroup containing $\Gamma(N)$ for some N. In what follows, we denote an arbitrary congruence subgroup of $\Gamma(1)$ by Γ; we are mostly interested in the cases $\Gamma = \Gamma(N)$ and $\Gamma = \Gamma_0(N)$.

We assume that Γ is equipped with a discrete topology, while H has a usual complex topology and hence is homeomorphic to an open disc.

PROPOSITION 5.1.4. *The topological space $\Gamma\backslash H$ is Hausdorff in the quotient topology.* □

The space $\Gamma\backslash H$ is canonically equipped with a structure of a noncompact Riemann surface. Indeed, let $N \geq 3$ and $\Gamma_N = \Gamma \cap \Gamma(N)$. According to Lemma 5.1.3, the group Γ_N contains no elements of finite order; therefore, for $N \geq 3$ the action of Γ_N on H is *free*, i.e., has no fixed points. Moreover, one can show that for any $v \in H$ there exists an open neighbourhood U_v homeomorphic to its image $V_v \subset \Gamma_N\backslash H$.

PROPOSITION 5.1.5. *There exists a unique structure of a Riemann surface on $\Gamma_N\backslash H$ such that homeomorphisms between U_v and V_v are complex analytic isomorphisms.* □

Since $\Gamma'_N = \Gamma/\Gamma_N$ is a finite group, we can define a Riemann surface $\Gamma\backslash H$ as a factor of $\Gamma_N\backslash H$ modulo the finite group Γ'_N.

LEMMA 5.1.6. *The Riemann surface structure on $\Gamma\backslash H$ thus defined does not depend on the choice of N.* □

A serious disadvantage of the Riemann surface $\Gamma\backslash H$ is that it is not compact. A canonical compactification $\Gamma\backslash H^*$ of the surface $\Gamma\backslash H$ is constructed as follows. Consider the set $\mathbb{P}^1(\mathbb{Q}) = \mathbb{Q} \cup \{\infty\}$ consisting of rational numbers and the symbol ∞. Each element of $\mathbb{P}^1(\mathbb{Q})$ can be represented in the form a/b with $a, b \in \mathbb{Z}$, where for $b = 0$ we put $a/b = \infty$.

The group $\Gamma(1)$ naturally acts on $\mathbb{P}^1(\mathbb{Q})$: if $\gamma = \begin{pmatrix} a & b \\ c & d \end{pmatrix}$, then

$$\gamma(r) = \frac{ar+b}{cr+d},$$

so that $\gamma(\infty) = a/c$ and $\gamma(r) = \infty$ if $cr + d = 0$.

Define $\Gamma\backslash H^* = (\Gamma\backslash H) \cup (\Gamma \backslash \mathbb{P}^1(\mathbb{Q}))$. Lemma 5.1.1(b) implies the following fact.

LEMMA 5.1.7. *The factor set $\Gamma \backslash \mathbb{P}^1(\mathbb{Q})$ is finite.* □

The finite set $\Gamma \backslash \mathbb{P}^1(\mathbb{Q})$ is called the set of cusps of the Riemann surface $\Gamma\backslash H^*$ or of the group Γ.

Let us define a complex structure on $\Gamma\backslash H^*$ such that its restriction to the open subset $\Gamma\backslash H$ coincides with that defined above, complex-analytic neighbourhoods of a point $r \in \mathbb{Q}$ are open discs tangent to the line $\mathrm{Im}(z) = 0$ at r, and neighbourhoods of ∞ are open half-planes of the form $\mathrm{Im}(z) > d$.

THEOREM 5.1.8. *The set $\Gamma\backslash H^*$ with the above-defined complex structure is a connected compact Riemann surface.* □

EXAMPLE 5.1.9. For $\Gamma = \Gamma(1)$, the surface $\Gamma\backslash H$ can be identified with $\mathbb{C}$, and $\Gamma\backslash H^*$ with the Riemann sphere $\mathbb{P}^1(\mathbb{C})$.

Consider a holomorphic function $f\colon H \to \mathbb{C}$ obtained as a composition

$$H \xrightarrow{\pi_{\Gamma(1)}} \Gamma(1)\backslash H \overset{\theta}{\xrightarrow{\sim}} \mathbb{C}$$

for some complex analytic isomorphism θ, where $\pi_{\Gamma(1)}\colon H \to \Gamma(1)\backslash H$ is the natural projection.

LEMMA 5.1.10. *The automorphism group of $\mathbb{C}$ as a Riemann surface consists of affine maps of the form $\{z \mapsto az+b\}$, $a \in \mathbb{C}^*$, $b \in \mathbb{C}$.* □

Thus, a function f is defined up to transformations $f \mapsto af+b$, $a \in \mathbb{C}^*$, $b \in \mathbb{C}$. Hence, there exists a unique function f of this form such that $f(i) = 1728$ and $f(\rho) = 0$, where $i = \sqrt{-1}$ and $\rho = \dfrac{-1+\sqrt{-3}}{2}$. We denote this function by j.

COROLLARY 5.1.11. *Let $z \in H$, let $\Lambda_z = \mathbb{Z}+\mathbb{Z}z$ be a lattice in $\mathbb{C}$, and let E_z be a complex elliptic curve such that $E_z(\mathbb{C}) = \mathbb{C}/\Lambda_z$. Then $j(z) = j(E_z)$ is the j-invariant of E_z.* □

EXAMPLE 5.1.12. For the group $\Gamma = \Gamma(2)$ the surface $\Gamma\backslash H$ can be identified with $\mathbb{C}\setminus\{0,1\}$: the identification is given by the *Legendre modulus* $\lambda = \lambda(z)$, where the complex elliptic curve E_z can be written as

$$y^2 = x(x-1)(x-\lambda(z));$$

one can easily calculate that

$$j(z) = \frac{256\left(1-\lambda(z)+\lambda(z)^2\right)^3}{\lambda(z)^2(1-\lambda(z))^2}.$$

Curves $X(N)$ and $X_0(N)$. If Γ is an arbitrary congruence subgroup of $\Gamma(1)$, then there exists a unique (up to isomorphism) smooth projective curve X_Γ over $\mathbb{C}$ such that X_Γ considered as a Riemann surface is isomorphic to $\Gamma\backslash H^*$. Moreover, the Riemann surface $\Gamma\backslash H$ is naturally isomorphic to the smooth affine curve Y_Γ over $\mathbb{C}$. The curves X_Γ and Y_Γ are called *modular curves*.

We are mostly interested in curves corresponding to the groups $\Gamma = \Gamma(N)$ and $\Gamma = \Gamma_0(N)$, which are denoted by

$$X(N) = X_{\Gamma(N)}, \qquad X_0(N) = X_{\Gamma_0(N)};$$
$$Y(N) = Y_{\Gamma(N)}, \qquad Y_0(N) = Y_{\Gamma_0(N)}.$$

Since $\Gamma(N)$ is a normal subgroup in $\Gamma(1)$, the group $\Gamma(1)/\Gamma(N) \cong \mathrm{PSL}_2(\mathbb{Z}/N\mathbb{Z})$ acts on the curve $X(N)$ (and also on $Y(N)$) according to the standard formula: if $\gamma = \begin{pmatrix} a & b \\ c & d \end{pmatrix} \in \Gamma(1)$ and $[z]$ is the image of $z \in H \cup \mathbb{P}^1(\mathbb{Q})$ in $X(N)$, then

$$\gamma\colon [z] \mapsto \left[\frac{az+b}{cz+d}\right].$$

Since $\Gamma(N)$ is normal in $\Gamma(1)$, the action is well defined. One can easily prove the following result.

LEMMA 5.1.13. *If $n \mid N$, then $X(N)/\overline{\Gamma}(n) = X(n)$, where $\overline{\Gamma}(n)$ is the image of the subgroup $\Gamma(n)$ in $\Gamma(1)/\Gamma(N)$.* □

The subgroup $\Gamma_0(N)$ is not normal in $\Gamma(1)$. Nevertheless, a direct computation implies the following facts.

LEMMA 5.1.14. (a) *Let $N = mn$. Then the element*

$$\tau_m = \frac{1}{\sqrt{m}}\begin{pmatrix} 0 & 1 \\ -m & 0 \end{pmatrix} \in \mathrm{SL}_2(\mathbb{R})$$

belongs to the normalizer of $\Gamma_0(N)$ in $\mathrm{SL}_2(\mathbb{R})$, i.e., $\tau_m^{-1}\Gamma_0(N)\tau_m = \Gamma_0(N)$.

(b) *The action of τ_m defines an automorphism w_m of the curve $X_0(N)$. All w_m are involutions (i.e., elements of order 2) and commute with each other.* □

Thus,

$$\mathrm{Aut}(X_0(N)) \supseteq (\mathbb{Z}/2\mathbb{Z})^{\sigma_0},$$

where $\sigma_0 = \sigma_0(N)$ is the number of divisors of N. The involutions w_m are called *Atkin–Lehner involutions.*

Tower of modular curves. Riemann surfaces of the form $\Gamma\backslash H^*$ constitute a natural tower: for two congruence subgroups $\Gamma' \subseteq \Gamma \subseteq \Gamma(1)$ there is a natural complex-analytic projection

$$\pi_{\Gamma',\Gamma} \colon \Gamma'\backslash H^* \to \Gamma\backslash H^*.$$

In particular, for any Γ there is a natural projection

$$\pi_{\Gamma,\Gamma(1)} = \pi_\Gamma \colon \Gamma\backslash H^* \to \Gamma(1)\backslash H^* = \mathbb{P}^1_{\mathbb{C}}.$$

The ramification structure of the projection

$$\pi_{\Gamma',\Gamma} \colon \Gamma'\backslash H^* \to \Gamma\backslash H^*$$

can easily be described. Indeed, let $a \in H \cup \mathbb{P}^1(\mathbb{Q}) = H^*$, and let $\Gamma_a \subset \Gamma(1)$ denote the stabilizer of a.

LEMMA 5.1.15. *Let $a \in H^*$, and let $\overline{a}$ be the image of a in $\Gamma'\backslash H^*$. Then the ramification index of the projection $\pi_{\Gamma',\Gamma}$ at $\overline{a}$ equals $[\Gamma_a : \Gamma'_a]$.* □

Thus, ramification points of the projection $\pi_{\Gamma',\Gamma}$ are precisely the images in $\Gamma'\backslash H^*$ of points $a \in H^*$ such that $\Gamma_a \neq \Gamma'_a$.

Let $S = \begin{pmatrix} 0 & -1 \\ 1 & 0 \end{pmatrix} \in \Gamma(1)$ and $T = \begin{pmatrix} 1 & 1 \\ 0 & 1 \end{pmatrix} \in \Gamma(1)$. The groups Γ_a for an arbitrary subgroup $\Gamma \subset \Gamma(1)$ can be computed using the following result, which describes the groups $\Gamma(1)_a$.

LEMMA 5.1.16. (a) *Let $z \in H$, $\rho = \dfrac{-1+\sqrt{-3}}{2}$, $i = \sqrt{-1}$.*
If $z \notin \Gamma(1)i \cup \Gamma(1)\rho$, then $\Gamma(1)_z = \{1\}$.
If $z = \gamma(i)$, $\gamma \in \Gamma(1)$, then $\Gamma(1)_z = \gamma\langle S\rangle\gamma^{-1}$ is a subgroup of order 2.
If $z = \gamma(\rho)$, $\gamma \in \Gamma(1)$, then $\Gamma(1)_z = \gamma\langle U\rangle\gamma^{-1}$ is a subgroup of order 3.

(b) *The action of $\Gamma(1)$ on $\mathbb{P}^1(\mathbb{Q})$ is transitive.*
If $a = \gamma(\infty)$, $a \in \mathbb{P}^1(\mathbb{Q})$, $\gamma \in \Gamma(1)$, then $\Gamma(1)_a = \gamma\langle T\rangle\gamma^{-1}$, where $\langle T\rangle$ is an infinite cyclic subgroup generated by T.

SKETCH OF THE PROOF. This is based on the following elementary facts, which can be verified by a simple brute-force computation: S is an element of order 2, and $U = ST = \begin{pmatrix} 0 & -1 \\ 1 & 1 \end{pmatrix}$ is of order 3 in $\Gamma(1)$. Moreover, $Si = i$ and $U\rho = \rho$. □

REMARK 5.1.17. In fact $\Gamma(1)$ is a free product of its cyclic subgroups $\langle S\rangle$ and $\langle U\rangle$. This means that any element $\gamma\in\Gamma(1)$ can be uniquely represented in the form $\gamma = S^aU^{b_1}SU^{b_2}\dots SU^{b_k}$, where $a\in\{0,1\}$, $b_k\in\{0,1,2\}$, and $b_j\in\{1,2\}$ for $j=1,\dots,k-1$.

Applying the results of the above lemmas together with the Hurwitz formula to the projection $\pi_\Gamma\colon X_\Gamma\to\mathbb{P}^1_{\mathbb{C}}$, one can compute the genus of X_Γ. Here is an answer for the most interesting curves $X(N)$ and $X_0(N)$.

PROPOSITION 5.1.18. (a) *Let $N\geq 3$. Then the genus $g(N)$ of $X(N)$ is*

$$g(N) = 1+\frac{(N-6)\mu_N}{12N},$$

where $\mu_N = [\Gamma(1):\Gamma(N)] = \dfrac{N^3}{2}\prod_{p|N}(1-p^{-2})$.

(b) *For the genus $g_0(N)$ of $X_0(N)$, we have*

$$g_0(N) = 1+\frac{\mu}{12}-\frac{\nu_2}{4}-\frac{\nu_3}{3}-\frac{\nu_\infty}{2},$$

where

$$\mu = [\Gamma(1):\Gamma_0(N)] = N\prod_{p|N}(1+p^{-1});$$

ν_2 is the number of $\Gamma_0(N)$-nonequivalent points of H that are $\Gamma(1)$-equivalent to $z=i$:

$$\nu_2 = \begin{cases} 0 & \text{if } 4\mid N \\ \prod_{p|N}\left(1+\left(\frac{-1}{p}\right)\right) & \text{if } 4\nmid N, \end{cases}$$

where

$$\left(\frac{-1}{p}\right) = \begin{cases} 0 & \text{for } p=2, \\ 1 & \text{for } p=1(\text{mod } 4), \\ -1 & \text{for } p=3(\text{mod } 4); \end{cases}$$

ν_3 is the number of $\Gamma_0(N)$-nonequivalent points of H that are $\Gamma(1)$-equivalent to ρ:

$$\nu_3 = \begin{cases} 0 & \text{if } 9\mid N, \\ \prod_{p|N}\left(1+\left(\frac{-3}{p}\right)\right) & \text{if } 9\nmid N, \end{cases}$$

where

$$\left(\frac{-3}{p}\right) = \begin{cases} 0 & \text{for } p=3, \\ 1 & \text{for } p=1(\text{mod } 3), \\ -1 & \text{for } p=2(\text{mod } 3); \end{cases}$$

and ν_∞ is the number of cusps of $X_0(N)$:

$$\nu_\infty = \sum_{\substack{d|N\\ d>0}}\varphi((d,N/d)),$$

where φ is the Euler totient function. □

COROLLARY 5.1.19. *Let $\ell \neq 11$ be either a prime or 1. Then $g_0(11\ell) = \ell$.* □

Below we need a description of ramification in the tower

$$X_0(11\ell) \xrightarrow{\varphi} X_0(11) \xrightarrow{\psi} \mathbb{P}^1,$$

where

$$\varphi = \pi_{\Gamma_0(11\ell),\Gamma_0(11)}, \qquad \psi = \pi_{\Gamma_0(11)}, \qquad \theta = \psi \circ \varphi = \pi_{\Gamma_0(11\ell)}.$$

Applying the above results to φ and ψ, one can derive the following facts.

PROPOSITION 5.1.20. (a) *Degrees of the projections φ, ψ, and θ are, respectively,*

$$\deg \varphi = \ell + 1, \qquad \deg \psi = 12, \qquad \deg \theta = 12(\ell + 1).$$

(b) *Outside the sets $\theta^{-1}(0)$, $\theta^{-1}(\infty)$, and $\theta^{-1}(1728)$, the projection θ is unramified. At the points lying over 0, 1728, and $\infty \in \mathbb{P}^1_{\mathbb{C}}$, we have the following:*

- *The set $\theta^{-1}(0)$ consists of $6(\ell+1)$ points $P_1, \ldots, P_{6(\ell+1)}$, each of them having ramification index 2.*
- *The set $\theta^{-1}(1728)$ consists of $4(\ell+1)$ points $R_1, \ldots, R_{4(\ell+1)}$, each of them having ramification index 3.*
- *The set $\theta^{-1}(\infty)$ consists of four points Q_1, Q_2, Q_3, Q_4 with ramification indices, respectively,*

$$e_1 = e_{Q_1} = 1, \qquad e_2 = e_{Q_2} = \ell, \qquad e_3 = e_{Q_3} = 11, \qquad e_4 = e_{Q_4} = 11\ell.$$

Ramification of ψ is described as follows:

- *The set $\psi^{-1}(0)$ consists of 4 points R'_1, R'_2, R'_3, R'_4, each of them having ramification index 3.*
- *The set $\psi^{-1}(1728)$ consists of 6 points $P'_1, \ldots, P'_6$, each of them having ramification index 2.*
- *The set $\psi^{-1}(\infty)$ consists of two points Q'_1 and Q'_2 with ramification indices $e'_1 = e_{Q'_1} = 1$ and $e'_2 = e_{Q'_2} = 11$.*

The projection φ is unramified outside $\theta^{-1}(0)$. Moreover, $\varphi^{-1}(Q'_1) = \{Q_1, Q_2\}$, the corresponding ramification indices of φ being $e''_1 = e_{Q_1} = 1$ and $e''_2 = e_{Q_2} = \ell$; and $\varphi^{-1}(Q'_2) = \{Q_3, Q_4\}$, with $e''_3 = 1$ and $e''_4 = \ell$.

Thus, for the ramification divisor B_θ of the projection θ, we have

$$B_\theta = \sum_{i=1}^{6(\ell+1)} P_i + 2 \sum_{m=1}^{4(\ell+1)} R_m + (\ell - 1)Q_2 + 10Q_3 + (11\ell - 1)Q_4.$$ □

REMARK 5.1.21. One can similarly describe the ramification structure in the towers

$$X_0(32\ell) \to X_0(32) \to \mathbb{P}^1,$$
$$X_0(2^n) \to X_0(32) \to \mathbb{P}^1,$$
$$X_0(3^m) \to X_0(27) \to \mathbb{P}^1,$$

where ℓ is a prime and where $n \geq 5$ and $m \geq 3$ are arbitrary integers, using the fact that the curves $X_0(32)$ and $X_0(27)$ are elliptic.

Weierstrass points. Since Weierstrass points play a role in the study of algebraic geometry codes, the question of Weierstrass points on curves of the form X_Γ is of interest. A complete description of Weierstrass points on such curves is known for several small values of N only; however, using Proposition 2.2.49 and Corollary 2.2.50, one can often find a rather representative set of Weierstrass points on such a curve. In this respect, cusps are especially convenient, since they quite often happen to be Weierstrass points.

Here we consider only the case $\Gamma = \Gamma_0(N)$, $X_\Gamma = X_0(N)$. Note that the canonical involution w_n interchanges 0 and $P = \infty$; therefore, ∞ is a Weierstrass point if and only if so is 0. Here we restrict ourselves to the case of the cusp $P = \infty$ (or 0, which is equivalent), since this point is especially often shown to be a Weierstrass point, its weight being sufficiently large. Moreover, we are mainly interested in the case of a curve of the form $X_0(p^2n)$, where p is a prime and $n \geq 2$.

PROPOSITION 5.1.22. (a) *The natural projection $X_0(p^2n) \to X_0(pn)$ is of degree p, and the ramification index of the point $0 \in X_0(p^2n)$ is p (i.e., 0 is totally ramified).*

(b) *If*

$$g(X_0(p^2n)) \geq p(g(X_0(pn)) + 1),$$

then 0 and ∞ are Weierstrass points on $X_0(p^2n)$.

SKETCH OF THE PROOF. The idea of the proof is as follows.

(a) This is obtained by considering a set of representatives of right cosets of $\Gamma_0(p^2n)$ in $\Gamma_0(pn)$.

(b) Apply the argument in the proof of Proposition 2.2.49. □

A laborious but straightforward argument also yields the following result.

PROPOSITION 5.1.23. *The point $\infty \in X_0(p^2n)$ is a Weierstrass point at least in the following cases:*

(a) *$p \geq 5$ and $p \mid n$, or*
$p = 3$ and $3 \mid n$, $n \neq 3$, $n \neq 9$, or
$p = 2$ and $2 \mid n$, $n \neq 2$, $n \neq 2\ell$, $n \neq 4\ell$, ℓ being a prime;

(b) *n is divisible by three different primes;*

(c) *n is divisible by m^2, where $p \nmid m$ and $p^2n \neq 36$;*

(d) *n is divisible by two different primes, and at least one of the congruences $x^2 + 1 \equiv 0 \pmod{pn}$ or $x^2 + x + 1 \equiv 0 \pmod{pn}$ has a solution;*

(e) *$p \equiv n \equiv 1 \pmod{12}$.* □

Thus, for curves $X_0(N)$ with N divisible by a square, ∞ is quite often a Weierstrass point. This effect is most important for curves of the form $X_0(2^n)$, $n \geq 7$. Hence, for a curve $X_0(2^n)$ one can indicate many nongaps that are not greater than the genus of the curve.

Another problem closely connected with Weierstrass points is that of hyperellipticity of the curves $X_0(n)$. One can easily write a number of small values of n for which $X_0(n)$ is hyperelliptic; for instance, $n = 37$.

However, for large n the curves $X_0(n)$ are never hyperelliptic. More precisely, there is the following fact.

PROPOSITION 5.1.24. *For $n \geq 108$ the curves $X_0(n)$ are not hyperelliptic.* □

Automorphic functions. Elements of the field of rational functions on a curve X_Γ, or, equivalently, meromorphic functions on a Riemann surface $\Gamma\backslash H^*$, can be considered as functions on H invariant under Γ such that their only singularities (both on $\mathbb{Q}\cup\{\infty\}$ and on H) are poles. These functions are called *automorphic functions* under Γ.

Let us start with the case $\Gamma=\Gamma(1)$. Here $\Gamma\backslash H^*=\mathbb{P}^1_{\mathbb{C}}$. Hence, the field of functions automorphic under $\Gamma(1)$ coincides with $\mathbb{C}(j)$. Any such function has a canonical Laurent series expansion at infinity, which is called its *q-expansion.* Since $T\in\Gamma(1)$, for any $f\in\mathbb{C}(j)$ and any $z\in H$ we have

$$F(Tz)=f(z+1)=f(z).$$

LEMMA 5.1.25. *The function $q=q_z=e^{2\pi iz}$ gives a local parameter at the point $\infty\in\mathbb{P}^1_{\mathbb{C}}$.* □

A local expansion over this parameter q of a function $\widetilde{f}$ induced on $\mathbb{P}^1_{\mathbb{C}}$ by f is called the q-expansion of f:

$$f=\sum_{m=-N}^{\infty}a_mq^m.$$

The number of terms with negative powers of q in this expansion is finite, since f is meromorphic.

The theory of elliptic functions yields the following fact.

PROPOSITION 5.1.26. *For the q-expansion of j, we have*

$$j(z)=q^{-1}\left(1+\sum_{m=0}^{\infty}c_mq^m\right),$$

where c_m are mutually coprime integers. □

Note that the normalization of j (i.e., the factor 1728) is motivated by Proposition 5.1.26.

Now let Γ be an arbitrary congruence subgroup such that $\Gamma\supseteq\Gamma(N)$. Then $T^N=\begin{pmatrix}1&N\\0&1\end{pmatrix}\in\Gamma$, so that for any automorphic function f with respect to the subgroup Γ we have

$$f(T^Nz)=f(z+N)=f(z),\quad \text{for } z\in H.$$

Therefore, any such function has a Laurent series expansion at infinity of the following form:

$$f(z)=\sum_{k=-M}^{\infty}a_kq^{k/N},\quad a_k\in\mathbb{C}.$$

This expansion is called the *q-expansion of f at infinity.* In fact, it is an expansion of the rational function $\widetilde{f}$ on X_Γ defined by f, in powers of a local parameter at the cusp of X_Γ corresponding to ∞. One can similarly define q-expansions of automorphic functions at other cusps, but we do not need them here.

Note that in the case $\Gamma=\Gamma_0(N)$ we have $T\in\Gamma$, and therefore functions automorphic under $\Gamma_0(N)$ have expansions in integral powers of q.

The field of automorphic functions with respect to $\Gamma_0(N)$ has a very simple form.

THEOREM 5.1.27. *The field of functions automorphic under $\Gamma_0(N)$ coincides with the field $\mathbb{C}(j(z),j(Nz))$.* □

In particular, the function $j_N(z) := j(Nz)$ is invariant under $\Gamma_0(N)$. The involution w_N on the curve $X_0(N)$ which permutes j and j_N induces an involution on the field $\mathbb{C}(j, j_N)$.

Modular equation. Since the field $\mathbb{C}(j(z), j_N(z))$ is the field of rational functions on an algebraic curve, the functions j and j_N satisfy a nontrivial relation of the form $F(j, j_N) = 0$, where F is a polynomial of two variables. Moreover, since the coefficients of q-expansions of j and j_N are integers, one can choose F with integral coefficients. A minimum relation $\Phi_N(x, j)$ of the form $F(j, x) = 0$ with integer coprime coefficients and positive leading term (i.e., $j^{\psi(N)}$; see below) satisfied by the functions $j = j(z)$ and $x = j_N(z)$ is called the *modular equation*. The following theorem gives its principal properties.

THEOREM 5.1.28. *Let*

$$\Phi_N(x, j) = \prod_{\alpha \in A} (x - j(\alpha(z))),$$

where α runs over the set A of all matrices of the form $\begin{pmatrix} a & b \\ 0 & d \end{pmatrix}$ with $a, b, d \in \mathbb{Z}$, $a > 0$, $d > 0$, $ad = N$, $0 \leq b < d$, $(a, b, d) = 1$, and $\alpha(z) = \dfrac{az + b}{d}$. Then

(a) $\Phi_N(x, j) \in \mathbb{Z}[x, j]$, *and*

$$\Phi_N(x, j) = x^{\psi(N)} + j^{\psi(N)} + \sum_{\substack{a < \psi(N) \\ b < \psi(N)}} c_{ab} x^a j^b,$$

where $\psi(N) = [\Gamma(1) : \Gamma_0(N)] = N \prod_{p|N} (1 + p^{(-1)})$; moreover,

$$\Phi_N(x, j) = \Phi_N(j, x),$$

so that

$$c_{ab} = c_{ba} \in \mathbb{Z}.$$

(b) *The polynomial $\Phi_N(x, j)$ is absolutely irreducible and satisfies the relation*

$$\Phi_N(j, j_N) = 0. \qquad \square$$

Thus, the modular equation $\Phi_N(x, j) = 0$ specifies an affine model of the curve $X_0(N)$. Moreover, since the coefficients of Φ_N are integers, the $X_0(N)$ can be considered as a curve defined over $\mathbb{Q}$. In what follows, we are mainly interested in the curve $X_0(N)$ over $\mathbb{Q}$; however, to study its properties, we often need to use the Riemann surface $\Gamma_0(N)\backslash H^*$.

Dedekind eta function. In many cases, the field $\mathbb{C}(j, j_N)$ can be explicitly constructed in terms of the Dedekind eta function, defined as the product

$$\eta(z) = q^{\frac{1}{24}} \prod_{n=1}^{\infty} (1 - q^n) = e^{\frac{\pi i z}{12}} \prod_{n=1}^{\infty} (1 - e^{2\pi i n z}),$$

which is a holomorphic function on H, though not invariant under the action of $\Gamma(1)$ or $\Gamma(N)$. It satisfies the following functional equations:

$$\eta(z + 1) = e^{\frac{\pi i}{12}} \eta(z),$$

$$\eta\left(-\frac{1}{z}\right) = \sqrt{-iz}\, \eta(z);$$

note that $\sqrt{-iz}$ is well defined as a holomorphic function on H.

For example,

$$\mathbb{C}(j(z), j(2z)) = \mathbb{C}(\xi), \quad \text{where} \quad \xi = \left(\frac{\eta(z)}{\eta(2z)}\right)^{24}.$$

Invariance of ξ under $\Gamma_0(2)$ easily follows from the above equations, which also imply that ξ has a simple pole at infinity and thus generates $F_2 = \mathbb{C}(j(z), j(2z))$. There exist many formulae of this kind; some of them are used in Sec. 5.3.1 below.

The *Weierstrass discriminant function*

$$\Delta(z) = \eta(z)^{24} = q \prod_{n=1}^{\infty} (1-q^n)^{24}$$

is very important in arithmetic of modular and elliptic curves. It satisfies the condition

$$\Delta(\gamma z) = (cz+d)^{12} \Delta(z) \quad \text{for} \quad \gamma = \begin{pmatrix} a & b \\ c & d \end{pmatrix} \in \Gamma(1).$$

Thus, $\Delta(z)$ is a unique (up to a constant factor) *modular form* of weight 12 which has a zero at infinity, i.e., the most simple *parabolic form* for $\Gamma(1)$. Note that if $y^2 = 4x^3 - g_2 x - g_3$ is the Weierstrass equation of the elliptic curve E_z such that $E_z(\mathbb{C}) \cong \mathbb{C}(\mathbb{Z}+z\mathbb{Z})$, then $\Delta = g_2^3 - 27g_3^2$.

5.1.2. Modular Curves over Finite Fields

We are mainly interested in modular curves over finite fields. To study these curves, one has to interpret modular curves as moduli varieties of elliptic curves. To obtain or even formulate the precise results, it would be necessary to use the theory of schemes, which we cannot do here. However, let us give some elementary formulations and rather rough explanations.

Modular curves as moduli varieties. We have mentioned above that the function $j\colon H \to \mathbb{P}^1_{\mathbb{C}}$, which defines the isomorphism $j\colon \Gamma(1)\backslash H^* \xrightarrow{\sim} \mathbb{P}^1_{\mathbb{C}}$, is in fact the absolute invariant of a variable elliptic curve $\mathbb{C}/(\mathbb{Z}+\mathbb{Z}z)$. Since equality of absolute invariants is a necessary and sufficient condition for complex elliptic curves to be isomorphic, we easily see that there is a bijection between $\Gamma(1)\backslash H = Y(1)$ and the set of isomorphism classes of complex elliptic curves; the bijection is defined by the absolute invariant j:

$$j\colon \text{elliptic curve } E \mapsto j(E).$$

This fact leads to interpretation of modular curves as moduli varieties of elliptic curves with extra structures. First, let us give two more examples.

Consider the curve $Y_0(N) = \Gamma_0(N)\backslash H$ and its natural projection onto the curve $Y(1) = \mathbb{A}^1$. The fibre of the projection $\pi_{\Gamma_0(N)}\colon Y_0(N) \to Y(1)$ over the point $j(\mathbb{C}/(\mathbb{Z}+\mathbb{Z}z))$ coincides (according to Lemma 5.1.1) with the set $\mathbb{P}^1(\mathbb{Z}/N\mathbb{Z})$ of cyclic subgroups of order N in the subgroup $E_N \cong (\mathbb{Z}/N\mathbb{Z})^2$ of points of order N on the curve $E = \mathbb{C}/(\mathbb{Z}+\mathbb{Z}z)$. Therefore, there is a natural bijection between $Y_0(N)$ and the set of isomorphism classes of pairs (E, C_N), where E is a complex elliptic curve and C_N is its cyclic subgroup of order N.

To state a similar result for the curve $Y(N)$, we need the following fact.

PROPOSITION 5.1.29. *Let $\psi\colon \mathbb{C}/(\mathbb{Z}+\mathbb{Z}z) \to E(\mathbb{C})$ be a holomorphic bijection; then $E_N = \psi\left(\left(\frac{1}{N}\Lambda\right)/\Lambda\right)$, where $\Lambda = \mathbb{Z}+\mathbb{Z}z$, and $s \in E_N$ is specified by a pair (a,b)*

of residues modulo N. *Define* $e(s,t)=e^{-2\pi i(ab'-a'b)/N}$ *for* $s=(a,b)$ *and* $t=(a',b')$. *Then the map*

$$E_N\times E_N\to\mu_N,\quad (s,t)\mapsto e(s,t)$$

defines a nondegenerate skew-symmetric pairing (*the* Weil pairing), *which yields a canonical identification* $\Lambda^2E_N=\mu_N$, *where* μ_N *is the group* $\{\zeta\in\mathbb{C}:\zeta^N=1\}$. □

The projection of $Y(N)$ onto $Y(1)$ shows that the points of $Y(N)$ are in a natural bijection with the set of pairs (E,α_N), where E is a complex elliptic curve and α_N is the so-called *structure of level* N *with* $\det\alpha_N=1$, i.e., an isomorphism

$$\alpha_N\colon E_N\xrightarrow{\sim}(\mathbb{Z}/N\mathbb{Z})^2$$

such that the isomorphism $\Lambda^2\alpha_N\colon\Lambda^2E_N=\mu_N\xrightarrow{\sim}\Lambda^2((\mathbb{Z}/N\mathbb{Z})^2)=\mathbb{Z}/N\mathbb{Z}$ takes $e^{2\pi i/N}$ to 1.

One can give as well a similar description of the projective curves $X(N)$ and $X_0(N)$ using an appropriate generalization of elliptic curves.

Good reductions of modular curves. Modular curves over finite fields are reductions of modular curves over $\mathbb{Q}$; now we explain what this means.

Let A be a commutative ring with the fraction field K (in the present section we need only rings of integers in some number field K, e.g., $A=\mathbb{Z}$, $K=\mathbb{Q}$), and let C be an absolutely irreducible projective smooth curve over K. Let $\mathfrak{p}\subset A$ be a maximal ideal. Then we say that C has a *good reduction* modulo $\mathfrak{p}$ if there exists an embedding $i\colon C\to\mathbb{P}^N(K)$ such that C is defined in the projective space $\mathbb{P}^N$ by a system of homogeneous polynomial equations $F_1=F_2=\ldots=F_r$ with coefficients in A and that the reduced modulo $\mathfrak{p}$ equations $\widetilde{F}_1=\widetilde{F}_2=\ldots=\widetilde{F}_r$ with coefficients in the residue field $k(\mathfrak{p})=A/\mathfrak{p}$ still define an absolutely irreducible projective smooth curve in $\mathbb{P}^N(k(\mathfrak{p}))$. It is possible to prove that all good reductions of C are isomorphic as curves over $k(\mathfrak{p})$; moreover, the genus of a good reduction equals that of C. If $\varphi\colon C\to C'$ is a K-morphism of algebraic curves which both have good reductions modulo $\mathfrak{p}$, then φ defines a $k(\mathfrak{p})$-morphism of algebraic curves $\varphi_{\mathfrak{p}}\colon C_{\mathfrak{p}}\to C'_{\mathfrak{p}}$; moreover, $(\varphi\circ\psi)_{\mathfrak{p}}=\varphi_{\mathfrak{p}}\circ\psi_{\mathfrak{p}}$. In particular, a good reduction modulc $\mathfrak{p}$ preserves automorphisms of C.

This definition is not directly applicable to modular curves, since they are by definition complex ones. However, their moduli interpretation allows us to prove that actually modular curves are defined over rather small number fields; moreover, they have good reductions modulo appropriate ideals.

THEOREM 5.1.30. *Let* $N\geq 3$. *Then*

1. *The curve* $X(N)$ *can be defined over the field* $\mathbb{K}_N$, *a unique quadratic subfield in* $\mathbb{Q}(\zeta_N)$. *Thus, there exists an absolutely irreducible projective smooth curve* $X_P(N)$ *over* $\mathbb{K}_N$ *such that the corresponding complex curve* $X_P(N)\otimes_{\mathbb{K}_N}\mathbb{C}$ *is canonically isomorphic to* $X(N)$. *Moreover, for any maximal ideal* $\mathfrak{p}$ *in the ring of integers* $B_N\subset\mathbb{K}_N$, *the reduction* $X_P(N)_{\mathfrak{p}}$ *is good if* $\mathfrak{p}$ *does not divide* N.

2. *The curve* $X_0(N)$ *can be defined over* $\mathbb{Q}$ *and has a good reduction* $X_0(N)_p$ *modulo any prime* $p\nmid N$. □

Note, however, that a "natural" field of definition for $X(N)$ and $Y(N)$ is $\mathbb{Q}(\zeta_N)$, $X_P(N)$ being its form over $\mathbb{K}_N$.

The same is true for the affine curves $Y(N),Y_0(N)$, and the corresponding reductions $Y_P(N)_{\mathfrak{p}},Y_0(N)_p$ are moduli varieties. More precisely, for any algebraically closed field $\Bbbk \supset k(\mathfrak{p})$ the set $Y_P(N)_{\mathfrak{p}}(\Bbbk)$ is in a natural bijection with the set

$$P(N)(\Bbbk) = \left\{\begin{matrix}\text{the set of isomorphism classes of pairs} \\ (E,\beta_N),\text{ where } E/\Bbbk \text{ is an elliptic curve} \\ \text{and } \beta_N \text{ is a projective structure of level } N\end{matrix}\right\},$$

where the *projective structure* β_N is a class of structures α_N of level N specified up to a factor $\lambda \in (\mathbb{Z}/N\mathbb{Z})^*$, and for any algebraically closed field $\Bbbk \supset \mathbb{F}_p$ the set $Y_0(N)_p(\Bbbk)$ is in a natural bijection with the set

$$F_0(N)(\Bbbk) = \left\{\begin{matrix}\text{the set of isomorphism classes of pairs} \\ (E,C_N),\text{ where } E/\Bbbk \text{ is an elliptic curve} \\ \text{and } C_N \text{ is a cyclic subgroup } E \text{ of order } N\end{matrix}\right\}.$$

Good reductions inherit some properties of the initial curves. In particular, $X_0(N)/p$ form towers similar to those described in Proposition 5.1.20, with the same degrees of maps and the same ramification structures. Moreover, the curve $X_0(N)/p$ inherits from $X_0(N)$ the set of involutions w_d, where $d \mid N$. Note also that hyperellipticity of $X_0(N)$ is equivalent to that of $X_0(N)/p$, since the reduction of a map defined by the canonical class gives a map defined by the canonical class of the reduction.

REMARK 5.1.31. The action of $\mathrm{PSL}_2(\mathbb{Z}/N\mathbb{Z})$ on $Y_P(N)$ and thus on $Y_P(N)/\mathfrak{p}$ can be described in the following way:

$$\gamma\colon (E,\beta_N) \mapsto (E,\beta_N \circ \gamma)$$

for $\gamma \in \mathrm{PSL}_2(\mathbb{Z}/N\mathbb{Z})$.

REMARK 5.1.32. The factor of the curve $X_p(N)/\mathfrak{p}$ modulo the subgroup $\mathrm{PSL}_2(\mathbb{Z}/N\mathbb{Z})$ of its automorphism group coincides with $\mathbb{P}^1$, and the projection $X_p(N)/\mathfrak{p} \to \mathbb{P}^1_{\Bbbk(\mathfrak{p})}$ is defined by the j-invariant (i.e., a pair (E,β_N) representing a point of $X_p(N)/\mathfrak{p}$ is mapped to $j(E)$).

$\mathbb{F}_{p^2}$-rational points. As we have seen, the curves $X(N)$ and $X_0(N)$ have good reductions "outside divisors of N". Moreover, these reductions have asymptotically maximal numbers of $\mathbb{F}_{p^2}$-rational points. Indeed, let us fix a prime p that is not a divisor of N. Then the curve $X_N = X_0(N)/p$ over $\mathbb{F}_p$ is smooth, as well as the curve $X'_N = X_p(N)/\mathfrak{p}$ over $\mathbb{F}_{p^2}$ where $\mathfrak{p}$ is a divisor of p in $\mathbb{K}_N$. Let $g_N = g(X_N) = g_0(N)$ and $g'_N = g(X'_N) = g(N)$, let R_N be the number of $\mathbb{F}_{p^2}$-rational points on X_N, and let R'_N be the number of $\mathbb{F}_{p^2}$-points on X'_N.

THEOREM 5.1.33. *We have*

$$\lim_{n\to\infty} R_N/g_N = p-1, \tag{5.1.1}$$

$$\lim_{n\to\infty} R'_N/g'_N = p-1. \tag{5.1.1$'$}$$

PROOF. First note that there is a tower of curves

$$X'_N \xrightarrow{\lambda_N} X_N \xrightarrow{\theta_N} \mathbb{P}^1,$$

where θ_N is a map of degree

$$\psi(N) = N\prod_{\ell\mid N}(1+\ell^{-1}),$$

defined by the j-invariant, and the composition $\pi_N = \theta_N \circ \lambda_N \colon X'_N \to \mathbb{P}^1$ is also defined by the j-invariant and is of degree equal to the order μ_N of the group $\mathrm{PSL}_2(\mathbb{Z}/N\mathbb{Z})$, which is $\dfrac{N^3}{2}\prod_{\ell|N}(1-\ell^{-2})$, so that λ_N is of degree

$$\deg \lambda_N = \deg \pi_N / \deg \theta_N = \frac{N^2}{2}\prod_{\ell|N}(1-\ell^{-1}).$$

Hence, it suffices to prove (5.1.1′), since we obviously have $R_N \geq R'_N/\deg\lambda_N$, and by the Hurwitz formula, $g_N \leq g'_N/\deg\lambda_N$, so that $R_N/g_N \geq R'_N/g'_N$. The inequality $\limsup_{N\to\infty} R_N/g_N \leq p-1$ follows from Theorem 4.1.30. To prove (5.1.1′), it suffices to appropriately bound R'_N from below. To this end, consider all points of X'_N lying over supersingular values of the j-invariant. Then it is sufficient to establish (a) and (b) as follows:

(a) The number of such points is $\mu_N \dfrac{p-1}{12}$.

(b) All such points are $\mathbb{F}_{p^2}$-rational.

Thus, $R_N \geq \dfrac{\mu_N(p-1)}{12}$, which implies (5.1.1′), since

$$g'_N = g(N) = 1 + \frac{\mu_N(N-6)}{12N}$$

(see Proposition 5.1.18).

Let us first check (a). The point $P \in X'_N$ is represented by a pair (E, β_N), where E is an elliptic curve and β_N is a projective structure of level N. On the fibre over the point $j(E) \in \mathbb{P}^1$ there acts the group $\mathrm{PSL}_2(\mathbb{Z}/N\mathbb{Z})$ of order μ_N, and the order of the isotropy group of this point equals $|\mathrm{Aut}\, E|/2$ (since isomorphic elliptic curves give one point on X'_N). Hence, the fibre over $j(E)$ contains $2\mu_N/|\mathrm{Aut}\, E|$ points, and the total number of points over supersingular values of j is (we use equation (2.4.8))

$$\sum_{j(E)} 2\mu_N |\mathrm{Aut}\, E|^{-1} = \mu_N(p-1)/12,$$

where the sum is over all supersingular values of $j(E)$.

Now we check (b). Let (E, β_N) be a supersingular elliptic curve over $\overline{\mathbb{F}}_p$ with a projective structure of level N. It suffices to show that the pair (E, β_N) is fixed under the action of the Frobenius endomorphism F over $\mathbb{F}_{p^2}$, which is a topological generator of the group $\mathrm{Gal}(\overline{\mathbb{F}}_p/\mathbb{F}_{p^2})$. Since the degrees of F and of the multiplication by p endomorphism coincide, and both morphisms are purely inseparable, we have $F = \varepsilon \circ p$ in the ring $\mathrm{End}_{\overline{\mathbb{F}}_p}(E)$, where ε is an automorphism of E. Indeed, since p annihilates the kernel of the $\mathbb{F}_p$-Frobenius morphism $F_0 \colon E \to E$, we have $p = f \circ F_0$, where $f \colon E' \to E$ is a morphism of degree p; note now that any morphism of degree p is of the form $\varepsilon \circ F_0$. If $|\mathrm{Aut}\, E| = 2$, then $\varepsilon = \pm 1$, $F = \pm p$, and F fixes β_N. If $|\mathrm{Aut}\, E| \neq 2$, we have to consider a representative of the point P on X'_N defined by a pair (E, β'_N) obtained from (E, β_N) by the action of ε. For this representative, the action of F coincides with $\pm p$, which is a central element. □

REMARK 5.1.34. In the notation of Theorem 5.1.33, we in fact have $R'_N = (p-1)g_n + c g_N^{2/3} + o(g_N^{2/3})$ for some $c > 0$. From the proof of Theorem 4.1.30 one can conclude that $N_1 \leq (\sqrt{q}-1)g + o(g/\log g)$, where N_1 is the number of $\mathbb{F}_q$-points on a curve of genus g. We do not know whether the difference $N_1-(\sqrt{q}-1)g$ can grow faster than $O(g^{2/3})$.

REMARK 5.1.35. The proof of Theorem 5.1.33 also implies that points of X_N lying over supersingular values of j are $\mathbb{F}_{p^2}$-rational and that there are at least $(p-1)g_0(N)$ of them.

5.1.3. Codes

Now we consider some properties of codes obtained from reductions of modular curves.

Parameters. We use the notation introduced in the proof of Theorem 5.1.33. Thus, $X_N = X_0(N)/p$ is an absolutely irreducible smooth projective curve of genus $g_N = g_0(N)$ over $\mathbb{F}_{p^2}$, and $X'_N = X_p(N)/g$ is an absolutely irreducible smooth projective curve of genus $g'_N = g(N)$ over either $\mathbb{F}_{p^2}$ or $\mathbb{F}_p$. By $\mathcal{P}$ we denote the set of supersingular points of X_N, $\mathcal{P} \subseteq X_N(\mathbb{F}_{p^2})$, and $\mathcal{P}'$ denotes the set of supersingular points of X'_N; we know that $\mathcal{P}' \subseteq X'_N(\mathbb{F}_{p^2})$; assume that $N \geq 3$.

PROPOSITION 5.1.36. (a) *Let D be a divisor of degree a on X_N defined over $\mathbb{F}_{p^2}$, and let* $\operatorname{Supp} D \cap \mathcal{P} = \varnothing$. *The $[n,k,d]_{p^2}$ code $C = (X_N, \mathcal{P}, D)_L$ has the following parameters:* $n = |\mathcal{P}| \in [1, n_0]$, *where*

$$n_0 = \frac{N(p-1)}{12} \prod_{\ell | N} (1+\ell^{-1}),$$
$$k \geq a - g_0(N) + 1,$$
$$d \geq n - a,$$

the product is over all prime divisors of N, and the value of $g_0(N)$ is given by Proposition 5.1.18(b).

(b) *Let D' be a divisor of degree a' on X'_N defined over $\mathbb{F}_{p^2}$, and let* $\operatorname{Supp} D' \cap \mathcal{P}' \neq \varnothing$. *The $[n',k',d']_{p^2}$ code $C' = (X'_N, \mathcal{P}', D')_L$ has the following parameters:* $n' = |\mathcal{P}'| \in [1, n'_0]$, *where*

$$n'_0 \geq \frac{N^3(p-1)}{24} \prod_{\ell | N} (1-\ell^{-2}),$$
$$k' \geq a' - \frac{N^2(N-6)}{24} \prod_{\ell | N} (1-\ell^{-2}),$$
$$d' \geq n' - a',$$

and the product is over all prime divisors of N.

PROOF. Immediately follows from Proposition 5.1.18 and Theorem 5.1.33. □

REMARK 5.1.37. Since the code length of C is at least $\dfrac{N(p-1)}{12} \prod_{\ell | N} (1+\ell^{-1})$, by shortening it if necessary, we may assume that

$$n = \left\lceil \frac{N(p-1)}{12} \prod_{\ell | N} (1+\ell^{-1}) \right\rceil.$$

The set of numbers of this form is rather dense on the number axis.

REMARK 5.1.38. The cardinality $\rho(n)$ of the set of integers of the form

$$m = \left\lceil \frac{N(p-1)}{12} \prod_{\ell \mid N} (1+\ell^{-1}) \right\rceil, \quad m \le n,$$

is asymptotically not less than $\frac{p-1}{12}\frac{n}{\ln n}$, i.e., $\rho(n) \ge f(n)$ with $\lim_{n\to\infty} \frac{12 f(n) \ln n}{(p-1)n}$ $= 1$. For the case of a prime N this follows from the prime distribution theorem.

For example, for $p = 7$, i.e., for codes over $\mathbb{F}_{49}$, one gets the following table giving the code length n versus N:

N	4	5	6	7	8	9	10	11	12	13	14	15	16	17	18	19	20	21	22
n	3	3	6	4	6	6	9	6	12	7	12	12	12	9	18	10	18	16	18

N	23	24	25	26	27	28	29	30	31	32	33	34	35	36	37	38	39	40
n	12	24	15	21	18	24	15	36	16	24	24	27	24	36	19	30	28	36

By using Weierstrass points, we can improve code parameters for small values of k. Now we shall consider only the point ∞ on curves $X_0(N)$ and restrict ourselves to the values $N = 2^m$, for which the effect of using Weierstrass points is the largest. From Proposition 5.1.18 it follows that the genus $g_N = g_0(N)$ of the curve $X_N = X_0(N)/p$ is given by

$$g_0(2^m) = 2^{m-3} + 1 - \delta_m,$$

where

$$\delta_m = \begin{cases} 3 \cdot 2^{m/2-2} & \text{for } m \text{ even,} \\ 2^{(m-1)/2} & \text{for } m \text{ odd.} \end{cases}$$

PROPOSITION 5.1.39. *Let $D = a\infty$ be a divisor of degree a on X_{2^m}, and let $C = (X_{2^m}, \mathcal{P}, D)_L$ be an $[n,k,d]_{p^2}$ code. Then for the parameters of C we have: $n = |\mathcal{P}| = 1, \ldots, n_0$, where*

$$n_0 \ge 2^{m-3}(p-1),$$
$$k \ge k(m,a),$$
$$d \ge n - a,$$

and $k(m,a)$ is defined as follows:

$$k(m,a) = \ell \quad \text{for} \quad 2^{m-3} - 2^{m-2-\ell} \le a \le 2^{m-3} - 2^{m-3-\ell}, \quad \ell = 1,2,\ldots,\lfloor m/2 \rfloor - 1.$$

Note that for these a the standard estimate $a - g + 1$ is nonpositive.

SKETCH OF THE PROOF. The only estimate that requires a proof is the estimate for k. Considering the projections $X_{2^m} \to X_{2^{2j}}$ and $X_{2^m} \to X_{2^{2j+1}}$ for $2j+1 \le m$ yields the following inequalities:

$$\ell(2^{m-2j}(2^{2j-3} - 3 \cdot 2^{j-2} + 1 + i)\infty) \ge i + 1;$$
$$\ell(2^{m-2j-1}(2^{2j-2} - 2^j + 1 + i)\infty) \ge i + 1.$$

The best of these bounds for $j = 1,2,\ldots,\lfloor m/2 \rfloor$ yields $k(m,a)$. □

Group properties. Since the group $\mathrm{PGL}_2(\mathbb{Z}/N\mathbb{Z})$ acts on the curve X'_N preserving the fibres of $X'_N \to \mathbb{P}^1$ and, in particular, preserves $\mathcal{P}'$ and the set of cusps of X'_N, it is natural to consider the codes on X'_N constructed from $\mathcal{P}'$ and the cusp divisors as group codes.

More precisely, let S_∞ be the set of points on X'_N that are mapped to $\infty \in \mathbb{P}^1$ under the natural projection $X'_N \to \mathbb{P}^1$. Each point $Q \in S_\infty$ is a ramification point of this projection with index N, so that

$$\nu_\infty = |S_\infty| = \mu_N/N = \frac{N}{2}\prod_{\ell|N}(1-\ell^{-2}).$$

Consider the divisor $F_\infty = \sum_{Q\in S_\infty} Q$ of degree $\deg F_\infty = \nu_\infty = |S_\infty|$. Since the point $\infty \in \mathbb{P}_1$ is $\mathbb{F}_p$-rational, the divisor F_∞ is defined over $\mathbb{F}_p$ (and hence over $\mathbb{F}_{p^2}$). Consider a divisor $D' = bF_\infty$ and the code $C' = (X'_N, \mathcal{P}', D')_L$ over $\mathbb{F}_{p^2}$.

The definition of D', the invariance of $\mathcal{P}'$ under the action of $\mathrm{PGL}_2(\mathbb{Z}/N\mathbb{Z})$, and Proposition 4.1.22 easily imply the following fact.

PROPOSITION 5.1.40. *The code C' has a natural action of* $\mathrm{PGL}_2(\mathbb{Z}/N\mathbb{Z})$. *The number of orbits under this action equals the number $s(p)$ of supersingular values of the j-invariant in characteristic p.* □

COROLLARY 5.1.41. *The code C' can be realized as a $\mathrm{PGL}_2(\mathbb{Z}/N\mathbb{Z})$-submodule in the algebra $\mathbb{F}_{p^2}[\mathrm{PGL}_2(\mathbb{Z}/N\mathbb{Z})]^{s(p)}$.* □

REMARK 5.1.42. If we restrict ourselves to the values of N such that $p \nmid (\ell^2-1)$ for all prime divisors $\ell|N$ (recall that we also assume that $p \nmid N$), then the algebra $\mathbb{F}_{p^2}[\mathrm{PGL}_2(\mathbb{Z}/N\mathbb{Z})]$ is semisimple, since in this case $p \nmid |\mathrm{PGL}_2(\mathbb{Z}/N\mathbb{Z})|$, which can considerably simplify the study of C' as a group code. Note that for $p \geq 5$ such N can be arbitrarily large.

PROBLEM 5.1.43. Give an explicit description of the code C' as a submodule of $\mathbb{F}_{p^2}[\mathrm{PGL}_2(\mathbb{Z}/N\mathbb{Z})]^{s(p)}$ and estimate its parameters without using the Riemann–Roch theorem.

REMARK 5.1.44. We cannot solve the above problem; however, the Riemann–Roch theorem provides some information on the structure of C' as a $\mathrm{PGL}_2(\mathbb{Z}/N\mathbb{Z})$-module. Put $a_0 = \lfloor N/6 \rfloor$. Then by the Riemann–Roch theorem, for $m \geq a_0$ we have

$$\ell(mF_\infty) = \mu_N(12mN+6)/12,$$

since $\deg(mF_\infty) > 2g'_N - 2$. Thus, $L((m+1)F_\infty)/L(mF_\infty)$ is a $\mathrm{PGL}_2(\mathbb{Z}/N\mathbb{Z})$-module of dimension

$$\mu_N/N = \frac{N^2}{2}\prod_{\ell|N}(1-\ell^{-2}).$$

Hence, C' considered as a $\mathrm{PGL}_2(\mathbb{Z}/N\mathbb{Z})$-module has a filtration of the form $C' = C_b \supset C_{b-1} \supset \ldots \supset C_{a_0}$, where C_m, $b \geq m \geq a_0$, is the image of $L(mF_\infty)$ in C', and all factors C_{m+1}/C_m, $m = a_0, \ldots, b-1$, are $\mathrm{PGL}_2(\mathbb{Z}/N\mathbb{Z})$-modules of dimension μ_N/N.

Polynomiality. Let us now briefly describe how to construct a generating matrix of an optimal algebraic geometry code $C = (X_N, \mathcal{P}, D)_L$ in polynomial time. Let us further assume that $N = 11\ell$ for a prime ℓ and take D in the form $D = t(Q_2 + Q_4)$ in the notation of Proposition 5.1.20.

The generator matrix which we want to compute is of the form

$$\left(f_\alpha(P_\beta)\right),$$

where f_α runs over a basis in $L(D)$ and $P_\beta \in \mathcal{P}$. Therefore, we have to compute f_α and P_β.

The main difficulty to construct $\{f_\alpha\}$ and $\{P_\beta\}$ is that though X_N is naturally a plane curve, it is very highly singular. First we compute its explicit plane equation, i.e., the polynomial $\Phi_N(x,j)$ (modulo p) for $N=11\ell$,

$$\Phi_N(x,j) = x^{12(\ell+1)} + j^{12(\ell+1)} + \sum_{m,n=0}^{12\ell+11} c_{mn} x^m j^n,$$

with coefficients $c_{ab} = c_{ba} \in \mathbb{F}_p$.

This can be done using the (polynomial-length) q-expansions of

$$j(q) = \frac{\left(1+240\sum\limits_{n=1}^{\infty}\sigma_3(n)q^n\right)^3}{q\prod\limits_{n=1}^{\infty}(1-q^n)^{24}} = \frac{1}{q} + 744 + 196\,844q + \ldots$$

and of

$$j(\alpha(z)) = \zeta^{-ba} t^{-a^2}\left(1+\sum_{m=1}^{\infty}\zeta^{mab} c_m t^{ma^2}\right), \quad z \in H,$$

for

$$\alpha = \begin{pmatrix} a & b \\ c & d \end{pmatrix} \in \mathrm{SL}_2(\mathbb{Z}), \qquad t = q^{1/11\ell}, \qquad \zeta = e^{2\pi i/11\ell}.$$

Then, using Theorem 5.1.28, one can explicitly compute $\Phi_N(x,j)$ with polynomial complexity; the computations are performed in the ring $\mathbb{Z}[\zeta_N]/\mathfrak{p}$, where $\mathfrak{p}$ is a divisor of p, but the resulting equation has its coefficients in $\mathbb{F}_p$.

The polynomial $\Phi_N(x,j)$ gives a highly singular model $\widetilde{X}_N$ of X_N, and to proceed we need to compute the coordinates of singular points of the plane model of $X_0(N)/p$ given by $\Phi_N(x,j)$. This can be done by computing the resultants

$$R_1(x) = \mathrm{Res}_j(\Phi_N, \partial\Phi_N/\partial x) \in \mathbb{F}_p[x], \qquad R_2(x) = \mathrm{Res}_j(\Phi_N, \partial\Phi_N/\partial j),$$

and their

$$\gcd(R_1(x), R_2(x)) = R(x).$$

The roots of $R(x)$ give the x-coordinates of singular points, which allows one to compute their y-coordinates in polynomial time.

The crucial point to construct $\{f_\alpha\}$ is to note that $L(D)$ can be realized by functions of the form $\dfrac{P(x,j)}{Q_0(x,j)}$, where $P(x,j)$ and $Q_0(x,j)$ are polynomials of degree $O(\ell t)$ and Q_0 is fixed for a fixed D. In this representation $L(D)$ becomes the common kernel of a system of linear functions on the space

$$V_N := \left\{ \frac{P(x,j)}{Q_0(x,j)} \;\middle|\; \deg P \le a(N) \right\},$$

where $a(N)$ is a constant, $a(N) \le O(\ell t)$. These functions depend on an explicit normalization of $\widetilde{X}_N$, which can be constructed polynomially using a sequence of standard blowing-ups; this normalization construction represents the points on X_N lying over a point $P=(a,b)\in\widetilde{X}_N$ as leaves of a certain normalization tree Γ_P describing the sequence of blowing-ups necessary to desingularize P. Note however that in general singular points are defined neither over $\mathbb{F}_p$ nor even over $\mathbb{F}_{p^2}$, so we need to consider some extension of $\mathbb{F}_p$, fortunately of a degree controlled polynomially by N.

Having constructed the whole forest $\bigcup_{P\in \mathrm{Sing}(\widetilde{X}_N)} \Gamma_P$ for all singular points $P \in \mathrm{Sing}(\widetilde{X}_N)$, one can write down the required linear functions on the space with elements $\dfrac{P(x,j)}{Q_0(x,j)}$ and thus construct $\{f_\alpha\}$ within this space.

The construction of $\mathcal{P} = \{P_\beta\}$ contains two steps: first, one constructs the image $\widetilde{\mathcal{P}}$ of $\mathcal{P}$ on $\widetilde{X}_N$ and then the set of leaves in the forest $\bigcup_{P\in\widetilde{\mathcal{P}}\cap \mathrm{Sing}(\widetilde{X}_N)} \Gamma_P$. Then one gets $\mathcal{P}$ as

$$\mathcal{P} = \mathrm{leaves}\Bigg(\bigcup_{P\in\widetilde{\mathcal{P}}\cap \mathrm{Sing}(\widetilde{X}_N)} \Gamma_P \Bigg) \cup \widetilde{\mathcal{P}} \setminus \mathrm{Sing}(\widetilde{X}_N).$$

To construct $\widetilde{\mathcal{P}}$ we note that the condition $P = (a,b) \in \widetilde{\mathcal{P}}$ is equivalent to the equation

$$\sum_{i=0}^{(p-1)/2} \binom{(p-1)/2}{i}^2 \lambda^i = 0,$$

which gives the supersingular values of λ and then the supersingular values of j by

$$a = j(P) = \frac{2^8(\lambda^2-\lambda+1)^3}{(1-\lambda)^2\lambda^2} \in \mathbb{F}_{p^2};$$

see the proof of Theorem 2.4.7. The second coordinate $b = j_N(P)$ is computed from the modular equation; note that it is automatically an element of $\mathbb{F}_{p^2}$.

This description of $\{f_\alpha\}$ and $\{P_\beta\}$ allows one to compute the matrix $\big(f_\alpha(P_\beta)\big)$. By thoroughly estimating all the above operations, one can obtain a polynomial upper bound on the complexity.

5.2. Drinfeld Modular Curves

Codes on reductions of classical modular curves yield good codes over $\mathbb{F}_{p^2}$. To obtain codes with similar properties over fields $\mathbb{F}_{q^2}$, where q is an arbitrary power of a prime p, we can consider Drinfeld modular curves. The latter have a number of advantages as compared to classical modular curves, since they are in many aspects simpler. However, this does not concern all aspects of modular curves; in some cases classical modular curves are more convenient, since they are reductions of curves in characteristic zero, and one can apply complex-analytical techniques.

In this section we give basic information on Drinfeld curves and codes on them. In Sec. 5.2.1 we discuss some properties of elliptic modules, which play the role of elliptic curves in the Drinfeld theory. Section 5.2.2 is devoted to the construction and properties of Drinfeld modular curves. In Sec. 5.2.3 we consider codes arising on these curves and briefly explain why they are polynomially constructible.

Throughout this section, we use the following notation: $A = \mathbb{F}_q[C']$, where $C' = C \setminus \{\infty\}$ is an affine absolutely irreducible curve over $\mathbb{F}_q$ obtained from a complete curve C over $\mathbb{F}_q$ by deleting an $\mathbb{F}_q$-rational point ∞. In fact, we are only interested in the case $C = \mathbb{P}^1$, $C' = \mathbb{A}^1$, $A = \mathbb{F}_q[T]$, and the reader would miss almost nothing if he restricts himself to this case. For $a \in A$ we use $|a| = |a|_\infty$ to denote the normalized absolute value corresponding to the point $\infty \in C$; thus, for $A = \mathbb{F}_q[T]$, we have $|a| = q^{-\deg a}$. By $\Bbbk$ we denote the fraction field of A, $\Bbbk = \mathbb{F}_q(C)$; the extension of the absolute value $|\cdot|$ to $\Bbbk$ is denoted by the same symbol: $\left|\frac{a}{b}\right| = \frac{|a|}{|b}$ for $a, b \in A$, $b \neq 0$. By $\Bbbk_\infty$ we denote the completion of $\Bbbk$ with respect to this absolute value; $\overline{\Bbbk}_\infty$ is the algebraic closure of $\Bbbk_\infty$. When reading this section, one should keep in mind the following list of analogies between objects of the classical theory and those of the Drinfeld theory:

Ring $\mathbb{Z}$	Ring A
Field $\mathbb{Q}$	Field $\Bbbk$
Absolute value $\lvert\cdot\rvert$	Absolute value $\lvert\cdot\rvert_\infty$
Field $\mathbb{R}$	Field $\Bbbk_\infty$
Field $\mathbb{C}$	Field $\overline{\Bbbk}_\infty$
Elliptic curve E	Elliptic module φ
Group $\mathrm{Hom}(E, E')$	A-module $\mathrm{Hom}(\varphi, \psi)$
Ring $\mathrm{End}_{\mathbb{Q}}(E)$	Ring $\mathrm{End}_{\Bbbk}(\varphi)$
Algebra $\mathrm{End}_{\mathbb{Q}}(E) \otimes \mathbb{Q}$	Algebra $\mathrm{End}_{\Bbbk}(\varphi) \otimes_A \Bbbk$
Lattice $\Lambda \in \mathbb{C}$	Lattice $\Gamma \subset \overline{\Bbbk}_\infty$
Positive integer N ("structure level")	Ideal $I \subset A$ ("structure level")
Group $\mathrm{PSL}_2(\mathbb{Z})$	Group $\mathrm{GL}_2(A)$
Complex modular curve $X(N)$	Drinfeld curve $\overline{M(I)} \otimes \overline{\Bbbk}_\infty$
Field $\mathbb{Q}(\zeta_N)$, $\zeta_N = e^{2\pi i}/N$	Field $\Bbbk(I)$
Groups $\Gamma_0(N)$ and $\Gamma_1(N)$	Groups $\Gamma_0(I)$ and $\Gamma_1(I)$
Reductions of modular curves $X_0(N)/p$ and $X_p(N)/\mathfrak{p}$	Curves $X_0(I)$ and $C(I)$ over $\mathbb{F}_q$
Supersingular elliptic curve E in characteristic p	"Supersingular" elliptic module φ of "characteristic" (T).

5.2.1. Elliptic Modules

Elliptic modules over a field. Let $\mathbb{K}$ be a field of characteristic p which is an A-algebra, and let $i\colon A \to \mathbb{K}$ be the structure morphism; i.e., $i(a) = a \cdot 1$. The prime ideal $\operatorname{Ker} i$ in A is said to be the "*characteristic*" of $\mathbb{K}$ and is denoted by "char" $\mathbb{K}$. If $\operatorname{Ker} i = (0)$, then $\mathbb{K}$ is said to have the "*general characteristic*", and if $\operatorname{Ker} i \neq (0)$, then $\mathbb{K}$ is a field of a "*finite characteristic*".

By $\mathbb{K}\{\tau\}$ we denote the ring of noncommutative polynomials of τ, i.e., of expressions of the form $\sum\limits_{i=0}^{n} a_i\tau^i$, $a_i \in \mathbb{K}$, with multiplication defined by the rule $\tau a = a^p\tau$ for $a \in \mathbb{K}$. By D we denote the natural projection

$$D\colon \mathbb{K}\{\tau\} \to \mathbb{K},$$

$$D\colon \sum_{i=0}^{n} a_i\tau^i \mapsto a_0.$$

The degree map $\deg\colon \mathbb{K}\{\tau\} \to \mathbb{Z}$ is defined by

$$\deg\left(\sum_{i=0}^{n} a_i\tau^i\right) = p^n \quad \text{if} \quad a_n \neq 0.$$

An *elliptic A-module over* $\mathbb{K}$ is a ring homomorphism

$$\varphi\colon A \to \mathbb{K}\{\tau\}$$

such that $\varphi(A)$ is not contained in $\mathbb{K}$ and the composition $D \circ \varphi$ coincides with i.

PROPOSITION 5.2.1. (a) *The homomorphism φ is injective.*

(b) *There exists a positive integer d such that for all $a \in A$ we have*

$$\deg\varphi(a) = |a|^d.$$

PROOF. (a) If $\operatorname{Ker}\varphi \neq (0)$, then $\varphi(A)$ is a subfield in $\mathbb{K}\{\tau\}$, and hence $\varphi(A) \subseteq \mathbb{K}$, which contradicts the definition of an elliptic module.

(b) Consider the map $a \mapsto \deg\varphi(a)$ from A to $\mathbb{Z}$. It is easy to check that this map can be extended to an absolute value on $\Bbbk$. Since for $a \in A$, $a \neq 0$, we have $\deg\varphi(a) \geq 1$, this absolute value cannot correspond to any prime ideal in A. Hence, $\deg\varphi(a) = |a|^d$ for some positive rational d; below we will see (Corollary 5.2.6) that d is in fact an integer. □

The positive integer d defined in Proposition 5.2.1 is called the *rank* of the elliptic A-module φ. Below we are interested in the case of $d = 2$ and $A = \mathbb{F}_q[T]$ only; thus, an elliptic A-module over $\mathbb{K}$ is specified by the element $\varphi(T) \in A$.

PROPOSITION 5.2.2. *For $d = 2$ we have*

$$\varphi(T) = i(T) + c_1F + c_2F^2, \quad c_i \in \mathbb{K}, \quad c_2 \neq 0,$$

where $F = \tau^m$ denotes the Frobenius element over $\mathbb{F}_q$, $q = p^m$.

SKETCH OF THE PROOF. This follows from commutativity of the ring $\varphi(A)$. □

Homomorphisms. A *homomorphism* from an elliptic A-module $\varphi\colon A \to \mathbb{K}\{\tau\}$ to an elliptic A-module $\psi\colon A \to \mathbb{K}\{\tau\}$ is an element $u \in \mathbb{K}\{\tau\}$ such that for all $a \in A$ we have

$$u\varphi(A) = \psi(A)u.$$

The set of homomorphisms from φ to ψ is denoted by $\mathrm{Hom}(\varphi,\psi)$. It has a natural structure of an A-module; its nonzero elements are called *isogenies*, and if $\mathrm{Hom}(\varphi,\psi)\neq(0)$, the modules φ and ψ are said to be *isogenous*.

LEMMA 5.2.3. *Isogenous modules are of the same rank.* □

The notion of isogeny makes it possible to define isomorphic elliptic modules. An element $u\in\mathbb{K}\{\tau\}$ defines an isomorphism of elliptic A-modules φ and ψ over $\mathbb{K}$ if $u\in\mathbb{K}^*$, and for any $a\in A$ we have $\psi(a)=u\varphi(a)u^{-1}$.

If $d=2$ and $A=\mathbb{F}_q[T]$, this notion becomes very explicit. Namely, one has the following result, which is obtained by simple direct computations.

LEMMA 5.2.4. (a) *Let*

$$\varphi(T)=i(T)+c_1F+c_2F^2,$$
$$\psi(T)=i(T)+d_1F+d_2F^2,$$

where φ and ψ are two elliptic A-modules over $\mathbb{K}$. Then φ is isomorphic to ψ if and only if for some $\lambda\in\mathbb{K}^*$ we have $d_1=\lambda^{q-1}c_1$ and $d_2=\lambda^{q^2-1}c_2$.

(b) *We have*

$$\mathrm{Aut}_{\mathbb{K}}(\varphi)=\begin{cases}\mathbb{F}_q^* & \text{if } c_1\neq 0,\\ \mathbb{F}_{q^2}^*\cap\mathbb{K} & \text{if } c_1=0.\end{cases}$$ □

Note that $j(\varphi)=c_1^{q+1}/c_2$ depends only on the isomorphism class of φ; it is an analogue of the classical j-invariant of an elliptic curve.

Let $\varphi\colon A\to\mathbb{K}\{\tau\}$ be an elliptic A-module, and let B be a $\mathbb{K}$-algebra. Define an action of A on B by the rule $a\cdot b=\varphi(a)b$; then we get an A-module $G_\varphi(B)$. Note that for $B=\mathbb{K}$ the A-module $G_\varphi(\mathbb{K})$ is different from the structure on $\mathbb{K}$ defined by i.

Points of finite order. Points of finite order of an elliptic A-module φ are A-torsion elements in the module $G_\varphi(\overline{\mathbb{K}})$, where $\overline{\mathbb{K}}$ is the algebraic closure of $\mathbb{K}$. The A-torsion submodule of $G_\varphi(\overline{\mathbb{K}})$ is denoted by $T(\varphi)$.

PROPOSITION 5.2.5. *There is an isomorphism*

$$T(\varphi)=\bigoplus_v(\Bbbk_v/A_v)^{d_v},$$

where the sum is over all maximal ideals v in A, A_v is the completion of A defined by v, $\Bbbk_v$ is its fraction field, d_v are integers, $d_v=d$ for $v\neq$ "char" $\mathbb{K}$, and $d_v<d$ for $v=$ "char" $\mathbb{K}$. In particular, if $\mathbb{K}$ has the "general characteristic", then $T(\varphi)=(\Bbbk/A)^d$.

PROOF. Since the A-module $T(\varphi)$ is divisible and is an A-torsion module, it is of the form $\bigoplus_v(\Bbbk_v/A_v)^{d_v}$. Since the class number of A is finite, there exists n such that $v^n=(a)$ for some $a\in A$. Let

$$M=\{x\in G_\varphi(\overline{\mathbb{K}})\mid ax=0\}.$$

Then $M=(A/(a))^{d_v}$; hence, $|M|=|A/(a)|^{d_v}=|a|^{d_v}$. On the other hand, it is clear that $|M|\leq\deg\varphi(a)=|a|^d$, and the condition $|M|=|a|^d$ is equivalent to the fact that the constant term of $\varphi(a)$ is nonzero, i.e., $v\neq$ "char" $\mathbb{K}$. □

COROLLARY 5.2.6. *The number d is an integer.* □

By considering points of finite order, one can prove the following fact.

PROPOSITION 5.2.7. *Let φ and ψ be elliptic A-modules over a field $\mathbb{K}$. Then* $\mathrm{Hom}(\varphi,\psi)$ *is a projective module of rank* $\le d^2$, *and the map*

$$\mathrm{Hom}(\varphi,\psi)\otimes_A A_v \to \mathrm{Hom}(T(\varphi),T(\psi))$$

is injective for all $v \neq \infty$, $v \neq$ "char" $\mathbb{K}$, where $T_v(\varphi)$ is a submodule in $T(\varphi)$ consisting of elements that are annihilated by some power of the ideal m_v. □

COROLLARY 5.2.8. *The ring* $\mathrm{End}(\varphi)\otimes_A \Bbbk$ *is a skew field. If $\mathbb{K}$ has the "general characteristic", then* $\mathrm{End}(\varphi)\otimes_A \Bbbk$ *is a field of degree at most d over $\Bbbk$.* □

The A-module $\mathrm{Hom}(\varphi,\psi)$ closely resembles the group $\mathrm{Hom}(X,Y)$, where X and Y are elliptic curves. Propositions 5.2.5 and 5.2.7 and Corollary 5.2.8 show that for $d=2$ structures of points of finite order and of the endomorphism ring of an elliptic module are completely analogous to those in the case of elliptic curves.

Lattices. As in the case of elliptic curves, which can be described in terms of lattices, there is a similar description for elliptic A-modules over a finite extension $\mathbb{L}$ of the field $\Bbbk_\infty$. The field $\mathbb{L}$ is complete with respect to the non-Archimedean absolute value which extends that on $\Bbbk_\infty$.

A *lattice* over $\mathbb{L}$ is a finitely generated discrete A-submodule Γ in the separable closure $\mathbb{L}^s$ of $\mathbb{L}$ which is invariant under the action of the Galois group $\mathrm{Gal}(\mathbb{L}^s/\mathbb{L})$. Such a module is projective over A; the *rank* of Γ is the rank of the corresponding projective module. A *morphism* from a lattice Γ_1 to a lattice Γ_2 is an element $\alpha \in \mathbb{L}$ such that $\alpha\Gamma_1 \subseteq \Gamma_2$. A composition of morphisms is by definition the multiplication in $\mathbb{L}$.

PROPOSITION 5.2.9. *The category of elliptic A-modules of rank d over $\mathbb{L}$ is equivalent to the category of lattices of rank d over $\mathbb{L}$.* □

Therefore, to any lattice Γ there corresponds an elliptic module φ_Γ, and to each morphism of lattices $\alpha\colon \Gamma_1 \to \Gamma_2$, a morphism of elliptic modules $u_\alpha\colon \varphi_{\Gamma_1} \to \varphi_{\Gamma_2}$; moreover, any elliptic A-module ψ is of the form φ_Γ for a suitable lattice Γ, and the map $\mathrm{Hom}_{\mathbb{L}}(\Gamma_1,\Gamma_2) \to \mathrm{Hom}_{\mathbb{L}}(\varphi_{\Gamma_1},\varphi_{\Gamma_2})$ is bijective.

We do not prove the proposition but only show how to construct an elliptic A-module φ_Γ from a lattice Γ.

Since Γ is discrete, the product

$$f_\Gamma(z) = z\cdot \prod_{\gamma\in\Gamma\setminus\{0\}} (1 - z\gamma^{-1})$$

converges and defines an *integral function* $f_\Gamma\colon \overline{\mathbb{L}} \to \overline{\mathbb{L}}$, i.e., a function of the form $\sum_{i=0}^{\infty} a_i z^i$ with $a_i \in \mathbb{L}$ and $z \in \overline{\mathbb{L}}$. Two main properties of f_Γ are as follows.

LEMMA 5.2.10. (a) *f_Γ is an additive function, i.e.,* $f_\Gamma(z+w) = f_\Gamma(z) + f_\Gamma(w)$.

(b) *Let $G_f(\overline{\mathbb{L}})$ be the A-module defined by the restriction of $f_\Gamma\colon \overline{\mathbb{L}} \to \overline{\mathbb{L}}$ to A. Then there exists a unique up to isomorphism elliptic A-module $\varphi = \varphi_\Gamma$ over $\mathbb{L}$ such that $G_f(\overline{\mathbb{L}})$ is isomorphic to $G_\varphi(\overline{\mathbb{L}})$ as an A-module.* □

5.2.2. Drinfeld Curves

Now we proceed to the construction and basic properties of Drinfeld modular curves. Here we continue to assume that $A = \mathbb{F}_q[T]$, $\Bbbk = \mathbb{F}_q(T)$, and $d = 2$.

Drinfeld curves over $\overline{\mathbb{k}}_\infty$. We start with curves over $\overline{\mathbb{k}}_\infty$, exactly as for the classical modular curves. A non-Archimedean analogue of the classical analytic theory, the theory of "rigid analytic spaces", establishes the existence of Drinfeld modular curves over $\mathbb{k}_\infty$. Complex modular curves appear as quotients $\Gamma\backslash H$. For Drinfeld modular curves, one should consider the "algebraist's upper half-plane"

$$\mathcal{H} := \overline{\mathbb{k}}_\infty \setminus \mathbb{k}_\infty$$

together with the natural action of the group $\mathrm{GL}_2(\mathbb{k}_\infty)$ on $\mathcal{H}$:

$$\begin{pmatrix} a & b \\ c & d \end{pmatrix} : z \mapsto \frac{az+b}{cz+d}, \qquad z \in \mathcal{H}.$$

For a discrete subgroup $\Gamma \subset \mathrm{GL}_2(\mathbb{k}_\infty)$, the quotient set $\Gamma\backslash\mathcal{H}$ possesses a natural structure of an algebraic curve over $\overline{\mathbb{k}}_\infty$. We are interested only in the case of $\Gamma \subset \mathrm{GL}_2(A)$ where Γ is a *congruence subgroup* of $\mathrm{GL}_2(A)$, which means that $\Gamma(I) \subset \Gamma$ for a nontrivial ideal $(0) \neq I \subset A$ and

$$\Gamma(I) = \left\{ \begin{pmatrix} a & b \\ c & d \end{pmatrix} \in \mathrm{GL}_2(A) \;\middle|\; \begin{pmatrix} a & b \\ c & d \end{pmatrix} \equiv \begin{pmatrix} 1 & 0 \\ 0 & 1 \end{pmatrix} \pmod{I} \right\} \subset \mathrm{GL}_2(A)$$

is the principal congruence subgroup of $\mathrm{GL}_2(A)$; thus, $\mathrm{GL}_2(A) = \Gamma(1)$.

In particular, the quotient set $\mathrm{GL}_2(A)\backslash\mathcal{H}$ is in a natural bijection with the affine line $\mathbb{A}^1$ over $\overline{\mathbb{k}}_\infty$, i.e., with $\overline{\mathbb{k}}_\infty$ itself. This bijection can be described as follows: any $\lambda \in \mathcal{H}$ gives a lattice $A + \lambda \cdot A$ corresponding (by Proposition 5.2.9) to a Drinfeld module φ_λ with

$$\varphi_\lambda(T) = T + c_1\tau + c_2\tau^2, \qquad j(\varphi_\lambda) = c_1^{q+1}/c_2;$$

then $\lambda \mapsto j(\varphi_\lambda)$ gives the desired bijection. This result can be expressed as a canonical isomorphism $Y((1)) \cong \mathbb{A}^1$ for the trivial ideal $I = (1) = A$. To construct a tower of Drinfeld modular curves, we need to use a *level structure*. Let $(0) \neq I \subset A$ be a proper ideal; let p_I be a monic polynomial generating this ideal, $I = p_I A$; and let $G = (\mathbb{L}, \varphi)$ be an elliptic module over a field $\mathbb{L}$ which we assume to be algebraically closed.

A *structure of level* I on the module φ is an isomorphism of A-modules

$$\lambda\colon (I^{-1}/A)^2 \to M_I,$$

where

$$M_I = \{x \in G_\varphi(\mathbb{L}) \mid p_I x = 0\}.$$

The quotient set $\Gamma(I)\backslash\mathcal{H}$ is in a natural bijection with the set

$$Y(I)(\overline{\mathbb{k}}_\infty) = \left\{ \begin{array}{l} \text{the set of isomorphism classes of pairs} \\ (\varphi, \lambda), \text{ where } \varphi \text{ is an elliptic } A\text{-module} \\ \text{over } \overline{\mathbb{k}}_\infty \text{ and } \lambda \text{ is a structure of level } I \end{array} \right\},$$

where $Y(I)$ is an affine algebraic curve corresponding to $\Gamma(I)\backslash\mathcal{H}$. This curve is equipped by a natural action of the group $\mathrm{GL}_2(A/I)$, since this group acts on the set $Y(I)(\overline{\mathbb{k}}_\infty)$ as follows: if φ is an elliptic module over $\overline{\mathbb{k}}_\infty$,

$$\lambda\colon (I^{-1}/A)^2 \xrightarrow{\sim} M_I$$

is a structure of level I, and $h \in \mathrm{GL}_2(A/I)$, then the action of h is given by

$$h\colon (\varphi, \lambda) \mapsto (\varphi, \lambda'),$$

where λ' is defined by the composition

$$\lambda' \colon (I^{-1}/A)^2 \xrightarrow{h^{-1}} (I^{-1}/A)^2 \xrightarrow{\lambda} M_I.$$

Note that here the subgroup $\mathbb{F}_q^* \subset \mathrm{GL}_2(A/I)$ acts on $Y(I)$ trivially, and thus in fact we have the action of $\mathrm{GL}_2(A/I)/\mathbb{F}_q^*$.

The affine curve has a natural compactification, i.e., a smooth projective curve $X(I) \supset Y(I)$. The finite set $\mathrm{Cusps}(I) := X(I) - Y(I)$ has the following description.

PROPOSITION 5.2.11. *The finite set* Cusps(I) *is canonically bijective to the quotient set* $\Gamma(I)\setminus\mathbb{P}^1(\Bbbk)$ *under the natural action of* $\Gamma(I)$ *on* $\mathbb{P}^1(\Bbbk)$:

$$\begin{pmatrix} a & b \\ c & d \end{pmatrix} : z \mapsto \frac{az+b}{cz+d}, \quad z \in \mathbb{P}^1(\Bbbk). \qquad \square$$

The theory of rigid analytic spaces allows us to find the genus of $X(I)$.

PROPOSITION 5.2.12. *The genus $g(I)$ of the curve $X(I)$ is*

$$1 + (q^2-1)^{-1} q^{2m} (q^m - q - 1) \prod_{r \mid p_I} (1 - q^{-2\deg r}),$$

where $m = \deg p_I$ and the product is over the set of irreducible monic polynomials r that divide p_I. $\square$

The curve $X(I)$ is defined over the field $\Bbbk(I)$, which is the maximal abelian extension of $\Bbbk$ of conductor I in which the point ∞ splits completely (see Sec. 6.3 below). This means that the ring $\Bbbk_\infty \otimes_k k(I)$ is isomorphic to a direct sum of several copies of $\Bbbk_\infty$ (the splitting condition), and the prime ideals that are completely splitting in $\Bbbk(I)$ are precisely prime ideals $\mathfrak{p}$ of the form (α), where $\alpha \in A$, $\alpha \equiv 1 \pmod{I}$. (The condition of complete splitting of $\mathfrak{p}$ means that the ring $\Bbbk(I) \otimes_\Bbbk \Bbbk_\mathfrak{p}$ is isomorphic to the direct sum of the fields $\Bbbk_\mathfrak{p}$, where $\Bbbk_\mathfrak{p}$ is the completion of $\Bbbk$ by the norm defined by the ideal $\mathfrak{p}$.) According to the class field theory, $\Bbbk(I)$ exists, it is unique, and there is a canonical isomorphism (the Artin symbol) of $(A(I))^*/\mathbb{F}_q^*$ onto the group $\mathrm{Gal}(\Bbbk(I)/\Bbbk)$; see Sec. 6.3. Therefore, the function field $\overline{\Bbbk}_\infty(X(I))$ coincides with the compositum $\overline{\Bbbk}_\infty \cdot \Bbbk(I)(M(I))$ for an absolutely irreducible smooth projective curve $M(I)$ defined over $\Bbbk(I)$, i.e., $\overline{\Bbbk}_\infty \cap \Bbbk(I)(M(I)) = \Bbbk(I)$. The curve $M(I)$ has a good reduction modulo any maximal ideal $P \subset B(I)$, where $B(I)$ is the integral closure of $A[I^{-1}]$ in the field $\Bbbk(I)$.

The action of $\mathrm{GL}_2(A/I)$ on the curve $X(I)$ extends to its action on the curve $M(I)$ and thus on $\Bbbk(I)$, and we have the following.

PROPOSITION 5.2.13. *For any $g \in \mathrm{GL}_2(A/I)$ the restriction to $\Bbbk(I)$ of the action of g on $\Bbbk(I)(M(I))$ is given by the image $\overline{\det g}$ of $\det g \in (A/I)^*$ in the group $(A/I)^*/\mathbb{F}_q^* = \mathrm{Gal}(\Bbbk(I)/\Bbbk)$.* $\square$

Modular curves over $\mathbb{F}_q$. In what follows we assume the ideal $I = (p_I)$ to be prime, so that the polynomial p_I is irreducible. We also assume the degree m of p_I to be odd and coprime with $q-1$. We are interested in modular curves over $\mathbb{F}_q$, i.e., curves obtained from the smooth complete (reducible) curve $X(I) = \overline{M(I)} \otimes_A \mathbb{F}_q$ over $\mathbb{F}_q$. Hereafter, expressions of the form $\otimes_A \mathbb{F}_q$ assume that the A-algebra structure on $\mathbb{F}_q$ is determined by the projection $\nu\colon A \to A/(T) = \mathbb{F}_q$. Thus, all elliptic A-modules considered below are of "characteristic" (T); Propositions 5.2.12 and 5.2.13 remain valid for the curve $X(I)$.

Consider the following subgroups of $\mathrm{GL}_2(A)$, which are analogous to the subgroups of the classical modular group $\mathrm{PSL}_2(\mathbb{Z})$ introduced above:

$$\Gamma_0(I) = \left\{ \begin{pmatrix} a & b \\ c & d \end{pmatrix} \in \mathrm{GL}_2(A) \mid c \in I \right\},$$
$$\Gamma_1(I) = \left\{ \begin{pmatrix} a & b \\ c & d \end{pmatrix} \in \mathrm{GL}_2(A) \mid c \in I,\ a-1 \in I \right\},$$
$$H(I) = \left\{ \begin{pmatrix} a & b \\ c & d \end{pmatrix} \in \mathrm{GL}_2(A) \mid b \in I,\ c \in I,\ a-d \in I \right\}.$$

Let $\overline{\Gamma_0(I)}$, $\overline{\Gamma_1(I)}$, and $\overline{H(I)}$ be, respectively, images of these subgroups in $\mathrm{GL}_2(A/I)$. Thus, $\overline{\Gamma_0(I)}$ is the subgroup of upper triangular matrices, i.e., those of the form $\begin{pmatrix} * & * \\ 0 & * \end{pmatrix}$, $\overline{\Gamma_1(I)}$ is the subgroup of matrices of the form $\begin{pmatrix} 1 & * \\ 0 & * \end{pmatrix}$, and $\overline{H(I)}$ coincides with the center of the group $\mathrm{GL}_2(A/I)$, which consists of matrices of the form $\begin{pmatrix} r & 0 \\ 0 & r \end{pmatrix}$, $r \in (A/I)^*$.

We define smooth complete curves $X_0(I)$, $X_1(I)$, and $C(I)$ as follows:

$$X_0(I) = X(I)/\overline{\Gamma_0(I)},$$
$$X_1(I) = X(I)/\overline{\Gamma_1(I)},$$
$$C(I) = X(I)/\overline{H(I)}.$$

The curves $X_0(I)$ and $X_1(I)$ are defined over $\mathbb{F}_q$; the curve $C(I)$ is defined over $\mathbb{F}_{q^2}$.

In our assumptions on the ideal I, we have the following.

PROPOSITION 5.2.14. (a) *The curves $X_0(I)$, $X_1(I)$, and $C(I)$ are absolutely irreducible.*

(b) *The curve $C(I)$ is a form of any irreducible component of $X(I)$.* □

The embeddings $\overline{H(I)} \subset \overline{\Gamma_0(I)}$ and $\overline{\Gamma_1(I)} \subset \overline{\Gamma_0(I)}$ define natural projections $C(I) \to X_0(I)$ and $X_1(I) \to X_0(I)$.

Note that there exists a canonical isomorphism $\overline{\Gamma_0(I)}/\overline{\Gamma_1(I)} \cdot \mathbb{F}_q^* = (A/I)^*/\mathbb{F}_q^*$.

PROPOSITION 5.2.15. *The action of the group $\overline{\Gamma_0(I)}/\overline{\Gamma_1(I)} \cdot \mathbb{F}_q^* = (A/I)^*/\mathbb{F}_q^*$ on the curve $X_1(I)$ is free.*

PROOF. This follows from Lemma 5.2.4(b), since our assumptions on I imply that the group $(A/I)^*/\mathbb{F}_q^*$ does not contain $\mathbb{F}_{q^2}^*/\mathbb{F}_q^*$. Indeed, since m is odd, we have

$$|(A/I)^*/\mathbb{F}_q^*| = \frac{q^m-1}{q-1} \equiv 1 \pmod{(q+1)}.$$ □

Since the curve $X_0(I)$ is identified with the factor of $X_1(I)$ modulo the action of $(A/I)^*/\mathbb{F}_q^*$, we obtain the following fact.

COROLLARY 5.2.16. *The morphism $X_1(I) \to X_0(I)$ is unramified of degree $(q^m-1)/(q-1)$.* □

From Proposition 5.2.11 we obtain a description of the set $\mathrm{Cusps}(I)$, the image of $\overline{M(I)}/M(I)$ in $X(I)$, and hence the following for the sets $\mathrm{Cusps}_0(I)$ and $\mathrm{Cusps}_1(I)$, i.e., the images of $\overline{M(I)}/M(I)$ in $X_0(I)$ and $X_1(I)$, respectively.

PROPOSITION 5.2.17. (a) *The set* $\mathrm{Cusps}_0(I)$ *consists of two points.*
(b) *The set* $\mathrm{Cusps}_1(I)$ *consists of* $2(q^m-1)/(q-1)$ *points.* □

A more detailed analysis using the action of Galois groups gives the following.

PROPOSITION 5.2.18. (a) *Any point of* $\mathrm{Cusps}_0(I)$ *is* $\mathbb{F}_q$*-rational.*
(b) *There exists a unique point of* $\mathrm{Cusps}_0(I)$ *such that all points of* $\mathrm{Cusps}_1(I)$ *lying over it are* $\mathbb{F}_q$*-rational.* □

In what follows we denote the point of $\mathrm{Cusps}_0(I)$ defined in Proposition 5.2.18(b) by Q_1 and use Q_2 for the other point of $\mathrm{Cusps}_0(I)$; note that no point of $\mathrm{Cusps}_1(I)$ lying over Q_2 is $\mathbb{F}_q$-rational.

Modular equation. The classical modular equation defines a plain (singular) model of the curve $X_0(N)$. In the case of Drinfeld curves, the curve $X_1(I)$ has a convenient plane model.

To write out the corresponding equation explicitly, one has to use an interpretation of $X_1(I)$ as a moduli variety over $\mathbb{F}_q$. More precisely, let $Y_1(I) = X_1(I)\setminus \mathrm{Cusps}_1(I)$ be the corresponding affine curve.

PROPOSITION 5.2.19. *For any extension L of $\mathbb{F}_q$ there exists a canonical bijection of sets*

$$Y_1(I)(L) \cong \left\{\begin{array}{l}\textit{the set of isomorphism classes of pairs}\\ (\varphi,c),\ \textit{where } \varphi \textit{ is an elliptic } A\textit{-module}\\ \textit{over } L \textit{ and } c\neq 0 \textit{ is a point of order } I\\ \textit{on } \varphi,\ \textit{i.e., } \varphi(p_I)(c)=0\end{array}\right\}. \qquad \square$$

This allows us to give an equation of a plane model of $X_1(I)$. Let a_k be coefficients of the polynomial p_I:

$$p_I(T) = \sum_{k=0}^{m} a_k T^k, \quad a_k \in \mathbb{F}_q, \quad a_m = 1.$$

Define the polynomials $P_k \in \mathbb{F}_q[x,y]$ by induction: let

$$P_0(x,y) = 1, \qquad P_{k+1}(x,y) = xP_k(x,y)^{q^2} + yP_k(x,y)^q \quad \text{for} \quad k \geq 0.$$

The following result can be deduced by induction on k.

LEMMA 5.2.20. *If $\varphi\colon A \to \overline{\mathbb{F}}_q\{\tau\}$ is an elliptic A-module over $\overline{\mathbb{F}}_q$ of characteristic (T),*

$$\varphi(T) = aF^2 + bF, \quad a,b \in \overline{\mathbb{F}}_q,$$

then the action of $\varphi(T^k)(1)$ on the element $1 \in \overline{\mathbb{F}}_q$ is given by

$$\varphi(T^k)(1) = P_k(a,b). \qquad \square$$

Let

$$F(x,y) = F_I(x,y) = \sum_{k=0}^{m} a_k P_k(x,y).$$

The equation $F_I(x,y)=0$ is an analogue of the modular equation for $X_1(I)$ rather than for $X_0(I)$. To state the corresponding result more precisely, we use the following notion. If $I=(p_I)$ is a prime ideal in A, we say that an ideal I' is *conjugate* to I (over the prime field $\mathbb{F}_p$) if I' is generated by a polynomial

$$p'_I = \sum_{k=0}^{m} a'_k T^k,$$

where $a'_k = \sigma a_k$ for some element $\sigma \in \mathrm{Gal}(\mathbb{F}_q/\mathbb{F}_p)$. In particular, if $a_k \in \mathbb{F}_p$ for $k = 0, \ldots, m$, then $I = I'$, $p_I = p'_I$.

PROPOSITION 5.2.21. *Let $\widetilde{X}_1(I) \in \mathbb{P}^2$ be a plane curve with affine equation $F_I(x,y) = 0$. Then*

(a) *The curve $\widetilde{X}_1(I)$ has exactly two singular points, namely $P_1 = (0:0:1)$ and $P_2 = (0:-1:1)$.*

(b) *The curve $\widetilde{X}_1(I)$ is birationally isomorphic over $\mathbb{F}_q$ to $X_1(I')$, where I' is the ideal conjugate to I over $\mathbb{F}_p$.*

PROOF. (a) Let us check that $\widetilde{X}_1(I)$ has no singularities in the finite part of the plane; i.e., the system of equations

$$\begin{cases} F(x,y) = 0, \\ F'_x(x,y) = 0, \\ F'_y(x,y) = 0 \end{cases}$$

has no solutions. Indeed, it is easy to check the identity

$$F - (xF'_x + yF'_y) = a_0.$$

Since p_I is irreducible and $a_0 \neq 0$, the system is inconsistent. Let $(Z_0 : Z_1 : Z_2)$ be homogeneous coordinates on $\mathbb{P}^2$ such that $x = Z_1/Z_0$ and $y = Z_2/Z_0$. The homogeneous equation $G(Z_0 : Z_1 : Z_2) = 0$ of the curve $\widetilde{X}_1(I)$ can be written as follows:

$$\begin{aligned} G(Z_0 : Z_1 : Z_2) &= Z_0^{(q^{2m}-1)(q^2-1)} F(Z_1/Z_0, Z_2/Z_0) \\ &= Z_1^{(q^{2m-2}-1)(q^2-1)} (Z_1 + Z_2)^{q^{2m-2}} + Z_0 H(Z_0 : Z_1 : Z_2), \end{aligned}$$

for a form $H(Z_0 : Z_1 : Z_2)$ with $\deg H = ((q^{2m}-1)/(q^2-1)) - 1$. Hence, $\widetilde{X}_1(I) \cap \{Z_0 = 0\}$ consists of exactly two points $\{P_1, P_2\}$, and it is easy to check that both are singular.

(b) Let us construct a map $\widetilde{v}\colon X_1(I) \to \widetilde{X}_1(I)$. To this end, we use Proposition 5.2.19 and define this map on $Y_1(I)$. Any point of $Y_1(I) = \mathbb{F}_q$ is defined by a pair (φ, c), where φ is an elliptic A-module over $\mathbb{F}_q$, $\varphi(T) = aF^2 + bF$, and $c \neq 0$ is a point of order p_I on φ, $a, b, c \in \overline{\mathbb{F}}_q$. We map the point $Y_1(I)$ defined by (φ, c) to the point

$$(c^{q^2-1}a, c^{q-1}b) \in \mathbb{A}^2.$$

Since $\varphi(p_I)(c) = 0$, one can easily deduce using Lemma 5.2.20 that

$$F_I(c^{q^2-1}a, c^{q-1}b) = 0,$$

so this point lies on $\widetilde{X}_1(I)$. It is also easy to check that $\widetilde{\nu}$ is well defined. Indeed, note that a pair (φ, c) is isomorphic to a pair (φ', c'), where

$$\varphi(T) = aF^2 + bF, \qquad \varphi'(T) = a'F^2 + b'F,$$

if and only if

$$a' = \lambda^{q^2-1}, \qquad b' = \lambda^{q-1}b, \qquad c' = \lambda^{-1}c$$

for some $\lambda \in \overline{\mathbb{F}}_q$. This also shows that the map

$$\widetilde{\nu}\colon Y_1(I)(\mathbb{F}_q) \xrightarrow{\sim} \widetilde{X}_1(I) \cap (\mathbb{A}^2 \setminus \{x=0\})$$

is bijective. It is also clear that the map $\widetilde{\nu}$ is defined over $\mathbb{F}_q$. Since this map is bijective on $\overline{\mathbb{F}}_q$-points, it is a composition of a birational isomorphism ν and some power of the Frobenius endomorphism, which implies (b). □

Thus, the curve $\widetilde{X}_1(I)$ is birationally isomorphic to $X_1(I')$ for an ideal of the same type as I. Note that the birational isomorphism $\nu\colon X_1(I') \to \widetilde{X}_1(I)$ is the normalization map for the curve $\widetilde{X}_1(I)$.

REMARK 5.2.22. If $(m,a)=1$, where $a=\log_p q$, then there exists a prime ideal $I \in A$ generated by a polynomial p_I of degree m whose coefficients lie in a prime field $\mathbb{F}_p$, so that $I=I'$.

The image of the set $\mathrm{Cusps}_1(I)$ on $\widetilde{X}_1(I)$ is described as follows.

COROLLARY 5.2.23. *The image of the set of $\mathbb{F}_q$-rational points of* $\mathrm{Cusps}_1(I)$ *coincides with $\{P_1,P_2\}$. The set of points of* $\mathrm{Cusps}_1(I)$ *that are not $\mathbb{F}_q$-rational is bijectively mapped onto the intersection of $\widetilde{X}_1(I)$ with the y axis, i.e., onto $\widetilde{X}_1(I) \cap \{x=0\}$.*

PROOF. Since $Y(I)(\overline{\mathbb{F}}_q)$ is in a bijection with $\overline{\mathbb{F}}_q$-points of

$$\widetilde{X}_1(I) \cap (\mathbb{A}^2 \setminus \{x=0\}),$$

$\nu(\mathrm{Cusps}_1(I))$ coincides with $(\widetilde{X}_1(I) \cap \{x=0\}) \cup \{P_1,P_2\}$. The set of ordinates of points of $\widetilde{X}_1(I) \cap \{x=0\}$ is given by the equation

$$F(y) = F_I(0,y) = \sum_{k=0}^{m} a_k y^{(q^k-1)/(q-1)} = 0.$$

This set contains $(q^m-1)/(q-1)$ elements, and none of them lies in $\mathbb{F}_q$ since $p_I(T)$ is irreducible. Therefore, it coincides with the image of nonrational points of $\mathrm{Cusps}_1(I)$, and $\{P_1,P_2\}$ with the image of rational ones. □

Let us give an explicit description of the action of the group $H=(A/I)^*/\mathbb{F}_q^*$ on the curve $\widetilde{X}_1(I)$.

COROLLARY 5.2.24. *Let*

$$r(T) = \sum_{k=0}^{m-1} r_k T^k \in A$$

be a polynomial of degree at most $m-1$ whose image in $(A/I)^*/\mathbb{F}_q^*$ is a generator θ of this group. Then the action of θ on the point $(x,y) \in \widetilde{X}_1(I)$ is given by

$$\theta\colon (x,y) \mapsto (x\lambda^{q^2-1}, y\lambda^{q-1}),$$

where

$$\lambda = \sum_{k=0}^{m-1} r_k P_k(x,y).$$

PROOF. The action of θ on triples of the form $(a,b,1)$ is given by

$$\theta\colon (a,b,1) \mapsto (a,b,c),$$

where

$$c = \sum_{k=0}^{m-1} r_k P_k(a,b),$$

since $\varphi(T^k)(1) = P_k(a,b)$, $\varphi(T) = aF^2 + bF$; this implies the result. □

$\mathbb{F}_{q^2}$-rational points. Similarly to the curves $X_0(N)/p$ and $X_p(N)/\mathfrak{p}$, the curves $X_0(I)$ and $C(I)$ have many $\mathbb{F}_{q^2}$-rational points.

PROPOSITION 5.2.25. *Let $N(I)$ be the number of $\mathbb{F}_{q^2}$-rational points of the curve $C(I)$, let $g(I)$ be its genus, and let $m(I)$ be the degree of an irreducible polynomial p_I generating the ideal I. Then*

$$\lim_{m(I)\to\infty} \frac{N(I)}{g(I)} = q-1.$$

PROOF. By Theorem 4.1.30, it suffices to show that

$$\lim \frac{N(I)}{g(I)} \geq q-1.$$

From Proposition 5.2.12 we get

$$g(I) = 1 + (q^{2m}-1)(q^m - q - 1)/(q^2-1),$$

where $m = m(I)$. Consider the projection

$$C(I) \to C(I)/PGL(2, A/I) = \mathbb{P}^1$$

of degree $q^m(q^{2m}-1) = |\mathrm{PGL}_2(A/I)|$. The points of $C(I)$ corresponding to pairs of the form (φ_0, λ), where φ_0 is an elliptic A-module such that $\varphi_0(T) = F^2$ (it is natural to call it a *supersingular* module), are ramification points of this projection with indices $q+1$, as follows from the description of the groups $\mathrm{Aut}(\varphi)$ for elliptic A-modules φ. Hence, the number of such points is

$$(q+1)^{-1}|\mathrm{PGL}_2(A/I)| = q^m(q^{2m}-1)/(q+1).$$

It suffices to prove that these points are $\mathbb{F}_{q^2}$-rational, since in this case

$$\frac{N(I)}{g(I)} \geq \frac{q^m(q^{2m}-1)/(q+1)}{1+(q^{2m}-1)(q^m-q-1)/(q^2-1)} \geq q-1.$$

Consider a point of $\overline{M(I)} \otimes \mathbb{F}_q$ defined by a pair (φ_0, λ), where λ is a structure of level I on φ_0. To show that (φ_0, λ) defines an $\mathbb{F}_{q^2}$-rational point $C(I)$, it suffices to show that under the action of the generator F^2 of the group $\mathrm{Gal}(\overline{\mathbb{F}}_q)/\mathbb{F}_{q^2}$ the structure λ is mapped to a structure λ' such that $\lambda = \lambda' \cdot h$ for some $h \in H(I)$. To this end, we note that F^2 lies in the center of $\mathrm{End}(\varphi_0)$, since, as one can easily check, $\mathrm{End}(\varphi_0) = \mathbb{F}_{q^2}\{F\}$ and $\mathrm{Cent}(\mathrm{End}(\varphi_0)) = \mathbb{F}_q[F^2]$. Hence, the image of F^2 in $\mathrm{GL}_2(A/I)$ under the action on the set of structures of level I belongs to the center of $\mathrm{GL}_2(A/I)$, i.e., to $H(I)$. □

REMARK 5.2.26. The proof of Proposition 5.2.25 is completely analogous to that of Theorem 5.1.33, and there exists only one supersingular elliptic module φ_0.

REMARK 5.2.27. The module φ_0 has no nontrivial points of order p, which justifies the term *supersingular*.

REMARK 5.2.28. Similarly to Remark 5.1.34, for the number $N(I)$ we have

$$N(I) \geq (q-1)g(I) + c \cdot g(I)^{2/3} + o(g(I)^{2/3})$$

for some $c > 0$.

The existence of a surjection $C(I) \to X_0(I)$ yields the following fact.

COROLLARY 5.2.29. *Let $N_0(I)$ be the number of $\mathbb{F}_{q^2}$-rational points of $X_0(I)$, and let $g_0(I)$ be the genus of $X_0(I)$. Then*

$$\lim_{m(I)\to\infty} \frac{N_0(I)}{g_0(I)} = q-1. \qquad \square$$

To obtain more precise information on $\mathbb{F}_{q^2}$-rational points of the curve $X_0(I)$, one can consider the modular equation $F_I(x,y)$. The equation $F_I(x,y) = 0$ gives a model of the curve $X_1(I')$ for some ideal I' conjugate to I, so we have to consider the curve $X_0(I')$ instead of $X_0(I)$. However, we do not lose generality when considering $X_1(I)$, since $I \mapsto I'$ is a bijection on the set of an ideal with given $m(I)$.

"Supersingular" points of $X_0(I)$, i.e., images of points of the form (φ_0, λ) on $C(I)$, are pulled up to points with $y = 0$ on $\widetilde{X}_1(I)$ whose abscissae satisfy the equation

$$F(x) = F_I(x,0) = \sum_{k=0}^{m} a_k x^{(q^{2k}-1)/(q^2-1)} = 0.$$

PROPOSITION 5.2.30. *The number of "supersingular" $\mathbb{F}_{q^2}$-points of $X_0(I)$ is $n = (q^m+1)/(q+1)$.*

PROOF. The polynomial $F(x)$ is separable; hence, the number of its roots is $\deg F(x) = (q^{2m}-1)/(q^2-1)$. These roots correspond to the points of $X_1(I)$ lying over "supersingular" points of $X_0(I)$; the number of them is

$$(q^{2m}-1)(q^2-1)^{-1}|(A/I)^*/\mathbb{F}_q^*|^{-1} = (q^m+1)/(q+1). \qquad \square$$

The action of the group $H = (A/I)^*/\mathbb{F}_q^*$ makes it possible to describe the decomposition of $F(x)$ into factors irreducible over $\mathbb{F}_{q^2}$. Let H_1 be the subgroup in H generated by the image of the element T, let $h_1 = |H_1|$ be its order, and let $h_2 = [H : H_1]$ be its index in H, $h = h_1 \cdot h_2$.

PROPOSITION 5.2.31. *The polynomial $F(x)$ is factorized over $\mathbb{F}_{q^2}$ into nh_2 irreducible factors, each of degree h_1.*

PROOF. Note that the image θ_1 of T in H acts on points of $\widetilde{X}_1(I)$ in the following way:

$$\theta_1 : (x,y) \mapsto \left(x(x+y)^{q^2-1}, y(x+y)^{q-1}\right),$$

as follows from Corollary 5.2.24. Hence, on "supersingular" points of $\widetilde{X}_1(I)$ with $y = 0$, the action of θ_1 coincides with the action of the topological generator F^2 of $\mathrm{Gal}(\overline{\mathbb{F}}_q/\mathbb{F}_{q^2})$. Since the action of H_1 on the set of points of the form $(x,0)$ is free, the proposition follows. $\square$

COROLLARY 5.2.32. *Let R be the set of roots of $\mathbb{F}(x)$. Then*

(a) *Each orbit of the action of H on R is of the form*

$$H\cdot r=\bigcup_{i=1}^{h_2} M_i,$$

where $r\in R$ and M_i is the set of roots of a polynomial F_i of degree h_1 which is irreducible over $\mathbb{F}_{q^2}$ and divides $F(x)$.

(b) *The set R lies in the field $\mathbb{F}_{q^{2h_1}}$.* □

5.2.3. Codes

Now we pass to a description of codes obtained from Drinfeld curves.

Parameters. We consider codes over $\mathbb{F}_{q^2}$ on the curves $C(I)$ and $X_0(I)$; we still assume the ideal I to satisfy the above assumptions.

PROPOSITION 5.2.33. *Let $C=(C(I),\mathcal{P},\mathscr{L})_H$ be an algebraic geometry $[n,k,d]_{q^2}$ code constructed from the curve $C(I)$, the set $\mathcal{P}$ of its "supersingular" $\mathbb{F}_{q^2}$-rational points, and a line bundle $\mathscr{L}$ of degree a. Then for the parameters of C we have*

$$\begin{aligned}
&n=q^m(q^{2m}-1)/(q+1),\\
&k\geq a-(q^{2m}-1)(q^m-q-1)/(q^2-1),\\
&d\geq q^m(q^{2m}-1)/(q+1).
\end{aligned}$$ □

REMARK 5.2.34. The set of possible lengths of codes obtained from curves of the type $C(I)$ and sets of their "supersingular" points is very sparse in the set of positive integers: when passing from a value m to $m+2$ (m must be odd), the length of such a code becomes approximately q^6 times as large. Taking into account that such codes are of interest mainly for $q\geq 7$ (i.e., $q^2\geq 49$), we see that n grows very rapidly with m. The first nontrivial code over $\mathbb{F}_{49}$, which corresponds to $m=5$ (for $m=1$ the genus of $C(I)$ is 0, and $m=3$ is not coprime to $q-1=6$), is of length greater than 6 billion. Therefore, the codes constructed from the curves $C(I)$ are of no avail for practical use.

PROPOSITION 5.2.35. *Parameters of an $[n,k,d]_{q^2}$ code $C=(X_0(I),\mathcal{P},\mathscr{L})_H$ constructed from the curve $X_0(I)$, the set $\mathcal{P}$ of its "supersingular" points, and a line bundle $\mathscr{L}$ of degree a satisfy the relations*

$$\begin{aligned}
&n=(q^m+1)/(q+1),\\
&k\geq a-(q^{2\ell}-1)/(q^2-1)+1, \quad \textit{where } \ell=\lfloor m/2\rfloor,\\
&d\geq (q^m+1)/(q+1)-a.
\end{aligned}$$

PROOF. It suffices to note that $g_0(I)\leq q(q^{2\ell}-1)/(q^2-1)$, as follows by the Hurwitz formula, the degree of the projection $C(I)\to X_0(I)$ being $q^m(q^m-1)$. □

REMARK 5.2.36. Remark 5.2.34 is valid, though to a lesser extent, for codes on curves $X_0(I)$. When passing from m to $m+2$, the length of such codes becomes q^2 times as large. Codes of relatively moderate lengths can be obtained for rather small values of m only.

EXAMPLE 5.2.37. The value $m=3$ is admissible for $q=2 \pmod 3$. In this case, the q^2-ary codes on $X_0(I)$ have the following parameters:

$$\begin{aligned} n &= q^2-q+1, \\ k &\geq a-q+1, \\ d &\geq q^2-q+1-a. \end{aligned}$$

Thus, parameters of these codes are much worse than those of Reed–Solomon codes.

EXAMPLE 5.2.38. The value $m=5$ is admissible for $q \not\equiv 1 \pmod 5$. For a q^2-ary code on $X_0(I)$, we have

$$\begin{aligned} n &= q^4-q^3+q^2-q+1, \\ k &\geq a-q(q^2+1)+1, \\ d &\geq q^4-q^3+q^2-q+1-a. \end{aligned}$$

The parameters of these codes are already not as bad as for $m=3$. Anyway, for small values of q the code length n lies in a reasonable range. For instance, for $q=7$ we have $n=2101$ and $k+d \geq 1852$ (for a code over $\mathbb{F}_{49}$); for $q=8$, $n=3301$ and $k+d \geq 2782$ (for a code over $\mathbb{F}_{64}$).

Group properties. Similarly to codes on reductions of $X(N)$, codes on $C(I)$ can be interpreted in terms of ideals in a group algebra. Moreover, since the "supersingular" elliptic module φ_0 is unique, the situation for $C(I)$ is simpler than for $X(N)$.

Let $\overline{\mathrm{Cusps}(I)}$ be the image of the set $\mathrm{Cusps}(I)$ in the curve $C(I)$; i.e., $\overline{\mathrm{Cusps}(I)}$ is the set of cusps of $C(I)$. Alternatively, $\overline{\mathrm{Cusps}(I)}$ can be defined as the inverse image of the point $\infty \in \mathbb{P}^1$ under the natural projection $j\colon C(I) \to \mathbb{P}^1$. Consider the divisor $F_\infty = \sum\limits_{P \in \overline{\mathrm{Cusps}(I)}} P$. The above interpretation of the set $\overline{\mathrm{Cusps}(I)}$ shows that F_∞ is an $\mathbb{F}_q$-rational divisor.

PROPOSITION 5.2.39. *The degree of the divisor* F_∞ *is* $q^{2m}-1$.

PROOF. The degree of F_∞ equals the number of cusps on $C(I)$. This set can be identified with $\Gamma(I) \backslash \mathbb{P}^1(\Bbbk)$. Since $\mathbb{P}^1(\Bbbk)$ is identified with $\mathrm{PGL}_2(A)/B$, where B is the subgroup of matrices of the form $\begin{pmatrix} 1 & a \\ 0 & 1 \end{pmatrix}$, $a \in A$, we have

$$\overline{\mathrm{Cusps}(I)} = \Gamma(I) \backslash \mathrm{PGL}_2(A)/B.$$

Noting that

$$\Gamma(I) \backslash \mathrm{PGL}_2(A)/B = \mathrm{PGL}_2(A/I)/\overline{B},$$

where $\overline{B}$ is the image of B in $\mathrm{PGL}_2(A/I)$, i.e., the subgroup of matrices of the form $\begin{pmatrix} 1 & \overline{a} \\ 0 & 1 \end{pmatrix}$, $\overline{a} \in A/I$, we obtain

$$\deg F_\infty = |\overline{\mathrm{Cusps}(I)}| = |\mathrm{PGL}_2(A/I)| \cdot |A/I|^{-1} = q^{2m}-1. \qquad \square$$

Since the projection $j\colon C(I) \to \mathbb{P}^1$ is merely factorizing over $\mathrm{PGL}_2(A/I)$, F_∞ is a $\mathrm{PGL}_2(A/I)$-invariant divisor. Let $D = a_1 F_\infty$, where a_1 is an arbitrary positive integer. Noting that the group $\mathrm{PGL}_2(A/I)$ acts transitively on the set $\mathcal{P}$ of "supersingular" $\mathbb{F}_{q^2}$-rational points of $C(I)$, we get the following result.

PROPOSITION 5.2.40. *The code C_{a_1} over $\mathbb{F}_{q^2}$ constructed from the curve $C(I)$, set $\mathcal{P}$, and divisor $D=a_1F_\infty$ has a natural transitive action of the group $G(I)=\mathrm{PGL}_2(A/I)$.* □

COROLLARY 5.2.41. *The code C_{a_1} can be realized as an ideal in the group algebra $\mathbb{F}_{q^2}[G(I)]$ lying in $\mathbb{F}_{q^2}[G(I)/H]=\mathbb{F}_{q^2}[G(I)]^H$, where H is the image of $\mathbb{F}_{q^2}^*$ in $G(I)$.* □

PROBLEM 5.2.42. Give an explicit description of C_{a_1} as an ideal in $\mathbb{F}_{q^2}[G(I)]$ and estimate its parameters without using the Riemann–Roch theorem.

REMARK 5.2.43. Though a solution of Problem 5.2.42 is not known, the Riemann–Roch theorem yields some information on the structure of this ideal. Let $a_0=\lfloor(2g(I)-2)/(q^{2m}-1)\rfloor+1=2q(q^{2\ell}-1)(q^2-1)^{-1}+1$, where $\ell=\left\lfloor\frac{m}{2}\right\rfloor$. Then, for $b\geq a_0$, by the Riemann–Roch theorem we have $\ell(bF_\infty)=b(q^{2m}-1)-g(I)+1$. Thus, $L((b+1)F_\infty)/L(bF_\infty)$ for $b\geq a_0$ is a $G(I)$-module of dimension $q^{2m}-1$. Hence, the code C_{a_1} as a $G(I)$-ideal has a filtration by $G(I)$-submodules of the form $C_{a_1}\supset C_{a_1-1}\supset\ldots\supset C_{a_0}$, where C_b for $a_1\geq b\geq a_0$ is the image of $L(bF_\infty)$ in C_{a_1}, and all factors C_{b+1}/C_b, $b=a_0,\ldots,a_1-1$, are of dimension $q^{2m}-1$. Unfortunately, the group algebra $\mathbb{F}_{q^2}[G(I)]$ is never semisimple, and thus the above filtration is not very useful.

REMARK 5.2.44. Corollary 5.2.41 can be interpreted in the following way: the asymptotically good code C_{a_1} is *transitive*, i.e., admits an automorphism subgroup of S_n, n being the code length, which acts transitively on the n coordinate positions of codewords. This is interesting, since in the case where such a transitive subgroup is cyclic, the code is cyclic, and the existence of asymptotically good cyclic codes is a well-known open problem.

Polynomiality. A polynomial construction for codes on Drinfeld modular curves can be given in the following setting:

$$C'=(X',\mathcal{P}',D')_L \quad \text{over } \mathbb{F}_{q^2}$$

for an ideal $I=(p_I)\subset A=\mathbb{F}_q[T]$, where p_I is a monic irreducible polynomial of an odd degree m with $\gcd(m,q-1)=1$, and $\mathcal{P}'$ is the set of supersingular $\mathbb{F}_{q^2}$-points of X', $D'=bQ_1$, Q_1 being the cusp of X' such that its inverse images in $X_1(I)$ are $\mathbb{F}_{q^2}$-rational. Here we have an explicit planar equation $F_I(x,y)=0$ for the curve $\widetilde{X}_1(I)$ birationally isomorphic to $X_1(I)$ and having two singular points P_1 and P_2.
Let $\pi\colon X=X_1(I)\to X_0(I)=X'$ be the natural projection.

The length of C' is $n=|\mathcal{P}'|=\dfrac{q^m+1}{q+1}$, and we need to construct a basis $\{f_\alpha\}$ of the space $L(D')$ and $\{P_\alpha\}=\mathcal{P}$ polynomially in n. The following remark allows one to realize $L(D')$ as a function space on $X_1(I)$ rather than on $X_0(I)=X'$, which is useful since $X_1(I)$ has an explicit equation $F_I(x,y)=0$ (Proposition 5.2.21). Recall that the projection π is merely a factorization over the natural faithful action of $H=(A/I)^*/\mathbb{F}_q$ on $X_1(I)$, i.e.,

$$X_0(I)\cong X_1(I)/H, \qquad |H|=\frac{q^n-1}{q-1}.$$

REMARK 5.2.45. The condition $f\in\pi^*(L(D'))$ is equivalent to $f=\sum\limits_{i=0}^{h-1}\theta^i g$ for a generator θ of H, where $h=|H|$.

This allows one to consider functions $\{f_\alpha\}$ as elements of $L(\pi^* D')$ invariant under θ.

On the other hand, one has the following result.

PROPOSITION 5.2.46. *Let V_m be the image in $\mathbb{F}_q(X_1(I))$ of the space*

$$W_m = \{P(x,y) \in \mathbb{F}_q[x,y] : \deg P \leq q^{4m}\}.$$

Then $L(\pi^* D') \subseteq V_m$. □

Therefore, elements of $L(\pi^* D')$ can be realized as polynomials in (x,y) of degree $\leq q^{4m} < n^5$.

Now we can construct desingularization trees $\Gamma_1 := \Gamma_{P_1}$ and $\Gamma_2 := \Gamma_{P_2}$ for the two singular points $\{P_1, P_2\}$ of $X_1(I)$, and this leads to constructing linear functionals on W_m such that the intersection of their kernels in W_m projects onto $L(\pi^* D')$ in V_m. Since the action of θ on $X_1(I)$ and thus on $L(\pi^* D')$ is given by an explicit formula in Corollary 5.2.24, this leads to constructing a basis $\{f_\alpha\}$ of $L(D')$.

Further, the elements $\{P_\beta\}$ of $\mathcal{P}$ are characterized as the images on $X_0(I)$ of the points $(x_\beta, 0) \in X_1(I)$, x_β being the roots of the equation $F(x) = F_I(x,0) = 0$; moreover, Proposition 5.2.31 gives a factorization of $F(x)$ into polynomials irreducible over $\mathbb{F}_{q^2}$. The set $\mathcal{P} = \{P_\beta\}$ can also be constructed explicitly, which completes the construction of the generator matrix $\big(f_\alpha(P_\beta)\big)$.

Since all the objects needed in this construction are bounded polynomially in q^m and thus in n, the construction is of polynomial complexity.

5.3. Elkies Explicit Towers

In the present section we describe how to write down explicit equations for some optimal towers of modular curves, both classical (Sec. 5.3.1) and Drinfeld (Sec. 5.3.2). Miraculously, the equations thus obtained coincide with the equations of recursive García–Stichtenoth towers; see Sec. 7.4 below.

5.3.1. Classical Modular Curve Towers

In this section we show how to describe certain optimal towers of modular curves quite explicitly. Below some complex modular curves $X_0(n)$ are written in terms of polynomial equations with integer coefficients; therefore, by considering their reductions given by the same equations over $\mathbb{F}_p$ we obtain optimal families over $\mathbb{F}_{p^2}$ for $p \nmid n$.

Curves $X_0(\ell^n)$. Fix a prime $\ell > 1$. The first family that we consider is given by the curves $X_0(\ell^n)$; then the affine curves $Y_0(\ell^n)$ parametrize elliptic curves with a cyclic ℓ^n-isogeny. Such an isogeny can equivalently be considered as a sequence of ℓ-isogenies

$$E_0 \to E_1 \to E_2 \to \ldots \to E_n \tag{5.3.1}$$

such that the composite isogeny $E_{j-1} \to E_{j+1}$ of degree ℓ^2 is cyclic for each j with $0 < j < n$. Then, as above, $X_0(\ell^n)$ is the smooth projective curve containing $Y_0(\ell^n)$, i.e., $X_0(\ell^n) = X_{\Gamma_0(\ell^n)}$, $Y_0(\ell^n) = Y_{\Gamma_0(\ell^n)}$.

Let $m \in [0, n]$. Then we get $n+1-m$ pairs of projections

$$\pi'_j : Y_0(\ell^n) \to Y_0(\ell^m),$$
$$\pi_j : X_0(\ell^n) \to X_0(\ell^m),$$

which are of degree ℓ^{n-m} for $m \geq 1$ and of degree $(\ell+1)\ell^{n-1}$ for $m = 0$. The projection π_j is given by extracting the isogeny $E_j \to E_{j+1}$ from (5.3.1). In particular, we have a tower of maps

$$Y_0(\ell^n) \xrightarrow{\pi'_0} Y_0(\ell^{n-1}) \xrightarrow{\pi'_0} Y_0(\ell^{n-2}) \xrightarrow{\pi'_0} \ldots \xrightarrow{\pi'_0} Y_0(\ell^2) \xrightarrow{\pi'_0} Y_0(\ell)$$

and thus a tower

$$X_0(\ell^n) \xrightarrow{\pi_0} X_0(\ell^{n-1}) \xrightarrow{\pi_0} X_0(\ell^{n-2}) \xrightarrow{\pi_0} \ldots \xrightarrow{\pi_0} X_0(\ell^2) \xrightarrow{\pi_0} X_0(\ell),$$

each map being of degree ℓ. Each $Y_0(\ell^n)$ and thus $X_0(\ell^n)$ also has an Atkin–Lehner involution $w_\ell = w_\ell^{(n)}$, taking a cyclic ℓ^n-isogeny to its dual isogeny, and the sequence (5.3.1) to the sequence

$$E_n \to \ldots \to E_2 \to E_1 \to E_0$$

of dual isogenies. Therefore,

$$w_\ell^{(m)} \circ \pi_j = \pi_{n-m-j} \circ w_\ell^{(n)}. \tag{5.3.2}$$

We regard $Y_0(N)$ as the quotient $H/\Gamma_0(N)$, and $X_0(N)$ as the quotient of the extended upper half-plane $(H \cup \mathbb{P}^1(\mathbb{Q}))/\Gamma_0(N)$; see Sec. 5.1.1. The $\Gamma_0(\ell^n)$ orbit of a point $\tau \in H$ parametrizes the isogeny between the elliptic curves with period lattices $\mathbb{Z}+\mathbb{Z}\tau$ and $\ell^{-n}\mathbb{Z}+\mathbb{Z}\tau$, the map π_j takes the $\Gamma_0(\ell^n)$ orbit of τ to the $\Gamma_0(\ell^m)$ orbit of $\ell^j\tau$, and the involution $w_\ell^{(n)}$ is represented by $\tau \leftrightarrow -1/\ell^n\tau$. It is induced by the matrix $\tau_{\ell^n} = \ell^{-n/2}\begin{pmatrix} 0 & 1 \\ -\ell^n & 0 \end{pmatrix}$ in the normalizer of $\Gamma_0(\ell^n)$ in Lemma 5.1.14.

We now observe that explicit formulae for $X_0(\ell)$ and $X_0(\ell^2)$, together with the involutions $w_\ell^{(1)}$ and $w_\ell^{(2)}$ of these curves and the map $\pi_0\colon X_0(\ell^2)\to X_0(\ell)$ between them, allow one to construct the whole tower explicitly. Namely, one has the following result.

PROPOSITION 5.3.1. *For $n\geq 2$ the product map*

$$\pi=\pi_0\times\pi_1\times\pi_2\times\ldots\times\pi_{n-2}\colon X_0(\ell^n)\to (X_0(\ell^2))^{n-1}$$

is a one-to-one map from $X_0(\ell^n)$ to the set of $(P_1,P_2,\ldots,P_{n-1})\in\big(X_0(\ell^2)\big)^{n-1}$ such that

$$\pi_0\big(w_\ell^{(2)}(P_j)\big)=w_\ell^{(1)}\big(\pi_0(P_{j+1})\big)\tag{5.3.3}$$

for each $j=1,2,\ldots,n-2$.

SKETCH OF THE PROOF. We will check this on the affine models $Y_0(\ell^n)$. The fact that the map is one-to-one to its image is clear, because a sequence (5.3.1) of ℓ-isogenies is determined by the ℓ^2-isogenies $E_{j-1}\to E_{j+1}$ parametrized by the jth coordinate of π $(0<j<n)$. Now $(P_1,\ldots,P_{n-1})$ is in the image of π if and only if the ℓ^2-isogenies parametrized by $P_1,\ldots,P_{n-1}$, regarded as sequences $E_0^j\to E_1^j\to E_2^j$ of ℓ-isogenies, fit together to form a sequence (5.3.1) with $E_i^j=E_{i+j}$, i.e., if and only if the isogenies $E_1^j\to E_2^j$ and $E_0^{j+1}\to E_1^{j+1}$ coincide for each $j=1,2,\ldots,n-2$. But these isogenies are represented by the points $\pi_1(P_j)$ and $\pi_0(P_{j+1})$ on $X_0(\ell)$. Thus, the necessary and sufficient condition is that

$$\pi_1(P_j)=\pi_0(P_{j+1})$$

for each $j=1,2,\ldots,n-2$; applying $w_\ell^{(1)}$ to both sides and then applying (5.3.2) to $w_\ell^{(1)}\big(\pi_1(P_j)\big)$ yields the equivalent form (5.3.3). □

Examples. The cases $\ell=2,3,5$. The above formulae are especially useful if the curves $X_0(\ell)$ and $X_0(\ell^2)$ are rational, which is the case for $\ell=2,3,5$ by Proposition 5.1.18(b). Therefore, we may regard $P_1,\ldots,P_{n-1}$ as $n-1$ rational coordinates on $X_0(\ell^n)$, and (5.3.3) as the $n-2$ algebraic relations on those coordinates that determine the curve $X_0(\ell^n)$. Let us give explicit formulae for $\ell=2,3,5$. Note also that the classical theory of modular functions gives explicit rational generators for the corresponding rational function fields. These explicit generators are analogous to but more complicated than that for $X_0(2)$, given at the end of Sec. 5.1.1.

PROPOSITION 5.3.2. (a) *The covers $X_0(4)\to X_0(2)$ and $X_0(9)\to X_0(3)$ are cyclic.*

(b) *We have*

$$\mathbb{C}(X_0(2))=\mathbb{C}(j(\tau),j(2\tau))=\mathbb{C}(h_2(\tau)),$$

where

$$h_2(\tau)=\left(\frac{\eta(\tau)}{\eta(2\tau)}\right)^{24}=q^{-1}-24+276q-2048q^2+11202q^3-\ldots,$$

and

$$\mathbb{C}(X_0(4))=\mathbb{C}(j(\tau),j(4\tau))=\mathbb{C}(\xi(\tau)),$$

where

$$\xi(\tau)=1+\frac{1}{8}\left(\frac{\eta(\tau)}{\eta(4\tau)}\right)^8=\frac{1}{8}(q^{-1}+20q-62q^3+216q^5-641q^7+\ldots).$$

Furthermore,

$$w_2^{(2)}(\xi(\tau))=\xi(-1/4\tau)=1+32\left(\frac{\eta(4\tau)}{\eta(\tau)}\right)^8=1+\frac{4}{\xi(\tau)-1}=\frac{\xi(\tau)+3}{\xi(\tau)-1}. \tag{5.3.4}$$

The map $X_0(4)\to X_0(2)$ is given by

$$h_2(\tau)=8\frac{(\xi(\tau)+1)^2}{(\xi(\tau)-1)}. \tag{5.3.5}$$

(c) *We have*

$$\mathbb{C}(X_0(3))=\mathbb{C}(j(\tau),j(3\tau))=\mathbb{C}(h_3(\tau)),$$

where

$$h_3(\tau)=\left(\frac{\eta(\tau)}{\eta(3\tau)}\right)^{12}=q^{-1}-12+54q-76q^2-243q^3+1188q^4\ldots,$$

and

$$\mathbb{C}(X_0(9))=\mathbb{C}(j(\tau),j(9\tau))=\mathbb{C}(\xi(\tau)),$$

where

$$\xi(\tau)=1+\frac{1}{3}\left(\frac{\eta(\tau)}{\eta(9\tau)}\right)^3=\frac{1}{3}(q^{-1}+5q-7q^5+3q^8+15q^{11}-32q^{14}\ldots).$$

Sketch of the proof. (a) This can be deduced from the fact that ±1 are the only units of $\mathbb{Z}/N\mathbb{Z}$ for $N=2$ and 3.

(b) Computing the map $\pi_0\colon X_0(4)\to X_0(2)$ amounts to writing h_2 as a rational function in ξ; necessarily that the function has degree 2 with a simple pole at the cusp $\xi=\infty$. But then $h_2-8\xi+24$ is a rational function of degree 1 in ξ with a simple zero at ∞, i.e., the inverse of a polynomial of degree 1. Comparing the q-expansions of $1/(h_2-8\xi+24)$ and ξ, we find that this polynomial is $(\xi+1)/32$ and recover formula (5.3.5).

(c) is slightly more involved but can be proved similarly. □

Proposition 5.3.3. (a) *Let $\ell=2$, and let x_j be the rational function $\xi(2^{j-1}\tau)$ on $X_0(2^n)$. Then the embedding $X_0(2^n)\hookrightarrow(\mathbb{P}^1)^{n-1}$ given by $(x_1,\ldots,x_{n-1})$ identifies $X_0(2^n)$ with the curve in $(\mathbb{P}^1)^{n-1}$ specified by the $n-2$ equations*

$$(x_j^2-1)(z_{j+1}^2-1)=1,\quad j=1,\ldots,n-2, \tag{5.3.6}$$

where

$$z_j=(x_j+3)/(x_j-1)$$

is obtained from x_j by the involution $w_2^{(2)}$.

(b) *Let $\ell=3$; then the curve $X_0(3^n)$ is isomorphic with the locus in $(\mathbb{P}^1)^{n-1}$ of $(x_1,\ldots,x_{n-1})$ satisfying*

$$(x_j^3-1)(z_{j+1}^3-1)=1,\quad j=1,\ldots,n-2, \tag{5.3.7}$$

where

$$z_j=(x_j+2)/(x_j-1). \tag{5.3.8}$$

SKETCH OF THE PROOF. (a) Note that x_j is the coordinate P_j of Proposition 5.3.1.

Using formula (5.3.4) for the involution $w_2^{(2)}$, from (5.3.5) we also obtain

$$h_2(\tau) = \frac{64}{\xi(-1/4\tau)^2 - 1}.$$

But by the functional equation

$$\eta(-1/\tau) = (\tau/i)^{1/2}\eta(\tau),$$

the involution $w_2^{(1)}$ acts on $X_0(2)$ by $h_2 \leftrightarrow 2^{12}/h_2$. Thus, $h_2(2\tau)$ is both $64(\xi(\tau)^2-1)$ and $64/(\xi(-1/8\tau)^2-1)$. Equating these two expressions yields an equation for the modular curve $X_0(8)$; more generally, we now deduce from Proposition 5.3.1 the explicit equations for the modular curve $X_0(2^n)$ for each $n>1$.

(b) Here the coordinate functions x_j on $X_0(3^n)$ are $\xi(3^{j-1}\tau)$, so ξ generates the field of rational functions on $X_0(9)$. The involution $w_3^{(2)}$ takes this ξ to $1+3/(\xi-1) = (\xi+2)/(\xi-1)$, whence (5.3.8); $h_3(\tau)$ goes to $3^6/h_3$ under $w_3^{(1)}$, and $h_3(3\tau)$ is both $27/(\xi(-1/27\tau)^3-1)$ and $27(\xi(\tau)^3-1)$, from which equations (5.3.7) and (5.3.8) for $X_0(3^n)$ follow thanks to Proposition 5.3.1. □

COROLLARY 5.3.4. (a) *For the reductions of* $X_0(2^k)$ *to characteristic* 3, *we get*

$$y_{j+1}^2 = y_j - y_j^2,$$

with $y_j = 1 - x_j^{-1}$.

(b) *For the reductions of* $X_0(3^k)$ *to characteristic* 2, *we get*

$$y_{j+1}^3 = y_j^3 + y_j^2 + y_j. \qquad \square$$

Now we consider the case $\ell = 5$.

PROPOSITION 5.3.5. (a) *We have*

$$\mathbb{C}(X_0(5)) = \mathbb{C}(j(\tau), j(5\tau)) = \mathbb{C}(h_5(\tau)),$$

where

$$h_5(\tau) = \left(\frac{\eta(\tau)}{\eta(5\tau)}\right)^6 = q^{-1} - 6 + 9q + 10q^2 - 30q^3 + 6q^4 - 25q^5 \ldots,$$

with $h_5(\tau)h_5(-1/5\tau) = 125$, *and*

$$\mathbb{C}(X_0(25)) = \mathbb{C}(j(\tau), j(25\tau)) = \mathbb{C}(\xi(\tau)),$$

where

$$\xi(\tau) = 1 + \frac{\eta(\tau)}{\eta(25\tau)} = q^{-1} - q + q^4 + q^6 - q^{11} - q^{14} + q^{21} + q^{24} - q^{26} \ldots.$$

(b) *Letting* $x_j = \xi(5^{j-1}\tau)$ *and* $z_j = w_\ell^{(2)}(x_j)$, *we have*

$$P(x_j)P(z_{j+1}) = 125, \quad j = 1, \ldots, n-2,$$

where

$$P(X) = X^5 + 5X^3 + 5X - 11, \qquad z_j = (x_j+4)/(x_j-1). \qquad \square$$

REMARK 5.3.6. The polynomial $P(X)$ is not as simple as the polynomials X^2-1 and X^3-1 occurring in (5.3.6), (5.3.7), because the cover $X_0(25) \to X_0(5)$ is not cyclic. It is, however, dihedral, as can be seen from the fact that $P(W-1/W) = W^5 - 11 - W^{-5}$.

The rationality of the curves $X_0(12)$ and $X_0(16)$, as well as ellipticity of $X_0(36)$, allows one to consider the following cases.

Cases $\ell = 4, 6$. In fact, the primality of ℓ is not necessary, and we can give formulae for the cases $\ell = 4, 6$.

In the first case $\ell = 4$, the curve $X_0(\ell^2) = X_0(16)$ is still rational.

PROPOSITION 5.3.7. (a) *The cover $X_0(16) \to X_0(4)$ is cyclic.*

(b) *We have*

$$\mathbb{C}(X_0(4)) = \mathbb{C}(j(\tau), j(4\tau)) = \mathbb{C}(h_4(\tau)),$$

where

$$h_4(\tau) = \left(\frac{\eta(\tau)}{\eta(4\tau)}\right)^8 = q^{-1} - 8 + 20q - 62q^3 + 216q^5 - 641q^7 + \dots,$$

and

$$\mathbb{C}(X_0(16)) = \mathbb{C}(j(\tau), j(16\tau)) = \mathbb{C}(\xi(\tau))$$

for

$$\xi(\tau) = 1 + \frac{1}{2}\frac{\eta^2(\tau)\eta(8\tau)}{\eta(2\tau)\eta^2(16\tau)} = \frac{1}{2}(q^{-1} + 2q^3 - q^7 - 2q^{11} + 3q^{15} + 2q^{19} \dots).$$

(c) *The curve $X_0(4^n)$ is isomorphic with the locus of $(x_1, \dots, x_{n-1})$ in $(\mathbb{P}^1)^{n-1}$ satisfying*

$$(x_j^4 - 1)(z_{j+1}^4 - 1) = 1, \quad j = 1, \dots, n-2,$$

where

$$z_j = (x_j + 1)/(x_j - 1),$$

the coordinate functions x_j on $X_0(4^n)$ being $\xi(4^{j-1}\tau)$. □

Indeed, $h_4(\tau)$ is mapped by $w_4^{(1)}$ to $4^4/h_4$, and we compute

$$h_4(4\tau) = \frac{16}{\xi(-1/64\tau)^4 - 1} = 16(\xi(\tau)^4 - 1).$$

The case of $\ell = 6$ is slightly more complicated because the curve $X_0(\ell^2)$ is no longer rational. It is, however, an elliptic curve with a simple Weierstrass equation.

PROPOSITION 5.3.8. (a) *We have*

$$\mathbb{C}(X_0(6)) = \mathbb{C}(j(\tau), j(6\tau)) = \mathbb{C}(h_6(\tau)),$$

where

$$h_6(\tau) = \frac{\eta^5(\tau)\eta(3\tau)}{\eta(2\tau)\eta^5(6\tau)} = q^{-1} - 5 + 6q + 4q^2 - 3q^3 - 12q^4 - 8q^5 + 12q^6 \dots.$$

(b) *The curve $X_0(6)$ is rational, and $X_0(36)$ is an elliptic curve given by the Weierstrass equation*

$$\gamma^2 = \xi^3 + 1$$

with

$$\xi(\tau)=\frac{\eta(12\tau)\eta^3(18\tau)}{\eta(6\tau)\eta^3(36\tau)}=q^{-2}+q^2+q^8-q^{14}-q^{20}+q^{26}+2q^{32}\dots,$$

$$\gamma(\tau)=\frac{\eta^4(12\tau)\eta^2(18\tau)}{\eta^2(6\tau)\eta^4(36\tau)}=q^{-3}+2q^3+q^9-2q^{15}-2q^{21}+2q^{27}+4q^{33}\dots.$$

(c) *The curve* $X_0(6^n)$ *is given by*

$$(x_j^3-8)(z_{j+1}^3-8)=72,\quad j=1,\dots,n-2,$$

where

$$z_j=\left(\frac{y_j+3}{x_j-2}\right)^2-x_j-2;$$

i.e.,

$$\mathbb{C}(X_0(6^n))=\mathbb{C}\big((x_1,y_1),\dots,(x_{n-1},y_{n-1})\big).$$ □

The tower $X_0(3\cdot 2^n)$. In this tower the first two curves $X_0(6)$ and $X_0(12)$ are still rational, and this leads to the following result.

PROPOSITION 5.3.9. (a) *The cover* $X_0(12)\to X_0(6)$ *is cyclic.*

(b) *The curve* $X_0(6)$ *is rational and can be represented as*

$$\mathbb{C}(X_0(6))=\mathbb{C}(j(\tau),j(6\tau))=\mathbb{C}(h_6(\tau)),$$

where

$$\widetilde{h}_6(\tau)=\left(\frac{\eta(2\tau)\eta^3(3\tau)}{\eta(\tau)\eta^3(6\tau)}\right)^3=q^{-1}+3+6q+4q^2-3q^3-12q^4-8q^5\dots$$

(*note that* $\widetilde{h}_6(\tau)$ *is different from* $h_6(\tau)$ *in Proposition* 5.3.8), *and*

$$\mathbb{C}(X_0(12))=\mathbb{C}(j(\tau),j(12\tau))=\mathbb{C}(\xi(\tau))$$

with

$$\xi(\tau)=\frac{\eta^4(4\tau)\eta^2(6\tau)}{\eta^2(2\tau)\eta^4(12\tau)}=q^{-1}+2q+q^3-2q^5-2q^7+2q^9+4q^{11}\dots.$$

(c) *We have*

$$\mathbb{C}(X_0(3\cdot 2^n))=\mathbb{C}(x_1,\dots,x_{n-1})$$

with

$$(x_j^2-1)(z_{j+1}^2-1)=-8,\quad j=1,\dots,n-2,$$

where

$$z_j=(3-x_j)/(1+x_j).$$ □

5.3.2. Explicit Towers of Drinfeld Modular Curves

Similar ideas can be applied to identify some Drinfeld modular curve towers with some García–Stichtenoth recursive towers (cf. Sec. 7.4).

Let $\Bbbk=\mathbb{F}_q$ and $\Bbbk_1=\mathbb{F}_{q^2}$. To construct the towers, it is convenient that instead of using $X_0(I)$, $X_1(I)$, and $X(I)$, we use other curves $\dot{X}_0(I)$, $\dot{X}_1(I)$, and $\dot{X}(I)$ defined as follows. We define smooth projective curves $\dot{X}_0(I)$, $\dot{X}_1(I)$, and $\dot{X}(I)$ as the curves over $\Bbbk$ with, respectively, the following rational function fields:

$$\Bbbk(\dot{X}_0(I)) = \mathbb{K}_0(I)(\theta),$$
$$\Bbbk(\dot{X}_1(I)) = \mathbb{K}_1(I)(\theta),$$
$$\Bbbk(\dot{X}(I)) = \mathbb{K}(I)(\theta),$$

where $\theta^{q+1} = -j$,

$$\mathbb{K}_0(I) = \mathbb{F}_q(X_0(I)),$$
$$\mathbb{K}_1(I) = \mathbb{F}_q(X_1(I)),$$
$$\mathbb{K}(I) = \mathbb{F}_q(X(I)).$$

We denote by $\dot{Y}_0(I)$, $\dot{Y}_1(I)$, and $\dot{Y}(I)$ the corresponding affine curves with $j(\varphi) \neq \infty$; the curves $\dot{X}_0(I)$, $\dot{X}_1(I)$, and $\dot{X}(I)$ are, respectively, their smooth projective models.

A Drinfeld module is said to be *normalized* if it satisfies

$$\varphi(T) = 1 + g\tau - \tau^2.$$

For a normalized module φ, one has $j(\varphi) = -g^{q+1}$, which implies the following result.

PROPOSITION 5.3.10. *The curves $\dot{Y}_0(I)$, $\dot{Y}_1(I)$, and $\dot{Y}(I)$ parametrize the corresponding isomorphism classes of normalized Drinfeld modules with the corresponding level structures.* □

Then define $\dot{X}'_0(T^n)$ as the curve parametrizing the triples (φ, x_1, G_n), where x_1 is a torsion point of φ and $G_n \ni x_1$ is a torsion group

$$G_n \cong A_0[t]/T^n.$$

PROPOSITION 5.3.11. (a) *The tower*

$$\ldots \to \dot{X}'_0(T^n) \to \dot{X}'_0(T^{n-1}) \to \ldots$$

can be identified with the García–Stichtenoth tower $\mathcal{W}_2$ of Section 7.4.2 (*equations* (7.4.3) *and* (7.4.4)) *given by*

$$x_j = x_{j-1}^{-1} z_j$$

with

$$z_j^q + z_j = x_j^{q+1},$$

i.e.,

$$\Bbbk(\dot{X}'_0(T^n)) = \Bbbk(x_1, \ldots, x_{n-1}).$$

Moreover, the set of supersingular points of $\dot{X}'_0(T^n)$ can be identified with the set of $\Bbbk$-rational points constructed by García and Stichtenoth (*see Proposition* 7.4.17(b)).

(b) *The tower*

$$\ldots \to X_0(T^n) \to X_0(T^{n-1}) \to \ldots,$$

where each $X_0(T^n)$ is the quotient of $\dot{X}'_0(T^n)$ by $\Bbbk_1^*$, can be identified with the tower given by

$$z_{j+1}(1+z_{j+1})^{q-1} = \frac{z_j^q}{(1+z_j)^{q-1}}, \quad j = 2, \ldots, n,$$

i.e.,

$$\Bbbk(X_0(T^n)) = \Bbbk(z_2, \dots, z_n),$$

which is the tower $\mathcal{W}_1$ *of Section* 7.4.2 (*equation* (7.4.2)). *The set of supersingular points of* $X_0(T^n)$ *coincides with the set*

$$\{\alpha \in \Bbbk_1 \mid \alpha^{q+1} = 1, \alpha \neq -1\}$$

of cardinality q^{n-1}.

SKETCH OF THE PROOF. First, we note that the curve $\dot{X}_1(T)$ parametrizing normalized Drinfeld modules φ with a T-torsion point is rational. Indeed, if

$$0 = T(x_1) = x_1 + g x_1^q - x_1^{q^2}, \tag{5.3.9}$$

i.e., x_1 is a nonzero torsion point of φ, then

$$g = x_1^{-q}(x_1^{q^2} - x_1).$$

Note also that $\dot{X}'_0(T) = \dot{X}'_1(T)$. Next, consider $\dot{X}'_0(T^2)$. It classifies the triples (φ, x_1, G_n), where

$$x_2 = x_1^{q-1} y_2 - y_2^q = P_{x_1}(y_2) \quad \text{for} \quad P_x(X) = x^{q-1}X - X^q.$$

Indeed, the choice of x_2 determines G_2. If we then define

$$t_x(X) := X + \frac{x^{q^2} - x}{x^q} X^q - X^{q^2},$$

we get

$$t_{x_1}(X) = P_{x_1}(Q_{x_1}(X)) \quad \text{for} \quad Q_x(X) = x^{1-q}X + X^q.$$

Thus, the condition $t_{x_1}(y_2) = x_1$ is equivalent to $Q_{x_1}(x_2) = x_1$, that is,

$$x_2 = x_1^{-1} z_2 \quad \text{for} \quad z_2^q + z_2 = x_1^{q+1}. \tag{5.3.10}$$

Thus, the curve $\dot{X}'_0(T^2)$ is just $z_2^q + z_2 = x_1^{q+1}$, which is the Hermitian curve over $\Bbbk_1$. Note that G_2 is the $\Bbbk$-vector space of zeroes of the linearized polynomial $P_{x_2} \circ P_{x_1}$. Moreover,

$$0 = P_{x_1}(Q_{x_1}(x_2)) = t'_{x_1}(x_2).$$

That is, x_2 is a T-torsion point on our T-isogenous module. It follows that $t'_{x_1} = t_{x_2}$. By induction, we can now determine the tower of curves $\dot{X}'_0(T^n)$ explicitly: the function field of $\dot{X}'_0(T^n)$ is generated by $x_1, x_2, \dots, x_{n-1}$ with relations $Q_{x_{j-1}}(x_j) = x_{j-1}$, or equivalently

$$x_j = x_{j-1}^{-1} z_j, \quad \text{where} \quad z_j^q + z_j = x_{j-1}^{q+1} \tag{5.3.11}$$

$(1 < j < n)$; the point $(x_1, x_2, \dots, x_{n-1})$ parametrizes the Drinfeld module with $\varphi_T = t_{x_1}$, with the T^n-torsion subgroup generated by any of the q^{n-1} solutions of

$$(P_{x_{n-1}} \circ P_{x_{n-2}} \circ \dots \circ P_{x_2} \circ P_{x_1})(y_n) = x_n.$$

This works because, by applying

$$Q_{x_j} \circ P_{x_j} = T_{x_j} = P_{x_{j-1}} \circ Q_{x_{j-1}}$$

$n-2$ times, we find

$$\begin{aligned} x_{n-1} = Q_{x_{n-1}}(x_n) &= (Q_{x_{n-1}} \circ P_{x_{n-1}} \circ P_{x_{n-2}} \circ \dots \circ P_{x_1})(y_n) = \dots \\ &= (P_{x_{n-2}} \circ \dots \circ P_{x_2} \circ P_{x_1} \circ T_{x_1})(y_n) = (P_{x_{n-2}} \circ \dots \circ P_{x_2} \circ P_{x_1})(T_{x_1}(y_n)); \end{aligned}$$

i.e., $T_{x_1}(y_n)$ satisfies the equation for y_{n-1}. As with G_2, we see that G_n is the $\Bbbk$-vector space of zeroes of the linearized polynomial

$$P_{x_n} \circ P_{x_{n-1}} \circ \ldots \circ P_{x_2} \circ P_{x_1}.$$

The set of $\Bbbk_1$-rational points on the tower (5.3.11), which coincides with the tower (7.4.3), can be described as follows. For each of the q^2-1 points on $\dot{X}_0'(T)$ with $x_1 \in \Bbbk_1^*$, we have $x_1^{q+1} \in \Bbbk_1^*$, so the q solutions z_2 of (5.3.10) are just the q elements of $\Bbbk_1$ whose trace to $\Bbbk$ is x_1^{q+1}. Clearly, none of these z_2 vanish, so $x_2 = x_1 z_2$ is again in $\Bbbk_1^*$. Inductively we see that x_1 lies under q^{n-1} rational points of $\dot{X}_0'(T^n)$ defined over $\Bbbk_1$. García and Stichtenoth use the resulting $(q^2-1)q^{n-1}$ rational points to confirm that the $\dot{X}_0'(T^n)$ form an asymptotically optimal tower over $\Bbbk_1$; see Secs. 3.4.3 and 7.4.2. From the Drinfeld modular viewpoint, we recognize $x_1 \in \Bbbk_1^*$ as the condition that a point on $\dot{X}_0'(T^n)$ is supersingular. Thus, the $(q^2-1)q^{n-1}$ points of $\dot{X}_0'(T^n)$ lying above $\Bbbk_1^*$ are precisely the supersingular points on $\dot{X}_0'(T^n)$.

The group $\Bbbk_1^*$ acts on $\dot{X}_0'(T^n)$ by

$$c(x_1,\ldots,x_n) = (cx_1, \bar{c}x_2, cx_3, \bar{c}x_4, \ldots, c^{q^{n-1}}x_n), \quad c \in \Bbbk_1^*, \quad \bar{c} := c^q.$$

If $c \in \Bbbk^*$, then this automorphism preserves φ (see (5.3.9)) and each G_j, but changes the generator x_1 of G_1 to cx_1. Thus we recover $\dot{X}_0(T^n)$ as the quotient of $\dot{X}_0'(T^n)$ modulo the action of $\Bbbk^*$, and the tower $\dot{X}_0(T^n)$ coincides with the tower (7.4.2). For arbitrary $c \in \Bbbk_1^*$, this automorphism multiplies g by the $(q+1)$st root of unity $c/\bar{c}$. This takes φ to a Drinfeld module φ_c with the same J-invariant but with a different choice of $(-J)^{1/(q+1)}$. The isomorphism c from φ to φ_c respects our choice of subgroups G_j. Thus, the quotient of $\dot{X}_0(T^n)$ modulo $\Bbbk_1^*/\Bbbk^*$, or equivalently of $\dot{X}_0'(T^n)$ modulo $\Bbbk_1^*$, is the Drinfeld modular curve $X_0(T^n)$.

Now let us verify that $\{X_0(T^n)\}$ coincides with (7.4.2). It is convenient to start the tower at $n = 2$. The curve $X_0(T)$ is the quotient of the x_1-line $\dot{X}_0'(T)$ modulo $\Bbbk_1^*$, so has genus zero and coordinate $x_1^{q^2-1}$. To obtain $X_0(T^2)$, raise both sides of equation (5.3.10) for $\dot{X}_0'(T^2)$ to the power $q-1$ to obtain

$$Z_2(1+Z_2)^{q-1} = x_1^{q^2-1}, \quad \text{where} \quad Z_2 := z_2^{q-1} = (x_1x_2)^{q-1}. \tag{5.3.12}$$

Now Z_2 is invariant under $\Bbbk_1^*$, and (5.3.12) shows that the Z_2-line is a degree-q cover of the $x_1^{q^2-1}$-line $X_0(T)$; since the cover $X_0(T^2)/X_0(T)$ is also of degree q, we conclude that Z_2 generates the function field of $X_0(T^2)$.

Thus, for $n \geq 2$ the function field of $X_0(T^n)$ is generated by

$$Z_j := z_j^{q-1} \quad (2 \leq j \leq n).$$

For each $j = 2,\ldots,n-1$, we have $Z_{j+1}(1+Z_{j+1})^{q-1} = x_j^{q^2-1}$ as in (5.3.12), which in turn equals

$$(z_j/x_{j-1})^{q^2-1} = Z_j^{q+1}/x_{j-1}^{q^2-1} = Z_j^q/(1+Z_j)^{q-1}.$$

Thus,

$$Z_{j+1}(1+Z_{j+1})^{q-1} = Z_j^q/(1+Z_j)^{q-1}.$$

This gives Z_{j+1} as an algebraic function of degree q in Z_j (and vice versa). Thus, these relations for $j = 2, \ldots, n-1$ determine the function field of $X_0(T^n)$. From our description of the supersingular points on $\dot{X}'_0(T^n)$ we see that the q^{n-1} supersingular points on $X_0(T^n)$ are the points for which each Z_j is in $\{Z \in \Bbbk_1 \mid Z^q = (1+Z)^{q-1}\}$. $\square$

Historical and Bibliographic Notes

Classical modular curves are one of those subjects which from the viewpoint of their origins are just a problem for a historian of mathematics. A reader interested in the history of this field can refer to Chapter VII of the book [**Die78**]; note however that even there he will not find a complete description. Some information on the history of modular curves and functions, and first of all on their connections with arithmetic, is contained in the book [**Vlă91**]; the reader will also find there a more detailed exposition of the theory of modular curves. Now we shall restrict ourselves to the principal landmarks of the history of modular curves and functions.

The initial point was the theory of elliptic integrals in the eighteenth century, where one of the main problems was that of transformation. In modern language this problem corresponds to that of search for an explicit expression for isogenies of elliptic curves. The problem of transformation was solved independently by N.H. Abel and C.G.J. Jacobi at the end of the 1820s. An equation for moduli of elliptic integrals related by transformation of order n is called the modular equation of order n. Examples of such equations for small n were written out by A.M. Legendre, N.H. Abel, and C.G.J. Jacobi. Modular functions were studied from different points of view by G.F.B. Riemann, L. Kronecker, C. Hermite, and many others. Later these functions proved to be rather a powerful instrument in many questions of analysis and number theory. In particular, E. Galois computed the Galois group of the modular equation. Up to the 1870s, the principal object to study were functions obtained from the Legendre modulus, the latter being a modular function for $\Gamma(2)$ and not for the complete modular group, which lead to a number of inconveniences in manipulating them. Due to R. Dedekind from the beginning of the 1870s the absolute invariant j was studied and widely used. A valuable contribution to the theory of modular functions was made by F. Klein; he found out that they give a uniformization of algebraic curves which were later called modular curves. In papers of L. Kronecker, H. Weber, R. Fueter, H. Basse, and M. Deuring modular functions were applied to construct class fields for complex quadratic fields (1880–1930). E. Hecke greatly contributed to the theory introducing operators now named after him. After Hecke the theory did not attract much attention up to the 1960s, when deep relations were found connecting modular functions to various fields of mathematics, first of all to infinite-dimensional representations of Lie groups (the Langlands program). The latter gave an explosion of activity in modular functions, which was reflected in its principle moments in the papers of conferences on modular functions [**Kuy73, DK73, KS73, BK75, SZ77a, SZ77b**].

The best reading on modular functions are the beautiful books by G. Shimura [**Shi71**] and S. Lang [**Lan76**]. Here the reader can find a detailed exposition of the material exposed in Sec. 5.1.1 except for the section on Weierstrass points where we borrowed from the paper by A.O.L. Atkin [**Atk67**] and for Proposition 5.1.24 due to R. Ogg [**Ogg74**]. The theory of moduli schemes of elliptic curves was developed by P. Deligne and M. Rapoport; it was essentially improved later by N. Katz and B. Mazur [**KM85**]. In Sec. 5.1.2 we follow [**DR73**]. Computation of the number of $\mathbb{F}_q$-points on reductions of modular curves, which is a simple exercise on application of Deligne–Rapoport results, is taken from [**TVZ82**]. The material of Sec. 5.1.3 is borrowed from [**TV91**].

In contrast to classical elliptic and modular curves, elliptic modules and their moduli varieties have a unique author (V.G. Drinfeld), and almost all results described in Secs. 5.2.1 and 5.2.2 are due to him. His motivation for inventing elliptic modules was their use in the proof of the Langlands conjecture for GL_2 over a function field. Note, however, that in the 1930s L. Carlitz [**Car38**] defined some functions which can be considered as an analytic counterpart of rank-1 Drinfeld modules and that those functions were applied by R. Hayes in the 1970s to construct an explicit class field theory for function fields; see Sec. 6.3.3. A great deal of Drinfeld's results are published in [**Dri74, Dri85**], while the rest were learned by the authors from their private conversations with him. The only exceptions are Propositions and Corollaries 5.2.15–5.2.32, which are just a simple application of Drinfeld's results. These results are due to S.G. Vlăduţ; see [**VM84**]. Section 5.2.3 is taken from [**TV91**].

For a thorough and accessible exposition of Drinfeld modules theory, see [**Gos96**]. Drinfeld modular curves are very well described in [**Gek01**]; see also [**Gek86**] and [**Gos80**].

The transitiveness in Remark 5.2.44 was essentially established in [**TV91**], though not stated as such. Another method for obtaining asymptotically good transitive codes was given in [**Sti06**], where the result was explicitly stated. However, in view of Section 5.3.2 it is quite probable that the construction of [**Sti06**] and that from Corollary 5.2.41 do essentially coincide.

The results of Sec. 5.3 belong completely to N.D. Elkies, namely Sec. 5.3.1 is borrowed from [**Elk97a**], and Sec. 5.3.2, from [**Elk01a**].

CHAPTER 6

Class Field Theory

The present chapter is mainly auxiliary. It contains some elements of the class field theory that we widely use below. Class field theory is the theory of abelian extensions of global fields. Global fields is a basic arithmetic notion, which is divided into two separate subclasses, function fields, i.e., fields of the form $\mathbb{F}_q(X)$ for a finite field $\mathbb{F}_q$ and an irreducible curve X over $\mathbb{F}_q$, and number fields, i.e., finite extensions of $\mathbb{Q}$. The results and technique of proofs are slightly different for these two cases, but have much more common features than distinctions.

In this chapter we expose basic definitions and constructions of class field theory, both for function fields and number fields. We do not present proofs (except for some most elementary ones in Sec. 6.1), which can be found in the books cited in the bibliographic notes below.

In Sec. 6.1 we define these fields and give some of their elementary properties. We also briefly discuss some parallels between the two cases. However, one should note that there exist much more parallels and analogies, which lead to the whole universe of profound results and conjectures.

The class field theory for global fields, which is our main topic in the chapter, is based on its local variant, i.e., the theory of abelian extensions of local fields. The latter are just completions of global fields. For local fields the corresponding theory, which is referred to as local class field theory, is simpler, and we give its main statements in Sec. 6.2.

Section 6.3 contains principal results of global class field theory, as well as its explicit version known for function fields. In Sec. 6.4 we give some specific but very important constructions of class field towers, which are largely used in Chapters 7 and 8.

We note, however, that our exposition of class field theory below is somewhat fragmentary and incomplete, in order to be as elementary as possible. We formulate main results that we use in subsequent chapters below, but methods of proofs are completely out of our exposition; we touch upon neither adeles nor cohomologies, which are indispensable to understand the general picture of the theory; L-functions are also briefly mentioned, despite their fundamental importance.

6.1. Global Fields

There are profound and important analogies between algebraic number fields, that is, finite extensions of $\mathbb{Q}$, and fields of rational functions on curves over finite fields, i.e., finite extensions of $\mathbb{F}_q(t)$. These two types of fields are called *global fields*. Their properties are parallel in many respects, and they can be studied by similar methods; we use them throughout this book. We start with their basic properties.

In Sec. 6.1.1 we consider function fields over finite fields; actually, here we merely reproduce the content of Secs. 2.5.1 and 2.5.2 of *Basic Notions*. In Sec. 6.1.2 we give basic definitions and state some basic properties of number fields and briefly discuss analogies between the two cases. Note that we define global fields in an *ad hoc* manner, but there is a natural and more conceptual definition, which, however, needs more theoretical background.

6.1.1. Function Fields

An *algebraic function field* $\mathbb{K}/\Bbbk$ *of one variable over* $\Bbbk$ (or, for short, simply a *function field* $\mathbb{K}/\Bbbk$) is any field $\mathbb{K} \supseteq \Bbbk$ which is a finite algebraic extension of $\Bbbk(x)$, where x is an element of $\mathbb{K}$ transcendental over $\Bbbk$ or, equivalently, a finitely generated extension of $\Bbbk$ of transcendence degree 1.

The set of elements of $\mathbb{K}$ that are algebraic over $\Bbbk$ is a subfield of $\mathbb{K}$, which is called the *field of constants* of $\mathbb{K}/\Bbbk$ and is denoted by $\widetilde{\Bbbk}$. We have $\Bbbk \subseteq \widetilde{\Bbbk} \subset \mathbb{K}$, and it is easily verified that $\mathbb{K}/\widetilde{\Bbbk}$ is also a function field over $\widetilde{\Bbbk}$. We say that $\Bbbk$ is *algebraically closed in* $\mathbb{K}$ (or is the *full constant field* of $\mathbb{K}$) if $\widetilde{\Bbbk} = \Bbbk$. Below we often assume the latter condition to be satisfied when saying that a function field $\mathbb{K}$ over $\Bbbk$ is given.

Recall that a *valuation ring* of the function field $\mathbb{K}/\Bbbk$ is a proper subring $\mathcal{O} \subset \mathbb{K}$ such that $\Bbbk \subset \mathcal{O}$, and for any $f \in \mathbb{K}$ at least one of the elements f and f^{-1} belongs to $\mathcal{O}$. The set $\mathcal{O}^* = \{f \in \mathbb{K} \mid f \in \mathcal{O}, f^{-1} \in \mathcal{O}\}$ is called the *group of units* of $\mathcal{O}$.

Properties of valuation rings are given in the following statement.

PROPOSITION 6.1.1. *Let $\mathcal{O}$ be a valuation ring of a function field $\mathbb{K}/\Bbbk$. Then*

(a) *$\mathcal{O}$ is a local ring; i.e., $\mathcal{O}$ has a unique maximal ideal P, and $P = \mathcal{O} \setminus \mathcal{O}^*$.*

(b) *If $f \in \mathbb{K}$, $f \neq 0$, then $f \in P$ if and only if $f^{-1} \notin \mathcal{O}$.*

(c) *$\widetilde{\Bbbk} \subseteq \mathcal{O}$ and $\widetilde{\Bbbk} \cap P = \{0\}$.*

(d) *P is a principal ideal.*

(e) *If $P = t\mathcal{O}$, then any nonzero element $f \in \mathbb{K}$ has a unique representation of the form $f = t^n u$, where $n \in \mathbb{Z}$ and $u \in \mathcal{O}^*$.*

(f) *$\mathcal{O}$ is a principal ideal domain; namely, if I is a nontrivial proper ideal in $\mathcal{O}$ and $P = t\mathcal{O}$, then $I = t^n\mathcal{O}$ for some $n \in \mathbb{N}$.* □

A ring having the above properties is called a *discrete valuation ring*.

Places of a function field. Now we are going to define the notion of a place, which replaces that of a point when we turn to the language of function fields.

A *place* P of a function field $\mathbb{K}/\Bbbk$ is the maximal ideal of some valuation ring $\mathcal{O}$ of $\mathbb{K}/\Bbbk$. Any element $t \in P$ such that $P = t\mathcal{O}$ is called a *local parameter at* P (also called a *prime element* or *uniformizing variable*). The set of all places of a function field $\mathbb{K}$ is denoted by $\mathbb{P}_{\mathbb{K}}$.

Since a valuation ring $\mathcal{O}$ is uniquely determined by its maximal ideal P, namely, $\mathcal{O} = \{f \in \mathbb{K} \mid f^{-1} \notin P\}$, $\mathcal{O} = \mathcal{O}_P$ is called the *valuation ring of the place* P.

REMARK 6.1.2. Theorem 2.1.50 states that for the case $\mathbb{K} = \Bbbk(X)$ and $\Bbbk = \overline{\Bbbk}$ there is a natural one-to-one correspondence between $\mathbb{P}_{\mathbb{K}}$ and $\{P \in X\}$.

Discrete valuation. A *discrete valuation* of $\mathbb{K}/\Bbbk$ is a function $v\colon \mathbb{K}^* \to \mathbb{Z}$ with the following properties:

(1) $v(fg) = v(f) + v(g)$ for any $f, g \in \mathbb{K}$;
(2) $v(f+g) \geq \min\{v(f), v(g)\}$ for any $f, g \in \mathbb{K}$;
(3) there exists an element $f \in \mathbb{K}$ with $v(f) = 1$;
(4) $v(c) = 0$ for any $c \in \Bbbk \setminus \{0\}$.

We extend v to the whole $\mathbb{K}$, putting by definition $v(0) = \infty$.

By Proposition 6.1.1(e), for each place $P \in \mathbb{P}_{\mathbb{K}}$, the choice of a local parameter t yields a unique representation of every nonzero element $f \in \mathbb{K}$ in the form $f = t^n u$, $u \in \mathcal{O}^*$.

Note that for any $f \neq 0$ the exponent n in this representation depends only on P but does not depend on the choice of a local parameter. Thus, the function $v_P\colon \mathbb{K} \setminus \{0\} \to \mathbb{Z}$ equal to the exponent n in this representation is well defined. We have the following result.

THEOREM 6.1.3. *Let $\mathbb{K}/\Bbbk$ be a function field.*

(a) *For any place $P \in \mathbb{P}_{\mathbb{K}}$, the function v_P is a discrete valuation. Moreover, we have*

$$\mathcal{O}_P = \{f \in \mathbb{K} \mid v_P(f) \geq 0\},$$
$$\mathcal{O}_P^* = \{f \in \mathbb{K} \mid v_P(f) = 0\},$$
$$P = \{f \in \mathbb{K} \mid v_P(f) > 0\},$$

and an element $t \in \mathbb{K}$ is a local parameter for P if and only if $v_P(t) = 1$.

(b) *Conversely, let v be a discrete valuation of the function field $\mathbb{K}/\Bbbk$. Then the set $P = \{f \in \mathbb{K} \mid v(f) > 0\}$ is a place of $\mathbb{K}$, and $\mathcal{O}_P = \{f \in \mathbb{K} \mid v(f) \geq 0\}$ is the corresponding valuation ring.*

(c) *Any valuation ring $\mathcal{O}$ of $\mathbb{K}/\Bbbk$ is a maximal proper subring of $\mathbb{K}$.* □

Value at a place. Since P is a maximal ideal, the residue class ring $\mathcal{O}_P/P$ is a field. This field $\Bbbk_P = \mathcal{O}_P/P$ is called the *residue class field* of P. For $f \in \mathcal{O}_P$ we denote the residue class of f modulo P by $f(P) = f + P \in \Bbbk_P$. For $f \in \mathbb{K} \setminus \mathcal{O}_P$ we put $f(P) = \infty$ by definition. Thus, there is a map $f \mapsto f(P)$ from $\mathbb{K}$ to $\Bbbk_P \cup \{\infty\}$. The image $f(P)$ of f under this map is called its *value at* P; elements $f \in \mathcal{O}_P$ are called *regular at* P.

Now it becomes clear why the fields in question are called function fields and their elements are often called *functions*.

REMARK 6.1.4. Note that values of the same element at different places belong to different fields. We can easily correct this shortcoming, observing that $\Bbbk_P \subseteq \overline{\Bbbk}$ for any P.

Degree of a place. From Proposition 6.1.1(c) we know that $\Bbbk \subseteq \mathcal{O}_P$ and $\Bbbk \cap P = \{0\}$, so the canonical residue class map $\mathcal{O}_P \to \mathcal{O}_P/P$ induces a canonical embedding $\Bbbk \hookrightarrow \Bbbk_P$. Therefore, we shall hereafter always consider $\Bbbk$ as a subfield of $\Bbbk_P$ (this also applies to the constant field $\widetilde{\Bbbk}$). We call $\deg P = [\Bbbk_P : \Bbbk]$ the *degree* of a place P. The degree of a place is always finite; moreover, we have the following statement.

PROPOSITION 6.1.5. *Let P be a place of a function field $\mathbb{K}/\Bbbk$, and let x be a nonzero element of P. Then*

$$\deg P \leq [\mathbb{K} : \Bbbk(x)].$$

□

Ramification, splitting, inertia. Let $\mathbb{K} \subset \mathbb{L}$ be a finite extension of function fields, and let P be a place of the field $\mathbb{K}$, corresponding to the local ring $\mathcal{O}_P$ and the maximal ideal P in it. We say that the place Q of $\mathbb{L}$ *lies over* P if the ideal $P \subseteq Q$; in this case we write $Q \mid P$.

PROPOSITION 6.1.6. *The following conditions are equivalent:*

- *The place Q lies over the place P.*
- $\mathcal{O}_P \subseteq \mathcal{O}_Q$.
- *There exists a positive integer $e \geq 1$ such that $v_Q(f) = e \cdot v_P(f)$ for any $f \in \mathbb{K}$.*

If these conditions are satisfied, then $P = Q \cap \mathbb{K}$ and $\mathcal{O}_P = \mathcal{O}_Q \cap \mathbb{K}$. □

From this it follows, in particular, that any place of $\mathbb{L}$ lies over some place of $\mathbb{K}$. Thus, a field extension gives a map from the set of places of the larger field to that of the smaller.

Further on we assume that the fields of constants of $\mathbb{K}$ and $\mathbb{L}$ coincide (this case corresponds to the coverings of curves). Then $e = e(Q \mid P)$ is called the *ramification index* of the place Q over P. If $e = 1$, we say that Q is *unramified* over P, and if $e > 1$, $Q \mid P$ is *ramified*. If all places over P are unramified, we say that P is *unramified in the extension* $\mathbb{L}/\mathbb{K}$. If all places of $\mathbb{K}$ are unramified, then the extension $\mathbb{L}/\mathbb{K}$ is called *unramified*.

A simple argument shows that if the place Q lies over P, then $\deg P \mid \deg Q$.

PROPOSITION 6.1.7. *Let Q run over all places lying over P. Then*

$$\sum_{Q \mid P} \deg Q \leq n \deg P,$$

where $n = [\mathbb{L} : \mathbb{K}]$ is the degree of the extension, and the equality holds if and only if the extension is unramified at P. □

Thus, over P there are at most n places. If the number of them is n, we say that P is *totally split*. If $\deg Q > \deg P$, we say that there is *inertia*. If $\deg Q = n \deg P$, then we say that P is *inert* in $\mathbb{L}/\mathbb{K}$; in this case it is, of course, unramified. If over P there is a unique place Q and $\deg Q = \deg P$, then P is *totally ramified*.

EXAMPLE 6.1.8. Let $\mathbb{K} = \Bbbk(t)$, $\mathbb{L} = \Bbbk(u)$, and $t = u^2$. Then all the places $P \in \mathbb{P}_{\mathbb{K}}$ except for $P_0 = \{t = 0\}$ and $P_\infty = \{t = \infty\}$ are unramified in $\mathbb{L}/\mathbb{K}$, while P_0 and P_∞ are totally ramified.

As we have seen before in Sec. 2.1.1, the field extension $\mathbb{L}/\mathbb{K}$ corresponds to a map of curves $Y \to X$, where $\mathbb{K} = \Bbbk(X)$, $\mathbb{L} = \Bbbk(Y)$. The place Q lies over P if and only if in the language of curves the point Q is taken to P by this map.

Different and the Hurwitz formula. As we have seen in Sec. 2.2 of *Basic Notions*, one of the most important results in the theory of algebraic curves is the Hurwitz formula; see Theorem 2.2.36 and Corollary 2.2.37. This result can also be readily formulated and proved in the language of function fields, and we are going to present this form of the Hurwitz formula. Note that many constructions of good curves, i.e., those having many rational points, use class field theory and are commonly and conveniently expressed in the function fields framework, making this version of the Hurwitz formula a very useful tool.

The main ingredient of the Hurwitz formula in its geometric version is the ramification divisor B_f of a map $f\colon X \to Y$ of smooth projective curves. Assume now that X and Y are defined over $\mathbb{F}_q$ and denote $\mathbb{K} := \mathbb{F}_q(X)$, $\mathbb{L} = \mathbb{F}_q(Y)$. We have a field embedding $f^*\colon \mathbb{K} \hookrightarrow \mathbb{L}$ and we identify $\mathbb{K}$ with its image $\mathbb{F}^*(\mathbb{K})$ in $\mathbb{L}$, thus obtaining a function field extension $\mathbb{L}/\mathbb{K}$ of a finite degree $n = \deg f$.

We define the *different* $\operatorname{Diff}(\mathbb{L}/\mathbb{K})$, which is just the divisor B_f written in the function fields language, by

$$\operatorname{Diff}(\mathbb{L}/\mathbb{K}) := \sum_{P\in\mathbb{P}_\mathbb{K}} \sum_{Q\mid P} d(Q\,|\,P)\,Q,$$

where the *different exponent* $d(Q\,|\,P)$ is defined by

$$d(Q\,|\,P) := -\upsilon_Q(t_P).$$

Note that all $d(Q \mid P)$, except for a finitely many, are zero.

Now the Hurwitz formula for function fields can be written as follows.

THEOREM 6.1.9. *Let $g_\mathbb{L}$ be the genus of $\mathbb{L}$ (i.e., the genus of Y) and $g_\mathbb{K}$ the genus of $\mathbb{K}$ $(= g(X))$. Then*

$$2g_\mathbb{L} - 2 = n(2g_\mathbb{K} - 2) + \deg \operatorname{Diff}(\mathbb{L}/\mathbb{K}),$$

where

$$\deg \operatorname{Diff}(\mathbb{L}/\mathbb{K}) = \sum_{P\in\mathbb{P}_\mathbb{K}} \sum_{Q\mid P} d(Q\,|\,P) \deg Q. \qquad \square$$

Zeroes and poles. Let $f \in \mathbb{K}$ and $P \in \mathbb{P}_\mathbb{K}$. We say that P is a *zero* of f if $\upsilon_P(f) > 0$; in this case $\upsilon_P(f)$ is called the *zero order*. A place P is called a *pole* of f if $\upsilon_P(f) < 0$; in this case, $-\upsilon_P(f)$ is the *pole order*.

PROPOSITION 6.1.10. *Let $f \in \mathbb{K}/\mathbb{k}$ be transcendental over $\mathbb{k}$. Then f has at least one zero and one pole.* $\square$

COROLLARY 6.1.11. *The set $\mathbb{P}_\mathbb{K}$ is nonempty.* $\square$

In view of the latter result, Proposition 6.1.5 immediately implies the following fact.

COROLLARY 6.1.12. *The constant field $\widetilde{\mathbb{k}}$ of a function field $\mathbb{K}/\mathbb{k}$ is a finite-degree extension of $\mathbb{k}$.* $\square$

An important property of discrete valuations of a function field is their independence in the following sense: for different discrete valuations $\upsilon_1, \dots, \upsilon_n$ of $\mathbb{K}/\mathbb{k}$, if for some $f \in \mathbb{K}$ we know $\upsilon_i(f)$, $i = 1, \dots, n-1$, we have no information on $\upsilon_n(f)$. More precisely, we have the following theorem.

THEOREM 6.1.13 (weak approximation theorem). *Let $\mathbb{K}/\mathbb{k}$ be a function field, $P_1, \dots, P_n$ be pairwise distinct places of $\mathbb{K}$, $f_1, \dots, f_n \in \mathbb{K}$, and $r_1, \dots, r_n \in \mathbb{Z}$. Then there exists some $f \in \mathbb{K}$ such that*

$$\upsilon_{P_i}(f - f_i) = r_i, \quad i = 1, \dots, n. \qquad \square$$

COROLLARY 6.1.14. *Any function field has infinitely many places.* $\square$

Using the weak approximation theorem, one can also prove the following result.

PROPOSITION 6.1.15. *Let $P_1, \dots, P_r$ be zeroes of an element f of a function field $\mathbb{K}/\mathbb{k}$. Then*

$$\sum_{i-1}^{r} v_{P_i}(f) \deg P_i \leq [\mathbb{K} : \mathbb{k}(f)]. \qquad \square$$

COROLLARY 6.1.16. *Any nonzero element of a function field has finitely many zeroes and poles.* $\square$

Ramification groups and Hilbert's formula. In many cases the Hurwitz formula reduces calculation of the genus to calculation of the different exponents $d(Q\,|\,P)$ for ramified places. In practice this last calculation in the case of a Galois extension $\mathbb{L}/\mathbb{K}$ often uses the *Hilbert different formula*, which we are going to explain. This formula is based on the study of *ramification groups*, which are also important in some other respects.

Let $\mathbb{L}/\mathbb{K}$ be a Galois extension of function fields having the same constant field $\mathbb{k} = \mathbb{F}_q$, with the Galois group $G := \mathrm{Gal}(\mathbb{L}/\mathbb{K})$ of order n. Let us fix a pair of places $Q \in \mathbb{P}_{\mathbb{L}}$, $P \in \mathbb{P}_{\mathbb{K}}$ such that $Q\,|\,P$. Then define the following subgroups of G:

$$\begin{aligned} G(Q/P) &:= \{\sigma \in G \mid \sigma(Q) = Q\}, \\ I(Q/P) &:= \{\sigma \in G \mid v_Q(\sigma t_Q - t_Q) > 0\}. \end{aligned} \qquad (6.1.1)$$

The group $G(Q/P)$ is called the *decomposition group* of Q over P, and $I(Q/P)$ is called its *inertia group*. The inertia group $I(Q/P)$ is nontrivial if and only if Q is ramified over P.

PROPOSITION 6.1.17. (a) *We have*

$$|G(Q/P)| = e(Q/P) \cdot f(Q/P),$$

where $f(Q/P)$ is the degree of the residue field extension $[\mathbb{k}_Q : \mathbb{k}_P]$.

(b) *We have*

$$|I(P/Q)| = e(P/Q).$$

(c) *If we define for any $\sigma \in G(Q/P)$ an element $\overline{\sigma} \in \mathrm{Gal}(\mathbb{k}_Q/\mathbb{k}_P)$ by $\overline{\sigma}(z(Q)) := \sigma(z)(Q)$ for any $z \in \mathcal{O}_Q$, we get a surjective homomorphism*

$$G(Q/P) \to \mathrm{Gal}(\mathbb{k}_Q/\mathbb{k}_P)$$

with the kernel $I(Q/P)$. $\square$

Now let $i \geq -1$ be an integer. Define the subgroups of $G(Q/P)$ by

$$G_i(Q/P) := \{\sigma \in G \mid v_Q(\sigma t_Q - t_Q) \geq i+1\}. \qquad (6.1.2)$$

The subgroups $G_i(Q/P)$ are called the *ramification groups* of Q over P; thus,

$$G_{-1}(Q/P) = G(Q/P), \qquad G_0(Q/P) = I(Q/P).$$

The group $G_1(Q/P)$ is a normal subgroup of G_0, called the *wild ramification* subgroup. Its order is a power of $p = \mathrm{char}\,\mathbb{F}_q$, and the order of G_0/G_1 is coprime to p.

Hilbert's different formula is as follows.

THEOREM 6.1.18. *We have*

$$d(Q\,|\,P) = \sum_{i \geq 0} (g_i(Q/P) - 1),$$

where $g_i(Q/P) = |G_i(Q/P)|$ is the order of the ith ramification group. $\square$

All the above definitions and results are valid for any constant field $\Bbbk$, but below we will only use them in the case $\Bbbk = \mathbb{F}_q$, where the function field $\mathbb{K}/\mathbb{F}_q$ is (by definition) a global field. Thus, if $\Bbbk$ is not finite, the function field $\mathbb{K}/\Bbbk$ is *not* a global field.

6.1.2. Number Fields

Another class of global fields is represented by number fields (of finite degree).

Algebraic integers. A finite extension $\Bbbk$ of the field $\mathbb{Q}$ of rational numbers is called an *algebraic number field*. Its *degree* $n = [\Bbbk : \mathbb{Q}]$ equals the dimension of $\Bbbk$ as a vector space over $\mathbb{Q}$. In the present subsection, $\Bbbk$ always denotes an algebraic number field.

If an element $x \in \Bbbk$ satisfies a relation

$$x^m + a_{m-1}x^{m-1} + \ldots + a_1 x + a_0, \quad a_i \in \mathbb{Z},$$

then x is called an *algebraic integer*, or an *integral element* of $\Bbbk$.

PROPOSITION 6.1.19. *Sums and products of algebraic integers are also algebraic integers.*

PROOF. First let us show that x is integral if and only if there exists a nontrivial finitely generated $\mathbb{Z}$-module $M \subset \Bbbk$ such that $xM \subseteq M$. Indeed, if x is integral, then we can set $M = \mathbb{Z} + x\mathbb{Z} + \ldots + x^{m-1}\mathbb{Z}$; conversely, if $xM \subseteq M$ and $v_1, \ldots, v_m$ is a basis in M over $\mathbb{Z}$, then multiplication by x is given in this basis by a matrix A_x, and hence by the Cayley–Hamilton theorem x is a root of the equation $\det(\lambda - A_x) = 0$, which is of degree m and with leading coefficient 1.

Now let x and y be integers in $\Bbbk$, and let M and N be finitely generated $\mathbb{Z}$-submodules in $\Bbbk$ such that $xM \subseteq M$ and $yN \subseteq N$. Set

$$L = MN = \{ab \mid a \in M,\ b \in N\}.$$

Clearly, L is also a finitely generated $\mathbb{Z}$-submodule of $\Bbbk$. Then $xyL \subseteq L$ and $(x \pm y)L \subseteq L$. □

COROLLARY 6.1.20. *The set of integral elements of $\Bbbk$ is a ring.* □

This ring $\mathcal{O}_\Bbbk$ is called the *ring of integers* of $\Bbbk$, or its *maximal order*. Any subring $\mathcal{O} \subset \mathcal{O}_\Bbbk$ of a finite index $[\mathcal{O}_\Bbbk : \mathcal{O}]$ is called an *order*.

PROPOSITION 6.1.21. *For any $z \in \Bbbk$ there exists $c \in \mathbb{Z}$ such that cz is an algebraic integer in $\Bbbk$.*

PROOF. Since z is an algebraic number, we have

$$a_m z^m + a_{m-1} z^{m-1} + \ldots + a_0 = 0$$

for some $a_i \in \mathbb{Z}$. Now it suffices to take $c = a_m$, since we have

$$(a_m z)^m + a_m a_{m-1} (a_m z)^{m-1} + \ldots + a_m^{m-1} a_0 = 0. \qquad \square$$

PROPOSITION 6.1.22. *The ring $\mathcal{O}_\Bbbk$ is finitely generated as a $\mathbb{Z}$-module.*

SKETCH OF THE PROOF. Let $\{w_1, \ldots, w_n\}$ be a basis of $\Bbbk$ over $\mathbb{Q}$. We may assume that $w_i \in \mathcal{O}_\Bbbk$ for all $i = 1, \ldots, n$. Consider the linear form ($\Bbbk/\mathbb{Q}$-trace)

$$\mathrm{Tr} = \mathrm{Tr}_{\Bbbk/\mathbb{Q}} \colon \Bbbk \to \mathbb{Q};$$

one can show that this form is nondegenerate (see Proposition 6.1.24 below). Let $\{w'_1,\dots,w'_n\}$ be the dual basis:

$$\mathrm{Tr}(w'_i w_j) = \delta_{ij} = \begin{cases} 1 & \text{for } i=j, \\ 0 & \text{for } i\neq j. \end{cases}$$

Let $c\in\mathbb{Z}$ be such that $cw'_i\in\mathcal{O}_{\Bbbk}$ for any $i=1,\dots,n$. Then for any $x\in\mathcal{O}_{\Bbbk}$ we have $cw_i x\in\mathcal{O}_{\Bbbk}$ and $\mathrm{Tr}(cw_i x)\in\mathbb{Z}$. Hence, $\mathcal{O}_{\Bbbk}\subseteq c^{-1}(\mathbb{Z}w_1+\dots+\mathbb{Z}w_n)$, and thus $\mathcal{O}_{\Bbbk}$ is finitely generated. □

COROLLARY 6.1.23. *Any order $\mathcal{O}$ in $\Bbbk$ is a free abelian group of rank n.*

PROOF. The rank of $\mathcal{O}$ equals n, since $\Bbbk/\mathcal{O}$ is a torsion group. □

Thus, there exists a basis $w_1,\dots,w_n$ of $\Bbbk$ over $\mathbb{Q}$ such that

$$\mathcal{O}_{\Bbbk}=\mathbb{Z}w_1+\dots+\mathbb{Z}w_n.$$

Such a basis is called a *fundamental basis* of $\Bbbk$.

Trace and norm. Now let $\Sigma=\{\sigma_1,\dots,\sigma_n\}$ be the set of distinct embeddings of $\Bbbk$ in $\mathbb{C}$. Since for any embedding σ_i such that $\sigma_i(\Bbbk)$ does not lie in $\mathbb{R}$, the complex conjugate embedding $\overline{\sigma}_i$ does not coincide with σ_i; such embeddings are present in Σ in pairs $(\sigma_i,\overline{\sigma}_i)$. Thus, if s is the number of embeddings $\sigma_i\colon\Bbbk\hookrightarrow\mathbb{C}$ with $\sigma_i(\Bbbk)\subset\mathbb{R}$ (such embeddings are said to be *real*) and t is the number of pairs $(\sigma_i,\overline{\sigma}_i)$ with $\sigma_i\neq\overline{\sigma}_i$ (such embeddings are said to be *complex*), then $s+2t=n$.

For any $x\in\Bbbk$ define

$$\mathrm{Tr}(x)=\mathrm{Tr}_{\Bbbk/\mathbb{Q}}(x)=\sum_{i=1}^{n}\sigma_i(x),$$

$$N(x)=N_{\Bbbk/\mathbb{Q}}(x)=\prod_{i=1}^{n}\sigma_i(x).$$

$\mathrm{Tr}(x)$ is called the $(\Bbbk/\mathbb{Q})$-*trace* of x, and $N(x)$ the $(\Bbbk/\mathbb{Q})$-*norm* of x.

PROPOSITION 6.1.24. (1) *If $a_m x^m+\dots+a_0=0$ is the minimal equation of x over $\mathbb{Q}$, then $m\mid n$, $\mathrm{Tr}(x)=-\dfrac{na_{m-1}}{ma_m}$, and $N(x)=(-1)^n\left(\dfrac{a_0}{a_m}\right)^{n/m}$.*

(2) *The bilinear form $\mathrm{Tr}(xy)$ is nondegenerate.*

(3) *We have $N(x)\in\mathbb{Z}$ if and only if $x\in\mathcal{O}_{\Bbbk}$.* □

Discriminant. Let $\Bbbk$ be an algebraic number field of degree n; let $\{w_1,\dots,w_n\}$ be its fundamental basis. The integer $D_{\Bbbk}=\det(\mathrm{Tr}(w_i w_j))$ is called the (absolute) *discriminant* of $\Bbbk$. One can check that this definition does not depend on the choice of a fundamental basis $\{w_1,\dots,w_n\}$.

The following fact is of importance.

THEOREM 6.1.25 (Hermite). *If $n>1$, i.e., $\Bbbk\neq\mathbb{Q}$, then $|D_{\Bbbk}|>1$.* □

COROLLARY 6.1.26. *There is no unramified extension of $\mathbb{Q}$.*

PROOF. Indeed, any p which divides $|D_{\Bbbk}|$ is ramified in $\Bbbk$; see Proposition 6.1.30 below. □

Another possible definition of $D_{\Bbbk}$ is as follows. Let s be the number of real embeddings σ_i and let t be the number of conjugate pairs $(\sigma_j, \overline{\sigma}_j)$ of complex embeddings. Let $A = \mathbb{R}^s \times \mathbb{C}^t$ be a commutative algebra over $\mathbb{R}$ of rank $n = s + 2t$, and let σ be the following ring embedding:

$$\sigma\colon \Bbbk \to \mathbb{R}^s \times \mathbb{C}^t, \quad a \mapsto (\sigma_1(a), \dots, \sigma_s(a); \sigma_{s+1}(a), \dots, \sigma_{s+t}(a)).$$

PROPOSITION 6.1.27. (1) *$\sigma(\Bbbk)$ generates A (over $\mathbb{R}$); $\sigma(\mathcal{O}_{\Bbbk})$ is a lattice in $A \cong \mathbb{R}^n$, i.e., a free abelian subgroup of $\mathbb{R}$ of full rank.*

(2) *We have*

$$|\det \sigma(\mathcal{O}_{\Bbbk})| = 2^{-t}\sqrt{|D_{\Bbbk}|}.$$

PROOF. Let $w_1, \dots, w_n$ be a fundamental basis of $\Bbbk$, let $\sigma_j(w_i) = x_{ij} \in \mathbb{R}$ for any i and any $j = 1, \dots, s$, and let $\sigma_{s+j}(w_i) = y_{ji} + \sqrt{-1}\,z_{ji}$ for any i and any $j = 1, \dots, t$, where $y_{ji}, z_{ji} \in \mathbb{R}$. Then

$$d = \det \sigma(\mathcal{O}_{\Bbbk}) = \det \begin{pmatrix} x_{11} & \dots & x_{s1} & y_{11} & z_{11} & \dots & y_{t1} & z_{t1} \\ \dots & \dots & \dots & \dots & \dots & \dots & \dots & \dots \\ x_{1n} & \dots & x_{sn} & y_{1n} & z_{1n} & \dots & y_{tn} & z_{tn} \end{pmatrix}.$$

It is clear that $d = d^*/(-2\sqrt{-1})^t$, where

$$d^* = \det \begin{pmatrix} x_{11} & \dots & x_{s1} & y_{11} + \sqrt{-1}\,z_{11} & y_{11} - \sqrt{-1}\,z_{11} & \dots \\ \dots & \dots & \dots & \dots & \dots & \dots \\ x_{1n} & \dots & x_{sn} & y_{1n} + \sqrt{-1}\,z_{1n} & y_{1n} - \sqrt{-1}\,z_{1n} & \dots \end{pmatrix};$$

i.e., $d^* = \det(\sigma_i(w_j))$, where $\{\sigma_1, \dots, \sigma_n\}$ is a full set of embeddings of $\Bbbk$ in $\mathbb{C}$ with $\sigma_{s+t+j} = \overline{\sigma}_{s+j}$, $j = 1, \dots, t$. Since by the definition of the trace we have

$$\operatorname{Tr}(w_i w_j) = \sum_{\ell=1}^{n} \sigma_\ell(w_i)\sigma_\ell(w_j), \quad 1 \le i, j \le n,$$

there is a matrix identity

$$(\operatorname{Tr}(w_i w_j)) = (\sigma_\ell(w_i))(\sigma_\ell(w_j))^T$$

(where ^{T} denotes transposition), whence

$$D_{\Bbbk} = \det(\operatorname{Tr}(w_i w_j)) = (d^*)^2,$$

as desired. □

Units. An element $a \in \mathcal{O}_{\Bbbk}$ is called a *unit* if $a^{-1} \in \mathcal{O}_{\Bbbk}$. Clearly, all units form an abelian group, which is denoted by $\mathcal{O}_{\Bbbk}^*$; the elements ζ of $\mathcal{O}_{\Bbbk}^*$ can be characterized by the condition $N(\zeta) = \pm 1$; torsion elements of $\mathcal{O}_{\Bbbk}^*$ are roots of unity, $\zeta^m = 1$ for some $m \in \mathbb{N}$.

The structure of the group $\mathcal{O}_{\Bbbk}^*$ is rather simple and is given by the following Dirichlet's unit theorem.

THEOREM 6.1.28. *The group $\mathcal{O}_{\Bbbk}^*$ is the product of its finite torsion subgroup and a free abelian group of rank $r = s + t - 1$.*

SKETCH OF THE PROOF. Consider the map

$$\log\colon \mathcal{O}_{\Bbbk}^* \to \mathbb{R}^{s+t}, \quad a \mapsto (\log|\sigma_1(a)|, \dots, \log|\sigma_s(a)|; \log|\sigma_{s+1}(a)|^2, \dots, \log|\sigma_{s+t}(a)|^2).$$

Its kernel is the torsion subgroup of $\mathcal{O}_{\Bbbk}^*$, and its image $\log(\mathcal{O}_{\Bbbk}^*)$ is contained in the hyperplane $H \subset \mathbb{R}^{s+t}$ defined by $x_1 + \dots + x_{s+t} = 0$. Indeed, if a lies in the kernel, then

$$|\sigma_1(a)\dots\sigma_s(a)||\sigma_{s+1}(a)|^2\dots|\sigma_{s+t}(a)|^2 = |N(a)| = 1$$

and $a^s = 1$ for some $s \in \mathbb{N}$. One can also show that $\log(\mathcal{O}_{\Bbbk}^*)$ is a (full-rank) lattice in H, which is implied by compactness of $H/\log(\mathcal{O}_{\Bbbk}^*)$, and the theorem follows. □

Now we define one more important notion. The *determinant* of this lattice,

$$R = R_{\Bbbk} = \det\begin{pmatrix} \log|\sigma_1(u_1)| & \dots & \log|\sigma_1(u_{s+t-1})| \\ \hdotsfor{3} \\ \log|\sigma_{s+t-1}(u_1)| & \dots & \log|\sigma_{s+t-1}(u_{s+t-1})| \end{pmatrix},$$

where $u_1, \dots, u_{s+t-1}$ is a basis of $\mathcal{O}_{\Bbbk}^*$ modulo torsion, is called the *regulator* of $\Bbbk$.

Places. To define places in the number field case, one should use a generalization of this notion given in Sec. 6.1.1. This generalization uses the language of absolute values. Let for a moment $\Bbbk$ be an arbitrary field. A map $\|\cdot\|\colon \Bbbk \to \mathbb{R}$ is called an *absolute value*, or a (*multiplicative*) *valuation*, if it satisfies the following conditions:

- $\|0\| = 0$, and $\|x\| > 0$ if $x \neq 0$.
- There exist $x, y \in \Bbbk$ such that $\|x\| \neq \|y\|$.
- $\|xy\| = \|x\|\|y\|$.
- There exists a positive real λ such that

$$\|x+y\| \leq \lambda(\|x\| + \|y\|).$$

Two absolute values $\|\cdot\|_1$ and $\|\cdot\|_2$ are *equivalent* if there exists a positive real θ such that $\|\cdot\|_1 = \|\cdot\|_2^{\theta}$. An equivalence class of absolute values is called a *place* of $\Bbbk$.

In the function field case it is easy to prove that the introduced notion is equivalent to that given in Sec. 6.1.1. Indeed, we can associate with any valuation ring $\mathcal{O}_P$, $P \in \mathbb{P}_{\Bbbk}$, a valuation $\|x\|_P := q^{v_P(x)}$, q being the cardinality of the field $\Bbbk_P = \mathcal{O}_P/P$. This gives a natural identification of places defined in Sec. 6.1.1 with those defined in this subsection.

One can give an explicit description of all places of a given number field, which is very close to that for a function field; one only needs to add so-called infinite places.

Let $\sigma\colon \Bbbk \hookrightarrow \mathbb{C}$ be an embedding of fields. Put $\|x\|_\sigma = |\sigma(x)|$ if σ is a real embedding (i.e., $\operatorname{Im}\sigma \in \mathbb{R}$), and $\|x\|_\sigma = |\sigma(x)|^2$ if σ is a complex embedding. These are absolute values. One can check that two such absolute values $\|\cdot\|_\sigma$ and $\|\cdot\|_{\sigma'}$ are equivalent if and only if either $\sigma' = \sigma$ or $\sigma' = \overline{\sigma}$. Thus we obtain s *real* and t *complex places* of $\Bbbk$. These places are said to be *infinite*, or *Archimedean*; the set of infinite places is denoted by S_∞.

Next, let $\mathfrak{p}$ be a maximal ideal of $\mathcal{O}_{\Bbbk}$. For $x \in \Bbbk^*$ define

$$\operatorname{ord}_{\mathfrak{p}}(x) = \max\{n \mid x \in \mathfrak{p}^n\}.$$

One easily checks that

$$\operatorname{ord}_{\mathfrak{p}}(xy) = \operatorname{ord}_{\mathfrak{p}}(x) + \operatorname{ord}_{\mathfrak{p}}(y)$$

for any $x, y \in \Bbbk^*$, and if also $x + y \in \Bbbk^*$, then

$$\operatorname{ord}_{\mathfrak{p}}(x+y) \geq \min\{\operatorname{ord}_{\mathfrak{p}}(x), \operatorname{ord}_{\mathfrak{p}}(y)\}.$$

Define the corresponding absolute value: for $x \in \Bbbk^*$ let

$$\|x\|_{\mathfrak{p}} = N(\mathfrak{p})^{-\operatorname{ord}_{\mathfrak{p}}(x)},$$

where $N(\mathfrak{p}) = |\mathcal{O}_{\Bbbk}/\mathfrak{p}|$. For such an absolute value (and for any equivalent to it), a stronger condition holds:

$$\|x+y\| \leq \lambda \max\{\|x\|, \|y\|\}$$

(which is not the case for Archimedean absolute values). Such absolute values are said to be *non-Archimedean*. If $\mathfrak{p} \neq \mathfrak{p}'$, then the corresponding absolute values are not equivalent; i.e., each maximal ideal corresponds to a place of $\Bbbk$. Such places are said to be *finite*, or *non-Archimedean*. Below we will sometimes refer to places of a number field as its *primes*.

It turns out that each place of a number field is either infinite or finite. If v is a place of $\Bbbk$, then the absolute values defined above are said to be *normalized* and are denoted by $\|\cdot\|_v$.

If there are no complex places, the number field is called *totally real*; if there are no real places, it is called *totally complex*.

Class group. Let $\mathfrak{a}$ be an ideal of $\mathcal{O}_{\Bbbk}$, and let $a \in \Bbbk^*$. The set $\mathfrak{c} = a^{-1}\mathfrak{a}$ is called a *fractional ideal*. The set of nonzero fractional ideals is a group with the composition law given by

$$\mathfrak{a} \cdot \mathfrak{b} = \{xy \mid x \in \mathfrak{a},\ y \in \mathfrak{b}\}.$$

Note that the inverse element is given by

$$\mathfrak{a}^{-1} = \{x \mid x^{-1} \in \mathfrak{a} \setminus \{0\}\} \cup \{0\}.$$

We call fractional ideals $\mathfrak{c}$ and $\mathfrak{c}_1$ *equivalent* if $\mathfrak{c}_1 = a\mathfrak{c}$ for some $a \in \Bbbk^*$. Equivalence classes of nonzero fractional ideals form a group $\mathrm{Cl}_{\Bbbk}$, which is called the (ideal) *class group* of $\Bbbk$.

THEOREM 6.1.29. *The group* $\mathrm{Cl}_{\Bbbk}$ *is finite for any algebraic number field* $\Bbbk$. $\square$

The order $h = h_{\Bbbk} = |\mathrm{Cl}_{\Bbbk}|$ of the class group is called the *class number* of $\Bbbk$. This is the most important invariant of a number field. For example, $h_{\Bbbk} = 1$ is equivalent to uniqueness of prime factorization in $\mathcal{O}_{\Bbbk}$, which holds, e.g., for $\Bbbk = \mathbb{Q}$, $\Bbbk = \mathbb{Q}(\sqrt{-1})$, $\Bbbk = \mathbb{Q}(\sqrt{2})$, and many other number fields, but does not hold, e.g., for $\Bbbk = \mathbb{Q}(\sqrt{-5})$ or $\Bbbk = \mathbb{Q}(\sqrt{-p})$ for a prime $p > 163$.

Extensions. Now consider a finite extension $\mathbb{K}$ of $\Bbbk$. Let $[\mathbb{K} : \Bbbk] = \dim_{\Bbbk} \mathbb{K} = m$, and let $\mathcal{O}_{\mathbb{K}}$ and $\mathcal{O}_{\Bbbk}$ be the corresponding rings of integers. Then any $x \in \mathbb{K}$ is a root of an equation of the form $P(x) = a_m x^m + \ldots + a_0 = 0$ irreducible over $\Bbbk$, where $a_i \in \mathcal{O}_{\Bbbk}$; i.e., $P(x)$ is a minimal polynomial of x.

Define the (*relative*) *trace* and *norm* as

$$\operatorname{Tr}_{\mathbb{K}/\Bbbk}(x) = -\frac{a_{m-1}}{a_m},$$
$$N_{\mathbb{K}/\Bbbk}(x) = (-1)^m \frac{a_0}{a_m}.$$

Different. Consider the following subset in $\mathbb{K}$:

$$\mathfrak{B}_{\mathbb{K}/\Bbbk} = \{x \in \mathbb{K} \mid \operatorname{Tr}_{\mathbb{K}/\Bbbk}(xy) \in \mathcal{O}_{\Bbbk}, \text{ for any } y \in \mathcal{O}_{\Bbbk}\}.$$

One can easily check that $\mathfrak{B}_{\mathbb{K}/\mathbb{k}}$ is an $\mathcal{O}_{\mathbb{k}}$-submodule in $\mathbb{K}$, which contains $\mathcal{O}_{\mathbb{K}}$. Hence, there is a unique ideal $\mathfrak{D}_{\mathbb{K}/\mathbb{k}}$ in $\mathcal{O}_{\mathbb{K}}$ such that $\mathfrak{D}_{\mathbb{K}/\mathbb{k}} \cdot \mathfrak{B}_{\mathbb{K}/\mathbb{k}} = \mathcal{O}_{\mathbb{K}}$, i.e., $\mathfrak{D}_{\mathbb{K}/\mathbb{k}} = \mathfrak{B}_{\mathbb{K}/\mathbb{k}}^{-1}$. The ideal $\mathfrak{D}_{\mathbb{K}/\mathbb{k}}$ is called the *different* of the extension $\mathbb{K}/\mathbb{k}$. The ideal

$$D_{\mathbb{K}/\mathbb{k}} = \{N_{\mathbb{K}/\mathbb{k}}(x) \mid x \in \mathfrak{D}_{\mathbb{K}/\mathbb{k}}\}$$

in $\mathcal{O}_{\mathbb{k}}$ is called the (*relative*) *discriminant* of the extension $\mathbb{K}/\mathbb{k}$.

This notion is consistent with the above definition of the absolute discriminant; i.e., the relative discriminant $D_{\mathbb{k}/\mathbb{Q}}$ coincides with the ideal in $\mathbb{Z}$ generated by the absolute discriminant $D_{\mathbb{k}}$.

Thus, $D_{\mathbb{k}}$ is determined by $D_{\mathbb{k}/\mathbb{Q}}$ up to a sign.

We have the following transitivity relation for differents and discriminants.

PROPOSITION 6.1.30. (1) *Let $\mathbb{L} \supset \mathbb{K} \supset \mathbb{k}$ be a tower of algebraic number fields. Then*

$$\mathfrak{D}_{\mathbb{L}/\mathbb{k}} = \mathfrak{D}_{\mathbb{L}/\mathbb{K}} \cdot \mathfrak{D}_{\mathbb{K}/\mathbb{k}}.$$

(2) *Let $[\mathbb{L} : \mathbb{K}] = \ell$. Then*

$$D_{\mathbb{L}/\mathbb{k}} = D_{\mathbb{K}/\mathbb{k}}^{\ell} \cdot N_{\mathbb{K}/\mathbb{k}}(D_{\mathbb{L}/\mathbb{K}}). \qquad \square$$

Unramified extensions. An algebraic extension $\mathbb{K}/\mathbb{k}$ is called *unramified* if its relative discriminant is trivial, i.e., $D_{\mathbb{K}/\mathbb{k}} = (1)$. Theorem 6.1.25 (together with Proposition 6.1.30) implies that $\mathbb{Q}$ has no unramified extensions.

The rule $\mathfrak{a} \mapsto \mathfrak{a}\mathcal{O}_{\mathbb{K}}$ defines a group homomorphism $\mathrm{Cl}_{\mathbb{k}} \to \mathrm{Cl}_{\mathbb{K}}$; the norm map defines a homomorphism $\mathrm{Cl}_{\mathbb{K}} \to \mathrm{Cl}_{\mathbb{k}}$. If an extension $\mathbb{K}/\mathbb{k}$ is unramified and abelian (i.e., normal with an abelian Galois group $\mathrm{Gal}(\mathbb{K}/\mathbb{k})$), then the global class field theory (see Theorem 6.3.3 below) yields the following result.

THEOREM 6.1.31. *The group* $\mathrm{Gal}(\mathbb{K}/\mathbb{k})$ *is isomorphic to the quotient group* $\mathrm{Cl}_{\mathbb{k}}/N_{\mathbb{K}/\mathbb{k}}(\mathrm{Cl}_{\mathbb{K}})$. $\square$

Moreover, there exists a maximal unramified abelian extension $\mathbb{K}_1$ of $\mathbb{k}$, which is called the *Hilbert* (or *absolute*) *class field* of $\mathbb{k}$. Its Galois group $\mathrm{Gal}(\mathbb{K}_1/\mathbb{k})$ is isomorphic to $\mathrm{Cl}_{\mathbb{k}}$.

Function fields and number fields. As we have seen, number fields and fields of rational functions on curves over finite fields have many features in common, and in many aspects they can be considered simultaneously. Above, we have seen some common features of both; let us outline some more analogies.

Let $\mathbb{k}$ be an algebraic number field, and let $\mathcal{O}_{\mathbb{k}}$ be its ring of integers. Let X be a curve over $\mathbb{F}_q$, $\mathbb{K} = \mathbb{F}_q(X)$; let F be a finite set of closed points of X, $U = X \setminus F$; and let $\mathcal{O}_F = \mathbb{F}_q[U]$ be the ring of rational functions that are regular on U.

For both rings $\mathcal{O}_{\mathbb{k}}$ and $\mathcal{O}_F$ any factor modulo a maximal ideal is a finite field. In the case of $\mathcal{O}_F$ all these fields contain $\mathbb{F}_q$ (the so-called "case of equal characteristics"); in the number field case, among these fields there is an extension of $\mathbb{F}_p$ for any prime p (the "case of mixed characteristics").

In the function field case we can choose various finite sets F and get various rings $\mathcal{O}_F$. In the number case we can choose a finite set S of maximal ideals of $\mathcal{O}_{\mathbb{k}}$ and consider the ring $\mathcal{O}_S$ which is obtained from $\mathcal{O}_{\mathbb{k}}$ by inverting nonzero elements of ideals from S; note that $\mathrm{Max}\,\mathcal{O}_S = \mathrm{Max}\,\mathcal{O}_{\mathbb{k}} \setminus S$, $\mathrm{Max}(A)$ being the set of maximal ideals in a commutative ring A, and $\mathcal{O}_{\varnothing} = \mathcal{O}_{\mathbb{k}}$. Rings of the form $\mathcal{O}_S$ or $\mathcal{O}_F$ can be characterized by a single conceptual algebraic condition, which we do not give here.

Number case problems are usually more difficult than their analogues for function fields. The main reason for this is the presence of infinite places, which are not determined by discrete valuation rings.

An analogue of $\mathbb{Q}$ is the field $\mathbb{F}_q(T)$, where T is a variable, since $\Bbbk = \mathbb{F}_q(X)$ (for a curve X over $\mathbb{F}_q$) is a finite extension of $\mathbb{F}_q(T)$; the ring $\mathbb{F}_q[T]$ is an analogue of $\mathbb{Z}$. Note nevertheless that there is no canonical embedding of $\mathbb{F}_q(T)$ in $\mathbb{F}_q(X)$, and hence we cannot say that $[\mathbb{F}_q(X):\mathbb{F}_q(T)]$ is an analogue of the degree of an algebraic number field.

However, one can suggest another analogue of the degree, namely, the number of $\mathbb{F}_q$-points of X. Indeed, if $|X(\mathbb{F}_q)| = N$, then the degree of a map $f\colon X \to \mathbb{P}^1$ (i.e., the degree of an extension $[\mathbb{F}_q(X):\mathbb{F}_q(T)]$) cannot be too small: $\deg f \geq \dfrac{N}{q+1}$, since any $\mathbb{F}_q$-point of X is mapped to an $\mathbb{F}_q$-point of $\mathbb{P}^1$ and any fibre of f contains at most $\deg f$ $\mathbb{F}_q$-rational points.

Let a map $f\colon X \to \mathbb{P}^1$ be fixed, and let us fix an $\mathbb{F}_q$-point ∞ on $\mathbb{P}^1$. Then we have $\mathbb{P}^1 \setminus \infty = \mathbb{A}^1$, $\mathbb{F}_q[\mathbb{A}^1] = \mathbb{F}_q[T]$, and we can regard the integral closure of $\mathbb{F}_q[T]$ in $\mathbb{F}_q(X)$ as an analogue of $\mathcal{O}_\Bbbk$. Note that this closure coincides with $\mathcal{O}_{F_\infty} = \mathbb{F}_q[X \setminus F_\infty]$, where $F_\infty = f^{-1}(\infty)$.

The ramification divisor B_f of a map f is an analogue of the different. The discriminant corresponds to the divisor $D = \sum e_p P$, where e_p is the ramification index of $P \in \mathbb{P}^1$. The Hurwitz formula (Theorem 2.2.36) corresponds to Proposition 6.1.30. The value $\log\sqrt{|D_\Bbbk|}$ can be thought of as an analogue of the genus of a curve (for a development of this idea, see Chapter 8 below).

A fractional ideal $\mathfrak{a} = \prod \mathfrak{p}^{n_\mathfrak{p}}$, where $\mathfrak{p}$ runs over prime ideals of $\mathcal{O}_\Bbbk$, corresponds to a divisor on X, $\mathfrak{a}^{-1}$ corresponds to $L(D)$, and the group $\operatorname{Pic} X$ is an analogue of $\mathrm{Cl}_\Bbbk$.

Note also that $\mathbb{F}_q(T)$ (just like $\mathbb{Q}$) has no unramified extensions.

Units $\mathcal{O}_S^*$ of the ring $\mathcal{O}_S$ correspond to units $\mathcal{O}_F^*$ of $\mathcal{O}_F$; the group $\mu_\Bbbk$ is an analogue of $\mathbb{F}_q^*$. Moreover, just as in the number field case, $\mathcal{O}_F^*/\mathbb{F}_q^*$ is a free abelian group of rank $|F|-1$ (note that an easy generalization of Theorem 6.1.28 says that $\mathcal{O}_S^* \cong \mathbb{Z}^{s+t+|S|-1}$ modulo torsion).

6.2. Local Class Field Theory

Local class field theory describes abelian extensions of local fields, and we are going to explain this notion and the main results on such extensions. We start with a field K and assume that there is an absolute value (or valuation) $\|\cdot\|\colon K \to \mathbb{R}_+$.

Equivalent valuations. A valuation $\|\cdot\|$ defines a metric on K with distance function

$$d(a,b) = \|a-b\|,$$

and hence a topology on K: for $a \in K$, the sets

$$U(a,\varepsilon) = \{x \in K \mid \|x-a\| < \varepsilon\}, \quad \varepsilon > 0,$$

form a fundamental system of open neighbourhoods of a. A set is open if and only if it is a union of sets of the form $U(a,\varepsilon)$.

For example, for the topology on $\mathbb{Q}$ defined by $\|\cdot\|_p$, a and b are close if their difference is divisible by a high power of p. In particular, the sequence

$$1, p, p^2, \dots, p^n, \dots$$

converges to 0 in this metric.

If $\mathfrak{p}$ is a place of a global field $\Bbbk$, then the topology defined by the $\mathfrak{p}$-adic valuation $\|\cdot\|_{\mathfrak{p}}$ is called the $\mathfrak{p}$-*adic topology* on $\Bbbk$.

PROPOSITION 6.2.1. *Let $\|\cdot\|_1$ and $\|\cdot\|_2$ be valuations on K, with $\|\cdot\|_1$ nontrivial. The following conditions are equivalent:*

(a) *$\|\cdot\|_1$ and $\|\cdot\|_2$ define the same topology on K;*
(b) *$\|\alpha\|_1 < 1 \Rightarrow \|\alpha\|_2 < 1$;*
(c) *$\|\cdot\|_2 = \|\cdot\|_1^a$ for some $a > 0$.* □

In fact, these conditions are equivalent to those given above, respectively, in Secs. 6.1.1 and 6.1.2 for number fields and function fields.

Two valuations are said to be *equivalent* if they satisfy the conditions of the proposition.

Completions. Let K be a field with a nontrivial valuation. A sequence (a_n) of elements in K is called a *Cauchy sequence* if, for every $\varepsilon > 0$, there is a number N such that

$$\|a_n - a_m\| < \varepsilon, \quad \text{for all } m, n > N.$$

The field K is said to be *complete* if every Cauchy sequence has a limit in K. Note that the limit is necessarily unique.

PROPOSITION 6.2.2. *If $\|\cdot\|$ is a discrete non-Archimedean valuation on K defined by a maximal ideal $\mathfrak{p} \subset \mathcal{O}_K$, then the corresponding completion $\widehat{K}_{\mathfrak{p}}$ is canonically isomorphic to the fraction field of the ring $\widehat{\mathcal{O}}_{\mathfrak{p}} = \varprojlim_{m\to\infty} \mathcal{O}_K/\mathfrak{p}^m$.* □

For a sequence of finite sets A_m with surjective maps $\pi_{m+1}\colon A_{m+1} \to A_m$, recall that the *inverse limit* (or *projective limit*) $\varprojlim$ is defined as a set of sequences $(a_1, a_2, \dots, a_i, \dots) \in \prod_{m=1}^{\infty} A_m$ such that $\pi_{m+1}(a_{m+1}) = a_i$ for any $i = 1, 2, \dots$. For instance, if $\mathfrak{p} = (p) \in \mathbb{Z}$, we have $\widehat{\mathcal{O}}_{\mathfrak{p}} = \mathbb{Z}_p$, the ring of integer p-adic numbers, and $\widehat{K}_{\mathfrak{p}} = \mathbb{Q}_p$, the field of p-adic numbers.

Local fields. A *local field* K is a field which is complete under a nontrivial valuation and is locally compact in the induced topology. In fact, this abstract definition can be reformulated more concretely.

PROPOSITION 6.2.3. *A field K is a local field if and only if it is one of the following*:

(a) *K is a finite extension of the field $\mathbb{Q}_p$ of p-adic numbers*;
(b) *K is a finite extension of the field of Laurent series $\mathbb{F}_p((T))$*;
(c) *$K = \mathbb{R}$ or $\mathbb{C}$.* □

Case (c) is referred to as Archimedean; cases (a) and (b) are said to be non-Archimedean; (a) and (c) correspond to completions of number fields; in case (b), K is a completion of a function field $\mathbb{F}_q(X)$.

Until the end of the section, we assume that K is local.

Extensions of valuations. One can extend a valuation of a local field to its finite extension.

THEOREM 6.2.4. *Let K be complete with respect to a discrete valuation $\|\cdot\|_K$, and let L be a finite separable extension of K of degree n. Then $\|\cdot\|_K$ extends uniquely to a discrete valuation $\|\cdot\|_L$ on L, and L is complete for the extended valuation. For all $\beta \in L$,*

$$\|\beta\|_L = \|N_{L/K}(\beta)\|_K^{1/n},$$

$N_{L/K}(\beta)$ being the determinant of the K-linear endomorphism of L given by multiplication by β. □

Let A be the discrete valuation ring in K, and let B be its integral closure in L. Let $\mathfrak{p}$ be the maximal ideal of A, and let $\mathfrak{P}$ be the unique maximal ideal in B such that $\mathfrak{P} \cap A = \mathfrak{p}$. Let k be the residue field of A.

Let $e \in \mathbb{N}$ be such that $\mathfrak{p}B = \mathfrak{P}^e$; then e is called the *ramification index* of $\mathfrak{p}$. We also denote by $f := [k_{\mathfrak{P}} : k_{\mathfrak{p}}]$ the degree of the residue field extension.

COROLLARY 6.2.5. *For K and L as in Theorem 6.2.4, one has $n = ef$.* □

One should compare this with Proposition 6.1.6 above.

If $e = n$, so that $\mathfrak{p}B = \mathfrak{P}^n$, then L is said to be *totally ramified* over K; if $f = n$, then L is said to be *unramified* over K. These notions are equivalent to those introduced above (in some sense to be specified below).

If K is non-Archimedean, then the residue field k is finite and thus perfect, i.e., having only separable extensions.

If L is an algebraic (possibly infinite) extension of K, we can still define

$$B = \{\alpha \in L \mid \|\alpha\| \leq 1\},$$
$$\mathfrak{p} = \{\alpha \in B \mid \|\alpha\| < 1\}$$

and call $B/\mathfrak{p}$ the residue field of L.

PROPOSITION 6.2.6. *There is a field $K_0 \in L$ containing all unramified extensions of K in L (called the largest unramified extension of K in L). In fact, it is obtained from K by adjoining all roots of 1 of order prime to the characteristic of k.* □

Let $\overline{K}$ and $\overline{k}$ be algebraic closures of K and k, respectively.

COROLLARY 6.2.7. *The residue field of* $\overline{K}$ *is* $\overline{k}$; *there is a subfield* K^{un} *of* $\overline{K}$ *such that a subfield* L *of* $\overline{K}$ *finite over* K *is unramified if and only if* $L \subset K^{\text{un}}$. □

Ramification groups. We can look at the ramification groups defined above in Section 6.1.1 from the local fields perspective. This leads to the following definitions. Let L be a finite Galois extension of K; then the Galois group $G := \text{Gal}(L/K)$ preserves the valuation on L. In particular, it preserves

$$B = \{\alpha \in L \mid \|\alpha\| \leq 1\}, \qquad \mathfrak{p} = \{\alpha \in L \mid \|\alpha\| < 1\}.$$

Let Π be a prime element of L (so that $\mathfrak{p} = (\Pi)$). We define a sequence of subgroups $G \supset G_0 \supset G_1 \supset \dots$ by the condition

$$\sigma \in G_i \iff \|\sigma\alpha - \alpha\| < \|\Pi\|^i, \quad \text{for all } \alpha \in B.$$

The group G_0 is called the *inertia group*, the group G_1 is called the (local) *ramification group*, and the groups G_i, $i > 1$, are called *higher ramification groups* of L over K.

LEMMA 6.2.8. *The groups* G_i *are normal subgroups of* G, *and for* i *large enough one has* $G_i = \{1\}$. □

THEOREM 6.2.9. *Let* L/K *be a Galois extension of local fields. Then*

(a) *The fixed field of* G_0 *is the largest unramified extension* K_0 *of* K *in* L, *and*

$$G/G_0 = \text{Gal}(K_0/K) = \text{Gal}(l/k).$$

(b) *For* $i \geq 1$, *one has*

$$G_i = \{\sigma \in G_0 \mid \|\sigma\Pi - \Pi\| < \|\Pi\|^i\}.$$ □

COROLLARY 6.2.10. *There is an exhaustive filtration* $G \supset G_0 \supset \dots$ *such that*

- $G/G_0 = \text{Gal}(l/k)$;
- $G_0/G_1 \hookrightarrow l^*$;
- $G_i/G_{i+1} \hookrightarrow l$.

Since k *is finite, the group* $\text{Gal}(L/K)$ *is solvable, i.e., admits a descending sequence of subgroups with abelian successive factors.* □

An extension L/K is said to be *wildly ramified* if $p \mid e$, where $p = \text{char}\, k$. Otherwise, it is said to be *tamely ramified*. Hence, for a Galois extension we have

$$L/K \text{ is unramified} \iff G_0 = \{1\}$$

and

$$L/K \text{ is tamely ramified} \iff G_1 = \{1\}.$$

Now let us briefly explain a relationship between the (local) ramification groups that we have just defined and the ramification groups for global fields defined in the preceding section by (6.1.1) and (6.1.2). In fact, they can be identified if we look at them properly. Indeed, for a finite global field extension $\mathbb{L}/\mathbb{K}$ and a pair of places $Q \mid P$, $Q \in \mathbb{P}_{\mathbb{L}}$, $P \in \mathbb{P}_{\mathbb{K}}$, we can define the local fields $K := \widehat{\mathbb{K}}_P$, $L = \widehat{\mathbb{L}}_Q$ as the respective completion of $\mathbb{K}$ and $\mathbb{L}$ with respect to the valuations defined by P and Q. Having done this, one gets the following result.

PROPOSITION 6.2.11. (i) *Any* $\sigma \in G(Q/P)$ *defines a unique* $\overline{\sigma} \in \text{Gal}(L/K)$ *by the condition* $\overline{\sigma}|_L = \sigma$, *and the correspondence* $\sigma \mapsto \overline{\sigma}$ *is an isomorphism of* $G(Q/P)$ *onto* $\text{Gal}(L/K)$.

(ii) *Under this bijection, the groups* $G_i(Q/P) \subset G(Q/P)$ *get identified with the groups* $G_i \subset \mathrm{Gal}(L/K)$. □

Galois groups. Before formulating the main results of the theory, we need some auxiliary information on Galois theory for infinite extensions.

Fix a field K. A field $\Omega \supset K$, not necessarily of finite degree, is said to be *Galois* over K if

(a) Ω is algebraic and separable over K, i.e., every element of Ω is a simple root of a polynomial with coefficients in K;
(b) Ω is normal over K, i.e., every irreducible polynomial with coefficients in K having a root in Ω splits into linear factors in $\Omega[X]$.

PROPOSITION 6.2.12. *A field* Ω *is Galois over* K *if and only if it is a union of finite Galois extensions of* K. □

If Ω is a Galois extension of K, then the *Galois group* $\mathrm{Gal}(\Omega/K)$ is defined to be the group of automorphisms of Ω fixing the elements of K, endowed with the topology for which the sets

$$\mathrm{Gal}(\Omega/E), \quad \Omega \supset E \supset K, \quad [E:K] < \infty,$$

form a fundamental system of neighbourhoods of 1. This means that two elements of $\mathrm{Gal}(\Omega/K)$ are close if they agree on some "large" field E, $\Omega \supset E \supset K$, $[E:K] < \infty$.

PROPOSITION 6.2.13. *The group* $\mathrm{Gal}(\Omega/K)$ *is compact and Hausdorff.* □

REMARK 6.2.14. On can show that

(1) The topology on $\mathrm{Gal}(\Omega/K)$ is discrete if and only if Ω is a finite extension of K.
(2) In this topology, $G = \mathrm{Gal}(\Omega/K)$ is a totally disconnected topological space, that is, has no nontrivial connected subspaces. Moreover,

$$G \cong \varprojlim_{\substack{[G:H]<+\infty \\ H \text{ closed}}} G/H;$$

that is, G is a *profinite* group.
(3) In general, there may exist finite-index subgroups $H \subset G$ that are not closed (and thus not open).

THEOREM 6.2.15. *Let* Ω *be a Galois extension of* K *with Galois group* G. *Then there is a one-to-one correspondence between the subfields of* Ω *and the closed subgroups of* G. *More precisely*:

(a) *For a subfield* E *of* Ω, $H := \mathrm{Gal}(\Omega/E)$ *is a closed subgroup of* G, *and* $E = \Omega^H$.
(b) *If* H *is a subgroup of* G, *then* $\mathrm{Gal}(\Omega/\Omega^H)$ *is the closure of* H *in* $\mathrm{Gal}(\Omega/K)$.

Moreover, the normal closed subgroups of G *correspond to the Galois extensions of* K, *and the open subgroups of* G *correspond to the finite extensions of* K. □

This result combined with claim (1) of Remark 6.2.14 implies that each of the discrete factors of H is finite.

Local class field theory. Local class field theory classifies abelian extensions of a local field, which we assume to be non-Archimedean, the cases of $\mathbb{R}$ and $\mathbb{C}$ being trivial since $\mathbb{C}$ is algebraically closed and $[\mathbb{C}:\mathbb{R}]=2$.

Thus, the valuation ring in K is denoted by $\mathcal{O}_K$, its maximal ideal by $\mathfrak{m}_K$ (or simply $\mathfrak{m}$), and its group of units by $\mathcal{O}_K^*$ or U_K. A generator of $\mathfrak{m}$ is called a prime element of K or a local uniformizing parameter for $\mathfrak{m}$. If π is a prime element of K, then every element of K^* can uniquely be written in the form $a=u\pi^m$ with $u\in\mathcal{O}_K^*$ and $m:=\operatorname{ord}_K(a)$. The residue field k of K has q elements, and its characteristic is p. The normalized valuation on K is defined by $\|a\|=q^{-\operatorname{ord}_K(a)}$.

We fix a separable algebraic closure $\overline{K}$ of K. Throughout the subsection, "extension of K" will mean "subfield of $\overline{K}$ containing K". Both ord_K and $\|\cdot\|$ have unique extensions to $\overline{K}$.

The composite of two finite abelian extensions of K is again a finite abelian extension of K. Therefore, the union K^{ab} of all finite abelian extensions of K is an infinite abelian extension whose Galois group is the quotient of $\operatorname{Gal}(\overline{K}/K)$ modulo the closure of its commutator subgroup.

Let L be a finite unramified extension of K. Then L is Galois over K, and there is a unique element $\sigma\in\operatorname{Gal}(L/K)$ such that $\sigma\alpha\equiv\alpha^q$ for all $\alpha\in\mathcal{O}_L$, i.e., such that σ induces the Frobenius automorphism on the residue field of L. This σ is called the *Frobenius element* of $\operatorname{Gal}(L/K)$ and is denoted by $\operatorname{Frob}_{L/K}$. It generates $\operatorname{Gal}(L/K)$.

The following two results form the essence of local class field theory.

THEOREM 6.2.16 (local reciprocity law). *For any non-Archimedean local field K there is a unique homomorphism*

$$\varphi_K\colon K^*\to\operatorname{Gal}(K^{\mathrm{ab}}/K)$$

with the following properties:

(a) *For any prime element π of K and any finite unramified extension L of K, $\varphi_K(\pi)|_L=\operatorname{Frob}_{L/K}$.*
(b) *For any finite abelian extension L of K, $N_{L/K}(L^*)$ is contained in the kernel of $a\to\varphi_K(a)|_L$, and φ_K induces an isomorphism*

$$\varphi_{L/K}\colon K^*/N_{L/K}(L^*)\to\operatorname{Gal}(L/K). \qquad \square$$

The maps φ_K and $\varphi_{L/K}$ are called the *local Artin maps* for K and L/K; they are also often called the *local reciprocity maps*, and $\varphi_{L/K}$ is often called the *norm residue map* or *symbol*.

THEOREM 6.2.17 (local existence theorem). *A subgroup N of K^* is of the form $N_{L/K}(L^*)$ for some finite abelian extension L of K if and only if it is of finite index and open.* □

As an immediate consequence, we get the following.

COROLLARY 6.2.18. *The map $L\mapsto N(L^*)$ is a bijection from the set of finite abelian extensions of K to the set of open subgroups of finite index in K^*. Moreover,*

$$\begin{aligned} L_1\subset L_2 &\iff N(L_1^*)\supset N(L_2^*);\\ N(L_1\cdot L_2) &= N(L_1)\cap N(L_2);\\ N(L_1\cap L_2) &= N(L_1)\cdot N(L_2). \end{aligned} \qquad \square$$

REMARK 6.2.19. Corollary 6.2.18 also trivially holds for Archimedean local fields. The case $K = \mathbb{C}$ is obvious, and the local Artin map in the real case

$$\mathbb{R}^*/\mathbb{R}^*_+ \to \mathrm{Gal}(\mathbb{C}/\mathbb{R})$$

sends r to the identity map or complex conjugation according to whether $r > 0$ or $r < 0$.

6.3. Global Class Field Theory

Global class field theory describes all the abelian extensions L/K for a global field K or, which is equivalent, the maximal abelian extension K^{ab} of K. This description being very precise and concise, is not however completely explicit in general. There are some cases where it is completely explicit, namely if $K = \mathbb{Q}$, or K is a complex quadratic field, or $\mathrm{char}(K) > 0$, i.e., K is a function global field.

For $K = \mathbb{Q}$, the class field theory is given by the following classical Kronecker–Weber theorem.

THEOREM 6.3.1 (Kronecker–Weber). *We have*

$$\mathbb{Q}^{\mathrm{ab}} = \bigcup_{n=1}^{\infty} \mathbb{Q}(\zeta_n), \quad \zeta_n := e^{2\pi i/n}. \qquad \square$$

For a complex quadratic number field, K^{ab} is given by the complex multiplication theory of elliptic curves, which do not touch upon here. In the function field case, K^{ab} can be described in terms of Drinfeld modules; see Sec. 6.3.3.

Below we give the main statements of the global class field theory in most elementary possible terms. First we give necessary formulations for number fields and then show how they should be modified for function fields.

6.3.1. Artin Map

As we have seen in the preceding section, in local class field theory the (local) Artin map is fundamental. Now let us define the corresponding global notion, also called the (global) Artin map.

Recall that for a number field K (see Sec. 6.1.2) we define a *prime* of K to be an equivalence class of nontrivial valuations of K. There are two types of primes: *finite primes*, which can be identified with the maximal ideals of $\mathcal{O}_K$, and *infinite primes*. A *real* infinite prime can be identified with an embedding of K into $\mathbb{R}$, and a *complex* infinite prime can be identified with a conjugate pair of embeddings of K into $\mathbb{C}$. We use $\mathfrak{p}$ or v to denote a prime, finite or infinite. We use S to denote a finite set of primes of K, and also the set of primes of a finite extension L of K lying over K. The set of infinite primes is denoted by S_∞.

Ideals prime to S. Let $I = I_K$ be the group of fractional ideals in K. For a finite set S of primes of K, we define I^S to be the subgroup of I generated by the prime ideals not in S. Each element $\mathfrak{a}$ of I^S factors uniquely as

$$\mathfrak{a} = \mathfrak{p}_1^{n_1} \dots \mathfrak{p}_s^{n_s}, \quad \mathfrak{p}_i \notin S, \quad n_i \in \mathbb{Z},$$

and so I^S can be identified with the free abelian group generated by the prime ideals not in S. Define

$$K^S = \{a \in K^* \mid (a) \in I^S\} = \{a \in K^* \mid \mathrm{ord}_{\mathfrak{p}}(a) = 0, \text{ for all finite } \mathfrak{p} \in S\}.$$

There is a natural map $i\colon K^S \to I^S$ sending an element a of K^S to the ideal $a\mathcal{O}_K$. For example, if $K = \mathbb{Q}$ and S is the set of prime numbers dividing n, then I^S can be identified with

$$\{r/s \mid r, s \in \mathbb{Z},\ r, s > 0,\ \gcd(r,n) = 1 = \gcd(s,n)\}$$

and

$$\mathbb{Q}^S = \{r/s \mid r, s \in \mathbb{Z},\ \gcd(r,n) = 1 = \gcd(s,n)\}.$$

In this case, the natural map $\mathbb{Q}^S \to I^S$ is surjective with kernel $\{\pm 1\}$.

LEMMA 6.3.2. *For any finite set S of prime ideals in $\mathcal{O}_K$, the sequence*

$$0 \to U_K \to K^S \to I^S \to C \to 0$$

is exact. (Here $U_K = \mathcal{O}_K^*$ and C is the full ideal class group $I/i(K^*)$.) □

Moduli. A *modulus* for K is a function

$$m\colon \{\text{primes of } K\} \to \mathbb{Z}$$

such that

(a) $m(\mathfrak{p}) \geq 0$ for all primes $\mathfrak{p}$, and $m(\mathfrak{p}) = 0$ for all but finitely many $\mathfrak{p}$.
(b) If $\mathfrak{p}$ is real, then $m(\mathfrak{p}) = 0$ or 1.
(c) If $\mathfrak{p}$ is complex, then $m(\mathfrak{p}) = 0$.

Traditionally, one writes

$$\mathfrak{m} = \prod_{\mathfrak{p}} \mathfrak{p}^{m(\mathfrak{p})}.$$

A modulus $\mathfrak{m} = \prod_{\mathfrak{p}} \mathfrak{p}^{m(\mathfrak{p})}$ is said to *divide* a modulus $\mathfrak{n} = \prod_{\mathfrak{p}} \mathfrak{p}^{n(\mathfrak{p})}$ if $m(\mathfrak{p}) \leq n(\mathfrak{p})$ for all $\mathfrak{p}$. In particular, a prime $\mathfrak{p}$ divides a modulus $\mathfrak{m}$ if and only if $m(\mathfrak{p}) > 0$.

A modulus $\mathfrak{m}$ can be written as

$$\mathfrak{m} = \mathfrak{m}_\infty \mathfrak{m}_0,$$

where $\mathfrak{m}_\infty$ is a product of real primes and $\mathfrak{m}_0$ is a product of positive powers of prime ideals, and hence can be identified with an ideal in $\mathcal{O}_K$.

Ray class group. For a modulus $\mathfrak{m}$, define $K_{\mathfrak{m},1}$ to be the set of $a \in K^*$ such that

$$\begin{cases} \operatorname{ord}_{\mathfrak{p}}(a-1) \geq m(\mathfrak{p}), & \text{for all finite } \mathfrak{p} \text{ dividing } \mathfrak{m}, \\ a_{\mathfrak{p}} > 0, & \text{for all real } \mathfrak{p} \text{ dividing } \mathfrak{m}. \end{cases}$$

Note that

$$\begin{aligned} \operatorname{ord}_{\mathfrak{p}}(a-1) \geq m(\mathfrak{p}) &\iff \pi^{m(\mathfrak{p})} \mid (a_{\mathfrak{p}} - 1) \\ &\iff a \mapsto 1 \text{ in } (\mathcal{O}_{\mathfrak{p}}/\mathfrak{p}^{m(\mathfrak{p})})^* = (\widehat{\mathcal{O}}_{\mathfrak{p}}/\widehat{\mathfrak{p}}^{m(\mathfrak{p})})^*, \end{aligned}$$

where π is a prime element in the completion $K_{\mathfrak{p}}$ at $\mathfrak{p}$. Let

$$S(\mathfrak{m}) = \{\text{primes dividing } \mathfrak{m}\}.$$

For any $a \in K_{\mathfrak{m},1}$ and prime ideal $\mathfrak{p}$ dividing $\mathfrak{m}$, $\operatorname{ord}_{\mathfrak{p}}(a-1) > 0 = \operatorname{ord}_{\mathfrak{p}}(1)$, and so

$$\operatorname{ord}_{\mathfrak{p}}(a) = \operatorname{ord}_{\mathfrak{p}}((a-1)+1) = 0.$$

Therefore, for any $a \in K_{\mathfrak{m},1}$, the ideal $i(a) := (a)$ lies in $I^{S(\mathfrak{m})}$. The quotient

$$C_{\mathfrak{m}} = I^{S(\mathfrak{m})}/i(K_{\mathfrak{m},1})$$

is called the *ray class group modulo* $\mathfrak{m}$.

We denote by $N(\mathfrak{a})$ the *absolute norm* of an ideal $\mathfrak{a} \subset \mathcal{O}_K$, i.e., $N(\mathfrak{a}) = [\mathcal{O}_K : \mathfrak{a}]$.

THEOREM 6.3.3. *For any modulus $\mathfrak{m}$ of K, there is an exact sequence*

$$0 \to U/U_{\mathfrak{m},1} \to K_{\mathfrak{m}}/K_{\mathfrak{m},1} \to C_{\mathfrak{m}} \to C \to 0$$

and canonical isomorphisms

$$K_{\mathfrak{m}}/K_{\mathfrak{m},1} \cong \prod_{\substack{\mathfrak{p} \text{ real} \\ \mathfrak{p}|\mathfrak{m}}} \{\pm\} \times \prod_{\substack{\mathfrak{p} \text{ finite} \\ \mathfrak{p}|\mathfrak{m}}} (\mathcal{O}_K/\mathfrak{p}^{m(\mathfrak{p})})^* \cong \prod_{\substack{\mathfrak{p} \text{ real} \\ \mathfrak{p}|\mathfrak{m}}} \{\pm\} \times (\mathcal{O}_K/\mathfrak{m}_0)^*,$$

where

$$K_{\mathfrak{m}} = K^{S(\mathfrak{m})} = \{\alpha \in K^* \mid \operatorname{ord}_{\mathfrak{p}}(\alpha) = 0 \ \ \textit{for all} \ \ \mathfrak{p} \mid \mathfrak{m}_0\},$$
$$U = \mathcal{O}_K^*, \textit{ the group of units in } K,$$
$$U_{\mathfrak{m},1} = U \cap K_{\mathfrak{m},1}.$$

Therefore, $C_{\mathfrak{m}}$ is a finite group of order

$$h_{\mathfrak{m}} = h \cdot (U : U_{\mathfrak{m},1})^{-1} \cdot 2^{r_0} \cdot N(\mathfrak{m}_0) \cdot \prod_{\mathfrak{p}|\mathfrak{m}_0} \left(1 - \frac{1}{N(\mathfrak{p})}\right),$$

where r_0 is the number of real primes dividing $\mathfrak{m}$ and h is the class number of K (the order of C). □

LEMMA 6.3.4. *Let S be a finite set of prime ideals of K. Then any element $\alpha \in K^S$ can be written as $\alpha = a/b$ with $a, b \in \mathcal{O}_K \cap K^S$.* □

Frobenius element. To define the Artin map, we need some specific elements in the Galois groups, called *Frobenius elements*. Let K be a number field, and let L be a finite Galois extension of K with group G. Let $\mathfrak{p}$ be an ideal of K, and let $\mathfrak{P}$ be an ideal of L lying over it. The decomposition group $D(\mathfrak{P})$ (or $G(\mathfrak{P})$) is defined to be

$$\{\tau \in G \mid \tau\mathfrak{P} = \mathfrak{P}\}.$$

Equivalently, it is the set of elements of G that act continuously for the $\mathfrak{P}$-adic topology, and so extend by continuity to an automorphism of the completion $L_{\mathfrak{P}}$. In this way we obtain an isomorphism

$$D(\mathfrak{P}) \to \operatorname{Gal}(L_{\mathfrak{P}}/K_{\mathfrak{p}}).$$

Assume that $\mathfrak{P}$ is unramified over $\mathfrak{p}$. Then the action of $\operatorname{Gal}(L_{\mathfrak{P}}/K_{\mathfrak{p}})$ on $\mathcal{O}_L$ induces an isomorphism

$$\operatorname{Gal}(L_{\mathfrak{P}}/K_{\mathfrak{p}}) \to \operatorname{Gal}(l/k),$$

where l and k are the residue fields.

The group $\operatorname{Gal}(l/k)$ is cyclic with a canonical generator, namely, the Frobenius element $x \mapsto x^q$, where q is the number of elements of k. Hence, $D(\mathfrak{P})$ is cyclic, and the generator of $D(\mathfrak{P})$ corresponding to the Frobenius element in $\operatorname{Gal}(l/k)$ is called the *Frobenius element* $(\mathfrak{P}, L/K)$ at $\mathfrak{P}$. It is the unique element σ of $\operatorname{Gal}(L/K)$ satisfying the following two conditions:

(a) $\sigma \in D(\mathfrak{P})$, i.e., $\sigma\mathfrak{P} = \mathfrak{P}$.
(b) For all $\alpha \in \mathcal{O}_L$, one has $\sigma\alpha \equiv \alpha^q \bmod \mathfrak{P}$, where q is the number of elements of the residue field $\mathcal{O}_K/\mathfrak{p}$, $\mathfrak{p} = \mathfrak{P} \cap K$.

LEMMA 6.3.5. *Let $\tau\mathfrak{P}$ be a second prime dividing $\mathfrak{p}$. Then $D(\tau\mathfrak{P}) = \tau D(\mathfrak{P})\tau^{-1}$ and*

$$(\tau\mathfrak{P}, L/K) = \tau(\mathfrak{P}, L/K)\tau^{-1}.$$ □

Function field case. For a function field K there is no ring $\mathcal{O}_K$, but one can choose and fix a point ∞ and define $\mathcal{O}_K := \mathbb{F}_q[X \setminus \infty]$, set I_K to be the group of fractional ideals of $\mathcal{O}_K$, and define I^S exactly as in the number field case. However, it is convenient to use divisors instead of fractional ideals and moduli, i.e.,

$$D = \sum_{i=1}^{s} n_i \mathfrak{p}_i \quad \Longleftrightarrow \quad \mathfrak{a} = \mathfrak{p}_1^{n_1} \dots \mathfrak{p}_s^{n_s}$$

for a fractional ideal $\mathfrak{a}$, and

$$\mathfrak{m} = \prod_{\mathfrak{p}} \mathfrak{p}^{m(\mathfrak{p})} \quad \Longleftrightarrow \quad D_{\mathfrak{m}} = \sum m(\mathfrak{p})\mathfrak{p}$$

for a modulus $\mathfrak{m}$. Thus, we simply switch from multiplicative notation to additive. The definition of the ray class group $C_{\mathfrak{m}}$ remains valid for $\mathfrak{m}$ with nonvoid $\operatorname{Supp} D_{\mathfrak{m}}$. The definition and all properties of the Frobenius element remain valid in the function field case.

Artin map. Now we are ready to define the global Artin map. Let L/K be an abelian extension with Galois group G.

For any finite set S of primes of K containing all primes that ramify in L, we have a homomorphism

$$\psi_{L/K} \colon I^S \to \operatorname{Gal}(L/K), \quad \mathfrak{p}_1^{n_1} \dots \mathfrak{p}_t^{n_t} \mapsto \prod (\mathfrak{p}_i, L/K)^{n_i},$$

called the *global Artin map* (or *reciprocity map*), where $(\mathfrak{p}_i, L/K) = (\mathfrak{P}_i, L/K)$ for any $\mathfrak{P}_i \mid \mathfrak{p}_i$; by Lemma 6.3.5 it is well defined.

PROPOSITION 6.3.6. *Let L be an abelian extension of K, and let K' be any intermediate field: $L \supset K' \supset K$. Then for any finite set S of prime ideals of K containing all those that ramify in L, S' being the set of primes of K' lying over a prime in S, we have*

$$\psi_{L/K}(N_{K'/K}(i)) = \psi_{L/K'}(i)$$

for any $i \in I_K^S$, where $N_{K'/K} \colon I_{K'}^{S'} \to I_K^S$ is the norm map. □

Taking $K' = L$ in this proposition, one obtains the following.

COROLLARY 6.3.7. *For any abelian extension L of K, we have*

$$N_{L/K}(I_L^S) \subset \operatorname{Ker}\left(\psi_{L/K} \colon I^S \to \operatorname{Gal}(L/K)\right).$$ □

Thus, the Artin map induces a homomorphism

$$\psi_{L/K} \colon I_K^S / N(I_L^S) \to \operatorname{Gal}(L/K)$$

whenever L/K is an abelian extension. Note that $I^S/N(I_L^S)$ is still an infinite group, and so $\psi_{L/K}$ cannot be injective.

6.3.2. Main Theorems

Now we formulate two principal results of global class field theory, namely the reciprocity law and the existence theorem.

Reciprocity law. Let S be a finite set of primes of K. We say that a homomorphism $\psi \colon I^S \to G$ *admits a modulus* if there exists a modulus $\mathfrak{m}$ with $S(\mathfrak{m}) \subset S$ such that $\psi(i(K_{\mathfrak{m},1})) = 0$. Thus, ψ admits a modulus if and only if it factors through $C_{\mathfrak{m}}$ for some $\mathfrak{m}$ with $S(\mathfrak{m}) \subset S$.

THEOREM 6.3.8 (reciprocity law). *Let L be a finite abelian extension of K, and let S be the set of primes of K ramifying in L. Then the Artin map $\psi\colon I^S \to \mathrm{Gal}(L/K)$ admits a modulus $\mathfrak{m}$ with $S(\mathfrak{m}) = S$, and it defines an isomorphism*

$$I_K^{S(\mathfrak{m})}/i(K_{\mathfrak{m},1}) \cdot N(I_L^{S(\mathfrak{m})}) \to \mathrm{Gal}(L/K). \qquad \square$$

A modulus as in the statement of the theorem is said to be a *defining modulus* for L.

It is not difficult to show that there is a minimal (with respect to divisibility) defining modulus of L. It is called the *conductor* of the extension L/K and is denoted by $\mathfrak{f}(L/K)$.

Existence theorem. Note that the reciprocity law does not imply that K has even a single nontrivial abelian extension. We write $I_K^{\mathfrak{m}}$ for the group of $S(\mathfrak{m})$-ideals in K, and $I_L^{\mathfrak{m}}$ for the group of $S(\mathfrak{m})'$-ideals in L, where $S(\mathfrak{m})'$ contains the primes of L lying over a prime in S. We call a subgroup H of $I_K^{\mathfrak{m}}$ a *congruence subgroup modulo* $\mathfrak{m}$ if

$$I_K^{\mathfrak{m}} \supset H \supset i(K_{\mathfrak{m},1}).$$

THEOREM 6.3.9 (existence theorem). *For any congruence subgroup H modulo $\mathfrak{m}$, there exists an abelian extension L/K such that $H = i(K_{\mathfrak{m},1}) \cdot N_{L/K}(I_L^{\mathfrak{m}})$.* $\square$

Note that, for H and L as in the theorem, the Artin map $\psi_{L/K}$ induces an isomorphism

$$I^{S(\mathfrak{m})}/H \to \mathrm{Gal}(L/K).$$

In particular, for each modulus $\mathfrak{m}$ there is a field $L_{\mathfrak{m}}$, called the *ray class field modulo* $\mathfrak{m}$ such that the Artin map defines an isomorphism $C_{\mathfrak{m}} \to \mathrm{Gal}(L_{\mathfrak{m}}/K)$. For a field $L \subset L_{\mathfrak{m}}$, set

$$N(C_{L,\mathfrak{m}}) = i(K_{\mathfrak{m},1}) \cdot N(I_L^{\mathfrak{m}}) \mod i(K_{\mathfrak{m},1}).$$

COROLLARY 6.3.10. *Fix a modulus $\mathfrak{m}$. Then the map $L \mapsto N(C_{L,\mathfrak{m}})$ is a bijection from the set of abelian extensions of K contained in $L_{\mathfrak{m}}$ to the set of subgroups of $C_{\mathfrak{m}}$. Moreover,*

$$\begin{aligned} L_1 \subset L_2 &\iff N(C_{L_1,\mathfrak{m}}) \supset N(C_{L_2,\mathfrak{m}}), \\ N(C_{L_1 \cdot L_2,\mathfrak{m}}) &= N(C_{L_1,\mathfrak{m}}) \cap N(C_{L_2,\mathfrak{m}}), \\ N(C_{L_1 \cap L_2,\mathfrak{m}}) &= N(C_{L_1,\mathfrak{m}}) \cdot N(C_{L_2,\mathfrak{m}}). \end{aligned} \qquad \square$$

As an immediate consequence of Theorem 6.3.9 we see that the maximal abelian extension K^{ab} of K is the union of ray class fields over all possible moduli.

Principal ideal theorem. The following theorem is an important complement to the existence theorem.

THEOREM 6.3.11 (principal ideal theorem). *Every ideal in K becomes principal in the Hilbert class field K_1 of K (recall that K_1 is the maximal unramified abelian extension of K).* $\square$

Function field case. All the above definitions and results of the present section remain valid for function field abelian extensions while replacing multiplicative notation of ideals with the additive notation of divisors, as was explained at the end of Sec. 6.3.1. Thus, the conductor $\mathfrak{f} = \mathfrak{f}(L/K)$ in this case becomes an effective divisor of K.

Dirichlet L-functions and densities. In Chapter 3 (see *Basic Notions*) we considered the zeta function in the case of algebraic curves, i.e., function fields. For a number field, we define the *Dirichlet L-functions*

$$L(\chi,s) := \prod_{\mathfrak{p}\in\mathrm{Max}(\mathcal{O}_K)} \frac{1}{1-\chi(\mathfrak{p})(N(\mathfrak{p}))^{-s}} = \sum_{\mathfrak{a}\subset\mathcal{O}_K} \chi(\mathfrak{a})/(N(\mathfrak{a}))^s$$

for a *Dirichlet character* χ modulo $\mathfrak{m}$, i.e., a homomorphism $\chi\colon C_{\mathfrak{m}} \to \mathbb{C}^*$ (its values are roots of 1); the sum is over all ideals $\mathfrak{a} \subset \mathcal{O}_K$. In particular, for $\chi = \chi_0 = \mathrm{id}$, we get the *Dedekind zeta function*

$$\zeta_k(s) := \prod_{\mathfrak{p}} \frac{1}{1-(N(\mathfrak{p}))^{-s}} = \sum_{\mathfrak{a}\subset\mathcal{O}_K} 1/(N(\mathfrak{a}))^s.$$

THEOREM 6.3.12. *The zeta function $\zeta_K(s)$ extends to a meromorphic function on the half-plane* $\mathrm{Re}(s) > 0$, *and*

$$\zeta_K(s) \sim \frac{2^{r_1}(2\pi)^{r_2}R_K}{w_K|D_{K/\mathbb{Q}}|^{1/2}} h_K \frac{1}{s-1} \quad \text{as } s\downarrow 1,$$

where r_1 and r_2 are the numbers of real and complex primes of K, respectively, R_K is the regulator of K, w_K is the number of roots of 1 in K, $D_{K/\mathbb{Q}}$ is the discriminant of $K/\mathbb{Q}$, and h_K is the class number. □

On applying log to the equality in the theorem, one finds that

$$\sum_{\mathfrak{p}} 1/(N(\mathfrak{p}))^s \sim \log\frac{1}{1-s} \quad \text{as } s\downarrow 1,$$

and one says that a set T of prime ideals in T has *Dirichlet density* δ if

$$\sum_{\mathfrak{p}\in T} 1/(N(\mathfrak{p}))^s \sim \delta\log\frac{1}{1-s} \quad \text{as } s\downarrow 1.$$

THEOREM 6.3.13. *For any ideal $\mathfrak{a}$ relatively prime to $\mathfrak{m}_0$, the prime ideals in $\mathcal{O}_K$ whose class in $C_{\mathfrak{m}}$ is $[\mathfrak{a}]$ have Dirichlet density $1/h_{\mathfrak{m}}$.* □

Applying L-functions, one can prove the following important result.

Chebotarev density theorem. Let L be a Galois extension of K with Galois group G. Recall that, for any prime ideal $\mathfrak{p}$ of K unramified in L,

$$(\mathfrak{p}, L/K) := \{(\mathfrak{P}, L/K) \mid \mathfrak{P}|\mathfrak{p}\}$$

is a conjugacy class in G.

THEOREM 6.3.14 (Chebotarev density theorem). *Let L/K be a finite extension of global fields with Galois group G, and let C be a conjugacy class in G. Then the set of prime ideals of K such that $(\mathfrak{p}, L/K) = C$ has density $|C|/|G|$ in the set of all prime ideals of K. In particular, if G is abelian, then, for a fixed $\tau\in G$, the set of prime ideals $\mathfrak{p}$ of K with $(\mathfrak{p}, L/K) = \tau$ has density $(G:1)$.* □

Applying the theorem to the splitting field of a polynomial $f(X)\in K[X]$, we get the following fact.

COROLLARY 6.3.15. *If a polynomial $f(X)\in K[X]$ splits into linear factors modulo $\mathfrak{p}$ for all but finitely many prime ideals $\mathfrak{p}$ in K, then it splits in $K[X]$.* □

For a finite extension L/K of number fields and a finite set S of primes of K, let $\mathrm{Spl}_S(L/K)$ be the set of primes of K not in S that split in L.

THEOREM 6.3.16. *If L and M are Galois extensions of K, then*

$$L \subset M \iff \mathrm{Spl}_S(L/K) \supset \mathrm{Spl}_S(M/K).$$

Hence,

$$L = M \iff \mathrm{Spl}_S(L/K) = \mathrm{Spl}_S(M/K). \qquad \square$$

In the function field case, the definition of the Artin map remains valid, as well as the reciprocity law and existence theorem.

6.3.3. Explicit Class Field Theory for Function Fields

For function fields, one can considerably refine the description of abelian extensions, giving more explicit results. There are two versions of such an explicit class field theory: the geometric one and the Hayes theory, based on rank-1 Drinfeld modules. We now formulate principal results of these versions of class field theory. Unfortunately, we have to be very concise, since a more detailed exposition needs much more elaborate techniques. However, the reader is at least assumed to understand two principal results, Theorems 6.3.23 and 6.3.28 below.

Geometric Lang–Rosenlicht theory. This theory is based on Jacobians and their generalizations. We start by recalling some facts on the Jacobians; cf. Sec. 2.1.4.

Let X be a projective, irreducible, and nonsingular algebraic curve, and let $f\colon X \to G$ be a rational map from X to a commutative algebraic group G. The set S of points of X where f is not regular is a finite set. If D is a divisor prime to S (i.e., of the form $D = \sum n_i P_i$ with $P_i \notin S$), $f(D)$ can be defined to be $\sum n_i f(P_i)$, which is an element of G.

When G is an abelian variety, $S = \varnothing$, and one knows that $f(D) = 0$ if D is the divisor (φ) of a rational function φ on X; in this case, $f(D)$ depends only on the linear equivalence class of D.

In the general case, we need to use "classes modulo S"; namely, for a *modulus* $\mathfrak{m}$ with support S, i.e., a positive divisor $\sum_{P_i \in S} n_i P_i$, $n_i > 0$, and for a rational function φ we say that φ is "congruent to 1 mod $\mathfrak{m}$" and write $\varphi \equiv 1 \bmod \mathfrak{m}$ if $v_i(1-\varphi) \geq n_i$ for all i, v_i denoting the valuation attached to the point P_i. Since all the n_i are positive, such a function is regular at the points P_i and takes the value 1 there; its principal divisor (φ) is thus prime to S.

THEOREM 6.3.17. *For every rational map $f\colon X \to G$ regular outside S there exists a modulus $\mathfrak{m}$ with support S such that $f(D) = 0$ for every divisor $D = (\varphi)$ with $\varphi = 1 \bmod \mathfrak{m}$.* $\square$

Conversely, given a modulus $\mathfrak{m}$, we can recover a "universal" group $J_\mathfrak{m}$ for the groups G with this last property. More precisely, we have the following.

THEOREM 6.3.18. *For every modulus $\mathfrak{m}$ there exists a commutative algebraic group $J_\mathfrak{m}$ and a rational map $f_\mathfrak{m}\colon X \to J_\mathfrak{m}$ such that the following property holds:*

For every rational map $f\colon X \to G$ satisfying the property of Theorem 6.3.17 with respect to $\mathfrak{m}$, there exists a unique rational (affine) homomorphism $\theta\colon J_\mathfrak{m} \to G$ such that $f = \theta \circ f_\mathfrak{m}$. $\square$

Note that for $\mathfrak{m} = 0$ we recover just the usual Jacobian $J = J_X$. More can be said about the structure of $J_\mathfrak{m}$. Indeed, let $C_\mathfrak{m}$ be the group of classes of divisors prime to S modulo those which can be written as $D = (\varphi)$ with $\varphi \equiv 1 \bmod \mathfrak{m}$, and let $C^0_\mathfrak{m}$ be the subgroup of $C_\mathfrak{m}$ formed by classes of degree 0. Denoting by C^0 the group of divisor classes of degree 0, we obtain a surjective homomorphism $C^0_\mathfrak{m} \to C^0$. The kernel $L_\mathfrak{m}$ of this homomorphism is formed by the classes in $C_\mathfrak{m}$ of divisors of the form (φ), with φ invertible at each point $P_i \in S$. But, for each $P_i \in S$, the invertible elements modulo those congruent to 1 mod $\mathfrak{m}$ form an algebraic group $R_{\mathfrak{m},i}$ of dimension n_i; let $R_\mathfrak{m}$ be the product of these groups. According to the weak approximation theorem for valuations, we can find a function corresponding to arbitrary given elements $r_i \in R_{\mathfrak{m},i}$. We conclude that $L_\mathfrak{m}$ gets identified with the quotient group $R_\mathfrak{m}/\boldsymbol{G}_m$, denoting by $\boldsymbol{G}_m$ the multiplicative group of constants embedded naturally in $R_\mathfrak{m}$. Putting $J = C^0$, we have an exact sequence of algebraic groups

$$0 \to R_\mathfrak{m}/\boldsymbol{G}_m \to C^0_\mathfrak{m} \to J \to 0.$$

Note that J has a natural structure of an algebraic group, since it is the Jacobian of X, and the same is true for $R_\mathfrak{m}/\boldsymbol{G}_m$. This extends to $C^0_\mathfrak{m}$ as follows.

THEOREM 6.3.19. *The map $f_\mathfrak{m}\colon X \to J_\mathfrak{m}$ defines, by extension to divisor classes, a bijection from $C^0_\mathfrak{m}$ to $J_\mathfrak{m}$. Identifying $C^0_\mathfrak{m}$ and $J_\mathfrak{m}$ by means of this bijection, the group $J_\mathfrak{m}$ becomes an extension of the group J by the group $R_\mathfrak{m}/\boldsymbol{G}_m$.* □

The groups $J_\mathfrak{m}$ are called the *generalized Jacobians* of the curve X.

Abelian coverings. Let G and G' be connected commutative algebraic groups, and let $\theta\colon G' \to G$ be an isogeny, i.e., a homomorphism of algebraic groups which is surjective with a finite kernel. We also assume that the corresponding field extension is separable, in which case we say that θ is separable. If $\mathfrak{g}$ denotes the kernel of θ, the group G is identified with the quotient $G'/\mathfrak{g}$, and G' is an unramified covering of G, with the abelian group $\mathfrak{g}$ as a Galois group.

Now let U be an algebraic curve, and let $f\colon U \to G$ be a regular map. One defines the *pull-back* $U' = f^{-1}(G')$ of G' by f as the subvariety of $U \times G'$ formed by the pairs (x, g') such that $f(x) = \theta(g')$. The projection $U' \to U$ makes U' an unramified covering of U with Galois group $\mathfrak{g}$.

More generally, let $f\colon X \to G$ be a rational map from an irreducible curve X to the group G, and let $X' \to X$ be a covering of X with Galois group $\mathfrak{g}$. If there exists a nonempty open neighbourhood U of X on which f is regular, and if the covering induced by X' on U is isomorphic to $f^{-1}(G')$, we will again say that X' is the *pull-back* of the isogeny $G' \to G$ by the map f.

With this convention, we have the following fact.

THEOREM 6.3.20. *Every abelian covering of an irreducible algebraic curve is the pull-back of a suitable isogeny.* □

Combining Theorem 6.3.20 with Theorems 6.3.17 and 6.3.18, we obtain the following.

COROLLARY 6.3.21. *Let $X' \to X$ be an abelian covering of an algebraic curve X. Then there exists a separable isogeny $\theta\colon G' \to J_\mathfrak{m}$, where $J_\mathfrak{m}$ is a generalized Jacobian of X, such that X' is isomorphic to $f_\mathfrak{m}^{-1}(G')$.* □

One can also prove the following results.

PROPOSITION 6.3.22. (a) *For fixed X' and $J_{\mathfrak{m}}$, the isogeny $\theta\colon G' \to J_{\mathfrak{m}}$ is* unique.

(b) *The modulus $\mathfrak{m}$ can be chosen so that its support S is exactly the set of ramification points of the given covering $X' \to X$.* □

In particular, unramified coverings correspond to isogenies of the Jacobian.

THEOREM 6.3.23. *If the abelian covering $X' \to X$ is defined and abelian over a finite field k, the isogeny $\theta\colon G' \to J_{\mathfrak{m}}$ of Corollary* 6.3.21 *can be defined over k.* □

Thus we get a construction of abelian extensions of the field $k(X)$ starting from k-isogenies of generalized Jacobians $J_{\mathfrak{m}}$ corresponding to moduli $\mathfrak{m}$ rational over k; moreover, this construction allows one to recover class field theory for the field $k(X)$.

Hayes theory. Hayes' explicit class field theory uses specific rank-1 Drinfeld modules to describe abelian unramified extensions of a function field $\Bbbk$.

Until the end of the section, we use the notation of Sec. 5.2.1, except that below we do not assume ∞ to be a rational place; we denote its degree over $\mathbb{F}_q$ by d_∞.

We begin with two auxiliary constructions valid for Drinfeld modules of any rank, namely normalized Drinfeld modules and the action of ideals on Drinfeld modules.

Normalized Drinfeld modules. Let $L/\Bbbk$ be a finite extension. Fix a Drinfeld module $\varphi\colon A \to L[\tau]$ of rank r. Then there exists a unique extension of φ to $\Bbbk_\infty \to L^{\mathrm{perf}}((\tau^{-1}))$, which we still denote by φ, where L^{perf} is the *perfectization* of L, i.e.,

$$L^{\mathrm{perf}} = \bigcup_{n=1}^{\infty} L^{(p^{-n})},$$

and $F((\tau^{-1}))$ for a perfect field F is the skew field of Laurent series, i.e., of expressions

$$\sum_{m=-N}^{\infty} a_i \tau^m, \quad \tau a = a^p \tau.$$

For each $x \in \Bbbk_\infty^*$, we define $\mu_\varphi(x) \in (L^{\mathrm{perf}})^*$ to be the first nonzero coefficient of the Laurent series $\varphi(x) \in L^{\mathrm{perf}}((\tau^{-1}))$. The first term of $\varphi(x)$ is $\mu_\varphi(x)\tau^{-rd_\infty \operatorname{ord}_\infty(x)}$. For a nonzero $x \in A$, one can also define $\mu_\varphi(x)$ as the leading coefficient of $\varphi(x)$ as a polynomial in τ.

For $x, y \in \Bbbk_\infty^*$, the value $\mu_\varphi(xy)$ is equal to the coefficient of

$$\mu_\varphi(x)\tau^{-rd_\infty \operatorname{ord}_\infty(x)} \cdot \mu_\varphi(y)\tau^{-rd_\infty \operatorname{ord}_\infty(y)},$$

and hence

$$\mu_\varphi(xy) = \mu_\varphi(x)\mu_\varphi(y)^{1/q^{rd_\infty \operatorname{ord}_\infty(x)}}.$$

With respect to the embedding $\mathbb{F}_\infty \hookrightarrow L$ arising from φ, one has $\mu_\varphi(x) = x$ for all $x \in \mathbb{F}_\infty^*$.

We say that φ is *normalized* if $\mu_\varphi(x)$ belongs to $\mathbb{F}_\infty^*$ for all $x \in \Bbbk_\infty^*$. If φ is normalized, then by the above equality the map $\mu_\varphi\colon \Bbbk_\infty^* \to \mathbb{F}_\infty^*$ is a group homomorphism that equals the identity map when restricted to $\mathbb{F}_\infty^*$; this is an example of a sign function, where a *sign function* for $\Bbbk_\infty$ is a group homomorphism $\varepsilon\colon \Bbbk_\infty^* \to \mathbb{F}_\infty^*$

with $\varepsilon(\mathbb{F}^*) = 1$. We say that φ is ε-*normalized* if it is normalized and $\mu_\varphi\colon \Bbbk_\infty^* \to \mathbb{F}_\infty^*$ is equal to ε composed with some k-automorphism of $\mathbb{F}_\infty$.

A sign function ε is trivial on $1+\mathfrak{m}_\infty$, so it is determined by its value $\varepsilon(\pi)$ for a fixed uniformizer π of $\Bbbk_\infty$. One can normalize an arbitrary Drinfeld module:

PROPOSITION 6.3.24. *Let ε be a sign function for $\Bbbk_\infty$, and let $\varphi'\colon A \to L[\tau]$ be a Drinfeld module. Then φ' is isomorphic over $\overline{L}$ to an ε-normalized Drinfeld module $\varphi\colon A \to \overline{L}[\tau]$.* □

The action of an ideal on a Drinfeld module. For a Drinfeld module $\varphi\colon A \to L[\tau]$ and a nonzero ideal $\mathfrak{a}$ of A, define $I_{\mathfrak{a},\varphi}$ to be the *left* ideal in $L[\tau]$ generated by the set $\{\varphi(a) : a \in \mathfrak{a}\}$. All left ideals of $L[\tau]$ are principal, so $I_{\mathfrak{a},\varphi} = L[\tau]\cdot\varphi_{\mathfrak{a}}$ for a unique monic polynomial $\varphi_{\mathfrak{a}} \in L[\tau]$. Using that $\mathfrak{a}$ is an ideal, we find that $I_{\mathfrak{a},\varphi}\varphi(x) \subseteq I_{\mathfrak{a},\varphi}$ for all $x \in A$. Thus for each $x \in A$, there is a unique polynomial $(\mathfrak{a} * \varphi)(x)$ in $L[\tau]$ that satisfies

$$\varphi_{\mathfrak{a}}\cdot\varphi(x) = (\mathfrak{a} * \varphi)(x)\cdot\varphi_{\mathfrak{a}}. \tag{6.3.1}$$

The map

$$\mathfrak{a} * \varphi\colon A \to L[\tau], \quad x \mapsto (\mathfrak{a} * \varphi)(x)$$

is also a Drinfeld module, and hence equation (6.3.1) shows that $\varphi_{\mathfrak{a}}$ is an isogeny from φ to $\mathfrak{a} * \varphi$. This action has the following properties.

PROPOSITION 6.3.25. (i) *Let $\mathfrak{a}$ and $\mathfrak{b}$ be nonzero ideals of A. Then $\varphi_{\mathfrak{a}\mathfrak{b}} = (\mathfrak{b} * \varphi)_{\mathfrak{a}}\cdot\varphi_{\mathfrak{b}}$ and $\mathfrak{a} * (\mathfrak{b} * \varphi) = (\mathfrak{a}\mathfrak{b}) * \varphi$.*

(ii) *Let $\mathfrak{a} = wA$ be a nonzero principal ideal of A. Then $\varphi_{\mathfrak{a}} = \mu_\varphi(w)^{-1}\cdot\varphi(w)$ and $(\mathfrak{a} * \varphi)(x) = \mu_\varphi(w)^{-1}\cdot\varphi(x)\cdot\mu_\varphi(w)$ for all $x \in A$.*

(iii) *Let $\sigma\colon L \hookrightarrow L'$ be an embedding of fields. Let $\sigma(\varphi)\colon A \to L'[\tau]$ be the Drinfeld module for which $\sigma(\varphi)(x) = \sigma(\varphi(x))$, where σ acts on the coefficients of $L[\tau]$. For each nonzero ideal $\mathfrak{a}$ of A, we have $\sigma(\mathfrak{a} * \varphi) = \mathfrak{a} * \sigma(\varphi)$ and $\sigma(\varphi_{\mathfrak{a}}) = \sigma(\varphi)_{\mathfrak{a}}$.* □

Hayes modules. Now we are ready to formulate the principal result of the Hayes theory.

Fix a sign function ε for $\Bbbk_\infty$. Let $\mathbb{C}_\infty$ be the completion of the algebraic closure $\overline{\Bbbk}_\infty$ of $\Bbbk_\infty$ with respect to the ∞-adic norm; it is both complete and algebraically closed (but not locally compact anymore).

A *Hayes module* for ε is a Drinfeld module $\varphi\colon A \to \mathbb{C}_\infty[\tau]$ of rank 1 that is ε-normalized and for which $D \circ \varphi\colon A \to \mathbb{C}_\infty$ is the inclusion map, D being the natural projection $\mathbb{K}\{\tau\} \to \mathbb{K}$ (see Sec. 5.2.1). Denote by X_ε the set of Hayes modules for ε.

The existence of Hayes modules follows from the analytic theory of Drinfeld modules together with Proposition 6.3.24.

Now let us take a Hayes module $\varphi \in X_\varepsilon$. Using that $\varphi_{\mathfrak{a}} \in \mathbb{C}_\infty[\tau]$ is monic along with (6.3.1), we see that the Drinfeld module $\mathfrak{a} * \varphi$ also belongs to X_ε. By Proposition 6.3.25(i) we find that the group $\mathcal{I}$ of fractional ideals of A acts on X_ε. Let $\mathcal{P}^+$ be the subgroup of principal fractional ideals generated by those $x \in \Bbbk^*$ that satisfy $\varepsilon(x) = 1$. The group $\mathcal{P}^+$ acts trivially on X_ε by Proposition 6.3.25(ii), and hence induces an action of the finite group $\mathrm{Pic}^+(A) := \mathcal{I}/\mathcal{P}^+$ on X_ε.

PROPOSITION 6.3.26. *The set X_ε is a principal homogeneous space for* $\mathrm{Pic}^+(A)$ *under the $*$ action (i.e., this action is free and transitive).* □

We now consider the arithmetic of the set X_ε. Take any $\varphi \in X_\varepsilon$ and choose a nonconstant $y \in A$. Let H_A^+ be the subfield of $\mathbb{C}_\infty$ generated by $\Bbbk$ and the coefficients of φ_y as a polynomial in τ. We call H_A^+ the *normalizing field* for the triple $(\Bbbk, \infty, \varepsilon)$.

LEMMA 6.3.27. *The extension $H_A^+/\Bbbk$ is finite and normal, and depends only on the triple $(\Bbbk, \infty, \varepsilon)$.* □

Thus, for every $\varphi \in X_\varepsilon$, we have $\varphi(A) \subseteq H_A^+[\tau]$. Let us regard φ as a Drinfeld module $\varphi\colon A \to H_A^+[\tau]$; then $\varphi(A) \subseteq B[\tau]$, where B is the integral closure of A in H_A^+. Since φ is normalized, we find that φ has *good reduction* at every place of H_A^+ not lying over ∞.

Thus, there is a natural action of the Galois group $\mathrm{Gal}(H_A^+/\Bbbk)$ on the set X_ε. With a fixed $\varphi \in X_\varepsilon$, Proposition 6.3.26 implies that there is a unique function $\psi\colon \mathrm{Gal}(H_A^+/\Bbbk) \hookrightarrow \mathrm{Pic}^+(A)$ such that $\sigma(\varphi) = \psi(\sigma) * \varphi$. Proposition 6.3.25 implies that ψ is a group homomorphism that does not depend on the initial choice of φ. The following important proposition implies that ψ is surjective, and hence an isomorphism.

THEOREM 6.3.28. *The extension $H_A^+/\Bbbk$ is unramified away from ∞. For each nonzero prime ideal $\mathfrak{p}$ of A, the class $\psi(\mathrm{Frob}_\mathfrak{p})$ of $\mathrm{Pic}^+(A)$ is the one containing $\mathfrak{p}$.* □

This proposition gives a complete explicit description of the maximal abelian extension $\Bbbk_\infty^{\mathrm{ab}}$ of $\Bbbk$ unramified away from ∞.

REMARK 6.3.29. Therefore, choosing another place $\infty' \neq \infty$ of $\Bbbk$ and taking the compositum $\Bbbk_\infty^{\mathrm{ab}}\Bbbk_{\infty'}^{\mathrm{ab}}$ of $\Bbbk_\infty^{\mathrm{ab}}$ and $\Bbbk_{\infty'}^{\mathrm{ab}}$, one constructs $\Bbbk^{\mathrm{ab}}$ explicitly. In fact, one can modify the construction appropriately to avoid the somewhat arbitrary choice of ∞'. However, we do not need this more elaborated construction here.

6.4. Class Field Towers

In this section we consider a specific but very important problem concerning infinite unramified extensions. Their very existence in the number field case is highly nontrivial and can be achieved by the class field tower construction, which will play a prominent role in Chapters 7 and 8.

6.4.1. Class Field Tower Problem

The simplest global fields $k = \mathbb{Q}$ and $\mathbb{F}_q(T)$ do not have unramified extensions.

Many other global fields have many finite unramified extensions, even abelian ones, as class field theory shows. However, it is still not immediately clear whether infinite unramified extensions do exist. In this section we give sufficient conditions for the existence of such extensions, first in the number field case and then in the function field case.

First, let k be a number field; then its maximal abelian unramified extension is the Hilbert class field k_1 of k. Its Galois group $\mathrm{Gal}(k_1/k)$ is canonically isomorphic to the class group Cl_k of k via the reciprocity isomorphism of class field theory. Theorem 6.3.11 says that every ideal (divisor) becomes principal in k_1, but there can be ideals in k_1 which are not principal, and thus k_2, the Hilbert class field of k_1, can be different from k_1. Continuing, we get a tower of extensions:

$$k \subset k_1 \subset k_2 \subset \ldots,$$

where each following field is the Hilbert class field of the preceding one. This tower is called the *Hilbert class field tower of* k. One puts $k_\infty = \bigcup_{i=1}^{\infty} k_i$; it is an unramified extension of k, but it is not a priori clear that it can be infinite.

First we note a simple criterion for k_∞ to be finite.

PROPOSITION 6.4.1. *The following two conditions are equivalent*:

1. *There exists an unramified extension* K/k *with* $h_K = 1$.
2. *The extension* k_∞/k *is finite* (*and thus* $h_{k_\infty} = 1$). □

Now we give some sufficient conditions for k_∞ to be infinite; moreover, these conditions concern only a p-extension for a fixed prime p. Therefore, we fix a prime p, and let K/k be a finite Galois extension. Then it is called a *p-extension* if its Galois group $G = \mathrm{Gal}(K/k)$ is a p-group, i.e., $|G| = p^m$ for some $m \geq 1$.

For a number field k, one can consider the maximal p-extension $k_1^{(p)} \subset k_1$ and then iterate this construction, thus obtaining the *Hilbert p-class field tower* of k:

$$k \subset k_1^{(p)} \subset k_2^{(p)} \subset \ldots \subset k_\infty^{(p)}.$$

In fact, one easily finds that $k_i^{(p)} \subset k_i$ and $k_\infty^{(p)} \subset k_\infty$. Thus, if the extension $k_\infty^{(p)}/k$ is infinite, the same is true for k_∞/k.

PROPOSITION 6.4.2. *The following conditions are equivalent*:

1. *There exists an unramified extension* K/k *with* $p \nmid h_K$.
2. *The extension* $k_\infty^{(p)}/k$ *is finite.* □

Now let us formulate a group-theoretic result, Theorem 6.4.3, which in many cases allows one to establish that the extension $k_\infty^{(p)}/k$ is infinite and thus to resolve the class field tower problem. This is possible thanks to the second main result below, Theorem 6.4.4, which shows that in many cases the conditions of Theorem 6.4.3 are satisfied.

Golod–Shafarevich theory. The following constructions give a solution to the class field tower problem. Let G be a finite group, and let G/p be the maximal abelian factor group of G of exponent p. Since the group G/p is annihilated by multiplying by p, G/p can be regarded as an $\mathbb{F}_p$-vector space. Denote by d_pG its dimension:

$$d_pG := \dim_{\mathbb{F}_p} G/p.$$

THEOREM 6.4.3. *For a number field k, one has*

$$d_p \operatorname{Cl}_k < 2+2\sqrt{r_1+r_2+\delta_k^p},$$

where

$$\delta_k^p = \begin{cases} 1 & \text{if } \zeta_p \in k, \\ 0 & \text{otherwise,} \end{cases}$$

r_1 and r_2 being as above. □

THEOREM 6.4.4. *Let $[k:\mathbb{Q}]=n$. Then*

$$d_p \operatorname{Cl}_k \geq t_k^p - (r_1+r_2)n,$$

where t_k^p is the number of ideals q in k with $p \mid e_k(q)$ for $e_k(q) = \gcd\{e(\mathfrak{q}) : \mathfrak{q} \mid q\}$, $\mathfrak{q}$ being a prime ideal in k. □

COROLLARY 6.4.5. *If k is an algebraic number field of degree n and*

$$t_k^p \geq 2+2\sqrt{r_1+r_2+\delta_k^p}+(r_1+r_2)n,$$

then the p-class field tower of k is infinite. □

COROLLARY 6.4.6. *Let $n>1$ and $p \mid n$. Then the p-class field tower of*

$$k := \mathbb{Q}(\sqrt[n]{q_1 \cdots q_N}), \quad N > 2+2\sqrt{r_1+r_2+\delta_k^p}+(r_1+r_2)n,$$

where $q_1, \dots, q_N$ are distinct primes, is infinite. □

Function field case. The existence of an unramified infinite extension for any function field $K/\mathbb{F}_q$ of genus >0, i.e., nonisomorphic to $\mathbb{F}_q(t)$, immediately follows from the results of Sec. 6.3.3. Indeed, the Jacobian of K is nontrivial, and by iterating the Hilbert class field construction one immediately obtains an infinite class tower.

However, in the function field case it is rather interesting to construct an infinite class field tower with additional conditions. The most natural and useful is the condition that some finite set S of places of K is split.

Let K be a function field over $\mathbb{F}_q$ and S be a finite nonempty set of places of K. Let $\mathcal{O}_S$ be the S-integral ring of K, i.e., $\mathcal{O}_S$ consists of all elements of K that have no poles outside S. Let Cl_S be the class group of $\mathcal{O}_S$. Let r be the greatest common divisor of the degrees of the places in S, and let ℓ be a prime number. The (S,ℓ)*-Hilbert class field* $H_{S,\ell}$ of K is the maximal unramified abelian extension of K with Galois group of type $(\ell,\ell,\dots,\ell)$ in which all places in S split totally. If ℓ divides r, it contains the constant field extension by $\mathbb{F}_{q^\ell}$, but if r and ℓ are prime to each other, there is no constant field extension. By global class field theory, the Galois group of the extension $H_{S,\ell}/K$ is isomorphic to $\mathrm{Cl}_S/\mathrm{Cl}_S^{\ell}$.

Now, let us define the *(S,ℓ)-Hilbert class field tower* of K. Let K_1 be the (S,ℓ)-Hilbert class field of K and S_1 be the set of places of K_1 over those in S. Recursively, we define K_i to be the (S_{i-1},ℓ)-Hilbert class field of K_{i-1} and S_i the set of places of K_i over those in S_{i-1}. Thus, we get a tower

$$K = K_0 \subseteq K_1 \subseteq K_2 \subseteq \dots.$$

Function field analogues of Theorems 6.4.3 and 6.4.4 and Corollary 6.4.5 give a sufficient condition for a p-class field tower to be infinite (that is, $K_i \neq K_{i-1}$ for all $i \geq 1$). The proofs remain valid in the function field case and yield the following result.

THEOREM 6.4.7. *Let K be a function field over $\mathbb{F}_q$ and S be a finite nonempty set of places of K. If $d_\ell \mathrm{Cl}_S \geq 1$ and*

$$|S| \leq \frac{(d_\ell \mathrm{Cl}_S)^2}{4} - d_\ell \mathrm{Cl}_S + \varepsilon,$$

where $\varepsilon = 0$ if $\ell \mid (q-1)$ and $\varepsilon = 1$ otherwise, then the (S,ℓ)-Hilbert class field tower of K is infinite. □

6.4.2. Applications to $A(q)$

Let us show how one can apply Theorem 6.4.7 to obtain lower bounds for $A(q)$. This is possible, since the ratio N/g does not change under an unramified extension.

Positivity of $A(q)$. The first result is strict positivity of $A(q)$ for any prime power q.

THEOREM 6.4.8. *We have*

$$A(q) > c \log q$$

for a positive constant c.

SKETCH OF THE PROOF. First assume that q is odd. Let $f = f_1 f_2 \dots f_6$, where $f_1, f_2, \dots, f_6$ are distinct monic irreducible polynomials in $\mathbb{F}_q[x]$ having even degrees. Then $\mathbb{F}_q(x, \sqrt{f_1}, \dots, \sqrt{f_6})$ is an unramified Galois extension of $\mathbb{F}_q(x, \sqrt{f})$, with Galois group $(\mathbb{Z}/2\mathbb{Z})^5$, in which both places of $\mathbb{F}_q(x, \sqrt{f})$ which contain $1/x$ are completely split. Applying Theorem 6.4.7, we obtain that $A(q) > 0$.

For odd q it remains only to prove the second statement for $q \gg 0$. Let $\alpha_1, \dots, \alpha_{r+1}$ be distinct elements of $\mathbb{F}_q$, and let $f(x) = (x-\alpha_1)\dots(x-\alpha_{r+1})$. Assume that r is even. Then $F := \mathbb{F}_q(\sqrt{x-\alpha_1}, \dots, \sqrt{x-\alpha_{r+1}})$ is an unramified Galois extension of $\mathbb{F}_q(x, \sqrt{f})$, with Galois group $(\mathbb{Z}/2\mathbb{Z})^r$. By the Hurwitz formula, $\mathbb{F}_q(x, \sqrt{f})$ has genus $g := r/2$ and F has genus $1 + 2^r(r/2-1)$. By the Weil bound, the number of rational places N on F satisfies

$$N \geq q + 1 - 2\sqrt{q} \cdot 2^r (r/2 - 1).$$

Exactly $N/2^r$ rational places of $\mathbb{F}_q(x, \sqrt{f})$ split completely in F. Choose r as the even integer nearest to $(\log_2 q)/3$. For $q \gg 0$, we have $r \geq 4$ and $N/2^r \geq (r-2)^2/4 - 1 \geq 0$, so it is possible to choose a subset S of these rational places with $|S| = \lfloor (r-2)^2/4 - 1 \rfloor$. By Theorem 6.4.7, the $(S,2)$-class field tower is infinite, so

$$A(q) \geq \frac{|S|}{g-1} = \frac{r}{2} + O(1) = c' \log q + O(1)$$

as $q \to \infty$, for some $c' > 0$.

The case of even q can be treated in a completely analogous manner, using Artin–Schreier extensions. □

Note that the bound of Theorem 6.4.8 is rather crude. Its main advantage is its uniformity over any q. However, similar ideas can give much better bounds, which we illustrate below for the case of $A(2)$. This can be achieved by applying bounds of Golod–Shafarevich type together with the technique of Hayes described in Sec. 6.3.3, which uses rank-1 Drinfeld modules.

Application to $A(2)$. Thus, we are going to give a class field construction of an asymptotically good family of function fields over $\mathbb{F}_2$ giving a good lower bound on $A(2)$. In fact, similar constructions also work for many fields $\mathbb{F}_p$, thus giving lower bounds on $A(p)$.

We start with the following simple result.

PROPOSITION 6.4.9. *Let $K/\mathbb{F}_q$ be a function field with an infinite (S,ℓ)-Hilbert class field tower. Then*

$$A(q) \geq \frac{|S|}{g-1},$$

g being the genus of K (i.e., the genus of a smooth absolutely irreducible projective curve $X/\mathbb{F}_q$ such that $\mathbb{F}_q(X) = K$).

SKETCH OF THE PROOF. This immediately follows from the Hurwitz genus formula, since if K_i/K is unramified and all places of S split totally in K_i, we get

$$2g(K_i) - 2 = [K_i : K](2g-2)$$

and

$$|S_i| \geq [K_i : K]|S|,$$

where S_i is the set of $\mathbb{F}_q$-rational places of an (absolutely irreducible smooth projective) curve X_i with $\mathbb{F}_q(X_i) = K_i$. □

To apply Theorem 6.4.7 to this result, one needs a good lower bound for the number $d_\ell \operatorname{Cl}_S$. Such an estimate can be given in the case of sufficiently many ramified places; in a sense, it is a refinement of Theorem 6.4.4.

PROPOSITION 6.4.10. *Let $K/\mathbb{F}_q$ be a global function field and F/K a finite abelian extension. Let $T \subsetneq P_K$ be such that $T' := \mathbb{P}_K \setminus T$ is finite and nonempty, and let S be the set of places of F lying over those in T. Then*

$$d_\ell \operatorname{Cl}_S \geq \sum_{P \in P_K} d_\ell G_P - (|T'| - 1 + \varepsilon) - d_\ell G,$$

where ε is as in Theorem 6.4.7, $G = \operatorname{Gal}(F/K)$, and G_P is the inertia group of a place P of K. □

To construct a global function field F satisfying the infinity criterion, one can use the Hayes theory, which leads to the following bound.

THEOREM 6.4.11. *We have*

$$A(2) \geq \frac{97}{376} = 0.257979\ldots. \qquad \square$$

To give a construction of such a global field K satisfying the conditions of Proposition 6.4.10, one can use the Hayes theory in the following situation. Let $\mathbb{k} = \mathbb{F}_2(x)$, and let f be the Drinfeld module

$$f\colon \mathbb{F}_2[x] \to \operatorname{End}_{\mathbb{F}_2} \overline{\mathbb{k}}, \quad f(x) \mapsto f(\varphi),$$

where $\varphi \in \operatorname{End}_{\mathbb{F}_2} \overline{\mathbb{k}}$ is given by

$$\varphi(u) = u^2 + xu, \quad \text{for all } u \in \overline{\mathbb{k}}.$$

It defines, as before, an $\mathbb{F}_2[x]$-module structure on $\overline{\mathbb{k}}$ by

$$(f(x))(u) = f(\varphi)u, \quad f(x) \in \mathbb{F}_2(x), \quad u \in \overline{\mathbb{k}}.$$

For a nonzero polynomial $M \in \mathbb{F}_2[x]$, let

$$\Lambda_M = \{z \in \overline{\mathbb{k}} : z^M = 0\}$$

be the set of M-torsion elements of $\overline{\mathbb{k}}$.

The Hayes theory yields the following result.

LEMMA 6.4.12. *Let $M \in \mathbb{F}_2[x]$ be a monic polynomial, and put $\mathbb{k}_M := \mathbb{k}(\Lambda_M)$. Then:*

(i) *$\mathbb{k}_M/\mathbb{k}$ is a finite abelian extension with the Galois group isomorphic to the multiplicative group $(\mathbb{F}_2[x]/(M))^*$ of the residue ring $\mathbb{F}_2[x]/(M)$. Thus, $[\mathbb{k}_M : \mathbb{k}] = \Phi(M) := |(\mathbb{F}_2[x]/(M))^*|$.*

(ii) *If $M = P^n$ for a monic irreducible polynomial P and an integer $n \geq 1$, then P is totally ramified in $\mathbb{k}_{P^n}/\mathbb{k}$, and all other places of $\mathbb{k}$ are unramified in $\mathbb{k}_{P^n}/\mathbb{k}$. Moreover, the pole place ∞ of x splits completely in $\mathbb{k}_{P^n}/\mathbb{k}$.* $\square$

Now, the field K in Proposition 6.4.10 is constructed as the compositum of two fields, K_1 and K_2, each arising from the Hayes theory. To construct K_1, first consider the polynomial $M = (x^4+x^3+x^2+x+1)^2 \in \mathbb{F}_2[x]$, and let K_1 be the subfield of the cyclotomic function field $\mathbb{k}_M$ fixed by the cyclic subgroup of $\operatorname{Gal}(\mathbb{k}_M/\mathbb{k}) = (\mathbb{F}_2[x]/(M))^*$ generated by $\overline{x}$, i.e., a subfield of $\mathbb{k}_M/\mathbb{k}$ with $\operatorname{Gal}(\mathbb{k}_M/K_1) = \langle\overline{x}\rangle$. It is clear that the only ramified place in $\mathbb{K}/\mathbb{k}$ is $(x^4+x^3+x^2+x+1)$, and it is totally ramified by Lemma 6.4.12(ii). The degree of the extension $K_1/\mathbb{k}$ is

$$|\operatorname{Gal}(\mathbb{k}_M/\mathbb{k})|/|\operatorname{Gal}(\mathbb{k}_M/K_1)| = 240/10 = 24.$$

One can show using the Hurwitz formula that $g(K_1) = 65$.

Then consider another cyclotomic field $\mathbb{k}_N$ with $N = x^4$, and let K_2 be the subfield of $\mathbb{k}_N$ fixed by the cyclic subgroup generated by $\overline{(x+1)^2}$. Then it is easy to see that $\operatorname{Gal}(L/\mathbb{k}) \cong (\mathbb{Z}/2\mathbb{Z})^2$, the only ramified place in $K_2/\mathbb{k}$ is (x), and it is totally ramified by Lemma 6.4.12(ii). A more thorough analysis allows one to compute the genus of $K_2/\mathbb{k}$.

The field K is defined as the compositum K_1K_2. It is clear that $K_1 \cap K_2 = \mathbb{k}$, since x is not ramified in $K_1/\mathbb{k}$. Hence, we have

$$\operatorname{Gal}(K/K_1) \cong \operatorname{Gal}(\mathbb{L}/\mathbb{k}) \cong (\mathbb{Z}/2\mathbb{Z})^2.$$

The only ramified places in K/K_1 are those places lying over x. There are 24 such places in total, and their inertia groups in K/K_1 are all isomorphic to $(\mathbb{Z}/2\mathbb{Z})^2$. Let T' be the set of the 24 places in $\mathbb{K}$ lying over ∞ and one place lying over x. Then $|T'|=25$. Put $T=\mathbb{P}_{K_1}\setminus T'$, let S be the set of the places of K lying over those places in T, and let $S'=\mathbb{P}_K\setminus S$. Then $|S'|=4\times 24+1=97$. Replacing F by K in Proposition 6.4.10, we get

$$d_2\,\mathrm{Cl}(\mathcal{O}_S)\geq\sum_P d_2G_P-(|T'|-1)-d_2G=24\times 2-24-2=22>2+2\sqrt{97}\approx 21.7.$$

Thus, the condition of Proposition 6.4.9 is satisfied, and

$$A(2)\geq\frac{|S'|}{g(K)-1}.$$

Applying the Hurwitz formula to K/K_1, one finds $g(K)-1=376$ and thus obtains the bound of Theorem 6.4.11.

Historical and Bibliographic Notes

The theory of global fields is a result of many centuries' development of number theory and algebraic geometry. Its history is a separate subject of many papers and several books. One can consult [**BS66**, **Lan70**, **Wei67**, **CF67**, **Wei84**, **IR90**, **Ros02**] and the references therein to have a glimpse of this long and complicated history. In particular, the analogy between number fields and function fields and their simultaneous treatment afterwards was one of the main sources of ideas during at least the last hundred years, which is partly reflected in these references. The same is true, albeit on a lesser time period, for the class field theory, which crowns the classical number theory. Some of its historical development is exposed in [**CF67**, Chapter XI].

There are many excellent expositions of this classical theory; see, e.g., the books by J.S. Milne [**Mil13**, **Mil14**], J.W.S. Cassels and A. Fröhlich [**CF67**], A. Weil [**Wei67**], and J. Neukirch [**Neu99**]. The explicit class field theory in its geometric version was developed in [**Ros54**, **Lan56**] and generalized and masterly exposed in [**Ser59**].

The Hayes theory was developed in [**Hay74**, **Hay79**]; see also [**Gos96**, Chapter 7] and [**NX01**, Chapter 3]. For Remark 6.3.29, see [**Zyw13**].

The class field tower problem was first proposed by P. Furtwängler and then popularized by H. Hasse in [**Has26**]. It was solved by E.S. Golod and I.R. Shafarevich in [**Sha63**, **GS64**]. A good exposition is contained in [**CF67**, Chapter IX] by P. Roquette; see also [**Tem01**] for Theorem 6.4.7.

The idea of applying the same method to obtain bounds for $A(q)$ belongs to J-P. Serre; see [**Ser83**, **Ser84**], where, in particular, Theorem 6.4.8 is proved. Further applications, in particular to $A(2)$, were given in [**Sch92**, **NX97**, **XY07**]. We mainly follow [**XY07**]; Proposition 6.4.10 is contained in [**NX01**]. Note that the current record for $A(2)$ is ≈ 0.3167 from [**DM13**].

CHAPTER 7

Curves with Many Points. II

The present chapter is devoted to curves with many rational points but is of a completely different spirit than Chapter 5, which treats a single, albeit very important, class of such curves, namely modular ones. On the contrary, in this chapter we try to address various questions on this rather large subject. However, it is impossible to touch upon all its facets, and we are forced to make some choices.

Recall that, speaking very generally, the subject of "curves with many points over finite fields" is studying the following quantity:

$$N_q(g) = \max_{C \text{ a curve of genus } g \text{ over } \mathbb{F}_q} |\{C(\mathbb{F}_q)\}|$$

for a given $g \geq 0$, $q = p^m$. Taken in that generality, the question of an explicit calculation (or a good estimation) of $N_q(g)$ is hopelessly difficult. Recall (see Sec. 3.4.4) that the answer is only known for $g \leq 2$ and for finitely many (≤ 20) values of q in the case of $g = 3$.

However, efforts of the four last decades have led to many results about $N_q(g)$. In particular, if we fix g and tend q to infinity, then

$$\lim_{q\to\infty} \frac{N_q(g) - q}{\sqrt{q}} = 2g,$$

which is a simple consequence of the Weil bound and some rather elementary constructions. For a fixed q and growing g, the Drinfeld–Vlăduţ bound (Theorem 3.2.1) gives

$$A(q) := \limsup_{g\to\infty} \frac{N_q(g)}{g} \leq \sqrt{q} - 1,$$

and families of various modular curves, e.g., those of Chapter 5, show that $A(q) = \sqrt{q} - 1$ if q is an even power of a prime (i.e., $\sqrt{q} - 1$ is an integer).

These asymptotic results can serve as an ideal for other questions concerning $N_q(g)$. However, there is no other definite result in this study but rather a large fan of different statements, sometimes with very difficult proofs, concerning $N_q(g)$ and related quantities. In this chapter we treat the following questions, which we regard to be essential and characteristic for the area (note however that a lot of very instructive results are not included here).

Section 7.1 contains the result of optimizing the explicit formula bound, which is called the Oesterlé bound; thus, it represents the best estimate for $N_q(g)$ that can be deduced from the explicit formula.

In Sec. 7.2 we consider curves that are absolutely maximal for their genera and, moreover, attain the Oesterlé bound, namely, the Deligne–Lusztig curves, which include the Ree and Suzuki curves (existing only in characteristics 2 and 3, respectively). We give a description of Deligne–Lusztig curves via class field theory, thus realizing the corresponding function fields as certain ray class fields.

Section 7.3 is devoted to some methods for constructing curves with many points for small genera: application of class field theory, fibre products, and quotients of Hermitian curves.

Section 7.4 contains a description of García–Stichtenoth towers, which are conjecturally related with modular curves; some confirmations of this conjecture are contained in Sec. 5.3. We also explain why codes on some of the towers can be constructed with a polynomial complexity much more reasonable than that of Secs. 5.1.3 and 5.2.2.

In Sec. 7.5 we briefly consider two other possible questions on curves with many points, namely, curves with a given number of points and curves for each genus.

7.1. Oesterlé Bound

Recall that the explicit formula from *Basic Notions* (Theorem 3.1.27 and Proposition 3.1.28) was used there to establish the Drinfeld–Vlăduţ bound, i.e., a bound on $\limsup\limits_{g\to\infty} \frac{N_q(g)}{g}$. However, each choice of the corresponding test polynomial results in a bound for each given pair (g,q). Moreover, it is possible to choose an (almost always) optimal polynomial for a given (g,q), and this result is known as the *Oesterlé bound*. Note however that it is formulated in another way, where $N_q(g)$ is an implicit function of (g,q).

Below we use the notation of Sec. 3.1.8; $\alpha_1,\dots,\alpha_g$ are the normalized inverse Frobenius roots, $\overline{\alpha}_1,\dots,\overline{\alpha}_g$ being their conjugates, and $\theta_j \in [0;2\pi)$ are given by $\alpha_j = e^{i\theta_j}$, $j=1,\dots,g$; $\psi(t) = \sum\limits_{r=1}^{m} c_r t^r$ is a real polynomial; $f(\theta) = 1 + 2\sum\limits_{r=1}^{m} c_r \cos(r\theta)$.

Recall that Proposition 3.1.28 says that if $c_r \ge 0$, $r=1,\dots,m$, and $f(\theta)\ge 0$ for any θ, then

$$N \le \frac{g}{\psi(q^{-1/2})} + \frac{\psi(q^{1/2})}{\psi(q^{-1/2})} + 1. \tag{7.1.1}$$

We are going to optimize the choice of ψ for a given N, thus obtaining the best possible lower bound for g in terms of N. We emphasize that here "the best possible" concerns bounds obtained by using the explicit formula; some other bounds can be (and are) better than this "best possible" bound. To formulate the result, we need some notation. Let φ_0 be defined as follows: for the integer m for which $q^{\frac{m}{2}} < N+1 \le q^{\frac{m+1}{2}}$, define

$$u := \frac{q^{\frac{m+1}{2}} - N + 1}{(N-1)\sqrt{q} - q^{m/2}} \in [0;1).$$

Then let φ_0 be a solution of the equation

$$\cos\Big(\frac{m+1}{2}\varphi\Big) + u\cos\Big(\frac{m-1}{2}\varphi\Big) = 0$$

in the interval $\Big[\frac{\pi}{m+1};\frac{\pi}{m}\Big)$; later we will see that φ_0 is unique.

THEOREM 7.1.1. *Let X be a curve of genus g over $\mathbb{F}_q$ with N points rational over $\mathbb{F}_q$. Then*

$$g \geq \sup_{\psi}\left\{(N-1)\psi(q^{-1/2})-\psi(q^{1/2})\right\} \geq \frac{(N-2)\sqrt{q}\cos\varphi_0+q-N+1}{q+1-2\sqrt{q}\cos\varphi_0},$$

where the supremum is taken over the polynomials $\psi(T)$ with nonnegative coefficients satisfying $\psi(t)+\psi(\overline{t})+1\geq 0$ for $t\in\mathbb{C}$, $|t|=1$. Moreover, if $q\geq 3$, the second inequality is actually an equality.

This result is called the *Oesterlé bound.*

To prove the theorem, given N and q ($q\geq 3$), we seek for the best choice of the $\{c_n\}$, i.e., the best possible lower bound on g, which can be made by choosing a measure μ on S^1 such that

$$\int_{S^1}\frac{1}{2}\,d\mu=(N-1)\sum_{n=1}^{\infty}c_nq^{-n/2}-\sum_{n=1}^{\infty}c_nq^{n/2}$$

and optimizing the $\{c_n\}$ for bounding g. Let us fix N and q; then inequality (7.1.1) can be rewritten as

$$g\geq (N-1)\sum_{n=1}^{\infty}c_nq^{-n/2}-\sum_{n=1}^{\infty}c_nq^{n/2}, \tag{7.1.2}$$

and we want to maximize the right-hand side.

For the measure μ on $S^1=\{z\in\mathbb{C}:|z|=1\}$ defined as

$$\mu=\sum_{j=1}^{g}\delta_{e^{i\theta_j}}+\delta_{e^{-i\theta_j}}$$

(in terms of Dirac measures δ), we have $\mu\geq 0$ and $\mu(S^1)=2g$. Formula (3.1.3), $N_n=q^n+1-q^{n/2}\sum_{i=1}^{2g}\alpha_i^n$, can be rewritten as

$$\frac{(1+q^n)-N_n}{q^{n/2}}=\sum_{j=1}^{g}2\cos(n\theta_j)=\sum_{j=1}^{g}e^{in\theta_j}+e^{-in\theta_j}=\int_{S^1}t^n\,d\mu=\int_{S^1}\frac{1}{2}(t^n+t^{-n})\,d\mu,$$

and our optimization task is equivalent to choosing a measure μ on S^1 such that

$$\int_{S^1}t^n\,d\mu\leq q^{n/2}-(N-1)q^{-n/2}:=\gamma_n \tag{7.1.3}$$

for all $n\geq 1$, as we will see below.

When varying μ and the $\{c_n\}$, let $g'(N,q)$ be the lower bound on $\frac{1}{2}\int_{S^1}d\mu$, and let $g(N,q)$ denote the maximum of the right-hand side in (7.1.2). Thus, our main purpose in this section is calculating $g(N,q)$. We begin with a couple of lemmas.

LEMMA 7.1.2. *If μ satisfies (7.1.3) and the $\{c_n\}$ satisfy (7.1.2), we have*

$$-\sum_{n=1}^{\infty}c_n\gamma_n\leq\int_{S^1}\frac{1}{2}\,d\mu. \tag{7.1.4}$$

In particular, $g(N,q)\leq g'(N,q)$.

PROOF. Indeed, since $f = 1 + \sum_n c_n(t^n + t^{-n})$ is positive and since

$$\mu(t^n) = \int_{S^1} t^n \, d\mu \le \gamma_n,$$

we have

$$0 \le \mu(f) = \mu(1) + 2\sum_{n=1}^{\infty} c_n \mu(t^n).$$

Hence,

$$\int_{S^1} \frac{1}{2}\, d\mu = \mu\Big(\frac{1}{2}\Big) \ge -\sum_{n=1}^{\infty} c_n \mu(t^n) \ge -\sum_{n=1}^{\infty} c_n \gamma_n. \qquad \square$$

LEMMA 7.1.3. *Equality in (7.1.4) takes place if and only if*

(a) *The support of μ on S^1 is contained in the zeroes of the function $f = 1 + \sum_n c_n(t^n + t^{-n})$ on S^1.*
(b) *There is equality in (7.1.3) for all n for which $c_n \neq 0$.*

SKETCH OF THE PROOF. In fact, we want $\mu(f) = 0$, so (a) is straightforward. Since we also want $\mu(t^n) = \gamma_n$ unless $c_n = 0$, (b) is also obvious. $\square$

Then we can obtain a measure-theoretic expression for $g(N,q)$.

PROPOSITION 7.1.4. *Let μ be as described in Lemma 7.1.3. Then*

$$g(N,q) = \int_{S^1} \frac{1}{2}\, d\mu = -\sum_{n=1}^{\infty} c_n \gamma_n.$$

PROOF. From Lemma 7.1.2 we have $g(N,q) \le g'(N,q)$, and by (7.1.4), we have $g'(N,q) \ge -\sum_n c_n \gamma_n$. By the assumption,

$$g'(N,q) \le \int_{S^1} \frac{1}{2}\, d\mu = -\sum_{n=1}^{\infty} c_n \gamma_n,$$

and thus

$$g(N,q) \le g'(N,q) \le g(N,q). \qquad \square$$

Let us give the first explicit corollary of the theory.

PROPOSITION 7.1.5. *For $q + 1 \le N \le q^{3/2} + 1$, the Weil bound is optimal.*

SKETCH OF THE PROOF. The choice $c_1 = 1/2$, $c_n = 0$, $n \ge 2$, obviously gives the Weil bound. For the optimality, we construct a measure μ on S^1 with the equality in (7.1.4), and then

$$g(N,q) = -\frac{1}{2}\gamma_1 = -\frac{1}{2}\left(q^{1/2} - (N-1)q^{-1/2}\right) = \frac{1}{2}\left((N-1)q^{-1/2} - q^{1/2}\right) \ge 0.$$

Thus, take $\mu = -\gamma_1 \delta_{-1}$. Then we have

- $1 + \cos(\theta) \ge 0$ and $c_n \ge 0$ (obvious);
- $\mathrm{Supp}(\mu) \subseteq \ker(1 + \cos(\theta))$ (obvious);
- $\mu(t) = \gamma_1$, since $\mu(t) = -\gamma_1(-1) = \gamma_1$ by the construction;

- $\mu(t^n) \le \gamma_n$ for $n \ge 2$: Indeed, for $n = 2$ we have
$$\begin{aligned}
&\mu(t^2) = 2\Big(-\frac{1}{2}\gamma_1\Big)(-1)^2 = -\gamma_1 \le \gamma_2 \\
&\iff -q^{1/2} + (N-1)q^{-1/2} \le q - (N-1)q^{-1} \\
&\iff (N-1)\left(q^{-1/2} + q^{-1}\right) \le q + q^{1/2} \\
&\iff (N-1)\left(1 - q^{1/2}\right) \le q^{3/2}\left(1 - q^{1/2}\right) \\
&\iff N - 1 \le q^{3/2},
\end{aligned}$$
and the last inequality is true by the condition on N. Next, $\mu(t^3) = -2(-\gamma_1/2) = \gamma_1$, and since $-2(-\gamma_1/2) \le 0$ and $\gamma_3 \ge 0$, we have $\mu(t^3) \le \gamma_3$. Next, $\mu(t^4) = 2(-\gamma_1/2) = -\gamma_1 \le \gamma_2 \le \gamma_4$, since γ_n grows for large n. This generalizes for larger n.

By Lemma 7.1.3 we now have equality in (7.1.4). □

Now we proceed with the general case and introduce the following notation: $\lambda = N - 1$, $\alpha = q^{1/2}$; define $m \in \mathbb{N}$ by $\alpha^m < \lambda \le \alpha^{m+1}$, i.e., $m = \left\lceil \frac{\log \lambda}{\log \alpha} \right\rceil - 1$. We may assume that $m \ge 2$, since $m = 1$ gives $q^{1/2} < N - 1 \le q$, which may be obtained by taking $g = 0$. Put
$$u = \frac{\alpha^{m+1} - \lambda}{\lambda\alpha - \alpha^m} \in [0, 1).$$

Using this notation, we may rewrite (7.1.3) as
$$\int_{S^1} t^n \, d\mu \le \alpha^n - \lambda\alpha^{-n}, \quad n \ge 1.$$

Now we are going to construct a measure μ on S^1 which satisfies this equation and also possesses the following properties:

(a) μ is concentrated in a symmetric set $T \subseteq S^1$ with $|T| = m - 1$.

(b) We have
$$\int_{S^1} t^n \, d\mu = \alpha^n - \lambda\alpha^{-n}, \quad n = 1, \ldots, m-1.$$

(c) μ can be written as
$$\mu = \sum_{t \in T} \nu_t \delta_t, \quad \nu_t > 0,$$
where $\nu_t = \nu_{\bar{t}}$.

Furthermore, let $\{c_n\}$ be such that
$$f(t) = 1 + \sum_{n=1}^{m-1} c_n(t^n + t^{-n}), \quad c_n \ge 0,$$
is zero on T and nonnegative on S^1. If this is possible, Proposition 7.1.4 implies
$$g(N, q) = \int_{S^1} \frac{1}{2} \, d\mu = -\sum_{n=1}^{m-1} c_n(\alpha^n - \lambda\alpha^{-n}).$$

The following three technical results allow one to obtain a definite answer for $q \geq 3$.

LEMMA 7.1.6. *Assume that $T \subseteq S^1$ satisfies the above conditions. Then T is contained in the set of solutions to*

$$s^{m+1} + 1 + u(s^m + s) = 0. \tag{7.1.5}$$

SKETCH OF THE PROOF. The conditions are equivalent to the following: for any polynomial ψ, $\deg(\psi) \leq m-1$, with constant term equal to zero, we have

$$\sum_{t \in T} \nu_t \psi(t) = \psi(\alpha) - \lambda\psi(\alpha^{-1}), \tag{7.1.6}$$

and a straightforward but lengthy computation leads to the conclusion. □

LEMMA 7.1.7. *Equation (7.1.5) has $m+1$ solutions in S^1 and exactly one of the form $t = e^{\pm i\varphi_0}$ with $\varphi_0 \in \left[\frac{\pi}{m+1}; \frac{\pi}{m}\right)$.*

PROOF. First note that if m is even, -1 is a solution, and if m is odd, this is not the case. Now put

$$F(\varphi) = \frac{\cos\left(\frac{m+1}{2}\varphi\right)}{\cos\left(\frac{m-1}{2}\varphi\right)}.$$

We see that $F(\varphi_0) = -u$ by the definition of φ_0. If $s = e^{i\varphi}$ with $F(\varphi) = -u$, we have

$$-u = F(\varphi) = \frac{s^{\frac{m+1}{2}\varphi} + s^{-\frac{m+1}{2}\varphi}}{s^{\frac{m-1}{2}\varphi} + s^{-\frac{m-1}{2}\varphi}},$$

and hence

$$-u = \frac{s^{m+1}+1}{s^m + s};$$

that is, s solves (7.1.5). This is also the case if $s = e^{-i\varphi}$. Now a given m corresponds to a $u \in [0; 1)$, and we must show that $F(\varphi) = -u$ has $m/2$ solutions if m is even and $(m+1)/2$ solutions if m is odd, which is elementary. □

Now let T be the complement of $e^{\pm i\varphi_0}$ in the set of solutions on S^1 to (7.1.6). Define $\nu_t \in \mathbb{C}$ by solving the $m-1$ equations in $m-1$ unknowns:

$$\sum_{t \in T} \nu_t t^n = \alpha^n - \lambda\alpha^{-n}, \quad n = 1, \ldots, m-1.$$

The following lemma requires long brute-force computations.

LEMMA 7.1.8. *We have*

(a) $\nu_t > 0$ *for all* $t \in T$.
(b) $\nu_t = \nu_{\bar{t}}$ *for all* $t \in T$.
(c) *For* $n \geq m$ *and* $\alpha \geq \sqrt{3}$, *we have*

$$\sum_{t \in T} \nu_t t^n \leq \alpha^n - \lambda\alpha^{-n}.$$

□

Now define a polynomial $P(X) = \prod_{t\in T}(X-t)$ and write

$$P(X)P(X^{-1}) = \sum_{n=-(m-1)}^{m-1} a_n X^n.$$

Since the ν_t are real, we have $P(z)P(z^{-1}) = P(z)P(\overline{z}) = P(z)\overline{P(z)} \geq 0$, and since $0 \notin T$, we have $a_0 > 0$. Therefore, me may define

$$f(t) = \frac{1}{a_0}P(t)P(t^{-1}) = 1 + \sum_{n=1}^{m-1} c_n(t^n + t^{-n}), \quad t \in S^1,$$

where $c_n = a_n/a_0$ for $n = 1, \dots, m-1$.

Now consider the measure

$$\mu = \sum_{t\in T} \nu_t \delta_t.$$

As a consequence of the above calculations, we obtain the following.

PROPOSITION 7.1.9. *The measure μ and function f defined above satisfy the condition*

$$\int_{S^1} t^n \, d\mu \leq \alpha^n - \lambda\alpha^{-n}, \quad n \geq 1,$$

and

$$g(N,q) = \int_{S^1} \frac{1}{2}\, d\mu = -\sum_{n=1}^{m-1} c_n(\alpha^n - \lambda\alpha^{-n})$$

for $q \geq 3$.

PROOF. By the construction, $f(t) = 0$ except on T and is nonnegative on S^1. Also, by the construction, the conditions on μ are satisfied. Now by Lemma 7.1.8 we may apply Proposition 7.1.4. □

Finally, we can state an explicit bound.

COROLLARY 7.1.10. *For $q \geq 3$, we have*

$$g \geq g(N,q) = \sum_{n=1}^{m-1} c_n(\lambda\alpha^{-n} - \alpha^n) = \frac{(\lambda-1)\alpha\cos(\varphi_0) + \alpha^2 - \lambda}{\alpha^2 - 2\alpha\cos(\varphi_0) + 1}.$$

PROOF. For $n = 0, \dots, m-1$ we find that

$$a_n = (m-n)\cos(n\varphi_0)\sin(\varphi_0) + \sin\big((m-n)\varphi_0\big),$$

and by substituting this in Proposition 7.1.9 we get the expression. □

REMARK 7.1.11. Note that for $q \geq 3$ the above bound is the best one, but for $\alpha = \sqrt{2}$, i.e., $q = 2$, this method does not necessarily give the optimal bound, because for some λ the inequality

$$\sum_{t\in T} \nu_t t^n \leq \alpha^n - \lambda\alpha^{-n}$$

does not hold for all $n \geq m$. For example, this is the case for the following values of N:

51–53, 70–77, 98–110.

7.2. Deligne–Lusztig Curves

There exist curves attaining the Oesterlé bound, but they are rather rare. As we have seen in Sec. 3.4.1, Hermitian curves attain the Weil bound and thus the Oesterlé bound. There are only two more families of curves attaining the Oesterlé bound, in characteristic 2 and 3 respectively. In fact, these families, called the Suzuki and Ree curves, are special examples of a very vast and important class of varieties over finite fields, namely the Deligne–Lusztig varieties. These varieties are associated with reductive algebraic groups over finite fields and provide very exact and profound information on representations of such groups.

In general, these varieties are of high dimensions, and there exist exactly three cases where they are of dimension 1, i.e., algebraic curves. The first case corresponds to groups of the type ${}^2A_2(q)$, i.e., (projective) unitary groups $\mathrm{PGU}_3(\mathbb{F}_q)$; these are exactly Hermitian curves. The two other cases are ${}^2B_2(q)$ (Suzuki) and ${}^2G_2(q)$ (Ree) groups, which lead to the Suzuki and Ree curves.

In the present section we give some basic information on these cases, including the group structure of the Hermitian curves, which we did not touch upon in Secs. 3.4.1 and 4.4.3 of *Basic Notions* while discussing Hermitian curves and codes. We are trying to be as elementary as possible and use neither the general theory of Deligne–Lusztig varieties nor the theory of reductive groups. This forces us to use some *ad hoc* definitions rather than natural but more elaborated ones.

Besides their purely theoretical interest, these curves possess some features which make them suitable for various applications in coding theory. The major advantage of (some) codes obtained from these curves is their automorphism groups and thus group-theoretic properties.

7.2.1. Group Codes on Hermitian Curves

First let us define the group $\mathrm{PGU}_3(q)$, $q=r^2$ being a square. Recall that the projective equation of the Hermitian curve $\mathcal{H}_q$ over $\mathbb{F}_r$,

$$X^{r+1}+Y^{r+1}+Z^{r+1}=0,$$

can be rewritten as

$$X\overline{X}+Y\overline{Y}+Z\overline{Z}=0, \tag{7.2.1}$$

where $a\mapsto\overline{a}\equiv a^r$ for $a\in\mathbb{F}_q$ is the Hermitian conjugation of $\mathbb{F}_q$ over $\mathbb{F}_r$. The group $\mathrm{U}_3(q)=\mathrm{U}_3(\mathbb{F}_q)$ is defined as the subgroup of $\mathrm{GL}_3(\mathbb{F}_q)$ leaving (7.2.1) invariant, and $\mathrm{PGU}_3(q)$ as the factor of $\mathrm{U}_3(q)$ modulo its center.

PROPOSITION 7.2.1. *The group* $\mathrm{PGU}_3(q)$ *is the full automorphism group* $G_H=\mathrm{Aut}(\mathcal{H}_q)$ *of* $\mathcal{H}_q$ *over* $\overline{\mathbb{F}}_q$, *and its order is*

$$|\mathrm{PGU}_3(q)|=qr(q-1)(qr-1)=q^{3/2}(q-1)(q^{3/2}-1). \qquad \square$$

REMARK 7.2.2. The order of $\mathrm{Aut}(\mathcal{H}_q)$ is extremely large and is in a sense exceptional among all algebraic curves. Indeed, recall first that for an (irreducible) algebraic curve X over any field k with $\mathrm{char}(k)=0$ we have the Hurwitz bound

$$|\mathrm{Aut}(X)|\le 84(g_X-1).$$

In a positive characteristic the Hurwitz inequality does not hold any more; however, we have the inequality

$$|\mathrm{Aut}(X)| \leq 16g_X^4 \tag{7.2.2}$$

with the only exception for Hermitian curves, which do not satisfy (7.2.2).

Now let us describe the action of $\mathrm{Aut}(\mathcal{H}_q)$ on the model $\mathcal{H}'_r$ of $\mathcal{H}_q$ given by $y^r + y = x^{r+1}$ with $x := \dfrac{\beta Z}{Y - \beta X}$, $y := Xx - \alpha Z$, where $\alpha, \beta \in \mathbb{F}_q$ satisfy $\alpha^r + \alpha = \beta^{r+1} = -1$ (cf. Sec. 3.4.1 in *Basic Notions*, Exercises 3.4.2 and 3.4.3); recall that this model is more convenient for applications related to algebraic geometry codes, and, in particular, the description of the Hermitian codes C_m in Sec. 4.4.3 uses this model.

The projective curve with affine equation

$$y^r + y = x^{r+1}$$

has a unique point $Q = P_\infty$ at infinity (i.e., the common pole of x and y), and this point is $\mathbb{F}_r$-rational. The stabilizer $G_Q \subset G_H = \mathrm{Aut}(\mathcal{H}_q)$ of Q can be described by its action on x and y as follows:

$$G_Q = \{\sigma \in G_H : \ \sigma(x) = x + b, \ \sigma(y) = a^{q+1}y + ab^q x + c\},$$

where $a, b, c \in \mathbb{F}_q$, $a \neq 0$, and $c^q + c = b^{q+1}$. Its order is

$$|G_Q| = qr(q-1) = q^{3/2}(q-1).$$

The unique p-Sylow subgroup $G_Q^{(1)} \subset G_Q$ (where $p = \mathrm{char}\,\mathbb{F}_q$) is given by

$$\begin{aligned} G_Q^{(1)} &= \{\sigma \in G_Q : \ \sigma(x) = x + b \text{ for some } b \in \mathbb{F}_q\} \\ &= \{\sigma \in G_Q : \ \sigma(x) = x + b, \ \sigma(y) = y + b^q x + c, \ c^q + c = b^{q+1}\}, \end{aligned}$$

and its order is $|G_Q^{(1)}| = q^{3/2}$. The factor group $G_Q/G_Q^{(1)}$ is cyclic of order $q-1$ and is generated by $\varepsilon \in G_Q$ with $\varepsilon(x) = ax$, $\varepsilon(y) = a^{q+1}y$ for any generator $a \in \mathbb{F}_q^*$.

PROPOSITION 7.2.3. *The group G is generated by G_Q and the involution*

$$\tau\colon \ x \mapsto \frac{x}{y}, \quad y \mapsto \frac{1}{y}. \qquad \square$$

Recall properties of the code $C_m = (\mathcal{H}'_r, \mathcal{P}, mQ)_L$ from Sec. 4.4.3, where $\mathcal{P} = \mathcal{H}'_r(\mathbb{F}_q) \setminus \{Q\}$, and also its group interpretation following from the fact that $G_Q^{(1)}$ stabilizes the infinite point P_∞.

PROPOSITION 7.2.4. (1) *Let $m \leq r^3 + r^2 - r - 2 =: t$. Then the codes C_m and C_{t-m} are dual.*

(2) *The space $L(mQ)$ has a basis given by*

$$\{x^i y^j : \ 0 \leq i, \ 0 \leq j \leq r-1, \ ir + j(r+1) \leq m\}.$$

(3) *Let $h := r^2 - r - 2$. Then the dimension k_m of C_m is given by*

$$k_m = \begin{cases} \nu(m) & \text{for } m \leq h, \\ m + 1 - \dfrac{q-r}{2} & \text{for } h < m < qr, \\ qr - \nu(t-m) & \text{for } m \geq qr, \end{cases}$$

where for $m \in \mathbb{N}$ we define an integer $\nu(m)$ as follows. Let $\widetilde{m}$ be the largest integer of the form $ir+j(r+1)$, $i \geq 0$, $0 \leq j \leq r-1$, not greater than m, and let $\widetilde{m} = ur+s$, $0 \leq s \leq r-1$. Then

$$\nu(m) := 1 + \frac{u(u+1)}{2} + \min\{u, s\}.$$

(4) *The code C_m is a group code for the group $G_r := G_Q^{(1)}$, i.e., can be represented as an ideal I_m in the group algebra $\mathbb{F}_q[G_r]$.* □

7.2.2. Suzuki Curves

Let us start with defining the Suzuki curves by elementary *ad hoc* formulae which allow one to study their properties without using an elaborated theory.

For $n \geq 1$, set $q = 2^{2n+1}$, $q_0 = 2^n$.

PROPOSITION 7.2.5. *The curve $X \subseteq \mathbb{P}^2(\overline{\mathbb{F}}_q)$ defined over $\mathbb{F}_q$ by the equation*

$$u^{q_0}(z^q + zu^{q-1}) = y^{q_0}(y^q + yu^{q-1})$$

is irreducible of genus $g = q_0(q-1)$. It has a unique point

$$P_\infty \in X \cap \{u = 0\}$$

at infinity.

PROOF. The irreducibility follows from Lemma 7.2.9 below; the genus is determined in Lemma 7.2.10. The claim about P_∞ is trivial. □

PROPOSITION 7.2.6. *The five rational functions on X*

$$f_0 = 1, \qquad f_q = \frac{y}{u}, \qquad f_{q+q_0} = \frac{z}{u},$$

$$f_{q+2q_0} = f_q^{2q_0+1} + f_{q+q_0}^{2q_0}, \qquad f_{q+2q_0+1} = f_q f_{q+q_0}^{2q_0} + f_{q+2q_0}^{2q_0}$$

are regular on $X \setminus \{P_\infty\}$. At P_∞ the function f_i has a pole of order i:

$$v_\infty(f_i) = -i, \quad i \in \{0, q, q+q_0, q+2q_0, q+2q_0+1\}.$$

PROOF. By definition, f_i is regular away from P_∞; the pole orders are determined in Lemma 7.2.9 below. □

Let $\mathscr{P}$ be the additive monoid generated by the orders of poles of the functions f_i:

$$\mathscr{P} = \langle 0, q, q+q_0, q+2q_0, q+2q_0+1 \rangle.$$

For each $n \in \mathscr{P}$ there is by definition an f_n such that f_n is regular on X except at P_∞, which is a pole of order n,

$$v_{P_\infty}(f_n) = -n,$$

namely

$$f_n = f_q^{a_1} f_{q+q_0}^{a_2} f_{q+2q_0}^{a_3} f_{q+2q_0+1}^{a_4},$$

where a_i are nonnegative integers such that

$$n = a_1 q + a_2(q+q_0) + a_3(q+2q_0) + a_4(q+2q_0+1).$$

One can calculate the cardinality of $\mathbb{N} \setminus \mathscr{P}$ directly.

PROPOSITION 7.2.7. *Let $\mathscr{P} = \langle 0, q, q+q_0, q+2q_0, q+2q_0+1\rangle$ be the additive monoid defined above. Then*

$$|\mathbb{N}\setminus\mathscr{P}| = q_0(q-1). \qquad \square$$

Thus, we get the following.

PROPOSITION 7.2.8. *Let f_n be chosen as above, one for each $n\in\mathscr{P}$. Then $L(dP_\infty)$ has as a basis the elements in the set*

$$\{f_n \mid n\in\mathscr{P},\ n\le d\}.$$

PROOF. The functions f_n are linearly independent, since they have mutually different pole orders at P_∞; note also that

$$\mathscr{P}\subseteq\{-v_{P_\infty}(f)\mid f \text{ is regular away from } P_\infty\}.$$

The complements of these two sets are finite of the same size $q_0(q-1)=g$, according to Propositions 7.2.5 and 7.2.7. We conclude that

$$\mathscr{P} = \{-v_{P_\infty}(f)\mid f \text{ is regular away from } P_\infty\}, \tag{7.2.3}$$

and that there is an integer d_0 such that for $d\ge d_0$,

$$\dim_{\mathbb{F}_q} L(dP_\infty) = d+1-g = d+1-q_0(q-1) = |\{n\in\mathscr{P}\mid n\le d\}|.$$

We claim that for all d,

$$\dim_{\mathbb{F}_q} L(dP_\infty) = |\{n\in\mathscr{P}\mid n\le d\}|.$$

Let d_1 be the largest integer for which the claim is false. Assume that $d_1>0$. Then

$$\begin{aligned}\dim_{\mathbb{F}_q} L((d_1+1)P_\infty) &= |\{n\in\mathscr{P}\mid n\le d_1+1\}|,\\ \dim_{\mathbb{F}_q} L(d_1P_\infty) &\ne |\{n\in\mathscr{P}\mid n\le d_1\}|.\end{aligned}$$

This implies that $d_1+1\notin\mathscr{P}$ and that d_1+1 is a pole order, in contradiction with (7.2.3). Therefore, the claim is true for all $d\ge 0$. For negative d the claim is trivially true. $\square$

In the affine plane $\mathbb{A}^2(\overline{\mathbb{F}}_q) = \mathbb{P}^2(\overline{\mathbb{F}}_q)\setminus\{u=0\}$ we get the corresponding affine equation:

$$X\setminus\{u=0\}:\ z^q+z = y^{q_0}(y^q+y).$$

Let F denote the function field of X over $\mathbb{F}_q$:

$$F=\mathbb{F}_q(y,z), \qquad z^q+z=y^{q_0}(y^q+y).$$

Let v_∞ denote the valuation corresponding to the point P_∞ at infinity.

Define $v, w\in F$ by

$$v = f_{q+2q_0}(z,y,1) = z^{2q_0}+y^{2q_0+1}, \qquad w = f_{q+2q_0+1}(z,y,1) = z^{2q_0}y+v^{2q_0}.$$

LEMMA 7.2.9. *In the above notation, we have*

$$F=\mathbb{F}_q(y,v), \qquad v^q+v=y^{2q_0}(y^q+y), \qquad v^{q_0}=z+y^{q_0+1}, \qquad w^{q_0}=v+zy^{q_0}.$$

The functions y, z, v, and w have no poles except at P_∞, where

$$\begin{aligned} v_\infty(y) &= -q, & v_\infty(z) &= -(q+q_0),\\ v_\infty(v) &= -(q+2q_0), & v_\infty(w) &= -(q+2q_0+1).\end{aligned}$$

The element $\pi = v/w$ is a prime element at P_∞.

The equation

$$z^q + z = y^{q_0}(y^q + y)$$

is irreducible, and $[F : \mathbb{F}_q(y)] = q$.

PROOF. The relations between y, z, v, and w follow by a brute force computation, which in turn imply that $F = \mathbb{F}_q(y, v)$.

Assume that v_∞ is a valuation of F such that $v_\infty(y) = -r < 0$. Then

$$\begin{aligned}
z^q + z = y^{q_0}(y^q + y) &\implies q v_\infty(z) = -r(q + q_0) &&\implies v_\infty(z) = -r\left(1 + \frac{q}{q_0}\right), \\
v^{q_0} = z + y^{q_0+1} &\implies q_0 v_\infty(v) = -r(1 + q_0) &&\implies v_\infty(v) = -r\left(1 + \frac{1}{q_0}\right), \\
w^{q_0} = v + z y^{q_0} &\implies q_0 v_\infty(w) = -r\left(1 + \frac{q_0}{q} + q_0\right) \\
&\implies v_\infty(w) = -r\left(1 + \frac{1}{q_0} + \frac{1}{q}\right) = -\frac{r}{q}\left(1 + \frac{q}{q_0} + q\right).
\end{aligned}$$

Since $(1 + (q/q_0) + q) \equiv 1 \bmod 2$, we conclude that $q \mid r$. Finally,

$$r \le [F : \mathbb{F}_q(y)] \le q \implies q = r \implies [F : \mathbb{F}_q(y)] = q.$$

This shows that v_∞ assumes the claimed values, the equation defining the curve is irreducible, and P_∞ is the only pole for y and thereby the only pole for z, v, and w. □

LEMMA 7.2.10. *The genus of* X *is* $q_0(q-1)$.

PROOF. According to Lemma 7.2.9 we have that $[F : \mathbb{F}_q(y)] = q$, the field extension is a Galois extension, and any automorphism of F over $\mathbb{F}_q(y)$ is of the form

$$\sigma : y \mapsto y, \qquad \sigma : z \mapsto z + \alpha, \quad \alpha \in \mathbb{F}_q.$$

For v and w defined above, we get

$$\sigma(v) = v + \alpha^{2q_0}, \qquad \sigma(w) = w + \alpha^{2q_0} y + \alpha^2.$$

For the prime element $\pi = v/w$, we obtain

$$v_\infty(\sigma(\pi) - \pi) = 2q_0 + 2, \quad \alpha \neq 0,$$

and thus the orders of the ramifications groups G_i of P_∞ in $F/\mathbb{F}_q(y)$ are

$$|G_i| = \begin{cases} q, & i = 0, 1, 2, \ldots, 2q_0 + 1, \\ 1, & i \ge 2q_0 + 2. \end{cases} \tag{7.2.4}$$

Since all the other places are unramified, by Hilbert's formula the different of $F/\mathbb{F}_q(y)$ is

$$(2q_0 + 2)(q - 1)P_\infty,$$

and the Hurwitz formula gives the result. □

Rational points. As we have mentioned above, these curves meet the Oesterlé bound; moreover, the structure of their rational points is rather simple. In fact, we have the following result.

PROPOSITION 7.2.11. *The curve* $X \subseteq \mathbb{P}^2(\overline{\mathbb{F}}_q)$ *defined over* $\mathbb{F}_q$ *by the equation*

$$u^{q_0}(z^q + zu^{q-1}) = y^{q_0}(y^q + yu^{q-1})$$

has $1+q^2$ $\mathbb{F}_q$*-rational points, namely one point* P_∞ *at infinity and all the* $\mathbb{F}_q$*-rational points in the affine plane* $\mathbb{A}^2(\overline{\mathbb{F}}_q) = \mathbb{P}^2(\overline{\mathbb{F}}_q) \setminus \{u=0\}$:

$$X(\mathbb{F}_q) = \{P_\infty\} \cup \mathbb{A}^2(\mathbb{F}_q).$$

The curve X *has the maximal number of points that a curve of genus* $q_0(q-1)$ *can have.*

PROOF. The rationality of these points is clear. To check the maximality, one can apply the Weil explicit formula with

$$\psi(t) = \frac{\sqrt{2}}{2}t + \frac{1}{4}t^2, \qquad a = \frac{4q}{4q_0+1}, \qquad b = 1 + \frac{2qq_0(2+q_0)}{4q_0+1}. \qquad \square$$

Group of automorphisms. Since the initial definition of the Suzuki curves comes from certain algebraic groups, it is but natural that they have very rich automorphism groups. Now we are going to describe these groups and their action on the curves.

Let $(z:y:u)$ be homogeneous coordinates on $\mathbb{P}^2(\overline{\mathbb{F}}_q)$. Consider the following subgroup of $\mathrm{GL}_3(\mathbb{F}_q)$:

$$S = \left\{ \begin{pmatrix} 1 & \alpha^{q_0} & \gamma \\ 0 & 1 & \alpha \\ 0 & 0 & 1 \end{pmatrix} \,\middle|\, \alpha, \gamma \in \mathbb{F}_q \right\},$$

which acts naturally on $\mathbb{P}^2(\overline{\mathbb{F}}_q)$.

PROPOSITION 7.2.12. *Let* S *be the group defined above. Then* $|S| = q^2$. *The action of* S *on* $\mathbb{P}^2(\overline{\mathbb{F}}_q)$ *restricts to the curve* X *so that* S *becomes a group of* $\mathbb{F}_q$*-rational automorphisms of* X. *The action of* S *on the* $\mathbb{F}_q$*-rational points* $X(\mathbb{F}_q)$ *has two orbits, namely* $\{P_\infty\}$ *and* $\mathbb{A}^2(\mathbb{F}_q)$. *The group* S *acts freely and transitively on the orbit* $\mathbb{A}^2(\mathbb{F}_q)$.

PROOF. The group clearly has order q^2. The action restricts to X, since the equation of X is invariant under any $\sigma \in S$. Consider the point $P_\infty = (1:0:0)$. Then $\sigma(P_\infty) = P_\infty$ for all $\sigma \in S$, so $\{P_\infty\}$ constitutes an orbit. Let $P \neq P_\infty$ be another $\mathbb{F}_q$-rational point; then P is fixed under σ if and only if $\sigma = \mathrm{id}$. Since the number of $\mathbb{F}_q$-rational points different from P_∞ equals the order of the group, the result follows. $\square$

The curve has more automorphisms; in fact, the Suzuki group ${}^2B_2(q)$ of order $q^2(q^2+1)(q-1)$ acts on X, and S is a Sylow 2-subgroup. Let us give an *ad hoc* description of the Suzuki group ${}^2B_2(q)$.

LEMMA 7.2.13. *Let* $q_0 = 2^s$, $s \geq 1$, *and* $q = 2q_0^2$. *For* $a, c \in \mathbb{F}_q$, *define the matrix*

$$\widetilde{T}_{a,c} := \begin{pmatrix} 1 & 0 & 0 & 0 \\ a & 1 & 0 & 0 \\ c & a^{2q_0} & 1 & 0 \\ ac + a^{2q_0+2} + c^{2q_0} & a^{2q_0+1} + c & a & 1 \end{pmatrix}.$$

Then the set $\widetilde{\boldsymbol{T}} = \{\widetilde{T}_{a,c} \mid a, c \in \mathbb{F}_q\}$ *is a group of exponent* 4 *and order* q^2. *For any* $b \in \mathbb{F}_q$ *with* $b \neq 0$, $\widetilde{\boldsymbol{T}}$ *is isomorphic to the* 2*-dimensional linear group over* $\mathbb{F}_q$

consisting of all linear transformation $(X,Y) \mapsto (X+a, b^{-1}a^{2q_0}X + Y + b^{-1}c)$ *with* a,c *running over* $\mathbb{F}_q$. *In particular, the center* $Z(\widetilde{\boldsymbol{T}})$ *is an elementary abelian group of order* 2^{2s+1} *whose nontrivial elements are those of order* 2 *of* $\widetilde{\boldsymbol{T}}$. □

Now define the following diagonal matrix group, which has a simple group structure (an easy direct calculation).

LEMMA 7.2.14. *For* $\alpha \in \mathbb{F}_q$, $\alpha \neq 0$, *let*

$$\widetilde{N}_\alpha := \begin{pmatrix} \alpha^{-q_0-1} & 0 & 0 & 0 \\ 0 & \alpha^{-q_0} & 0 & 0 \\ 0 & 0 & \alpha^{q_0} & 0 \\ 0 & 0 & 0 & \alpha^{q_0+1} \end{pmatrix}.$$

The set $\widetilde{\boldsymbol{N}} = \{\widetilde{N}_\alpha \mid \alpha \in \mathbb{F}_q, \alpha \neq 0\}$ *is a group of order* $q-1$ *isomorphic to* $\mathbb{F}_q^*$. □

Also define

$$\widetilde{W} := \begin{pmatrix} 0 & 0 & 0 & 1 \\ 0 & 0 & 1 & 0 \\ 0 & 1 & 0 & 0 \\ 1 & 0 & 0 & 0 \end{pmatrix}.$$

The Suzuki group ${}^2B_2(q)$ is defined as the 4-dimensional linear group generated by $\widetilde{\boldsymbol{T}}$, $\widetilde{\boldsymbol{N}}$, and $\widetilde{W}$. It is a simple group, and therefore embeds in $\mathrm{PGL}_4(\mathbb{F}_q)$. One can show that ${}^2B_2(q)$ preserves the ovoid O_3 of $\mathbb{P}^3(\mathbb{F}_q)$ consisting of the points $(0:0:0:1)$ and $(1:x:y:xy+x^{2q_0+1}+y^{2q_0})$ with x,y ranging over $\mathbb{F}_q$; moreover, it can be characterized by this last property.

Codes on $\boldsymbol{X}$. Let $P_1,\dots,P_{q^2}$ be the $\mathbb{F}_q$-rational points $\mathbb{A}^2(\mathbb{F}_q)$ on the curve X, $\mathbb{A}^2(\mathbb{F}_q) = \{P_1,\dots,P_{q^2}\}$, and let $m \geq 0$ be an integer. Consider the algebraic geometry code $C_m := (X, \mathbb{A}^2(\mathbb{F}_q), mP_\infty)_L$. Let $\mathscr{P}$ be the monoid $\langle 0, q, q+q_0, q+2q_0, q+2q_0+1\rangle \subseteq \mathbb{N}$ as above.

PROPOSITION 7.2.15. (1) *The code* C_m *is a* $[q^2, k, d \geq q^2 - m]$ *code with*

$$k = \dim(C_m) = |\{n \mid n \in \mathscr{P},\ n \leq m\}|, \quad m < q^2$$

(*as usual, up to a possible spoiling*). *In particular, if* $2q_0(q-1) - 2 < m < q^2$, *then*

$$k = m + 1 - q_0(q-1), \qquad k + d \geq q^2 + 1 - q_0(q-1).$$

(2) *The evaluation map*

$$\mathrm{Ev}_{\mathbb{A}^2(\mathbb{F}_q)} \colon L(mP_\infty) \to \mathbb{F}_q[S], \quad f \mapsto (f(P_1), \dots, f(P_{q^2})),$$

identifies the code C_m *as a left ideal in* $\mathbb{F}_q[S]$. □

Now we can describe the dual code using Theorems 4.1.25 and 4.1.35. Namely, the code $C_m^\perp$ dual to $C_m = (X, \mathbb{A}^2(\mathbb{F}_q), mP_\infty)_L$ is $C_m^\perp = (X, \mathbb{A}^2(\mathbb{F}_q), P_1 + \ldots + P_{q^2} - mP_\infty + (\eta))_L$, where η is a differential form with simple poles at $P_1,\dots,P_{q^2}$ with $\mathrm{Res}_{P_i}(\eta) = 1$.

PROPOSITION 7.2.16. *Let* $C_m = (X, \mathbb{A}^2(\mathbb{F}_q), mP_\infty)_L$ *be the code associated with* mP_∞. *Then* $C_m^\perp$ *is of the same type*; *namely,*

$$C_m^\perp = C_{q^2+2q_0(q-1)-2-m}.$$

PROOF. Let $\eta = dx/x$ with $x = y^q + y$. Then all the P_i are simple poles of η

and

$$(\eta) = \left(\frac{dx}{x}\right) = (dx) - (P_1 + \ldots + P_{q^2}) + q^2 P_\infty$$
$$= (2g-2)P_\infty - (P_1 + \ldots + P_{q^2}) + q^2 P_\infty;$$

hence,

$$P_1 + \ldots + P_{q^2} - mP_\infty + (\eta) = (q^2 + 2q_0(q-1) - 2 - m)P_\infty. \qquad \square$$

7.2.3. Ree Curves

For these curves we also give a direct elementary description as for the Suzuki curves. Unfortunately, details are here more cumbersome than in the previous case, and we mainly state the results with almost no justification.

Now let $s \geq 1$, $q = 3^{2s+1}$, $q_0 = 3^s$. Define $F_R = \mathbb{F}_q(x, y_1, y_2)$, where

$$\begin{aligned} y_1^q - y_1 &= x^{q_0}(x^q - x), \\ y_2^q - y_2 &= x^{q_0}(y_1^q - y_1). \end{aligned} \tag{7.2.5}$$

Throughout this subsection, we always consider F_R as an extension of $\mathbb{F}_q(x)$. Let Q_∞ be the place at infinity of $\mathbb{F}_q(x)$, and let P_∞ be a place of F_R lying above Q_∞. We denote the discrete valuation at P_∞ by v_∞ and the ramification index of P_∞ with respect to Q_∞ by e_∞.

Define the following functions $w_i \in F_R$, $i = 1, \ldots, 8$:

$$w_1 = x^{3q_0+1} - y_1^{3q_0}, \qquad w_2 = xy_1^{3q_0} - y_2^{3q_0}, \qquad w_3 = xy_2^{3q_0} - w_1^{3q_0},$$
$$w_4 = xw_2^{q_0} - y_1 w_1^{q_0}, \qquad w_5 = y_1 w_3^{q_0} - y_2 w_2^{q_0},$$
$$w_6 = v^{3q_0} - w_2^{3q_0} + xw_4^{3q_0}, \quad \text{where} \quad v = xw_3^{q_0} - y_2 w_1^{q_0},$$
$$w_7 = y_1 w_2^{q_0} - xw_3^{q_0} - w_6^{3q_0}, \qquad w_8 = w_5^{3q_0} + xw_7^{3q_0}.$$

THEOREM 7.2.17. *We have*

(i) $e_\infty = q^2$, *and* P_∞ *is the only place of* F_R *lying above* Q_∞.
(ii) *The number of* $\mathbb{F}_q$*-rational places of* F_R *is* $q^3 + 1$.
(iii) F_R *has genus* $g_R = \frac{3}{2} q_0(q-1)(q + q_0 + 1)$.

SKETCH OF THE PROOF. (i) One can show that

$$v_\infty(w_8) = -\left(1 + \frac{1}{q_0} + \frac{2}{q} + \frac{1}{q_0 q} + \frac{1}{q^2}\right) e_\infty,$$

implying $e_\infty = eq^2$, $e \in \mathbb{N}$. On the other hand, $e_\infty \leq [F_R : \mathbb{F}_q(x)] \leq q^2$, and therefore $e = q^2$. Consequently, P_∞ is the only place of F_R lying above Q_∞.

(ii) From (i) we have that P_∞ is an $\mathbb{F}_q$-rational place. Let Q_β, $\beta \in \mathbb{F}_q$, be an $\mathbb{F}_q$-rational place of $\mathbb{F}_q(X)$ outside infinity. Then there are q^2 $\mathbb{F}_q$-rational places $P_{\beta\gamma\delta}$, $\gamma, \delta \in \mathbb{F}_q$, of F_R lying above Q_β (here $P_{\beta\gamma\delta}$ denotes the place corresponding to the solution $x = \beta$, $y_1 = \gamma$, $y_2 = \delta$ of (7.2.5)). Hence, the number of $\mathbb{F}_q$-rational places of F_R is $q^3 + 1$.

(iii) P_∞ is the only place of F_R which ramifies with respect to $\mathbb{F}_q(x)$, and we have

$$2g_R - 2 = -2q^2 + v_\infty\left(\frac{dx^{-1}}{du}\right), \tag{7.2.6}$$

where $u \in F_R$ is a prime element of P_∞. A careful computation shows that $u = w_6/w_8$ is a prime element. Then we obtain

$$\frac{dx^{-1}}{du} = \frac{w_8^2}{x^2\left(w_6 w_7^{3q_0} - w_4^{3q_0} w_8\right)}$$

and thus

$$v_\infty\left(\frac{dx^{-1}}{du}\right) = 3q_0q^2 + 3q^2 - q - 3q_0 - 2,$$

which together with (7.2.6) proves (iii). $\square$

The two equations (7.2.5) make it possible to consider $\mathcal{X}_R$ as a curve in $\mathbb{P}_q^3$. However, $\mathcal{X}_R$ is not a complete intersection defined by

$$f_1(X,Y,Z) = f_2(X,Y,Z) = 0$$

with

$$f_1 = Z^{q_0}(Y_1^q - Y_1 Z^{q-1}) - X^{q_0}(X^q - XZ^{q-1}),$$
$$f_2 = Z^{q_0}(Y_2^q - Y_2 Z^{q-1}) - X^{q_0}(Y_1^q - Y_1 Z^{q-1}),$$

since the latter is a reducible curve. In fact, we need to add the third equation,

$$f_3(X,Y,Z) = 0,$$

with

$$f_3 = (Y_1^q - Y_1 Z^{q-1})^2 - (X^q - XZ^{q-1})(Y_2^q - Y_2 Z^{q-1}),$$

which gives a space model of $\mathcal{X}_R$ nonsingular outside a unique point P_∞ at infinity $Z = 0$ (at P_∞, the ideal (f_1, f_2, f_3) generates the square of the corresponding maximal ideal $\mathfrak{m}_{P_\infty}$).

In fact, it is possible to represent $\mathcal{X}_R$ as a plane curve. Indeed, the function w_2 satisfies $w_2^q - w_2 = y_1^{3q_0}(x^q - x)$, and by using this equation and the first equation in (7.2.5), we obtain

$$f(x, w_2) = w_2^{q^2} - [1 + (x^q - x)^{q-1}]w_2^q + (x^q - x)^{q-1}w_2 - x^q(x^q - x)^{q+3q_0} = 0,$$

which is the desired plane affine model of $\mathcal{X}_R$, since

$$y_1 \mapsto \left(\frac{w_2^q - w_2}{x^q - x}\right)^{q_0} - x^{q_0}(x^q - x), \qquad y_2 \mapsto x^{q_0}\left(\frac{w_2^q - w_2}{x^q - x}\right)^{q_0} - x^{2q_0}(x^q - x) - w_2^{q_0}$$

gives an isomorphism.

Automorphism group. Now we describe the automorphism group $G_R = \mathrm{Aut}(F_R/\overline{\mathbb{F}}_q)$; it turns out that $G_R = \mathrm{Aut}(F_R/\mathbb{F}_q)$, i.e., all the automorphisms are $\mathbb{F}_q$-rational. The group G_R acts as a permutation group on the set of $\overline{\mathbb{F}}_q$-points of $\mathcal{X}_R$.

THEOREM 7.2.18. *The group G_R is the Ree group ${}^2G_2(q)$ of order $|G_R| = q^3(q-1)(q^3+1)$.* $\square$

We do not give a proof of this result or even a definition of the Ree group ${}^2G_2(q)$; it is a simple finite group of the order given in the theorem. However, we give some information on its structure.

LEMMA 7.2.19. (1) *The stabilizer of the place P_∞ is*

$$G_R(P_\infty) = \{\psi_{\alpha\beta\gamma\delta} \mid \alpha \in \mathbb{F}_q^*,\ \beta, \gamma, \delta \in \mathbb{F}_q\},$$

where

$$\psi_{\alpha\beta\gamma\delta}(x) = \alpha x + \beta, \qquad \psi_{\alpha\beta\gamma\delta}(y_1) = \alpha^{q_0+1}y_1 + \alpha\beta^{q_0}x + \gamma,$$
$$\psi_{\alpha\beta\gamma\delta}(y_2) = \alpha^{2q_0+1}y_2 - \alpha^{q_0+1}\beta^{q_0}y_1 + \alpha\beta^{2q_0}x + \delta;$$

(2) *$G_R(P_\infty)$ acts transitively on $\mathcal{X}_R(\mathbb{F}_q)\setminus\{P_\infty\}$.* □

Then define $\varphi \in G_R$ by the following explicit formulae:

$$\varphi(x) = w_6/w_8, \qquad \varphi(y_1) = w_{10}/w_8, \qquad \varphi(y_2) = w_9/w_8,$$

for

$$w_9 = w_2^{q_0}w_4 - y_1 w_6^{q_0}, \qquad w_{10} = y_2 w_6^{q_0} - w_3^{q_0}w_4.$$

One can check by a direct (yet cumbersome) computation that

$$\varphi(y_1)^q - \varphi(y_1) = \varphi(x)^{q_0}(\varphi(x)^q - \varphi(x)),$$
$$\varphi(y_2)^q - \varphi(y_2) = \varphi(x)^{q_0}(\varphi(y_1)^q - \varphi(y_1)),$$

and thus φ is indeed an automorphism, in fact an involution, $\varphi^{-1} = \varphi$. Moreover, the order of the subgroup $\widetilde{G}_R \subset \mathrm{Aut}(\mathcal{X}_R)$ generated by $G_R(P_\infty)$ and φ is exactly $q^3(q-1)(q^3+1)$, and this subgroup acts transitively on $\mathcal{X}_R(\mathbb{F}_q)$. The main point in the proof of Theorem 7.2.18 is to show that $\widetilde{G}_R$ cannot be a proper subgroup of $\mathrm{Aut}(\mathcal{X}_R)$, since otherwise the group $\mathrm{Aut}(\mathcal{X}_R)$ would violate the bound $|\mathrm{Aut}(\mathcal{X}_R)| \le 16g_{\mathcal{X}_R}^4$.

Zeta function and uniqueness. The structure of Ree curves and their rational points is extremely rigid. As we have mentioned above, they attain the Oesterlé bound; moreover, one has complete information on their zeta functions. Namely, one has the following.

THEOREM 7.2.20. (i) *F_R has the maximal number of $\mathbb{F}_q$-rational places possible for a function field of genus g_R.*

(ii) *The zeta function of F_R is*

$$Z(F_R,\mathbb{F}_q)(t) = \frac{(1+3q_0t+qt^2)^{q_0(q^2-1)}(1+qt^2)^{q_0(q-1)(q+3q_0+1)/2}}{(1-t)(1-qt)}.$$

(iii) *The number of placed of degree one in $F_R/\mathbb{F}_{q^m}$ is*

$$N_m = q^m + 1 - q_0\sqrt{q}(q-1)[(q+3q_0+1)\cos m\pi/2 + 2(q+1)\cos 5m\pi/6].$$

In particular, $F_R/\mathbb{F}_{q^m}$ is a maximal function field if and only if $m \equiv 6 \pmod{12}$. □

One can prove uniqueness of the field F_R (and thus of the curve $\mathcal{X}_R$).

THEOREM 7.2.21. *Let $q = 3^{2s+1}$ and $q_0 = 3^s$, $s \ge 1$. Let $F/\mathbb{F}_q$ be a function field over the finite field $\mathbb{F}_q$ satisfying*

(i) *$F/\mathbb{F}_q$ has genus $g = \dfrac{3}{2}q_0(q-1)(q+q_0+1)$.*
(ii) *The automorphisms in $G = \mathrm{Aut}(F/\mathbb{F}_q)$ are $\mathbb{F}_q$-rational, and G is a Ree group of order $q^3(q-1)(q^3+1)$.*
(iii) *$F/\mathbb{F}_q$ has $1+q^3$ $\mathbb{F}_q$-rational places, on which G acts as a permutation group.*

Then F is $\mathbb{F}_q$-isomorphic to $\mathbb{F}_q(x,y_1,y_2)$, where

$$y_1^q - y_1 = x^{q_0}(x^q - x), \qquad y_2^q - y_2 = x^{q_0}(y_1^q - y_1).$$ □

The proof of Theorem 7.2.21 is based on a precise description of ramification groups at infinity, namely,

$$G_0 = G_1 = G_2 = \ldots = G_{3q_0+1}, \qquad |G_0| = q^2,$$
$$G_{3q_0+2} = \ldots = G_{q+3q_0+1}, \qquad |G_{3q_0+2}| = q, \qquad G_{q+3q_0+2} = \{1\}.$$

7.2.4. Deligne–Lusztig Curves via Class Field Theory

Deligne–Lusztig curves can also be represented as rather simple coverings corresponding to some ray class fields of a very clear nature, and below we give this description.

THEOREM 7.2.22. *The function fields of the Hermitian, Suzuki, and Ree curves are isomorphic to the largest ray class fields of $\mathbb{F}_r(x)$ with conductor $D = kP_\infty$ in which all places $Q \neq P_\infty$ of degree one split completely, where*

$$k = \begin{cases} p^{\lceil f/2 \rceil} + 2 & \text{if } q = p^f \text{ is a square or } p = 2, \\ p^{\lceil f/2 \rceil} + 3 & \text{if } q = p^f \text{ is not a square or } p = 3. \end{cases} \qquad \square$$

Let us briefly expose the main points of this construction.

Hermitian curves. Consider the function field $\mathbb{F} = \mathbb{F}_r(H) = \mathbb{F}_r(x,y)$ with $y^q + y = x^{q+1}$, as an extension of $\mathbb{F}_r(x)$. Recall that

(a) The field degree is $[\mathbb{F} : \mathbb{F}_r(x)] = r$.
(b) P_∞ is the only place of $\mathbb{F}_r(x)$ which is ramified in $\mathbb{F}$, and it is totally ramified in $\mathbb{F}$.
(c) All rational places P_a, $a \in \mathbb{F}_r$, split completely in $\mathbb{F}$.
(d) $\mathbb{F}/\mathbb{F}_r(x)$ is abelian; its Galois group G is given by the substitutions $y \to y + b$ with $b^q + b = 0$.

PROPOSITION 7.2.23. *The conductor of the abelian extension $\mathbb{F}/\mathbb{F}_r(x)$ is $\mathfrak{f} = (r+2)P_\infty$.*

SKETCH OF THE PROOF. By property (b) above, the conductor is a multiple of P_∞, i.e., $\mathfrak{f} = kP_\infty$ for some positive integer k. The ramification theory of abelian extensions implies that

$$k = 1 + \frac{1}{g}(g_1 + g_2 + \ldots + g_s), \tag{7.2.7}$$

where $g = |G| = r$, $g_i = |G_i|$, s being the largest i with $g_i > 1$. Then formula (7.2.4) in the proof of Lemma 7.2.10 shows that $k = 1 + s = r + 2$. $\square$

Thus, the main theorem for Hermitian fields takes the following form.

THEOREM 7.2.24. *The field $\mathbb{F}$ can be characterized as the largest abelian extension of $\mathbb{F}_r(x)$ which has conductor $(r+2)P_\infty$ and in which all finite places P_a, $a \in \mathbb{F}_r$, split completely.*

SKETCH OF THE PROOF. Let $\mathbb{F}'$ be the largest extension of $\mathbb{F}_r(x)$ with the properties as announced. Then $\mathbb{F} \subset \mathbb{F}'$. Let $[\mathbb{F}' : \mathbb{F}] = m$, so that $[\mathbb{F}' : \mathbb{F}_r(x)] = mr$. We claim that $m = 1$.

Let us count the number N' of $\mathbb{F}_r$-rational places of $\mathbb{F}'$. In $\mathbb{F}_r(x)$ there are $q = r^2$ finite $\mathbb{F}_r$-rational places P_a, $a \in \mathbb{F}_r$. Every one of them splits completely in $\mathbb{F}'$, which gives a total of mrq $\mathbb{F}_r$-rational places of $\mathbb{F}'$.

It is sufficient to get an upper bound for the genus of $\mathbb{F}'$, which will guarantee the maximality of $\mathbb{F}'$, since then the uniqueness of the Hermitian curve will imply $m = 1$. This can be achieved by applying the Hurwitz formula together with a complementary argument (the conductor-discriminant formula), implying that $d' \leq (mr-1)(r+2)$ for the degree d' of the different of $\mathbb{F}'/\mathbb{F}_r(x)$. Therefore, $g' = g_{\mathbb{F}'} \leq \frac{1}{2}r(mr-1)$, implying the maximality of $\mathbb{F}'$; thus, $\mathbb{F} = \mathbb{F}'$, which completes the proof. □

COROLLARY 7.2.25. *Let $k = r+2$. Then*

$$\left|(\mathbb{F}_r[T]/T^k)^*/\mathbb{F}_r^*/\langle 1-\alpha T \mid \alpha \in \mathbb{F}_r^*\rangle\right| = r.$$

Furthermore, this quotient is trivial if $k < r+2$. □

Suzuki curves. We use the notation of Sec. 7.2.2 and consider the extension $\mathbb{F}_q(x,y)/\mathbb{F}_q(x)$, where $y^q + y = x^{q_0}(x^q + x)$, $q = 2^{2m+1}$, $q_0 = 2^m$. For this extension $\mathbb{F}$, we have the following (see above):

(a) $[\mathbb{F} : \mathbb{K}(x)] = q$.
(b) The place $P_\infty = Q$ is the only place of $\mathbb{F}_q(x)$ which is ramified in $\mathbb{F}$, and it is totally ramified in $\mathbb{F}$.
(c) The genus of $\mathbb{F}$ is $g = q_0(q-1)$, and the number of rational places is q^2+1.
(d) Every finite rational place splits completely in $\mathbb{F}$.
(e) $\mathbb{F}/\mathbb{F}_q(x)$ is abelian; its Galois group G is given by the substitutions $y \to y+\alpha$ with $\alpha \in \mathbb{K}$.
(f) The conductor of the abelian extension $\mathbb{F}/\mathbb{F}_q(x)$ is $\mathfrak{f} = (2q_0+2)Q$.

The main theorem for the function fields of the Suzuki curves is then as follows.

THEOREM 7.2.26. *The extension $\mathbb{F}$ can be characterized as the largest abelian extension of $\mathbb{F}_q(x)$ which has conductor $(2q_0+2)Q$ and in which all finite places P_a, $a \in \mathbb{K}$, split completely.*

SKETCH OF THE PROOF. Again let $\mathbb{F}'$ be the largest extension of $\mathbb{F}_q(x)$ with the properties as announced; then $\mathbb{F} \subset \mathbb{F}'$. Let $[\mathbb{F}' : \mathbb{F}] = m$, so that $[\mathbb{F}' : \mathbb{F}_q(x)] = mq$. By the same argument as in the Hermitian case above, $\mathbb{F}'$ has $N' = q^2m+1$ rational places. Using the trigonometric polynomial

$$h_1(\theta) = 1 + 2\left(\frac{\sqrt{2}}{2}\cos(\theta) + \frac{1}{4}\cos(2\theta)\right)$$

in the explicit formula, we find that

$$N' \leq \frac{4q}{4q_0+1}g' + \frac{4q_0q+q^2}{4q_0+1} + 1. \tag{7.2.8}$$

As above, the Hurwitz formula and an additional argument give

$$g' \leq q_0(mq-1),$$

and thus (7.2.8) becomes

$$q^2m + 1 \leq q^2\frac{4q_0m+1}{4q_0+1} + 1,$$

implying that $m = 1$. □

COROLLARY 7.2.27. *Let $\mathbb{K} = \mathbb{F}_q$ be the finite field with $q = 2^{2m+1} = 2q_0^2$ elements. Let $k = 2q_0 + 2$. Then*

$$\left|(\mathbb{K}[T]/T^k)^*/\mathbb{K}^*/\langle 1 - \alpha T \mid \alpha \in \mathbb{K}^* \rangle\right| = q.$$

Furthermore, this quotient is trivial if $k < 2q_0 + 2$. □

Ree curves. The function field $F_R = \mathbb{K}(x, y_1, y_2)$ with

$$y_1^q - y_1 = x^{q_0}(x^q - x), \qquad y_2^q - y_2 = x^{q_0}(y_1^q - y_1)$$

is considered as an extension of $\mathbb{F}_q(x)$, and again we have the following facts; see above.

(a) The field degree is $[F_R : \mathbb{F}_q(x)] = q^2$.
(b) The place P_∞ is the only place of $\mathbb{F}_q(x)$ which is ramified in F_R, and it is totally ramified in F_R.
(c) The genus of F_R is $g = \dfrac{3}{2} q_0(q-1)(q + q_0 + 1)$, and the number of rational places is $q^3 + 1$. The field F_R has the maximum possible number of rational places for its genus. The trigonometric polynomial which is chosen to show that the Ree curves are maximal is

$$h_2(\theta) = 1 + 2\sum c_n \cos(n\theta),$$

where

$$c_1 = \frac{\sqrt{3}}{2}, \qquad c_2 = \frac{7}{12}, \qquad c_3 = \frac{\sqrt{3}}{6}, \qquad c_4 = \frac{1}{12}, \qquad c_i = 0, \quad i > 4.$$

(d) Every finite rational place P_a, $a \in \mathbb{F}_q$, splits completely in F_R.
(e) The Galois group of $F_R/\mathbb{F}_q(x)$ is abelian; it is a subgroup of the full Ree group.
(f) The conductor of the abelian extension $F_R/\mathbb{F}_q(x)$ is $\mathfrak{f} = (3q_0 + 3)P_\infty$.

The main theorem for the function fields of the Ree curves is then as follows; its proof is a direct modification of those for Hermitian and Suzuki curves.

THEOREM 7.2.28. *The function field $\mathbb{F}_R$ can be characterized as the largest abelian extension of $\mathbb{F}_q(x)$ of conductor $(3q_0 + 3)P_\infty$ in which all finite places P_a, $a \in \mathbb{F}_q$, split completely.* □

7.3. Some Curves of Small Genera

Recall (Sec. 4.4.4) that values of $N_q(g)$ are known for $g \leq 2$. For any fixed $g \geq 3$, values of $N_q(g)$ are known for a finite set of q only, and their determination is usually a difficult problem for any given g and q. In this section we review some methods for constructing curves, which give some lower bounds for the quantities $N_q(g)$. We describe methods from class field theory, fibre products of Kummer and Artin–Schreier curves, and quotients of curves with many points. Note however that there are many more methods for constructing curves of small genera with many points; some of them are also mentioned in the Historical and Bibliography Notes to the present section.

7.3.1. Curves with Many Points from Ray Class Fields

As we have seen above in Sec. 7.2.3, the function fields of Deligne–Lusztig curves possessing many rational places can be naturally represented as ray class fields of a very simple structure. This allows one to hope that ray class fields of a similar type could be a source of curves with many rational points, and indeed there are many cases where this really does work; we are going to describe some examples of this nature. We consider two constructions: the first is almost the simplest possible, and the second uses the rank-1 Drinfeld module techniques as in Sec. 6.4.2.

Simple construction. Let $\mathbb{K} = \mathbb{F}_q(t)$. We take a point P_d of degree d and consider the module $D = kP_d$ for a natural number k. For the ray class field $\mathbb{K}_D$ with conductor D, we have the following.

LEMMA 7.3.1. (i) $G_D = \mathrm{Gal}(\mathbb{K}_D/\mathbb{K}) \cong \left(\mathbb{F}_q[T]/(P_d(T))^k\right)^*/\mathbb{F}_q^*$.
(ii) *Let $k = 1$. Then the Artin conductor of a nontrivial character χ of G_D is d.* □

The characterization in Lemma 7.3.1(i) can be used to compute the finite quotients obtained by splitting places of $\mathbb{K}$; one splits places of degree one by taking the quotient of this group modulo the cyclic groups generated by the polynomials of the form $T + \nu$.

For $k = 1$, we get the following result.

PROPOSITION 7.3.2. *Let $\mathbb{K} = \mathbb{F}_q(t)$, $q = p^m$, $D = P_d$, $d \geq 3$. Then the degree of the extension obtained by splitting just one place of degree one of $\mathbb{K}$ in the ray class field of conductor D is $(q^d - 1)/(q - 1)$, the number of places of degree one in the extension field is at least $(q^d - 1)/(q - 1)$, and all nontrivial characters have conductor d, so the genus is given by the formula*

$$g = \frac{(d-2)(q^{d-1} + \ldots + q)}{2}. \qquad \square$$

EXAMPLE 7.3.3. When $q = 4$ and $d = 3$, we get a curve over $\mathbb{F}_4$ of genus 10 with at least 21 points. This is not so good, since over $\mathbb{F}_4$ there exists a curve of genus 10 with 27 points.

EXAMPLE 7.3.4. For $q = 8$ and $d = 3$, the curve has $N \geq 73$, $g = 36$. This is not so good either, since over $\mathbb{F}_8$ there exists a curve of genus 36 with 105 points.

When splitting more points, it is necessary to compute case by case the quotient group of $(\mathbb{F}_q[T]/(P_d(T)))^*/\mathbb{F}_q^*$ modulo degree-one polynomials. Note however that with the use of Lemma 7.3.1 one can easily obtain the following result.

PROPOSITION 7.3.5. *Let $D=2P_d$, $q=2^m$. Then splitting one point gives an extension of degree $q^d(q^d-1)$ and genus*

$$1+q^d n_d(d-1)-\frac{d(n_d+1)}{2},$$

where $n_d=q^{d-1}+\ldots+q+1$. □

COROLLARY 7.3.6. *For $d=2$, we get a curve of genus $(q^2-1)(q+1)$ having $q^2(q+1)$ points for each q.* □

EXAMPLE 7.3.7. Let $q=2$, $d=2$, split one point: $N=12$, $g=9$, and this is optimal.
Let $q=2$, $d=3$, split one point: $N=56$, $g=101$.
Let $q=4$, $d=2$, split one point: $N=80$, $g=75$.

Now let $d=1$, k arbitrary, i.e., $D=kP_1$, $k\in\mathbb{N}$.

EXAMPLE 7.3.8. Let $q=4$, $k=4$. Then splitting all four of the other rational points of $\mathbb{K}$, one obtains an extension of degree 2 in which the nontrivial character has conductor 4. This is an elliptic curve which is maximal over F_4: $g=1$, $N=9$.

EXAMPLE 7.3.9. Let $q=4$, $k=6$. Then splitting all four of the other rational points of $\mathbb{K}$, one obtains an extension of degree 8 in which one character has conductor 4 and six characters have conductor 6. This extension corresponds to a curve of genus 13 with 33 points, which is the maximum possible according to the Oesterlé bound, so $N_4(13)=33$.

EXAMPLE 7.3.10. Let $q=4$, $k=7$. Then splitting all four of the other rational points of $\mathbb{K}$, one obtains an extension of degree 16 in which one character has conductor 4, six characters have conductor 6, and eight characters have conductor 7. This gives a curve of genus 33 with 65 points. Since the Oesterlé bound is 66 in this case, we obtain $N_4(33)=65$ or 66. For $k=8$, one gets $g=177$, $N=257$.

EXAMPLE 7.3.11. For $q=4$, $k=6$, one obtains an example of a curve of genus 27 with 49 points, which is rather good.

EXAMPLE 7.3.12. Splitting four points over $\mathbb{F}_8$ with $D=4P_1$, one gets a curve of genus 14 with 65 points, which is in fact a Suzuki curve.

Using rank-1 Drinfeld modules. It is natural that the technique of rank-1 Drinfeld modules applied to $A(2)$ in Sec. 6.4.2 can also be employed to construct good curves of small genera. We give some examples of such constructions for $q=3$, 4, and 5. The following two theorems are based on this technique.

THEOREM 7.3.13. *Let $F=\mathbb{F}_q(x)$, and let L/F be an extension of degree $r\geq 2$ with $g(L)\geq 1$. Assume that there are two different rational places P and P_∞ of L which are totally ramified over F, with P_∞ lying over ∞, and assume that $1,2,\ldots,r-1$ are gap numbers of P. Then there exists a global function field K with full constant field $\mathbb{F}_q$ such that*

$$g(K)=\frac{h(L)}{r}(g(L)-1)+1 \qquad \text{and} \qquad N(K)=\frac{2h(L)}{r}.$$ □

THEOREM 7.3.14. *Let the global function field L be a constant field extension of the global function field L_1 with $N(L_1) \geq 1$. Then there exists a global function field K with the same full constant field as L such that*

$$g(K) = \frac{h(L)}{h(L_1)}(g(L)-1)+1 \qquad \text{and} \qquad N(K) = \frac{h(L)N(L_1)}{h(L_1)}. \qquad \square$$

These theorems allow one to generate numerous good examples of curves with many $\mathbb{F}_q$-rational places, and now we give some of them for small values of q.

Case $q = 3$.

EXAMPLE 7.3.15. $g(K) = 1$, $N(K) = 7$, $K = \mathbb{F}_3(x, y)$ with

$$y^2 = x^3 - x + 1.$$

In the Kummer extension K/F, the places x, $x+1$, and $x-1$ split completely, and the place ∞ is totally ramified. The only other ramified place of K is lying over $x^3 - x + 1$. This is an optimal elliptic curve.

EXAMPLE 7.3.16. $g(K) = 2$, $N(K) = 8$, $K = \mathbb{F}_3(x, y)$ with

$$y^2 = x^6 - x^2 + 1.$$

All rational places of F split completely in the Kummer extension K/F. The only ramified place of K lies over $x^6 - x^2 + 1$. This is an optimal curve.

EXAMPLE 7.3.17. $g(K) = 3$, $N(K) = 10$, $K = \mathbb{F}_3(x, y)$ with

$$y^3 - y = x^4 - x^2.$$

In the Artin–Schreier extension K/F, the places x, $x+1$, and $x-1$ split completely, and the place ∞ is totally ramified. The curve is optimal.

EXAMPLE 7.3.18. $g(K) = 4$, $N(K) = 12$, $K = \mathbb{F}_3(x, y)$ with

$$y^3 - y = \frac{x^3 - x}{(x^2+1)^2}.$$

All rational places of F split completely in the Artin–Schreier extension K/F. The only ramified place of K lies over $x^2 + 1$. The curve is optimal.

EXAMPLE 7.3.19. $g(K) = 5$, $N(K) = 12$, $K = \mathbb{F}_3(x, y_1, y_2)$ with

$$y_1^2 = x^3 - x + 1, \qquad y_2^2 = -x^4 + x^2 + 1.$$

In this tower of Kummer extensions, the places x, $x+1$, and $x-1$ split completely in K/F. The first Kummer extension is the one from Example 7.3.15. This is suboptimal; the optimal value is 13.

EXAMPLE 7.3.20. $g(K) = 7$, $N(K) = 16$. Consider the irreducible polynomial $P = x^4 + x^3 + x^2 + x + 1$ over $\mathbb{F}_3$ and the corresponding cyclotomic function field F_P. The Galois group $\mathrm{Gal}(F_P/F) = B_P^*$ has order 80. Let K be the subfield of F_P/F fixed by the subgroup $H = \mathbb{F}_3^*\langle\overline{x}\rangle$ of B_P^*, where $\langle\overline{x}\rangle$ is the cyclic subgroup generated by $\overline{x} \in B_P^*$. Then $|H| = 10$ and $[K : F] = 8$. One can show that P is the unique ramified place in K/F and that it is totally and tamely ramified. Moreover, the place ∞ splits completely in K/F. Since $\overline{x} \in H$, we see that x also splits completely in K/F by considering the Artin symbol of x. The genus of K follows from the Hurwitz formula, which yields $2g(K) - 2 = -2 \cdot 8 + 4 \cdot (8-1)$, i.e., $g(K) = 7$. The curve is optimal.

Case $q=4$. We write α for an element of $\mathbb{F}_4$ satisfying $\alpha^2+\alpha+1=0$.

EXAMPLE 7.3.21. $g(K)=1$, $N(K)=9$, $K=\mathbb{F}_4(x,y)$ with

$$y^2+y=x^3.$$

The places x, $x+1$, $x+\alpha$, and $x+\alpha+1$ split completely in the Artin–Schreier extension K/F, and the place ∞ is totally ramified in K/F. This is just a Hermitian curve and thus is optimal.

EXAMPLE 7.3.22. $g(K)=2$, $N(K)=10$, $K=\mathbb{F}_4(x,y)$ with

$$y^2+y=\frac{x}{x^3+x+1}.$$

All rational places of F split completely in the Artin–Schreier extension K/F. The only ramified place of K lies over x^3+x+1. This curve is optimal.

EXAMPLE 7.3.23. $g(K)=3$, $N(K)=14$. Consider the irreducible polynomial $P=x^3+\alpha$ over $\mathbb{F}_4$ and the corresponding cyclotomic function field F_P. The Galois group $\mathrm{Gal}(F_P/F)=B_P^*$ has order 63. Let K be the subfield of F_P/F fixed by the subgroup of B_P^* generated by $\overline{x}$. Then $[K:F]=7$. One can easily check that the place ∞ splits completely in K/F, and x also splits completely in K/F by the Artin symbol argument. Note also that P is the unique ramified place in K/F and it is totally and tamely ramified, and the Hurwitz formula gives $2g(K)-2=-2\cdot 7+3\cdot(7-1)$, i.e., $g(K)=3$. This is optimal.

EXAMPLE 7.3.24. $g(K)=4$, $N(K)=15$, $K=\mathbb{F}_4(x,y_1,y_2)$ with

$$y_1^2+y_1=x^3, \qquad y_2^3=(x^2+x)y_1+x^4+1.$$

Let $L=\mathbb{F}_4(x,y_1)$ be the field in Example 7.3.21. Then the four places of L given by $(0,0)$, $(0,1)$, (α,α), and $(\alpha+1,\alpha+1)$ split completely in the Kummer extension K/L. The place of L lying over ∞ and the two places of L lying over $x+1$ are totally ramified in K/L. This curve is optimal.

EXAMPLE 7.3.25. $g(K)=5$, $N(K)=17$. Consider the cyclotomic function field F_M with $M=x^4\in\mathbb{F}_4[x]$ and Galois group $\mathrm{Gal}(F_M/F)=B_M^*$ of order 192. Let K be the subfield of F_M/F fixed by the subgroup $H=\mathbb{F}_4^*\cdot(\mathbb{F}_2[x]/(M))^*$ of B_M^*. Then $|H|=24$ and $[K:F]=8$. The places $x+1$ and ∞ split completely in K/F, and the place x is totally ramified in K/F. This curve is optimal.

Case $q=5$.

EXAMPLE 7.3.26. $g(K)=1$, $N(K)=10$, $K=\mathbb{F}_5(x,y)$ with

$$y^2=3(x^4+2).$$

All rational places of F except ∞ split completely in the Kummer extension K/F. The only ramified place of K lies over x^4+2. This elliptic curve is optimal.

EXAMPLE 7.3.27. $g(K)=2$, $N(K)=12$, $K=\mathbb{F}_5(x,y)$ with

$$y^2=(x^2+2)(x^4+3x^2+3).$$

All rational places of F split completely in the Kummer extension K/F. The only ramified places of K are those lying over x^2+2 or x^4+3x^2+3. This curve is optimal.

EXAMPLE 7.3.28. $g(K)=4$, $N(K)=18$. Consider the cyclotomic function field $E=F_{P_1P_2}$ with the irreducible polynomials $P_1=x^2+2$ and $P_2=x^2+3$ in $F_5[x]$ and the subfields $L_1=F_{P_1}$ and $L_2=F_{P_2}$ of E/F. For $i=1,2$, let K_i be the subfield of L_i/F fixed by the cyclic subgroup $\langle\overline{x}\rangle$ of $\mathrm{Gal}(L_i/F)=B^*_{P_i}$; then $[K_i:F]=3$. The places x and ∞ split completely in both K_1/F and K_2/F. Let K be the composite field of K_1 and K_2; then $[K:F]=9$, since L_1 and L_2 are linearly disjoint, as follows from the Hayes theory. Furthermore, the places x and ∞ split completely in K/F; thus, $N(K)\geq 18$. The only ramified places in K/F are those lying over x^2+2 or x^2+3, each with ramification index 3 by Abhyankar's lemma (Theorem 7.4.10 below), and the Hurwitz formula gives $2g(K)-2=-2\cdot 9+3\cdot 2\cdot(3-1)+3\cdot 2\cdot(3-1)$, that is, $g(K)=4$. Since $N_5(4)=18$, we must have $N(K)=18$. This curve is optimal.

7.3.2. Fibre Products

In many cases a fibre product of good curves also yields a good curve. We illustrate this techniques with some simple examples of fibre products of Kummer and Artin–Schreier curves.

Kummer curves. Let $n_1,n_2\geq 2$ be integers with $\gcd(n_1,q)=\gcd(n_2,q)=1$. Let $h_1(x),h_2(x)\in\mathbb{F}_q(x)$ and $E=\mathbb{F}_q(x,y_1,y_2)$ be the algebraic function field with

$$y_1^{n_1}=h_1(x),\qquad y_2^{n_2}=h_2(x).$$

Let P_∞ be the place of $\mathbb{F}_q(x)$ corresponding to the pole of $x\in\mathbb{F}_q(x)$.

REMARK 7.3.29. For a rational place P of $\mathbb{F}_q(x)$ distinct from P_∞, there exists a uniquely determined $v\in\mathbb{F}_q$ such that P corresponds to the zero of $x-v\in\mathbb{F}_q(x)$.

Let P_0 be a rational place of $\mathbb{F}_q(x)$ distinct from P_∞. Using Remark 7.3.29, let $u\in\mathbb{F}_q$ be such that P_0 corresponds to the zero of $x-u\in\mathbb{F}_q(x)$. Let a_1 and a_2 be integers, and let $f_1(x),f_2(x)\in\mathbb{F}_q(x)$ be such that

$$h_1(x)=(x-u)^{a_1}f_1(x),\qquad h_2(x)=(x-u)^{a_2}f_2(x),$$

and $v_{P_0}(f_1(x))=v_{P_0}(f_2(x))=0$. We note that the integers a_1,a_2 and the rational functions $f_1(x),f_2(x)\in\mathbb{F}_q(x)$ are uniquely determined by the rational place P_0. Let $\overline{n}_1=\gcd(n_1,a_1)$, $\overline{n}_2=\gcd(n_2,a_2)$, and $m=\gcd\left(\frac{n_1}{\overline{n}_1},\frac{n_2}{\overline{n}_2}\right)$. We note that $\overline{n}_i$ is positive, while a_i need not be positive, for $i=1,2$.

Let $f_1(u)$ and $f_2(u)$ be evaluations of $f_1(x)$ and $f_2(x)$ at P_0.

PROPOSITION 7.3.30. *Under the above notation, assume that $\mathbb{F}_q$ is the full constant field of E and $[E:\mathbb{F}_q(x)]=n_1n_2$. Moreover, assume that the integers $\overline{n}_1$, $\overline{n}_2$, and m divide $q-1$. If a rational place of E over P_0 exists, then there are exactly $\overline{n}_1\overline{n}_2m$ rational places of E over P_0. This happens if and only if the following conditions 1–4 hold simultaneously:*

(1) *$f_1(u)$ is an $\overline{n}_1$-power in $\mathbb{F}_q$.*
(2) *$f_2(u)$ is an $\overline{n}_2$-power in $\mathbb{F}_q$.*
(3) $(m\,\mathrm{lcm}(\overline{n}_1,\overline{n}_2))\mid(q-1)$.
(4) *Under the assumptions of 1 and 2, let α and β be elements of $\mathbb{F}_q$ with $\alpha^{\overline{n}_1}=f_1(u)$ and $\beta^{\overline{n}_2}=f_2(u)$. Let A and B be integers satisfying*

$$A\frac{n_1}{\overline{n}_1}+B\frac{a_1}{\overline{n}_1}=1;$$

then

$$\frac{\beta}{\alpha^{(a_2/\overline{n}_2)B}} \text{ is an } m\text{-power in } \mathbb{F}_q. \qquad \square$$

Next we consider the rational places of E over P_∞.

Let $f_{1,1}(x)$, $f_{1,2}(x)$, $f_{2,1}(x)$, $f_{2,2}(x)$ be monic polynomials in $\mathbb{F}_q[x]$, and let $c_1, c_2 \in \mathbb{F}_q$ be such that $\gcd(f_{1,1}(x), f_{1,2}(x)) = 1$, $\gcd(f_{2,1}(x), f_{2,2}(x)) = 1$,

$$h_1(x) = c_1 \frac{f_{1,1}(x)}{f_{1,2}(x)}, \qquad \text{and} \qquad h_2(x) = c_2 \frac{f_{2,1}(x)}{f_{2,2}(x)}.$$

We note that the monic polynomials $f_{1,1}(x), f_{1,2}(x), f_{2,1}(x), f_{2,2}(x) \in \mathbb{F}_q[x]$ and $c_1, c_2 \in \mathbb{F}_q$ are uniquely determined. Let $d_{1,1}, d_{1,2}, d_{2,1}, d_{2,2}$ be the degrees of the polynomials $f_{1,1}(x), f_{1,2}(x), f_{2,1}(x), f_{2,2}(x)$, respectively. Let $d_1 = d_{1,1} - d_{1,2}$ and $d_2 = d_{2,1} - d_{2,2}$. Also, let $\overline{n}_1 = \gcd(n_1, d_1)$, $\overline{n}_2 = \gcd(n_2, d_2)$, and $m = \gcd\left(\frac{n_1}{\overline{n}_1}, \frac{n_2}{\overline{n}_2}\right)$. As in the case of Proposition 7.3.30, we note that $\overline{n}_i$ is positive, while d_i need not be positive, for $i = 1, 2$.

PROPOSITION 7.3.31. *Under the notation as above, assume that $\mathbb{F}_q$ is the full constant field of E and $[E : \mathbb{F}_q(x)] = n_1 n_2$. Moreover, assume that the integers $\overline{n}_1$, $\overline{n}_2$, and m divide $q-1$. If a rational place of E over P_∞ exists, then there are exactly $\overline{n}_1 \overline{n}_2 m$ rational places of E over P_∞. This happens if and only if the following conditions 1–4 hold simultaneously:*

(1) *c_1 is an $\overline{n}_1$-power in $\mathbb{F}_q$.*
(2) *c_2 is an $\overline{n}_2$-power in $\mathbb{F}_q$.*
(3) $(m \operatorname{lcm}(\overline{n}_1, \overline{n}_2)) \mid (q-1)$.
(4) *Under the assumptions of 1 and 2, let α and β be elements of $\mathbb{F}_q$ with $\alpha^{\overline{n}_1} = c_1$ and $\beta^{\overline{n}_2} = c_2$. Let A and B be integers satisfying*

$$A \frac{n_1}{\overline{n}_1} + B \frac{a_1}{\overline{n}_1} = 1;$$

then

$$\frac{\beta}{\alpha^{\frac{a_2}{\overline{n}_2} B}} \text{ is an } m\text{-power in } \mathbb{F}_q. \qquad \square$$

EXAMPLE 7.3.32. Let $E = \mathbb{F}_8(x, y_1, y_2)$ be the function field over $\mathbb{F}_8$ given by the following equations:

$$y_1^7 = w^3 (x+1)^4 (x+w)^2,$$
$$y_2^7 = \frac{(x+1)^4 (x+w)}{x + w^6},$$

where $w^3 + w + 1 = 0$. The genus of E is $g(E) = 36$, and $N(E) = 112$. This is the best known value.

EXAMPLE 7.3.33. Let $E = \mathbb{F}_{16}(x, y_1, y_2)$ be the function field over F_{16} given by the following equations:

$$y_1^3 = \frac{w^3 x (x+1)}{x + w^{10}},$$
$$y_2^5 = x^3 (x+1)^3 (x^6 + x^5 + x^3 + x + 1),$$

where $w^4 + w + 1 = 0$. The genus of E is $g(E) = 20$, and $N(E) = 127$. This is the best known value.

EXAMPLE 7.3.34. Let $E = \mathbb{F}_9(x, y_1, y_2)$ be the function field over $\mathbb{F}_9$ given by the following equations:

$$y_1^2 = \frac{x^2 + w^7x + w^5}{x + w^3},$$
$$y_2^2 = \frac{x^4 + w^5x^3 + x^2 + w^3x + w^3}{x + w^3},$$

where $w^2 + 2w + 2 = 0$. The genus of E is $g(E) = 5$, and $N(E) = 32$. This is the best known value.

EXAMPLE 7.3.35. Let $E = \mathbb{F}_9(x, y_1, y_2)$ be the function field over $\mathbb{F}_9$ given by the following equations:

$$y_1^8 = -(x^6 + x^5 + wx^4 + 2x^3 + w^7x^2 + x + 2),$$
$$y_2^2 = \frac{x^4 + w^6x^3 + w^7x + w^5}{x + w^3},$$

where $w^2 + 2w + 2 = 0$. The genus of E is $g(E) = 9$, and $N(E) = 48$. This is the best known value.

EXAMPLE 7.3.36. Let $E = \mathbb{F}_{27}(x, y_1, y_2)$ be the function field over $\mathbf{F}_{27}$ given by the following equations:

$$y_1^2 = (x-1)^6(x^3 + w^{11}x^2 + w^{11}x + w^{15}),$$
$$y_2^2 = \frac{x^3 + w^{11}x + w^{12}}{x^2},$$

where $w^3 + 2w + 1 = 0$. The genus of E is $g(E) = 4$, and $N(E) = 64$. This is optimal.

EXAMPLE 7.3.37. Let $E = \mathbb{F}_{81}(x, y_1, y_2)$ be the function field over $\mathbf{F}_{81}$ given by the following equations:

$$y_1^{10} = x(x+1),$$
$$y_2^2 = \frac{x^2 + w^3x + w^6}{x},$$

where $w^4 + 2w^3 + 2 = 0$. The genus of E is $g(E) = 8$, and $N(E) = 226$. This function field is maximal.

Artin–Schreier curves and class groups. An Artin–Schreier cover C_f of a curve $C/\mathbb{F}_q$ of genus g is given by the equation

$$y^p - y = f(x), \quad f(x) \in \mathbb{F}_p(C).$$

To get f for which C_f will have many rational points provided that C has many points, one can use the class group of C.

EXAMPLE 7.3.38. Let $C/\mathbb{F}_2$ be the elliptic curve defined by $y^2 + y = x^3 + x$. It has five rational points $P_0 = \infty$, $P_1 = (0,0)$, $P_2 = (1,0)$, $P_3 = (1,1)$ and $P_4 = (0,1)$. Since we have an isomorphism $C(\mathbb{F}_2) \cong \mathbb{Z}/5\mathbb{Z}$ of abelian groups such that under the isomorphism P_i corresponds to $i \pmod 5$, the divisor $\sum a_iP_i$ is of degree 0 and linearly equivalent to 0 if and only if

$$\sum_{i=0}^{4} a_i = 0 \qquad \text{and} \qquad \sum_{i=0}^{4} a_i i \equiv 0 \pmod 5.$$

Thus, there exists a function $f \in \mathbb{F}_2(C)$ with divisor (f) equal to a given divisor $\sum a_i P_i$ if and only if the 5-tuple $(a_0, a_1, \ldots, a_4) \in \mathbb{Z}^5$ satisfies the above equation.

For brevity, let us denote a relation for linear equivalence $\sum_{i=0}^{4} a_i P_i \sim 0$ simply by $[a_0, a_1, \ldots, a_4]$. Two examples of such relations for the present curve are $[-3, -1, 2, 1, 1]$ and $[-1, -3, 1, 1, 2]$, which give functions f_1 and f_2 and lead to curves C_{f_1} and C_{f_2} of genera $g = 4$ with $N = 8$ rational points. Note that $N_2(4) = 8$, so these curves realize the maximum for genus 4. The fibre product $C_{f_1} \times_C C_{f_2}$ has genus $g = 11$ with $N = 14$ rational points, and this is also optimal.

EXAMPLE 7.3.39. Let $C/\mathbb{F}_2$ be the curve of genus 2 defined by the affine equation $y^2 + y = (x^2 + x)/(x^3 + x^2 + 1)$. It has six rational points P_i $(i = 1, \ldots, 6)$ and class group $J_C(\mathbb{F}_2) \cong \mathbb{Z}/19\mathbb{Z}$. Let $\varphi_C \colon C \to J_C$ be the Abel–Jacobi map given by $Q \mapsto Q - \deg(Q) P_1$. For a suitable numbering of the P_i, the images of the P_i in $J_C(\mathbb{F}_2) \cong \mathbb{Z}/19\mathbb{Z}$ are as follows:

i	1	2	3	4	5	6
$\varphi(P_i)$	0	1	14	6	16	4

From this table we see that we have the following linear equivalence:

$$-3P_1 - 3P_2 + 2P_3 + 2P_4 + P_5 + P_6 \sim 0.$$

Let $f \in \mathbb{F}_q(C)$ be the function with $(f) = -3P_1 - 3P_2 + 2P_3 + 2P_4 + P_5 + P_6$; then the curve C_f obtained as the double cover of C defined by $w^2 + w = f$ has genus $g = 7$ and 10 rational points. This is optimal. Similarly, the relation

$$-9P_1 + 2P_2 + 2P_3 + P_4 + 2P_5 + 2P_6 \sim 0$$

gives rise to a curve with $g = 8$ and 11 rational points, again an optimal curve over $\mathbb{F}_2$. Let Q be the divisor of zeroes of the polynomial $x^3 + x^2 + 1$ on C; then $-Q + \sum_{i=1}^{6} P_i \sim 0$, which gives us a curve of genus 9 with 12 points, an optimal curve.

EXAMPLE 7.3.40. Consider the elliptic curve $C/\mathbb{F}_3$ defined by the equation $y^2 = x^3 + 2x + 1$. It has 7 points P_i with $i = 0, \ldots, 7$ such that $P_i \mapsto i \pmod 7$ defines an isomorphism $C(\mathbb{F}_3) \cong \mathbb{Z}/7\mathbb{Z}$. The relations $[-4, -1, 0, 1, 2, 1, 1]$ and $[-4, -2, 1, 2, 1, 1, 1]$ give a fibre product of genus $g = 30$ with 38 points, the best known value.

EXAMPLE 7.3.41. The curve $C/\mathbb{F}_3$ of genus 2 given by $y^2 = x^5 + x^3 + x + 1$ has zeta function $(9t^4 + 9t^3 + 5t^2 + 3t + 1)/(1 - t)(1 - 3t)$. The curve has 7 rational points and class group $\mathbb{Z}/27\mathbb{Z}$, and the 7 rational points map to $0, 1, 26, 17, 10, 23, 4$ in the group. We have the relations $[1, 1, 1, 1, -2, 2, -4]$ and $[1, 1, 1, 2, -1, 1, -5]$, and the corresponding fibre product has $g = 44$ with 47 points, the best known value.

7.3.3. Maximal Curves Covered by a Hermitian Curve

The Hermitian curve $H_q : y^q + y = z^{q+1}$ is maximal over $\mathbb{F}_{q^2}$. Therefore, any curve covered by H_q is also maximal over $\mathbb{F}_{q^2}$. In particular, this is true for

$$C_{q,m} : \quad y^q + y = z^m, \quad \text{with} \quad m \mid (q+1),$$

of genus $g=(m-1)(q-1)/2$, which gives, e.g., the maximal curves

$$\begin{array}{lllll}
C_{5,6}: & y^5+y=z^6 & \text{over } \mathbb{F}_{5^2}, & g=10, & N_q(g)=126,\\
C_{7,4}: & y^7+y=z^4 & \text{over } \mathbb{F}_{7^2}, & g=9, & N_q(g)=176,\\
C_{11,3}: & y^{11}+y=z^3 & \text{over } \mathbb{F}_{11^2}, & g=10, & N_q(g)=342,\\
C_{13,2}: & y^{13}+y=z^2 & \text{over } \mathbb{F}_{13^2}, & g=6, & N_q(g)=326,\\
C_{17,2}: & y^{17}+y=z^2 & \text{over } \mathbb{F}_{17^2}, & g=8, & N_q(g)=562,
\end{array}$$

etc.

Another example of a curve covered by H_q is

$$z^n=t(t+1)^{q-1}, \quad \text{with} \quad n \mid (q^2-1),$$

of genus $g=(n-\delta)/2$, where $\delta=\gcd(n,q-1)$, which gives, e.g., the maximal curves

$$\begin{array}{llll}
z^{10}=t(t+1)^{48} & \text{over } \mathbb{F}_{7^4}, & g=4, & N_q(g)=2794,\\
z^{16}=t(t+1)^{120} & \text{over } \mathbb{F}_{11^4}, & g=4, & N_q(g)=15\,610,\\
z^{21}=t(t+1)^{12} & \text{over } \mathbb{F}_{13^2}, & g=9, & N_q(g)=404,\\
z^{12}=t(t+1)^{16} & \text{over } \mathbb{F}_{17^2}, & g=4, & N_q(g)=426,\\
z^{15}=t(t+1)^{18} & \text{over } \mathbb{F}_{19^2}, & g=6, & N_q(g)=590.
\end{array}$$

7.4. Recursive Towers

In Sec. 3.4.3 of *Basic Notions* we mentioned some asymptotically optimal towers, namely the García–Stichtenoth towers. These towers are of a very special nature: they are *recursive* in a sense specified below (are defined by essentially the same equations at every step). This property greatly simplifies the difficult problem of computing genera in infinite towers; their theory, developed also by García and Stichtenoth, makes possible a systematic treatment of many recursive towers, thus leading to numerous examples of asymptotically good and optimal towers. Moreover, conjecturally, in the case of optimality these towers are connected with modular towers of the type considered in Chapter 5. Another very pleasant feature of these towers is the possibility of constructing the corresponding algebraic geometry codes with a reasonable polynomial complexity (at least for some of them).

This section is devoted to a brief review of this theory. In this section (except for Sec. 7.4.5) we use the language of function fields rather than that of algebraic curves.

First we give some basic facts on recursive towers, and then some most interesting examples of such towers over fields of square cardinality. We briefly describe a polynomial algorithm for constructing codes on one of those towers. We give a description of a very good tower over finite fields of nonsquare cardinality, the construction of which is rather close in spirit to that of optimal recursive towers over fields with square cardinality and is related to higher-rank Drinfeld modules. The section ends with a very recent construction of a good recursive tower over any field $\mathbb{F}_q$, $q \geq 4$, in particular over any prime field $\mathbb{F}_p$, $p \geq 5$.

All towers described in the present section can be called García–Stichtenoth towers in a wide sense; however, sometimes this term is used more specifically to refer to the most useful towers $\mathcal{W}_1$ and $\mathcal{W}_2$ from Sec. 7.4.2.

7.4.1. Some Basic Facts on Towers

A *tower* $\mathcal{F}$ *over* $\mathbb{F}_q$ (or an $\mathbb{F}_q$*-tower*) is an infinite sequence $\mathcal{F} = (F_0, F_1, F_2, \ldots)$ of function fields $F_i/\mathbb{F}_q$ such that

(i) $F_0 \subsetneqq F_1 \subsetneqq F_2 \subsetneqq \ldots \subsetneqq F_n \subsetneqq \ldots$.
(ii) Each extension F_{n+1}/F_n is finite and separable.
(iii) The genera satisfy $g(F_n) \to \infty$ as $n \to \infty$.

Clearly, this notion (which will also be used in Sec. 8.1) is equivalent to a tower of algebraic curves over $\mathbb{F}_q$

$$X_0 \leftarrow X_1 \leftarrow X_2 \leftarrow \ldots,$$

where $F_i = \mathbb{F}_q(X_i)$, which is a *projective* system of curves, corresponding to an *inductive* system of fields

$$F_0 \hookrightarrow F_1 \hookrightarrow F_2 \hookrightarrow \ldots.$$

We will consider explicit towers only, i.e., those given by explicit polynomial equations. Moreover, we are interested in recursive towers, which we are going to define.

Let $\mathcal{F} = (F_0, F_1, F_2, \ldots)$ be a tower of function fields over $\mathbb{F}_q$, where $F_0 = \mathbb{F}_q(x_0)$ is the rational function field. We say that the tower $\mathcal{F}$ is *recursive* if there exist a polynomial $f(X,Y) \in \mathbb{F}_q[X,Y]$ and functions $x_n \in F_n$ such that:

(i) $f(X,Y)$ is separable in both variables X and Y;

(ii) $F_{n+1} = F_n(x_{n+1})$ with $f(x_n, x_{n+1}) = 0$, for all $n \geq 0$;

(iii) $[F_{n+1} : F_n] = \deg_Y f(X,Y)$, for all $n \geq 0$.

We also say that the tower $\mathcal{F}$ is given by the equation $f(X,Y) = 0$ or that $\mathcal{F}$ is defined recursively by the polynomial $f(X,Y)$. Sometimes a tower is recursively given by an equation of the form

$$g(X,Y) = h(X,Y)$$

with rational functions $g(X,Y), h(X,Y) \in \mathbb{F}_q(X,Y)$. It is obvious that, after clearing denominators, such an equation can be transformed into the form $f(X,Y) = 0$ with a polynomial $f(X,Y) \in \mathbb{F}_q[X,Y]$. For example, the defining equation

$$Y^\ell - Y = X^\ell/(X^{\ell-1} + 1)$$

(cf. Sec. 7.4.2 below) can be transformed into the polynomial equation

$$f(X,Y) = (X^{\ell-1} + 1)(Y^\ell - Y) - X^\ell = 0;$$

however, it is often more convenient to use the initial form of the equation.

For a recursive tower $\mathcal{F} = (F_0, F_1, F_2, \ldots)$, much information about it is already contained in the field $F_1 = \mathbb{F}_q(x_0, x_1)$. Thus, we introduce the following definition.

Let $\mathcal{F}$ be a recursive tower over $\mathbb{F}_q$ given by a polynomial equation $f(X,Y) = 0$. Then its *basic function field* is defined as

$$F = \mathbb{F}_q(x,y), \quad \text{with the relation} \quad f(x,y) = 0.$$

We associate with a recursive tower the following fundamental parameters: its *limit*

$$\lambda(\mathcal{F}) := \lim_{n\to\infty} N(F_n)/g(F_n), \quad 0 \leq \lambda(\mathcal{F}) \leq A(q) \leq \sqrt{q} - 1;$$

its *splitting rate*

$$\nu(\mathcal{F}) := \lim_{n\to\infty} N(F_n)/[F_n : F_0], \quad 0 \leq \nu(\mathcal{F}) \leq N(F_0);$$

and its *genus*

$$\gamma(\mathcal{F}) := \lim_{n\to\infty} g(F_n)/[F_n : F_0].$$

A tower $\mathcal{F}$ is said to be *asymptotically good* if it has a positive limit $\lambda(\mathcal{F}) > 0$. If $\lambda(\mathcal{F}) = 0$, then $\mathcal{F}$ is said to be *asymptotically bad*. If $\lambda(\mathcal{F}) = A(q)$, the tower is *asymptotically optimal*. Existence of these limits easily follows from the Hurwitz formula; namely, one has the following (the same arguments are used in Lemmas 8.1.9 and 8.1.10 below).

LEMMA 7.4.1. *Let $F_0 \subseteq F \subseteq E$ be finite separable extensions of algebraic function fields over $\mathbb{F}_q$. Then we have*

$$\frac{N(F)}{[F:F_0]} \geq \frac{N(E)}{[E:F_0]} \qquad \textit{and} \qquad \frac{g(F)-1}{[F:F_0]} \leq \frac{g(E)-1}{[E:F_0]}.$$

In particular, if $g(F) \geq 2$, then

$$\frac{N(F)}{g(F)-1} \geq \frac{N(E)}{g(E)-1}. \qquad \square$$

PROPOSITION 7.4.2. *Given a tower $\mathcal{F} = (F_0, F_1, F_2, \ldots)$ over $\mathbb{F}_q$, the limits $\nu(\mathcal{F})$ and $\gamma(\mathcal{F})$ do exist.*

PROOF. By Lemma 7.4.1, the sequence $(N(F_n)/[F_n:F_0])_{n\geq 0}$ is decreasing, and hence is convergent in $\mathbb{R}$. On the other hand, we have that the sequence $((g(F_n)-1)/[F_n:F_0])$ is increasing, hence convergent in $\mathbb{R}\cup\{\infty\}$. Since

$$\lim_{n\to\infty}(g(F_n)-1)/[F_n:F_0]=\lim_{n\to\infty}g(F_n)/[F_n:F_0],$$

the proposition follows. □

It follows from Proposition 7.4.2 that the sequence $N(F_n)/g(F_n)$ is convergent, since we have

$$\frac{N(F_n)}{g(F_n)}=\frac{N(F_n)/[F_n:F_0]}{g(F_n)/[F_n:F_0]}\to\frac{\nu(\mathcal{F})}{\gamma(\mathcal{F})},\quad \text{as}\quad n\to\infty.$$

Therefore, we obtain the following fact.

COROLLARY 7.4.3. *For a tower $\mathcal{F}$ over $\mathbb{F}_q$ one has*

$$\lambda(\mathcal{F})=\nu(\mathcal{F})/\gamma(\mathcal{F}).$$

Moreover, the following statements are equivalent:

(i) *The tower $\mathcal{F}$ is asymptotically good.*
(ii) *The genus $\gamma(\mathcal{F})$ is finite, and the splitting rate $\nu(\mathcal{F})$ is strictly positive.* □

It can be shown that for a recursive tower $\mathcal{F}$ with positive limit $\lambda(\mathcal{F})>0$ (which we are only interested in) one has

$$\deg_X f(X,Y)=\deg_Y f(X,Y).$$

For the corresponding basic function field $F=\mathbb{F}_q(x,y)$, this condition means that

$$[F:\mathbb{F}_q(x)]=[F:\mathbb{F}_q(y)]\quad\text{if}\quad \lambda(\mathcal{F})>0.$$

Let $\mathcal{E}=(E_0,E_1,E_2,\ldots)$ and $\mathcal{F}=(F_0,F_1,F_2,\ldots)$ be two towers over $\mathbb{F}_q$. We call $\mathcal{E}$ a subtower of $\mathcal{F}$ if for any E_n there exists some F_m such that $E_n\subseteq F_m$.

One can easily prove the following result.

PROPOSITION 7.4.4. *If $\mathcal{E}$ is a subtower of $\mathcal{F}$, then $\lambda(\mathcal{E})\geq\lambda(\mathcal{F})$. In particular, if the tower $\mathcal{F}$ is asymptotically good (respectively, optimal), then any subtower $\mathcal{E}$ of $\mathcal{F}$ is also asymptotically good (respectively, optimal).* □

Let $\mathcal{F}=(F_0,F_1,F_2,\ldots)$ be a tower over $\mathbb{F}_q$, and let P be a place of F_0. We say that P is *ramified in the tower* $\mathcal{F}$ if for some $n\geq 1$ there exists a place Q of F_n lying above P such that $Q|P$ is ramified; i.e., the ramification index satisfies $e(Q|P)>1$. If there exists an index $n\geq 1$ and a place Q of F_n above P such that $Q|P$ is wildly ramified (i.e., the characteristic of $\mathbb{F}_q$ divides the ramification index $e(Q|P)$), then P is said to be *wildly ramified in the tower* $\mathcal{F}$. Otherwise, the place P is said to be *tame* in $\mathcal{F}$. The set

$$V(\mathcal{F}):=\{P\mid P \text{ is a place of } F_0 \text{ which is ramified in } \mathcal{F}\}$$

is called the *ramification locus* of $\mathcal{F}$.

We will only consider towers with finite ramification loci.

PROPOSITION 7.4.5. *Let $\mathcal{F}=(F_0,F_1,F_2,\ldots)$ be a tower with a finite ramification locus $V(\mathcal{F})$. Assume that for each place $P\in V(\mathcal{F})$ there exists a real constant $c_P>0$ such that, for all $n\geq 1$ and for all places Q of F_n lying above P, we have*

$$d(Q|P)\leq c_P\cdot e(Q|P),$$

$d(Q|P)$ being the different exponent and $e(Q|P)$ the ramification index of the place Q over P. Then the genus of the tower is finite and satisfies

$$\gamma(\mathcal{F}) \le g(F_0) - 1 + \frac{1}{2} \sum_{P \in V(\mathcal{F})} c_P \cdot \deg P < \infty. \qquad \square$$

This proposition is a consequence of the Hurwitz formula. An important special case of Proposition 7.4.5 is the case of tame towers.

COROLLARY 7.4.6. *Let $\mathcal{F} = (F_0, F_1, F_2, \ldots)$ be a tower with a finite ramification locus $V(\mathcal{F})$, and assume that all places $P \in V(\mathcal{F})$ are tame in $\mathcal{F}$. Then*

$$\gamma(\mathcal{F}) \le g(F_0) - 1 + \frac{1}{2} \sum_{P \in V(\mathcal{F})} \deg P. \qquad \square$$

In fact, we will consider even more specific towers and assume that the tower $\mathcal{F} = (F_0, F_1, F_2, \ldots)$ is recursively defined by the equation

$$\varphi(Y) = \psi(X),$$

where $\varphi(T), \psi(T) \in \mathbb{F}_q(T)$ are rational functions. As above, let $F = \mathbb{F}_q(x,y)$ be the corresponding basic function field which is given by the equation $\varphi(y) = \psi(x)$, and define

$$V_0 := \{P \mid P \text{ is a place of } \mathbb{F}_q(x) \text{ which ramifies in } F/\mathbb{F}_q(x)\}.$$

The set V_0 is finite, since the extension $F/\mathbb{F}_q(x)$ is separable. We also define

$$R_0 := \{x(P) \mid P \in V_0\}.$$

Clearly, this set R_0 is a finite subset of $\overline{\mathbb{F}}_q \cup \{\infty\}$.

PROPOSITION 7.4.7. *Let $\mathcal{F} = (F_0, F_1, F_2, \ldots)$ be a tower over $\mathbb{F}_q$ which is recursively defined by $\varphi(Y) = \psi(X)$. Assume that there exists a finite subset $R \subseteq \overline{\mathbb{F}}_q \cup \{\infty\}$ such that the following two conditions hold:*

(a) *The set R contains $R_0 = \{x(P) \mid P \in V_0\}$.*

(b) *If $\beta \in R$ and $\alpha \in \overline{\mathbb{F}}_q \cup \{\infty\}$ satisfy the equation $\varphi(\beta) = \psi(\alpha)$, then $\alpha \in R$.*

Then the ramification locus of the tower $\mathcal{F}$ satisfies

$$V(\mathcal{F}) \subseteq \{P \mid P \textit{ is a place of } F_0 \textit{ with } x_0(P) \in R\};$$

in particular, $V(\mathcal{F})$ is finite, and moreover,

$$\sum_{P \in V(\mathcal{F})} \deg P \le |R|. \qquad \square$$

Now we want to establish a criterion which implies a positive splitting rate $\nu(\mathcal{F}) > 0$, for a particular class of recursive towers.

Let $\mathcal{F} = (F_0, F_1, F_2, \ldots)$ be a tower over $\mathbb{F}_q$, and let P be a rational place of F_0 (i.e., $\deg P = 1$). We say that P *splits completely in the tower $\mathcal{F}$* if P splits completely in all extensions F_n/F_0; i.e., there are exactly $[F_n : F_0]$ places of F_n above the place P (and they are rational places of F_n). The set

$$Z(\mathcal{F}) := \{P \mid P \text{ is a rational place of } F_0 \text{ which splits completely in } \mathcal{F}\}$$

is called the *splitting locus of the tower $\mathcal{F}$*. It is obvious that $Z(\mathcal{F}) \cap V(\mathcal{F}) = \varnothing$.

A simple argument in the spirit of those in the proofs of Proposition 7.4.2 and Lemmas 8.1.9 and 8.1.10 shows the following.

LEMMA 7.4.8. *Let $\mathcal{F}$ be a tower over $\mathbb{F}_q$. Then*

$$\nu(\mathcal{F}) \geq |Z(\mathcal{F})|. \qquad \square$$

7.4.2. Some Specific García–Stichtenoth Towers and the Elkies Conjecture

Now we give some examples of recursive asymptotically good towers and formulate their conjectural relation with modular towers. We begin with the tame case.

Tame towers. First let us consider recursive tame towers having a finite ramification locus and a nonempty splitting locus $Z(\mathcal{T})$. Then we get from the above the following lower bound on the limit $\lambda(\mathcal{T})$ of the tower $\mathcal{T}$.

LEMMA 7.4.9. *Assume that $\mathcal{T}$ is a recursive tame tower defined by $\varphi(Y) = \psi(X)$ and satisfying the assumptions of Proposition 7.4.7. Then its limit $\lambda(\mathcal{T})$ satisfies the estimate*

$$\lambda(\mathcal{T}) \geq \frac{2|Z(\mathcal{T})|}{|R|-2},$$

where R is the finite set defined in Proposition 7.4.7.

PROOF. We have the inequalities $\nu(\mathcal{T}) \geq |Z(\mathcal{T})|$ (see Lemma 7.4.8) and also $\gamma(\mathcal{T}) \leq (-2+|R|)/2$ (see Corollary 7.4.6 and Proposition 7.4.7). Then the assertion follows immediately, since $\lambda(\mathcal{T}) = \nu(\mathcal{T})/\gamma(\mathcal{T})$ (see Corollary 7.4.3). $\square$

The main tool to work with such towers is the following famous Abhyankar's lemma.

THEOREM 7.4.10 (Abhyankar's lemma). *Let E/F be a function field extension, and let E_1 and E_2 be two intermediate fields with $E = E_1 \cdot E_2$ (i.e., E is the composite field of E_1 and E_2). Let Q be a place of E, and denote by Q_1, Q_2, and P its restrictions to E_1, E_2, and F. If $Q_1 \,|\, P$ or $Q_2 \,|\, P$ is tame, then*

$$e(Q \,|\, P) = \operatorname{lcm}\big(e(Q_1 \,|\, P), e(Q_2 \,|\, P)\big),$$

where lcm *stands for the least common multiple.* $\square$

In many tame cases, repeatedly using this result, one can compute the required parameters of a tower.

Tower $\mathcal{T}_1$. Let ℓ be any prime power, and let $q = \ell^r$, where $r \in \mathbb{N}$ and $r \geq 2$. Consider the tower $\mathcal{T}_1$ over $\mathbb{F}_q$ which is given recursively by the equation

$$Y^m = (X+1)^m - 1, \quad \text{with} \quad m = (q-1)/(\ell-1). \tag{7.4.1}$$

One can easily show that the place $P_0 = (x_0 = 0)$ of the function field $F_0 = \mathbb{F}_q(x_0)$ is totally ramified in all extensions F_n/F_0, so that equation (7.4.1) does define a recursive tower $\mathcal{T}_1 = (F_0, F_1, F_2, \ldots)$. Indeed, let P_i be the zero of the function $x_i \in \mathbb{F}_q(x_i)$. Then $e(P_i \,|\, P_0) = 1$ for all $i \geq 0$ (as follows from the equation $x_i^m = (x_{i-1}+1)^m - 1$). Therefore, equation (7.4.1) does define a tower, and the place P_0 is totally ramified in all extensions F_n/F_0.

Let $F = \mathbb{F}_q(x,y)$ with $y^m = (x+1)^m - 1$ be the basic function field corresponding to the tower $\mathcal{T}_1$, and let V_0, as above, be the set of places of $\mathbb{F}_q(x)$ which ramify in $F/\mathbb{F}_q(x)$. The set $R_0 = \{x(P) \mid P \in V_0\}$ is here given by

$$R_0 = \{\beta \in \overline{\mathbb{F}}_q \mid (\beta+1)^m = 1\}.$$

We claim that the set $R := \mathbb{F}_q$ satisfies the conditions of Proposition 7.4.7. In fact, we have $R_0 \subseteq \mathbb{F}_q$ (since $m = (q-1)/(\ell-1)$ is the norm exponent of the extension $\mathbb{F}_q/\mathbb{F}_\ell$), and for $\beta \in \mathbb{F}_q$ and $\alpha \in \overline{\mathbb{F}}_q$ with $\beta^m = (\alpha+1)^m - 1$ it follows that $(\alpha+1)^m = 1+\beta^m \in \mathbb{F}_\ell$; hence, $\alpha \in \mathbb{F}_q$. By Proposition 7.4.7, the ramification locus $V(\mathcal{T}_1)$ is finite and satisfies

$$V(\mathcal{T}_1) \subseteq \{P \mid P \text{ is a place of } F_0 \text{ with } x_0(P) \in \mathbb{F}_q\}.$$

Next we show that the place $P_\infty = (x_0 = \infty)$ of the rational function field $F_0 = \mathbb{F}_q(x_0)$ splits completely in the tower $\mathcal{T}_1$. We have

$$F_1 = F_0\left(\frac{x_1}{x_0+1}\right) \quad \text{and} \quad \left(\frac{x_1}{x_0+1}\right)^m = 1 - \left(\frac{1}{x_0+1}\right)^m.$$

The right-hand side of the last equality above takes the value 1 at the place P_∞, and since the exponent $m = (q-1)/(\ell-1)$ is the norm exponent of $\mathbb{F}_q/\mathbb{F}_\ell$, we conclude that P_∞ splits completely in the extension F_1/F_0. Let Q_∞ be a place of F_1 above P_∞; then we have from the equation $x_1^m = (x_0+1)^m - 1$ that $v_{Q_\infty}(x_1) = v_{Q_\infty}(x_0) = -1$. Since

$$F_2 = F_1\left(\frac{x_2}{x_1+1}\right) \quad \text{and} \quad \left(\frac{x_2}{x_1+1}\right)^m = 1 - \left(\frac{1}{x_1+1}\right)^m,$$

it follows as above that the place Q_∞ splits completely in the extension F_2/F_1. Repeating this argument, we find that P_∞ splits completely in the tower $\mathcal{T}_1$; thus, $|Z(\mathcal{T}_1)| \geq 1$. From Lemma 7.4.1 we then get a positive limit

$$\lambda(\mathcal{T}_1) \geq 2/(q-2) > 0;$$

i.e., the tower $\mathcal{T}_1$ over $\mathbb{F}_q$ is asymptotically good.

Tower $\mathcal{T}_2$. Now let p be an odd prime number, and let $q = p^2$. Consider the equation

$$Y^2 = \frac{X^2+1}{2X}$$

over the field $\mathbb{F}_q$. Similarly to the case of the tower $\mathcal{T}_1$, one sees that this equation does recursively define a tower $\mathcal{T}_2 = (F_0, F_1, F_2, \ldots)$ over $\mathbb{F}_q$ and that the place $P_\infty = (x_0 = \infty)$ of the rational function field $F_0 = \mathbb{F}_q(x_0)$ is totally ramified in all extensions F_n/F_0.

Let $a \in \mathbb{F}_q$ be such that $a^2 = -1$ (it exists, since $q = p^2$). The set R_0 in Proposition 7.4.7 is here given by $R_0 = \{0, \infty, \pm a\}$. We claim that the set

$$R = \{0, \infty, \pm a, \pm 1\}$$

satisfies condition (b) in Proposition 7.4.7. This follows from the elementary facts:

- If $\beta = 0$ and $\beta^2 = (\alpha^2+1)/(2\alpha)$, then $\alpha = \pm a$.
- If $\beta = \infty$ and $\beta^2 = (\alpha^2+1)/(2\alpha)$, then $\alpha \in \{0, \infty\}$.
- If $\beta = \pm a$ and $\beta^2 = (\alpha^2+1)/(2\alpha)$, then $\alpha = -1$.
- If $\beta = \pm 1$ and $\beta^2 = (\alpha^2+1)/(2\alpha)$, then $\alpha = 1$.

Further, there are at least $2(p-1)$ places of F_0 that split completely over $\mathbb{F}_{p^2}$; i.e., $|Z(\mathcal{T}_2)| \geq 2(p-1)$. Therefore, we get from Lemma 7.4.9 that

$$\lambda(\mathcal{T}_2) \geq \frac{2 \cdot 2(p-1)}{6-2} = p-1,$$

and thus the tower $\mathcal{T}_2$ attains the Drinfeld–Vlăduţ bound over $\mathbb{F}_{p^2}$; in particular, $|Z(\mathcal{T}_2)| = 2(p-1)$.

For some more examples of optimal tame towers, see Theorem 7.4.19 below.

The $\mathcal{T}_2$ tower is optimal for $q = p^2$, but the great majority of known optimal towers are wild ones, and we are going to give the two most useful examples.

Wild towers. In the case of wild towers, Abhyankar's lemma is not valid but can often be replaced with the following result.

LEMMA 7.4.11. *Let $F/\mathbb{F}_q$ be a function field, and let E_1/F and E_2/F be distinct Galois extensions of F such that $[E_1 : F] = [E_2 : F] = p = \mathrm{char}(\mathbb{F}_q)$. Denote by $E = E_1 \cdot E_2$ the composite field of E_1 and E_2. Let Q be a place of E, and denote by Q_1, Q_2, and P its restrictions to the subfields E_1, E_2, and F. Assume that we have $d(Q_i \mid P) = 2(e(Q_i \mid P) - 1)$ for $i = 1, 2$. Then we also have*

$$d(Q \mid Q_i) = 2(e(Q \mid Q_i) - 1) \quad \textit{for} \quad i = 1, 2. \qquad \square$$

Tower $\mathcal{W}_1$. Let $q = r^2$, where r is a power of a prime number p. We consider the tower $\mathcal{W}_1$ over $\mathbb{F}_q$ which is recursively given by the equation

$$Y^r + Y = \frac{X^r}{X^{r-1}+1}. \tag{7.4.2}$$

First one has to show that this equation actually defines a tower.

LEMMA 7.4.12. *Let $\mathcal{W}_1 = (F_0, F_1, F_2, \ldots)$, where $F_0 = \mathbb{F}_q(x_0)$ is the rational function field and $F_{n+1} = F_n(x_{n+1})$ with the relation $x_{n+1}^r + x_{n+1} = x_n^r/(x_n^{r-1}+1)$, for each $n \geq 0$. Then we have the the following:*

(i) *For each $n \geq 0$, the extension F_{n+1}/F_n has degree $[F_{n+1} : F_n] = r$, and the field $\mathbb{F}_q$ is algebraically closed in F_{n+1}.*
(ii) *The pole P_∞ of x_0 in F_0 is totally ramified in all extensions F_{n+1}/F_0. If Q_{n+1} denotes the unique place of F_{n+1} above P_∞, then Q_{n+1} is a simple pole of the function x_{n+1}.* $\square$

Next, one has to determine the ramification locus $V(\mathcal{W}_1)$ of the tower $\mathcal{W}_1$. Recall that $(x_0 = \alpha)$ denotes the place of the rational function field F_0 which is the zero of the function $x_0 - \alpha$. Using Proposition 7.4.7, one can prove the following.

LEMMA 7.4.13. *The ramification locus $V(\mathcal{W}_1)$ of the tower $\mathcal{W}_1$ is given by*

$$V(\mathcal{W}_1) = \{P_\infty\} \cup \{(x_0 = \alpha) \mid \alpha^r + \alpha = 0\}. \qquad \square$$

The splitting locus $Z(\mathcal{W}_1)$ of the tower $\mathcal{W}_1$ can be found by using Lemma 7.4.8.

LEMMA 7.4.14. *The splitting locus $Z(\mathcal{W}_1)$ of the tower $\mathcal{W}_1$ is given by*

$$Z(\mathcal{W}_1) = \{(x_0 = \alpha) \mid \alpha \in \mathbb{F}_q \text{ and } \alpha^r + \alpha = 0\}.$$

In particular, the splitting rate $\nu(\mathcal{W}_1)$ satisfies the inequality $\nu(\mathcal{W}_1) \geq r^2 - r$. $\square$

An argument using Lemma 7.4.11 allows one to compute the different and prove the following result.

THEOREM 7.4.15. *Let $q = r^2$ be a square, and let $\mathcal{W}_1 = (F_0, F_1, F_2, \ldots)$ be the tower over $\mathbb{F}_q$ which is recursively defined by the equation $Y^r + Y = X^r/(X^{r-1}+1)$. Then we have*

$$\gamma(\mathcal{W}_1) = r, \qquad \nu(\mathcal{W}_1) = r^2 - r, \qquad \lambda(\mathcal{W}_1) = r - 1.$$

In particular, the tower $\mathcal{W}_1$ attains the Drinfeld–Vlăduţ bound and is therefore asymptotically optimal over $\mathbb{F}_q$. □

For more details on this tower, see Sec. 7.4.3 below, where the tower is used to construct good algebraic geometry codes.

Tower $\mathcal{W}_2$. We are still assuming that $q = r^2$ is a square. For each divisor $m \geq 1$ of $r+1$, here we construct an asymptotically optimal tower $\mathcal{W}_2$ over $\mathbb{F}_q$ as follows.

We start with the tower $\mathcal{E} = (E_0, E_1, E_2, \ldots)$ over $\mathbb{F}_q$ which is defined as follows:

$$\begin{aligned} &E_0 = \mathbb{F}_q(z_0) \text{ is the rational function field,} \\ &E_1 = E_0(z_1) = \mathbb{F}_q(z_1) \text{ with } z_1^r + z_1 = z_0, \\ &E_{n+1} = E_n(z_{n+1}) \text{ with } z_{n+1}^r + z_{n+1} = z_n^r/(z_n^{r-1}+1), \quad \text{for all } n \geq 1. \end{aligned} \tag{7.4.3}$$

It is clear that the tower $(E_1, E_2, E_3, \ldots)$ is just the tower $\mathcal{W}_1$ discussed above, and that $E_0 \subseteq E_1$ is the subfield $E_0 = \mathbb{F}_q(z_1^r + z_1) \subseteq \mathbb{F}_q(z_1) = E_1$. The place $(z_0 = 0)$ of E_0 splits completely in E_1/E_0 into the places $(z_1 = \alpha)$ of E_1 with $\alpha^r + \alpha = 0$. The place $(z_0 = \infty)$ is the only ramified place in the extension E_1/E_0; its extension to E_1 is the pole $(z_1 = \infty)$ of z_1, and the different exponent d of $(z_1 = \infty)$ over $(z_0 = \infty)$ is given by $d = 2(r-1)$. Therefore, the ramification locus of the tower $\mathcal{E}$ is (see Lemma 7.4.13)

$$V(\mathcal{E}) = \{(z_0 = 0), (z_0 = \infty)\}.$$

For a place $P_0 \in V(\mathcal{E})$, an integer $n \geq 1$, and a place Q_0 of the field E_n lying above P_0, we have

$$d(Q_0 \,|\, P_0) = 2(e(Q_0 \,|\, P_0) - 1).$$

This follows from Lemma 7.4.11 and the transitivity of different exponents applied to the extensions $E_0 \subseteq E_1 \subseteq E_n$. Now we fix an integer $m \geq 1$ which is a divisor of $r+1$ and consider the fields

$$\begin{aligned} &F_0 := \mathbb{F}_q(x_0) \text{ with } x_0^m = z_0, \\ &F_n := F_0 \cdot E_n, \quad \text{for all } n \geq 1. \end{aligned} \tag{7.4.4}$$

It is obvious that we thus obtain a tower

$$\mathcal{W}_2 := (F_0, F_1, F_2, \ldots)$$

over $\mathbb{F}_q$; clearly, the above tower $\mathcal{W}_1$ is a subtower of $\mathcal{W}_2$.

The defining equations of the tower $\mathcal{W}_2$ are given by $F_{n+1} := F_n(z_{n+1})$ with

$$z_{n+1}^q + z_{n+1} = x_n^{q+1}, \quad x_n := z_n/x_{n-1} \in F_n, \quad n \geq 1.$$

THEOREM 7.4.16. *Let $q = r^2$ be a square, and let $m \geq 1$ be a divisor of $r+1$. Then the tower $\mathcal{W}_2$ over $\mathbb{F}_q$ which is defined by (7.4.3) and (7.4.4) is asymptotically optimal; i.e., its limit is $\lambda(\mathcal{W}_2) = r - 1$, and it therefore attains the Drinfeld–Vlăduţ bound.* □

One can give explicit formulae for the genus $g_n = g(F_n)$ and the number $N_n = N(F_n)$ of $\mathbb{F}_q$-rational points.

PROPOSITION 7.4.17. (a) *The genus g_n is given by the following formula:*

$$g_n = \begin{cases} q^{n+1} + q^n - q^{\frac{n+2}{2}} - 2q^{\frac{n}{2}} + 1, & n \text{ even}, \\ q^{n+1} + q^n - \frac{1}{2}q^{\frac{n+3}{2}} - \frac{3}{2}q^{\frac{n+1}{2}} - q^{\frac{n-1}{2}} + 1, & n \text{ odd}. \end{cases}$$

(b) *All the places of*

$$S = \{(x_0 = \alpha), \alpha \neq 0\}$$

split completely in F_n/F_0, and thus

$$N_n \geq (q-1)r^n.$$

□

In Sec. 12.5 we will also need the following specific result.

LEMMA 7.4.18. *Let $(x_n)_0^{(n)}$ be the divisor of zeroes of the function x_n on F_n. Then $(x_n)_0^{(n)} = r^n Q_n$, where Q_n is the unique common zero of the functions $x_0, z_1, \dots, z_n$.* □

The Elkies conjecture. As we have seen in Sec. 5.3 above, some optimal modular towers, both for classical modular curves and Drinfeld modular curves, have explicit equations identifying the corresponding towers with some recursive towers, which are tame for classical modular curves and wild for Drinfeld modular curves. This lead Elkies to conjecture that any asymptotically optimal recursive tower should be modular (the *Elkies conjecture*). Moreover, all new optimal recursive towers constructed afterwards were found to be modular, which strongly supports the conjecture.

Here we give some more examples of optimal recursive towers over small fields supporting the conjecture (we do not repeat those of Sec. 5.3, which are of the same type).

THEOREM 7.4.19. *The following (tame) asymptotically optimal recursive towers are isomorphic with modular towers associated with the following congruence subgroups G of $\mathrm{PSL}_2(\mathbb{Z})$ (n being the index of the field F_n):*

- $G = \Gamma_1(5) \cap \Gamma_0(2^n)$ *for*
$$F(X,Y) = 2XY^2 + (X^2 + X + 1)Y + X^2 + X + 2$$
over $\mathbb{F}_9$;
- $G = \Gamma_1(12) \cap \Gamma_0(2^{n+1})$ *for*
$$F(X,Y) = (4X+1)Y^2 + (X^2 + X + 2)Y + X + 3$$
over $\mathbb{F}_{25}$;
- $G = \Gamma_1(5) \cap \Gamma_0(2^n)$ *for*
$$F(X,Y) = (X^2 + 6)Y^2 + XY + X^2 + 4$$
over $\mathbb{F}_{49}$;
- $G = \Gamma_1(9) \cap \Gamma_0(3^{n+1})$ *for*
$$F(X,Y) = X^2Y^3 + (X^3 + X^2 + X)Y^2 + (X+1)Y + X^3 + X$$
over $\mathbb{F}_4$. □

REMARK 7.4.20. These polynomials were found by computer search, which showed that, quite naturally, optimal recursive towers are extremely rare. The technique for proving the optimality is a development of the above technique for tame towers, and that for proving the modularity is a generalization of the technique in Sec. 5.3.1 with some highly nontrivial refinements.

7.4.3. Polynomially Constructible Codes from the $\mathcal{W}_1$ Tower

As we have seen in Sections 5.1 and 5.2, some algebraic geometry codes coming from modular curves (either classical or Drinfeld) can be constructed by polynomial algorithms. However, as was mentioned above, those algorithms (though being polynomial) cannot be transformed to algorithms of reasonably low complexity due to their complicated structure.

For codes coming from García–Stichtenoth towers, the situation is somewhat better. Namely, certain codes on curves in the $\mathcal{W}_1$ tower of Sec. 7.4.2 can be constructed with reasonably low complexity. We are going to describe the main technical points leading to the corresponding algorithm.

Thus, we consider the tower $\mathcal{W}_1$ recursively defined by

$$Y^r + Y = \frac{X^r}{X^{r-1}+1}$$

for $q = r^2$, r being a prime power.

Let us give more detail on this tower. Thus, let

$$F_m := \mathbb{F}_q(x_0, \ldots, x_m) \quad \text{for} \quad m \geq 0,$$

i.e., $F_0 = \mathbb{F}_q(x_0)$, $F_k = F_{k-1}(x_k)$, and

$$x_k^r + x_k = \frac{x_{k-1}^r}{x_{k-1}^{r-1}+1} \quad \text{for} \quad 1 \leq k \leq m.$$

Let

$$\Omega := \{\alpha \in \mathbb{F}_q \mid \alpha^r + \alpha = 0\}, \qquad \Omega^* := \Omega \setminus \{0\} = \{\alpha \in \mathbb{F}_q \mid \alpha^{r-1} = -1\}.$$

First we consider the first stage of the tower.

PROPOSITION 7.4.21. *The following assertions hold:*

(i) $[F_1 : F_0] = [F_1 : \mathbb{F}_q(x_1)] = r$.

(ii) *The function x_0 has a unique pole P_∞ in F_1, and this place P_∞ is totally ramified in F_1/F_0.*

(iii) *For any $\alpha \in \Omega^*$, the function $(x_0 - \alpha)$ has a unique zero P_α in F_1, and this place P_α is totally ramified in F_1/F_0.*

(iv) *For any $\gamma \in \Omega$, there is a unique common zero Q_γ of x_0 and $x_1 - \gamma$ in F_1.*

(v) *The principal divisors in F_1 of the functions $(x_0 - \alpha)$ and $(x_1 - \gamma)$, $\alpha, \gamma \in \Omega$, are as follows:*

$$(x_0) = \sum_{\gamma \in \Omega} Q_\gamma - rP_\infty,$$

$$(x_0 - \alpha) = rP_\alpha - rP_\infty, \quad \text{for} \quad \alpha \in \Omega^*,$$

$$(x_1 - \gamma) = rQ_\gamma - P_\infty - \sum_{\alpha \in \Omega^*} P_\alpha, \quad \text{for} \quad \gamma \in \Omega.$$

(vi) *The places of F_1 that are ramified over F_0 are exactly the places P_∞ and P_α, $\alpha \in \Omega^*$. Their different exponents with respect to the extension F_1/F_0*

are

$$d(P_\infty) = d(P_\alpha) = 2(r-1).$$

(vii) *The places of F_1 that are ramified over $\mathbb{F}_q(x_1)$ are exactly the places Q_γ, $\gamma \in \Omega$.* □

An exact description of the successive different is given in the following proposition.

PROPOSITION 7.4.22. (a) *For any $n \geq 1$ we have*

$$\deg \operatorname{Diff}(F_n/F_{n-1}) = 2(r-1)r^{\lfloor \frac{n+1}{2} \rfloor}.$$

(b) *Therefore, the genus $g_n = g(F_n)$ is given by*

$$g(F_n) = \begin{cases} \left(r^{\frac{n+1}{2}} - 1\right)^2, & n \text{ odd}, \\ \left(r^{\frac{n}{2}} - 1\right)\left(r^{\frac{n+2}{2}} - 1\right), & n \text{ even}. \end{cases}$$ □

PROPOSITION 7.4.23. (a) *Let*

$$S := \{P_\alpha \mid \alpha \in \mathbb{F}_q \setminus \Omega\} \subset \mathbb{P}_{F_0}.$$

Then any $P \in S$ splits completely in all extensions F_n/F_0.

(b) *Therefore, the number of rational places $N_n = N(F_n)$ satisfies*

$$N(F_n) \geq r^n(r^2 - r).$$ □

Clearly, formulae of Propositions 7.4.22(b) and 7.4.23(b) together imply the optimality of the tower, since

$$\frac{N(F_n)}{g(F_n)} \geq \frac{r^{n+1}(r-1)}{g(F_n)} \to r-1 \quad \text{as} \quad n \to \infty.$$

Hereinafter in this section we assume that $r = 2^m$, $m \geq 2$. Let us consider algebraic geometry codes

$$C_a(X_n) := (X_n, \mathcal{P}, aP_\infty^{(n)})_L, \quad a \geq 2g_n - 1,$$

where $\mathbb{F}_q(X_n) = F_n$,

$$\mathcal{P} = \bigcup_\theta \{P_\theta^{(n)}, \theta \in \mathbb{F}_q \setminus \Omega\} = S^{(n)},$$

$P_\theta^{(n)}$ being the set of places (which are rational by Proposition 7.4.23(a)) of F_n lying over P_θ, $\theta \in \mathbb{F}_q \setminus \Omega$, and $P_\infty^{(n)}$ being a unique place of F_n lying over $P_\infty \in \mathbb{P}_{F_1}$. The main result of these codes is as follows.

THEOREM 7.4.24. *Let $r = 2^s \geq 4$. Then the generator matrix of $C_a(X_n)$ can be computed using no more than $n^3N^3/(r-1)^2$ operations in $\mathbb{F}_q$, where $N = r^n(r^2 - r)$ is the length of the code.* □

COROLLARY 7.4.25. *If $n^3 < (r-1)^2$ (which is the only case of practical interest for a reasonably large q), then this complexity is less than N^3.* □

Now let us briefly describe some structures and results which allow one to construct such an algorithm.

Define

$$R_n = \text{integral closure of } F_0 \text{ in } F_n.$$

Then R_n is the ring of functions that have pole only at $P_\infty^{(n)}$ and has an alternative description:

$$R_n = \bigcup_{a \geq 0} L\big(aP_\infty^{(n)}\big).$$

We will refer to elements of R_n as *regular functions.*

Finding a suitable basis in the Riemann–Roch space allowing one to compute a code generator matrix is based on a thorough study of R_n and its structure.

The main step in this study is constructing a specific basis of R_n as an $R_0 = F_0$ module. More precisely, for any n we look for a vector space basis $\{z_i\}$ of F_n over F_0 such that

$$\sum_i R_0 z_i \subseteq R_n \subseteq \sum_i R_0 z_i^*,$$

where $\{z_i^*\}$ is the dual basis, i.e., $\mathrm{Tr}_{F_n/F_0}(z_i z_j^*) = \delta_{ij}$.

For $0 \leq k \leq n$, put

$$u_{k,e} = \begin{cases} x_k^e & \text{if } 0 \leq e < r-1, \\ x_k^{q-1}+1 & \text{if } e = q-1, \end{cases}$$

$$g_k = x_k^{q-1}+1, \qquad \pi_k = g_0 g_1 \dots g_k.$$

PROPOSITION 7.4.26. *The following is a pair of dual bases for F_k/F_{k-1}, $1 \leq k \leq n$:*

$$\{1, x_k, \dots, x_k^{q-1}\} \longleftrightarrow \{u_{k,q-1}, u_{k,q-2}, \dots, u_{k,0}\}. \qquad \square$$

Now let

$$E = \{\mathbf{e} = (e_1, \dots, e_n) \mid 0 \leq e_i \leq q-1 \text{ for all } i\}.$$

We get the following result.

THEOREM 7.4.27. *The n-fold products*

$$Z := \{z_{\mathbf{e}}\} = \left\{ \prod_{j=1}^n \pi_{j-1} u_{j,e_j} \mid (e_1, \dots, e_n) \in E \right\}$$

form a basis for F_n/F_0 having dual $\{z_e^*\}$ given by

$$Z^* := \{z_{\mathbf{e}}^*\} = \left\{ \prod_{j=1}^n \frac{1}{\pi_{j-1}} x_j^{e_j} \mid (e_1, \dots, e_n) \in E \right\}. \qquad \square$$

The next two results allow one to prove Theorem 7.4.24.

PROPOSITION 7.4.28. *Every function in R_n has an expression of the form*

$$x_0^\ell \left(\sum_{e_0=0}^{(n-1)q+1} \sum_{e_1=0}^{q-1} \dots \sum_{e_n=0}^{q-1} c_{\mathbf{e}} g_0 \frac{x_0^{e_0} x_1^{e_1} \dots x_n^{e_n}}{\pi_1 \dots \pi_{n-1}} \right),$$

where $\ell \geq 0$ and $c_{\mathbf{e}} \in \mathbb{F}_q$. $\square$

A pole cancellation procedure allows us to obtain a somewhat more convenient expression. Let E_1 and E_2 denote, respectively, the set of n-tuples $(e_1, \dots, e_n) \in E$ satisfying

$$\sum_{j=1}^n q^{n-j} e_j + q^n + q^{n-1} + \dots + q \leq 2q^n$$

and its complement in E. An alternative expression for the functions in R_n is as follows.

PROPOSITION 7.4.29. *Every function in R_n has an expression of the form*

$$x_0^\ell\left(\sum_{e_0=0}^{(n-1)q+1}\sum_{e\in E_1} c_e g_0 \frac{x_0^{e_0}x_1^{e_1}\dots x_n^{e_n}}{\pi_1\dots\pi_{n-1}}\right)+x_0^\ell\left(\sum_{e_0=0}^{(n-1)q+1}\sum_{e\in E_2} c_e g_0^2 \frac{x_0^{e_0}x_1^{e_1}\dots x_n^{e_n}}{\pi_1\dots\pi_{n-1}}\right),$$

where $\ell\geq 0$ and $c_e\in\mathbb{F}_q$. □

The main point following from this description of R_n is the fact that all the functions involved in the computation of a generator matrix of $C_a(X_n)$ are regular outside $P_\infty^{(n)}$, and the minimum order of poles of those functions at $P_\infty^{(n)}$ is not greater than

$$\nu_{\max}=r^{n+1}-r^{\lfloor n/2\rfloor+1}+r^n.$$

Clearly, this gives a decisive advantage against the algorithm (say) of Sec. 5.1, since in that algorithm one needs to keep tract of pole orders at all the leaves of the desingularization tree rather than at the single place $P_\infty^{(n)}$ here.

Then a minute analysis gives the desired complexity estimate of Theorem 7.4.24.

7.4.4. A Very Good Tower for $q=p^{2m+1}$, $m\geq 1$

Let us now touch upon a very good tower over $\mathbb{F}_{p^{2m+1}}$, which comes rather close to the Drinfeld–Vlăduţ bound for small p and large m. The tower is also especially interesting, since it can be interpreted in terms of Drinfeld modules of higher ranks.

We begin with the main result on this tower.

THEOREM 7.4.30. *Let q be a prime power. For any integer $n>1$ and $1\leq k<n$ with $\gcd(k,n-k)=1$ there exists a tower $\mathcal{N}=(N_1\subset N_2\subset\dots)$ over $\mathbb{F}_\ell$, where $\ell=q^n$, such that*

(i) *$N_1=\mathbb{F}_\ell(z_1)$ is a rational function field.*
(ii) *For each $i\geq 2$, the extension N_i/N_1 is a Galois extension having as Galois group an extension of a subgroup of $\mathrm{GL}_{n-1}(\mathbb{F}_q)$ by a p-group. The extension N_i/N_2 is a p-extension.*
(iii) *The place $[z_1=-1]$ of N_1 splits completely in $\mathcal{N}$, i.e., it splits completely in each extension N_i/N_1.*
(iv) *The only places of N_1 which are ramified in $\mathcal{N}$ are $P_0:=[z_1=0]$ and $P_\infty:=[z_1=\infty]$, and they are weakly ramified, i.e., their second ramification groups are trivial.*
(v) *For each $i>1$, the extension N_i/N_2 is 2-bounded; more precisely, for any place P of N_2 the ramification index $e(P)$ and different exponent $d(P)$ of P in the extension N_i/N_2 satisfy*

$$d(P)=2(e(P)-1).$$

(vi) *Let $e_i(P_0)$ and $e_i(P_\infty)$ denote the ramification indices in the extension N_i/N_1 of the places P_0 and P_∞, respectively, and assume that $i>1$. We have*

$$e_i(P_0)=(q^k-1)q^{(i-1)(n-k)-k}p^{\varepsilon_1(i)}$$

and

$$e_i(P_\infty) = (q^{n-k}-1)q^{(i-1)(n-k)}p^{\varepsilon_2(i)},$$

with $\varepsilon_1(i), \varepsilon_2(i) \geq 0$.

(vii) *The limit of the tower satisfies*

$$\lambda(\mathcal{N}) \geq 2\left(\frac{1}{q^k-1}+\frac{1}{q^{n-k}-1}\right)^{-1}. \qquad \square$$

Choosing $k=m$ and $n-k=m+1$, we obtain the following bound.

COROLLARY 7.4.31. *We have*

$$A(p^{2m+1}) \geq \frac{2(p^{m+1}-1)}{p+1+\varepsilon}$$

with $\varepsilon = \dfrac{p-1}{p^m-1}$. $\square$

Note that the Drinfeld–Vlăduţ bound reads in this case as

$$A(p^{2m+1}) \leq p^m\sqrt{p}-1.$$

For a fixed p and $m \to \infty$, the ratio of the bound in Corollary 7.4.31 to the Drinfeld–Vlăduţ bound tends to $\frac{2\sqrt{p}}{p+1}$; in particular, for $p=2$ we get

$$\frac{2\sqrt{p}}{p+1} = \frac{2\sqrt{2}}{3} \approx 0.9428,$$

getting very close to 1.

In fact, the tower of Theorem 7.4.30 is the Galois closure of the following recursive tower.

Let $n = 2m+1 = j+k$ with $k \geq 1$, $j = n-k \geq 1$, and coprime k and j: $\gcd(k,j)=1$, and let Tr_a for any $a \geq 1$ be the trace

$$\mathrm{Tr}_a(s) = \mathrm{Tr}_{\mathbb{F}_{q^a}/\mathbb{F}_q}(s) = \sum_{i=0}^{a-1} s^{q^i}$$

from $\mathbb{F}_{q^a}$ to $\mathbb{F}_q$.

Define the (rational) function

$$f(X,Y) := \mathrm{Tr}_j\left(\frac{Y}{X^{q^k}}\right) + \mathrm{Tr}_k\left(\frac{Y^{q^j}}{X}\right) - 1$$

and consider the recursive tower $\mathcal{F} = (F_0, F_1, \ldots)$ defined by $f(X,Y)$; note that here the equation $f(X,Y)=0$ cannot be written in a separated form $g(X)=h(Y)$ anymore. This tower, however, also satisfies the recursive equation

$$\frac{Y^{q^n}-Y}{Y^{q^j}} = \frac{X^{q^n}-X}{X^{q^n-q^k+1}},$$

but this last equation is reducible.

For this tower one can prove the following result.

THEOREM 7.4.32. *The recursive tower* $\mathcal{F}$ *satisfies*

$$\lambda(\mathcal{F}) \geq 2\left(\frac{1}{q^j-1}+\frac{1}{q^k-1}\right)^{-1}. \qquad \square$$

Note that the parameter n in the construction of the tower $\mathcal{F}$ of Theorem 7.4.32 actually need not be odd; for instance, for $k=j=1$, i.e., $n=2$, we obtain another description of the $\mathcal{W}_2$ tower.

The proof of Theorem 7.4.32 uses a far-reaching development of the methods of Sec. 7.4, i.e., the theory of recursive towers.

Another very interesting feature of the tower $\mathcal{F}$ is its relation to Drinfeld modules of higher (>2) ranks. More precisely, let us consider the following subtower $\mathcal{H} \subseteq \mathcal{F}$, $\mathcal{H} = (H_1, H_2, \ldots)$, where $H_i = \mathbb{F}_{p^n}(z_1, \ldots, z_i)$ with

$$z_i = -x_i^{q^n - 1}$$

for $F_i = \mathbb{F}_{p^n}(x_i, x_{i+1})$ in its recursive representation.

The one can show that H_i satisfies the separated (reducible) recursion

$$\frac{(Y+1)^{N_n}}{Y^{N_j}} = \frac{(X+1)^{N_n}}{X^{q^k N_j}}$$

for $N_r = \theta_{r-1} = \dfrac{q^r - 1}{q-1}$ $(r \geq 1)$. The last equation can be rewritten in the form

$$\mathrm{J}(\varphi) = \mathrm{J}(\psi)$$

for two normalized n-rank Drinfeld modules over $\mathbb{F}_{p^n}[T]$ with

$$\begin{aligned} \varphi(T) &= -\tau^n + g\tau^j + 1, \\ \psi(T) &= -\tau^n + h\tau^j + 1, \end{aligned}$$

which are related by an isogeny of the form $\tau^k - X^{q^k - 1}$, where $\mathrm{J}(\varphi) := g^{N_n}$ and $\mathrm{J}(\psi) := h^{N_n}$.

Thus, the equation defining the tower generalizes those of Sec. 5.3.2 and is related to modular varieties for higher-rank (>2) Drinfeld modules similarly to the Elkies conjecture which relates the Drinfeld modular curves (which are modular varieties of rank-2 modules) with optimal towers.

7.4.5. Good Recursive Towers over $\mathbb{F}_q$, $q \geq 4$

All of the above constructions in the present section do not concern prime fields. However, good recursive towers over prime fields $\mathbb{F}_p$ do exist for $p \geq 5$.

Let us first reformulate the definition of a recursive tower in geometric (rather than algebraic) terms, those of fiber products. A recursive tower $\mathcal{T} = (X_m)_{m \in \mathbb{N}}$ can be described by two morphisms $f, g \colon C_0 \to C_{-1}$, where C_0 and C_{-1} are curves defined over $\mathbb{F}_q$. This recursively defines $(C_m)_{m \geq 0}$ by the fiber product

$$\begin{array}{ccc} C_{m+1} & \longrightarrow & C_0 \\ {\scriptstyle \pi_m}\big\downarrow & & \big\downarrow{\scriptstyle g} \\ C_m & \xrightarrow[f_m]{} & C_{-1} \end{array}$$

with $f_m(P_0, \ldots, P_m) = f(P_m)$ and $\pi_m(P_0, \ldots, P_{m+1}) = (P_0, \ldots, P_m)$. In other terms, we have

$$C_m = \left\{(P_0, \ldots, P_m) \in C_0^{m+1} : g(P_i) = f(P_{i-1}),\, 1 \leq i \leq m\right\};$$

we denote by X_m the normalization of C_m and write $\pi_m \colon X_{m+1} \to X_m$ for the induced cover.

The idea of the construction is as follows. Starting from a good choice of a cover $f\colon C_0 \to C_{-1}$, construct the morphism g by carefully choosing an automorphism ψ (respectively, φ) of C_0 (respectively, C_{-1}) and define $g = \varphi \circ f \circ \psi$. More precisely, let $\mathcal{R}$ be a subset of $C_0(\overline{\mathbb{F}}_q)$ containing the ramification points of f, and $\mathcal{S} \subset C_0(\mathbb{F}_q)$ be the inverse image under f of the set of totally split points of f. We assume that there exists an automorphism ψ of C_0 (respectively, φ of C_{-1}) preserving $\mathcal{R}$ and $\mathcal{S}$ (respectively, $f(\mathcal{R})$ and $f(\mathcal{S})$). If we can guarantee that f,g define a tower $\mathcal{T} = (X_m)_{m\in\mathbb{N}}$, then $\mathcal{R}$ (respectively, $\mathcal{S}$) contains the ramification locus (respectively, splitting locus) of the tower, and the results of Sec. 7.4.1 give that $\lambda(\mathcal{T}) \geq \dfrac{2|\mathcal{S}|}{|\mathcal{R}|-2}$.

In fact we use a particular cyclic Galois cover $f\colon \mathbb{P}^1 \to \mathbb{P}^1$ of degree $q+1$ given by the following construction. The ramification points $Q, \overline{Q}$ of f are defined over $\mathbb{F}_{q^2}$, whereas the point at infinity is totally split, i.e., $f^{-1}(\infty) = \mathbb{P}^1(\mathbb{F}_q)$. We let $\mathcal{R} = \mathbb{P}^1(\mathbb{F}_{q^2}) \setminus \mathbb{P}^1(\mathbb{F}_q)$ and $\mathcal{S} = \mathbb{P}^1(\mathbb{F}_q)$; then there exist automorphisms ψ and φ preserving these sets and their image under f. Moreover, when $q > 3$, we can impose that $g(Q) = Q$ and $Q, \overline{Q}$ are not ramification points of g. This implies that the curves X_m are absolutely irreducible and are of growing genus; hence, $\mathcal{T}$ is a good tower.

More precisely, let $q = p^n$ be a prime power, and for $x \in \mathbb{F}_{q^2}$ denote by $\overline{x}$ its Galois conjugate over $\mathbb{F}_q$. Let $Q = (\theta : 1) \in \mathbb{P}^1(\mathbb{F}_{q^2}) \setminus \mathbb{P}^1(\mathbb{F}_q)$. The automorphism group of $\mathbb{P}^1$ over $\mathbb{F}_q$ is $\mathrm{PGL}_2(\mathbb{F}_q)$, the isotropy group G of Q is a cyclic subgroup of $\mathrm{PGL}_2(\mathbb{F}_q)$ of order $q+1$, and the point $\overline{Q} = (\overline{\theta} : 1)$ has the same isotropy group G.

The group G defines a tame Galois cover $f\colon \mathbb{P}^1 \to \mathbb{P}^1$ of degree $q+1$, where Q and $\overline{Q}$ are totally ramified and, by the Hurwitz formula, are the only ramified points; thus, G acts transitively on $\mathbb{P}^1(\mathbb{F}_q)$. We choose a coordinate x on $\mathbb{P}^1$ such that $f(Q) = Q$, $f(\overline{Q}) = \overline{Q}$, and $f(\infty) = \infty$, where $x(\infty) = 0$. We then have $f^{-1}(\infty) = \mathbb{P}^1(\mathbb{F}_p)$, and therefore ∞ is totally split, whereas Q and $\overline{Q}$ are the two branched points.

A simple direct calculation proves the following fact.

LEMMA 7.4.33. *We have*

$$f(\mathcal{R}) = \{\gamma : \gamma \in \mathbb{F}_{q^2} \text{ with } \mathrm{Tr}_{\mathbb{F}_{q^2}/\mathbb{F}_q}(\gamma) = \mathrm{Tr}_{\mathbb{F}_{q^2}/\mathbb{F}_q}(\theta)\}. \qquad \square$$

Since an automorphism $x \mapsto cx + d$ preserves

$$f(\mathcal{R}) = \{\gamma : \gamma \in \mathbb{F}_{q^2} \text{ with } \mathrm{Tr}_{\mathbb{F}_{q^2}/\mathbb{F}_q}(\gamma) = a\}$$

if and only if $\mathrm{Tr}_{\mathbb{F}_{q^2}/\mathbb{F}_q}(c\gamma + d) = a$, i.e., $(1-c)a = 2d$, we get the following.

PROPOSITION 7.4.34. *An automorphism $\varphi\colon \mathbb{P}^1 \to \mathbb{P}^1$ of the form $x \mapsto cx + d$ and defined over $\mathbb{F}_q$ preserves $f(\mathcal{R})$ if and only if $(1-c)a = 2d$ with $c \neq 0$ and $a = \mathrm{Tr}_{\mathbb{F}_{q^2}/\mathbb{F}_q}(\theta)$.* $\square$

Now let $g = \varphi \circ f \circ \psi$ for a choice of an automorphism ψ (respectively, φ) defined over $\mathbb{F}_q$ and preserving $\mathcal{R}$ and $\mathcal{S}$ (respectively, $f(\mathcal{R})$ and $f(\mathcal{S})$). Consider the correspondence $f, g\colon \mathbb{P}^1 \to \mathbb{P}^1$ and the curves $(X_m)_{m\in\mathbb{N}}$ that they define.

The tower is asymptotically good.

THEOREM 7.4.35. *Assume that $q > 3$. Then for $\mathcal{T} = (X_m)_{m\in\mathbb{N}}$ over $\mathbb{F}_q$ we have*

$$\lambda(\mathcal{T}) \geq \frac{2}{q-2}.$$

SKETCH OF THE PROOF. First we get that $\mathcal{T} = (X_m)_{m\in\mathbb{N}}$ is indeed a tower, since all the $\pi_{m+1}\colon X_{m+1} \to X_m$ are ramified, because $g(Q) = Q$ and Q is not a ramification point for g, i.e., $\varphi(Q) \notin \{Q, \overline{Q}\}$. For $q > 3$, there always exist $c, d \in \mathbb{F}_q$ with $c \neq 0$ and $(1-c)a = 2d$ such that $\varphi(x) = cx + d$ does not map Q on itself ($c = 1$, $d = 0$) or on $\overline{Q}$ ($c = -1$, $d = a$). Picking such (c,d) and the corresponding φ and setting $S = \varphi^{-1}(Q)$, we see that $f^{-1}(S) \subseteq \mathcal{R}$, since φ stabilizes $f(\mathcal{R})$ and $Q \in f(\mathcal{R})$. Then pick an automorphism ψ defined over $\mathbb{F}_q$ to ensure $g(Q) = Q$, and by Sec. 7.4.1 we get

$$\lambda(\mathcal{T}) \geq \frac{2|\mathcal{S}|}{|\mathcal{R}| - 2} = \frac{2}{q-2}. \qquad \square$$

Note that for $q = 4$ the tower is optimal, whereas for $q = 5$ it is rather good; however, the function $\dfrac{2}{q-2}$ decreases with q, and for large q the bound becomes worse than $\dfrac{1}{96}\log q$ obtained from class field towers.

One can check that for $q \geq 7$ the choice

$$f(x) = \frac{x^{q+1} + b}{x^q - x} \quad \text{and} \quad g(x) = \frac{2b(x^{q+1} + n)}{(b+n)(x^q - x)}$$

with $-n, -b \in \mathbb{F}_q^*$ two nonsquares, $n \neq \pm b$, is good for the above construction, whereas for $q = 5$ one can take

$$f(x) = \frac{x^6 + x + 2}{x^5 - x} \quad \text{and} \quad g(x) = \frac{x^6 + x^5 + 2x + 3}{x^5 - x}.$$

7.5. Two Nonstandard Problems

In the preceding sections, we considered the problem of finding good curves in its "standard" setting: find curves with as many rational points as possible under certain conditions on the genus (which can either be fixed, or tend to infinity, etc.). However, there are other possible settings for the problem of curves with many points, and in the present section we briefly discuss two less standard problems where we rather have to choose the genus g in compliance with some conditions on the number N of rational points. In general, this kind of question is very difficult. In its most general (and definitely hopeless) form, it is to characterize all the triples (q,g,N) for which there exists a curve $X/\mathbb{F}_q$ of genus g with $N(X)=N$. However, there are two highly nontrivial results in this direction, which we would like to expose.

The first is whether for a given N there exists a curve C of genus g over $\mathbb{F}_q$ with $|C(\mathbb{F}_q)|=N$; clearly, the question only makes sense for triples (q,g,N) satisfying the consequences of Serre–Weil and Oesterlé bounds.

The second question is of asymptotic nature and is exactly of whether there are curves over $\mathbb{F}_q$ of every genus with many rational points.

7.5.1. Curves with a Prescribed Number of Points

To formulate the result, it is convenient to consider the set

$$\mathcal{G}(q,N) := \{g \mid \text{there exists a curve } X \text{ over } \mathbb{F}_q \text{ of genus } g \text{ with } N(X)=N\}.$$

Thus, we ask for what values of g a curve over $\mathbb{F}_q$ of genus g with N rational points exists. A necessary condition for a number g to be in $\mathcal{G}(q,N)$ is

$$g \geq \frac{|N-(q+1)|}{\lfloor 2\sqrt{q}\rfloor},$$

as follows from the Serre bound. However, we do not get an upper bound for the set $\mathcal{G}(q,N)$; moreover, we have the following result.

THEOREM 7.5.1. *For any finite field $\mathbb{F}_q$ and any integer $N\geq 0$ there is an integer $g_0\geq 0$ such that for every $g\geq g_0$ there exists a curve X over $\mathbb{F}_q$ of genus $g(X)=g$ having exactly N rational points.*

In other words, the set $\mathbb{N}\setminus\mathcal{G}(q,N)$ is finite for all q and N.

To prove the theorem, we recall two basic facts.

LEMMA 7.5.2. *Given a function field $E/\mathbb{F}_q$, there is a constant $C\geq 0$ such that for every integer $k\geq C$ there exists a place P of E with $\deg P=k$.* □

LEMMA 7.5.3. *Let $E/\mathbb{F}_q$ be a function field and $z\in E\setminus\mathbb{F}_q$ be a nonconstant element with pole divisor $(z)_\infty = r_1Q_1+\ldots+r_sQ_s$. Assume that $r_1,\ldots,r_s$ are relatively prime to q. Let $F=E(y)$, where y satisfies the equation*

$$y^q - y = z.$$

Then we have the following:

(i) *The extension F/E is Galois of degree $[F:E]=q$.*
(ii) *The places $Q_1,\ldots,Q_s$ are totally ramified in F/E, and all other places of E are unramified in F/E.*

(iii) *The different exponent of* Q_j *in the different* $\mathfrak{D}_{F/E}$ *of* F/E *is* $d(Q_j) = (r_j+1)(q-1)$, *for* $1 \le j \le s$. *Hence, we have*

$$\deg \mathfrak{D}_{F/E} = \sum_{j=1}^{s} (r_j+1)(q-1)\deg Q_j.$$

(iv) *Assume that* Q *is a rational place of* $E/\mathbb{F}_q$ *which is a zero of* z. *Then* Q *splits completely in* F/E; *i.e., there are* q *rational places of* F *lying above* Q.

(v) *Assume that* Q *is a rational place of* $E/\mathbb{F}_q$ *which is a zero of* $z-a$ *for* $a \in \mathbb{F}_q \setminus \{0\}$. *Then* Q *is inert in* F/E; *i.e., the places of* F *lying above* Q *are not rational.* □

We also need the following lemma.

LEMMA 7.5.4. *For any nonnegative integers* j *and* N *there exists a function field* $E/\mathbb{F}_q$ *with* $g(E) \equiv j \bmod (q-1)$ *and* $N(E) \ge N$.

PROOF. Start with a quadratic extension $E_0/\mathbb{F}_q(x)$ of the rational function field $\mathbb{F}_q(x)$ with $g(E_0) = j$ and $N(E_0) \ge 2$ (it is an easy exercise to show that such a field E_0 exists). We proceed by induction over N.

For $N \le 2$ we take $E := E_0$, and we are done. Now assume that $N > 2$. By the induction hypothesis there exists a function field $H/\mathbb{F}_q$ with

$$g(H) \equiv j \mod (q-1) \qquad \text{and} \qquad N(H) \ge N-1.$$

Choose rational places $P_1, \dots, P_{N-1}$ of H and a place $P \in \mathbb{P}_H$ with $\deg P \ge g(H) + N - 1$, which is possible by Lemma 7.5.2. Since $\deg(P - (P_1 + \ldots + P_{N-1})) \ge g(H)$, the Riemann–Roch theorem implies that $\ell(P - (P_1 + \ldots + P_{N-1})) > 0$. Next, choose a nonzero element $z \in L(P - (P_1 + \ldots + P_{N-1}))$; then P is the only pole of z, and it is a simple pole. Moreover, the places $P_1, \dots, P_{N-1}$ are zeroes of z. Consider the extension $E := H(y)$ with $y^q - y = z$. By Lemma 7.5.3, E/H is Galois of degree $[E:H] = q$, the full constant field of E is $\mathbb{F}_q$, the place P is the only place of H which is ramified in E/H, and the degree of the different of E/H is $2(q-1)\deg P$; therefore,

$$2g(E) - 2 = q(2g(H)-2) + 2(q-1)\deg P,$$

and hence, $g(E) \equiv g(H) \equiv j \bmod (q-1)$. The places $P_1, \dots, P_{N-1}$ split completely in the extension E/H by Lemma 7.5.3. Therefore, $N(E) \ge q(N-1) > N$, since $N > 2$. □

PROOF OF THEOREM 7.5.1. Let q and N be given. In order to prove the theorem, it is sufficient to show that for every $j = 0, \dots, q-2$ there is a constant $g_0^{(j)}$ such that for every $g \ge g_0^{(j)}$ with $g \equiv j \bmod (q-1)$ there exists a function field $F/\mathbb{F}_q$ with $g(F) = g$ and $N(F) = N$.

First we choose (by Lemma 7.5.4) a function field $E/\mathbb{F}_q$ with

$$g(E) \equiv j \mod (q-1) \qquad \text{and} \qquad N(E) \ge N.$$

By Lemma 7.5.2 there is an integer $C \ge 0$ such that we have the following:

($*$) $C > 2g(E) + N(E) - N$ and

($**$) for all $t \ge C$ there is a place P of E with $\deg P = t$.

We set

$$g_0^{(j)} := g(E) + (q-1)(g(E) - 1 + C + N).$$

Clearly, $g_0^{(j)}$ satisfies the condition $g_0^{(j)} \equiv g(E) \equiv j \bmod (q-1)$. Let $g \geq g_0^{(j)}$ be an integer such that $g \equiv j \bmod (q-1)$, say $g = g_0^{(j)} + r(q-1)$. Denote by $P_1, \ldots, P_N$ and $Q_1, \ldots, Q_s$ all rational places of E, so that $s = N(E) - N$. Choose a place P of E with $\deg P = C + r$ (which is possible by $(**)$). It follows from $(*)$ that

$$\deg(P - (Q_1 + \ldots + Q_s)) = C + r - (N(E) - N) > 2g(E).$$

By the Riemann–Roch theorem we can find nonzero elements $u, x_1, \ldots, x_N$ with

$$u \in L(P - (Q_1 + \ldots + Q_s)),$$
$$x_i \in L(P + P_i - (Q_1 + \ldots + Q_s)) \setminus L(P - (Q_1 + \ldots + Q_s))$$

for $1 \leq i \leq N$. We set

$$x := \begin{cases} x_1 + \ldots + x_N & \text{if } P \text{ is a pole of } x_1 + \ldots + x_N, \\ x_1 + \ldots + x_N + u & \text{otherwise.} \end{cases}$$

Then the pole divisor of x is

$$(x)_\infty = P + P_1 + \ldots + P_N,$$

and the places $Q_1, \ldots, Q_s$ are zeroes of x. Define $F := E(y)$ with

$$y^q - y = x + 1.$$

Lemma 7.5.3 implies that F/E is a Galois extension of degree $[F : E] = q$, the places $P, P_1, \ldots, P_N$ are totally ramified in F/E with a different exponent $d = 2(q-1)$, all other places of E are unramified in F, and the places $Q_1, \ldots, Q_s$ are inert in F/E. We conclude that $N(F) = N$, and the genus of F satisfies

$$2g(F) - 2 = q(2g(E) - 2) + 2(q-1)(C + r + N);$$

hence,

$$g(F) = g(E) + (q-1)(g(E) - 1 + C + N) + r(q-1) = g_0^{(j)} + r(q-1) = g. \qquad \square$$

Thus, we see that the sets

$$\mathcal{N}(q, g) := \{N \in \mathbb{N} \mid \text{there exists a curve } X \text{ over } \mathbb{F}_q \text{ of genus } g \text{ with } N(X) = N\}$$

contain many consecutive integers.

COROLLARY 7.5.5. *Given a prime power q and an integer $N \geq 0$, there exists a constant $g_1 = g_1(q, N) \geq 0$ with the following property: For every integer $g \geq g_1$ and every $S \in \{0, \ldots, N\}$ there is a curve X over $\mathbb{F}_q$ with $g(X) = g$ having exactly S rational points.* $\square$

REMARK 7.5.6. The above proof of Theorem 7.5.1 yields an explicit bound on g_0, but it is worse than the bound following from asymptotically good towers.

7.5.2. Curves for Every Genus

Now we are interested in the quantity

$$A^-(q) := \liminf_{g \to \infty} \frac{N_q(g)}{g},$$

which translates the question of whether *there are curves with many points for every given genus*. Note that it is not clear at all whether this quantity is positive, i.e., whether there are curves of every genus with the number of points growing linearly in g.

However, this is true; moreover, $A^-(q)$ and $A(q)$ are related in a simple manner.

THEOREM 7.5.7. *We have*

$$A^-(q) \geq A(q)/4,$$

and thus

(i) *For any prime power* q,
$$A^-(q) \geq d \log q$$
for an absolute constant $d > 0$.

(ii) *For any square prime power* q,
$$A^-(q) \geq (\sqrt{q}-1)/4.$$

The proof is based on the following result.

PROPOSITION 7.5.8. *Let* C *be a curve of genus* g *over* $\mathbb{F}_q$, *and let* $g \geq 3$. *For every integer* $h \geq 4g$, *there exists a genus* h *curve* $B/\mathbb{F}_q$ *that admits a degree* 2 *covering map* $B \to C$ *defined over* $\mathbb{F}_q$.

PROOF. The assertion is clear when $g = 0$, so we assume that $g \geq 1$.

First assume that q is even. If $|C(\mathbb{F}_q)| > 0$, then let R be a rational point of C. Set $m = 2h - 4g + 1$; by the Riemann–Roch theorem, there exists a function f in $\mathbb{F}_q(C)$ with $(f)_\infty = mR$. Let $B \to C$ be the cover of C obtained by adjoining a root of $X^2 + X + f$ to $\mathbb{F}_q(C)$; this cover has degree two, and its different is $(m+1)R'$, where R' is the point of B lying over R. By the Hurwitz formula, the genus of B is h, and we are done. We are left with the case where $|C(\mathbb{F}_q)| = 0$. Since every genus 1 curve over a finite field has a rational point, we must have $g \geq 2$. Then an easy counting argument implies the existence of a place P of degree $d := h - 2g + 1$. By the Riemann–Roch theorem, there exists $f \in \mathbb{F}_q(C)$ with $(f)_\infty = P$. If we form B as before, the different of the cover $B \to C$ is $2P'$, where P' is the place of B lying above P. Again the genus of B is h, and we are done.

Now assume that q is odd. Let $J(\mathbb{F}_q)$ be the group of $\mathbb{F}_q$-rational points on the Jacobian of C. Let $d = h - 2g + 1$, and let T_d denote the set of places on C of degree d. Consider the map $T_d \to J(\mathbb{F}_q)/2J(\mathbb{F}_q)$ defined by $P \mapsto [P - P_0]$, where P_0 is any fixed degree d divisor on C. Assume that that $|T_d| > |J(\mathbb{F}_q)/2J(\mathbb{F}_q)|$. Then there are distinct places $P, P' \in T_d$ with the same image in $J(\mathbb{F}_q)/2J(\mathbb{F}_q)$, so there is a function f on C with $(f) = P - P' + 2D$ for some divisor D. Adjoining $\sqrt{f}$ to the function field of C produces a degree 2 cover $B \to C$ over $\mathbb{F}_q$, with the genus of B equal to $2g + d - 1 = h$. Thus, we need only to show that $|T_d| > |J(\mathbb{F}_q)/2J(\mathbb{F}_q)|$. But $|J(\mathbb{F}_q)/2J(\mathbb{F}_q)|$ is equal to the number of 2-torsion points in $J(\mathbb{F}_q)$, and this is at most 2^{2g}.

Now one only needs to note that if q is odd and $d > 2g$, then $B_d(C) \geq 2^{2g}$, with equality if and only if $q = d = 3$, $g = 1$, and $|C(\mathbb{F}_3)| = 6$, which is a simple consequence of the formulas

$$dB_d(C) = \sum_{m|d} \mu(d/m)|C(\mathbb{F}_{q^m})|$$

and

$$dB_d(C) \geq q^d + 1 - 2gq^{d/2} - \sum_{\substack{m|d \\ 1 \leq m < d}} (q^m + 1 + 2gq^{m/2}),$$

following from the Weil bound. □

COROLLARY 7.5.9. *Let C be a curve over $\mathbb{F}_q$ of genus g. For every integer $h \geq 4g$, we have $N_q(h) \geq |C(\mathbb{F}_q)|$.*

PROOF. By Proposition 7.5.8, there is a genus h curve B over $\mathbb{F}_q$ that admits a degree 2 covering map $B \to C$ over $\mathbb{F}_q$. The result follows, since either B or its quadratic twist over C has at least as many $\mathbb{F}_q$-rational points as does C. □

COROLLARY 7.5.10. *If $h \geq 4g$, then $N_q(h) \geq N_q(g)$.* □

This corollary implies Theorem 7.5.7. Indeed, using Theorem 6.4.8, we obtain Theorem 7.5.7(i) with $d := c/4$, and using the equality $A(q) = \sqrt{q} - 1$ for a square q, we get Theorem 7.5.7(ii).

REMARK 7.5.11. In fact, using more elaborate arguments, one can prove that

$$A^-(q) \geq \begin{cases} \dfrac{\sqrt{q}-1}{2 + (\log 2/\log q)} & \text{if } q \text{ is an even square,} \\ \dfrac{\sqrt{q}-1}{2 + (\log 4/\log q)} & \text{if } q \text{ is an odd square,} \end{cases}$$

thus almost doubling the above bound in the square case.

Historical and Bibliographic Notes

The original proof of the bound by J. Oesterlé was never published; in Sec. 7.1 we follow [**Han95**, **Sch90**].

The general Deligne–Lusztig varieties varieties were introduced in [**DL76**]. Afterwards, their properties were widely investigated. In Secs. 7.2.1–7.2.3 we follow [**HS90**, **Ped92**, **HP93**]. The description of ${}^2B_2(q)$ in Lemmas 7.2.13 and 7.2.14 follows [**GKT06**]. The results of Sec. 7.2.4 are borrowed from [**Lau99**].

Constructions for curves of small genera with many rational points are extremely diversified. One can list the following approaches:

I. Methods from general class field theory.
II. Methods from class field theory based on Drinfeld modules of rank 1.
III. Fibre products of Artin–Schreier and Kummer curves.
IV. Towers of curves with many points.
V. Miscellaneous methods such as:
   (1) formulae for $N_q(1)$ and $N_q(2)$;
   (2) explicit curves, e.g., Hermitian curves, Klein's quartic, Artin–Schreier curves, Kummer extensions, complete intersections, or curves obtained by computer search;
   (3) elliptic modular curves $X(n)$ associated to the full congruence subgroups $\Gamma(n)$;
   (4) quotients of curves with many points.

All these methods are used in the online table of the best known curves maintained at `manypoints.org` by G. van der Geer, E.W. Howe, K.E. Lauter, and C. Ritzentaler, which contains all the values for $g \le 50$ and $q = 2^a$, $a \in [1,7]$; $q = p^b$, $3 \le p \le 19$, $b \in [1,5]$; $q = p$ prime, $3 \le q \le 97$. For more details, see [**vdGvdV00**].

In Sec. 7.3 we describe a tiny part of those methods. The ray class field technique (the "simple construction" of Sec. 7.3.1) and the corresponding examples are borrowed from [**Lau96**]; the results and examples concerning rank-1 Drinfeld modules are contained in [**NX97**].

Section 7.3.2 is based on [**ÖT07**, **vdG09**]. Some examples listed in Sec. 7.3.2 were previously obtained by other methods; see references in [**NX97**]. If in an example in this section we write "this is the best known value", this means that it corresponds to an entry in the table from `manypoints.org` at the time of writing. If we write "this is optimal", this means that the corresponding entry at `manypoints.org` has a proof of its optimality (which can be well outside the scope of the present book). Section 7.3.3 is based on [**RI10**].

The initial idea of G.L. Feng (see [**FR95**]) to use recursive tower constructions was developed by R. Pellikaan, who proposed a tower over $\mathbb{F}_4$ defined recursively by $x_{j+1}^2 x_j + x_j^2 + x_{j+1} = 0$, $j = 1, 2, \ldots$, which has many $\mathbb{F}_4$-rational points, but was unable to compute the genus. The last question was answered by A. García and H. Stichtenoth in [**GS95**], containing the optimal tower $\mathcal{W}_2$; the $\mathcal{W}_1$ tower appeared in [**GS96b**]. For the $\mathcal{T}_1$ and $\mathcal{T}_2$ towers, see also [**GS96b**]. For Abhyankar's lemma, see [**Sti09**, Theorem 3.9.1].

The theory of recursive towers was developed in [**GS05**, **GS07**]. Theorem 7.4.19 is proved in [**LMSE02**]. The Elkies conjecture was proposed in [**Elk01a**, **Elk01b**] and confirmed by examples in [**LMSE02**]. The results of Sec. 7.4.3 are contained in [**SAK$^+$01**]. The towers of Sec. 7.4.4 were constructed in [**BBGS14**, **BBGS15**]. In Sec. 7.4.5 we follow [**BR18**].

The results of Sec. 7.5.1 are borrowed from [**Sti11**], and those of Sec. 7.5.2, from [**EHK$^+$04**].

CHAPTER 8

Infinite Global Fields

This chapter is devoted to a theory which was initially motivated by application to algebraic geometry codes (see Chapter 4) and sphere packings (Chapter 10 below), but can be of independent value. In fact, it is of a somewhat different nature than the preceding chapters, which expose well-established results and theories. Here we try to develop an approach to a theory of infinite-degree extensions of global fields. We cannot state that this theory is fully successful, but as far as we are aware, it is the only known technique to treat the extensions that lead to some nontrivial (albeit incomplete) conclusions.

With a global field, i.e., an infinite extension of either $\mathbb{Q}$ or $\mathbb{F}_q(t)$, we can associate a countable set (sequence) of numerical invariants. The fact that these invariants are not always zero is highly nontrivial; however, the Golod–Shafarevich theory gives many examples of such extensions. Moreover, these parameters satisfy some natural inequalities, which can be considered as refinements of asymptotic forms of the Odlyzko–Serre, Stark, and Drinfeld–Vlăduţ bounds; these inequalities, in turn, can be concisely written in terms of a certain limit zeta function. Passing to the limit in Weil's explicit formula gives a limit explicit formula for the zeta function thus constructed. In our opinion, those results, exposed in Sec. 8.1, form just the beginning of a theory of infinite global fields to be developed in the future.

Our approach also allows us to answer the question of the necessity of conditions in the Brauer–Siegel theorem, which was open for several decades. Moreover, it gives quantitative results on the Brauer–Siegel ratio, which is 1 in the classical situation, establishing some bound for its value. These results form the subject of Sec. 8.2. In Sec. 8.3 we present numerous examples when this ratio is different from 1 and varies in some interval around 1. Finally, Sec. 8.4 contains many open questions raised by the theory and essential for its eventual development.

Note also that here we are forced to use some means of analytic number theory which were not mentioned in the preceding chapters. This remark concerns the number field case, the function field one being more elementary.

8.1. Invariants and Basic Inequalities

Recall that a global field K is a finite algebraic extension either of the field $\mathbb{Q}$ of rational numbers or of the field $\mathbf{Q}_r = \mathbb{F}_r(t)$ of rational functions. An *infinite global field* $\mathcal{K}$ is either an infinite algebraic extension of $\mathbb{Q}$ or an infinite algebraic extension of $\mathbf{Q}_r$ such that $\mathcal{K} \cap \overline{\mathbb{F}}_r = \mathbb{F}_r$. In the first case, we call $\mathcal{K}$ an *infinite number field*, and in the second, an *infinite function field over* $\mathbb{F}_r$; note that here the word *infinite* refers to the degree of extension, a usual global field being of infinite cardinality.

We produce a series of invariants, and introduce and study a kind of zeta function of such a field.

Some of our results give generalizations of Odlyzko–Serre bounds, taking into account non-Archimedean places. Let us recall these bounds in the classical setting. Our first results in Sec. 8.1.1 generalize asymptotic versions of these bounds.

To start with, an infinite global field is always the (direct) limit, i.e., a union, of a tower of finite ones:

$$\mathcal{K} = \operatorname{ind}\lim_{i\to\infty} K_i = \bigcup_{i=1}^{\infty} K_i, \quad \text{where} \quad K_1 \subset K_2 \subset K_3 \subset \dots .$$

This tower is, of course, not unique, and we are looking for invariants of $\mathcal{K}$, i.e., for parameters of $K_1 \subset K_2 \subset K_3 \subset \dots$ that do not depend on the tower, but only on its limit.

We use the following notation: for a sequence $\{K_i\}$, $i = 1, 2, \dots$, of global fields, either number or function, we set

$$g_i = \operatorname{genus}(K_i)$$

in the function field case, and

$$g_i = g_{K_i} = \log\sqrt{|D_i|}$$

in the number field case; we call it the *genus* of a number field.

Attention: In this chapter we use the following agreement. In the number field case, the notation log means the natural logarithm $\log_e$. In the function field case over $\mathbb{F}_r$, the same notation log means $\log_r$; as we shall see below, this can be justified by the uniformity of our results.

Let us recall the classical inequality involving the Minkowski constant.

THEOREM 8.1.1 (Minkowski bound). *Let K be a number field, $n = [K:\mathbb{Q}]$, $n = r_1 + 2r_2$, where r_1 is the number of real embeddings and r_2 is the number of pairs of complex embeddings. Then*

$$|D_K|^{1/2} \geq C_K^{-1},$$

where

$$C_K := \frac{n!}{n^n}\left(\frac{4}{\pi}\right)^{r_2}$$

is called the Minkowski constant for the field K. □

This classical result can be strengthened a little, which leads to so-called Odlyzko–Serre bounds; see Theorem 8.1.2 below.

Using this result and the geometry of numbers, one can prove that the set of number fields of bounded discriminant is finite.

One of the reasons to suppose that our definition of genus for the number field case is natural is that for any given g_0 there are only finitely many number fields K whose genus does not exceed g_0. The same is true for function fields with a given constant field.

We call a sequence $\{K_i\}$ of global fields a *family* if K_i is nonisomorphic to K_j for $i \neq j$. A family is called *tower* if also $K_i \subset K_{i+1}$ for any i. In any family $g_i \to \infty$ for $i \to \infty$. In the function field case we always assume that the constant field of all K_i is one and the same field $\mathbb{F}_r$.

In the number field case, let

$$n_i = [K_i : \mathbb{Q}] = r_1(K_i) + 2r_2(K_i),$$

where r_1 and r_2 stand for the numbers of real and (pairs of) complex embeddings. We assume also that $g_i > 0$ for any i, i.e., $K_i \neq \mathbb{Q}$ in the number field case, and K_i is nonisomorphic to $\mathbb{F}_r(T)$ in the function field case. This assumption does not restrict the generality of our considerations.

We consider the set $A = \{\mathbb{R}, \mathbb{C}; 2, 3, 4, 5, 7, 8, 9, \dots\}$ of all prime powers plus two auxiliary symbols $\mathbb{R}$ and $\mathbb{C}$ as the set of indices. The parameters that we are going to present will be indexed by elements $\alpha \in A$. In the function field case over $\mathbb{F}_r$, the set A is reduced to $A_r = \{r, r^2, r^3, \dots\}$, meaning that for $\alpha \in A \setminus A_r$ the parameters vanish.

For a prime power q we set

$$N_q(K_i) := |\{v \in P(K_i) : \mathrm{Norm}(v) = q\}|,$$

where $P(K_i)$ is the set of non-Archimedean places of K_i. We also put $N_{\mathbb{R}}(K_i) = r_1(K_i)$ and $N_{\mathbb{C}}(K_i) = r_2(K_i)$.

By h_i we denote the class number of K_i (which equals the number of $\mathbb{F}_r$-rational points on the Jacobian of K_i in the function field case); R_i denotes the regulator of K_i in the number field case and equals 1 in the function field case.

For an infinite global field $\mathcal{K} = \bigcup K_i$ and all $\alpha \in A$, let us introduce the following quantities:

$$\varphi_\alpha = \varphi_\alpha(\mathcal{K}) := \lim_{i \to \infty} \frac{N_\alpha(K_i)}{g_i}.$$

Of course, we need to prove that these limits exist and do not depend on the tower.

Note that $\varphi_{\mathbb{R}}$ and $\varphi_{\mathbb{C}}$ are finite, if they exist, since the ratio $n/g_i = r_1(K_i)/g_i + 2r_2(K_i)/g_i$ is bounded on the set of all number fields $\neq \mathbb{Q}$ by the Minkowski bound (or Minkowski constant) for number fields, which implies in particular that

$$\sqrt{|D_K|} \geq \left(\frac{\pi}{4}\right)^{n/2} \frac{n^n}{n!}$$

for a number field K of a finite degree $n = [K : \mathbb{Q}]$.

More generally, we call a family $\mathcal{K} = \{K_i\}$, $i = 1, 2, \dots$, of global fields *asymptotically exact* if and only if for any $\alpha \in A$ there exist the limit

$$\varphi_\alpha = \varphi_\alpha(\mathcal{K}) := \lim_{i \to \infty} \frac{N_\alpha(K_i)}{g_i}.$$

We call the family $\mathcal{K}$ *asymptotically good* (respectively, *bad*) if there exists $\alpha \in A$ with $\varphi_\alpha > 0$ (respectively, $\varphi_\alpha = 0$ for any $\alpha \in A$).

It is important to point out that, by abuse of notation, $\mathcal{K}$ is used both for an infinite global field and for an asymptotically exact family. This is reasonable, since below we prove that for an infinite global field all our definitions and results do not depend on the choice of the tower.

The notion of an asymptotically exact family is much more general than that of a tower. In particular, a simple diagonal argument shows that any family contains an asymptotically exact subfamily; namely, we choose a subfamily having φ_{α_1}, then its subfamily having φ_{α_2}, etc., and then take the diagonal subfamily.

The quantities $\varphi = \{\varphi_\alpha\}$ give rise to the following definition. The *limit zeta function* of an asymptotically exact family is defined by the product

$$\zeta_{\mathcal{K}}(s) = \zeta_\varphi(s) = \prod_q (1 - q^{-s})^{-\varphi_q},$$

where q runs over all prime powers. Here and in what follows, raising to a complex power a function in s defined for $\operatorname{Re} s > a \geq 0$ and such that its values are real positive for real $s > a$ is understood to be the unique analytic continuation of its branch which is real positive for real s. The *completed zeta function* is defined in the number field case as

$$\widetilde{\zeta}_{\mathcal{K}}(s) = \widetilde{\zeta}_{\varphi}(s) = e^s 2^{-\varphi_{\mathbb{R}}} \pi^{-s\varphi_{\mathbb{R}}/2} (2\pi)^{-s\varphi_{\mathbb{C}}} \Gamma\left(\frac{s}{2}\right)^{\varphi_{\mathbb{R}}} \Gamma(s)^{\varphi_{\mathbb{C}}} \prod_q (1-q^{-s})^{-\varphi_q}.$$

In the function field case, consistent with our convention, we set

$$\widetilde{\zeta}_{\mathcal{K}}(s) = \widetilde{\zeta}_{\varphi}(s) = r^s \prod_{m=1}^{\infty} (1-r^{-ms})^{-\varphi_{r^m}}.$$

The products defining zeta functions $\zeta_\varphi(s)$ and $\widetilde{\zeta}_\varphi(s)$ converge absolutely for $\operatorname{Re}(s) \geq 1$. These functions depend only on $\varphi = \{\varphi_\alpha\}$ and do not depend on the particular sequence of global fields. Therefore, $\zeta_{\mathcal{K}}(s)$ and $\widetilde{\zeta}_{\mathcal{K}}(s)$ are defined for any infinite global field $\mathcal{K}$.

The zeta function of a family is thus the limit of the gth roots of the usual zeta functions of the fields K_i. Moreover, for $\operatorname{Re} s \geq 1+\varepsilon$ the convergence is uniform.

It is also true that a family is asymptotically exact if and only if the limit $\lim \zeta_{K_i}(s)^{1/g_i}$ exists.

Now let us denote in the number field case

$$\begin{aligned}\xi_{\mathcal{K}}(s) = \xi_\varphi(s) &= (\log \widetilde{\zeta}_\varphi)' = \widetilde{\zeta}'_\varphi / \widetilde{\zeta}_\varphi \\ &= 1 - \frac{\varphi_{\mathbb{R}}}{2} \log \pi - \varphi_{\mathbb{C}} \log 2\pi + \frac{1}{2} \varphi_{\mathbb{R}} \psi\left(\frac{s}{2}\right) + \varphi_{\mathbb{C}} \psi(s) - \sum_q \varphi_q \frac{\log q}{q^s - 1},\end{aligned}$$

where $\psi(s) = \dfrac{\Gamma'}{\Gamma}(s)$, and in the function field case,

$$\xi_{\mathcal{K}}(s) = \xi_\varphi(s) = (\log_r \widetilde{\zeta}_\varphi)' = 1 - \sum_{m=1}^{\infty} \frac{m\varphi_{r^m}}{r^{ms} - 1}.$$

Recall the generalized Riemann hypothesis (GRH).

Generalized Riemann Hypothesis. *Let K be a number field (of a finite degree over $\mathbb{Q}$). Then all the zeroes s of the zeta function $\zeta_K(s)$ in the critical strip $0 < \operatorname{Re}(s) < 1$ have* $\operatorname{Re}(s) = 1/2$.

Studying the number field case, we often assume the GRH to hold for number fields in question, but most of our results also have an unconditional (weaker) formulation. To distinguish between the two, we always write GRH in relevant cases. Note that the function field case does not need it, the GRH being proved.

Now let us recall Odlyzko–Serre bounds, which relate the degree of a number field with its discriminant and improve on Theorem 8.1.1. Some of the results below on invariants of infinite global fields can be considered as refinements of asymptotic versions of these bounds. Note that these bounds improve (considerably) the Minkowski bound mentioned above.

To formulate the bounds, it is convenient to define the *root discriminant* of a number field K with $n = n_K = [K : \mathbb{Q}] < \infty$ by

$$\mathrm{rd} = \mathrm{rd}_K = |D_K|^{1/n} = \exp(2g_K/n).$$

As usual, let $\gamma \approx 0.577\ldots$ denote the Euler constant.

THEOREM 8.1.2 (Odlyzko–Serre bounds). *We have*

1. (*unconditionally*)
$$\mathrm{rd} \geq \left(4\pi e^{1+\gamma}\right)^{r_1/n}\left(4\pi e^{\gamma}\right)^{2r_2/n} - O(n^{-2/3});$$
2. (*under the GRH*)
$$\mathrm{rd} \geq \left(8\pi e^{\gamma+\pi/2}\right)^{r_1/n}\left(8\pi e^{\gamma}\right)^{2r_2/n} - O\left((\log n)^{-2}\right). \qquad \square$$

Taking the logarithms, one can easily rewrite the asymptotic fcrms of these inequalities as follows.

COROLLARY 8.1.3. *We have*

1. (*unconditionally*)
$$g_K \geq \alpha'_{\mathbb{R}} r_1 + \alpha'_{\mathbb{C}} r_2 - o(n),$$
or equivalently
$$1 \geq \alpha'_{\mathbb{R}}\frac{r_1}{g} + \alpha'_{\mathbb{C}}\frac{r_2}{g} - o(1),$$
for
$$\alpha'_{\mathbb{R}} = \frac{\gamma}{2} + \frac{1}{2} + \log 2\sqrt{\pi}, \qquad \alpha'_{\mathbb{C}} = \gamma + \log 4\pi;$$
2. (*under the GRH*)
$$g_K \geq \alpha_{\mathbb{R}} r_1 + \alpha_{\mathbb{C}} r_2 - o(n),$$
or equivalently
$$1 \geq \alpha_{\mathbb{R}}\frac{r_1}{g} + \alpha_{\mathbb{C}}\frac{r_2}{g} - o(1),$$
for
$$\alpha_{\mathbb{R}} = \log 2\sqrt{2\pi} + \frac{\pi}{4} + \frac{\gamma}{2}, \qquad \alpha_{\mathbb{C}} = \log 8\pi + \gamma. \qquad \square$$

Proofs of these results use the explicit formula of Proposition 8.1.17 below.

In fact, these inequalities are a development of a more elementary Stark inequality, which reads as follows.

PROPOSITION 8.1.4 (Stark's inequality, unconditional). *We have*
$$\mathrm{rd}_K \geq \left(4\pi e^{\gamma}\right)^{r_1/n}\left(2\pi e^{\gamma}\right)^{2r_2/n} e^{o(n)}. \qquad \square$$

Its proof is based on the following Stark formula:
$$\log|D| = r_1(\log\pi - \psi(s/2)) + 2r_2(\log(2\pi) - \psi(s)) - \frac{2}{s} - \frac{2}{s-1} + 2{\sum_{\rho}}' \frac{1}{s-\rho} + 2\sum_{P}\sum_{m=1}^{\infty} N(P)^{-ms}\log N(P),$$
where in the first sum ρ runs over the zeroes of $\zeta_K(s)$ in the critical strip $0 < \mathrm{Re}(s) < 1$, and $\sum'$ means that the ρ and $\overline{\rho}$ are to be grouped together, while the external sum in the double sum is taken over all prime divisors P of K.

Actually, the Odlyzko–Serre bounds are obtained by applying the explicit formula to the Stark formula.

The logarithmic version of Stark's inequality reads as follows.

COROLLARY 8.1.5. *We have*

$$g_K \geq \alpha''_{\mathbb{R}} r_1 + \alpha''_{\mathbb{C}} r_2 - o(n),$$

or equivalently

$$1 \geq \alpha''_{\mathbb{R}} \frac{r_1}{g} + \alpha''_{\mathbb{C}} \frac{r_2}{g} - o(1),$$

for

$$\alpha''_{\mathbb{R}} = \log 2\sqrt{\pi} + \frac{\gamma}{2}, \qquad \alpha''_{\mathbb{C}} = \log 2\pi + \gamma.$$ □

8.1.1. Invariants of Infinite Global Fields

Now we give some basics on infinite global fields and asymptotically exact families showing that these notions are worth studying.

LEMMA 8.1.6. *For any given g_0 there are only finitely many number fields K whose genus does not exceed g_0. The same is true for function fields over a given constant field (considered up to an isomorphism).* □

We do not give a proof of this result, since, though being direct and simple, it requires a certain technique that lies beyond the scope of this volume; however, we mention that this is a consequence of the Minkowski bound together with some auxiliary arguments, and in the function field case this follows from the existence of modular varieties classifying algebraic curves.

LEMMA 8.1.7. *Every family of global fields $\{K_i\}$ contains an asymptotically exact subfamily.*

PROOF. In the number field case, for any q we have $N_q(K_i) \leq n(K_i)$, where $n(K) = [K : \mathbb{Q}]$. On the other hand, since $K_i \neq \mathbb{Q}$ there exists a universal constant C such that $|D_{K_i}|^{1/n} \geq C$. This follows from the Minkowski constant argument and Lemma 8.1.6. Thus, $n(K_i)/g(K_i) \leq 1/\log C$, and all limit points of $N_q(K_i)/g(K_i)$ lie between 0 and $1/\log C$.

Therefore, for each individual q there is a limit over some subsequence of $\mathcal{K} = \{K_i\}$. The same is true for $r_\alpha(K_i)/g(K_i)$, $\alpha = \mathbb{R}, \mathbb{C}$.

Choose a subsequence $\mathcal{K}_0$ such that $\varphi_{\mathbb{R}}$ exists. Take its subsequence $\mathcal{K}_1$ where $\varphi_{\mathbb{C}}$ exists, then its subsequence $\mathcal{K}_2$ where φ_2 exists, and so on. Now take a diagonal sequence, i.e., such that $K_1 \in \mathcal{K}_1$, $K_2 \in \mathcal{K}_2$, etc. For this sequence, φ_α exists for any α.

The function field case is treated similarly; this diagonal argument works as well. □

REMARK 8.1.8. This proof only uses the fact that the space of positive measures with mass 1 is a compact metric space.

Therefore, the study of any family can be reduced in a certain sense to that of asymptotically exact families, and from now on we assume (in this chapter) all our families to be asymptotically exact.

LEMMA 8.1.9. *Let $L \supseteq K$. Then, in the number field case, for $g(K) \geq 1$,*

$$\frac{N_{\mathbb{R}}(L)}{g(L)} + 2\frac{N_{\mathbb{C}}(L)}{g(L)} \leq \frac{N_{\mathbb{R}}(K)}{g(K)} + 2\frac{N_{\mathbb{C}}(K)}{g(K)},$$

and, for any prime p and any $n \geq 1$,

$$\sum_{m=1}^{n} \frac{mN_{p^m}(L)}{g(L)} \leq \sum_{m=1}^{n} \frac{mN_{p^m}(K)}{g(K)}.$$

In the function field case, for $g(K) \geq 2$ and for any $n \geq 1$,

$$\sum_{m=1}^{n} \frac{mN_{r^m}(L)}{g(L)-1} \leq \sum_{m=1}^{n} \frac{mN_{r^m}(K)}{g(K)-1}.$$

PROOF. In the number field case, for an extension $L \supset K$ we have by the transitivity of the different (Proposition 6.1.30)

$$|D_L| \geq |D_K|^{[L:K]},$$

and in the function field case we have by the Hurwitz formula

$$g(L) - 1 \geq [L:K](g(K)-1).$$

On the other hand, if a place v of K is decomposed into a set $\{v_1, v_2, \dots\}$ of places of L, then

$$\prod \operatorname{Norm} v_i \leq (\operatorname{Norm} v)^{[L:K]}.$$

Therefore,

$$\sum_{m=1}^{n} mN_{p^m}(L) \leq [L:K] \sum_{m=1}^{n} mN_{p^m}(K)$$

for any prime p and any $n \geq 1$. Dividing, we get the required monotonicity.

The Archimedean inequality and that for function fields are similar. □

LEMMA 8.1.10. *Any infinite tower $K_0 \subset K_1 \subset K_2 \subset \dots$ is an asymptotically exact family.*

PROOF. First consider the number field case. For a given prime p, by the second inequality of Lemma 8.1.9, the sequence $\sum_{m=1}^{n} \frac{mN_{p^m}(K_i)}{g(K_i)}$, $i = 1, 2, \dots$, is non-increasing for any fixed n, and hence has a limit. (In the function field case we divide by $g(K_i) - 1$ rather than $g(K_i)$.) Taking $n = 1$, we see that φ_p exists; setting $n = 2$, we derive the existence of φ_{p^2}, etc.

Using the first inequality of Lemma 8.1.9, we prove the existence of $\varphi_{\mathbb{R}}$ and then of $\varphi_{\mathbb{C}}$.

The function field case is treated similarly. □

Let $\mathcal{K}$ be an infinite global field; then $\mathcal{K} = \bigcup_{i=1}^{\infty} K_i$, where $K_1 \subset K_2 \subset K_3 \subset \dots$, and we can define the corresponding parameters φ_α, $\alpha \in A = \{\mathbb{R}, \mathbb{C}; 2, 3, 4, 5, 7, 8, 9, \dots\}$.

LEMMA 8.1.11. *For an infinite global field $\mathcal{K}$, the parameters φ_α do not depend on the choice of the tower $K_1 \subset K_2 \subset K_3 \subset \dots$.*

PROOF. Let $\mathcal{K} = \bigcup_{i=1}^{\infty} K_i = \bigcup_{i=1}^{\infty} L_i$ be two representations. Since each K_i is generated by finitely many elements of $\mathcal{K}$, it is contained in some L_j. Using the inequalities of Lemma 8.1.9 we see that the limit of the ratio in question for the tower $\{L_j\}$ is less than or equal to the corresponding value for the tower $\{K_i\}$.

As in the proof of Lemma 8.1.10, we have to treat φ_p first, and then use induction over the degree. We get $\varphi_\alpha(\mathcal{L}) \le \varphi_\alpha(\mathcal{K})$ and, by symmetry, vice versa. □

LEMMA 8.1.12. *Let $\mathcal{K} \subset \mathcal{L}$ be infinite global fields. Then*

$$\varphi_{\mathbb{R}}(\mathcal{K}) + 2\varphi_{\mathbb{C}}(\mathcal{K}) \ge \varphi_{\mathbb{R}}(\mathcal{L}) + 2\varphi_{\mathbb{C}}(\mathcal{L}),$$

and for any prime p and any $n \ge 1$,

$$\sum_{m=1}^{n} m\varphi_{p^m}(\mathcal{K}) \ge \sum_{m=1}^{n} m\varphi_{p^m}(\mathcal{L}).$$

In particular,

$$\varphi_p(\mathcal{K}) \ge \varphi_p(\mathcal{L}).$$

PROOF. We follow the same lines as above. Consider a tower $\{L_j\}$ filtering $\mathcal{L}$. Then $\{K_j = \mathcal{K} \cap L_j\}$ filters $\mathcal{K}$. Using the monotonicity of Lemma 8.1.9 for the pair $K_j \subset L_j$, we get the result. □

An infinite global field $\mathcal{K}$ (or an asymptotically exact family $\mathcal{K} = \{K_i\}$) is called *asymptotically bad* if and only if all $\varphi_\alpha = 0$. If at least one of the parameters $\varphi_\alpha > 0$, it is called *asymptotically good*.

For example, any sequence of global fields of bounded degree is asymptotically bad.

Families $\{K_i\}$ of number fields satisfying the condition

$$n/\log\sqrt{|D|} \to 0$$

are asymptotically bad. Indeed, this condition clearly implies $\varphi_{\mathbb{R}} = \varphi_{\mathbb{C}} = 0$ and then all $\varphi_q = 0$ as well, since one has the following fact.

LEMMA 8.1.13. *For any asymptotically exact family $\mathcal{K} = \{K_i\}$ of number fields and for any prime p one has*

$$\sum_{m=1}^{\infty} m\varphi_{p^m} \le \varphi_{\mathbb{R}} + 2\varphi_{\mathbb{C}}.$$

PROOF. Indeed, all places of a number field K whose norm is a power of p lie over p in $\mathbb{Q}$ and the product of their norms is not greater than p^n, where $n = r_1 + 2r_2$ is the degree of K. □

Using a conductor-discriminant formula, one can readily show that abelian fields are bad.

PROPOSITION 8.1.14. *Let L be a (finite) global field and L^{ab} be its maximal abelian extension. If for an infinite global field $\mathcal{K}$ the field $\mathcal{K} \cap L^{\mathrm{ab}}$ is also infinite, then $\mathcal{K}$ is asymptotically bad. In particular, if $\mathcal{K}$ is abelian over a finite global field, then $\mathcal{K}$ is asymptotically bad.* □

Now let K_1 be a number field with the infinite Hilbert class field tower $\{K_i\}$ or any other unramified tower. Then the family $\{K_i\}$ is asymptotically good, since the ratio n_i/g_i is constant in such a tower, and thus at least one of $\varphi_{\mathbb{R}}$ and $\varphi_{\mathbb{C}}$ is nonzero. In fact, in the number field case we do not know essentially different methods for constructing asymptotically good families (however, one can take composita with a fixed number field, one can mix different class field towers, etc.).

For the function field case, the families constructed above in Chapter 5 and Sec. 7.4 are clearly asymptotically good.

8.1.2. Basic Inequalities

One can note that the Odlyzko–Serre bounds presented in their asymptotic logarithmic form are parallel to the Drinfeld–Vlăduţ inequality: they bound linear forms in some specific parameters of an asymptotically exact family, namely $\varphi_{\mathbb{C}}$ and $\varphi_{\mathbb{R}}$ in the number field case and φ_r in the function field case. Indeed, Stark's inequality and the Odlyzko–Serre bounds in this setting become

0. $\alpha''_{\mathbb{R}}\varphi_{\mathbb{R}} + \alpha''_{\mathbb{C}}\varphi_{\mathbb{C}} \le 1,$
1. (unconditionally) $\alpha'_{\mathbb{R}}\varphi_{\mathbb{R}} + \alpha'_{\mathbb{C}}\varphi_{\mathbb{C}} \le 1,$
2. (GRH) $\alpha_{\mathbb{R}}\varphi_{\mathbb{R}} + \alpha_{\mathbb{C}}\varphi_{\mathbb{C}} \le 1,$

while the Drinfeld–Vlăduţ bound reads

$$(\sqrt{r}-1)\varphi_1 \le 1.$$

Now we are going to establish basic inequalities that generalize the Drinfeld–Vlăduţ inequality in the function field case and the Odlyzko–Serre bounds in the number field case and provide a basis for all other developments in this chapter.

Function field case. Let us prove an inequality which generalizes the Drinfeld–Vlăduţ inequality taking into account points of all higher degrees.

PROPOSITION 8.1.15 (basic inequality in the function field case). *For an infinite function field, as well as for any asymptotically exact family of function fields, we have*

$$\sum_{m=1}^{\infty} \frac{m\varphi_{r^m}}{r^{m/2}-1} \le 1.$$

To start with, let us introduce some measures on the unit circle. We normalize the circle to be $\mathbb{R}/2\pi\mathbb{Z}$ represented by $(-\pi,\pi]$. For a zero ρ of the zeta function $\zeta_{K_j}(s)$, let $t(\rho)$ be defined by

$$t(\rho) = \frac{1}{i}\Big(\rho - \frac{1}{2}\Big) \pmod{2\pi}.$$

Let

$$\Delta_j := \frac{\pi}{g_j} \sum_{\zeta_{K_j}(\rho)=0} \delta_{t(\rho)},$$

$\delta_{t(\rho)}$ being the Dirac measure supported at $t(\rho)$. Then Δ_j is a measure of total mass 2π on $\mathbb{R}/2\pi\mathbb{Z}$, and Δ_j is symmetric with respect to $t \mapsto -t$.

Now we establish a preliminary result (cf. Sec. 7.1).

LEMMA 8.1.16. *Let μ be a positive measure on $T = \mathbb{R}/2\pi\mathbb{Z}$ of total mass 2π. Assume that all of its Fourier coefficients*

$$a_n = \mu(2\cos n\theta) = \int_0^{2\pi} 2\cos n\theta \, d\mu, \quad n = 1,2,\ldots,$$

satisfy $a_n \leq 0$. *Then*

$$\sum_{n=1}^{\infty} |a_n| \leq 1.$$

PROOF. Let

$$P_m(\theta) = \frac{1}{2m+1}(1+2\cos\theta+2\cos 2\theta+\ldots+2\cos m\theta)^2.$$

It is easy to verify that

$$P_m = \sum_n 2b_{m,n}\cos(n\theta)$$

with

$$b_{m,n} = \begin{cases} 1-\dfrac{n}{2m+1} & \text{for } 0 \leq n \leq 2m, \\ 0 & \text{for } n > 2m. \end{cases}$$

Since $P_m(\theta) \geq 0$, we get

$$\mu(P_m) = \sum_{n\geq 0} b_{m,n}a_n \geq 0, \quad \text{for any } m \geq 1.$$

Since $a_0 = 1$, $b_{m,0} = 1$, and $a_n = -|a_n|$ for $n \geq 1$, we get

$$\sum_{n\geq 1} b_{m,n}|a_n| \leq 1, \quad \text{for any } m \geq 1.$$

Since all the $b_{m,n}$ and $|a_n|$ are nonnegative, we may pass to the limit as $m \to \infty$, which yields the inequality. □

PROOF OF PROPOSITION 8.1.15. To get a proof of the proposition, one notes that

$$\sum_{m=1}^{\infty} \frac{m\varphi_{r^m}}{r^{m/2}-1} = \sum_{m=1}^{\infty} \nu_m r^{-m/2}$$

with

$$\nu_m := \sum_{k|m} k\varphi_{r^k}.$$

Now it suffices to note that

$$\nu_m r^{-m/2} = \Delta(2\cos m\theta)$$

for the measure

$$\Delta = \frac{F}{2\pi}\,d\theta$$

with

$$F = 1 - \sum_m r^{-m/2}\nu_m \cos m\theta.$$

This measure is positive, since

$$\Delta = \lim_{j\to\infty} \Delta_j,$$

and each Δ_j is positive. □

This result generalizes the Drinfeld–Vlăduţ theorem, saying that for the number N of points of degree one on algebraic curves over a finite field $\mathbb{F}_r$ we have

$$N \le (\sqrt{r}-1)g + o(g),$$

as g tends to infinity, i.e., that $\varphi_r \le \sqrt{r}-1$. Indeed, it is sufficient to omit all terms in the sum except for the first one.

Number field case. In the number field case, proofs are more elaborate and use "explicit formulae" for the zeta functions.

Let K be a number field of degree n and discriminant $D = D_K$ with r_1 real and r_2 pairs of complex embeddings, and let $\zeta_K(s)$ be its zeta function; by $P(K)$ we denote the set of prime divisors of K which can be identified with the set of non-Archimedean places of K.

We use the following form of the Guinand–Weil explicit formula.

PROPOSITION 8.1.17. *Let $F(x)$ be a differentiable even real positive function defined on the whole real line $\mathbb{R}$ such that $F(0)=1$ and there exist positive real constants c and ε such that*

$$F(x), F'(x) \le ce^{-(1/2+\varepsilon)|x|} \quad \text{as} \quad |x| \to \infty.$$

Define

$$\Phi(s) := \int_{-\infty}^{\infty} F(x) e^{(s-1/2)x}\, dx.$$

Then we have the following formula:

$$\begin{aligned}\log|D| = r_1\frac{\pi}{2} + n(\gamma + \log 8\pi)\\ -n\int_0^\infty \frac{1-F(x)}{2\sinh\frac{x}{2}}\,dx - r_1\int_0^\infty \frac{1-F(x)}{2\cosh\frac{x}{2}}\,dx - 4\int_0^\infty F(x)\cosh\frac{x}{2}\,dx\\ + {\sum_\rho}' \Phi(\rho) + 2\sum_P \sum_{m=1}^\infty N(P)^{-m/2} F(m\log N(P)) \log N(P),\end{aligned}$$

where in the first sum ρ runs over the zeroes of $\zeta_K(s)$ in the critical strip $0 < \mathrm{Re}(s) < 1$, $\sum'$ meaning that the ρ and $\overline{\rho}$ terms are to be grouped together, the outer sum in the last term is taken over all prime divisors $P \in P(K)$, and in the inner sum $N(P)$ denotes the absolute norm of P. □

GRH basic inequality. Let us now apply this formula to the case of an asymptotically exact family $\mathcal{K} = \{K_i\}$ of number fields.

PROPOSITION 8.1.18 (GRH basic inequality). *For an asymptotically exact family of number fields one has*

$$\sum_q \frac{\varphi_q \log q}{\sqrt{q}-1} + \alpha_{\mathbb{R}}\varphi_{\mathbb{R}} + \alpha_{\mathbb{C}}\varphi_{\mathbb{C}} \le 1,$$

the sum being over all prime powers q.

PROOF. Let us apply the above explicit formula to the field $K = K_i$ from our sequence, substituting $F(x) = e^{-yx^2}$ for a real positive y. We get

$$\log|D| = r_1\frac{\pi}{2} + n(\gamma+\log 8\pi)$$
$$-n\int_0^\infty \frac{1-e^{-yx^2}}{2\sinh\frac{x}{2}}\,dx - r_1\int_0^\infty \frac{1-e^{-yx^2}}{2\cosh\frac{x}{2}}\,dx - 4\int_0^\infty e^{-yx^2}\cosh\frac{x}{2}\,dx$$
$$+\operatorname{Re}\sum_t{}' \int_{-\infty}^\infty e^{itx-yx^2}\,dx + 2\sum_P\sum_{m=1}^\infty N(P)^{-m/2}e^{-ym^2\log^2 N(P)}\log N(P),$$

where in the first sum t runs over all reals such that $\zeta_K(1/2+it)=0$, and the outer sum in the last term is taken over all prime divisors P of K.

Dividing the last equation by $2g = \log|D|$ and using the relation $n = r_1 + 2r_2$, we get

$$1 = \frac{r_1}{g}\left(\frac{\pi}{4}+\frac{\gamma}{2}+\frac{\log 8\pi}{2}\right) + \frac{r_2}{g}(\gamma+\log 8\pi)$$
$$-\frac{n}{g}\int_0^\infty \frac{1-e^{-yx^2}}{4\sinh\frac{x}{2}}\,dx - \frac{r_1}{g}\int_0^\infty \frac{1-e^{-yx^2}}{4\cosh\frac{x}{2}}\,dx - \frac{2}{g}\int_0^\infty e^{-yx^2}\cosh\frac{x}{2}\,dx$$
$$+\operatorname{Re}\frac{1}{2g}\sum_t \int_{-\infty}^\infty e^{itx-yx^2}\,dx + \sum_q \frac{N(q)}{g}\sum_{m=1}^\infty q^{-m/2}e^{-ym^2\log^2 q}\log q,$$

where $N(q)$ is the number of prime divisors in K of norm q.

Thus, $1 = T_1+T_2-T_3-T_4-T_5+T_6+T_7$ is presented as a sum of seven terms T_j, $j=1,\ldots,7$. Now we set $y = 1/\log g$ and tend g to infinity. Then y tends to zero, and we are going to show that $T_1+T_2+T_7$ tends to the left-hand side of the basic inequality, while $T_3+T_4+T_5$ tends to zero, T_6 being nonnegative, which proves the theorem.

Indeed, it is clear that T_1 tends to $\alpha_{\mathbb{R}}\varphi_{\mathbb{R}}$, and T_2 tends to $\alpha_{\mathbb{C}}\varphi_{\mathbb{C}}$. Then, for T_7 we have

$$T_7 = \sum_q \frac{N(q)}{g}\sum_{m=1}^\infty q^{-m/2}e^{-ym^2\log^2 q}\log q \le \sum_q \frac{N(q)}{g}\sum_{m=1}^\infty q^{-m/2}\log q,$$

since $e^{-ym^2\log^2 q}\le 1$. On the other hand, since $e^{-ym^2\log^2 q} \ge 1 - ym^2\log^2 q$, we get

$$T_7 \ge \sum_{q\le\log g}\frac{N(q)}{g}\sum_{m=1}^{[\log^{1/4} g]} q^{-m/2}(1-ym^2\log^2 q)\log q$$
$$\ge \left(1-\frac{(\log\log g)^2}{\sqrt{\log g}}\right)\sum_{q\le\log g}\frac{N(q)}{g}\sum_{m=1}^{[\log^{1/4} g]} q^{-m/2}\log q.$$

These inequalities show that T_7 tends to

$$\sum_q \varphi_q \sum_{m=1}^\infty q^{-m/2}\log q = \sum_q \frac{\varphi_q\log q}{\sqrt{q}-1}$$

as $g\to\infty$.

Now let us estimate T_3, T_4, and T_5. Since

$$0 \le \int_0^\infty \frac{1-e^{-yx^2}}{4\cosh\frac{x}{2}}\,dx \le \int_0^\infty \frac{1-e^{-yx^2}}{4\sinh\frac{x}{2}}\,dx,$$

if we show that T_3 tends to zero, then T_4 also tends to 0. We write

$$\int_0^\infty \frac{1-e^{-yx^2}}{\sinh\frac{x}{2}}\,dx = \int_0^1 \frac{1-e^{-yx^2}}{\sinh\frac{x}{2}}\,dx + \int_1^\delta \frac{1-e^{-yx^2}}{\sinh\frac{x}{2}}\,dx + \int_\delta^\infty \frac{1-e^{-yx^2}}{\sinh\frac{x}{2}}\,dx$$

for any $\delta > 1$, all three integrals being positive. Since $e^{-yx^2} \ge 1 - yx^2$,

$$\int_0^1 \frac{1-e^{-yx^2}}{\sinh\frac{x}{2}}\,dx \le \int_0^1 \frac{yx^2}{\sinh\frac{x}{2}}\,dx = O(y).$$

For the same reason,

$$\int_1^\delta \frac{1-e^{-yx^2}}{\sinh\frac{x}{2}}\,dx \le (\delta-1)\max_{1\le x\le\delta}\frac{1-e^{-yx^2}}{\sinh\frac{x}{2}} \le (\delta-1)\frac{1-e^{-y\delta^2}}{\sinh\frac{1}{2}} \le \frac{\delta-1}{\sinh\frac{1}{2}}y\delta^2.$$

For the third integral we have

$$\int_\delta^\infty \frac{1-e^{-yx^2}}{\sinh\frac{x}{2}}\,dx \le \int_\delta^\infty \frac{1}{\sinh\frac{x}{2}}\,dx = O(e^{-\delta}),$$

as δ tends to infinity. If δ tends to infinity in such a way that $y\delta^3$ tends to zero (e.g., put $\delta = y^{-1/4} = \log^{1/4} g$), these inequalities show that T_3 (and thus T_4) tends to zero.

For T_5 we have

$$0 \le T_5 = \frac{2}{g}\int_0^\infty e^{-yx^2}\cosh\frac{x}{2}\,dx \le \frac{2}{g}\int_0^\infty e^{-yx^2+\frac{x}{2}}\,dx \le \frac{2}{g}\int_{-\infty}^\infty e^{-yx^2+\frac{x}{2}}\,dx$$
$$= \frac{2}{g}\sqrt{\frac{\pi}{y}}e^{\frac{1}{16y}} = 2g^{-15/16}\sqrt{\pi\log g},$$

which shows that it also tends to zero.

Then it is sufficient to note that all terms in the sum T_6 are positive; indeed,

$$\int_{-\infty}^\infty e^{itx-yx^2}\,dx = e^{-t^2/4y}\int_{-\infty}^\infty e^{-y(x-\frac{it}{2y})^2}\,dx = \sqrt{\frac{\pi}{y}}e^{-t^2/4y} > 0,$$

which completes the proof. □

One can ask how large the difference $\delta = \delta_{\mathcal{K}}$ between the right- and left-hand sides of Propositions 8.1.15 and 8.1.18 can be.

We call $\delta_{\mathcal{K}}$ the *deficiency* of the family. Of course, $0 \le \delta_{\mathcal{K}} \le 1$, and for asymptotically bad families $\delta_{\mathcal{K}} = 1$. In the function field case, at least when r is an even prime power, families defined by modular curves and García–Stichtenoth towers provide examples with $\delta_{\mathcal{K}} = 0$. In the number field case we do not know such a family. In fact, all known examples are those of unramified class field towers, often with extra splitting conditions (see Secs. 8.3 and 8.2), which do not reach the $\delta = 0$ value; namely, the class field tower of the Martinet field

$$\mathbb{Q}\Big(\cos\frac{2\pi}{11}, \sqrt{2}, \sqrt{-23}\Big)$$

has $\delta_{\mathcal{K}} \leq 0.1601\ldots$, which was the best known example for many years. For slight improvements of this result, see Sec. 8.3.2.

Unconditional basic inequalities. Let us now give some partial results that can be obtained without assuming the GRH, which are albeit much weaker.

PROPOSITION 8.1.19 (basic inequality′). *For any asymptotically exact family of number fields, one has*

$$2\sum_q \varphi_q \log q \sum_{m=1}^{\infty} \frac{1}{q^m+1} + \alpha'_{\mathbb{R}}\varphi_{\mathbb{R}} + \alpha'_{\mathbb{C}}\varphi_{\mathbb{C}} \leq 1,$$

the first sum being over all prime powers q.

Note that the Archimedean coefficients are

$$\alpha'_{\mathbb{R}} := \gamma/2 + 1/2 + \log 2\sqrt{\pi} \approx 2.054\ldots \qquad \text{and} \qquad \alpha'_{\mathbb{C}} := \gamma + \log 4\pi \approx 3.108\ldots,$$

whereas in Proposition 8.1.18 they are

$$\alpha_{\mathbb{R}} = \gamma/2 + \pi/4 + \log 2\sqrt{2\pi} \approx 2.686\ldots \qquad \text{and} \qquad \alpha_{\mathbb{C}} = \gamma + \log 8\pi \approx 3.801\ldots.$$

This result improves the asymptotic form of the unconditional Odlyzko–Serre inequality.

SKETCH OF THE PROOF. The method of the proof is exactly the same as for the basic inequality (GRH Proposition 8.1.18), with the only difference being in the choice of the function $F(x)$; without the GRH we choose $F(x) = \dfrac{e^{-yx^2}}{\cosh(x/2)}$ to have $\operatorname{Re}\Phi(s)$ positive on the whole critical strip, which is checked by an elementary computation using the maximum principle. □

The following simple result improves upon the asymptotic Stark's inequality.

PROPOSITION 8.1.20 (basic inequality″). *For any asymptotically exact family of number fields, one has*

$$\sum_q \frac{\varphi_q \log q}{q-1} + \alpha''_{\mathbb{R}}\varphi_{\mathbb{R}} + \alpha''_{\mathbb{C}}\varphi_{\mathbb{C}} \leq 1.$$

Note that Archimedean coefficients are

$$\alpha''_{\mathbb{R}} := \gamma/2 + \log 2\sqrt{\pi} \approx 1.554\ldots \qquad \text{and} \qquad \alpha''_{\mathbb{C}} := \gamma + \log 2\pi \approx 2.415\ldots.$$

PROOF. Apply Stark's formula (see p. 145) to $K = K_j$ for $s = 1 + \dfrac{1}{\sqrt{g_j}}$, where, as usual, $g_j = \log\sqrt{|D_j|}$, and divide it by $2g_j$. We get

$$1 = \frac{r_1}{2g_j}(\log\pi - \psi(s/2)) + \frac{r_2}{g_j}(\log 2\pi - \psi(s)) - \frac{1}{\sqrt{g_j}(1+\sqrt{g_j})} - \frac{1}{\sqrt{g_j}} + \frac{1}{g_j}\sum_\rho{}' \frac{1}{s-\rho} + \frac{1}{g_j}\sum_P \sum_{m=1}^{\infty} N(P)^{-ms}\log N(P).$$

When g_j grows, the first two terms tend, respectively, to $\alpha''_{\mathbb{R}}\varphi_{\mathbb{R}}$ and $\alpha''_{\mathbb{C}}\varphi_{\mathbb{C}}$, since $\psi\left(\frac{1}{2}\right) = -\gamma - \log 4$ and $\psi(1) = -\gamma$; the third and forth terms tend to zero, the fifth being positive, and the last term tends to $\sum_q \frac{\varphi_q \log q}{q-1}$, which completes the proof. $\square$

8.1.3. Limit Zeta Function

We define the *limit zeta function* of an asymptotically exact family $\mathcal{K}$ as

$$\zeta_{\mathcal{K}}(s) = \zeta_\varphi(s) := \prod_q (1-q^{-s})^{-\varphi_q},$$

and its *completed limit zeta function* as

$$\widetilde{\zeta}_{\mathcal{K}}(s) = \widetilde{\zeta}_\varphi(s) := e^s 2^{-\varphi_{\mathbb{R}}} \pi^{-s\varphi_{\mathbb{R}}/2} (2\pi)^{-s\varphi_{\mathbb{C}}} \Gamma\left(\frac{s}{2}\right)^{\varphi_{\mathbb{R}}} \Gamma(s)^{\varphi_{\mathbb{C}}} \prod_q (1-q^{-s})^{-\varphi_q}$$

in the number field case, and

$$\widetilde{\zeta}_{\mathcal{K}}(s) = \widetilde{\zeta}_\varphi(s) := r^s \prod_q (1-q^{-s})^{-\varphi_q}$$

in the function field case; q runs over all prime powers in the number field case and over powers $q = r^m$ in the function field case. An expression $(1-q^{-s})^{-\varphi_q}$ is understood here and in what follows in the sense given by the binomial series.

PROPOSITION 8.1.21 (GRH). *For any asymptotically exact family of global fields, the product defining ζ_φ (and $\widetilde{\zeta}_\varphi$) converges absolutely in the closed half-plane* $\operatorname{Re}(s) \geq 1/2$ *and defines an analytic function in the open half-plane* $\operatorname{Re}(s) > 1/2$. *In the function field case, the result is unconditional.*

PROOF. The product converges absolutely if and only if the series

$$\sum_q \varphi_q \log\left|\frac{1}{1-q^{-s}}\right|$$

does, but for $\operatorname{Re}(s) \geq 1/2$ one has

$$\sum_q \varphi_q \log\left|\frac{1}{1-q^{-s}}\right| \leq \sum_q \varphi_q \log\frac{1}{1-q^{-1/2}},$$

which in its turn converges, since, starting from some q, we have

$$\varphi_q \log\frac{1}{1-q^{-1/2}} \leq \frac{\varphi_q \log q}{\sqrt{q}-1},$$

and the series

$$\sum_q \frac{\varphi_q \log q}{\sqrt{q}-1}$$

converges by GRH Proposition 8.1.18. $\square$

REMARK 8.1.22. In fact, since the coefficients of the Dirichlet series corresponding to ζ_φ are positive and $\zeta_\varphi(1/2)$ is finite, the product converges absolutely for $\operatorname{Re}(s) > \frac{1}{2} - \varepsilon(\varphi)$, where $\varepsilon(\varphi)$ depends on how large φ_q are; it can happen that $\widetilde{\zeta}_\varphi(s)$ is an entire function. For asymptotically bad families we have $\widetilde{\zeta}_\varphi(s) = e^s$ or r^s, depending on the case under consideration. In fact, we do not know a single example of an infinite global field for which the product does not converge in the half-plane $\mathrm{Re}(s) > 0$; note also that the Archimedean factors are analytic in this half-plane.

Now recall that

$$\begin{aligned}\xi_\varphi(s) &:= (\log \widetilde{\zeta}_\varphi)' = \widetilde{\zeta}_\varphi'/\widetilde{\zeta}_\varphi \\ &= 1 - \frac{\varphi_\mathbb{R}}{2}\log\pi - \varphi_\mathbb{C}\log 2\pi + \frac{1}{2}\varphi_\mathbb{R}\psi\Big(\frac{s}{2}\Big) + \varphi_\mathbb{C}\psi(s) - \sum_q \varphi_q \frac{\log q}{q^s - 1}\end{aligned}$$

in the number field case, and

$$\xi_\varphi(s) := \big(\log_r \widetilde{\zeta}_\varphi\big)' = \frac{1}{\log r}\left(\frac{\widetilde{\zeta}_\varphi'}{\widetilde{\zeta}_\varphi}\right) = 1 - \sum_{m=1}^{\infty} \frac{m\varphi_{r^m}}{r^{ms}-1}$$

in the function field case.

Then one can express the basic inequality (GRH Proposition 8.1.18) as

$$\xi_\mathcal{K}(1/2) \geq 0.$$

Unconditionally, we have the following.

PROPOSITION 8.1.23. (a) *For any asymptotically exact family of global fields the product defining ζ_φ (and $\widetilde{\zeta}_\varphi$) converges absolutely in the closed half-plane* $\mathrm{Re}\, s \geq 1$ *and defines an analytic function in* $\mathrm{Re}\, s > 1$.

(b) *For* $\mathrm{Re}\, s > 1$ *we have pointwise limits*

$$\zeta_\varphi^0(s) = \lim_{j\to\infty} \zeta_{K_j}(s)^{1/g_j},$$

$$\widetilde{\zeta}_\varphi(s) = \lim_{j\to\infty} \widetilde{\zeta}_{K_j}(s)^{1/g_j},$$

where $\widetilde{\zeta}_{K_j}(s)$ is the completed zeta function defined by

$$\widetilde{\zeta}_{K_j}(s) = |D_j|^{s/2} 2^{-r_1(K_j)} \pi^{-sr_1(K_j)/2} (2\pi)^{-sr_2(K_j)} \Gamma(s/2)^{r_1(K_j)} \Gamma(s)^{r_2(K_j)} \zeta_{K_j}(s)$$

in the number field case, and by

$$\widetilde{\zeta}_{K_j}(s) = r^{s(g_j-1)} \zeta_{K_j}(s)$$

in the function field case. For any $\varepsilon > 0$ the convergence is uniform in the half-plane $\mathrm{Re}\, s > 1+\varepsilon$ *(and thus on compact subsets in the half-plane* $\mathrm{Re}\, s > 1$*).*

(c) *Conversely, let $s_0 > 1$, $s_0 \neq 2, 4$, be a real number such that the limit*

$$\lim_{j\to\infty} \widetilde{\zeta}_{K_j}(s_0)^{1/g_j}$$

exists for some family $\{K_j\}$. Then the family is asymptotically exact.

PROOF. Let us prove (b) in the number field case, the proof in the function fields case being essentially the same but simpler. Let K_j be a field from our family. Note that it is sufficient to consider the case of real s, since $\big|\zeta_{K_j}(x)^{1/g}/\zeta_\varphi(x) - 1\big| \leq \big|\zeta_{K_j}(s)^{1/g}/\zeta_\varphi(s) - 1\big|$ if $x = s + it$ with $s > 1$ (look at the corresponding Dirichlet series). For a real $s > 1$ we have

$$\zeta_{K_j}(s)^{1/g}/\zeta_\varphi(s) = \prod_q (1 - q^{-s})^{N_q(K_j)/g_j - \varphi_q}.$$

Let g_0 be a positive integer such that for any $q \leq M$ (here M is a positive integer to be specified below) one has $|\varphi_q - N_q(K_j)/g_j| \leq \delta_1$ for $g_j \geq g_0$, where a real positive δ_1 is also to be specified. Then we have

$$\prod_{q\le M}(1-q^{-s})^{\delta_1}\le\prod_{q\le M}(1-q^{-s})^{N_q(K_j)/g_j-\varphi_q}\le\prod_{q\le M}(1-q^{-s})^{-\delta_1}.$$

For any $s\ge 1+\varepsilon$ the product $\prod_{q\le M}(1-q^{-s})$ satisfies the inequalities

$$F_M(\varepsilon)^{-1}\le\prod_{q\le M}(1-q^{-s})\le F_M(\varepsilon)$$

for some real $F_M(\varepsilon)>1$. Thus

$$F_M(\varepsilon)^{-\delta_1}\le\prod_{q\le M}(1-q^{-s})^{N_q(K_j)/g_j-\varphi_q}\le F_M(\varepsilon)^{\delta_1}.$$

Let us now estimate the "tail" $\prod_{q\ge M+1}(1-q^{-s})^{N_q(K_j)/g_j-\varphi_q}$.

We show first that always $|N_q(K_j)/g_j-\varphi_q|\le 3q$ for sufficiently large q. Indeed, in the function field case we have

$$N_q(K_j)/g_j\le\frac{q+1+2g_jq^{1/2}}{g_j}\le 2q$$

and $\varphi_q\le(q^{1/2}-1)/m<q^{1/2}$ by the basic inequality, which proves the assertion in this case. In the number field case, $N_q(K_j)\le n_j=\deg K_j\le\mathrm{const}\cdot g_j$, and thus $N_q(K_j)/g_j$ is bounded by a constant; on the other hand, the unconditional basic inequality implies that $\varphi_q<q$, which proves the assertion (with much room to spare) in the number field case.

The assertion implies that the tail lies between

$$G(\varepsilon,M)=\prod_{q\ge M+1}^{\infty}\left(1-q^{-(1+\varepsilon)}\right)^{3q}$$

and its inverse.

Since $\left(1-q^{-(1+\varepsilon)}\right)^{3q}=\left((1-q^{-(1+\varepsilon)})^{3q^{(1+\varepsilon)}}\right)^{q^{-\varepsilon}}$ and $\left(1-q^{-(1+\varepsilon)}\right)^{3q^{(1+\varepsilon)}}$ tends to e^{-3} as $q\to\infty$, we get the tail to be between $C\exp\left(\sum\limits_{q\ge M+1}^{\infty}q^{-\varepsilon}\right)$ and its inverse for any $C>1$, $s\ge 1+\varepsilon$, and M sufficiently large. Choosing C and δ_1 such that $F_M(\varepsilon)^{\delta_1}\le\sqrt{1+\delta}$ and M such that $C\exp\left(\sum\limits_{q\ge M+1}^{\infty}q^{-\varepsilon}\right)\le\sqrt{1+\delta}$, we get the result.

Let us prove (c). Let us assume that the family is not asymptotically exact; it means that there exists α such that the sequence $N_\alpha(K_j)/g_j$ has at least two different limit points, which we denote φ'_α and φ''_α. We can choose two asymptotically exact subfamilies $\mathcal{K}'$ and $\mathcal{K}''$ of our family such that $\varphi_\beta(\mathcal{K}')=\varphi_\beta(\mathcal{K}'')$ for all $\beta\ne\alpha$ and $\varphi_\alpha(\mathcal{K}')=\varphi'_\alpha$, $\varphi_\alpha(\mathcal{K}'')=\varphi''_\alpha$. Then using (b) we see that our condition implies $(f_\alpha(s_0))^{\varphi'_\alpha}=(f_\alpha(s_0))^{\varphi''_\alpha}$, where f_α is the factor in the product defining $\widetilde{\zeta}_\varphi$ that corresponds to α. Since $f_\alpha(s_0)\ne 1$ for $s_0\ne 2,4$ (for non-Archimedean factors, this is true for any s_0, but for the gamma factors one needs to throw out the values $s_0=2,4$), we get the result. □

REMARK 8.1.24. Proposition 8.1.20 can be rewritten as $\xi_\varphi(1)\ge 0$.

REMARK 8.1.25. Part (b) of Proposition 8.1.23 shows in which sense the sequence of zeta functions of an asymptotically exact family of global fields tends to the limit zeta function.

REMARK 8.1.26. The definition of $\widetilde{\zeta}_\varphi(s)$ in the number field case is chosen on one hand so as to write the basic inequality and the generalized Brauer–Siegel theorem in the shortest possible way, and on the other hand so that it is a natural asymptotic analogue of the function

$$\widetilde{\zeta}_K(s) = |D|^{s/2} 2^{-r_1} \pi^{-sr_1/2} (2\pi)^{-sr_2} \Gamma(s/2)^{r_1} \Gamma(s)^{r_2} \zeta_K(s),$$

which is invariant under $s \mapsto 1-s$.

8.1.4. Limit Explicit Formula

Function field case. In this section we give a formula explaining the defect of a family as a certain "limit density" of zeta zeroes.

PROPOSITION 8.1.27 (explicit formula and basic equality in the function field case). *In the function field case, the limit*

$$\Delta_{\mathcal{K}} = \lim_{j\to\infty} \Delta_j$$

exists in the weak topology on the space of measures on $\mathbb{R}/2\pi\mathbb{Z}$. *Moreover, the measure* $\Delta_{\mathcal{K}}$ *has a continuous density* $M_{\mathcal{K}}$, *namely*

$$M_{\mathcal{K}}(\theta) = \operatorname{Re}\left(\xi_{\mathcal{K}}\left(\frac{1}{2} + \frac{i}{\log_e r}\theta\right)\right) = 1 - \sum_{m=1}^{\infty} m\varphi_{r^m} h_m(\theta)$$

for

$$h_m(\theta) = \frac{r^{m/2}\cos(m\theta) - 1}{r^m + 1 - 2r^{m/2}\cos(m\theta)},$$

which depends only on the family $\varphi = \{\varphi_{r^m}\}$, *and the following basic equality holds*:

$$\xi_{\mathcal{K}}\left(\frac{1}{2}\right) = 1 - \sum_{m=1}^{\infty} \frac{m\varphi_{r^m}}{r^{m/2} - 1} = M_{\mathcal{K}}(0).$$

PROOF. This is a simple corollary of the proof of Proposition 8.1.15. Indeed, the formula for $\xi_{\mathcal{K}}(1/2)$ is its immediate consequence; the general formula follows from the identity

$$h_m(\theta) = \sum_{n=1}^{\infty} r^{-mn/2}\cos(mn\theta). \qquad \square$$

Number field case. We are going to study the asymptotic distribution of zeroes of ζ_K for g_K tending to infinity, assuming the GRH to hold.

Recall some notation. Let $\mathcal{D}'$ be the space of distributions dual to $\mathcal{D} = C_0^\infty(\mathbb{R})$, and let $\mathcal{S}'$ be that of tempered distributions, which is dual to the Schwartz space $S(\mathbb{R}) \subset C^\infty(\mathbb{R})$. The space $\mathcal{M} \subset \mathcal{D}'$ of measures contains the cone $\mathcal{M}_+$ of positive measures, i.e., measures whose value at a positive real-valued function is positive, and for the cone $\mathcal{D}'_+$ of positive distributions satisfying the same condition, we have $\mathcal{D}'_+ = \mathcal{M}_+$. The space of measures of slow growth is $\mathcal{M}_{\mathrm{sl}} = \mathcal{M} \cap \mathcal{S}'$, and a measure is of slow growth if and only if the integral

$$I_k = \int_{-\infty}^{\infty} (x^2+1)^{-k}\, d\mu$$

converges for some positive integer k.

We are going to associate some measure of slow growth with an asymptotically exact family $\mathcal{F} = \{K_j\}$ of number fields. For each K_j we define the measure

$$\Delta_{K_j} := \frac{\pi}{g_j} \sum_{\zeta_{K_j}(\rho)=0} \delta_{t(\rho)},$$

where $g_j := g_{K_j}$, $t(\rho) = \left(\rho - \frac{1}{2}\right)/i$, and ρ runs over all nontrivial zeroes of the zeta function $\zeta_{K_j}(s)$. In virtue of the GRH, $t(\rho)$ is real, and Δ_{K_j} is a discrete measure on $\mathbb{R}$, which is of slow growth. The latter fact follows, e.g., from the Weil explicit formula; see the proof of GRH Theorem 8.1.28 below.

Now we are ready to formulate the main result of this section, expressing the limit distribution of zeta zeroes in terms of the parameters $\varphi = \{\varphi_\alpha\}$ of the asymptotically exact family.

THEOREM 8.1.28 (GRH explicit formula). *For an asymptotically exact family $\mathcal{K}$, the weak limit*

$$\Delta = \Delta_{\mathcal{K}} := \lim_{j\to\infty} \Delta_{K_j}$$

exists in the space of measures of slow growth on $\mathbb{R}$. Moreover, the measure Δ has a continuous density M_φ given by

$$M_\varphi(t) = \operatorname{Re}\xi_\varphi\left(\frac{1}{2}+it\right) = 1 - \sum_q \varphi_q h_q(t)\log q + \frac{1}{2}\varphi_{\mathbb{R}} \operatorname{Re}\psi\left(\frac{1}{4}+\frac{it}{2}\right)$$
$$+\varphi_{\mathbb{C}} \operatorname{Re}\psi\left(\frac{1}{2}+it\right) - \frac{\varphi_{\mathbb{R}}}{2}\log\pi - \varphi_{\mathbb{C}}\log 2\pi,$$

where

$$h_q(t) = \frac{\sqrt{q}\cos(t\log q) - 1}{q+1-2\sqrt{q}\cos(t\log q)}.$$

REMARK 8.1.29. This density (and thus the measure Δ) depends only on the parameters $\varphi = \{\varphi_\alpha\}$.

COROLLARY 8.1.30 (GRH basic equality). *We have*

$$\sum_q \frac{\varphi_q \log q}{\sqrt{q}-1} + \varphi_{\mathbb{R}}\left(\log\sqrt{8\pi} + \frac{\pi}{4} + \frac{\gamma}{2}\right) + \varphi_{\mathbb{C}}(\log 8\pi + \gamma) = 1 - M_\varphi(0);$$

therefore, for the deficiency of the family we have

$$\delta_{\mathcal{K}} = \xi_{\mathcal{K}}\left(\frac{1}{2}\right) = M_\varphi(0).$$

PROOF. Put $t = 0$ in the formula for $M_\varphi(t)$. □

PROOF OF GRH THEOREM 8.1.28. The plan of the proof is as follows. First we are going to prove that Δ exists as an element of $\mathcal{S}'$; therefore, being positive, it lies in $\mathcal{M}_{sl,+} = \mathcal{S}'_+$. The next point is to show that this measure is absolutely continuous with respect to the Lebesgue measure dt, i.e., is of the form $F(t)\,dt$; to do this, we have to prove that neither a skyscraper nor singular component occurs. Then we shall compare this measure $\Delta = F(t)\,dt$ with the measure $\Delta_0 = M_\varphi(t)\,dt$ of the theorem. We show that they coincide on the set of some specific test functions $H_{y,a}(t)$, which will suffice to deduce the conclusion.

We prove that Δ is well defined as a tempered distribution using the Weil explicit formula. To formulate it, we assume in the notation of Sec. 8.1.2 that $F \in \mathcal{S}(\mathbb{R})$ and that

$$(*) \qquad F(x), F'(x) \leq ce^{-(1/2+\varepsilon)|x|} \quad \text{as} \quad |x| \to \infty.$$

Note that for $s = \frac{1}{2} + it$ we have

$$\Phi(s) := \int_{-\infty}^{\infty} F(x)e^{(s-1/2)x}\, dx = \widehat{F}(t),$$

$\widehat{F}(t)$ being the Fourier transform of F.

Then the Weil explicit formula reads as follows.

THEOREM 8.1.31. *Let K be a number field. Then the limit*

$$\sum{}' \Phi(\rho) := \lim_{T\to\infty} \sum_{|\rho|<T} \Phi(\rho) = \lim_{T\to\infty} \sum_{\substack{|\frac{1}{2}+it|<T \\ \zeta_K(\frac{1}{2}+it)=0}} \widehat{F}(t)$$

exists, where ρ in the sum runs over the set of zeroes of $\zeta_K(s)$ on the critical line $\mathrm{Re}(s) = 1/2$ (which is assumed to be the set of all critical zeroes of $\zeta_K(s)$), and we have the following formula:

$$\begin{aligned}\sum{}' \Phi(\rho) - \Phi(0) - \Phi(1) &= F(0)(\log|D_K| - r_1 \log \pi - 2r_2 \log 2\pi) \\ &- \sum_P \sum_{m=1}^{\infty} N(P)^{-m/2}[F(m \log N(P)) + F(-m \log N(P))] \log N(P) \\ &+ \frac{r_1}{2\pi} \int_{-\infty}^{\infty} \frac{\widehat{F}(t) + \widehat{F}(-t)}{2} \mathrm{Re}\,\psi\Big(\frac{1}{4} + \frac{it}{2}\Big)\, dt \\ &+ \frac{r_2}{\pi} \int_{-\infty}^{\infty} \frac{\widehat{F}(t) + \widehat{F}(-t)}{2} \mathrm{Re}\,\psi\Big(\frac{1}{2} + it\Big)\, dt,\end{aligned}$$

where the outer sum is taken over all primes P of K, and $N(P)$ denotes the absolute norm of P. □

We apply the formula only to functions from $\mathcal{S}(\mathbb{R})$, which guarantees that all expressions in the proof are correctly defined.

Replacing any function F from $\mathcal{S}(\mathbb{R})$ with $\dfrac{F(t) + F(-t)}{2}$, we can apply the result to F.

Let us rewrite the sum

$$\sum_P \sum_{m=1}^{\infty} N(P)^{-m/2}[F(m \log N(P)) + F(-m \log N(P))] \log N(P)$$

as

$$\sum_q N_q(K) \sum_{m=1}^{\infty} q^{-m/2}[F(m \log q) + F(-m \log q)] \log q,$$

the sum being over all prime powers q.

Let $\widehat{\mathcal{D}} \subset \mathcal{S}$ be the Fourier dual of $\mathcal{D}$, i.e., the set $\widehat{F}$ for $F \in \mathcal{D}$; it is a dense subspace of $\mathcal{S}$ (since $\mathcal{D}$ is dense in $\mathcal{S}$); and let $K = K_j$ be a field from our family. For any $f \in \widehat{\mathcal{D}}$ put $f = \widehat{F}$ with $F \in \mathcal{D}$. Then we have $f(t) = \Phi\left(\frac{1}{2}+it\right)$. The function F satisfies the above condition $(*)$, and, dividing by $2g_{K_j} = \log|D_{K_j}|$, we get

$$\begin{aligned}
\Delta_{K_j}(f) &= \frac{\pi}{g_{K_j}} \sum_{\zeta_K(\frac{1}{2}+it)=0}' f(t) \\
&= 2\pi\left(\frac{\Phi(0)+\Phi(1)}{2g_{K_j}} + F(0)\left(1 - \frac{r_1}{2g_{K_j}}\log\pi - \frac{r_2}{g_{K_j}}\log 2\pi\right)\right) \\
&\quad - 2\pi\left(\sum_q \frac{N_q(K_j)}{2g_{K_j}} \sum_{m=1}^{\infty} q^{-m/2}[F(m\log q) + F(-m\log q)]\log q\right) \\
&\qquad + \frac{r_1}{2g_{K_j}} \int_{-\infty}^{\infty} \frac{f(t)+f(-t)}{2} \operatorname{Re}\psi\left(\frac{1}{4}+\frac{it}{2}\right) dt \\
&\qquad\quad + \frac{r_2}{g_{K_j}} \int_{-\infty}^{\infty} \frac{f(t)+f(-t)}{2} \operatorname{Re}\psi\left(\frac{1}{2}+it\right) dt.
\end{aligned}$$

We fix f and tend j to infinity; then the right-hand side tends to

$$\begin{aligned}
\Delta(f) := 2\pi F(0)&\left(1 - \frac{\varphi_{\mathbb{R}}}{2}\log\pi - \varphi_{\mathbb{C}}\log 2\pi\right) \\
&- \pi \sum_q \varphi_q \sum_{m=1}^{\infty} q^{-m/2}[F(m\log q) + F(-m\log q)]\log q \\
&\quad + \frac{\varphi_{\mathbb{R}}}{2} \int_{-\infty}^{\infty} \frac{f(t)+f(-t)}{2} \operatorname{Re}\psi\left(\frac{1}{4}+\frac{it}{2}\right) dt \\
&\qquad + \varphi_{\mathbb{C}} \int_{-\infty}^{\infty} \frac{f(t)+f(-t)}{2} \operatorname{Re}\psi\left(\frac{1}{2}+it\right) dt,
\end{aligned}$$

since the family is asymptotically exact, and this implies the existence of the limit

$$\Delta(f) = \lim_{j\to\infty} \Delta_{K_j}(f)$$

for any $f \in \widehat{\mathcal{D}}$. Since the map $f \mapsto \Delta(f)$ is linear and since $\widehat{\mathcal{D}}$ is dense in $\mathcal{S}$, in order to prove that Δ is a tempered distribution, it suffices to verify that Δ is continuous in the topology of $\mathcal{S}$. Now let $\{f_i\}$, $i = 1, 2, \ldots$, be a sequence with $f_i \in \widehat{\mathcal{D}}$ tending to zero in the topology of $\mathcal{S}$; then the sequence $\{F_i\}$, where $f_i = \widehat{F}_i$, tends to zero as well, since the Fourier transform is an isometry of $\mathcal{S}$. In particular, both sequences $\{f_i\}$ and $\{F_i\}$ tend to zero uniformly. Let us then show that Δ is continuous. Indeed, we have

$$\Delta(f_i) = T_1(f_i) - T_2(f_i) + T_3(f_i) + T_4(f_i),$$

where

$$T_1(f_i) = 2\pi F_i(0)\left(1 - \frac{\varphi_{\mathbb{R}}}{2}\log\pi - \varphi_{\mathbb{C}}\log 2\pi\right) = \delta(\widehat{f_i})\left(1 - \frac{\varphi_{\mathbb{R}}}{2}\log\pi - \varphi_{\mathbb{C}}\log 2\pi\right),$$

$$T_2(f_i) = \pi\sum_q \varphi_q \sum_{m=1}^{\infty} q^{-m/2}[F_i(m\log q) + F_i(-m\log q)]\log q,$$

$$T_3(f_i) = \frac{\varphi_{\mathbb{R}}}{2}\int_{-\infty}^{\infty} \frac{f_i(t)+f_i(-t)}{2}\operatorname{Re}\psi\Big(\frac{1}{4}+\frac{it}{2}\Big)\,dt,$$

$$T_4(f_i) = \varphi_{\mathbb{C}}\int_{-\infty}^{\infty} \frac{f_i(t)+f_i(-t)}{2}\operatorname{Re}\psi\Big(\frac{1}{2}+it\Big)\,dt.$$

The term $T_1(f_i)$ is clearly continuous as well as the terms $T_3(f_i)$ and $T_4(f_i)$, since the measures $\operatorname{Re}\psi\Big(\frac{1}{2}+it\Big)\,dt$ and $\operatorname{Re}\psi\Big(\frac{1}{4}+\frac{it}{2}\Big)\,dt$ are of slow growth. The continuity of $T_2(f_i)$ follows from the inequality

$$T_2(f_i) \le 2\pi\sup_{x\in\mathbb{R}}|F_i(x)|,$$

valid in virtue of the GRH version of the basic inequality, and the continuity of Δ is proved.

Thus, Δ is a tempered distribution, $\Delta \in \mathcal{S}'$, and it is positive by its very definition. Therefore, Δ is a measure of slow growth, and we have to show that $\Delta = M_\varphi\,dx$; for this, we need the following lemmas.

For any $a\in\mathbb{R}$ and any $y>0$ let us define the function $H_{y,a}(x)\in\mathcal{S}(\mathbb{R})$ by

$$H_{y,a}(x) := \frac{1}{2\sqrt{\pi y}}\exp\Big(\frac{-(a-x)^2}{4y}\Big)$$

so that

$$F_{y,a}(x) = \frac{\widehat{H}_{y,a}(x)}{2\pi} := \frac{1}{2\pi}\exp(-yx^2+iax).$$

LEMMA 8.1.32. *We have*

$$\lim_{y\to+0}\Delta(H_{y,a}) = M_\varphi(a). \qquad\square$$

Note that for y tending to zero the function $H_{y,a}$ tends to δ_a in the sense of distributions.

LEMMA 8.1.33. *Let μ be a positive measure on $\mathbb{R}$ such that for any $a\in\mathbb{R}$ one has*

$$\lim_{y\to+0}\mu(H_{y,a}) = M(a)$$

with a function M continuous on $\mathbb{R}$. Then $\mu = M(x)\,dx$. $\square$

The theorem follows from these two lemmas, and their proofs are rather standard exercises; for Lemma 8.1.33, it uses the decomposition of μ as a sum of an absolutely continuous, singular, and Dirac measures. $\square$

REMARK 8.1.34. Theorem 8.1.28 implies that zeta zeroes are asymptotically uniformly distributed for any asymptotically bad family, e.g., for a family of number fields of fixed absolute degree.

8.2. The Generalized Brauer–Siegel Theorem

In this section we show that our approach allows us to generalize and improve the classical Brauer–Siegel theorem, which reads as follows.

THEOREM 8.2.1 (Brauer–Siegel). *If K_i is a sequence of number fields with $n_i = [k_i : \mathbb{Q}] < \infty$ such that*

$$\frac{n_i}{\log |D_{K_i}|} \to 0$$

and one of the following holds:

(a) *GRH or*

(b) *K_i are normal over $\mathbb{Q}$,*

then

$$\lim_{i\to\infty} \frac{\log(h_i R_i)}{g_i} = 1,$$

where h_i denotes the class number of K_i. □

Note that this theorem is naturally a result on an asymptotically bad family of number fields. A natural question of whether this last assumption is necessary was not answered for several decades. Our setting and techniques allow us to establish this necessity and give a precise form of the theorem for any asymptotically exact family.

8.2.1. Main Result

For an asymptotically exact family $\mathcal{K} = \{K_i\}$ of global fields consider the limits

$$\mathrm{BS}(\mathcal{K}) = \lim_{i\to\infty} \frac{\log(h_i R_i)}{g_i}$$

(the Brauer–Siegel ratio) and

$$\varkappa(\mathcal{K}) = \lim_{i\to\infty} \frac{\log \varkappa_i}{g_i}.$$

Here h_i is the class number, R_i the regulator, and $\varkappa_i$ the zeta residue at $s = 1$. We are going to show that these limits exist and depend only on the parameters $\varphi = \{\varphi_\alpha\}$, and therefore $\mathrm{BS}(\mathcal{K})$ is well defined for an infinite global field $\mathcal{K}$.

Function field case. We begin with an elementary setting for function fields. We set $R = 1$, and, of course, φ_α can be nonzero only for $\alpha = r^m$, $m = 1, 2, 3, \ldots$.

First of all we have the following generalized Brauer–Siegel theorem:

For an asymptotically exact family of function fields over $\mathbb{F}_r$ we have

$$\mathrm{BS}(\mathcal{K}) = \lim_{i\to\infty} \frac{\log_r h_i}{g_i} = 1 + \sum_{m=1}^{\infty} \varphi_{r^m} \log_r \frac{r^m}{r^m - 1}.$$

The function field case analogue of the classical Brauer–Siegel theorem corresponds to the asymptotically bad case (i.e., $\varphi_\alpha = 0$ for any α), for which $\mathrm{BS}(\mathcal{K}) = 1$.

Next we have the following bounds.

PROPOSITION 8.2.2. *For any family of function fields over $\mathbb{F}_r$ we have*

$$1 \le \liminf_{i\to\infty} \frac{\log_r h_i}{g_i} \le \limsup_{i\to\infty} \frac{\log_r h_i}{g_i} \le 1 + (\sqrt{r} - 1) \log_r \frac{r}{r-1}.$$

PROOF. We use the following classical formula for h_i:

$$h_i = \prod_{j=1}^{g_i}(1-\omega_{ij}) = r^{g_i}\prod_{j=1}^{g_i}(1-\overline{\omega}_{ij}),$$

where ω_{ij}, $j=1,\dots,g_i$, are the Frobenius eigenvalues of K_i. Let

$$a_{ij} := \frac{\omega_{ij}}{\sqrt{r}}, \quad |\alpha_{ij}| = 1, \quad j=1,\dots,g_i.$$

Thus we get

$$\begin{aligned}\log h_i = \sum_{j=1}^{g_i}\log(1-\omega_{ij}) &= g_i + \sum_{j=1}^{g_i}\log(1-\alpha_{ij}/\sqrt{r}) \\ &= g_i - \sum_{j=1}^{g_i}\sum_{m=0}^{\infty}\frac{\alpha_{ij}^m}{r^{m/2}} = g_i - \sum_{m=0}^{\infty} r^{-m/2}\sum_{j=1}^{g_i}\alpha_{ij}^m.\end{aligned}$$

Hence,

$$\frac{\log h_i}{g_i} = 1 - \sum_{m=0}^{\infty} r^{-m/2}\frac{1}{g_i}\sum_{j=1}^{g_i}\alpha_{ij}^m.$$

Since

$$\frac{r^{-m/2}}{g_i}\sum_{j=1}^{g_i}\alpha_{ij}^m = \frac{-N_{r^m}+1+r^m}{g_i r^{m/2}}$$

tends to φ_{r^m} for a fixed j and $g_i\to\infty$, we get the conclusion. □

We also have some partial existence results. In particular, both bounds 1 and $1+(\sqrt{r}-1)\log_r\frac{r}{r-1}$ are attainable; the lower bound 1 is attained for any asymptotically bad family, while the value $1+(\sqrt{r}-1)\log_r\frac{r}{r-1}$ is attained for any asymptotically maximal family, i.e., such that $\varphi_r=\sqrt{r}-1$ and $\varphi_{r^m}=0$ for all $m\neq 1$, e.g., for those from Chapter 5 or Sec. 7.4.

Another limit parameter $\varkappa(\mathcal{K})$ gives no new information in the function field case, since $\varkappa(\mathcal{K}) = \mathrm{BS}(\mathcal{K})-1$ for any $\mathcal{K}$.

Now we are going to present number field analogues of these results, which happen to be much more involved.

THEOREM 8.2.3 (generalized Brauer–Siegel inequality). *For an asymptotically exact family of number fields one has*

$$\limsup_{i\to\infty}\frac{\log(h_iR_i)}{g_i} \le 1+\sum_q\varphi_q\log\frac{q}{q-1} - \varphi_{\mathbb{R}}\log 2 - \varphi_{\mathbb{C}}\log 2\pi,$$

$$\limsup_{i\to\infty}\frac{\log\varkappa_i}{g_i} \le \sum_q\varphi_q\log\frac{q}{q-1},$$

the sum being over all prime powers q.

This result is unconditional. Assuming the GRH, we get the equality.

THEOREM 8.2.4 (GRH generalized Brauer–Siegel theorem). *For an asymptotically exact family $\mathcal{K}$ of number fields the limits* $\mathrm{BS}(\mathcal{K})$ *and* $\varkappa(\mathcal{K})$ *exist and we have*

$$\mathrm{BS}(\mathcal{K}) := \lim_{i\to\infty} \frac{\log(h_i R_i)}{g_i} = 1 + \sum_q \varphi_q \log\frac{q}{q-1} - \varphi_{\mathbb{R}}\log 2 - \varphi_{\mathbb{C}}\log 2\pi,$$

$$\varkappa(\mathcal{K}) := \lim_{i\to\infty} \frac{\log \varkappa_i}{g_i} = \sum_q \varphi_q \log\frac{q}{q-1},$$

the sum being over all prime powers q.

One can express the generalized Brauer–Siegel theorem (GRH Theorem 8.2.4 and Theorem 8.2.5 below) either as

$$\lim_{i\to\infty} \frac{\log h_i R_i}{g_i} = \log\widetilde{\zeta}_\varphi(1)$$

or as

$$\lim_{i\to\infty} \frac{\log \varkappa_i}{g_i} = \log\zeta_\varphi(1),$$

where $\varkappa_i$ is the residue of $\zeta_{K_i}(s)$ at 1. The function field case of the generalized Brauer–Siegel theorem reads either as

$$\lim_{i\to\infty} \frac{\log_r h_i}{g_i} = \log_r(\widetilde{\zeta}_\varphi(1))$$

or as

$$\lim_{i\to\infty} \frac{\log_r \varkappa_i}{g_i} = \log_r(\zeta_\varphi(1)).$$

If we restrict our attention to the case of *almost normal towers*, we can prove the same unconditionally. To formulate the result we need one more definition.

Let K be a number field. We call K *almost normal* if there exists a finite tower of number fields $\mathbb{Q} = K_0 \subset K_1 \subset \ldots \subset K_m = K$ such that all the extensions K_i/K_{i-1} are normal. A family is called almost normal if all its fields are, and an infinite number field is called almost normal if it is a limit of an almost normal tower.

THEOREM 8.2.5 (unconditional generalized Brauer–Siegel theorem). *For an asymptotically good tower $\mathcal{K} = \{K_i\}$, $K_1 \subset K_2 \subset \ldots$, of almost normal number fields, the limits* $\mathrm{BS}(\mathcal{K})$ *and* $\varkappa(\mathcal{K})$ *do exist and we have*

$$\mathrm{BS}(\mathcal{K}) := \lim_{i\to\infty} \frac{\log(h_i R_i)}{g_i} = 1 + \sum_q \varphi_q \log\frac{q}{q-1} - \varphi_{\mathbb{R}}\log 2 - \varphi_{\mathbb{C}}\log 2\pi,$$

$$\varkappa(\mathcal{K}) := \lim_{i\to\infty} \frac{\log \varkappa_i}{g_i} = \sum_q \varphi_q \log\frac{q}{q-1},$$

the sum being over all prime powers q.

Now we are going to prove these results.

PROOF OF THEOREMS 8.2.3, 8.2.4, AND 8.2.5. We begin with the inequality of Theorem 8.2.3, which does not require additional conditions. Passing to a subfamily, we can assume that there exits the limit (maybe infinite)

$$\lim_{i\to\infty} \frac{\log(h_i R_i)}{g_i}.$$

For any real $s>1$ and any K we have

$$\zeta_K(s)=\frac{\varkappa_K}{s-1}F_K(s),$$

$\varkappa_K$ being the residue of $\zeta_K(s)$ at 1, and $F_K(s)$ being an analytic function in a neighbourhood of 1 with $F_K(1)=1$.

Let us first remark that

$$\frac{\log \varkappa_{K_j}}{g_j} \longrightarrow \lim_{j\to\infty}\frac{\log(h_jR_j)}{g_j}-1+\varphi_{\mathbb{R}}\log 2+\varphi_{\mathbb{C}}\log 2\pi.$$

Indeed, we have

$$\varkappa_K=\frac{2^{r_1}(2\pi)^{r_2}(h_KR_K)}{w_K\sqrt{|D_K|}},$$

and taking the logarithm for $g_j=\log\sqrt{|D_{K_j}|}\to\infty$, i.e., for $j\to\infty$, we get the desired formula, since $\log w_{K_j}/\log|D_{K_j}|\to 0$ by virtue of the inequality $w_{K_j}\le cn_{K_j}^2$ for an absolute constant c.

Let us put $s=1+\theta$ with $\theta=\theta_j>0$ being a small positive real number, its dependence on j to be specified later. Taking the logarithm of

$$\zeta_{K_j}(s)=\frac{\varkappa_{K_j}}{s-1}F_{K_j}(s)$$

and dividing by g_j we get

$$\frac{\log\zeta_{K_j}(1+\theta_j)}{g_j}=\frac{\log\varkappa_{K_j}}{g_j}+\frac{\log F_{K_j}(1+\theta_j)}{g_j}-\frac{\log\theta_j}{g_j}.$$

Now it suffices to show that for a proper choice of θ_j for $j\to\infty$ the following relations hold:

(i) $$\frac{\log\zeta_{K_j}(1+\theta_j)}{g_j}\longrightarrow\sum_q\varphi_q\log\frac{q}{q-1};$$

(ii) $$\frac{\log\theta_j}{g_j}\longrightarrow 0;$$

(iii) $$\liminf\frac{\log F_{K_j}(1+\theta_j)}{g_j}\ge 0,$$

which can be proved using standard analytic techniques.

LEMMA 8.2.6 (GRH). *For any asymptotically exact family of number fields the function*

$$Z_j(s):=\frac{-(\log(\zeta_{K_j}(s)))'-1/(s-1)}{g_j}$$

tends for $j\to\infty$ to

$$Z_\varphi(s):=\sum_q\varphi_q\frac{\log q}{q^s-1}$$

uniformly on $\mathrm{Re}(s)\ge 1/2+\delta$ *for any* $\delta>0$. □

The proof is a rather tedious exercise in analytic number theory.

END OF PROOF OF GRH THEOREM 8.2.4. From Lemma 8.2.6 it follows, in particular, that $Z_j(s)$ tends to $Z_\varphi(s)$ for $j\to\infty$ uniformly on $\mathrm{Re}(s)\ge 1$. Therefore, for small enough ε, large enough j, and any $\theta > 0$, we have

$$|(\log F_{K_j}(1+\theta)/g_j)'| = |Z_j(1+\theta)| \leq |Z_\varphi(1)| + \varepsilon \leq 1,$$

since $|Z_\varphi(1)| < 1$ because of Proposition 8.1.20. Thus

$$\log F_{K_j}(1+\theta_j)/g_j \leq \theta_j.$$

This proves (iii) and the theorem. □

REMARK 8.2.7. In fact, one notes that Corollary 8.2.10 remains valid under the assumption that there are no zeta zeroes with $\mathrm{Re}(s) \geq 1-\delta$ for arbitrary fixed $\delta > 0$, so that we do not need the full strength of the GRH.

To prove the opposite inequality of Theorem 8.2.5, it is sufficient to show that for an asymptotically good almost normal tower $\{K_j\}$ and for a proper choice of θ_j we have the following:

(i)″
$$\liminf f_j(\theta_j) \geq \sum_q \varphi_q \log \frac{q}{q-1} = f(0);$$

(ii)″
$$\frac{\log \theta_j}{g_j} \longrightarrow 0;$$

(iii)″
$$\limsup \log F_{K_j}(1+\theta_j)/g_j \leq 0.$$

We should stress that the choice of θ_j in the proof of Theorem 8.2.5 is completely different from that in the proofs of Theorems 8.2.3 and 8.2.4. We need the following definition.

Let K be a number field (of a finite degree n). A real number ρ is called an *exceptional zero* of $\zeta_K(s)$ if $\zeta_K(\rho) = 0$ and

$$1 - (4\log|D_K|)^{-1} \leq \rho < 1;$$

an exceptional zero ρ of $\zeta_K(s)$ is called its *Siegel zero* if

$$1 - (16\log|D_K|)^{-1} \leq \rho < 1.$$

It is known that for any K there exists at most one exceptional zero of $\zeta_K(s)$.

We are going to show that under some conditions asymptotically exact families have no Siegel zero. We begin with the following fundamental property of Siegel zeroes due to Heilbronn and Stark.

LEMMA 8.2.8. *Let K be an almost normal number field, and let ρ be a Siegel zero of $\zeta_K(s)$. Then there is a quadratic subfield k of K such that $\zeta_k(\rho) = 0$.* □

LEMMA 8.2.9. *Let $\mathcal{K} = \{K_i\}$ be an asymptotically good family of almost normal number fields. Then there exists a positive integer I such that $\zeta_{K_i}(s)$ has no Siegel zero for any $i \geq I$. In other words, in such a family almost all fields have no Siegel zero.*

PROOF. In view of Lemma 8.2.8 it is sufficient to prove that the set $Q(\mathcal{K})$ of quadratic fields k contained in at least one of the fields K_i is finite. Indeed, if this is the case, let

$$\beta = \max\{\rho \in \mathbb{R} : \zeta_k(\rho) = 0 \text{ for } k \in Q(\mathcal{K})\}.$$

Since $Q(\mathcal{K})$ is finite, the maximum exists and $\beta < 1$. Now if

$$g_{K_i} > \frac{1}{16(1-\beta)},$$

then $\zeta_{K_i}(s)$ has no Siegel zeroes by Lemma 8.2.8.

Let us verify the finiteness of $Q(\mathcal{K})$. The ratio n/g is nonincreasing in extensions, since

$$|D_K| \geq |D_k|^{[K:k]}.$$

Moreover, $n_{K_i}/g_{K_i} \to \varphi_\infty = \varphi_\mathbb{R} + 2\varphi_\mathbb{C} > 0$, the family $\mathcal{K}$ being asymptotically good; therefore, there exists a positive real number ε such that $n_{K_i}/g_{K_i} \geq \varepsilon$ for any i. If $k \in Q(\mathcal{K})$, $k \subseteq K_i$, then

$$2/g_k = n_k/g_k \geq n_{K_i}/g_{K_i} \geq \varepsilon,$$

and it follows that $g_k \leq 2/\varepsilon$ for any $k \in Q(\mathcal{K})$, and $|D_k| \leq e^{4/\varepsilon}$, which implies the finiteness of $Q(\mathcal{K})$. □

COROLLARY 8.2.10. *Let K be a number field with infinite Hilbert class field tower $\{K_i\}$. Then almost all fields K_i have no Siegel zero.* □

Note that for any $\theta > 0$

$$\left(\frac{\log F_{K_j}(1+\theta)}{g_j}\right)' = Z_j(1+\theta)$$

for

$$Z_j(s) := \frac{-(\log(\zeta_{K_j}(s)))' - 1/(s-1)}{g_j}.$$

LEMMA 8.2.11. *There exist absolute constants C_0 and $C > 0$ such that for any number field K which has no Siegel zero we have*

$$|Z(1+\theta)| \leq Cg^6$$

for

$$Z(s) := \frac{-(\log(\zeta_K(s)))' - 1/(s-1)}{g},$$

any real $\theta \in (0,1)$, and any $g = g(K) > C_0$. □

The proof uses Lagarias–Odlyzko's estimate of the Chebyshev step function and standard techniques of analytic number theory.

Lower bounds for regulators. As an application of the generalized Brauer–Siegel theorem, we can obtain a lower bound for regulators of number fields in asymptotically good families which is better than the known general bound.

THEOREM 8.2.12 (GRH). *For an asymptotically good family of number fields, we have*

$$\liminf_{i\to\infty} \frac{\log R_i}{g_i} \geq (\log\sqrt{\pi e} + \gamma/2)\varphi_\mathbb{R} + (\log 2 + \gamma)\varphi_\mathbb{C}.$$

PROOF. We begin with an estimate for the class numbers of fields in question, which could be of independent interest.

PROPOSITION 8.2.13. *For an asymptotically exact family of number fields, we have*

$$\limsup_{i\to\infty}\frac{\log h_i}{g_i}\le 1-(\log 2\sqrt{\pi}+\frac{\gamma+1}{2})\varphi_{\mathbb{R}}-(\log 4\pi+\gamma)\varphi_{\mathbb{C}}+\sum\varphi_q\log\frac{q}{q-1}.$$

PROOF. Let $\mathcal{K}=\{K_i\}$ be an asymptotically exact family of number fields and let

$$\zeta_{K_i}(s)=\sum_{n=1}^{\infty}D_n^{(i)}n^{-s}$$

be the corresponding zeta functions. We shall use the following result on the asymptotic behaviour of the coefficients $D_n^{(i)}$.

LEMMA 8.2.14. *Let n_i, $i=1,2,\ldots$, be a sequence of positive integers such that the limit*

$$\nu:=\lim_{i\to\infty}\frac{n_i}{g_i}$$

exists. Then

$$\limsup_{i\to\infty}\frac{\log D_{n_i}^{(i)}}{g_i}=\nu\limsup_{i\to\infty}\frac{\log D_{n_i}^{(i)}}{n_i}\le\nu+\sum_q\varphi_q\log\frac{q}{q-1},$$

where the sum is over all prime powers.

PROOF. Indeed, from the Euler product for $\zeta_{K_i}(s)$ we see the following:

(a) The function $n\to D_n^{(i)}$ is multiplicative, i.e., $D_{nn'}^{(i)}=D_n^{(i)}D_{n'}^{(i)}$ for coprime n and n'. In particular,

$$D_n^{(i)}=\prod D_{p_j^{m_j}}^{(i)},$$

$n=\prod p_j^{m_j}$ being the prime factorization of n.

(b) We have

$$D_{p^m}^{(i)}=\sum_{\substack{(b_1,\ldots,b_m)\\ b_1+2b_2+\ldots+mb_m=m}}\prod_{s=1}^{m}\binom{N_{p^s}(K_i)+b_s-1}{b_s},$$

the sum being over all partitions of m, $b_i\in\mathbb{Z}$, $b_i\ge 0$.

Let $n_i=\prod p_j^{m_{ij}}$. Then

$$\log D_{n_i}^{(i)}=\sum_j\log D_{p_j^{m_{ij}}}^{(i)},$$

$$D_{p_j^{m_{ij}}}^{(i)}=\sum_{\substack{(b_1,\ldots,b_{m_{ij}})\\ b_1+2b_2+\ldots+m_{ij}b_{m_{ij}}=m_{ij}}}\prod_{s=1}^{m_{ij}}\binom{N_{p_j^s}(K_i)+b_s-1}{b_s}.$$

This implies

$$\log D_{p_j^{m_{ij}}}^{(i)}\le\log\boldsymbol{p}(m_{ij})+\max_{\substack{(b_1,\ldots,b_{m_{ij}})\\ b_1+2b_2+\ldots+m_{ij}b_{m_{ij}}=m_{ij}}}\left(\sum_{s=1}^{m_{ij}}\log\binom{N_{p_j^s}(K_i)+b_s-1}{b_s}\right),$$

where $\boldsymbol{p}(x)$ is the partition function. Now the argument used in the proof of Lemma 4.5.12 shows that

$$\log D^{(i)}_{p_j^{m_{ij}}} \le O(\sqrt{g}) + \log p_j^{m_{ij}} + \varphi_{p_j^{m_{ij}}} \log \frac{p_j^{m_{ij}}}{p_j^{m_{ij}} - 1},$$

which proves the lemma. □

END OF PROOF OF PROPOSITION 8.2.13. Zimmert's theorem on twin classes states that for $\lambda_{\mathbb{R}} = \log 2\sqrt{\pi} + \frac{\gamma+1}{2}$, $\lambda_{\mathbb{C}} = \log 4\pi + \gamma$, and any class $\mathcal{C}$ of ideals of a number field k, we have

$$\frac{n_{\inf}(\mathcal{C}) + n_{\inf}(\mathcal{C}^*)}{2} \le g - \lambda_{\mathbb{R}} r_1 - \lambda_{\mathbb{C}} r_2$$

and thus

$$\frac{\nu_{\inf}(\mathcal{C}) + \nu_{\inf}(\mathcal{C}^*)}{2} \le 1 - \lambda_{\mathbb{R}}\varphi_{\mathbb{R}} - \lambda_{\mathbb{C}}\varphi_{\mathbb{C}},$$

where $\nu_{\inf}(\mathcal{C}) = n_{\inf}(\mathcal{C})/g$, $n_{\inf}(\mathcal{C})$ being the minimum norm of an ideal from $\mathcal{C}$, and $\mathcal{C}^*$ being the twin class of the class $\mathcal{C}$ (i.e., the class $\mathcal{D}\mathcal{C}^{-1}$, where $\mathcal{D}$ is the class of the different). This implies that the class number h_i of K_i is not greater than two times the number of "small norm" ideals, i.e., those counted in $D^{(i)}_{n_i'}$ for $n_i' \le \widetilde{n}_i = g_i(1 - \lambda_{\mathbb{R}}\varphi_{\mathbb{R}} - \lambda_{\mathbb{C}}\varphi_{\mathbb{C}}) + o(g_i)$. By Lemma 8.2.14, we get

$$\begin{aligned}
\limsup_{i\to\infty} \frac{\log h_i}{g_i} &\le \limsup_{i\to\infty} \frac{1}{g_i} \log(2 \sum_{n_i' \le \widetilde{n}_i} D^{(i)}_{n_i'}) \\
&\le \limsup_{i\to\infty} \frac{1}{g_i} \log(\max_{n_i' \le \widetilde{n}_i} D^{(i)}_{n_i'}) \\
&\le 1 - \lambda_{\mathbb{R}}\varphi_{\mathbb{R}} - \lambda_{\mathbb{C}}\varphi_{\mathbb{C}} + \sum \varphi_q \log \frac{q}{q-1}.
\end{aligned}$$

□

END OF PROOF OF THEOREM 8.2.12. To finish the proof, one compares the generalized Brauer–Siegel theorem with Proposition 8.2.13. □

Passing to an asymptotically exact subfamily we easily deduce the following.

COROLLARY 8.2.15 (GRH). *Let $\mathcal{S}$ be any family of number fields which does not contain an asymptotically bad subfamily. Then there exists a strictly positive $A = A(\mathcal{S})$ such that for any $K \in \mathcal{S}$ and its regulator R_K we have*

$$R_K \ge A(\sqrt{\pi}e^{(\gamma+1)/2})^{r_1}(2e^{\gamma})^{r_2}.$$

□

REMARK 8.2.16. One can compare Theorem 8.2.12 with a result by Zimmert which in our notation reads

$$\liminf_{i\to\infty} \frac{\log R_i}{g_i} \ge (\log 2 + \gamma)\varphi_{\mathbb{R}} + 2\gamma\varphi_{\mathbb{C}}.$$

The numerical values of Zimmert's coefficients are $\log 2 + \gamma \approx 1.270\ldots$ and $2\gamma \approx 1.154\ldots$, while $\log\sqrt{\pi e} + \gamma/2 \approx 1.361\ldots$ and $\log 2 + \gamma \approx 1.270\ldots$, respectively.

Applying the same argument to the case of an asymptotically good tower of almost normal number fields we get an unconditional version of Theorem 8.2.12.

THEOREM 8.2.17. *For an asymptotically good tower of almost normal number fields, we have*

$$\liminf_{i\to\infty} \frac{\log R_i}{g_i} \geq (\log\sqrt{\pi e}+\gamma/2)\varphi_{\mathbb{R}}+(\log 2+\gamma)\varphi_{\mathbb{C}}. \qquad \square$$

8.2.2. Bounds for the Brauer–Siegel Ratios

The generalized Brauer–Siegel theorem implies that the quantity $\lim \dfrac{\log(h_i R_i)}{g_i}$ can be different from the classical value 1. Moreover, it allows one to find some upper and lower bounds for this ratio, which we are going to explain. This requires some elementary application of the linear programming technique.

Linear programming problem. To get optimal bilateral estimates for the limit points of $\dfrac{\log(h_i R_i)}{g_i}$ and $\dfrac{\log(\varkappa_i)}{g_i}$, we come to the following linear programming problem.

Let q run over all prime powers and let there be given two sets of nonnegative coefficients $\{a_q\}$ and $\{b_q\}$ as well as nonnegative a_0, b_0, a_1, b_1 with the properties that if $a_i = 0$, then $b_i = 0$ for all $i = 0, 1, q$. Assume that

(1) For any $m \geq n$, we have

$$\frac{m}{n} \geq \frac{a_{p^m}}{a_{p^n}}.$$

(2) For any $q_1 \leq q_2$ such that $a_{q_1} \neq 0$, $a_{q_2} \neq 0$, we have

$$\frac{b_{q_1}}{a_{q_1}} \geq \frac{b_{q_2}}{a_{q_2}}.$$

(3) If $a_0 \neq 0$ and $a_1 \neq 0$, then $a_1 \geq a_0$, $b_1 \geq b_0$, and

$$\frac{b_0}{a_0} \leq \frac{b_1}{a_1}.$$

(4) We have

$$\sum_q b_q = \sum_q a_q = \infty.$$

Consider the following linear programming problem:

Find the maximum and the minimum of

$$F(x) = \sum_q b_q x_q - b_0 x_0 - b_1 x_1$$

under the conditions

(i) *for any* $i = 0$, 1, *or* q,

$$x_i \geq 0;$$

(ii) *for any prime* p,

$$\sum_{m=1}^{\infty} m x_{p^m} \leq x_0 + 2x_1;$$

(iii) *we have*

$$\sum_q a_q x_q + a_0 x_0 + a_1 x_1 \leq 1;$$

(iv) *if for some $i=0,1,q$ we have $a_i=0$, then it is assumed that the corresponding variables $x_i=0$.*

We consider this problem in two versions, one being when for all $i=0,1,q$ such that $a_i\neq 0$ the corresponding x_i are variables, and the other being when x_0 and x_1 are fixed and x_q vary. We assume that either x_0 or x_1 is nonzero, since otherwise $\max F(x)=\min F(x)=0$.

PROPOSITION 8.2.18. *If $a_1\neq 0$, then*

$$\min F(x) = -\frac{b_1}{a_1}.$$

If $a_1=0$ and $a_0\neq 0$, then

$$\min F(x) = -\frac{b_0}{a_0}.$$

The same problem for fixed x_0 and x_1 has

$$\min_{\substack{x_0,x_1\\ \text{fixed}}} F(x) = -b_0x_0-b_1x_1.$$

PROOF. The assertion is almost obvious. Indeed, given a vector x with a nonzero x_q for some q, change it, putting $x_q=0$. The value of $F(x)$ then diminishes, leaving all the conditions satisfied. Therefore the minimum is attained when $x_q=0$ for every q. Then we see that the minimum of $-b_0x_0-b_1x_1$ under the conditions $x_i\geq 0$ and $a_0x_0+a_1x_1\leq 1$ is attained for $a_0x_0+a_1x_1=1$ and one of the two x_i being 0, namely, that with the smaller ratio b_i/a_i. □

PROPOSITION 8.2.19. *Assume that x_0 and x_1 are fixed. Then*

$$\max_{\substack{x_0,x_1\\ \text{fixed}}} F(x) = (x_0+2x_1)\Big(\sum_{p<p'} b_p+\alpha b_{p'}\Big)-b_0x_0-b_1x_1,$$

where p' and $\alpha\in(0,1]$ satisfy the condition

$$\sum_{p<p'} a_p+\alpha a_{p'} = \frac{1-a_0x_0-a_1x_1}{x_0+2x_1}.$$ □

PROPOSITION 8.2.20. *Assume that x_0, x_1, and all x_q vary. Then*

$$\max F(x) = \frac{\sum\limits_{p\leq p_0} b_p - b}{\sum\limits_{p\leq p_0} a_p + a},$$

where

- $a=\dfrac{a_1}{2}$, $b=\dfrac{b_1}{2}$ *if $a_0=0$ and $a_1\neq 0$;*
- $a=a_0$, $b=b_0$ *if $a_0\neq 0$ and $a_1=0$;*
- *and if both $a_0\neq 0$ and $a_1\neq 0$, we have to compare two possibilities: $a=\dfrac{a_1}{2}$, $b=\dfrac{b_1}{2}$; and $a=a_0$, $b=b_0$.*

The prime p_0 is defined as follows:

$$p_0=\max\left\{\text{prime } p' :\ 0\leq \frac{\sum\limits_{p\leq p'} b_p - b}{\sum\limits_{p\leq p'} a_p + a}\leq \frac{b_{p'}}{a_{p'}}\right\}.$$ □

Proofs of these results are rather easy and can be obtained either directly or by writing out the dual linear problem.

Bounds. Applying the above linear programming problems to basic inequalities, one gets the following results by brute-force calculations.

THEOREM 8.2.21 (GRH bounds). *For any family of number fields we have*

$$\mathrm{BS}_{\mathrm{lower}} \le \liminf_{i\to\infty} \frac{\log(h_i R_i)}{g_i} \le \limsup_{i\to\infty} \frac{\log(h_i R_i)}{g_i} \le \mathrm{BS}_{\mathrm{upper}},$$

$$0 \le \liminf_{i\to\infty} \frac{\log \varkappa_i}{g_i} \le \limsup_{i\to\infty} \frac{\log \varkappa_i}{g_i} \le \varkappa_{\mathrm{upper}},$$

where

$$\mathrm{BS}_{\mathrm{lower}} = 1 - \frac{\log 2\pi}{\gamma + \log 8\pi} \approx 0.5165\ldots,$$

$$\mathrm{BS}_{\mathrm{upper}} = 1 + \frac{\log\frac{3}{2} + \log\frac{5}{4} + \log\frac{7}{6}}{\frac{\gamma}{2} + \frac{\pi}{4} + \log 2\sqrt{2\pi} + \frac{\log 2}{\sqrt{2}-1} + \frac{\log 3}{\sqrt{3}-1} + \frac{\log 5}{\sqrt{5}-1} + \frac{\log 7}{\sqrt{7}-1}} \approx 1.0938\ldots,$$

$$\varkappa_{\mathrm{upper}} = \frac{\log 2 + \log\frac{3}{2}}{\frac{\gamma}{2} + \log 2\sqrt{2\pi} + \frac{\log 2}{\sqrt{2}-1} + \frac{\log 3}{\sqrt{3}-1}} \approx 0.2164\ldots.$$

Moreover, if all the fields in the family are totally real, then

$$\liminf_{i\to\infty} \frac{\log(h_i R_i)}{g_i} \ge \mathrm{BS}_{\mathbb{R},\mathrm{lower}}$$

where

$$\mathrm{BS}_{\mathbb{R},\mathrm{lower}} = 1 - \frac{\log 2}{\frac{\gamma}{2} + \frac{\pi}{4} + \log 2\sqrt{2\pi}} \approx 0.7419\ldots,$$

and

$$\liminf_{i\to\infty} \frac{\log(\varkappa_i)}{g_i} \le \varkappa_{\mathbb{R},\mathrm{upper}}$$

where

$$\varkappa_{\mathbb{R},\mathrm{upper}} = \frac{\log 2 + \log\frac{3}{2}}{\frac{\gamma}{2} + \log 2\sqrt{2\pi} + \frac{\pi}{4} + \frac{\log 2}{\sqrt{2}-1} + \frac{\log 3}{\sqrt{3}-1}} \approx 0.1874\ldots.$$

If all the fields are totally complex, then

$$\limsup_{i\to\infty} \frac{\log(h_i R_i)}{g_i} \le \mathrm{BS}_{\mathbb{C},\mathrm{upper}},$$

where

$$\mathrm{BS}_{\mathbb{C},\mathrm{upper}} = 1 + \frac{\sum\limits_{\substack{p=2\\ \text{prime}}}^{13} \log\frac{p}{p-1} - \frac{1}{2}\log 2\pi}{\frac{\gamma}{2} + \log 2\sqrt{2\pi} + \sum\limits_{\substack{p=2\\ \text{prime}}}^{13} \frac{\log p}{\sqrt{p}-1}} \approx 1.0764\ldots. \qquad \square$$

THEOREM 8.2.22 (unconditional upper bound). *For any family of number fields, we have*

$$\limsup_{i\to\infty}\frac{\log(h_iR_i)}{g_i}\le 1+\frac{\sum\limits_{\substack{p=3\\ \text{prime}}}^{23}\log\dfrac{p}{p-1}}{\dfrac{\gamma}{2}+\dfrac{1}{2}+\log 2\sqrt{\pi}+2\sum\limits_{\substack{p=2\\ \text{prime}}}^{23}\log p\sum\limits_{m=1}^{\infty}\dfrac{1}{p^m+1}}\approx 1.1588\ldots,$$

$$\limsup_{i\to\infty}\frac{\log\varkappa_i}{g_i}\le\varkappa_{\text{unc,upper}}=\frac{\sum\limits_{\substack{p=3\\ \text{prime}}}^{5}\log\dfrac{p}{p-1}}{\dfrac{\gamma}{2}+\log 2\sqrt{\pi}+2\sum\limits_{\substack{p=2\\ \text{prime}}}^{5}\log p\sum\limits_{m=1}^{\infty}\dfrac{1}{p^m+1}}\approx 0.3151\ldots.$$

Moreover, if all the fields are totally complex, then

$$\limsup_{i\to\infty}\frac{\log(h_iR_i)}{g_i}\le 1+\frac{\sum\limits_{\substack{p=2\\ \text{prime}}}^{179}\log\dfrac{p}{p-1}-\log\sqrt{2\pi}}{\dfrac{\gamma}{2}+\log 2\sqrt{\pi}+2\sum\limits_{\substack{p=2\\ \text{prime}}}^{179}\log p\sum\limits_{m=1}^{\infty}\dfrac{1}{p^m+1}}\approx 1.1309\ldots,$$

and if all the fields are totally real, then

$$\limsup_{i\to\infty}\frac{\log\varkappa_i}{g_i}\le\frac{\sum\limits_{\substack{p=3\\ \text{prime}}}^{5}\log\dfrac{p}{p-1}}{\dfrac{\gamma}{2}+\log 2\sqrt{\pi}+\dfrac{1}{2}+2\sum\limits_{\substack{p=2\\ \text{prime}}}^{5}\log p\sum\limits_{m=1}^{\infty}\dfrac{1}{p^m+1}}\approx 0.2816\ldots.\qquad\square$$

THEOREM 8.2.23 (unconditional lower bound). *If for a tower of almost normal number fields there exists $\alpha>0$ such that $\inf\dfrac{N_i}{g_i}\ge\alpha$, then*

$$\liminf_{i\to\infty}\frac{\log(h_iR_i)}{g_i}\ge 1-\frac{\log 2\pi}{\gamma+\log 4\pi}\approx 0.4087\ldots.$$

If, in addition, the fields are totally real, then

$$\liminf_{i\to\infty}\frac{\log(h_iR_i)}{g_i}\ge 1-\frac{2\log 2}{\gamma+1+\log 4\pi}\approx 0.6625\ldots.\qquad\square$$

8.3. Class Field Tower Examples

Now we give some examples in the number field case which show that the Brauer–Siegel ratio can indeed vary significantly; all these examples are class field towers with controlled ramification, either trivial or close to that.

8.3.1. Unramified Towers with Splitting Conditions

Because of the Brauer–Siegel theorem, we need families of fields for which $[K:\mathbb{Q}]/g_K$ does not tend to 0. One's thought immediately turns to unramified towers, and the only infinite examples we know are Hilbert class field towers satisfying some extra conditions. Below we will use the notation of Sec. 6.4.1. Theorems 6.4.3 and 6.4.4 and Corollary 6.4.5 and some of their refinements allow one to deduce, in an unobvious way, the following results with small values of the root discriminant rd_K.

THEOREM 8.3.1. *The fields*

$$\mathbb{Q}(\sqrt{3\cdot 5\cdot 13\cdot 29\cdot 61}),$$
$$\mathbb{Q}(\sqrt{2},\sqrt{3\cdot 5\cdot 7\cdot 23\cdot 29}),$$
$$\mathbb{Q}(\sqrt{-3\cdot 5\cdot 17\cdot 19}),$$
$$\mathbb{Q}\Big(\cos\frac{2\pi}{11},\sqrt{2},\sqrt{-23}\Big)$$

have infinite unramified class field 2-towers. □

We shall also need unramified towers with some extra splitting conditions. To study them, let us fix some notation. Let S_∞ be the set of Archimedean places of K. Fix a finite set S of prime ideals of K. In what follows, if S is empty, we omit S from the corresponding notation. Let $O_{K,S}$ be the ring of S-integers, $E_{K,S}=O_{K,S}^*=U_{K,S}\cap K^*$ the group of S-units in K, $I_{K,S}$ the group of fractional ideals coprime with prime ideals of S, I_K^S the group of ideals generated by prime ideals in S (of course, $I_K=I_{K,S}\oplus I_K^S$), $P_{K,S}$ the image of the principle ideal group P_K in $I_{K,S}$ under $P_K\subset I_K\to I_{K,S}$, $I_K\to I_{K,S}$ being the natural projection, $\mathrm{Cl}_{K,S}=I_{K,S}/P_{K,S}=\mathrm{Cl}_K/\mathrm{Im}(I_K^S)$ the group of S-classes, and W_K the group of roots of unity lying in K.

THEOREM 8.3.2. *Let k be a number field. Let $P=\{p_1,\dots,p_t\}$ and $Q=\{q_1,\dots,q_r\}$ be disjoint sets of prime ideals of k, and let t_0 be the number of principal ideals in P. Consider a number field K/k of a prime degree ℓ ramified exactly at Q. Let S be the set of prime ideals in K lying over P, $s=|S|$. If*

$$r\geq s-t_0+r_1+r_2+\delta_\ell+2-\rho+2\sqrt{\ell(r_1+r_2-\rho/2)+\delta_\ell+s},$$

then the field K has an infinite unramified class field ℓ-tower $K=K_0\subset K_1\subset K_2\subset\dots$ where S splits completely.

To prove this theorem, we need some auxiliary results.

The following proposition is a number field analogue of Theorem 6.4.7.

PROPOSITION 8.3.3. *If*

$$d_\ell(\mathrm{Cl}_{K,S})\geq 2+2\sqrt{d_\ell(E_{K,S})+1},$$

then K has an infinite class field ℓ-tower $K=K_0\subset K_1\subset K_2\subset\dots$, where S splits completely. □

LEMMA 8.3.4. *We have*

$$\begin{aligned} d_\ell(E_{K,S}) &= r_1(K)+r_2(K)+\delta_\ell(K)-1+s\\ &= \ell(r_1+r_2-\rho/2)+\delta_\ell-1+s. \end{aligned}$$

PROOF. By Dirichlet's unit theorem, we have $E_{K,S}=W_K\oplus\mathbb{Z}^{r_1(K)+r_2(K)+s-1}$, and in our case $r_1(K)=\ell(r_1-\rho)$, $r_2(K)=\ell r_2+\ell\rho/2$, and $\delta_\ell(K)=\delta_\ell$, since no new ℓ-root can appear in an ℓ-extension. □

LEMMA 8.3.5. *We have*

$$d_\ell(\mathrm{Cl}_K)\geq r-r_1-r_2+\rho-\delta_\ell.$$ □

LEMMA 8.3.6. *We have*

$$d_\ell(\mathrm{Cl}_{K,S})\geq d_\ell(\mathrm{Cl}_K)-s+t_0.$$

PROOF. Let P_0 be the set of principal ideals lying in P. Let $\varphi\colon I_K^S\to\mathrm{Cl}_K$ be the composition of natural maps $I_K^S\to I_K$ and $I_K\to\mathrm{Cl}_K$. By the definition, $\mathrm{Cl}_{K,S}=\mathrm{Cl}_K/\operatorname{Im}\varphi$. We have $I_K^S=\prod\limits_{p\in P}(\prod\limits_{w|p}w^{\mathbb{Z}})\simeq\mathbb{Z}^s$. Since for any $p\in P_0$ we have $\prod\limits_{w|p}w\in\operatorname{Ker}\varphi$, it follows that $\mathrm{rk}_{\mathbb{Z}}\operatorname{Ker}\varphi\geq t_0$; therefore, $\mathrm{rk}_{\mathbb{Z}}\operatorname{Im}\varphi\leq s-t_0$, and it remains to apply d_ℓ. □

PROOF OF THEOREM 8.3.2. By Lemmas 8.3.4, 8.3.5, and 8.3.6 and by the inequality in the theorem statement, we get

$$\begin{aligned} d_\ell(\mathrm{Cl}_{K,S}) &\geq d_\ell(\mathrm{Cl}_K)-s+t_0\\ &\geq r-r_1-r_2+\rho-\delta_\ell-s+t_0\\ &\geq s-t_0+r_1+r_2+\delta_\ell+2-\rho+2\sqrt{\ell(r_1+r_2-\rho/2)+\delta_\ell+s}\\ &\quad -r_1-r_2+\rho-\delta_\ell-s+t_0\\ &=2+2\sqrt{\ell(r_1+r_2-\rho/2)+\delta_\ell+s}\\ &=2+2\sqrt{d_\ell(E_{K,S})+1}. \end{aligned}$$

Applying Proposition 8.3.3 completes the proof. □

COROLLARY 8.3.7. *Let $P=\{p_1,\dots,p_t\}$ and $Q=\{q_1,\dots,q_r\}$ be disjoint sets of rational primes. Consider a quadratic number field $K/\mathbb{Q}$ ramified exactly at Q. Let σ be the number of primes in P that split in K, and $s=t+\sigma$ the total number of prime ideals in K lying over P. Assume that either K is complex quadratic and*

$$r\geq 3+\sigma+2\sqrt{2+s},$$

or K is real quadratic and

$$r\geq 4+\sigma+2\sqrt{3+s}.$$

Then K has an infinite unramified class field 2-tower totally splitting all prime ideals over P.

PROOF. Indeed, here $k=\mathbb{Q}$, $\ell=2$, $\delta_2=1$, $r_1=1$, $r_2=0$, $\rho=1$ for the complex quadratic case and 0 for the real quadratic one, $t_0=t$. □

COROLLARY 8.3.8. *The field*

$$\mathbb{Q}(\sqrt{11\cdot 13\cdot 17\cdot 19\cdot 23\cdot 29\cdot 31\cdot 37\cdot 41\cdot 43\cdot 47\cdot 53\cdot 59\cdot 61\cdot 67})$$

has an infinite unramified class field 2-tower in which nine prime idecls lying over 2, 3, 5, 7, *and* 71 *split completely.*

PROOF. A straightforward check shows that 2, 3, 5, 7 split in $K/\mathbb{Q}$ and 71 is inert. Then we apply Corollary 8.3.7. □

COROLLARY 8.3.9. *The field*

$$\mathbb{Q}(\sqrt{-13\cdot 17\cdot 19\cdot 23\cdot 29\cdot 31\cdot 37\cdot 41\cdot 43\cdot 47\cdot 53\cdot 59\cdot 61\cdot 73\cdot 79})$$

has an infinite unramified class field 2-tower in which ten prime ideals lying over 2, 3, 5, 7, *and* 11 *split completely.* □

We shall use these examples below. We also get the following.

COROLLARY 8.3.10. *The field*

$$\mathbb{Q}(\sqrt{-3\cdot 5\cdot 7\cdot 11\cdot 13\cdot 17\cdot 23\cdot 31})$$

has an infinite unramified class field 2-tower in which both prime ideals lying over 2 *split completely.* □

Small ratios. We begin with examples where the Brauer–Siegel ratio is considerably less than 1.

THEOREM 8.3.11 (GRH). *Consider the Martinet field*

$$K=\mathbb{Q}\Big(\cos\frac{2\pi}{11},\sqrt{2},\sqrt{-23}\Big)$$

of degree 20 *over* $\mathbb{Q}$. *We have*

$$D_K=2^{30}11^{16}23^{10},$$

$$g=g(K)=\log\sqrt{D_K}\approx 45.2578\ldots.$$

This field has an infinite unramified 2*-tower* $\mathcal{K}$, *and we have*

$$1-\frac{10\log(2\pi)}{g}=\mathrm{BS}_{\mathrm{lower}}(\mathcal{K})\le \mathrm{BS}(\mathcal{K})\le \mathrm{BS}_{\mathrm{upper}}(\mathcal{K}),$$

$$0\le\varkappa(\mathcal{K})\le\varkappa_{\mathrm{upper}}(\mathcal{K}),$$

where

$$\mathrm{BS}_{\mathrm{upper}}(\mathcal{K})=\mathrm{BS}_{\mathrm{lower}}(\mathcal{K})+\frac{(\sqrt{23}-1)\log\frac{23}{22}}{\log 23}\Big(1-\frac{10(\gamma+\log 8\pi)}{g}\Big),$$

$$\varkappa_{\mathrm{upper}}(\mathcal{K})=\frac{(\sqrt{23}-1)\log\frac{23}{22}}{\log 23}\Big(1-\frac{10(\gamma+\log 8\pi)}{g}\Big),$$

which numerically reads as follows:

$$0.5939\ldots\le \mathrm{BS}(\mathcal{K})\le 0.6025\ldots,$$

$$0\le\varkappa(\mathcal{K})\le 0.0086\ldots.$$

REMARK 8.3.12. The deficiency $\delta(\mathcal{K})$ of this tower is at most

$$1-\frac{10(\gamma+\log 8\pi)}{g}\approx 0.1601\ldots.$$

PROOF OF THEOREM 8.3.11. First recall (Theorem 8.3.1) that this field has an infinite class field tower.

Let $K_0 = \mathbb{Q}\left(\cos\frac{2\pi}{11}\right)$, $K_{11} = \mathbb{Q}(\sqrt[11]{1})$, $k = \mathbb{Q}\left(\cos\frac{2\pi}{11}, \sqrt{2}\right)$, $F_{23} = \mathbb{Q}(\sqrt{-23})$, and $F_2 = \mathbb{Q}(\sqrt{2})$. We have

$$D_{K_{11}} = 11^9,$$

since the discriminant of a cyclotomic field $\mathbb{Q}(\zeta_p)$ equals p^{p-2} for a prime p. Hence, $D_{K_{11}}/\mathbb{Q}$ and $D_{K_0}/\mathbb{Q}$ are unramified outside of 11, and since 11 is totally ramified in $D_{K_{11}/\mathbb{Q}}$, it is also totally ramified in $K_0/\mathbb{Q}$; therefore, $D_{K_0} = 11^4$. The field K is the composite of K_0, F_2, and F_{23}, and therefore we get

$$D_K = D_{K_0}^4 D_{F_2}^{10} D_{F_{23}}^{10} = 11^{16} 2^{30} 23^{10}.$$

First let us prove that in $K/\mathbb{Q}$ we have the following decomposition of small primes:

v	2	3	5	7	11	13	17	19	23
e_v	2	1	1	1	5	1	1	1	2
f_v	5	10	20	10	4	10	10	20	1
n_v	2	2	1	2	1	2	2	1	10

where e_v is the ramification index, f_v the inertia one, and n_v is the number of places over v.

In $K_{11}/\mathbb{Q}$ only 11 is ramified (totally) and for a prime p the inertia index f_p equals the smallest f such that $p^f \equiv 1 \pmod{11}$. We easily check that $f_3 = f_5 = 5$, $f_2 = f_7 = f_{13} = f_{17} = f_{19} = 10$, and $f_{23} = 1$. Since 2 does not divide 5, K_0 being an index 2 subfield of K_{11}, we see that in $K_0/\mathbb{Q}$ all the primes of our table except 11 and 23 are inert, i.e., $f_p = 5$, that 11 is totally ramified and 23 is totally split. In $F_2/\mathbb{Q}$ only 2 is ramified, and p is split if and only if 2 is a square modulo p, i.e., if and only if $p \equiv \pm 1 \pmod 8$; such primes are 7, 17, and 23, and the rest (3, 5, 11, 13, and 19) are inert. In $F_{23}/\mathbb{Q}$ the only ramified prime is 23, since $-23 \equiv 1 \pmod 4$, and thus is a square $\bmod 2$. Thus, 2, 3, and 13 are split, and 5, 7, 11, 17, and 19 are inert.

Summing up this information, we get the above table. Note also that in $k/\mathbb{Q}$ there are 10 prime ideals over 23, and they all ramify in K/k.

Looking at the decomposition table above, we see that the smallest norm for which there exists a prime ideal of K is 23.

Now apply the linear programming method to find the minimum and maximum of $\mathrm{BS}(\mathcal{K})$. We set $a_0 = 0$, as well as $a_q = 0$ for $q < 23$, recalling that if $a_i = 0$ we have also $b_i = 0$ and we do not optimize over x_i for these i. Then the right-hand side of GRH Theorem 8.2.4 becomes

$$1 - \frac{10}{g}\log 2\pi + F(x) = \mathrm{BS}_{\mathrm{lower}}(\mathcal{K}) + F(x),$$

where

$$F(x) = \sum_{q \geq 23} b_q x_q, \quad b_q = \log\frac{q}{q-1}, \quad x_q = \varphi_q \quad \text{for} \quad q \geq 23,$$

under the following constraints: (i) $x_q \geq 0$, (ii) $\sum\limits_{m=1}^{\infty} m x_{p^m} \leq \frac{20}{g}$ for any p, and (iii) $\frac{10}{g}(\gamma + \log 8\pi) + \sum\limits_{q \geq 23} a_q x_q \leq 1$, where $a_q = \frac{\log q}{\sqrt{q} - 1}$.

The minimum of $F(x)$ is clearly 0, and it is easy to check that the maximum is attained for all $x_q = 0$ except for

$$x_{23} = \frac{\sqrt{23}-1}{\log 23}\left(1 - \frac{10}{g}(\gamma + \log 8\pi)\right).$$

Indeed, $x_{23} \approx 0.19\ldots < \frac{20}{g}$, which checks (ii), and x_{23} is chosen so that (iii) becomes an equality.

As for $\varkappa(\mathcal{K})$, we have

$$\varkappa(\mathcal{K}) = \mathrm{BS}(\mathcal{K}) - 1 + \varphi_{\mathbb{C}} \log(2\pi).$$

□

REMARK 8.3.13. Since the difference $\mathrm{BS}(\mathcal{K}) - \varkappa(\mathcal{K})$ is known, we do not give the value of $\varkappa(\mathcal{K})$ in some cases below.

THEOREM 8.3.14 (GRH). *Consider the real Martinet field*

$$K = \mathbb{Q}\left(\sqrt{2}, \sqrt{3\cdot 5\cdot 7\cdot 23\cdot 29}\right)$$

of degree 4 *over* $\mathbb{Q}$. *We have*

$$D_K = 2^8 \cdot (3\cdot 5\cdot 7\cdot 23\cdot 29)^2,$$
$$g = g(K) = \log\sqrt{D_K} \approx 13.9293\ldots.$$

This field has an infinite unramified 2*-tower* $\mathcal{K}$, *and*

$$\mathrm{BS}(\mathcal{K}) \in (\mathrm{BS}_{\mathrm{lower}}(\mathcal{K}), \mathrm{BS}_{\mathrm{upper}}(\mathcal{K})),$$

where

$$\mathrm{BS}_{\mathrm{lower}}(\mathcal{K}) = 1 - \frac{4\log 2}{g}$$

while

$$\mathrm{BS}_{\mathrm{upper}}(\mathcal{K}) = \mathrm{BS}_{\mathrm{lower}}(\mathcal{K}) + \frac{\log 2}{g} + \frac{\sqrt{7}-1}{g\log 7}\left(g - 2\gamma - \pi - 2\log 8\pi - \frac{\log 2}{\sqrt{2}-1}\right)\log\frac{7}{6},$$

i.e., approximately in the interval

$$(0.8009\ldots, 0.8648\ldots).$$

The deficiency of this tower is at most

$$1 - \frac{2\gamma + \pi + 2\log 8\pi}{g} \approx 0.2286\ldots.$$

PROOF. We proceed as in the proof of Theorem 8.3.11. Our field has an infinite unramified tower by Theorem 8.3.1. Let $K_1 = \mathbb{Q}(\sqrt{2})$, $K_2 = \mathbb{Q}(\sqrt{3\cdot 5\cdot 7\cdot 23\cdot 29})$, and $K = K_1 \cdot K_2$. Since $3\cdot 5\cdot 7\cdot 23\cdot 29 = 70035 \equiv 3 \pmod 4$, we have $D_{K_2} = 4\cdot 70035$, and 2, 3, 5, 7, 23, and 29 are ramified in K_2. Since $D_K = 2^8 \cdot 70035^2$, we see that the ideal lying over 2 is also ramified in K/K_2, i.e., 2 is totally ramified in $K/\mathbb{Q}$. In $K_1/\mathbb{Q}$ only 2 is ramified, $D_{K_1} = 8$. Since 2 is congruent to a square modulo 7, and noncongruent to a square modulo 3 and 5, we see that 7 splits, but 3 and 5 remain inert in $K_1/\mathbb{Q}$. Thus, in $K/\mathbb{Q}$ there is only one ideal of norm 2, no ideals of norm 3 and 5, and two ideals of norm 7. There are 4 real places and no complex ones.

Using, as above, the linear programming approach, with $a_1 = a_3 = a_5 = 0$ and $x_2 \le \frac{1}{g}$, we get

$$\mathrm{BS}_{\mathrm{lower}}(K) = 1 - \frac{4\log 2}{g}$$

and

$$\mathrm{BS}_{\mathrm{upper}}(K) = \mathrm{BS}_{\mathrm{lower}}(K) + \max F(x),$$

where $F(x) = \sum_{q \neq 3,5} b_q x_q$.

The maximum is attained for $x_2 = \frac{1}{g}$, $x_q = 0$ for $q = 4$ and $q > 7$, and the value of x_7 is chosen so that (iii) becomes an equality. □

REMARK 8.3.15. The other two fields of Corollary 8.3.1 give the following numerical results. For

$$K = \mathbb{Q}(\sqrt{-3 \cdot 5 \cdot 17 \cdot 19})$$

we GRH-have

$$g(K) \approx 4.9359\ldots,$$
$$\delta(\mathcal{K}) \le 0.2298\ldots,$$

and

$$\mathrm{BS}_{\mathrm{lower}}(\mathcal{K}) \approx 0.6276\ldots, \qquad \mathrm{BS}_{\mathrm{upper}}(\mathcal{K}) \approx 0.7228\ldots.$$

For

$$K = \mathbb{Q}(\sqrt{3 \cdot 5 \cdot 13 \cdot 29 \cdot 61})$$

we GRH-have

$$g(K) \approx 7.0687\ldots,$$
$$\delta(\mathcal{K}) \le 0.2400\ldots,$$

and

$$\mathrm{BS}_{\mathrm{lower}}(\mathcal{K}) \approx 0.8038\ldots, \qquad \mathrm{BS}_{\mathrm{upper}}(\mathcal{K}) \approx 0.9080\ldots.$$

Large ratios. Now we give examples where the Brauer–Siegel ratio is considerably greater than 1.

THEOREM 8.3.16 (GRH). *Consider the totally real quadratic field*

$$K = \mathbb{Q}(\sqrt{11 \cdot 13 \cdot 17 \cdot 19 \cdot 23 \cdot 29 \cdot 31 \cdot 37 \cdot 41 \cdot 43 \cdot 47 \cdot 53 \cdot 59 \cdot 61 \cdot 67})$$

of genus $g \approx 25.9882\ldots$. *It has an infinite unramified* 2*-tower* $\mathcal{K}$ *in which nine prime ideals lying over* 2, 3, 5, 7, *and* 71 *split completely,* $\mathrm{BS}(\mathcal{K}) \in (\mathrm{BS}_{\mathrm{lower}}(\mathcal{K}),$

$\mathrm{BS}_{\mathrm{upper}}(\mathcal{K})$), *and* $\varkappa(\mathcal{K}) \in (\varkappa_{\mathrm{lower}}(\mathcal{K}), \varkappa_{\mathrm{upper}}(\mathcal{K}))$, *where*

$$\mathrm{BS}_{\mathrm{lower}}(\mathcal{K}) = 1 + \frac{2\log\frac{3}{2} + 2\log\frac{5}{4} + 2\log\frac{7}{6} + \log\frac{5041}{5040}}{g},$$

$$\mathrm{BS}_{\mathrm{upper}}(\mathcal{K}) = \mathrm{BS}_{\mathrm{lower}}(\mathcal{K}) + \frac{1}{g}\sum_{p=11}^{47}\log\frac{p}{p-1}$$

$$+\frac{\sqrt{53}-1}{g\log 53}\left(g - \gamma - \frac{\pi}{2} - \log 8\pi - 2\sum_{p=2}^{7}\frac{\log p}{\sqrt{p}-1} - \frac{\log 71^2}{70} - \sum_{p=11}^{47}\frac{\log p}{\sqrt{p}-1}\right)\log\frac{53}{52},$$

$$\varkappa_{\mathrm{lower}}(\mathcal{K}) = \frac{2\log 2 + 2\log\frac{3}{2} + 2\log\frac{5}{4} + 2\log\frac{7}{6} + \log\frac{5041}{5040}}{g},$$

$$\varkappa_{\mathrm{upper}}(\mathcal{K}) = \mathrm{BS}_{\mathrm{upper}}(\mathcal{K}) - 1 + \frac{2\log 2}{g},$$

the sums being over prime p. *Numerically,*

$$\mathrm{BS}(\mathcal{K}) \in (1.0602\ldots, 1.0798\ldots),$$
$$\varkappa(\mathcal{K}) \in (0.1135\ldots, 0.1331\ldots).$$

PROOF. Let $d = 11\cdot 13\cdot 17\cdot 19\cdot 23\cdot 29\cdot 31\cdot 37\cdot 41\cdot 43\cdot 47\cdot 53\cdot 59\cdot 61\cdot 67$. A straightforward, though tedious, check shows that d is congruent to 1 modulo 8, and it is a square modulo 3, 5, and 7, but not modulo 71. Hence, 2, 3, 5, and 7 split in $K/\mathbb{Q}$, and 71 is inert. Corollary 8.3.8 shows that there is an unramified tower splitting the nine ideals over 2, 3, 5, 7, and 71. Applying the same linear programming method, we obtain $\mathrm{BS}_{\mathrm{lower}}(K)$, $\mathrm{BS}_{\mathrm{upper}}(K)$, $\varkappa_{\mathrm{lower}}(\mathcal{K})$, and $\varkappa_{\mathrm{upper}}(\mathcal{K})$. □

THEOREM 8.3.17 (GRH). *Consider the totally complex quadratic field*

$$K = \mathbb{Q}\left(\sqrt{-13\cdot 17\cdot 19\cdot 23\cdot 29\cdot 31\cdot 37\cdot 41\cdot 43\cdot 47\cdot 53\cdot 59\cdot 61\cdot 73\cdot 79}\right)$$

of genus $g \approx 27.0169\ldots$. *It has an infinite unramified* 2*-tower* $\mathcal{K}$ *in which ten prime ideals lying over* 2, 3, 5, 7, *and* 11 *split completely,* $\mathrm{BS}(\mathcal{K}) \in (\mathrm{BS}_{\mathrm{lower}}(\mathcal{K})$, $\mathrm{BS}_{\mathrm{upper}}(\mathcal{K}))$, *and* $\varkappa(\mathcal{K}) \in (\varkappa_{\mathrm{lower}}(\mathcal{K}), \varkappa_{\mathrm{upper}}(\mathcal{K}))$, *where*

$$\mathrm{BS}_{\mathrm{lower}}(K) = 1 - \frac{1}{g}\log 2\pi + \frac{2}{g}\left(\log 2 + \log(3/2) + \log(5/4) + \log(7/6) + \log(11/10)\right),$$

$$\mathrm{BS}_{\mathrm{upper}}(\mathcal{K}) = \mathrm{BS}_{\mathrm{lower}}(\mathcal{K}) + \frac{1}{g}\sum_{p=13}^{61}\log\frac{p}{p-1}$$

$$+\frac{\sqrt{67}-1}{g\log 67}\left(g - \gamma - \log 8\pi - 2\sum_{p=2}^{11}\frac{\log p}{\sqrt{p}-1} - \sum_{p=13}^{61}\frac{\log p}{\sqrt{p}-1}\right)\log\frac{67}{66},$$

$$\varkappa_{\mathrm{lower}}(K) = \frac{2}{g}\left(\log 2 + \log(3/2) + \log(5/4) + \log(7/6) + \log(11/10)\right),$$

$$\varkappa_{\mathrm{upper}}(K) = \mathrm{BS}_{\mathrm{upper}}(\mathcal{K}) - 1 + \frac{1}{g}\log 2\pi,$$

the sums being over prime p. Numerically,

$$(\mathrm{BS}_{\mathrm{lower}}(\mathcal{K}), \mathrm{BS}_{\mathrm{upper}}(\mathcal{K})) = (1.0482\ldots, 1.0653\ldots),$$
$$(\varkappa_{\mathrm{lower}}(\mathcal{K}), \varkappa_{\mathrm{upper}}(\mathcal{K})) = (0.1162\ldots, 0.1333\ldots).$$

PROOF. Along the same lines as in the proofs of Theorems 8.3.16, 8.3.14, and 8.3.11, using Corollary 8.3.9. □

REMARK 8.3.18. Ihara's example of Corollary 8.3.10,

$$K = \mathbb{Q}(\sqrt{-3\cdot 5\cdot 7\cdot 11\cdot 13\cdot 17\cdot 23\cdot 31}),$$

with two divisors of 2 splitting in the tower, has $g \approx 9.5097\ldots$, its deficiency δ is at most $0.2483\ldots$, and $(\mathrm{BS}_{\mathrm{lower}}(\mathcal{K}), \mathrm{BS}_{\mathrm{upper}}(\mathcal{K})) = (0.9525\ldots, 1.010\ldots)$; therefore, this field does not give a counterexample to the classical Brower–Siegel theorem.

Unconditional results. Let us see what can be obtained without the GRH. We consider the same fields as in GRH Theorems 8.3.11 and 8.3.14.

THEOREM 8.3.19. *Consider the Martinet field*

$$K = \mathbb{Q}\Big(\cos\frac{2\pi}{11}, \sqrt{2}, \sqrt{-23}\Big)$$

of degree 20 *over* $\mathbb{Q}$. *This field has an infinite unramified* 2*-tower* $\mathcal{K}$, *and* $\mathrm{BS}(\mathcal{K}) \in (\mathrm{BS}_{\mathrm{lower}}(\mathcal{K}), \mathrm{BS}_{\mathrm{unc,upper}}(\mathcal{K}))$, *where*

$$\mathrm{BS}_{\mathrm{lower}}(\mathcal{K}) = 1 - \frac{10\log(2\pi)}{g}$$

and

$$\mathrm{BS}_{\mathrm{unc,upper}}(\mathcal{K}) = \mathrm{BS}_{\mathrm{lower}}(\mathcal{K}) + \frac{10\log\frac{23}{22}}{g} + \frac{2\log\frac{32}{31}}{g} + \frac{20}{g}\sum_{p=37}^{97}\log\frac{p}{p-1},$$

i.e., approximately in the interval

$$(0.5939\ldots, 0.7108\ldots).$$

SKETCH OF THE PROOF. Proceed along the same lines as before. Having no GRH at hand, instead of GRH Theorem 8.2.4 we use Theorem 8.2.5 (the tower being almost normal, as any 2-tower over a normal field), and instead of GRH Proposition 8.1.18 we use either Proposition 8.1.19 or Proposition 8.1.20. □

REMARK 8.3.20. Using Proposition 8.1.19 instead of Proposition 8.1.20 we can do better. We can also use further information on the prime decomposition in $K/\mathbb{Q}$. In particular, the only possible norms between 37 and 1000 are in fact 121, 353, 439, 463, 593, 967, and 991. This makes the constant better, and we unconditionally get $\mathrm{BS}(\mathcal{K}) \leq 0.62077\ldots$.

THEOREM 8.3.21. *Consider the real Martinet field*

$$K = \mathbb{Q}(\sqrt{2}, \sqrt{3\cdot 5\cdot 7\cdot 23\cdot 29})$$

of degree 4 *over* $\mathbb{Q}$. *Its genus equals*

$$g = g(K) = \log\sqrt{|D_K|} \approx 13.9293\ldots,$$

it has an infinite unramified 2-tower $\mathcal{K}$, *and* $\mathrm{BS}(\mathcal{K}) \in (\mathrm{BS}_{\mathrm{lower}}(\mathcal{K}), \mathrm{BS}_{\mathrm{unc,upper}}(\mathcal{K}))$, *where*

$$\mathrm{BS}_{\mathrm{unc,upper}}(\mathcal{K}) = \mathrm{BS}_{\mathrm{lower}}(\mathcal{K}) + \frac{\log 2 + 2\log\frac{7}{6} + 4\log\frac{11}{10} + 4\log\frac{13}{12}}{g}$$
$$+ \frac{1}{2gA_{17}}(g - 2\gamma - \pi - 2\log 8\pi - A_2 - 2A_7 - 4A_{11} - 4A_{13})\log\frac{17}{16},$$

where $A_p = 2\log p \sum_{m=1}^{\infty}(p^m+1)^{-1}$, *i.e., approximately in the interval*

$$(0.8009\ldots, 0.8648\ldots).$$

□

8.3.2. Hajir–Maire Tame Towers

So far, we have worked with unramified towers. F. Hajir and C. Maire have noticed that if one does not insist that towers should be unramified but rather allows for some tame (or close to tame) ramification, then one gains some flexibility in applying the Golod–Shafarevich criterion and can obtain better asymptotic results. In this subsection we briefly expose some of these constructions.

Tame Towers. In all examples above, the base field K is taken as a Galois extension of degree 2 of another number field k in which many places are ramified; furthermore, the set of these places is stable under the Galois group of $K/\mathbb{Q}$. Moreover, $K = F \cdot k$ is abelian for a quadratic extension $F/\mathbb{Q}$.

The method of Hajir and Maire uses two observations: first is that requiring the set of ramifying primes in K/k to be Galois-stable over $\mathbb{Q}$ is too restrictive, and sometimes asymmetric constructions yield better results. The second observation is that although unramified towers are nice, having constant relative genus, all that is needed is that the relative genus remain bounded, which is also true for tamely ramified towers. In this case, the number of generators does not decrease, while the number of relations decreases by 1, and this is often just enough for the application of the Golod–Shafarevich test.

To apply the above results—essentially, Proposition 8.3.3 and Lemmas 8.3.4–8.3.6—to the tamely ramified situation, one only needs to give a bound for the genus of number fields contained in a tamely ramified tower.

LEMMA 8.3.22. *Let* ℓ *be a prime, and assume that* T *is a finite set of prime ideals of a number field* K *not dividing* ℓ. *Let* $F \subset K_T$ *be a finite* ℓ*-extension of* K *unramified outside* T, *let* $m = [F : K]$, *and let*

$$\mu = \left(\sum_{\mathcal{P}\in T} \ln N_{F/\mathbb{Q}}(\mathcal{P})\right) / [K : \mathbb{Q}].$$

Then

$$g_F \le g_K + \mu\left(1 - \frac{1}{m}\right) < \mu + g_K,$$

with equality in the left-hand inequality if and only if every $\mathcal{P} \in T$ *is totally ramified in* F.

SKETCH OF THE PROOF. The extension F/K is tamely ramified, and therefore the $\mathcal{P}$-valuation of the norm in F/K of the different $\mathfrak{D}_{F/K}$ is $m(1 - 1/e_{\mathcal{P}})$ for each prime ideal $\mathcal{P}$ of K, where $e_{\mathcal{P}}$ is the ramification index of $\mathcal{P}$ in F/K. Note that, for each $\mathcal{P}$, $e_{\mathcal{P}}$ is a divisor of m, equal to m if and only if $\mathcal{P}$ is totally ramified in F/K and to 1 if $\mathcal{P} \notin T$; and Proposition 6.1.30 completes the proof. □

PROPOSITION 8.3.23. *Let $k = \mathbb{Q}(\xi)$ be a totally real field, where ξ is a root of $f = x^4 - x^3 - 4x^2 + 2x + 1$ of discriminant $3981 = 3 \cdot 1327$. Let $K = k(\sqrt{\eta})$, where $\eta = -29\xi^3 - 23\xi^2 + 25\xi - 65$ is a totally negative integral element of k. Let $\mathfrak{P}_3$ be the unique ideal in $\mathcal{O}_K$ of absolute norm 3, and let $T = \{\mathfrak{P}_3\}$. Then K has an infinite T-tamely unramified 2-tower, and its root discriminant is bounded by*

$$\mathrm{rd} \leq 3981^{1/4} \cdot (3 \cdot 5 \cdot 9 \cdot 13 \cdot 23 \cdot 37 \cdot 41)^{1/8} \cdot 3^{1/8} \approx 85.706. \qquad \square$$

In this example four infinite and seven finite places of k are ramified in K.

PROPOSITION 8.3.24. *Let $k = \mathbb{Q}(\xi)$ be a totally real field, where ξ is a root of $f = x^4 - 2x^3 - 4x^2 + 3x + 3$ of discriminant $7053 = 3 \cdot 2351$. Let $K = k(\sqrt{\eta})$, where $\eta = -25\xi^3 - 8\xi^2 + 30\xi - 17$ is a totally negative integral element of k. Let π_3 be the unique prime element of k of absolute norm 3, and let $T = \{T_1, T_2\}$ be the set of prime ideals in K dividing π_3. Then K has an infinite T-tamely unramified 2-tower with*

$$\mathrm{rd} \leq 7053^{1/4} \cdot (9 \cdot 7 \cdot 13^2 \cdot 17 \cdot 31)^{1/8} \cdot 9^{1/8} \approx 84.140. \qquad \square$$

In this example four infinite and six finite places of k are ramified in K.

PROPOSITION 8.3.25. *Let $k = \mathbb{Q}(\xi)$ be a quartic field of signature $(r_1, r_2) = (2, 1)$, where ξ is a root of $f = x^4 - x^3 + x^2 + x - 1$ of discriminant -331. Let $K = k(\sqrt{\eta})$ for a totally negative integral element $\eta = -127\xi^3 - 139\xi^2 - 181\xi - 146$. Let π_5 be the unique prime element of k of absolute norm 5, and let $T = \{T_1, T_2\}$ be the set of prime ideals in K dividing π_5. Then K has an infinite T-tamely unramified 2-tower with root discriminant bounded by*

$$\mathrm{rd} \leq 331^{1/4} \cdot (13^2 \cdot 17 \cdot 19 \cdot 23^2 \cdot 31)^{1/8} \cdot 25^{1/8} \approx 83.885.$$

Moreover, the deficiency of the corresponding infinite number field satisfies

$$\delta_K \leq 1 - \frac{\gamma + \log(8\pi)}{\log(\mathrm{rd})} = 0.141\ldots. \qquad \square$$

In this example both infinite places as well as seven finite places of k ramify in K. In all these examples, 2 is ramified in $K/\mathbb{Q}$.

Proofs of these propositions use massive algebraic computer computations.

8.4. Further Theory and Open Questions

In this section we give some directions of further study of infinite global fields, and more generally, of asymptotically exact families. Most of the following problems look very difficult, but some of them might be easier than others, and one hopes to obtain at least some partial answers.

The section is divided into two subsections of a slightly different nature. In Sec. 8.4.1 we collect questions common for the number field and function field cases. On the contrary, in Sec. 8.4.2 we pose some questions specific for function fields and their generalizations such as abelian varieties; actually, we have no idea of how their number-theoretic counterpart could look.

8.4.1. Common Questions

Structure of the parameter set. For an infinite global field $\mathcal{K}$ we have defined the sequence $\varphi_{\mathcal{K}}$ of parameters:

$$\varphi_{\mathcal{K}} = (\varphi_{\mathbb{R}}, \varphi_{\mathbb{C}}, \varphi_2, \dots)$$

for the number field case, and

$$\varphi_{\mathcal{K}} = (\varphi_r, \varphi_{r^2}, \varphi_{r^3}, \dots)$$

for the function field one. We have shown that these sequences are sometimes nontrivial. Then it is natural to ask about the structure of possible parameters. We define the sets Φ and Φ_r in $\mathbb{R}^\infty$ by

$$\begin{aligned} \Phi &= \{\varphi_{\mathcal{K}} : \mathcal{K} \text{ an infinite number field}\}, \\ \Phi_r &= \{\varphi_{\mathcal{K}} : \mathcal{K} \text{ an infinite function field over } \mathbb{F}_r\}. \end{aligned}$$

We also introduce the sets $\widetilde{\Phi} \subset \mathbb{R}^\infty$ and $\widetilde{\Phi}_r \subset \mathbb{R}^\infty$ defined exactly as Φ and Φ_r, but for all asymptotically exact families. Clearly, $\{0\} \in \Phi \subseteq \widetilde{\Phi}$, and $\{0\} \in \Phi_r \subseteq \widetilde{\Phi}_r$.

The above considerations show that neither of these sets reduce to $\{0\}$. However, their structure remains mysterious. Let us put some natural questions on this structure, for brevity, only in the case of Φ; exactly the same questions are equally interesting for the other three sets.

PROBLEM 8.4.1. Is Φ closed in some natural topology on $\mathbb{R}^\infty$?

One can propose a natural class of topologies to consider. Let $a = (a_{\mathbb{R}}, a_{\mathbb{C}}, a_2, \dots)$ be a sequence of positive real numbers indexed exactly as those of the sequences φ. Then one can define the weighted spaces $\ell_{p,a}$ with $1 \le p \le \infty$ using the norm

$$\|x\|_{p,a} := (\sum_\alpha a_\alpha |x|^p)^{1/p}.$$

Our basic inequality says that $\Phi \subseteq \ell_{1,a}$ with an appropriate sequence a depending on the version of the basic inequality used.

PROBLEM 8.4.2. Is Φ (relatively) compact in some natural topology on $\mathbb{R}^\infty$?

PROBLEM 8.4.3. Does Φ contain a nonempty open set in $\mathbb{R}^\infty$? Is it convex? Is it a restricted cone, i.e., does $\varphi_0 \in \Phi$ imply $\mu\varphi_0 \in \Phi$ for $\mu \in [0,1]$? Is it true that $\Phi \subset \ell_\infty$?

PROBLEM 8.4.4. Show that the cardinality of $\operatorname{Supp}\varphi$ is unbounded on Φ, where $\operatorname{Supp}\varphi$ is the set of indices α with $\varphi_\alpha \neq 0$. Does there exist $\varphi \in \Phi$ with infinite $\operatorname{Supp}\varphi$?

The last question is also related to the unramified Fontaine–Mazur conjecture and to the convergence abscissa for $\zeta_\varphi(s)$; cf. Remark 8.1.22.

The deficiency problem. Since a complete description of the sets $\widetilde{\Phi}$ and Φ is, most probably, very difficult, one can also be interested in possible values of the deficiency, which has his own importance.

PROBLEM 8.4.5. Does there exist an infinite global field $\mathcal{K}$ (an asymptotically exact family) with zero deficiency $\delta_{\mathcal{K}}$?

If the answer is positive, one would like to have an explicit construction of such a family. Note that the positive answer is known for function global fields over $\mathbb{F}_r$ with a square r.

More generally, one can put the following.

PROBLEM 8.4.6. Describe the set of possible values of the deficiency for infinite global fields (asymptotically exact families).

One can also be interested in properties of this set: whether it is closed, convex (i.e., an interval), of positive measure, etc.

At the moment these problems seem to be inaccessible and we would like to put forward a more modest question concerning the refinement of existing estimates for δ_K.

PROBLEM 8.4.7. Produce an example of an infinite global field $\mathcal{K}$ (or of an asymptotically exact family) with the value $\delta_{\mathcal{K}}$ as small as possible.

The best example known in the number field case is with $\delta_{\mathcal{K}} \leq 0.141\ldots$.

In particular one should consider families with $\varphi_q \neq 0$ for at least one prime power q. It looks promising to search for ramified towers $\mathcal{K}$ of number fields with good values of $\delta_{\mathcal{K}}$ to replace the class field towers of Sec. 8.3 above.

Connection with the unramified Fontaine–Mazur conjecture. Let K be a number field, p a prime, let T be a finite set of primes of K, none above p, and let $G_{K,T}^{(p)} = \operatorname{Gal}(\mathcal{K}_T^{(p)}/K)$ be the Galois group of the maximal algebraic pro-p extension $\mathcal{K}_T^{(p)}$ of K unramified outside T. Then the unramified Fontaine–Mazur conjecture reads as follows:

Any continuous representation

$$\rho\colon G_{K,T}^{(p)} \to \mathrm{GL}_n(\mathbb{Z}_p)$$

has a finite image.

One can consider a *just-infinite* subextension $\mathcal{L}/K$ of $\mathcal{K}_T^{(p)}$, i.e., an extension which contains no proper infinite subextensions of K. Then the unramified Fontaine–Mazur conjecture is equivalent to the finiteness of the image for any representation

$$\rho\colon \operatorname{Gal}(\mathcal{L}/K) \to \mathrm{GL}_n(\mathbb{Z}_p)$$

for all just-infinite extensions $\mathcal{L}/K$.

One says that a pro-p group is *torsion-riddled* if all its open subgroups have torsion. N. Boston put forth the following conjecture:

The Galois group $\mathrm{Gal}(\mathcal{L}/K)$ *is torsion-riddled for any just-infinite* $\mathcal{L}$.

This conjecture would imply the unramified Fontaine–Mazur conjecture and it is ultimately connected with the following.

PROBLEM 8.4.8. Does there exist an infinite number field $\mathcal{K}$ for which the set $S_{\mathcal{K}}$ of prime powers q such that $\varphi_q > 0$ is infinite? If so, characterize such families.

Boston's conjecture would be implied by the following.

CONJECTURE. *Let* $\mathcal{K}$ *be a number field just-infinite over* K. *Then* $S_{\mathcal{K}}$ *is infinite.*

Indeed, if it is the case, there exist in $\mathrm{Gal}(\mathcal{K}/K)$ infinitely many Frobenius elements of finite order.

Let us also note that Ihara conjectured the existence of an unramified extension $\mathcal{K}$ with $\delta_{\mathcal{K}} = 0$ and infinite $S_{\mathcal{K}}$.

Around the asymptotic explicit formula. One easily sees that almost all GRH results of our paper have their unconditional counterparts, with one notable exception, namely GRH Theorem 8.1.28 and its consequence, GRH Corollary 8.1.30. It is but natural to pose the following.

PROBLEM 8.4.9. What are unconditional analogues of GRH Theorem 8.1.28 and Corollary 8.1.30?

At the moment we have no approach to this problem.

Around the generalized Brauer–Siegel theorem. Now we discuss some open questions concerning the generalized Brauer–Siegel theorem. First of all, in the proof of the generalized Brauer–Siegel theorem we do not really use the whole strength of the GRH; moreover, under some mild conditions we have totally dispensed with GRH (Theorem 8.2.5). Therefore, it is but natural to ask whether one really needs the GRH to prove the result, which leads to the following.

PROBLEM 8.4.10. Prove GRH Theorem 8.2.4 unconditionally.

Let us now discuss some problems connected with the Brauer–Siegel ratio introduced and studied above. First of all, its very existence (i.e., the existence of the corresponding limit) is proved only under the GRH or for almost normal asymptotically good infinite global fields, which leads to the following.

PROBLEM 8.4.11. Prove unconditionally that for any asymptotically exact family $\mathcal{K}$ of number fields the Brauer–Siegel ratio $\mathrm{BS}(\mathcal{K})$ is well defined.

The following problem is connected with the fact that for an arbitrary asymptotically exact family unconditionally we have only an *upper* bound for $\mathrm{BS}(\mathcal{K})$; cf. Theorems 8.2.3, 8.2.5, and 8.2.22.

PROBLEM 8.4.12. Give an unconditional lower bound for the Brauer–Siegel ratio $\mathrm{BS}(\mathcal{K})$ for *any* asymptotically exact family.

Note that for towers of normal number fields this results from Theorems 8.2.5 and 8.2.22. One can hope that this problem can be solved if one refines the technique of the proof of the Brauer–Siegel theorem, i.e., estimates in the adelic integral representation of the zeta function.

There also is the question of how good are our bounds and examples.

PROBLEM 8.4.13. Improve the bounds of GRH Theorem 8.2.21 and/or of Theorem 8.2.22.

PROBLEM 8.4.14. Construct examples of class field towers (or other asymptotically exact families) with $\mathrm{BS}(\mathcal{K})$ GRH-smaller than those of GRH Theorems 8.4.17 and 8.3.14 or GRH-greater than those of GRH Theorems 8.3.16 and 8.3.17.

Our results in Secs. 8.1 and 8.2 are of an asymptotic nature. However, it is clear that a good part of them can be made effective, which leads to the following challenge.

PROBLEM 8.4.15. Give effective versions of the above results with the remainder terms as good as possible.

8.4.2. Results Specific for the Function Field Case and Corresponding Problems

In the function field case we have some specific results which do not yet have their number field counterparts. Let us discuss some of them.

Asymptotics of zeta function coefficients in families. In this case we can get rather good estimates for the coefficients of zeta functions. More precisely, let

$$Z_K(t) = \sum_{m=1}^{\infty} D_m t^m$$

for a function field K; one knows that $D_m = D_m(K)$ is the number of positive divisors of degree m. Then we have the following.

THEOREM 8.4.16. *Let $\mathcal{K} = \{K_j\}$ be an asymptotically exact family of function fields over $\mathbb{F}_r$, and let*

$$\mu_0 = \sum_{m=1}^{\infty} \frac{m\varphi_{r^m}}{r^m - 1} = 1 - \xi_\varphi(1).$$

Then for any real $\mu > 0$

$$\lim_{i\to\infty} \frac{D_{[\mu g_i]}}{g_i} = \mu \log \Lambda + \sum_{m=1}^{\infty} \varphi_{r^m} \log \frac{\Lambda^m}{\Lambda^m - 1},$$

where $\Lambda = \Lambda(\mu)$ equals r for $\mu \geq \mu_0$, and is defined from the equation

$$\sum_{m=1}^{\infty} \frac{m\varphi_{r^m}}{\Lambda^m - 1} = \mu$$

for $\mu \leq \mu_0$. □

Moreover, one can get an even more precise result for the ratio D_m/h. Let $h_j = h(K_j)$ be the class number. Then we have the following.

THEOREM 8.4.17. *Let $\mathcal{K} = \{K_j\}$ be an asymptotically exact family. Then for any $\varepsilon > 0$ and any m with $m/g \geq \mu_1 + \varepsilon$, we have*

$$\frac{D_m(K_j)}{h_j} = \frac{r^{m-g+1}}{r-1}(1 + o(1))$$

for $j \to \infty$, $o(1)$ being uniform in m. Here $\mu_1 = \mu_1(\varphi)$ is defined as the largest of the two roots of the equation

$$\frac{\mu}{2} + \mu \log_r \frac{\mu}{2} + (2-\mu)\log_r(1-\frac{\mu}{2}) = -2\log_r \zeta_\varphi(1). \qquad \square$$

Note that the estimate of Theorem 8.4.17 is much more precise than that of Theorem 8.4.16 (there we have an exponential $o(1)$ instead of the multiplicative one of Theorem 8.4.17).

PROBLEM 8.4.18. What is the number field counterpart of the above results on the number of positive divisors?

It does not look likely that Theorem 8.4.17 has a proper number field analogue. On the other hand, one can hope to obtain an analogue of Theorem 8.4.16.

One can also consider the asymptotic behaviour of $w_m(\mathcal{K})$, the number of classes of positive divisors of degree m. More precisely, let $\mathcal{K} = \{K_j\}$ be an asymptotically exact family, and let

$$w(\mathcal{K},\mu)_{\inf} := \liminf (w_{[\mu g]}(K_j))^{1/g_j},$$
$$w(\mathcal{K},\mu)_{\sup} := \limsup (w_{[\mu g]}(K_j))^{1/g_j}$$

for a real number $\mu \in (0,1)$. Clearly,

$$w(\mathcal{K},\mu)_{\inf} \le w(\mathcal{K},\mu)_{\sup} \le d(\mathcal{K},\mu) = \lim \frac{D_{[\mu g]}(K_j)}{h_j}.$$

It is known that for $\mu \in (0,1/r)$ one has $w(\mathcal{K},\mu)_{\inf} = w(\mathcal{K},\mu)_{\sup} = d(\mathcal{K},\mu)$, and for $\mu > 1/r$ the ratio $w(\mathcal{K},\mu)_{\inf}/d(\mathcal{K},\mu)$ is bounded from below by $r^{-\varphi_r R_r(1-\mu/\varphi_r)}$ for any asymptotic upper bound R_r for r-ary linear codes (recall that $R_r(\delta)$ is a decreasing continuous function on $\left[0, \frac{r-1}{r}\right]$ with $R_r(0) = 1$, $R_r\left(\frac{r-1}{r}\right) = 0$).

PROBLEM 8.4.19. Is it true that one has $w(\mathcal{K},\mu)_{\inf} = w(\mathcal{K},\mu)_{\sup} = d(\mathcal{K},\mu)$ for any $\mu \in (0,1)$?

The above results use geometric arguments, in particular, the construction of algebraic geometry codes and have no evident number theory counterparts.

Integrality. All of the asymptotic results that we have seen above were of analytic nature, i.e., in fact we worked with *real Weil systems*, namely with sets of inverse roots of polynomials $P(t)$ with real coefficients such that $P(0) = 1$ and all inverse roots are of absolute value $\sqrt{q}$. This is quite natural; a priori in the asymptotics we cannot trace the integrality of the coefficients $P(t)$, since we divide by the growing genus g and pass to the limit.

We can define another measure $\mu(x)$ on the interval $[-2\sqrt{q}, 2\sqrt{q}]$ corresponding to the curve X, namely

$$\mu_X(x) = \frac{1}{g_X}\sum_{i=1}^{g} \delta_{x_i}(x)$$

with $x_i = \omega_i + \overline{\omega}_i$, where $\{\omega_1,\ldots,\omega_g,\overline{\omega}_i,\ldots,\overline{\omega}_g\}$ is the set of inverse roots of $P_X(t)$.

One can prove the following fact (true not only for curves but for abelian varieties as well).

PROPOSITION 8.4.20. *We have*

$$\mu_X(x) = \frac{1}{2\pi}\alpha_*\Delta_X(t),$$

where the measure Δ_X *is defined in Sec. 8.1.2 and* $\alpha\colon (-\pi,\pi] \to [-2\sqrt{q}, 2\sqrt{q}]$ *is given by* $x = 2\sqrt{q}\cos t$.

THEOREM 8.4.21. *Let* $H(x)$ *be any polynomial from* $\mathbb{Z}[x]$. *Define* $\log^\bullet(x)$ *as the function equal to* $\log(x)$ *for* $x \neq 0$ *and to* 0 *for* $x = 0$. *Then for the limit measure* $\mu(x)$ *on the interval* $[-2\sqrt{q}, 2\sqrt{q}]$ *corresponding to an asymptotically exact family of curves over* $\mathbb{F}_q$, *we have*

$$\langle \log^\bullet |H(x)|, \mu(x)\rangle \geq 0.$$

The same is true for the limit measure $\mu(x)$ *corresponding to an asymptotically exact family of abelian varieties.*

PROOF. Let us start with a fixed curve X. Fix a polynomial $H(x)$. Consider the set $\{x_i\}$ of algebraic integers with $x_i = \omega_i + \overline{\omega}_i$. This set splits into orbits under the action of the group $\operatorname{Gal}\overline{\mathbb{Q}}/\mathbb{Q}$:

$$\{x_1, \ldots, x_g\} = \bigcup O_k.$$

Then

$$\langle \log^\bullet |H(x)|, \mu_X(x)\rangle = \frac{1}{g_X}\sum_{i=1}^{g} \log^\bullet |H(x_i)| = \frac{1}{g_X}\sum_k \sum_{O_k} \log^\bullet |H(x_i)|.$$

If the x_i are roots of $H(x)$, then the sum over the orbit $\sum_{O_k} \log^\bullet |H(x_i)|$ is zero by the definition of $\log^\bullet$. Otherwise,

$$\sum_{O_k} \log^\bullet |H(x_i)| = \log \prod_{O_k} |H(x_i)| = \log |\operatorname{Norm} H(x_1)|^a,$$

a being a positive integer. Indeed, $H(x_i)$ is also an algebraic integer and O_k is a union of its orbits. The latter is nonnegative, since the absolute value of the norm of a nonzero algebraic integer is at least 1.

For the limit measure $\mu(x)$ the integral is the limit of integrals and thus is also nonnegative. Note however that passing to the limits requires a certain carefulness. □

We get a funny consequence, which is formulated in terms of elementary calculus.

COROLLARY 8.4.22. *Let* $C = 2\sqrt{q}$, q *being a prime power. Then for any polynomial* $H(x)$ *with integer coefficients*

$$\int_{-1}^{1} \log |H(Ct)| \frac{dt}{\sqrt{1-t^2}} \geq 0.$$

SKETCH OF THE PROOF. It suffices to use an asymptotically bad family of curves. □

Now a problem in calculus: (1) Is Corollary 8.4.22 true for any $C \geq C_0$? Find C_0. (2) Is Corollary 8.4.22 true for any monic $H(x)$ with real coefficients?

Which measures correspond to curves and abelian varieties? One of the most interesting questions of asymptotic theory is to find out which measures correspond to actual families of abelian varieties and of curves.

Since the measure is expressible in terms of $\{\varphi_q\}$, it is the same as asking which families $\{\varphi_q\}$ are realizable by abelian varieties and which by curves, or asking this question for the limit zeta functions.

Let $\mathcal{M}$ be the set of measures μ on $[-2\sqrt{q}, 2\sqrt{q}]$ such that $\langle 1, \mu\rangle = 4\sqrt{q}$, and let $\mathcal{M}_0 \subset \mathcal{M}$ consist of measures μ corresponding to the sequence $\{\varphi_q\}$ satisfying the basic inequality.

Define $\mathcal{A} \subseteq \mathcal{M}$ as the set of measures corresponding to asymptotically exact families of abelian varieties, and $\mathcal{C} \subseteq \mathcal{M}_0$ of those corresponding to curves.

PROBLEM 8.4.23. Characterize $\mathcal{A}$ and $\mathcal{C}$. At least find some conditions distinguishing them from $\mathcal{M}$ and $\mathcal{M}_0$.

We have but some partial results in this direction. The picture is rather curious. While measures from $\mathcal{M}_0$ have continuous densities and other nice properties (in particular, the support of such a measure is always the whole segment), measures from $\mathcal{A}$ can have rather unexpected properties. For instance, a measure from $\mathcal{A}$ can be atomic. Even if the measure has no atoms (i.e., the measure of any point is zero), it can be supported on a proper subset of the segment. In particular, we have the following.

PROPOSITION 8.4.24. *For any segment $S \subset [-2\sqrt{q}, 2\sqrt{q}]$ of length greater than 4, there exists a measure from $\mathcal{A}$ with no atoms whose support equals S. If the length of S is smaller than 4, this is impossible.* □

Historical and Bibliographic Notes

The theory of infinite global fields originated from different sources. First, the study of asymptotic invariants of such fields is motivated by applications to algebraic geometry and similar number-theoretic codes. On the other hand, many classical results in number theory—such as Golod–Shafarevich towers, the Brauer–Siegel theorem, and the Odlyzko–Serre bound—fit perfectly into this framework; moreover, there exist purported connections with more advanced and recent topics such as the (unramified) Fontaine–Mazur conjecture. We believe that the theory which is in its infant stage of development can grow out to a rich and interesting subject.

Let us give some bibliographic remarks on the chapter.

The theory was mainly developed in [**TV97**] for the function field case and in [**TV02**] for the number field case; see, however, also [**Tsf92**] for some earlier results in the same vein. Similar ideas in the case of unramified towers can be found in [**Iha83**].

The existence of the Minkowski constant can be found in [**Lan70**]. Theorem 8.1.2 and Proposition 8.1.4 can be found in [**Odl90**]; see also [**Sta74, Sta75, Odl75, Odl76, Odl77a, Odl77b, Ser75, Poi77b**].

Lemma 8.1.6 in the number field case can be proved by the geometry of numbers (cf. [**Lan70**, Theorem V.4.5]). In the function field case it follows from the existence of moduli spaces for genus g curves. Indeed, those are varieties over the ground field which is finite, and thus they have but finitely many points defined over it.

Remark 8.1.8 is due to J-P. Serre (personal communication). Proposition 8.1.17 (as well as its unconditional version) can be found in [**Odl90**]. Propositions 8.1.18 and 8.1.20 were proved in [**Iha83**] in the particular case of unramified towers. The result of Theorem 8.1.28 in the case of asymptotically bad families (Remark 8.1.34) is essentially proved in [**Lan71**]. The version of the explicit formula used in the proof of Theorem 8.1.28 (i.e., Theorem 8.1.31) can be found in [**Poi77b**]. In this proof we use the standard properties of distributions from [**Sch66**, Chapters I and VII]. For the conductor-discriminant formula, see [**Neu99**, Chapter VII, Sec. 11, p. 534].

The problem of necessary conditions for the Brauer–Siegel theorem was widely discussed at the end of the 1950s at a seminar led by E. Artin; cf. [**Lan70**]. The inequality $w_{K_j} \leq cn_{K_j}^2$ used in the proof of Theorem 8.2.3 can be found in [**Lan70**, Sec. XVI.1, proof of Lemma 1]. Lemma 8.2.8 was discovered in [**Hei73**] and improved in [**Sta74**]. Lagarias–Odlyzko's estimate mentioned just after Lemma 8.2.11 is given in [**LO77**]. The inequality from Remark 8.2.16 is essentially established in [**Zim81**], as well as Zimmert's theorem on twin classes; cf. [**Oes85**].

Theorem 8.3.1 is proved by J. Martinet [**Mar78**]. Proposition 8.3.3 can be found in [**Roq67**]; cf. [**Mai96**]. Lemma 8.3.5 is proved in [**Mar78**]. Corollary 8.3.10 belongs to Y. Ihara [**Iha83**]. The formula for the discriminant of a cyclotomic field used in the proof of Theorem 8.3.11 can be found, e.g., in [**Lan70**, Sec. IV.1], and the formula for $E_{K,S}$ used in the proof of Lemma 8.3.4, in [**Lan70**, Sec. V.1].

Hajir–Maire tame towers are discovered in [**HM01**]; see also [**HM02**] for some generalizations. The example mentioned just after Problem 8.4.7 can also be found in [**HM01**]. The Fontaine–Mazur conjecture was proposed in [**FM95**], and

N. Boston's conjecture in [**Bos98**]. Ihara's conjecture on $\delta_{\mathcal{K}}$ was stated in [**Iha83**]. Theorems 8.4.16 and 8.4.17 are proved in [**TV97**];further inequalities on $w(\mathcal{K},\mu)_{\mathrm{inf}}$ were proved in [**TV97, Vla98b**]. The technique mentioned just after stating Problem 8.4.12 is considered in [**Lan70**, Sec. XVI.2, Lemma 3].

For the results on integrality in the second half of Sec. 8.4.2, see [**Tsf18**]. Theorem 8.4.21 was orally communicated to M. Tsfasman by J-P. Serre in the mid-1990s. Positive answers to the problem stated after Corollary 8.4.22 was obtained by V.A. Kleptsyn; see [**Tsf18**].

Proposition 8.4.24 (see [**Ser18**]) is based on old results by R.M. Robinson; in fact, a much stronger result is true: S can be any compact set of capacity greater than 1.

CHAPTER 9

Decoding: Some Examples

In the present chapter we touch upon the decoding problem for algebraic geometry codes, which is the most important from the practical point of view. However, since this domain is becoming more and more vast, we are obliged to make some choice of what we can expose in this subject. Recall that Sec. 4.1.4 of *Basic Notions* contains a formulation of the problem in the setting of a so-called *minimum distance decoder*, together with a basic decoding algorithm. By developing this algorithm, one can give a satisfactory construction of the minimum distance decoder; see Historical and Bibliographic Notes to the present chapter. However, we have chosen another decoding technique for our exposition, that of a *list decoder*. The reason is that in this setting the algebraic geometry codes can be decoded, under certain conditions, to an almost incredible error rate, approaching the (upper!) Singleton bound.

More precisely, we describe the Guruswami–Sudan decoding algorithm for algebraic geometry codes and some of its developments. Its natural framework is not just the decoding problem in the sense of Sec. 4.1.3 but another method, in some cases more efficient in practice and also having theoretical advantages, namely *list decoding*. We begin the chapter with a description of this concept and its principal properties in Sec. 9.1. Then in Sec. 9.2 we describe the Guruswami–Sudan list decoding algorithm, first for Reed–Solomon codes and then for general algebraic geometry codes. Section 9.3 contains a particular example of applying these techniques to codes on a Hermitian curve. The last section, Sec. 9.4, is devoted to an even more ambitious task of approaching the Singleton bound for decoding codes on curves from the García–Stichtenoth tower $\mathcal{W}_1$.

9.1. List Decoding

The problem of list decoding a code C of length n up to a fraction p of errors (or radius pn) is the following: given a received word, output the list of all codewords $c \in C$ within Hamming distance pn from it. It is clear that this could be of interest only if the output list size is reasonable, say bounded by a polynomial in the length, irrespective of which word is received. This motivates the following definition.

For $0 < p < 1$ and an integer $L \geq 1$, a code $C \subseteq \mathbb{F}_q^n$ is said to be *list decodable up to a fraction p of errors with list size L*, or simply *(p, L)-list-decodable*, if for every $y \in \mathbb{F}_q^n$, the number of codewords $c \in C$ within Hamming distance pn from y is at most L. For a function $\widehat{L}\colon \mathbb{Z}^+ \to \mathbb{Z}^+$ and $0 < p < 1$, a family of codes is said to be *$(p, \widehat{L})$-list-decodable* if every code C in the family is $(p, \widehat{L}(n))$-list-decodable, where n is the length of C. If $\widehat{L}(n) = L$ for all $n \in \mathbb{Z}^+$, we simply say that the family is (p, L)-list-decodable.

Note that a code being $(p,1)$-list-decodable is equivalent to saying that its relative distance δ is strictly greater than $2p$.

9.1.1. Johnson Bound

If a code has distance d, every Hamming ball of radius less than $d/2$ has at most one codeword, and thus the list decoding radius for a list size of 1 equals $\left\lfloor \frac{d-1}{2} \right\rfloor$. The Johnson bound, implicitly used in the proof of the Bassalygo–Elias bound (Theorem 1.1.50), shows that up to a radius that we call the *Johnson radius*, which is much larger than $d/2$, the number of codewords within some Hamming ball remains reasonably small.

In what follows, $d(a,b)$ denotes the Hamming distance between a and b, and $B(r,e)$ or $B_q(r,e)$ denotes the Hamming ball of radius e centered at r in the space $\mathbb{F}_q^n$.

THEOREM 9.1.1 (Johnson bound). *Assume that $r \in \mathbb{F}_q^n$, and let $S \subseteq \mathbb{F}_q^n$ be any subset of cardinality $M = |S|$. Let d be the average pairwise Hamming distance between distinct elements of S, and let e be the average distance between r and elements of S, namely*

$$d = \mathop{\mathbf{E}}_{\substack{x,y \in S \\ x \neq y}} d(x,y) = \frac{2}{M(M-1)} \sum_{\substack{x,y \in S \\ x \neq y}} d(x,y),$$

$$e = \mathop{\mathbf{E}}_{x \in S} d(r,x) = \frac{1}{M} \sum_{x \in S} d(r,x).$$

Then

$$M \leq \frac{\frac{q}{q-1}\frac{d}{n}}{\left(1 - \frac{q}{q-1}\frac{e}{n}\right)^2 - \left(1 - \frac{q}{q-1}\frac{d}{n}\right)},$$

provided that the denominator is positive.

COROLLARY 9.1.2. *Let C be any q-ary code of length n and minimum distance $d = \left(1 - \frac{1}{q}\right)(1-\delta)n$ for some $\delta \in (0,1)$. Let $e = \left(1 - \frac{1}{q}\right)(1-\gamma)n$ for some $\gamma \in (0,1)$ be an integer, and let $\gamma^2 > \delta$. Then, for all $r \in \mathbb{F}_q^n$, we have*

$$|B_q(r,e) \cap C| \leq \frac{1-\delta}{\gamma^2 - \delta}.$$

PROOF. Let $S = B_q(r,e) \cap C$. Let

$$\mathop{\mathbf{E}}_{\substack{x,y \in S \\ x \neq y}} d(x,y) = \left(1 - \frac{1}{q}\right)(1-\delta')n,$$

$$\mathop{\mathbf{E}}_{x \in S} d(r,x) = \left(1 - \frac{1}{q}\right)(1-\gamma')n.$$

We then have $\delta' \leq \delta < \gamma^2 \leq \gamma'^2$, and by Theorem 9.1.1,

$$|B_q(r,e) \cap C| \leq \frac{1-\delta'}{\gamma'^2 - \delta'} = 1 + \frac{1-\gamma'^2}{\gamma'^2 - \delta'} \leq 1 + \frac{1-\gamma^2}{\gamma^2 - \delta} = \frac{1-\delta}{\gamma^2 - \delta}.$$ □

The quantity $n\left(1-\frac{1}{q}\right)\left(1-\sqrt{1-\frac{q}{q-1}\frac{d}{n}}\right)$ is called the (q-ary) *Johnson radius*, and in every q-ary code of relative distance d/n, every Hamming ball of this radius therefore has few codewords.

SKETCH OF THE PROOF OF THEOREM 9.1.1. To keep the notation simple, we will assume that r is the all-zero vector. Pick distinct codewords $x=(x_1,\dots,x_n)$, $y=(y_1,\dots,y_n)\in S$, uniformly at random. We will obtain a lower bound (in terms of e and M) on the expected number of coordinates where x and y agree. We know that this expectation is $n-d$, and the theorem will follow by comparing these two quantities.

For $1\le i\le n$ and $\alpha\in\mathbb{F}_q$, let $\mu_i(\alpha)=|\{x\in S : x_i=\alpha\}|$. Then

$$\Pr(x_i=y_i)=\binom{M}{2}^{-1}\sum_{\alpha\in\mathbb{F}_q}\binom{\mu_i(\alpha)}{2}\ge\binom{M}{2}^{-1}\left[\binom{\mu_i(0)}{2}+(q-1)\binom{\frac{M-\mu_i(0)}{q-1}}{2}\right]$$

by Jensen's inequality. Therefore, the expected number of coordinates where x and y agree equals

$$\sum_{i=1}^{n}\Pr(x_i=y_i)\ge\binom{M}{2}^{-1}n\left[\binom{\mu}{2}+(q-1)\binom{\frac{M-\mu}{q-1}}{2}\right],$$

using Jensen's inequality again, where $\mu=\frac{1}{n}\sum_i\mu_i(0)=M\left(\frac{n-e}{n}\right)$. This expectation is exactly $n-d$, and thus we get

$$\frac{n-d}{n}\binom{M}{2}\ge\binom{\mu}{2}+(q-1)\binom{\frac{M-\mu}{q-1}}{2}.$$

Substituting $1-\frac{e}{n}$ for $\frac{\mu}{M}$ and simplifying, we obtain the desired result. □

We can also state the following alphabet-independent version of the Johnson bound.

THEOREM 9.1.3. *Let C be a q-ary code of length n and distance d. Let $r\in\mathbb{F}_q^n$ and assume that $(n-e)^2>n(n-d)$. Then*

$$|B_q(r,e)\cap C|\le\frac{nd}{(n-e)^2-n(n-d)}.$$

In particular, if $(n-e)^2\ge n(n-d)+1$, then $|B(r,e)\cap C|\le n^2$.

SKETCH OF THE PROOF. One can transform the denominator of the Johnson bound (Theorem 9.1.1) and then use the obvious inequality $\frac{q}{q-1}\frac{e^2}{n^2}\ge\frac{e^2}{n^2}$. □

Theorem 9.1.3 says that a code of relative distance δ is supposed to be list decoded up to a fraction $(1-\sqrt{1-\delta})$ of errors with polynomial size lists. Note that for $\delta\to 1$, the fraction of errors approaches 1, whereas with unique decoding, we can never correct more than a fraction $1/2$ of errors.

REMARK 9.1.4. The reader may wonder whether the Johnson bound is tight. It turns out that purely as a function of the distance, the Johnson bound is optimal; i.e., there exist codes which have super-polynomially many codewords in a ball of radius just slightly bigger than the Johnson radius.

However, it is known that for most codes, the Johnson bound is not tight and the list decoding radius for polynomial-sized lists far exceeds the Johnson radius; this follows from random coding (Gilbert–Varshamov) arguments used in the proofs of Theorems 9.1.6 and 9.1.7 below.

REMARK 9.1.5. It is easy to see that the Singleton bound is always a natural upper bound for list decoding, i.e., any ball of radius δn contains exponentially many codewords whenever $R+\delta>1$.

9.1.2. Capacity of List Decoding

We now turn to determining the "capacity" of list decoding, which means the trade-off between the rate of a code and its list decoding radius. Throughout this subsection, p is the fractional error-correction radius.

Recall that we have (cf. the Hamming bound)

$$q^{H_q(p)n-o(n)} \le |B_q(0,pn)| \le q^{H_q(p)n}. \tag{9.1.1}$$

We begin with an elementary upper bound on the list decoding radius.

THEOREM 9.1.6. *Let $q\ge 2$ be an integer, $0<p<1-1/q$, and $\varepsilon>0$. Then for all n large enough, if C is a q-ary code of length n and rate $1-H_q(p)+\varepsilon$, then there exists an $r\in\mathbb{F}_q^n$ such that $|B_q(r,pn)\cap C|\ge q^{\varepsilon n/2}$. In other words, C cannot be list decoded up to a fraction p of errors with polynomial-sized lists.*

PROOF. Pick a random vector $r\in\mathbb{F}_q^n$ and consider the random variable $Z=|B_q(r,pn)\cap C|$. Clearly, we have $\mathbf{E}\,Z=|C|\dfrac{|B_q(0,pn)|}{q^n}\ge q^{\varepsilon n-o(n)}$ by (9.1.1). Therefore, there must exist a center r such that $|B_q(r,pn)\cap C|\ge q^{\varepsilon n-o(n)}\ge q^{\varepsilon n/2}$. □

It turns out that the above simple bound is in fact tight and a rate approaching $1-H_q(p)$ can in fact be attained for correcting a fraction p of errors, as is shown below.

THEOREM 9.1.7. *For integers $q,L\ge 2$ and every $p\in(0,1-1/q)$, there exists a family of (p,L)-list-decodable q-ary error-correcting codes of rate R satisfying $R\ge 1-H_q(p)-1/L$.*

SKETCH OF THE PROOF. We use a standard expurgation argument. Fix a large enough length n; for simplicity assume that $e=pn$ is an integer. The idea is to pick (with a replacement) a random code of length n consisting of M codewords, where M is a parameter that will be fixed later in the proof. We will show that with high probability the resulting code will be (p,L)-list-decodable.

The probability that a fixed set of $L+1$ codewords all lie in a fixed Hamming sphere of radius pn equals $(|B_q(0,pn)|/q^n)^{L+1}$. By (9.1.1), this probability is at most $q^{-(L+1)(1-H_q(p))n}$, and therefore, by the union bound, the probability that some tuple of $L+1$ codewords all lie in some Hamming ball of radius pn is at most

$$\binom{M}{L+1}q^n q^{-(L+1)(1-H_q(p))n}.$$

If $M=q^{rn}$ for $r=1-H_q(p)-1/(L+1)$, then the above probability is at most $1/(L+1)!<1/3$. Also, among the M chosen codewords there are at least $M/2$

distinct codewords with probability at least $1/2$. Hence, there exists a (p,L)-list-decodable code with $q^{(1-H_q(p)-1/(L+1))n}/2 \geq q^{(1-H_q(p)-1/L)n}$ distinct codewords, or in other words with rate at least $1-H_q(p)-1/L$, as claimed. □

The above two results imply that, over alphabets of size q, the optimal rate possible for list decoding to radius p is $1-H_q(p)$, and the proof is easily carried over to the case of linear codes. In both cases, the results are purely existential, with no hint at constructivity.

THEOREM 9.1.8. *For every prime power $q \geq 2$, every $p \in (0,1-1/q)$, and every integer $L \geq 2$, there exists a family of (p,L)-list-decodable q-ary linear error-correcting codes of rate*

$$R \geq 1-H_q(p) - \frac{1}{\log_q(L+1)}.$$ □

Large alphabets. Let us consider the behaviour of the list decoding capacity $1-H_q(p)$ for growing q. We have

$$1-H_q(p) = 1-p+p\log_q\left(\frac{q}{q-1}\right) - \frac{H(p)}{\log q}.$$

In the limit of large q, the list decoding capacity thus approaches $1-p$. In other words, we can list decode a code of rate R up to a fraction of errors approaching $1-R$. Since $1-H_q(p) \geq 1-p-\frac{1}{\log q}$, we can get within ε of capacity, i.e., achieve a rate of $1-p-\varepsilon$ for list decoding up to a fraction p of errors, over an alphabet of size $\exp(O(1/\varepsilon))$. Moreover, a list size of $O(1/\varepsilon)$ suffices for this result. Conversely, if we fix p and let $\varepsilon \to 0$, then an alphabet size of $\exp(\Omega(1/\varepsilon))$ is necessary in order to achieve a rate of $1-p-\varepsilon$ for list decoding up to a fraction p of errors. (Recall that $f = \Omega(\alpha)$ for a positive function $f = f(\alpha)$ with $\alpha > 0$ means that $\limsup_{\alpha\to+\infty} f/\alpha > 0$.) Therefore, we have the following.

COROLLARY 9.1.9. *Let $0 < R < 1$, and let $\varepsilon > 0$ be sufficiently small. Then there exists a family of error-correcting codes of rate R over an alphabet of size $\exp(O(1/\varepsilon))$ that are $(1-R-\varepsilon, O(1/\varepsilon))$-list-decodable. Conversely, for a fixed p and $\varepsilon \to 0$, if there exists a family of q-ary $(p-\varepsilon)$-list-decodable codes of rate $1-p$, then $q \geq \exp(\Omega(1/\varepsilon))$ (here the constant in $\Omega(1/\varepsilon)$ depends on p).* □

One would like to achieve this result constructively, i.e., give explicit codes of rate R and efficient algorithms to list decode them up to a fraction $1-R-\varepsilon$ of errors; this is a challenge of fundamental importance.

9.2. Guruswami–Sudan Algorithm

Now let us explain the main ideas and some technical points of the Guruswami–Sudan algorithm, which gives list decoding for algebraic geometry codes up to a very satisfactory (and in fact rather astonishing) rate. We start with the main ideas of the algorithm for Reed–Solomon codes, which is called the Sudan algorithm, and then pass to more general algebraic geometry codes. However, we only treat a very special (but sufficient in many respects) case of a one-point code on a curve X with an evaluation set disjoint from the support of the divisor mP_∞. Our description in this case is split into two parts: the first describes some problems and algorithms for their solution parallel to the steps of the Guruswami–Sudan algorithm for Reed–Solomon codes provided that some basic operations in the function field $\mathbb{K} = \mathbb{F}_q(X)$ can be done effectively. The last subsection analyzes this last issue, which uses some specific representation of algebraic geometry codes under consideration and allows one to give an algorithm which is polynomial in the strict sense of the word.

9.2.1. Decoding of Reed–Solomon Codes

The list decoding problem for Reed–Solomon codes can be reduced to the following "curve-fitting" problem over a field $\mathbb{F}$: Given n points $(x_i,y_i)_{i=1}^n$, $x_i,y_i \in \mathbb{F}$, a degree parameter k, and an agreement parameter t, find all univariate polynomials p of degree at most k such that $y_i = p(x_i)$ for at least t values of $i \in \{1,\dots,n\}$; here the parameter t is more convenient to use than the number of errors $e = n-t$.

The Johnson bound gives constraints on possible values of k and t.

PROPOSITION 9.2.1. *For n distinct points $(x_i,y_i) \in \mathbb{F}_q \times \mathbb{F}_q$, $i = 1,\dots,n$, there are at most $\dfrac{n(t-k)}{t^2-kn}$ polynomials $p \in \mathbb{F}_q[x]$ of degree $\le k$ satisfying the condition $|\{i \mid p(x_i) = y_i\}| \ge t$ for $t > \sqrt{kn}$.* □

This result follows from Theorem 9.1.3 applied to the corresponding Reed–Solomon code.

Thus, the goal is to get a solution to the curve-fitting problem for $t > \sqrt{kn}$. Let us first assume for simplicity that $t > \sqrt{2kn}$ and expose the main ideas of the Sudan algorithm.

Description of the Sudan algorithm. The Sudan algorithm consists of two steps. In the first step we find a polynomial $Q(x,y)$ in two variables satisfying $Q(x_i,y_i) = 0$ for all $i = 1,\dots,n$. Then in the second step we consider Q as a univariate polynomial $Q(y)$ with coefficients in $\mathbb{F}_q[x]$ and factorize it; the aim is to find all small-degree roots of $Q(y)$, i.e., all polynomials p of degree at most k such that $y - p(x)$ is a factor of $Q(x,y)$. All these polynomials p form the output of the algorithm.

We have to check two points:

(i) The existence of Q.
(ii) If for $p(x)$ with $\deg p(x) \le k$ one has $p(x_i) = y_i$ for at least t values of i, then $p(x)$ is a root of $Q(y)$.

For (i) note that, allowing $Q(x,y)$ to have sufficiently large partial degrees $\deg_x Q$ and $\deg_y Q$, one easily finds Q by solving a homogeneous linear system.

For (ii) we need the degree of Q to be sufficiently small, and then for any $p(x)$ with $\deg p(x) \leq k$ such that $p(x_i) = y_i$ for at least t values of i, the degree of the polynomial $g(x) = Q(x,p(x))$ is also bounded from above. If this degree is less than t, then $g(x)$ must be identically zero, and we are done.

It turns out that the optimal choice of the parameters is as follows. Let the $(1,k)$-*weighted degree* $\deg_{(1,k)} Q$ be at most $D = \sqrt{2nk}$, where

$$\deg_{(1,k)} Q := \max\left\{i + kj : Q = \sum q_{ij} x^i y^j,\, q_{ij} \neq 0\right\}.$$

One can easily check the following statements.

PROPOSITION 9.2.2. *For any set $\{(x_i,y_i) \mid 1 \leq i \leq n\}$ of n points in $\mathbb{F}_q \times \mathbb{F}_q$, there exists a nonzero polynomial Q with $\deg_{(1,k)} \leq \sqrt{2nk}$ such that $Q(x_i,y_i) = 0$ for all $i = 1,\ldots,n$.* □

PROPOSITION 9.2.3. *If a polynomial $Q(x,y)$ with $\deg_{(1,k)} Q = D$ and a polynomial $y - p(x)$ with $\deg p \leq k$ have more than D common zeroes, then $y - p(x)$ divides $Q(x,y)$.* □

Since both solving a system of linear equations and factoring multivariate polynomials have very efficient algorithms, one can deduce the following result.

THEOREM 9.2.4. *The Sudan algorithm solves the list decoding problem for $[n,k+1,n-k]_q$ Reed–Solomon codes using $O(n^2 \log n)$ operations over $\mathbb{F}_q$ provided that $t > \sqrt{2kn}$. Moreover, the list size is at most $\sqrt{2n/k}$.* □

Improvement. One can improve the Sudan algorithm, thus obtaining what is called the *Guruswami–Sudan algorithm*. The improvement will only be in the number of corrected errors; its complexity grows but remains polynomially bounded.

The idea of the improvement is to force Q to have multiple points, i.e., to pass through each point (x_i,y_i) at least r times. In this case the degree D of Q grows, but since now $p(x)$ agrees with t *multiple* points of Q, the condition $t > D$ transforms into $rt > D$ (see Lemma 9.2.6 below), which is profitable. Growing r yields the result.

Now we give a formal description of the procedure. We begin with multiple points of Q, and we say that a polynomial $Q(x,y)$ passes through the point $(0,0)$ at least r times if it has no monomials of total degree less than r. Shifting $(0,0)$ to (α,β), we get the corresponding notion for any $P \in \mathbb{F}_q \times \mathbb{F}_q$.

Since the coefficients $\{q_{ij}\}$ of Q and the coefficients $\{(q_{\alpha,\beta})_{ij}\}$ of $Q_{\alpha,\beta} = Q(x+\alpha,y+\beta)$ are related by a linear equation

$$(q_{\alpha,\beta})_{ij} = \sum_{i' \geq i} \sum_{j' \geq j} \binom{i'}{i} \binom{j'}{j} q_{i',j'} \alpha^{i'-i} \beta^{j'-j},$$

finding Q passing through a set of n given points at least r times is still a linear system problem.

The algorithm is based on the following main lemma.

LEMMA 9.2.5. *If $Q(x,y)$ passes through a point (α,β) at least r times and $\beta = p(\alpha)$ for a polynomial p, then $(x-\alpha)^r$ divides the polynomial $g(x) := Q(x,p(x))$.*

PROOF. Consider the polynomial $p_1(x) = p(x+\alpha) - \beta$ and note that $p_1(0) = 0$. Hence, $p_1(x) = xp_2(x)$ for some polynomial $p_2(x)$. Now, let $g_1(x) := Q_{\alpha,\beta}(x,p_1(x))$;

then we have

$$g(x)=Q(x,p(x))=Q_{\alpha,\beta}(x-\alpha,p(x)-\beta)=Q_{\alpha,\beta}(x-\alpha,p_1(x-\alpha))=g_1(x-\alpha).$$

On the other hand, $Q_{\alpha,\beta}$ has no monomials of total degree less than r. Thus, by substituting $xp_2(x)$ for $p_1(x)$, we get a polynomial g_1 such that x^r divides $g_1(x)$. Therefore, $(x-\alpha)^r$ divides $g_1(x-\alpha)=g(x)$. □

LEMMA 9.2.6. *Let $Q(x,y)$ be a polynomial with $\deg_{(1,k)} Q=D$ passing through each of n distinct points (x_i,y_i) at least r times. If $p(x)$ is a polynomial of degree at most k such that $y_i=p(x_i)$ for at least t values of $i=1,\dots,n$ and $rt>D$, then $y-p(x)$ divides $Q(x,y)$.*

PROOF. Consider the polynomial $g(x)=Q(x,p(x))$; then $\deg g\le D$, and by Lemma 9.2.5, for every i such that $y_i=p(x_i)$, we know that $(x-x_i)^r$ divides $g(x)$. Thus, if S is the set of i such that $y_i=p(x_i)$, then $\prod_{i\in S}(x-x_i)^r$ divides $g(x)$. Since $|S|\ge t$, we have a polynomial of degree at least rt dividing g, but $\deg g=D<rt$; therefore, g is identically zero and $p(x)$ is a root of $Q(y)$. □

Now we will show that a required polynomial Q exists under a natural condition.

LEMMA 9.2.7. *If $n\binom{r+1}{2}<\dfrac{D(D+2)}{2k}$, then there exists a polynomial Q with $\deg_{(1,k)} Q=D$ passing through each of n given distinct points at least r times.*

PROOF. As we have seen, finding the coefficients of Q is solving a homogeneous linear system. The number of equations is $n\binom{r+1}{2}$. The number of unknowns equals the number of monomials of $(1,k)$-weighted degree at most D, and this number is

$$\sum_{j_2=0}^{\lfloor \frac{D}{k}\rfloor}\sum_{j_1=0}^{D-kj_2}1=\sum_{j_2=0}^{\lfloor \frac{D}{k}\rfloor}(D+1-kj_2)=(D+1)\left(\left\lfloor\frac{D}{k}\right\rfloor+1\right)-\frac{k}{2}\left\lfloor\frac{D}{k}\right\rfloor\left(\left\lfloor\frac{D}{k}\right\rfloor+1\right)$$
$$\ge\left(\left\lfloor\frac{D}{k}\right\rfloor+1\right)\left(D+1-\frac{D}{2}\right)\ge\frac{D}{k}\frac{D+2}{2}.$$ □

Setting $D=rt-1$, the condition $n\binom{r+1}{2}<\dfrac{D(D+2)}{2k}$ becomes a quadratic inequality in r, and if $t^2>kn$, then this inequality has positive integer solutions. Therefore, Lemmas 9.2.6 and 9.2.7 imply the following result.

THEOREM 9.2.8. *The Guruswami–Sudan algorithm solves the list decoding problem for Reed–Solomon codes in polynomial time provided that $t>\sqrt{kn}$.* □

Algorithm. Let us give a formal description of the Gusruswami–Sudan algorithm for Reed–Solomon codes.

ALGORITHM 9.2.9.

Input: the number of points $n=$ the length of the code, degree k of the polynomial $=\dim C-1$, agreement parameter t, and a set of points $\{(x_i,y_i)\}$, $i=1,\dots,n$, where $x_i,y_i\in\mathbb{F}_q$.

0. Set
$$r := 1 + \left\lfloor \frac{kn + \sqrt{k^2n^2 + 4(t^2 - kn)}}{2(t^2 - kn)} \right\rfloor,$$
$$D := rt - 1.$$

1. Find a nonzero polynomial $Q(x,y) = \sum_{j_1 + kj_2 \leq D} q_{j_1,j_2} x^{j_1} y^{j_2}$ of $(1,k)$-weighted degree at most D for which all coefficients of the shifted polynomial Q_{x_i,y_i} of total degree less than r are zero; more precisely,
$$(q_{x_i,y_i})_{j_1,j_2} := \sum_{j_1' \geq j_1} \sum_{j_2' \geq j_2} \binom{j_1'}{j_1} \binom{j_2'}{j_2} q_{j_1',j_2'} x_i^{j_1'-j_1} y_i^{j_2'-j_2} = 0$$
for all $i = 1, \ldots, n$ and all $j_1, j_2 \geq 0$ such that $j_1 + j_2 < r$.
2. Find all polynomials $p \in \mathbb{F}_q[x]$ of degree at most k such that $y - p(x)$ is a factor of $Q(x,y)$. For each such polynomial p check whether $p(x_i) = y_i$ for at least t values of $i = 1, \ldots, n$; if so, include p in the output list.

Output: The list of p's.

REMARK 9.2.10. One can also consider a weighted version of the Guruswami–Sudan algorithm. In the weighted version, the polynomial Q is forced to pass through different points a different number r_i times, where r_i is proportional to the weight w_i of the point.

Complexity of the algorithm. As well as the Sudan algorithm, the Guruswami–Sudan algorithm can be implemented in polynomial time. This follows by the same arguments: the first step can be accomplished by solving a homogeneous linear system, and the second step can be performed in polynomial time by using known bivariate polynomial factorization algorithms. The final result is as follows.

THEOREM 9.2.11. (a) *For any integer $L \geq 1$, an $[n, k+1, n-k]_q$ Reed–Solomon code can be list decoded to a radius of e for $e < n - \frac{r+1}{2(L+1)}n - \frac{L}{2r}k$ using lists of size L in $O(n^2L^5 \log q)$ field operations over $\mathbb{F}_q$, where*
$$r = \min\{r \in \mathbb{Z}, 1 \leq r \leq L : kL(L+1) \leq nr(r+1)\}.$$

(b) *An $[n, k+1, n-k]_q$ Reed–Solomon code can be list decoded to a radius of e for $e < n - \sqrt{kn}$ in $O(n^{12} \log q)$ operations over $\mathbb{F}_q$ using lists of size $O(n^2)$.*

(c) *If $e \leq n - \sqrt{(1+\varepsilon)kn}$, then one can perform list decoding in $O(n^2\varepsilon^{-5} \log q)$ operations over $\mathbb{F}_q$ using lists of size $O(\varepsilon^{-1}\sqrt{n/k})$.* □

Note that field operations over $\mathbb{F}_q$ can be implemented in $O(\log q \log^{O(1)} \log q)$ time.

9.2.2. Decoding of Algebraic Geometry Codes

We now describe the extension of the Guruswami–Sudan algorithm to a class of algebraic geometry codes. More precisely, we consider one-point algebraic geometry codes of the form $C = (X, \mathcal{P}, mP_\infty)_L$ for $\mathcal{P} = \{P_1, \ldots, P_n\}$, where $P_1, \ldots, P_n, P_\infty$ are distinct $\mathbb{F}_q$-rational points of a (smooth absolutely irreducible) curve X over $\mathbb{F}_q$; we also assume that $m \geq 2g - 1$.

Results of the previous section are extended to the case of algebraic function fields, yielding an approach to the list decoding problem for algebraic geometry codes. In particular, the decoding problem is reduced to some basis computations in an algebraic function field and to a factorization problem over an algebraic function field. Note that neither of these tasks is known to be efficiently solvable given only a generator (or a parity-check) matrix of the linear code. However, by precomputing a polynomial amount of additional information about the linear code and the underlying algebraic structures, one can solve both parts efficiently.

As in the case of Reed–Solomon codes, we begin with a description of the data points $\{(P_i, y_i)\}$ by some polynomial Q. We let Q be a polynomial in y with coefficients in the function field $\mathbb{K} = \mathbb{F}_q(X)$, $Q(y) \in \mathbb{K}[y]$. Then, given a value of $y_i \in \mathbb{F}_q$, $Q(y_i)$ yields an element of $\mathbb{K}$, which can be evaluated at $P_i \in X$, and we require $Q(P_i, y_i) = (Q(y_i))(P_i)$ to be zero. Moreover, as in the Guruswami–Sudan algorithm for Reed–Solomon codes, we require that (P_i, y_i) "behave" like a zero of multiplicity r of Q. We also insist that $Q(y)$ has a bounded number of poles (say, at most ℓ) at P_∞ for any substitution of y with a function in $L(mP_\infty)$. Having found such a Q, we then look for roots $f \in L(mP_\infty)$ of Q and output those that satisfy $f(P_i) = y_i$ for at least t values of i, $1 \le i \le n$.

Now let us give some details.

Notation. We denote by $v_i = v_{P_i}$ the discrete valuation on $\mathbb{F}_q(X)$ corresponding to the point P_i, $i = 1, \dots, n, \infty$.

Since $m \ge 2g-1$, $\ell(mP_\infty) = m - g + 1$ by Riemann–Roch, and by $\varphi_1, \dots, \varphi_{m-g+1}$, we denote a basis in $L(mP_\infty)$ with

$$-v_\infty(\varphi_1) < -v_\infty(\varphi_2) < \dots < -v_\infty(\varphi_{m-g+1}).$$

The following result shows the existence of bases where the zero multiplicity of a given point different from P_∞ is increasing; moreover, this result is effective.

PROPOSITION 9.2.12. *Let P_i, $1 \le i \le n$, be a point and let $m \ge 2g - 1$. Then there exists a basis $\varphi_{1,i}, \dots, \varphi_{m-g+1,1}$ of $L(mP_\infty)$ such that*

$$v_i(\varphi_{1,i}) < v_i(\varphi_{2,i}) < \dots < v_i(\varphi_{m-g+1,i}).$$

In what follows, such a basis will be called an *increasing-zero basis* with respect to the point P_i.

SKETCH OF THE PROOF. Take a basis $B = \{\varphi_1, \dots, \varphi_{m-g+1}\}$ of $L(mP_\infty)$. Let $B_i := 0$ and proceed as follows:

Let $e := \min\{v_i(f) \mid f \in B\}$ and $A := \{f \in B \mid v_i(f) = e\}$.

If $|A| = 1$, then let

$$B := B \setminus A,$$
$$B_i := B_i \cup A.$$

If $|A| > 1$, then let $A = \{a_1, \dots, a_{|A|}\}$ be a list and write each a_j in the form

$$a_j = t_i^e u_j, \quad j = 1, \dots, |A|,$$

where t_i is a local parameter and $u_j \in \mathcal{O}_{P_i}^*$. Now let

$$B := (B \setminus A) \cup \{t_i^e(u_j(P_i)u_1 - u_1(P_i)u_j) \mid j = 2, \dots, |A|\},$$
$$B_i := B_i \cup \{a_1\};$$

this ensures that $v_i(f) > e$ for all $f \in B$.

The process above is repeated $m-g+1$ times until $B=\varnothing$, and then the obtained set B_i is an increasing-zero basis of $L(mP_\infty)$ with respect to P_i, where two of its elements cannot have the same valuation at P_i. □

For $t \in \mathbb{N}$, denote by $g(t)$ the number of gaps $\leq t$ at P_∞, therefore

$$g(t) := t+1-\dim(L(tP_\infty)), \tag{9.2.1}$$

and let R denote the following vector space:

$$R := \bigcup_{i=0}^{\infty} L(iP_\infty).$$

If R is spanned by $\{\varphi_i \mid i \geq 1\}$ with strictly increasing pole orders, then the space $R[z] \subset \mathbb{K}[z]$ is spanned by $\{\varphi_i z^j \mid i \geq 1,\, j \geq 0\}$. We assign to z its weight $\rho(z)$ and define the weight

$$\rho(\varphi_i z^j) = -v_\infty(\varphi_i) + j\rho(z)$$

for $\varphi_i z^j$ in $R[z]$. For any function $f \in R[z] \setminus \{0\}$ with $f = \sum_{i,j} f_{i,j}\varphi_i z^j$, we define its weight as

$$\rho(f) = \max\{\rho(\varphi_i z^j) \mid f_{i,j} \neq 0\}.$$

We can also define a lexicographic ordering $\prec$ on the basis functions $\varphi_i z^j$:

$$\varphi_i z^j \prec \varphi_a z^b \quad\Longleftrightarrow\quad \rho(\varphi_i z^j) < \rho(\varphi_a z^b) \quad \text{or} \quad \rho(\varphi_i z^j) = \rho(\varphi_a z^b) \text{ and } j < b.$$

The following lemma describes the weights of the basis functions.

LEMMA 9.2.13. *Assume that the basis functions $\{Q_0, Q_1, \ldots\}$ of $R[z]$ are listed in ascending lexicographic ordering induced by $\rho(z) \geq 2g-1$, i.e.,*

$$\rho(Q_0) \leq \rho(Q_1) \leq \ldots.$$

Then the weight of Q_j is given by

$$\rho(Q_j) = (s-1)\rho(z) + t,$$

where s and t are positive integers defined by the following conditions:

$$\binom{s}{2}\rho(z) - (s-1)g \leq j < \binom{s+1}{2}\rho(z) - sg,$$

$$ts - g(t) \leq j - \left(\binom{s}{2}\rho(z) - (s-1)g\right) < (t+1)s - g(t+1),$$

g being the genus of X and $g(t)$ being the gap function defined in (9.2.1).

PROOF. Group the basis functions into disjoint sets

$$M_i = \{Q_\ell \mid (i-1)\rho(z) \leq \rho(Q_\ell) < i\rho(z)\}.$$

Note that for each t' with $0 \leq t' < \rho(z)$, the number of functions in M_i with weight $t' + (i-1)\rho(z)$ is exactly i if t' is a nongap and $i-1$ if t' is a gap. Thus, $|M_i| = i(\rho(z)-g) + (i-1)g = i\rho(z) - g$ and $|M_1| + |M_2| + \ldots + |M_{i-1}| = \binom{i}{2}\rho(z) - (i-1)g$. By the definition of s, we get $Q_j \in M_s$, and by the definition of t, $\rho(Q_j) = (s-1)\rho(z) + t$. □

Now we can define the zero multiplicity for a function $f \in R[z]$; namely, for $z_0 \in \mathbb{F}_q$, a pair (P_i, z_0) is said to be a *zero of multiplicity* r of $f = \sum_{j=0}^{\deg f} f_j(z-z_0)^j$ if

$$v_i(f_j) > r-j \quad \text{for all } j \le r \qquad \text{and} \qquad v_i(f_j) = r-j \quad \text{for some } j \le r.$$

One can easily deduce the following.

LEMMA 9.2.14. *If (P_i, z_0) is a zero of multiplicity r of f, then $v_i(f(t+z_0)) \ge r$ for any $t \in R$ with $v_i(t) \ge 1$.* □

Let $\varphi_{1,i}, \dots, \varphi_{m-g+1,i}$ be an increasing-zero basis with respect to P_i. Then a polynomial $f \in R[z]$ can be represented as

$$f(z) = \sum_{j=0}^{u} \sum_{k} f_{j,k} \varphi_{k,i} z^j$$

with $u \ge \deg f$. We give a sufficient condition for (P_i, z_0) to be a zero of multiplicity r of a nonzero polynomial $f \in R[z]$, as easily follows from the above definition.

LEMMA 9.2.15. *If for all $j+k \le r$ the coefficients*

$$f'_{j,k} := \sum_{\ell=j}^{u} f_{\ell,k} \binom{\ell}{j} z_0^{\ell-j}$$

are zero, then (P_i, z_0) is a zero of multiplicity at least r of f. □

Decoding algorithm. Now we are ready to describe the Guruswami–Sudan decoding algorithm for algebraic geometry codes. Here is a formal description.

ALGORITHM 9.2.16.

Input: A code $C = (X, \mathcal{P}, mP_\infty)_L$, $\mathcal{P} = \{P_1, \dots, P_n\}$, where $P_1, \dots, P_n, P_\infty$ are distinct $\mathbb{F}_q$-rational points of a (nonsingular absolutely irreducible) curve X of genus g, $n > m \ge 2g-1$; a received word $w = (w_1, \dots, w_n) \in \mathbb{F}_q^n$; and a parameter $r \ge 1$.

0. Set $\rho(z) := v_\infty(\varphi_k^{-1})$ and compute s_r and t so that

$$\binom{s_r}{2}\rho(z) - (s_r-1)g \le n\binom{r+1}{2} < \binom{s_r+1}{2}\rho(z) - s_r g,$$

$$ts_r - g(t) \le n\binom{r+1}{2} - \left(\binom{s_r}{2}\rho(z) - (s_r-1)g\right) < (t+1)s_r - g(t+1),$$

   where $g(t)$ is given by (9.2.1). Now set

$$e_r = n - \left\lfloor \frac{(s_r-1)\rho(z)+t}{r} \right\rfloor - 1.$$

1. Compute $Q(z) \in \mathbb{F}_q(X)[z] \setminus \{0\}$ so that
   - For all i, Q has (P_i, w_i) as a zero of multiplicity at least r;
   - $\rho(Q) \le (s_r-1)\rho(z) + t$.
2. Factorize Q into irreducible factors.
3. If $z-f$ divides Q for $f \in L(mP_\infty)$, and $d\big((f(P_1), \dots, f(P_n)), w\big) \le e_r$, then add the vector $(f(P_1), \dots, f(P_n))$ to the output list.

In what follows, any polynomial Q satisfying the two conditions in the algorithm will be called a *Q-polynomial* (with zero multiplicity r). Correctness of this algorithm must be shown by proving the existence of a Q-polynomial and proving that such a polynomial has the right factors. First, we have the following theorem.

THEOREM 9.2.17. *A Q-polynomial does exist.*

PROOF. By Lemma 9.2.15 the first condition on a Q-polynomial can be written as $n\binom{r+1}{2}$ homogeneous linear equations in $n\binom{r+1}{2}+1$ unknowns. Lemma 9.2.13 ensures that a polynomial that satisfies the first condition on a Q-polynomial can be constructed in such a way that the weight does not exceed $(s_r-1)\rho(z)+t$. □

The fact that a Q-polynomial has all the factors corresponding to codewords in a ball of radius e_r centered at the received word w follows from the results below.

LEMMA 9.2.18. *Let $f\in L(mP_\infty)$ and assume that $f(P_i)=w_i$. Then*

$$v_i(Q(f))\geq r.$$

PROOF. Let $p=f-w_i$. Then $p(P_i)=0$, so $v_i(p)\geq 1$. Since (P_i,w_i) is a zero of Q of multiplicity at least r, Lemma 9.2.14 implies that $v_i(Q(p+w_i))\geq r$. □

THEOREM 9.2.19. *Let Q be a Q-polynomial. Let a function $f\in L(mP_\infty)$ be such that $d\big((f(P_i),\dots,f(P_n)),w\big)\leq e_r$. Then $(z-f)\mid Q$.*

PROOF. Let $h=Q(f)$. Then $\rho(h)\leq(s_r-1)\rho(z)+t$. By Lemma 9.2.18 we have $v_i(h)\geq r$ for each value of i where $f(P_i)=w_i$. This happens at least $n-e_r$ times, and therefore the total number of zero multiplicities of h is at least

$$\sum_{i=1}^{n}v_i(h)\geq r(n-e_r)=r\left(\left\lfloor\frac{(s_r-1)\rho(z)+t}{r}\right\rfloor+1\right)>(s_r-1)\rho(z)+t\geq v_\infty(h^{-1}).$$

Thus, h is the zero function, since otherwise the number of its zeroes would exceed the number of its poles; therefore, $z-f$ divides Q. □

REMARK 9.2.20. Note that by Lemma 9.2.13 the degree of Q must be less than s_r, and thus the list size L satisfies $L<s_r$.

9.2.3. Representations of Algebraic Geometry Codes

The algorithm described above uses some steps the depend on computations in $\mathbb{K}=\mathbb{F}_q(X)$. To be really effective, these computations should be performed effectively, and thus there arise several issues on how to represent elements from the function field $\mathbb{K}$ and its places. Specifically, we would like to perform the following operations efficiently:

(i) Given two elements $x,y\in\mathbb{K}$, compute their sum and product in $\mathbb{K}$.
(ii) Given $f\in\mathbb{K}$ and $P\in\mathbb{P}_{\mathbb{K}}$, compute the zero or pole order of f at P, and compute $f(P)$.
(iii) Given a divisor $D\geq 0$, compute a basis for the vector space $L(D)$ over $\mathbb{F}_q$.

Note that, $\mathbb{K}$ being an infinite set, we cannot assume that operations in $\mathbb{K}$ are unit operations. Instead one must fix a representation for the elements and one must give explicit algorithms to perform these operations that are efficient with respect to the size of the representation of an element. A natural representation of elements of $\mathbb{K}$ is a ratio of two homogeneous multivariate polynomials, and for this representation, the field operations in $\mathbb{K}$ can be done in time polynomial in the sum of degrees of the respective polynomials. However, for question (iii) above, it is not known whether for general function fields we can find a basis for $L(D)$ over $\mathbb{F}_q$ with a short representation as the ratio of polynomials.

For applications to decoding, one does not need to work with all of $\mathbb{K}$ and instead focuses their attention only on elements in $L(D)$ for some divisor $D \geq 0$; this allows one to represent elements of $L(D)$ as vectors in $\mathbb{F}_q^{\dim L(D)}$ with respect to some *fixed* basis of $L(D)$ over $\mathbb{F}_q$, and below we use this representation in a somewhat augmented form. The basic step of the above algorithm is the root-finding task; we begin with formally stating an algorithm which solves this task and then discussing the arising representation issues.

Efficient root-finding.

PROCEDURE *ROOT-FIND*$_{(K,D)}(H)$.

Input: A degree-m polynomial $H = \sum_{i=0}^{m} a_i T^i \in \mathbb{K}[T]$ where each $a_i \in L(D)$.

Output: All roots of H that lie in $L(D)$.

1. "Reduce" H modulo a place $R \in \mathbb{P}_{\mathbb{K}}$ of degree $r = \deg R$ large enough, i.e. compute $b_i = a_i(R)$ for $0 \leq i \leq m$ and take the polynomial $P = \sum_{i=0}^{m} b_i y^i \in \mathbb{F}_{q^r}[y]$.
2. Compute the roots, say $\alpha_1, \ldots, \alpha_t$, of P that lie in $\mathbb{F}_{q^r}$, using a root-finding algorithm for finite fields.
3. For each α_j, $1 \leq j \leq t$, "find" $\beta_i \in L(D)$, if any, such that $\beta_i(R) = \alpha_t$.

The difficulty in the above algorithm is that we need to evaluate an $f \in L(D)$ at some place $R \in \mathbb{P}_{\mathbb{K}}$ of interest, and to this end we represent a place R by the values $\varphi_i(R)$, for $1 \leq i \leq p$, where $p = \dim L(D)$ and $\varphi_1, \ldots, \varphi_p$ is a basis of $L(D)$ over $\mathbb{F}_q$. Together with the representation of any element of $L(D)$ as a linear combination of the basis functions φ_i, this clearly enables us to evaluate any element of $L(D)$ at R; since $\varphi_i(R) \in \mathbb{F}_{q^r}$, $i = 1, \ldots, p$, where $r = \deg R$, we can represent R as an element of $\mathbb{F}_{q^r}^p$ when evaluating $f(R)$ for $f \in L(D)$.

Given these representations, we can provide efficient algorithms.

Now let us describe more details, taking all representation issues into account. This description will imply that the algorithm complexity is polynomial given the representation.

In fact, for the polynomial $H \in \mathbb{K}[T]$ of degree m having all its coefficients in $L(D)$, we will solve the problem of finding its roots that lie in $L(D)$ for some divisor $D \geq 0$; actually, we will need $D = mP_\infty$.

Instead of finding only roots that lie in $L(D)$, we will more generally find all roots $\alpha \in \mathbb{K}$ such that α has no poles outside $\operatorname{Supp} D$. We first prove a simple lemma that limits the number of poles that any such root can have at a place in $\operatorname{Supp} D$.

LEMMA 9.2.21. *Let $H \in \mathbb{K}[z]$ be a nonzero polynomial all of whose coefficients lie in $L(D)$ for some divisor $D \geq 0$, and let $\alpha \in \mathbb{K}$ be a root of H such that α has no poles outside* $\operatorname{Supp} D$. *Then in fact $\alpha \in L(D')$, where $D' = \sum_{P \in \operatorname{Supp} D} \frac{\deg D}{\deg P} P$. (Note that* $\deg D' = (\deg D)|\operatorname{Supp} D|$.*)*

PROOF. Let α be a root of H. We prove that for any $P \in \operatorname{Supp} D$ we have $v_P(\alpha) \leq \frac{\deg D}{\deg P}$, which implies the result. Let $H(z) = a_m z^m + \ldots + a_1 z + a_0$, where each $a_j \in L(D)$, and let $D = \sum_{R \in \operatorname{Supp} D} e_R R$, where each $e_R > 0$; then $v_P(a_j\alpha^j) \leq e_P + jv_P(\alpha)$, since $v_P(a_j) \leq e_P$. If $v_P(\alpha) \leq 0$, we are done, so assume that $v_P(\alpha) > 0$, and we get

$$v_P(H(\alpha) - a_m\alpha^m) = v_P\left(\sum_{j=0}^{m-1} a_j\alpha^j\right) \leq e_P + (m-1)v_P(\alpha). \tag{9.2.2}$$

We now upper bound $v_P(a_m)$. Since $\deg(a_m) = 0$ and $a_m \in L(D)$, we have

$$\sum_{R \in \operatorname{Supp} D} v_R(a_m) \deg R \geq 0,$$

and this together with $v_R(a_m) \leq e_R$ for every R gives

$$v_P(a_m) \deg P \geq -\sum_{R \in (\operatorname{Supp} D)\setminus P} e_R \deg R = -\deg D + e_P \deg P.$$

Thus we have

$$v_P(a_m\alpha^m) \geq -\frac{\deg D}{\deg P} + e_P + mv_P(\alpha), \tag{9.2.3}$$

and since $H(\alpha) = 0$, we obtain $v_P(a_m\alpha^m) = v_P(H(\alpha) - a_m\alpha^m)$. Using equations (9.2.2) and (9.2.3), this gives $e_P + (m-1)v_P(\alpha) \geq -\frac{\deg D}{\deg P} + e_P + mv_P(\alpha)$, which implies $v_P(\alpha) \leq \frac{\deg D}{\deg P}$. □

REMARK 9.2.22. The correctness of the algorithm is based on the following two simple facts. First, recall that for any function field $\mathbb{K}$, there exists a place of degree m in $\mathbb{P}_K$ for every large enough integer m. The second fact is that if $f_1, f_2 \in L(A)$ for some divisor $A \geq 0$ and $f_1(R) = f_2(R)$ for some place R with $\deg R > \deg A$, then $f_1 = f_2$.

We are now ready to formally present the root-finding algorithm.

ALGORITHM 9.2.23 (Algorithm *ROOT-FIND*$_{(K,D)}(H)$).

Nonuniform input which depends only on D and is independent of the actual input: A place $R \in \mathbb{P}_{\mathbb{K}}$ such that $\deg R = r > (\deg D)|\operatorname{Supp} D|$. The place R is represented as an s-tuple $(\zeta_1^R, \ldots, \zeta_s^R)$ over $\mathbb{F}_{q^r}$ comprised of evaluations of the basis functions $\mathcal{B}' = \{\varphi_1, \ldots, \varphi_p, \varphi_{p+1}, \ldots, \varphi_s\}$ of $L(D')$ at R. Here $D' = \sum_{P \in \operatorname{Supp} D} \frac{\deg D}{\deg P} P$, $s = \dim L(D)$, and the basis $\mathcal{B}'$ extends a basis $\mathcal{B} = \{\varphi_1, \ldots, \varphi_p\}$ of $L(D)$.

Input: A polynomial $H = a_0 + a_1 z + \ldots + a_m z^m$ of degree m in $\mathbb{K}[z]$, where each $a_i \in L(D)$ for some divisor D of $\mathbb{K}$. Each a_i is presented as a p-tuple $(a_{i1}, \ldots, a_{ip})$ over $\mathbb{F}_q$, where $p = \dim L(D)$, $a_{ij} \in \mathbb{F}_q$, i.e., $a_i = \sum_{j=1}^{p} a_{ij}\varphi_j$ for a basis $\mathcal{B} = \{\varphi_1, \ldots, \varphi_p\}$ of $L(D)$ over $\mathbb{F}_q$.

Output: All the roots of H that lie in $\mathbb{K}$ and have no poles outside $\operatorname{Supp} D$; by Lemma 9.2.21, all such roots lie in $L(D')$.

1. Reduce H modulo R to obtain a polynomial $h \in \mathbb{F}_{q^r}[z]$; i.e., set $h(z) = b_0 + b_1 z + \ldots + b_m z^m$ for

$$b_i = a_i(R) = \sum_{j=1}^{p} a_{ij}\zeta_j^R \in \mathbb{F}_{q^r}, \quad 0 \leq i \leq m.$$

2. Find all the (distinct) roots $\alpha_1,\ldots,\alpha_t$ of h that lie in $\mathbb{F}_{q^r}$ using a standard root finding algorithm for finite fields, e.g., Berlekamp's algorithm.
3. (Recovering the original roots): For each $\alpha_i \in \mathbb{F}_{q^r}$ such that $h(\alpha_i) = 0$, "find" a *unique* $\beta_i \in L(D')$ such that $\beta_i(R) = \alpha_i$, in terms of its coefficients $c_{ij} \in \mathbb{F}_q$ with respect to the basis $\mathcal{B}'$ of $L(D')$. (Recall that $D' = \sum_{P\in \mathrm{Supp}\, D} \frac{\deg D}{\deg P} P$ and thus $\deg D' = (\deg D)|\mathrm{Supp}\, D|$, so a β_i, if any, is unique; see Remark 9.2.22.) For each i, the coefficients c_{ij} can be found by solving $\sum_{j=1}^{s} c_{ij}\zeta_j^R = \alpha_i$, which can be viewed as a linear system of equations over $\mathbb{F}_q$ (by fixing some representation of elements of $\mathbb{F}_{q^r}$ over $\mathbb{F}_q$).
4. Output the list of roots $\{\beta_1,\ldots,\beta_t\}$.

It is clear in view of Lemma 9.2.21 (and Remark 9.2.22) that the algorithm correctly finds all roots of H that have no poles outside $\mathrm{Supp}\, D$. Moreover, the algorithm complexity is clearly polynomial given the nonuniform input, and we get the following result.

THEOREM 9.2.24. *There is an efficient root-finding algorithm that, for any function field* $\mathbb{K}$ *and any divisor* $D \geq 0$, *given an "advice" that depends only on* D *and is of size polynomial in* $\deg D$, *finds, in* $\mathrm{poly}(m, \deg D)$ *time, the roots of any degree-*m *polynomial in* $\mathbb{K}[z]$ *whose coefficients all lie in* $L(D)$. □

Since root-finding is the main algorithmic step in the Guruswami–Sudan list decoding algorithm and the other steps in the decoding algorithm can be performed efficiently under this representation, we conclude the main result.

THEOREM 9.2.25. *For every algebraic-geometric code* $C = (X, \mathcal{P}, \alpha Q)_L$ *of block length* n *and designed distance* d, *there is a representation of the code of size polynomial in* n, *such that, given this representation, list decoding from up to* $n - \sqrt{n(n-d)}$ *errors can be done in polynomial time.* □

9.3. Example: Hermitian Curves

In the present section we give an example of applying the Guruswami–Sudan algorithm to codes on Hermitian curves, paying attention to implementation issues. More precisely, we describe an efficient implementation of the Guruswami–Sudan algorithm, presented in Sec. 9.2.2, whose main ingredients are a fast way of computing increasing-zero bases, an efficient interpolation algorithm for finding a Q-polynomial, and a reduction of the problem of factoring the Q-polynomial to the problem of factoring a univariate polynomial over a large finite field.

Recall that Hermitian codes over $\mathbb{F}_q$ with $q = q_0^2$ are codes defined using the Hermitian curve X (see Sec. 3.4.1),

$$x^{q_0+1} - y^{q_0} - y = 0,$$

having q_0^3 affine $\mathbb{F}_{q_0^2}$-rational points and genus $\frac{q_0(q_0-1)}{2}$, the point P_∞ being the point at infinity. As in Sec. 9.2.2, we consider one-point algebraic geometry codes $C = (X, \mathcal{P}, mP_\infty)_L$, $\mathcal{P} = \{P_1, \dots, P_n\}$, where $P_1, \dots, P_n, P_\infty$ are distinct $\mathbf{F}_q$-rational points of X.

9.3.1. Bases and Interpolation

Computing increasing-zero bases. Proposition 9.2.12 ensures the existence of an increasing-zero base described in this proposition. However, it is not trivial to compute the standard representation of a function and evaluate the unit; here we show how to solve this problem in the case of the Hermitian function field.

To compute the standard representation of a polynomial $f \in \mathcal{O}_{P_i}$ with respect to the point $P_i = (x_i, y_i)$, $i \in \{1, \dots, n\}$, we can use $x - x_i$ as a local parameter in P_i. Indeed,

$$x^{q_0+1} - y^{q_0} - y = (x-x_i)^{q_0+1} - (y-y_i)^{q_0} + x_i(x-x_i)^{q_0} + x_i^{q_0}(x-x_i) - (y-y_i),$$

and $x - x_i$ is a local parameter if and only if $x_i \neq 0$.

Since f is regular in P_i, the valuation of f in P_i can be computed by repeatedly dividing f by $x - x_i$ until a unit is obtained; thus, we need an explicit method to divide by $x - x_i$ and to determine if a function is a unit.

A basis for the ring $R = \mathbb{F}_q[x,y]/(x^{q_0+1} - y^{q_0} - y) \subset \mathbb{F}_q(X)$ of polynomials is given by

$$\{(x-x_i)^\alpha (y-y_i)^\beta \mid 0 \le \alpha < q_0 + 1,\ \beta \ge 0\}, \quad i \in \{1, \dots, n\}.$$

Thus, dividing a basis element by $x - x_i$ is easy if it contains a factor $(x - x_i)$. Otherwise, one needs to compute $\frac{y-y_i}{x-x_i}$, but then we note that

$$\frac{y-y_i}{x-x_i} = \frac{(x-x_i)^{q_0} + x_i(x-x_i)^{q_0-1} + x_i^{q_0}}{(y-y_i)^{q_0-1} + 1}, \tag{9.3.1}$$

and the denominator on the right-hand side is obviously a unit; we denote $e := (y-y_i)^{q_0-1} + 1$.

Now we want to obtain a standard representation for

$$f = \sum_{\alpha=0}^{q_0} \sum_{\beta \in \mathbb{N}} f_{\alpha,\beta} (x-x_i)^\alpha (y-y_i)^\beta,$$

where $f_{\alpha,\beta} = \sum_{j\in\mathbb{Z}} f_{\alpha,\beta,j}e^j$ with $f_{\alpha,\beta,j} \in \mathbb{F}_{q_0^2}$. Note that such a representation of f is not unique, but this causes no difficulties, and f can be divided by $x - x_i$ with a result of the same form.

For $(\alpha,\beta) \neq (0,0)$ it is already shown how to divide $f_{\alpha,\beta}(x-x_i)^\alpha(y-y_i)^\beta$ by $x - x_i$; it remains to show how to compute $\frac{f_{0,0}}{x-x_i}$ in the case where f is not a unit, i.e., $\sum_{j\in\mathbb{Z}} f_{0,0,j} = 0$.

Let $s = \min\{j \mid f_{0,0,j} \neq 0\}$; then

$$f_{0,0} = e^s \sum_{j\in\mathbb{Z}} f_{0,0,j}e^{j-s},$$

where $\sum_{j\in\mathbb{Z}} f_{0,0,j}e^{j-s}$ is a polynomial in y which is divisible by $y - y_i$. Put $h := \left(\sum_{j\in\mathbb{Z}} f_{0,0,j}e^{j-s}\right)/(y-y_i)$ and note that

$$\frac{f_{0,0}}{x-x_i} = e^s \frac{y-y_i}{x-x_i}h = e^{s-1}\big((x-x_i)^{q_0} + x_i(x-x_i)^{q_0-1} + x_i^{q_0}\big)h. \tag{9.3.2}$$

In this way we obtain the following algorithm for the standard representation of a polynomial.

ALGORITHM 9.3.1.

Input: Polynomial f and point $P_i = (x_i, y_i)$.

Output: u and m such that $m = v_i(f)$ and $f = u(x-x_i)^m$.

1. Initialization: $f^{(0)} := f = \sum_{\alpha,\beta} f_{\alpha,\beta}(x-x_i)^\alpha(y-y_i)^\beta$, $m := 0$.
2. If $\sum_{j\in\mathbb{Z}} f_{0,0,j} \neq 0$, then stop.
3. Use equations (9.3.2) and (9.3.1) to compute

$$\begin{aligned} f^{(m+1)} &:= \frac{f^{(m)}}{x-x_i} \\ &= \frac{f^{(m)}_{0,0}}{x-x_i} + \sum_{\alpha\geq 1,\,\beta} f^{(m)}_{\alpha,\beta}(x-x_i)^{\alpha-1}(y-y_i)^\beta + \sum_{\beta\geq 1} f^{(m)}_{0,\beta}(y-y_i)^{\beta-1}\frac{y-y_i}{x-x_i}, \end{aligned}$$

$$m := m+1.$$

4. Go to Step 2.

Note that we use this algorithm for polynomials only, but it can also be applied to any rational function $h = \frac{f}{g}$ by treating f and g separately.

The above algorithm allows one to implement the method for determining increasing-zero bases in the proof of Proposition 9.2.12, since the unit of the standard representation is given in a form which can be directly evaluated.

To make the implementation more efficient, it may be observed that the increasing-zero basis is computed only for the use in the interpolation step, and this implies that in practice a full increasing-zero basis is not needed. It is sufficient to have a basis where a given point P_i is a zero of increasing multiplicity for the first $r+1$ basis elements (where r is the parameter of the decoding algorithm).

The other basis elements should just have P_i as a zero of multiplicity greater than r, which allows one to speed up computations, as well as save computer memory, since many of the basis elements are just shifted monomials and can be computed whenever they are needed.

Interpolation. Now we want to determine a valid Q-polynomial, which exists in the vector space

$$\langle Q_0,\ldots,Q_\ell\rangle = \mathbb{F}_q Q_0 \oplus \mathbb{F}_q Q_1 \oplus \ldots \oplus \mathbb{F}_q Q_\ell, \quad \ell = \binom{r+1}{2} n,$$

$Q_0,\ldots,Q_\ell$ being as in Lemma 9.2.13; a Q-polynomial can be found by solving a system of linear equations expressing its very definition (see p. 200), which will be referred to as *zero-conditions* in what follows. Moreover, the system is of a special structure, which allows one to make the computations more efficient, and now we give a method for this improvement.

Let ord: $R[z] \to \mathbb{N} \cup \{-\infty\}$ be given by $\operatorname{ord}(0) = -\infty$, $\operatorname{ord}(\langle Q_0\rangle \setminus \{0\}) = 0$, and

$$\operatorname{ord}(\langle Q_0,\ldots,Q_i\rangle \setminus \langle Q_0,\ldots,Q_{i-i}\rangle) = i,$$

for $i > 0$. If $f \in R[z]$, then $\operatorname{ord}(f)$ is called the order of the function f.

We want to find a polynomial that satisfies the zero-conditions and is minimal with respect to ord, which gives Q, and it suffices to look for it in the vector space

$$V = \big\{ f z^j \mid f \in R,\ 0 \le j \le \max\{\deg(Q_j) \mid 0 \le j \le \ell\}\big\}.$$

The idea of the algorithm is to make a partition of V and then consider the zero-conditions one by one, while maintaining a list of polynomials—one from each partition class—which all satisfy the zero-conditions considered so far and all are minimal within the given partition class.

To this end, we need for each point an element of $R[z]$ that has a given pair (P_i, W_i) as a zero of multiplicity at least r and which, whenever multiplied by a polynomial, gives a result within the same partition class as that polynomial. To do this, the order of these elements must be the same for all $i \in \{1,\ldots,n\}$. Therefore, choose t to be the smallest integer such that

$$\operatorname{ord}(\varphi_{j(i),i}) = t \quad \text{and} \quad v_i(\varphi_{j(i),i}) \ge r, \quad i = 1,\ldots,n.$$

Now construct a partition of V as follows.

Let

$$A := \operatorname{ord}(V) \setminus \big\{ i \in \operatorname{ord}(V) \bigm| \exists f \in \operatorname{ord}^{-1}(i),\ \exists h \in \operatorname{ord}^{-1}(t) : h \mid f \big\}.$$

Furthermore, let $h \in \operatorname{ord}^{-1}(t)$ and define

$$T := \bigcup_{i=0}^{\infty} \operatorname{ord}^{-1}(\operatorname{ord}(h^i)),$$

which does not depend on the choice of h.

Let $A = \{a_1,\ldots,a_{|A|}\}$, and let

$$G_j := \operatorname{ord}^{-1}\big(\operatorname{ord}(\{f_j g \mid f_j \in \operatorname{ord}^{-1}(a_j),\ g \in T\})\big).$$

Now $G_1,\ldots,G_{|A|}$ is a partition of V; furthermore, if a polynomial in some G_j is multiplied by a polynomial in $\operatorname{ord}^{-1}(t)$, then the result will remain in G_j.

Finally, for $f = \sum_{j,k} f_{j,k}\varphi_{k,i}z^j \in R[z]$, we put

$$\mathrm{coef}(f, \varphi_{k,i}z^j) = f_{j,k}.$$

This leads to the following algorithm.

ALGORITHM 9.3.2.

Input: Pairs $(P_1, w_1), \ldots, (P_n, w_n)$ and a required zero multiplicity r.

1. Initialize by setting
$$G := \{g_1, \ldots, g_{|A|}\},$$
so that $\mathrm{ord}(g_j) = \min \mathrm{ord}(G_j)$ for $j = 1, \ldots, |A|$;
2. For $i = 1, \ldots, n$ do:
   For each pair $(j,k) \in \mathbb{N}^2$ with $j + k \leq r$ and $k \geq 1$ do:
   Let $G^* := \{g \in G \mid \mathrm{coef}(g(z+w_i), \varphi_{k,i}z^j) \neq 0\}$;
   If $G^* \neq \varnothing$, then choose $f \in G^*$ so that $\mathrm{ord}(f) = \min \mathrm{ord}(G^*)$ and let
$$G^* = \{\varphi_{j(i),i}f\} \cup \{\mathrm{coef}\big(f(z+w_i), \varphi_{k,i}z^j\big)g - \mathrm{coef}\big(g(z+w_i), \varphi_{k,i}z^j\big)f\};$$
3. After this, the result is given by the polynomial in G which is smallest with respect to ord.

Consecutive analysis of the preceding steps yields the following conclusion.

PROPOSITION 9.3.3. *The algorithm works correctly.* □

REMARK 9.3.4. In this algorithm, an upper bound on the order of the solution is known, and we can discard polynomials exceeding this upper bound.

9.3.2. Factorization

To finish, we need the factorization step. Note, however, that since the Q-polynomial is a polynomial in $R[z]$, the factorization is not obvious. Note also that fairly simple and efficient methods exist for factoring univariate polynomials over a finite field. It is possible to transform our task of factoring the Q-polynomial into a problem of factoring a univariate polynomial over a large finite field.

Indeed, we have $R = \mathbb{F}_{q_0^2}[x,y]/(x^{q_0+1} - y^{q_0} - y) = \langle \{x^\alpha y^\beta \mid 0 \leq \alpha \leq q_0, \beta \geq 0\} \rangle$; the polynomials in $L(mP_\infty)$ have degree in x smaller than $q_0 + 1$, and the degree in y is also bounded by some $c \in \mathbb{N}$. Now let $f(y) \in \mathbb{F}_q[y]$ with $\deg(f) \geq c$ and

$$D_1 = \mathbb{F}_q[T]/(f(T));$$

furthermore, let $g(x) := x^{q_0+1} - y^{q_0} - y \bmod (f)$ and

$$D_2 = D_1[T]/(g(T)).$$

Consider the maps $\varphi\colon \mathbb{F}_q[x,y][z] \to D_1[x][z]$ and $\theta\colon D_1[x][z] \to D_2[z]$ given by

$$\varphi(h_1) = h_1 \bmod (f),$$
$$\theta(h_2) = h_2 \bmod (g),$$

and their composition $\theta\varphi$. Assume that $h \in L(mP_\infty)$ and $(z-h) \mid Q$; then $\theta\varphi(z-h) \mid \theta\varphi(Q)$ and, furthermore, reducing $z-h$ modulo f and g will not change $z-h$, since the degree in y of f and the degree in x of g is higher than the degrees of h in y and x, respectively. Therefore, it will be sufficient to factorize $\theta\varphi(Q)$.

This is especially useful, since D_1 is a finite field, and if $g(x)$ is irreducible over D_1, then D_2 is also a finite field; for an irreducible f we have $D_1 \cong \mathbb{F}_q^{c_1}$ with $c_1 = \deg(f)$, and $g(x) = x^{q_0+1} - (y^{q_0} + y)$ is a binomial in $\mathbb{F}_q^{c_1}[x]$.

Now we can apply the following theorem.

THEOREM 9.3.5. *Let $\omega \in \mathbb{F}_q^c \setminus \{0\}$, and let e denote the order of ω. Then $x^{q_0+1} - \omega$ is irreducible over $\mathbb{F}_q^c$ if and only if each prime factor of q_0+1 divides e but not $\dfrac{q^c-1}{e}$.* □

Note that, since q_0+1 divides $q^c - 1$, the theorem states that $x^{q_0+1} - \omega$ is irreducible if and only if $\gcd\left(q_0+1, \dfrac{q^c-1}{e}\right) = 1$, where e is the order of ω in $\mathbb{F}_q^c \setminus \{0\}$.

Using the last result, one can obtain a polynomial f as follows.

THEOREM 9.3.6. *Let $c_1 \geq 1$ be an integer. Then there exists a polynomial $f(y)$ such that $\deg(f) \geq c_1$ and the order e of $y^{q_0} + y$ in $\mathbb{F}_q[y]/(f(y))$ satisfies the condition*

$$\gcd\left(q_0+1, \frac{q^c-1}{e}\right) = 1.$$

SKETCH OF THE PROOF. Take $f_1(y) \in \mathbb{F}_q[y]$ to be an arbitrary irreducible polynomial of degree c_1, let $\mathbb{F}_Q = \mathbb{F}_q[y]/(f_1(y))$, where $Q = q^{c_1}$, and set $h(x) = x^{q_0} + x - \alpha \in \mathbb{F}_Q[x]$ for a primitive $\alpha \in \mathbb{F}_Q$. Let γ be a root of $h(x)$ in an extension of $\mathbb{F}_Q$.

One easily deduces that the difference of two roots of $h(x)$ is an element of $\mathbb{F}_q \subseteq \mathbb{F}_Q$.

Now define $f(y)$ as the minimal polynomial of γ over $\mathbb{F}_q$, and let $\mathbb{F}_q(\gamma) \cong \mathbb{F}_q[y]/(f(y)) = \mathbb{F}_Q$. Then a direct computation yields the conclusion. □

The two explicit procedures of interpolation and factorization make it possible to apply the Guruswami–Sudan algorithm to one-point codes on Hermitian curves.

Example. Let us give a particular example of list decoding a code over $\mathbb{F}_4 = \mathbb{F}_2[\alpha]$, $\alpha^2 + \alpha + 1 = 0$. The Hermitian curve $x^3 - y^2 - y = 0$ over $\mathbb{F}_4$ is elliptic and has the following affine $\mathbb{F}_4$-points:

$$\mathcal{P} = \{P_1, \dots, P_8\} = \{(0,0),(0,1),(1,\alpha),(1,\alpha^2),(\alpha,\alpha),(\alpha,\alpha^2),(\alpha^2,\alpha),(\alpha^2,\alpha^2)\}.$$

We take the $[8,4,4]$ code $(X, \mathcal{P}, 4P_\infty)_L$ and use the parameters $r = 6$, $e_6 = 2$, and $s_6 = 9$.

The transmitted codeword in this example is the all-zero word. We assume that two errors occur, and the received word is

$$(\alpha^2, 0, 0, \alpha^2, 0, 0, 0, 0).$$

In this case, increasing-zero bases are needed for $L(36P_\infty)$ with respect to each point. For example, several first elements of an increasing-zero basis with respect to $P_4 = (1, \alpha^2)$ are as follows for $e := 1 + y$:

$$\begin{aligned}
\varphi_{1,4} &:= 1 = (x-1)^0, \\
\varphi_{2,4} &:= (x-1) = (x-1)^1, \\
\varphi_{3,4} &:= (x-1)^2 = (x-1)^2, \\
\varphi_{4,4} &:= (x-1) + (y - \alpha^2) \\
&= (x-1)^3\big((e^{-3} + e^{-2} + e^{-1}) + (e^{-3} + e^{-2})(x-1) + e^{-3}(x-1)^2\big).
\end{aligned}$$

Furthermore, $\mathbb{F}_{4^3}$ is constructed as $D_1 = \mathbb{F}_4[y]/(y^3+\alpha^2 y+1)$, and $\mathbb{F}_{4^{3\cdot 3}}$ as $D_2 = \mathbb{F}_{4^3}[x]/(x^3-(y^2+y))$.

We compute the Q-polynomial using the interpolation method, and the result after reducing modulo $y^3+\alpha^2y+1$ and $x^3-(y^2+y)$ is the following polynomial in $\mathbb{F}_{4^{3\cdot 3}}[z]$:

$$\begin{aligned}Q_{\mathrm{red}} = \big(&(\alpha+\alpha y)+(\alpha y+\alpha^2y^2)x+(\alpha y+y^2)x^2\big)z+\big((\alpha+\alpha^2y)+(\alpha+\alpha^2y)x\\ &+(1+\alpha y^2)x^2\big)z^2+\big((\alpha^2y+\alpha^2y^2)+(\alpha+\alpha y+\alpha^2y^2)x+(\alpha^2+\alpha y\\ &+\alpha^2y^2)x^2\big)z^3+\big((\alpha^2+y+\alpha y^2)+(\alpha^2+\alpha^2y+\alpha y^2)x+(\alpha y+y^2)x^2\big)z^4\\ &+\big((\alpha+y)+(\alpha+\alpha y^2)x+(1+\alpha y)x^2\big)z^5+\big((y+\alpha y^2)+(1+\alpha y+y^2)x\\ &+(\alpha+\alpha^2y^2)x^2\big)z^6+\big((1+\alpha^2y^2)+(\alpha^2+\alpha y^2)x+(\alpha+y^2)x^2\big)z^7+z^8.\end{aligned}$$

It turns out that this polynomial has three linear factors,

$$\big((\alpha^2+\alpha y+y^2)+(\alpha y+\alpha^2y^2)x+(1+\alpha^2y+\alpha y^2)x^2\big)+z,$$
$$(\alpha^2+\alpha^2x+\alpha^2x^2)+z, \qquad z,$$

but the first of them does not lie in $L(4P_\infty)[z]$ and is ignored. The last two factors give the following list of codewords:

$$(\alpha^2,\alpha^2,\alpha^2,\alpha^2,0,0,0,0), \qquad (0,0,0,0,0,0,0,0);$$

both of these codewords have distance 2 to the received word, forming a desired list in this case.

9.4. Approaching the Singleton Bound

The Guruswami–Sudan algorithm in its asymptotic setting allows one to list decode up to $(1-\sqrt{R})n$ errors. However, one could ask whether it is possible to approach the Singleton bound of $(1-R)n$ errors for list decoding. Surprisingly, this is possible for certain subcodes of algebraic geometry codes, in particular those coming from the García–Stichtenoth tower $\mathcal{W}_1$. Clearly, this advance requires some new ideas and techniques, and in fact the resulting constructions do not yield a constant size list but give a list of very slowly increasing cardinality; nevertheless, here we expose some principal points of the corresponding algorithms, giving their extraordinary performance.

We begin with formulating the result and then giving some details of the corresponding constructions and algorithms.

THEOREM 9.4.1. *For every $R \in (0,1)$ and $\varepsilon > 0$ there exists a deterministic polynomial time constructible family of error-correcting codes of rate R over an alphabet of size $\exp(O(\varepsilon^{-2}\log^2(1/\varepsilon)))$ that can be list decoded in polynomial time from a fraction $1-R-\varepsilon$ of errors, outputting a list of size at most*

$$\exp_{1/\varepsilon}\big(\exp_{1/\varepsilon}(\exp(O(\log^* N)))\big),$$

where N is the block length of the code.

Here $\log^* N$ denotes the minimum m such that the mth iterated logarithm $\log^{(m)} N = \underbrace{\log(\log(\ldots(\log N)\ldots))}_{m \text{ times}}$ satisfies $\log^{(m)} N \le 1$.

Note that these codes are some subcodes of the algebraic geometry codes on García–Stichtenoth curves, specially constructed for validity of Theorem 9.4.1.

A complete proof of the theorem is rather involved, and below we only give three of its main points. Sections 9.4.1 and 9.4.2 describe algebraic constructions serving to list decode the codes of Theorem 9.4.1. In Sec. 9.4.3 we describe the algorithm for the general algebraic geometry codes, and finally, in Sec. 9.4.4, we give a construction of the codes coming from the $\mathcal{W}_1$ tower together with a version of the algorithm applied to those codes. Note that the parameters of list decoding in Theorem 9.4.1 are optimal, which explains our attention to it.

9.4.1. Periodic Affine Subspaces

The first algebraic tool for the list decoders used in the proof of Theorem 9.4.1 is as follows.

For a vector $\boldsymbol{y} = (y_1, y_2, \ldots, y_m) \in \mathbb{F}_q^m$ and positive integers $t_1 \le t_2 \le m$, we denote its projection onto coordinates t_1 through t_2 by $\mathrm{proj}_{[t_1,t_2]}(\boldsymbol{y}) \in \mathbb{F}_q^{t_2-t_1+1}$, i.e., $\mathrm{proj}_{[t_1,t_2]}(\boldsymbol{y}) = (y_{t_1}, y_{t_1+1}, \ldots, y_{t_2})$. When $t_1 = 1$, we use $\mathrm{proj}_t(\boldsymbol{y})$ to denote $\mathrm{proj}_{[1,t]}(\boldsymbol{y})$. These notions are extended to subsets of vectors in an obvious way: $\mathrm{proj}_{[t_1,t_2]}(S) = \{\mathrm{proj}_{[t_1,t_2]}(\boldsymbol{x}) \mid \boldsymbol{x} \in S\}$.

For positive integers r, b, and Δ, an affine subspace $H \subset \mathbb{F}_q^{b\Delta}$ is said to be (r,Δ,b)-*periodic* if there exists a subspace $S \subseteq \mathbb{F}_q^{\Delta}$ of dimension at most r such that for every $j = 1, 2, \ldots, b$ and every "prefix" $\boldsymbol{a} \in \mathbb{F}_q^{(j-1)\Delta}$, the projected affine subspace of $\mathbb{F}_q^{\Delta}$ defined as

$$\{\mathrm{proj}_{[(j-1)\Delta+1,j\Delta]}(\boldsymbol{x}) \mid \boldsymbol{x} \in H \text{ and } \mathrm{proj}_{(j-1)\Delta}(\boldsymbol{x}) = \boldsymbol{a}\}$$

is contained in an affine subspace $S+\boldsymbol{v}_{\boldsymbol{a}} \subseteq \mathbb{F}_q^{\Delta}$ for some vector $\boldsymbol{v}_{\boldsymbol{a}} \in \mathbb{F}_q^{\Delta}$ dependent on $\boldsymbol{a}$.

In the list decoding below, for a code over $\mathbb{F}_{q^m}$ of dimension k with an auxiliary parameter s governing the quality of the decoder relative to the Singleton bound, we use an $(s-1,m,k)$-periodic affine subspace of $\mathbb{F}_q^{mk}$.

The following properties of periodic affine spaces immediately follow from the definition.

PROPOSITION 9.4.2. *Let H be an (r,Δ,b)-periodic affine subspace. Then for each $j=1,2,\ldots,b$, we have the following:*

1. *The projection $\mathrm{proj}_{j\Delta}(H)=\{\mathrm{proj}_{j\Delta}(\boldsymbol{x}) \mid \boldsymbol{x}\in H\}$ of H onto the first j blocks of Δ coordinates satisfies $\dim \mathrm{proj}_{j\Delta}(H)\leq jr$.*
2. *For each $\boldsymbol{a}\in\mathbb{F}_q^{(j-1)\Delta}$, there are at most q^r extensions $\boldsymbol{y}\in\mathrm{proj}_{j\Delta}(H)$ with $\mathrm{proj}_{(j-1)\Delta}(\boldsymbol{y})=\boldsymbol{a}$.* □

The *canonical representation of an (r,Δ,b)-periodic subspace H* consists of a matrix $B\in\mathbb{F}_q^{\Delta\times\Delta}$ such that $\ker(B)$ has dimension at most r, and of vectors $a_i\in\mathbb{F}_q^{\Delta}$ and matrices $A_{i,j}\in\mathbb{F}_q^{\Delta\times\Delta}$, for $1\leq i\leq b$ and $1\leq j<i$, such that $\boldsymbol{x}\in H$ if and only if the equality

$$a_i+\left(\sum_{j=1}^{i-1} A_{i,j}\cdot\mathrm{proj}_{[(j-1)\Delta+1,j\Delta]}(\boldsymbol{x})\right)+B\cdot\mathrm{proj}_{[(i-1)\Delta+1,i\Delta]}(\boldsymbol{x})=0$$

holds for every $i=1,2,\ldots,b$.

There is an even stronger property that holds for the subspaces output by the linear-algebraic list decoder. An affine subspace H of $\mathbb{F}_q^{\varkappa}$ is said to be (r,Δ)-*ultra periodic* if for every integer ℓ, $1\leq\ell\leq\frac{\varkappa}{\Delta}$, we have that $\mathrm{proj}_{b_\ell\cdot\ell\Delta}(H)$ is $(\ell r,\ell\Delta,b_\ell)$-periodic for $b_\ell=\left\lfloor\frac{\varkappa}{\ell\Delta}\right\rfloor$. These subspaces look periodic in all "scales" simultaneously.

9.4.2. Subspace Designs

Further algebraic objects that will be used to prove Theorem 9.4.1 are the so-called *algebraic designs*, defined as follows.

Let Λ be a positive integer and q a prime power. For positive integers $r<\Lambda$ and d, an (r,d)-*subspace design* in $\mathbb{F}_q^{\Lambda}$ is a collection of subspaces of $\mathbb{F}_q^{\Lambda}$ such that for every r-dimensional subspace $W\subset\mathbb{F}_q^{\Lambda}$, we have

$$\sum_{H\in\mathcal{H}}\dim(W\cap H)\leq d.$$

The cardinality of a subspace design $\mathcal{H}$ is the number of subspaces in its collection, i.e., $|\mathcal{H}|$. If all subspaces in $\mathcal{H}$ have the same dimension t, then we refer to t as the *dimension* of the subspace design $\mathcal{H}$.

Note that the inequality in the above definition in particular implies that for every r-dimensional subspace W, at most d of the subspaces in an (r,d)-subspace design nontrivially intersect it.

The following key result shows how periodic subspaces interrelate with subspace designs.

PROPOSITION 9.4.3. *Assume that* $H_1, H_2, \ldots, H_b$ *are subspaces in an* (r,d)-*subspace design in* $\mathbb{F}_q^\Lambda$, *and* T *is an* (r,Λ,b)-*periodic affine subspace of* $\mathbb{F}_q^{\Lambda b}$ *with underlying subspace* S. *Then the set*

$$\mathcal{T} = \{(\boldsymbol{f}_1, \boldsymbol{f}_2, \ldots, \boldsymbol{f}_b) \in T \mid \boldsymbol{f}_j \in H_j \text{ for } j = 1,2,\ldots,b\}$$

is an affine subspace of $\mathbb{F}_q^{\Lambda b}$ *of dimension at most* d. *Also, the underlying subspace of* $\mathcal{T}$ *is contained in* $\mathcal{S} := S \cap (H_1 \times H_2 \times \ldots \times H_b)$. □

The task of explicitly constructing subspace designs is not easy; however, we have the following.

THEOREM 9.4.4. *For every* $\eta > 0$, *integers* r *and* Λ *with* $r \le \eta\Lambda/4$, *and prime powers* q *satisfying* $q^{\eta\Lambda/(2r)} > 2r/\eta$, *for any* $b \le q^{\eta\Lambda/(4r)}$ *there exists an explicit* $(r, r^2/\eta)$-*subspace design of cardinality* b *and dimension* $(1-\eta)\Lambda$ *that can be constructed deterministically in* $\mathrm{poly}(b,q)$ *time. In the case where* $q > \Lambda$, *one can explicitly construct an* $(r, 2r/\eta)$-*subspace design with the same parameters.* □

In fact, to apply the technique of subspace designs to algebraic geometry codes from García–Stichtenoth towers, we need an even more complicated notion, since the length of those codes grows unrestrictedly, whereas the possible size of subspace designs inside a fixed space $\mathbb{F}_q^m$ is naturally bounded. This leads to the following definition of subspace designs of increasing length.

Let l be a positive integer. For positive integers $r_0 \le r_1 \le \ldots \le r_l$ and $m_0 \le m_1 \le \cdots \le m_l$ such that $m_{i-1} \mid m_i$ for $1 \le i \le l$, an $(r_0, r_1, \ldots, r_l)$-*cascaded subspace design* with *length vector* $(m_0, m_1, \ldots, m_l)$ and *dimension vector* $(d_0, d_1, \ldots, d_{l-1})$ is a collection of l subspace designs, specifically an (r_{i-1}, r_i)-subspace design in $\mathbb{F}_{q^{m_{i-1}}}$ of cardinality m_i/m_{i-1} and dimension d_{i-1} for each $i = 1,2,\ldots,l$.

Note that the $l = 1$ case of the above definition corresponds to an (r_0, r_1)-subspace design in $\mathbb{F}_q^{m_0}$ of dimension d_0 and cardinality m_1/m_0.

Generalizing the space $\mathcal{S}$ from Proposition 9.4.3, we get the following object, a canonical space.

Let $\mathcal{M}$ be a cascaded subspace design with length vector $(m_0, m_1, \ldots, m_l)$ such that the ith subspace design in $\mathcal{M}$ has subspaces

$$H_1^{(i)}, H_2^{(i)}, \ldots, H_{m_i/m_{i-1}}^{(i)} \subset \mathbb{F}_q^{m_{i-1}}, \quad \text{for } 1 \le i \le l.$$

The *canonical subspace* associated with such a cascaded subspace design, denoted by $U(\mathcal{M})$, is a subspace of $\mathbb{F}_q^{m_l}$ defined as follows: A vector $\boldsymbol{x} \in \mathbb{F}_q^{m_l}$ belongs to $U(\mathcal{M})$ if and only if for every $i \in \{1,2,\ldots,l\}$, each of the m_i-sized blocks of $\boldsymbol{x}$ given $\mathrm{proj}_{[jm_i+1,(j+1)m_i]}(\boldsymbol{x})$ for $0 \le j < m_l/m_i$ belongs to $H_1^{(i)} \times H_2^{(i)} \times \ldots \times H_{m_i/m_{i-1}}^{(i)}$.

Finally, one can construct a canonical subspace that has low-dimensional intersection with ultra-periodic subspaces based on the explicit subspace designs of Theorem 9.4.4.

THEOREM 9.4.5. *Let* $q \ge 4$ *be a prime power. Let* $\eta \in (0,1)$ *and integers* $\Lambda, r \ge 2$ *satisfy* $\Lambda \ge cr\eta^{-1}\log(r/\eta)$ *for a large enough (absolute) constant* $c > 0$. *For all large enough multiples* $\varkappa$ *of* Λ, *we can construct a subspace* U *of* $\mathbb{F}_q^\varkappa$ *of dimension at least* $(1-\eta)\varkappa$ *such that for every* (r,Λ)-*ultra periodic affine subspace* $T \subset \mathbb{F}_q^\varkappa$, *the dimension of the affine subspace* $U \cap T$ *is at most* $(r/\eta)^{2^{O(\log^* \varkappa)}}$. *The subspace* U *can be constructed deterministically in* $\mathrm{poly}(\varkappa, q)$ *time.* □

9.4.3. Case of the General Algebraic Geometry Codes

Here is a description of the decoding procedure in the case of the general algebraic geometry codes. Note however that those "general" algebraic geometry codes are not simply algebraic geometry codes but some special subcodes of them. More precisely, let $X/\mathbb{F}_q$ be a (nonsingular absolutely irreducible) curve of genus g, and let $K/\mathbb{F}_q$ be the corresponding function field. Let $P_\infty, P_1, P_2, \ldots, P_N$ be its $N+1$ distinct $\mathbb{F}_q$-rational places. Let $\sigma \in \mathrm{Gal}(\mathbb{F}_{q^m}/\mathbb{F}_q)$ be the Frobenius automorphism; we extend σ to an automorphism in $\mathrm{Gal}(K_m/K)$, where $K_m = \mathbb{F}_{q^m}(X)$, and then $P^\sigma = P$ for any place of K.

For a divisor $D = lP_\infty$ on X, we set $L_m(D) = L(D) \otimes \mathbb{F}_{q^m} \subset K_m^*$. Consider the one-point algebraic geometry code

$$C(m,\ell) = (X/\mathbb{F}_{q^m}, \mathcal{P}, \ell P_\infty)_L, \quad \mathcal{P} = \{P_1, \ldots, P_n\}.$$

Thus, the code $C(m,\ell)$ is a code over F_{q^m} obtained from $\mathbb{F}_q$-rational points of X, and we immediately get the following.

LEMMA 9.4.6. *The code $C(m,\ell)$ is an $\mathbb{F}_{q^m}$-linear code over $\mathbb{F}_{q^m}$ of rate at least $(\ell - g + 1)/N$ and minimum distance at least $N - \ell$.* □

We now present a list decoding algorithm for the above codes. Assume that a codeword encoding $f \in L_m((k + 2g - 1)P_\infty)$ is transmitted and received as $\boldsymbol{y} = (y_1, y_2, \ldots, y_N)$.

Given such a received word, we will interpolate a nonzero linear polynomial over K_m,

$$Q(Y_1, Y_2, \ldots, Y_s) = A_0 + A_1 Y_1 + A_2 Y_2 + \ldots + A_s Y_s, \tag{9.4.1}$$

where $A_i \in L_m(DP_\infty)$ for $i = 1, 2, \ldots, s$ and $A_0 \in L_m((D + k + 2g - 1)P_\infty)$ with the degree parameter D chosen to be

$$D = \left\lfloor \frac{N - k + (s-1)g + 1}{s+1} \right\rfloor. \tag{9.4.2}$$

If we fix a basis of $L_m(DP_\infty)$ and extend it to a basis of $L_m((D + k + 2g - 1)P_\infty)$, then the dimension of the corresponding subspace of A_0 is at least $D + k + g$, and that of A_i is at least $D - g + 1$ for $i \geq 1$. Thus, the total number of coefficients in the polynomial Q is

$$s(D - g + 1) + D + k + g = (s+1)(D+1) - (s-1)g - 1 + k > N$$

for the above choice (9.4.2) of D. The interpolation conditions on $Q \in K_m[Y_1, \ldots, Y_s]$ are as follows:

$$Q(y_i, y_i^\sigma, \ldots, y_i^{\sigma^{s-1}}) = A_0(P_i) + A_1(P_i)y_i + A_2(P_i)y_i^\sigma + \ldots + A_s(P_i)y_i^{\sigma^{s-1}} = 0 \tag{9.4.3}$$

for $i = 1, 2, \ldots, N$. Thus, we have N equations in total. Therefore, a nonzero linear function $Q \in K_m[Y_1, \ldots, Y_s]$ of the form (9.4.1) satisfying (9.4.3) can be found by solving a homogeneous linear system over $\mathbb{F}_{q^m}$ with $\leq N$ equations and $\geq s(D - g + 1) + D + k + g$ variables.

The message functions $f \in L_m((k + 2g - 1)P_\infty)$ that are of interest in list decoding must satisfy the following condition.

LEMMA 9.4.7. *If f is a function in $L_m((k + 2g - 1)P_\infty)$ whose encoding agrees with the received word $\boldsymbol{y}$ in at least t positions with $t > D + k + 2g - 1$, then*

$$Q(f, f^\sigma, \ldots, f^{\sigma^{s-1}}) = A_0 + A_1 f + A_2 f^\sigma + \ldots + A_s f^{\sigma^{s-1}} = 0. \tag{9.4.4}$$

PROOF. The proof proceeds by comparing the number of zeroes of the function $Q(f, f^{\sigma}, \dots, f^{\sigma^{s-1}}) = A_0 + A_1 f + A_2 f^{\sigma} + \dots + A_s f^{\sigma^{s-1}} \in L_m((D+k+2g-1)P_{\infty})$ with $D+k+2g-1$. If position i of the encoding of f agrees with $\boldsymbol{y}$, then

$$\begin{aligned} 0 &= A_0(P_i) + A_1(P_i)y_i + A_2(P_i)y_i^{\sigma} + \dots + A_s(P_i)y_i^{\sigma^{s-1}} \\ &= A_0(P_i) + A_1(P_i)f(P_i) + A_2(P_i)(f(P_i))^{\sigma} + \dots + A_s(P_i)(f(P_i))^{\sigma^{s-1}} \\ &= A_0(P_i) + A_1(P_i)f(P_i) + A_2(P_i)f^{\sigma}(P_i) + \dots + A_s(P_i)f^{\sigma^{s-1}}(P_i) \\ &= (A_0 + A_1 f + A_2 f^{\sigma} + \dots + A_s f^{\sigma^{s-1}})(P_i), \end{aligned}$$

i.e., P_i is a zero of $Q(f, f^{\sigma}, \dots, f^{\sigma^{s-1}})$. Thus, there are at least t zeroes for all the agreeing positions. Hence, $Q(f, f^{\sigma}, \dots, f^{\sigma^{s-1}})$ must be the zero function when $t > D+k+2g-1$. □

Let P be a rational place in K, and let $T \in K$ be a local parameter of P. Then $T^{\sigma} = T$, and there are two possibilities: either $P = P_{\infty}$ or $P \neq P_{\infty}$. Here we only consider the case of $P = P_{\infty}$ for the García–Stichtenoth tower.

Assume that $f \in L_m((k+2g-1)P_{\infty})$ has a local expansion at P as follows:

$$f = \sum_{j=0}^{\infty} f_j T^j \tag{9.4.5}$$

with $f_j \in \mathbb{F}_{q^m}$. Then f is uniquely determined by $(f_0, f_1, \dots, f_{k+2g-1})$, because $\deg(f)_{\infty} \geq 1-2g-k$.

LEMMA 9.4.8. *The set of solutions $f \in L_m((k+2g-1)P_{\infty})$ to equation* (9.4.3),

$$A_0 + A_1 f + A_2 f^{\sigma} + \dots + A_s f^{\sigma^{s-1}} = 0,$$

when at least one A_i is nonzero has size at most $q^{(s-1)(k+2g-1)}$. Moreover, the coefficients $(f_0, f_1, \dots, f_{k+2g-1})$ of the local expansion of f at P belong to an $(s-1, m)$-ultra periodic affine subspace of $\mathbb{F}_q^{(k+2g-1)m}$.

PROOF. Let $u = \min\{v_{P_{\infty}}(A_i) : i = 1, 2, \dots, s\}$. Then it is clear that $u \geq 0$ and $v_P(A_0) \geq u$. Any A_i for $i = 0, 1, \dots, s$ has a local expansion at P as follows:

$$A_i = T^u \sum_{j=0}^{\infty} a_{i,j} T^j.$$

If f has a local expansion (9.4.5) at P, then the local expansion of f^{σ^i} at P is as follows:

$$f^{\sigma^i} = \sum_{j=0}^{\infty} f_j^{q^i} T^j.$$

For $\ell = 0, 1, \dots,$ define the linearized polynomial

$$B_{\ell}(X) := a_{1,\ell} X + a_{2,\ell} X^q + \dots + a_{s,\ell} X^{q^{s-1}}$$

and note that $B_0(X) \neq 0$. Equating the coefficient of T^{d+u} in $A_0 + A_1 f + A_2 f^{\sigma} + \dots + A_s f^{\sigma^{s-1}}$ to 0, we get the condition

$$a_{0,d} + B_d(f_0) + B_{d-1}(f_1) + \dots + B_0(f_d) = 0. \tag{9.4.6}$$

Let $W = \{\alpha \in \mathbb{F}_{q^m} : B_0(\alpha) = 0\}$; then W is an $\mathbb{F}_q$-subspace of $\mathbb{F}_{q^m}$ with $\dim W \leq s-1$, since B_0 is a nonzero linearized polynomial of q-degree at most $s-1$. For each fixed $f_0, f_1, \ldots, f_{d-1}$, the coefficient f_d must belong to a coset of the subspace W. This implies that the coefficients $(f_0, f_1, \ldots, f_{k+2g-1})$ belong to an $(s-1, m, k+2g-1)$-periodic subspace of $\mathbb{F}_q^{m(k+2g-1)}$. In particular, there are at most $q(s-1)(k+2g-1)$ solutions $f \in L_m((k+2g-1)P_\infty)$ to (9.4.3).

Equation (9.4.6) also shows that each group of ℓ successive coefficients $f_{d-\ell+1}, f_{d-\ell+2}, \ldots, f_d$ belong to cosets of the same underlying $\ell(s-1)$-dimensional subspace of $\mathbb{F}_q^{m\ell}$. This implies that $(f_0, f_1, \ldots, f_{k+2g-1})$ in fact belong to an $(s-1, m)$-ultra periodic subspace. □

Restricting message functions using local expansions. Lemma 9.4.8 will be used to recover the message in terms of the coefficients of its local expansion at P; however, this needs to restrict the subspace of possible solutions, which is possible via choosing a subcode that corresponds to restricting the coefficients to a carefully constructed subset. This requires one to index message functions in terms of the local expansion coefficients, but not all $k+2g-1$ tuples over $\mathbb{F}_{q^m}$ arise in the local expansion of functions in the k-dimensional subspace $L_m((k+2g-1)P_\infty)$. The problem can be solved by finding a k-dimensional subspace of $L_m((k+2g-1)P_\infty)$ such that the top k local expansion coefficients give rise to all k-tuples over $\mathbb{F}_{q^m}$.

More precisely, the following result serves as a base for all further developments.

LEMMA 9.4.9. *In the space $L_m((k+2g-1)P_\infty)$ there exist a set of functions $\{g_1, g_2, \ldots, g_k\}$ such that the $k \times k$ matrix A formed by taking the ith row of A to be the first k coefficients in the local expansion* (9.4.5) *for g_i at P is nonsingular.*

PROOF. Let $\{\psi_1, \psi_2, \ldots, \psi_g\}$ be a basis of $L_m((k+2g-1)P_\infty - kP)$. Extend this basis to a basis $\{\psi_1, \psi_2, \ldots, \psi_g, g_1, g_2, \ldots, g_k\}$ of $L_m((k+2g-1)P_\infty)$; then the functions $\{g_1, g_2, \ldots, g_k\}$ satisfy the condition.

Indeed, assume by contradiction that the matrix A obtained from the functions g_i is singular. This implies that there exist elements $\{\lambda_i\}_{i=1}^k$ such that the function $\sum_{i=1}^{k} \lambda_i g_i$ has expansion $\sum_{i=k}^{\infty} a_i T^i$ at P for some $a_i \in \mathbb{F}_{q^m}$. Therefore, the function $\sum_{i=1}^{k} \lambda_i g_i$ belongs to the space $L_m((k+2g-1)P_\infty - kP)$, and $\sum_{i=1}^{k} \lambda_i g_i$ is a linear combination of $\psi_1, \psi_2, \ldots, \psi_g$. This forces all λ_i to be zero, since $\{\psi_1, \ldots, \psi_g, g_1, g_2, \ldots, g_k\}$ are linearly independent. □

Now we can describe our algebraic geometry code in a manner adapted for using the possible local expansion coefficients. Note that, since those codes are not usual algebraic geometry codes, they need a specific encoding procedure.

Encoding. Take a subset $\{g_1, g_2, \ldots, g_k\}$ of functions of $L_m((k+2g-1)P_\infty)$ as in Lemma 9.4.9. Then we may assume that A is the $k \times k$ identity matrix, applying, if necessary, the elementary operations of Lemma 9.4.9; thus, we assume that each function g_i, $1 \leq i \leq k$, has expansion $T^{i-1} + \sum_{j=k}^{\infty} \lambda_{ij} T^j$ for some $\lambda_{ij} \in \mathbb{F}_{q^m}$.

Now for any subset $M \subseteq \mathbb{F}_{q^m}^k$, we may assume that our messages belong to M and encode each message $(a_1, a_2, \ldots, a_k) \in M$ to the codeword of the form

$(f(P_1), f(P_2), \ldots, f(P_N))$ for $f = \sum_{i=1}^{k} a_i g_i$. Thus, the actual code is a subcode of a code $C(m, k+2g-1)$ defined as follows:

$$C(m, k+2g-1 \mid M)$$
$$:= \left\{ (f(P_1), f(P_2), \ldots, f(P_N)) : f = \sum_{i=1}^{k} a_i g_i, \ (a_1, a_2, \ldots, a_k) \in M \right\}.$$

Decoding. To decode the thus obtained codes, we first use equation (9.4.4) and solve it to find the subspace of possible first k coefficients $f_0, f_1, \ldots, f_{k-1}$ in the local expansion of the function $f = \sum_{i=1}^{k} a_i g_i$ at P. Note that the message tuple $(a_1, a_2, \ldots, a_k)$ belongs to this subspace.

LEMMA 9.4.10. *Let* $f = \sum_{i=1}^{k} f_{i-1} g_i$ *be the local expansion of* f *at* P. *Then* $f_i = a_{i+1}$ *for* $i = 0, \ldots, k-1$.

PROOF. Since g_i has local expansion $T^{i-1} + \sum_{j=k}^{\infty} \lambda_{ij} T^j$, it is clear that the local expansion of f is $\sum_{i=0}^{k-1} a_{i+1} T^i + \sum_{j=k}^{\infty} a_{j+1} T^j$ for some $a_{k+1}, a_{k+2}, \ldots$ in $\mathbb{F}_{q^m}$, and hence the first k coefficients of the local expansion of f are $a_1, a_2, \ldots, a_k$. □

Combining Lemmas 9.4.7, 9.4.8, and 9.4.10 and recalling the choice of D in (9.4.2), we get the following.

COROLLARY 9.4.11. *For the code* $C(m, k+2g-1 \mid \mathbb{F}_{q^m}^k)$, *one can find an* $(s-1, m)$*-ultra periodic subspace of* $\mathbb{F}_q^{mk}$ *that includes all messages whose encoding differs from a received word* $\boldsymbol{y} \in \mathbb{F}_{q^m}^N$ *in at most* $\frac{s}{s+1}(N-k) - \frac{3s+1}{s+1} g$ *positions.* □

One can get for our codes a trade-off between the rate and list decoding radius in a form convenient for improvements to the list size using subspace designs.

THEOREM 9.4.12. *Let* q *be an even power of a prime. Let* $1 \le s \le m$ *be integers, and let* $R \in (0,1)$. *Then for infinitely many* N *(all integers of the form* $q^{e/2}(\sqrt{q}-1)$*) there is a deterministic polynomial-time construction of an* $\mathbb{F}_{q^m}$*-linear code* $\mathrm{GS}^{(q,m)}[N,k]$ *of block length* N *and dimension* $k = RN$ *that can be list decoded in* $\mathrm{poly}(N, m, \log q)$ *time from* $\frac{s}{s+1}(N-k) - \frac{3N}{\sqrt{q}-1}$ *errors, pinning down the messages to one of* $q^{O(mN)}$ *possible* $(s-1, m)$*-ultra periodic* $\mathbb{F}_q$*-affine subspaces of* $\mathbb{F}_q^{mk}$. □

The claim about the number of possible solution subspaces follows since the subspace is determined by $A_0, A_1, \ldots, A_s$, and for our choice of parameter D, there are at most $q^{O(mN)}$ choices of those.

9.4.4. Codes from the $\mathcal{W}_1$ Tower

Now we can adapt the above algorithm to the $\mathcal{W}_1$ tower, and put $K = F_e$, $e \ge 1$, to be the eth function field in the tower $\mathcal{W}_1$ as in Lemma 7.4.12. Its genus $g_e = g(F_e)$ is given in Proposition 7.4.22(b); recall that $q = r^2$. The encoding and decoding is almost identical to the general algebraic geometric code described in Sec. 9.3.2 except that here we use P_∞ for local expansion.

Encoding. As in Lemma 9.4.9, we can find a set of functions $\{h_1, h_2, \ldots, h_k\}$ in $L_m((k+2g_e-1)P_\infty)$ such that the $k \times k$ matrix A formed by taking the ith row of A to be the first k coefficients in the expansion (9.4.7) for h_i at P_∞ is nonsingular; note however that in this case the local expansion starts from $T^{-(k+2g_e-1)}$, while in Sec. 9.3.2 the local expansion starts from T^0. Furthermore, we may assume that the functions h_i, $1 \le i \le k$, have expansions $T^{-(k+2g_e-1)}\left(T^{i-1} + \sum_{j=k}^{\infty} \lambda_{ij} T^j\right)$ for some $\lambda_{ij} \in \mathbb{F}_{q^m}$.

We encode a message k-tuple $(a_1, a_2, \ldots, a_k) \in \mathbb{F}_{q^m}^k$ by the codeword of the form $(f(P_1), f(P_2), \ldots, f(P_N))$, where $f = \sum_{i=1}^{k} a_i h_i$ and $P_1, P_2, \ldots, P_N \neq P_\infty$ are arbitrary $\mathbb{F}_q$-rational points. The block length N can be any integer satisfying $k \le N \le r^e(r^2 - r)$, and as above, we can consider the subcode obtained by only encoding tuples in M:

$$C_{\mathrm{GS}}(m, k+2g_e-1 \mid M) := \left\{ (f(P_1), f(P_2), \ldots, f(P_N)) : \ f = \sum_{i=1}^{k} a_i h_i, \ (a_1, a_2, \ldots, a_k) \in M \right\}$$

for any subset $M \subseteq \mathbb{F}_{q^m}^k$. According to Sec. 7.4.3, the code C_{GS} can be polynomially constructed.

List decoding. In order to list decode, we can write a functional equation $A_0 + A_1 f + A_2 f^\sigma + \ldots + A_s f^{\sigma^{s-1}} = 0$, and to solve for f from it, we consider the local expansions of the message functions f at P_∞. Let $T \in F_e$ be a local expansion of P_∞ and assume that a function $f \in L_m((k+2g_e-1)P_\infty)$ has a local expansion at P_∞:

$$f = T^{-(k+2g_e-1)} \sum_{j=0}^{\infty} f_j T^j \tag{9.4.7}$$

for some $f_j \in \mathbb{F}_{q^m}$. Furthermore, if $f = \sum_{i=1}^{k} a_i h_i$, then the top k coefficients $f_0, f_1, \ldots, f_{k-1}$ in the above local expansion equal $a_1, a_2, \ldots, a_k$. Thus, we can determine such f uniquely by finding $f_0, f_1, \ldots, f_{k-1}$. The following lemma is similar to Lemma 9.4.8 and shows that the coefficients belong to an ultra-periodic subspace.

LEMMA 9.4.13. *Assume that $f \in L_m((k+2g_e-1)P_\infty)$ satisfies the equation*

$$A_0 + A_1 f + A_2 f^\sigma + \ldots + A_s f^{\sigma^{s-1}} = 0$$

when at least one A_i is nonzero. Then the coefficients $(f_0, f_1, \ldots, f_{k-1})$ of the local expansion (9.4.7) of f at P_∞ belong to an $(s-1, m)$-ultra periodic affine subspace of $\mathbb{F}_q^{km}$.

PROOF. Let $u = \min\{\upsilon_{P_\infty}(A_i) : i = 1, 2, \ldots, s\}$; then it is clear that $u \ge -D$ and $\upsilon_{P_\infty}(A_0) \ge u - (k+2g_e-1)$. Each A_i has a local expansion at P_∞ as follows:

$$A_0 = T^{u-(k+2g_e-1)} \sum_{j=0}^{\infty} a_{0,j} T^j \quad \text{and} \quad A_i = T^u \sum_{j=0}^{\infty} a_{i,j} T^j \quad \text{for} \quad i = 1, 2, \ldots, s.$$

Assume that the function f has a local expansion (9.4.5) at P_∞; then f^{σ^i} has a local expansion at P as follows:

$$f^{\sigma^i} = \sum_{j=0}^{\infty} f_j^{q^i} T^j.$$

For $\ell = 0,1,\ldots$, define the linearized polynomial

$$B_\ell(X) := a_{1,\ell}X + a_{2,\ell}X^q + \ldots + a_{s,\ell}X^{q^{s-1}}.$$

The definition of u implies that $B_0(X) \neq 0$. Then the coefficient of $T^{d+u-(k+2g_e-1)}$ in $A_0 + A_1 f + A_2 f^\sigma + \ldots + A_s f^{\sigma^{s-1}}$ gives the condition

$$a_{0,d} + B_d(f_0) + B_{d-1}(f_1) + \ldots + B_0(f_d) = 0,$$

and arguing as in Lemma 9.4.8, this implies that $(f_0, f_1, \ldots, f_{k-1})$ belongs to an $(s-1,m)$-ultra periodic subspace of $\mathbb{F}_q^{mk}$. □

Good list-decodable subcodes. We now combine our code constructions in Section 9.4.3 with a pre-coding step that restricts coefficients to belong to a subspace design, and thereby obtain subcodes that are list decodable with smaller list size in polynomial time.

For a finite field $\mathbb{F}_q$, constant $\varepsilon > 0$, and integers s and m satisfying $1 \le s \le O(\varepsilon m/\log(1/\varepsilon))$ and $m \ge \Omega(1/\varepsilon^2)$, we will define subcodes of $\mathrm{GS}^{(q,m)}[N,k]$ guaranteed by Theorem 9.4.12. Note that the message space of this code can be identified with $\mathbb{F}_q^{mk}$.

Define the subcode $\widehat{GS}$ of $\mathrm{GS}^{(q,m)}[N,k]$ consisting of the encodings of a subspace $U \subseteq \mathbb{F}_q^{mk}$ guaranteed by Theorem 9.4.5, when applied with the parameter choices

$$\eta = \varepsilon, \qquad r = s-1, \qquad \Lambda = m, \qquad \varkappa = km.$$

Note that $\widehat{GS}$ is an $\mathbb{F}_q$-linear code over the alphabet $\mathbb{F}_{q^m}$ of rate $(1-\varepsilon)k/N$. Also, it can be constructed in $\mathrm{poly}(k,m,q)$ time by virtue of the construction complexity of U.

LEMMA 9.4.14. *Given an input string $\boldsymbol{y} \in \mathbb{F}_{q^m}^N$, a basis of an affine subspace of dimension at most*

$$(s/\varepsilon)^{2^{O(\log^*(km)}}$$

that includes all codewords of the above subcode within Hamming distance

$$\frac{s}{s+1}(N-k) - \frac{3N}{\sqrt{q}-1}$$

from $\boldsymbol{y}$ can be found in deterministic $\mathrm{poly}(n,\log q,m)$ *time.* □

By taking $q = \Theta(1/\varepsilon^2)$ and choosing $s = \Theta(1/\varepsilon)$ and $m = \Theta(\varepsilon^{-2}\log(1/\varepsilon))$ in the above lemma, we get Theorem 9.4.1.

Historical and Bibliographic Notes

The traditional decoding approach to algebraic geometry codes was very briefly described in Sec. 4.1.3. Its main feature is that it deals with the worst possible error pattern. Recall that the idea of the basic decoding algorithm due to J. Justesen, K.J. Larsen, H.E. Jensen, A. Havemose, and T. Høholdt appeared in [**JLJ$^+$89**] and was developed and formalized by A.N. Skorobogatov and S.G. Vlăduţ in [**SV90**] and independently by V.Yu. Krachkovsky in [**Kra88**]. This initial basic algorithm does not achieve the natural bound $\left\lfloor \frac{d^*-1}{2} \right\rfloor$ for the number of corrected errors, where d^* is the designed minimum distance of the code.

An effective algorithm which corrects $(d^*-1)/2$ errors was given by D. Ehrhard [**Ehr93**]. An elegant solution of the decoding problem for algebraic geometry codes by a majority vote for unknown syndromes was proposed by G.L. Feng and T.R.N. Rao [**FR93**]. They showed in collaboration with I.M. Duursma [**Duu93b**, **Duu93a**] that the algorithm corrects $(d^*-1)/2$ errors. The origin of these methods stems from the decoding of cyclic codes beyond the BCH error-correcting capacity by G.L. Feng and K.K. Tzeng [**FT94**]. As a result of the majority voting, it was noticed that sometimes one can even correct beyond half the Goppa designed minimum distance [**FR94**].

The Berlekamp–Massey algorithm [**Ber68**, **Mas69**] on *linear recurring relations* in one variable was generalized by S. Sakata [**Sak88**, **Sak90**] to several variables (the BMS algorithm). This algorithm was applied by J. Justesen, K.J. Larsen, H.E. Jensen, and T. Høholdt [**JLJH92**] and B. Shen [**She93**, **She92**] to get faster implementations of the modified algorithm, and later the majority voting was also incorporated by S. Sakata, J. Justesen, Y. Madelung, H.E. Jensen, and T. Høholdt [**Mad93**, **SJM$^+$95a**, **SJH95**, **SJM$^+$95b**]. Using the block-Hankel structure of codes on plane curves, G.L. Feng, V.K. Wei, T.R.N. Rao, and K.K. Tzeng [**FWRT94**] could lower the complexity of the majority voting scheme as well. For more details, consult [**HP95**]. For an exposition of further progress in this direction and of publications devoted to the minimum distance decoders, see also [**BH08**] and the references therein.

However, considering minimum distance decoders severely limits the maximum fraction of errors that the above-mentioned algorithms can tolerate. In turn, this was the source of a big hiatus between the error-correction performance known for probabilistic noise models (pioneered by C.E. Shannon) and what was thought to be the limit for the more powerful, worst-case noise models (suggested by R. Hamming).

The main progress to fill this gap was done by a model called *list decoding*. It was introduced independently by P. Elias [**Eli57**] and J.M. Wozencraft [**Woz58**] as a relaxation of the "classical" notion of decoding by allowing the decoder to output a list of codewords as answers. The decoding is considered successful as long as the correct message is included in the list. List decoding was also used by C.E. Shannon, R.G. Gallager, and E.R. Berlekamp [**SGB67**] in investigating low rate average error bounds for general discrete memoryless channels, and R. Ahlswede [**Ahl73**] showed that it enables one to determine the capacity of a wide class of communication channels.

The basic question raised during next two decades was: How many errors could still be recovered with lists of small size? Two basic parameters are thus the number of errors and the allowed size of the output list. These parameters are usually studied as a function of some of the more classical parameters of error-correcting codes. How large can the rate of a code be if we want small list sizes for a certain number of errors? And how do codes of large minimum distance perform with respect to list decoding? These problems were largely investigated in the works of V.V. Zyablov and M.S. Pinsker [**ZP81**], V.M. Blinovsky [**Bli86**, **Bli97**], and P. Elias [**Eli91**]. (The paper [**Eli91**] also gives a very good summary of the prior work and history.) Applying the list decoding approach to algebraic geometry codes (initiated in [**Sud97**] for Reed–Solomon codes) had given major results in developing efficient algorithms for such decoding [**Sud97**, **GS99**, **SW99**, **GS00**], which also allowed some rather astonishing advances such as approaching the Singleton bound (described in Sec. 9.4).

The Johnson bound is a classical bound in coding theory and is at the heart of the Bassalygo–Elias bound on rate as a function of relative distance. The Johnson bound was stated and proved in the context of list decoding, in a form similar to that presented here, in [**GRS00**, Sec. 4.1]. A simpler geometric proof of the Johnson bound appeared in [**Gur04**, Chapter 3]. The combinatorial proof sketched in Sec. 9.1.1 is borrowed from [**Gur06**] and belongs to Jaikumar Radhakrishnan; see also, e.g., [**Sud00**, Chapter 1].

For the claim of Remark 9.1.4 for general codes, see [**GRS00**, Sec. 4.3]; for linear codes, the result was shown for the binary case in [**Gur02**]. Theorem 9.1.7 is taken from [**Eli91**]. Theorem 9.1.8 first appeared implicitly in the work of V.V. Zyablov and M.S. Pinsker [**ZP81**]; the reader may find a detailed proof in [**Gur04**, Theorem 5.6]. In general, for Sec. 9.1.1, see [**Gur06**]. The Sudan and the Guruswami–Sudan algorithms are exposed in [**Sud97**, **GS99**, **Sud01**, **Gur04**]; see also [**SW99**]. The Berlekamp algorithm for finding roots in a finite field was given in [**Ber67**].

For the material of Sec. 9.2.3, see [**GS01**]; a discussion concerning question (iii) at the beginning of Sec. 9.2.3 is contained in [**SW99**]. The results of Sec. 9.3 are borrowed from [**HN99**]; our presentation in Sec. 9.2.2 also follows this paper. The fundamental iterative algorithm used in Sec. 9.3.1 was presented in [**FT91**], and its application in this section follows the lines of [**Köt96**, Chapter 4]. For the factorization issues in Sec. 9.3.2, one can consult [**LN97**, Chapter 4]; for Theorem 9.3.5, see [**LN97**, Theorem 3.75].

The main theorem of Sec. 9.4 (Theorem 9.4.1) is the result of successive efforts by V. Guruswami, C. Xing, and others; see [**GX12**, **GX13**, **GX17**] and the references therein. Note that [**GX17**] also contains another version of the main result, with a fixed list size but without an explicit polynomial construction (the construction is of a Monte Carlo type).

CHAPTER 10

Sphere Packings

This chapter is mostly devoted to another kind of problem which greatly resembles code issues, namely, to a classical problem of the dense packing of equal nonoverlapping spheres in $\mathbb{R}^N$. It turns out that both the direct application of algebraic geometry codes and the use of intuition developed while studying them are quite useful.

In Sec. 10.1 we give basic definitions, put the packing problem, both for lattice and nonlattice cases, consider examples, and discuss asymptotic results. Section 10.2 is devoted to some constructions of sphere packings using codes; algebraic geometry codes happen to be rather useful, especially for asymptotic problems. In particular, these methods allow one to give the first construction of lattice packings with exponentially large kissing numbers. In Sec. 10.3 we consider various constructions of packings which use global fields, both number and functional ones; these constructions are somewhat modified analogues of algebraic geometry codes. They also lead to numerous asymptotically good families of codes and lattices. In the last section we consider another class of lattices coming from algebraic geometry, namely Mordell–Weil lattices. They do not yield good asymptotic families, but nevertheless they are the best known in many dimensions between 24 and 4096; in particular, the famous Leech lattice also belongs to this class.

10.1. Definitions, Examples, and Constructions

In this section we give basic definitions for the problem of densely packing equal nonoverlapping spheres in an N-dimensional Euclidean space $\mathbb{R}^n$. We are especially interested in lattice packings and in the behaviour of the packing parameters as $n \to \infty$. We also consider the case of random packings, which is a kind of analogue of the Gilbert–Varshamov expurgation procedure.

10.1.1. Parameters and Some Basic Examples

Packings. Consider the classical problem of packing equal nonoverlapping open spheres in $\mathbb{R}^N$. Any such packing is determined by the set P of sphere centers; let

$$d = d(P) = \inf_{\substack{v,u \in P \\ v \neq u}} |u - v|$$

be the *minimum distance* of the packing, which equals the maximum possible diameter of nonoverlapping spheres centered at P.

The *density* of P is the fraction of $\mathbb{R}^N$ covered by spheres; more precisely, it is defined as

$$\Delta(P) = \limsup_{u\to\infty} \frac{v(S\cap B_u)}{v(B_u)},$$

where v is the standard volume in $\mathbb{R}^N$,

$$S = \{x \in \mathbb{R}^N \mid \exists u \in P: |x-u| < d/2\}$$

is the set of points inside spheres, and

$$B_u = \{x \in \mathbb{R}^N \mid |x| \le u\}$$

is a ball of radius u.

Another quantity of interest for a given P is its *kissing number* (or *Newton number*)

$$\tau(P) = \max_{x\in P} |\{y \in P: |x-y| = d\}|.$$

Lattices. If P is an additive subgroup of $\mathbb{R}^N$, we call P a *lattice packing* (or simply a *lattice*; in this case we use the notation L rather than P). In what follows we always assume that the rank of L is N, since otherwise $\Delta(L) = 0$. If L is a lattice, any choice of a basis $e_1,\dots,e_n$ defines a map $\mathbb{Z}^N \to \mathbb{R}^N$; the matrix of this map is called a *generator matrix* of the lattice.

PROPOSITION 10.1.1. *The volume of the fundamental domain*

$$F = \left\{ \sum_{i=1}^{N} x_i a_i \mid 0 \le x_i < 1 \right\} \subseteq \mathbb{R}^N$$

equals the absolute value of the determinant $\det L$ *of the generator matrix.* □

This volume is called the *determinant* of the lattice, and the determinant of the Gram matrix $((e_i,e_j))_{i,j=1,\dots,n}$ is called the *discriminant* of the lattice, $\operatorname{discr} L$; they are closely related as follows.

PROPOSITION 10.1.2. *We have*

$$\operatorname{discr} L = (\det L)^2.$$ □

The minimum distance of a lattice is clearly $d(L) = \min_{v\in L\setminus\{0\}} |v|$, and the density of a lattice packing is computed is follows.

PROPOSITION 10.1.3. *We have*

$$\Delta(L) = \frac{d(L)^N V_N}{2^N \det L},$$

$V_N = \dfrac{\pi^{N/2}}{\Gamma(N/2+1)}$ *being the volume of a unit sphere in* $\mathbb{R}^N$. □

By Stirling's formula we have

$$\log_2 V_N = \frac{N}{2}\log_2\left(\frac{2\pi e}{N}\right) - \log_2\sqrt{\pi N} + o(1)$$

as $N \to \infty$, and therefore

$$\frac{1}{N}\log_2 V_N \sim \frac{1}{2}\log_2\left(\frac{2\pi e}{N}\right).$$

Thus, as $N \to \infty$, we obtain

$$-\frac{1}{N}\log_2 \Delta(L_N) \sim -\log_2\sqrt{\frac{\pi e}{2}} + \log_2\sqrt{N} - \log_2 d(L) + \frac{1}{N}\log_2(\det L).$$

Other parameters. One can also define other parameters of packings, namely

$$\begin{aligned} \delta(P) &= \Delta(P)/V_N && \text{(center density)},\\ \lambda(P) &= -(\log_2 \Delta(P))/N && \text{(density exponent)},\\ \nu(P) &= \log_2 \delta(P); \end{aligned}$$

sometimes they are more convenient to use than Δ. In particular, $\delta(P)$ is more suitable for root lattices, and $\nu(P)$ is more useful to compute densities of lattices obtained by constructions of Sec. 10.2.1. The density exponent $\lambda(P)$ is specially adopted to asymptotic problems, since we have

$$\Delta(P) = 2^{-\lambda(P)N};$$

note that $\Delta(P) \le 1$, and thus we have $\lambda(P) \ge 0$ for any packing.

Densest packings. Define

$$\lambda(N) = \inf_{P\subset\mathbb{R}^N} \lambda(P), \qquad \Delta(N) = \sup_{P\subset\mathbb{R}^N} \Delta(P), \qquad \ldots.$$

A natural problem of finding the densest possible sphere packing in a given dimension can be split into two problems:

A. Find the exact value of $\lambda(N)$ (or, equivalently, of $\Delta(N)$, $\delta(N)$, ...).
B. Specify a packing P with $\lambda(P) = \lambda(N)$.

These problems are completely solved for $N = 1$, 2, 3, 8, and 24.

For $N = 1$, the answer is obvious: equal segments cover the whole line, and hence for this packing L_1 we have $\Delta(L_1) = 1$; i.e., $\Delta(1) = 1$ and $\lambda(1) = 0$.

For $N = 2$, the problem is not so simple; however, one can prove that $\Delta(2) = \pi/(2\sqrt{3})$ and that the lattice $L_2 \subset \mathbb{R}^2 = \mathbb{C}$ generated by 1 and $\dfrac{1+\sqrt{-3}}{2}$ (the so-called *hexagonal lattice*) has density $\Delta(2)$, i.e., is optimal.

The value of $\Delta(3) = \pi/(3\sqrt{2})$ is the subject of the famous Kepler conjecture, completely settled quite recently. The so-called face-centered cubic packing is optimal in this case. For $N = 8$ and 24 the optimal packings are given by the lattice E_8 and the Leech lattice, respectively (see pp. 233–234 for their description).

Densest lattices. For lattice packings, more is known. Define

$$\lambda_\ell(N) = \inf_{L\subset\mathbb{R}^N} \lambda(L), \qquad \Delta_\ell(N) = \sup_{L\subset\mathbb{R}^N} \Delta(L), \ldots$$

and therefore

$$\lambda_\ell(N) \ge \lambda(N), \qquad \Delta_\ell(N) \le \Delta(N), \ldots.$$

Recall that a lattice L is said to be *unimodular* if $\det L = 1$. In this case, the *dual lattice*

$$L^\perp = \{x \in \mathbb{R}^N \mid (x,\ell) \in \mathbb{Z} \text{ for all } \ell\}$$

is also unimodular. Recall also that the inner product in $\mathbb{R}^N$ induces a positive definite bilinear form on L.

Now let $\varphi(x,y)$ be a positive definite $\mathbb{R}$-valued bilinear form in N variables, and let $f(x) = \varphi(x,x)$ be the corresponding quadratic form. Assume also that φ is unimodular, i.e., $\operatorname{discr}\varphi = 1$. Such forms are in bijection with unimodular lattices $L \subset \mathbb{R}^N$. Define

$$\gamma(L) = \gamma(\varphi) = \min_{x \in \mathbb{Z}^N \setminus \{0\}} f(x);$$

in terms of lattices, this is the squared length of the shortest nonzero vector, and an easy check yields the following result.

PROPOSITION 10.1.4. *We have*

$$\Delta(L) = V_N \left(\frac{\gamma(L)}{4}\right)^{N/2}, \qquad \gamma(L) = 4\left(\frac{\Delta(L)}{V_N}\right)^{2/N}. \qquad \square$$

The definition of $\gamma(\varphi)$ naturally extends to the nonunimodular case as follows:

$$\gamma(L) = \gamma(\varphi) = \min_{x \in \mathbb{Z}^N \setminus \{0\}} f(x)/\sqrt[N]{4 \operatorname{discr}\varphi}.$$

Define

$$\gamma(N) = \max_{\varphi} \gamma(\varphi),$$

where the maximum is over all positive definite bilinear forms in N variables (or, equivalently, over all lattices of rank N). The parameters $\gamma(N)$ and $\Delta_\ell(N)$ are related by formulae similar to those of Proposition 10.1.4.

Note that for $N \to \infty$ these formulae take the form

$$\lambda(N) \sim \log_2 \sqrt{\frac{2N}{\pi e \gamma(N)}},$$

$$\log_2(\gamma(N)) \sim \log_2\left(\frac{2N}{\pi e}\Delta^{2/N}\right) = -2\lambda(N) + \log_2\left(\frac{2N}{\pi e}\right).$$

Precise values of $\Delta_\ell(N)$ are known for $N = 1, 2, \ldots, 8, 24$; see Table A.1 in the Appendix to this chapter, where the values of all the above parameters are presented.

Note that within the range of this table $\Delta_\ell(N)$ monotonically decreases and $\gamma(N)$ increases. It would be interesting to know whether this is the case for any N.

In Table A.1, all $\delta(N)^{-2}$ are integers, and we do not not know whether this holds for any N. Note, however, that the densest lattice packing of a given rank can always be generated by a matrix with rational entries, which implies the rationality of $\delta(N)^{-2}$.

Some lattices. Now let us give some basic examples. Namely, we shall describe the densest lattices for $N \leq 8$ and introduce some interesting infinite families of lattices.

We construct families of lattices $L \subset \mathbb{R}^N$ and give the values $d(L)$ and $\det(L)$ for them. From this, one readily calculates their other density parameters.

The simplest family is

$$\mathbb{Z}^N \subset \mathbb{R}^N, \quad \text{with} \quad d(\mathbb{Z}^N) = 1, \quad \det(\mathbb{Z}^N) = 1.$$

The next three lattice classes, A, D, and E, are the so-called root lattices; they are ubiquitous in all parts of mathematics.

***A*- and *D*-lattices.** The lattices A_N and D_N of rank N are given by

$$A_N = \left\{ \sum_{i=1}^{N+1} a_i e_i \mid a_i \in \mathbb{Z},\ \sum a_i = 0 \right\} \subset \mathbb{R}^{N+1},$$

$$D_N = \left\{ \sum_{i=1}^{N} a_i e_i \mid a_i \in \mathbb{Z},\ \sum a_i \equiv 0 \pmod 2 \right\} \subset \mathbb{R}^{N+1},$$

where $\{e_i\}$ is a standard basis in $\mathbb{R}^{N+1}$.

PROPOSITION 10.1.5. (a) *The lattice A_N is generated by the vectors*

$$\alpha_1 = e_1 - e_2, \quad \alpha_2 = e_2 - e_3, \quad \dots, \quad \alpha_N = e_N - e_{N+1},$$

and we have

$$d(A_N) = \sqrt{2}, \qquad \det(A_N) = \sqrt{N+1}.$$

(b) *The lattice D_N is generated by the vectors*

$$\alpha_1 = e_1 - e_2, \quad \alpha_2 = e_2 - e_3, \quad \dots, \quad \alpha_{N-1} = e_{N-1} - e_N, \quad \alpha_N = e_{N-1} + e_N,$$

and we have

$$d(D_N) = \sqrt{2}, \qquad \det(A_N) = 2.$$

□

***E*-lattices.** The next remarkable lattices exist only for $N = 4,5,6,7,8$; define the lattice E_N in $\mathbb{R}^8$ by its basis:

$$\alpha_1 = \frac{1}{2}(e_1 + e_8) - \frac{1}{2}(e_2 + \ldots + e_7),$$
$$\alpha_2 = e_1 + e_2, \qquad \alpha_i = e_i + e_{i-1}, \quad i = 3, \dots, N.$$

PROPOSITION 10.1.6. *The lattices E_8, E_7, and E_6 can be given by*

$$E_8 = \left\{ \sum_{i=1}^{8} a_i e_i \mid 2a_i \in \mathbb{Z},\ a_i - a_j \in \mathbb{Z},\ \sum_{i=1}^{8} a_i \in 2\mathbb{Z} \right\},$$

$$E_7 = \left\{ x = \sum_{i=1}^{8} a_i e_i \mid x \in E_8,\ a_7 = -a_8 \right\}, \qquad \operatorname{rk} E_7 = 7,$$

$$E_6 = \left\{ x = \sum_{i=1}^{8} a_i e_i \mid x \in E_7,\ a_6 = a_7 \right\}, \qquad \operatorname{rk} E_6 = 6,$$

and we have

$$d(E_N) = \sqrt{2}, \qquad \det(E_N) = 9 - N.$$

□

Note also that $A_1 = \mathbb{Z}$, $D_3 = A_3$, $E_4 = A_4$, $E_5 = D_5$.

Γ-lattices. Now let $N \geq 8$, $N \equiv 0 \pmod 4$. Define

$$\Gamma_N = \left\{ \sum_{i=1}^{N} a_i e_i \mid 2a_i \in \mathbb{Z},\ a_i - a_j \in \mathbb{Z},\ \sum_{i=1}^{N} a_i \in 2\mathbb{Z} \right\}.$$

PROPOSITION 10.1.7. *The lattice Γ_N is generated by the vectors $e_i + e_j$ and $\frac{1}{2}\sum_{i=1}^{N} e_i$, and we have*

$$d(\Gamma_N) = \sqrt{2}, \qquad \det(\Gamma_N) = 1.$$

□

Note that $\Gamma_8 = E_8$.

The lattices $A_1 = \mathbb{Z}$, A_2, $A_3 = D_3$, D_4, D_5, E_6, E_7, and $E_8 = \Gamma_8$ are the densest lattices in their dimensions. A proof of this fact is based on the reduction theory of quadratic forms, and we do not give it here.

Note that for all the described infinite families of lattices we have

$$\lambda(L_N) \sim \log_2 \sqrt{N} \to \infty \quad \text{as} \quad n \to \infty.$$

We shall soon show that there are lattices which are asymptotically much better.

For $N \geq 9$, precise values of $\lambda_\ell(N)$ are not known except for $N = 24$; we only have some bounds. As in the case of codes, it is natural to refer to upper bounds on $\Delta(N)$ and $\Delta_\ell(N)$ as *possibility bounds*, and to lower ones as *existence bounds* (note, however, that for $\lambda(N)$, on the contrary, possibility bounds are lower ones, while existence bounds are upper ones).

Leech lattice. In dimension 24 there exists an exceptionally good lattice, which is a unique integral even unimodular lattice that has no vector of length $\sqrt{2}$ (recall that a lattice is *even* if the scalar square of any of its vectors is even). Moreover, it gives the best packing in this dimension.

This lattice is called the *Leech lattice* and is denoted by Λ_{24}. It is closely related to the $[24, 12, 8]_2$ Golay code C_{24}, and its first construction used this code. There are many ways to construct it, and here we give one of the simplest.

The lattice Λ_{24} is generated by vectors

$$v_{i,c} = \frac{1}{\sqrt{8}} u_{i,c}, \quad i = 1, 2, \ldots, 24, \quad c \in C_{24},$$

where $u_{i,c}$ has ∓ 3 in the ith position and ± 1 in all other positions; the upper sign is chosen for the positions where the codeword c has 1.

The vectors $v_{i,c}$ are the shortest vectors of the lattice Λ_{24}; their length is 2 (and there are $2^{12} \cdot 24$ of them). There are two more types of the shortest vectors, namely $t_{c,\varepsilon}$ and $w_{i,j,\delta}$, defined as follows:

$$t_{c,\varepsilon} = \frac{1}{\sqrt{8}} s_{c,\varepsilon}, \quad c \in C_{24}, \quad \mathrm{wt}(c) = 8, \quad \varepsilon \in (\mathbb{Z}_2)^7,$$

where the vector $s_{c,\varepsilon}$ has ± 2 in the positions where c has 1, the number of components equal to -2 is even, and the distribution of signs is defined by ε; in all other positions, $s_{c,\varepsilon}$ has zeroes (there are $2^7 \cdot 759$ vectors of this type). The vectors $w_{i,j,\delta}$ are given by

$$w_{i,j,\delta} = \frac{1}{\sqrt{8}} x_{i,j,\delta}, \quad 1 \leq i, j \leq 24, \quad i \neq j, \quad \delta \in (\mathbb{Z}_2)^2,$$

where $x_{i,j,\delta}$ has ± 4 in positions i and j, with arbitrary distribution of signs, which is defined by δ, and has zeroes in all other positions (there are 1104 vectors of this type). These three types of vectors are denoted, respectively, by $(3^1 1^{23})$, $(2^8 0^{16})$, and $(4^2 0^{22})$. In total, there are 196 560 shortest lattice vectors; i.e., each sphere in the corresponding packing kisses exactly 196 560 other spheres.

Using the fact that there are no vectors of length $\sqrt{2}$ in Λ_{24}, one can prove the following result.

PROPOSITION 10.1.8. *We have*

$$\det(\Lambda_{24}) = 1, \qquad d(\Lambda_{24}) = 2,$$

and hence

$$\delta(\Lambda_{24}) = 1, \qquad \nu(\Lambda_{24}) = 0, \qquad \gamma(\Lambda_{24}) = 4,$$
$$\Delta(\Lambda_{24}) \approx 0.001932, \qquad \lambda(\Lambda_{24}) \approx 0.376. \qquad \square$$

The *covering radius* of the Leech lattice is $2\sqrt{2}$; i.e., balls of radius $2\sqrt{2}$ centered at points of Λ_{24} cover the whole space. One can also describe "deep holes", i.e., points with distance $2\sqrt{2}$ from the nearest lattice point.

The Leech lattice has an enormous automorphism group called $\mathcal{C}o_0$; namely,

$$|\mathcal{C}o_0| = 8\,315\,553\,613\,086\,720\,000,$$

and the largest sporadic group, the Fischer–Griess Monster, can be realized as the automorphism group of an algebra closely connected with the Leech lattice.

Analogy with codes. There is a system of beautiful analogies between codes and packings. Indeed, one can consider an $[n, k, d]_q$ code $C \subseteq \mathbb{F}_q^n$ as the set of centers of a sphere packing (of radius $t = \left\lceil \frac{d-1}{2} \right\rceil$) in the Hamming metric. The minimum distance of a code corresponds to the diameter $d(L)$ of a sphere packing.

Linear codes correspond to lattice packings, since a linear code is a subset in $\mathbb{F}_q^n$ which is closed under addition and under multiplication by elements of $\mathbb{F}_q$, and a lattice is closed under addition and multiplication by integers. Strictly speaking, we can consider "quasilinear" codes, i.e., subsets which are closed under addition and under multiplication by elements of $\mathbb{F}_p$ (rather than $\mathbb{F}_q$) as an analogue of lattices, but we do not pursue this idea here.

Let C be a linear $[n, k, d]_q$ code. Then the volume (i.e., cardinality) of the factor space $\mathbb{F}_q^n/C$ is q^{n-k}. For a lattice $L \subset \mathbb{R}^N$, the volume of the factor space $\mathbb{R}^N/L$ equals $\det L$; i.e., $\log(\det L)$ is an analogue of the code codimension $n - k$. To be definite, we shall assume that log in the expression $\log(\det L)$ means the binary logarithm.

There are two possible analogues for the dimension N of a lattice: the length, n, and dimension, k, of a code. We use the first one; nevertheless, we think that the second can also be of some use.

The density of a packing corresponds to the density of a packing in the Hamming metric. Note that the density of a lattice packing equals the volume of the ball of radius d divided by $\det L$. For the density of a packing in the Hamming metric, an analogous statement is also true if we assume that the ball volume is normalized:

$$\text{ball volume} = (\text{number of points in the ball})/q^n.$$

However, an analogy between code and lattice parameters is not complete, since the density of packing does not change under a dilation $L \mapsto aL$. Therefore, we can assume that $d(L) = 1$ (or $\det(L) = 1$), and thus a packing has two essential parameters, N and Δ, whereas a code has three essential parameters, n, k, and d. Thus, a single asymptotic parameter λ is an analogue of the pair of code asymptotic parameters (δ, R).

An asymptotically good packing family (i.e., with $\lambda < \infty$ as $n \to \infty$) is an analogue of an asymptotically good code family (i.e., with $R\delta > 0$ as $n \to \infty$).

The Gilbert–Varshamov bound corresponds to the Minkowski bound (see Theorem 10.1.10), and the Hamming bound to the condition $\lambda \geq 0$. It is not clear whether there exists a reasonable analogue of the Plotkin bound, and this could be an interesting question. The Kabatiansky–Levenshtein bound corresponds to the McEliece–Rodemich–Rumsey–Welch bound.

Packings on a sphere correspond to constant-weight codes.

An interesting question about analogies between concrete code families and lattice families is mostly open; for instance, parity check codes correspond to either lattices A_N or D_N.

The θ-function of a lattice L, which is defined as a formal series $\sum_{x\in L} q^{\|x\|^2}$, corresponds to the code enumerator; this analogy is quite useful.

Unimodular lattices correspond to self-dual codes.

We are also interested in analogues of algebraic geometry codes, and we describe some of them in Sec. 10.3 below. These analogues are closely connected to a deep analogy between algebraic curves over finite fields and algebraic number fields, considering them both as global fields; cf. Sec. 6.1.

10.1.2. Asymptotic Problems

Now let us discuss the asymptotic behaviour of density parameters as $N \to \infty$. Introduce the parameter

$$\widetilde{\lambda} = \liminf_{N\to\infty} \lambda(N).$$

Computing $\widetilde{\lambda}$, i.e., finding what is the maximum asymptotic (as $N \to \infty$) density $\widetilde{\Delta} = 2^{-\widetilde{\lambda}N}$ of a high-dimensional packing, is a very difficult problem, and we will be interested in bounds for these parameters. Note that $\widetilde{\lambda}$ is an analogue of the parameter $\alpha_q(\delta)$ in coding theory.

A *family of packings* $\{P_N\}$ is a set of packings $P_N \subset \mathbb{R}_N$ where N runs over an infinite subset of $\mathbb{N}$.

Let

$$\lambda(\{P_N\}) = \liminf_{N\to\infty} \lambda(P_N);$$

we refer to families of packings with $\lambda(\{P_N\}) < \infty$ as (asymptotically) *good families* (they are analogues of good families of codes).

PROPOSITION 10.1.9. *We have* $\inf_{\{P_N\}} \lambda(\{P_N\}) = \widetilde{\lambda}$. □

Similarly, for lattices we define

$$\widetilde{\lambda}_\ell = \liminf_{N\to\infty} \lambda_\ell(N) = \inf_{\{L_N\}} \lambda(\{L_N\}).$$

Bounds. The best known estimates for $\widetilde{\lambda}$ are as follows.

THEOREM 10.1.10. *We have* $1 \geq \widetilde{\lambda}_\ell \geq \widetilde{\lambda} \geq 0.599$. □

The upper bound, which is the existence bound, is called the Minkowski bound, the lower one is the Kabatiansky–Levenshtein bound. Note that the Minkowski bound is a random choice bound and thus is parallel to the Gilbert–Varshamov bound; however, chronologically it precedes the Gilbert–Varshamov bound. A family of (nonlattice) packings with $\lambda(\{P_N\}) = 1$ is constructed similarly to Exercise 1.1.61 (cf. Theorem 10.1.13); the inequality $\widetilde{\lambda}_\ell \leq 1$ can be proved by computing the mean value of $\lambda(L_N)$ on a sufficiently large set of lattices. As in the case of codes (see Remark 1.3.18), almost all lattice families have $\lambda(\{L_N\}) = 1$.

REMARK 10.1.11. The bound $\widetilde{\lambda}_\ell \le 1$ is asymptotic, and the corresponding lattices are "nonconstructive". However, if $N \ge 2^{10}$, then all known constructions of lattices $L_N \subset \mathbb{R}^N$ lead to worse values of $\lambda(L_N)$, i.e., $\lambda(L_N) > 1$.

Here is another bound, called the Rogers bound.

PROPOSITION 10.1.12. *We have*

$$\Delta(N) \le \sigma_N,$$

where σ_N *is the ratio of the volume of the intersection* $\left(\bigcup_{i=1}^{N+1} B_i \cap \Sigma_N\right)$ *to the volume of* Σ_N, *where* Σ_N *is the regular simplex of edge length* 2 *and* $B_i, \dots, B_{N+1}$ *are unit balls centered at the vertices of* Σ_N. □

This bound is asymptotically worse; it gives $\widetilde{\lambda} \ge 0.5$. However, it is quite useful for moderate values of N.

By the Minkowski bound we know that there exist lattices of density $\Delta \sim 2^{-N}$, but we do not know how to construct them explicitly, and the problem of explicit construction of dense packings naturally arises. Formalization of the "explicity" requirement leads to the following definition.

We call a packing family $\{P_N\}$ *polynomial* if there exists an algorithm which requires a polynomial in N number of operations over real numbers and for an output has an algorithm of consecutive construction of corresponding sphere centers (for lattices its output can be just the lattice basis).

Let

$$\widetilde{\lambda}^{\text{pol}} = \inf_{\{P_N\} \text{ polynomial}} \lambda(\{P_N\});$$

similarly one defines $\widetilde{\lambda}_\ell^{\text{pol}}$ for lattices, and also $\widetilde{\lambda}^{\text{exp}}$ and $\widetilde{\lambda}_\ell^{\text{exp}}$ (for *exponential* families).

Let us look for upper bounds for these parameters; note that it is not even immediately clear that they are finite.

THEOREM 10.1.13. *We have*

$$\widetilde{\lambda}^{\text{exp}} \le 1.$$

SKETCH OF THE PROOF. For a given N, consider a cube K with edge length $N^{1/4}$ centered at the origin and the cube K_1 of edge length $N^{1/4}+2$ homothetic to K. Let us pack the cube K by unit spheres centered in $N^{-3/4}\mathbb{Z}$. To do this, arrange all the obtained centers in some order; note that there are approximately N^N of them. Using this order, we can apply expurgation to construct the desired packing. Then we cover $\mathbb{R}^N$ by translations of K_1. □

One can show that this algorithm is exponential in N.

The situation with $\widetilde{\lambda}_\ell^{\text{exp}}$ is more complicated. One can prove the following fact.

THEOREM 10.1.14. *We have*

$$\widetilde{\lambda}_\ell^{\text{exp}} \le 1 + 8 \cdot 10^{-9}.$$

We do not prove this result here, we only give a brief description of the method.

METHOD OF THE PROOF. Identify the field $\mathbb{F}_p$, $p \ne 2$, with the set

$$I_p = \left\{-\frac{p-1}{2}, \dots, 0, \dots, \frac{p-1}{2}\right\} \subset \mathbb{Z}.$$

For a positive integer σ and for $x \in \mathbb{F}_p^N$, set $\|x\|_\sigma = \sum |x_i|^\sigma$, where $|a|$ for $a \in \mathbb{F}_p$ denotes the absolute value of a as an integer in I_p. Let $C \subseteq \mathbb{F}_p^n$ be a linear subspace and let $k = \dim C$, $w = \min \|c\|_\sigma$, where the minimum is over all $c \in C \setminus \{0\}$; C is called an $[n,k,w]_p$ "σ-code". Usual linear codes correspond to the degenerate case $\sigma = 0$. A proof of Theorem 10.1.14 can be obtained by proving the following three claims:

(1) For "σ-codes" there exists an analogue of the Gilbert–Varshamov bound.
(2) By applying the "simplest construction" of Sec. 10.2.1 (see below) to "2-codes", one obtains
$$\det L = p^{n-k}, \qquad d(L) = \min\{p, \sqrt{w}\}.$$
(3) One can apply the result of claim (2) to "2-codes" lying on the analogue of the Gilbert–Varshamov bound (for $n = w = p^2 \to \infty$). □

Note, however, that one can improve Theorem 10.1.14 and get the following result.

THEOREM 10.1.15. *We have*
$$\widetilde{\lambda}_\ell^{\exp} \leq 1. \qquad \square$$

10.1.3. Random Packings

In this section we describe the behaviour of the density of *typical* lattice packings and of *typical* random packings. Here, the word "typical" refers to the natural probability distributions, defined below.

Random lattices. Let L be a lattice. Clearly, its geometric density remains the same after multiplication by a scalar. Therefore, we can consider here only unimodular lattices, i.e., those with $\det(L) = 1$. The space of unimodular lattices in $\mathbb{R}^n$ is naturally isomorphic to the homogeneous space $\Lambda_n = \mathrm{SL}_n(\mathbb{R})/\mathrm{SL}_n(\mathbb{Z})$, which is equipped with the Haar measure μ_n with $\mu_n(\Lambda_n) < +\infty$, and we can normalize it so that $\mu_n(\Lambda_n) = 1$. After such a choice, the density $\Delta(L)$ becomes a random variable. As the following theorem shows, in order to get a nontrivial limiting distribution for it, one should also normalize it by a factor of 2^n.

THEOREM 10.1.16. *Let*
$$F_n(x) = \mu_n\{L \in \Lambda_n \mid 2^n \Delta(L) \leq x\}$$
be the distribution function of the random variable $2^n\Delta(L)$. *Then*
$$\lim_{n\to\infty} F_n(x) = 1 - e^{-x/2}.$$

This result follows from a theorem of Schmidt.

THEOREM 10.1.17. *Let* $n \geq 13$. *Let* S *be a Borel set in* $\mathbb{R}^n$ *with* $S \cap (-S) = \varnothing$, *and let* $\mathrm{vol}(S) \leq n-1$. *Then*
$$\mu_n\{L \in \Lambda_n \mid L \cap S = \varnothing\} = e^{-\mathrm{vol}(S)}(1 - R_n),$$
where
$$|R_n| < 6 \cdot \left(\frac{3}{4}\right)^{n/2} e^{4\,\mathrm{vol}(S)} + \mathrm{vol}(S)^{n-1} n^{-n+1} e^{\mathrm{vol}(S)+n}. \qquad \square$$

PROOF OF THEOREM 10.1.16. To derive the theorem from Theorem 10.1.17, take $S = S(d)$ to be "one half of the ball" $B(0,d)$, so as to satisfy the conditions $S \cap (-S) = \varnothing$ and $S \cup (-S) = B(0,d) \setminus \{0\}$. Then, for a lattice L, the condition $L \cap S = \varnothing$ is equivalent to the condition $d(L) > d$; therefore,

$$\begin{aligned}\mu_n\{L \in \Lambda_n \mid d(L) > d\} &= \mu_n\{L \in \Lambda_n \mid L \cap S(d) = \varnothing\} \\ &= e^{-\operatorname{vol}(S(d))}(1 - R_n).\end{aligned}$$

Let us put $x(L) = v_n d(L)^n$; then $2^n \Delta(L) = x(L)$, since $\det(L) = 1$. Put $d = (x/v_n)^{1/n}$. Then

$$\begin{aligned}F_n(x) &= \mu_n\{L \in \Lambda_n \mid 2^n\Delta(L) \le x\} = \mu_n\{L \in \Lambda_n \mid x(L) \le x\} \\ &= 1 - \mu_n\{L \in \Lambda_n \mid d(L) > d\} = 1 - \mu_n\{L \in \Lambda_n \mid L \cap S(d) = \varnothing\}.\end{aligned}$$

Since $\operatorname{vol}(S(d)) = x/2$, we get

$$\begin{aligned}F_n(x) &= 1 - \mu_n\{L \in \Lambda_n \mid L \cap S(d) = \varnothing\} \\ &= 1 - e^{-\operatorname{vol}(S(d))}(1 - R_{n,x}) = 1 - e^{-x/2}(1 - R_{n,x}),\end{aligned}$$

where for $n \to \infty$ we have

$$|R_{n,x}| < 6 \cdot \left(\frac{3}{4}\right)^{n/2} e^{4x/2} + \left(\frac{x}{2}\right)^{n-1} n^{-n+1} e^{x/2+n} \to 0. \qquad \square$$

Random packings. Now let us consider configurations σ of point random fields and discuss the same questions as we considered for lattices above.

A *point random field* in $\mathbb{R}^n$ is understood to be a probability measure $\mathbf{P}$ on the set $\mathcal{S}$ of all countable locally finite subsets of $\mathbb{R}^n$. To define such a measure, we first introduce, for every $V \subset \mathbb{R}^n$, the notation $\mathcal{S}_V$ for the set of all locally finite subsets of V, and for every $\sigma \in \mathcal{S}$ we denote the intersection $\sigma \cap V$ by $\sigma_V \in \mathcal{S}_V$.

A random field $\mathbf{P}^\lambda$ is called a *Poisson random field with intensity* $\lambda > 0$ if

(i) For every two disjoint subsets $V, W \subset \mathbb{R}^n$, $V \cap W = \varnothing$, the random configurations σ_V and σ_W are independent.
(ii) For every finite subset $V \subset \mathbb{R}^n$, the conditional distribution of σ_V under the condition that $|\sigma_V| = m$ is the Lebesgue measure on V^m normalized by the factor of $1/(\operatorname{vol}(V))^m$, while the probability of the event $|\sigma_V| = m$ is given by

$$\mathbf{P}\{\sigma \mid |\sigma_V| = m\} = e^{-\lambda \operatorname{vol}(V)} \frac{(\lambda \operatorname{vol}(V))^m}{m!}.$$

In what follows, by $\mathbf{P}$ we denote the Poisson random field with intensity $\lambda = 1$, and we omit λ from our notation.

Let σ be our point random field; then it is easy to see that for every $\varepsilon > 0$ we have

$$\mathbf{P}(\Delta_N(\sigma) > \varepsilon) \to 0 \quad \text{as} \quad N \to \infty.$$

Since our goal is to construct subsets with positive geometric density, this is unsatisfactory, and thus we allow ourselves to decimate a configuration σ by considering only "a fraction ν" of our collection of points σ, where $\nu \in (0,1)$. To this end, we introduce the quantity

$$\Delta_{N,\nu}(\sigma) = \max\left\{\Delta_N(\tau) \mid \tau \subset \sigma_N,\ \nu|\sigma_N| - |\sigma_N|^{1/2+\varepsilon} \le |\tau| \le \nu|\sigma_N| + |\sigma_N|^{1/2+\varepsilon}\right\},$$

where ε is some fixed small number. This fraction is designed so that for any τ satisfying the restrictions in the above definition of $\Delta_{N,\nu}(\sigma)$ we have $|\tau|/|\sigma_N| \to \nu$ as $N \to \infty$ for typical σ, which allows us to improve the situation.

THEOREM 10.1.18. *With* **P**-*probability* 1, *the limit*

$$\mathcal{D}(\nu) = \liminf_{N\to\infty} \Delta_{N,\nu}(\sigma)$$

does not depend on σ *for any* $\nu \in (0,1)$.

(i) *Introduce the function* $\nu_1(d) = 1 - \frac{1}{2}v_n(d)$. *Then for any* $d > 0$ *we have*

$$\mathcal{D}(\nu_1(d)) \geq 2^{-n} v_n(d)\nu_1(d), \tag{10.1.1}$$

provided that $\nu_1(d) > 0$. *In particular, if* $v_n(d_n) = 1$, *we get* $\nu_1(d_n) = \frac{1}{2}$, *and therefore*

$$\mathcal{D}\Big(\frac{1}{2}\Big) \geq 2^{-n-1}.$$

(ii) *For the function* $\nu_2(d) = 1 - \frac{1}{2}v_n(d)e^{-2v_n(d)}$, *we have*

$$\mathcal{D}(\nu_2(d)) \leq 2^{-n} v_n(d)\nu_2(d) \tag{10.1.2}$$

for any $d > 0$.

(iii) *For the function* $\nu_3(d) = 1 - \frac{1}{3}v_n(d) - \frac{1}{6}v_n(d)e^{-v_n(d)}$, *we have*

$$\mathcal{D}(\nu_3(d)) \geq 2^{-n} v_n(d)\nu_3(d) \tag{10.1.3}$$

for any $d > 0$. □

Relations (10.1.1)–(10.1.3) give implicit bounds for the function $\mathcal{D}(\nu)$; in particular, (10.1.1) can be rewritten as $\mathcal{D}(\nu) \geq 2^{-n+1}\nu(1-\nu)$. The other relations cannot be rewritten so easily; (10.1.3) improves upon (10.1.1); see Fig. 10.1.

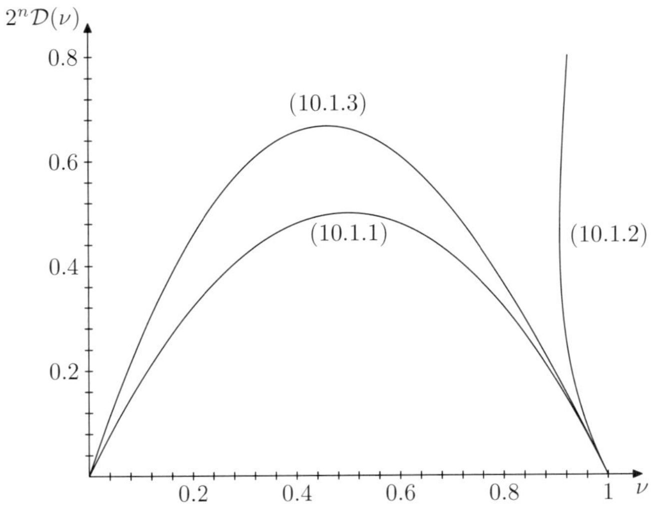

FIGURE 10.1. Upper (10.1.2) and lower (10.1.1), (10.1.3) bounds on $\mathcal{D}(\nu)$.

10.2. Asymptotically Dense Packings

In the preceding section we have seen that there are numerous analogies between linear codes and lattices; moreover, the Leech lattice construction uses the Golay code. Here we continue to use codes to obtain packings. In Sec. 10.2.1 we describe some constructions which give various dense packings using various code families and consider asymptotically good packings obtained in this way. Section 10.2.2 is devoted to spherical codes obtained from algebraic geometry codes and to bounds on kissing numbers that follow from their constructions. In Sec. 10.2.3 we describe how one can obtain lattices with exponentially many minimum vectors from algebraic geometry codes.

10.2.1. Constructions of Dense Packings

First we list constructions of lattices using codes.

The simplest construction (Construction A). Let p be a prime. Let us embed $\mathbb{F}_p$ in $\mathbb{Z}$ as the subset $\{0,1,\dots,p-1\}$. For a linear $[N,k,d]_p$ code C, its image under the composition map $\mathbb{F}_p^N \hookrightarrow \mathbb{Z}^N \subset \mathbb{R}^N$ will also be denoted by C; then the lattice $C\mathbb{Z} = \{ac \mid a \in \mathbb{Z}, c \in C\}$ is of rank k. Put

$$L = C\mathbb{Z} + p\mathbb{Z}^N \quad \text{or} \quad L = \{x \in \mathbb{Z}^N \mid x \pmod{p} \in C\}.$$

The length of a vector $x \in L$ is at least p (for $x \in p\mathbb{Z}^N$), or at least $\sqrt{d}$ (for $x \notin p\mathbb{Z}^N$, since in this case x has at least d nonzero integer entries), and therefore we get

$$d(L) \geq \min\{p, \sqrt{d}\},$$

with equality in the case $p = 2$.

Let us compute $\det L$; since $\det(p\mathbb{Z}^N) = p^N$ and $L/p\mathbb{Z}^N \cong C \subseteq \mathbb{F}_p^N \cong \mathbb{Z}^N/p\mathbb{Z}^N$, we have $|L/p\mathbb{Z}^N| = p^k$ and

$$\det L = p^{N-k}.$$

EXAMPLE 10.2.1. (1) Let C be the single-parity-check $[N, N-1, 2]_2$ code; then the construction gives the lattice D_N.

(2) Let C be the $[8,4,4]_2$ code obtained from the Hamming code by adding the overall parity check; then $L \cong E_8$.

Note, however, that the described simplest construction cannot yield asymptotically good packing families. To get such families, we need more elaborated techniques, which are as follows.

***T*-lattices.** Let $L \subset \mathbb{R}^m$ be a lattice, and let T be the composition of an orthogonal transformation of $\mathbb{R}^m$ and a dilation with coefficient $t = \|T\|$. We call L a *(p,b)-lattice* if the following conditions hold:

(a) T maps L to $\dfrac{1}{p}L$;
(b) T^{-1} maps L to L;
(c) $pT = F(T^{-1})$ for some $F \in \mathbb{Z}[x]$;
(d) $b = -m\log_p t$ is an integer.

REMARK 10.2.2. Any lattice $L \subset \mathbb{R}^m$ is a (p,m)-lattice for $t = p^{-1}$.

EXAMPLE 10.2.3. For $\mathbb{Z}^2$ we can set $T = \frac{1}{2}\begin{pmatrix} 1 & 1 \\ 1 & -1 \end{pmatrix}$. Hence, $2T = T^{-1} = \begin{pmatrix} 1 & 1 \\ 1 & -1 \end{pmatrix}$, $p = 2$, $m = 2$, and $b = 1$.

Generalizing this example, we call a $(2,k)$-lattice for $m = 2k$ a *T-lattice*. Below we shall see that D_4, E_8, and Λ_{24} are T-lattices.

REMARK 10.2.4. For T-lattices one can define the values $\widetilde{\lambda}_T$, $\widetilde{\lambda}_T^{\mathrm{exp}}$, and $\widetilde{\lambda}_T^{\mathrm{pol}}$ in a way similar to that of Sec. 10.1.2. There are obvious inequalities illustrated by the following diagram (where $\to$ stands for $\geq$):

$$\begin{array}{ccccccccc}
 & & & & \widetilde{\lambda} & & & & \\
 & & & \nearrow & & \nwarrow & & & \\
 & & \widetilde{\lambda}_\ell & & & & \widetilde{\lambda}^{\mathrm{exp}} & & \\
 & \nearrow & & \nwarrow & & \nearrow & & \nwarrow & \\
\widetilde{\lambda}_T & & & & \widetilde{\lambda}_\ell^{\mathrm{exp}} & & & & \widetilde{\lambda}^{\mathrm{pol}}. \\
 & \nwarrow & & \nearrow & & \nwarrow & & \nearrow & \\
 & & \widetilde{\lambda}_T^{\mathrm{exp}} & & & & \widetilde{\lambda}_\ell^{\mathrm{pol}} & & \\
 & & & \nwarrow & & \nearrow & & & \\
 & & & & \widetilde{\lambda}_T^{\mathrm{pol}} & & & &
\end{array}$$

To obtain good lattice families, we use the following construction.

Basic construction. Let us consider a (p,b)-lattice $L \subset \mathbb{R}^m$. It is clear that L has a sublattice K of rank b with

$$q = |K/(T^{-1}L \cap K)| = p^b.$$

Define $\psi_0 \colon \mathbb{F}_q \to \mathbb{R}^m$ for $q = p^b$ as the composition

$$\mathbb{F}_q \xrightarrow{\sim} \mathbb{F}_p^b \xrightarrow{\lambda^b} \mathbb{Z}^b \xrightarrow{\beta} K \hookrightarrow \mathbb{R}^m$$

for the embedding λ given by

$$\lambda \colon \mathbb{F}_p \xrightarrow{\sim} \{0,1,\dots,p-1\} \subset \mathbb{Z}$$

and β corresponding to a fixed basis in K. Set $\psi_i = T^i\psi_0$ and assume (rescaling if necessary) that $\rho_0 = d(L)^2 \geq 1$.

Take a family of nested $[n,k_i,d_i]_q$ codes $\mathbb{F}_q^n \supset C_1 \supset \ldots \supset C_a$ such that $d_i t^{2i}\rho_0 \geq 1$ and $C_i = \mathbb{F}_p\langle c_1,\dots,c_{bk_i}\rangle$. Set $L_0 = L$ and

$$L_i = \mathbb{Z}\langle L_{i-1}, \psi_i(c_1),\dots,\psi_i(c_{bk_i})\rangle, \quad \text{for } i = 1,\dots,a.$$

Let us compute the parameters of these lattices using the parameter $\nu(L) = \log_2 \delta(L)$.

THEOREM 10.2.5. *For the lattices L_i obtained from a lattice L and a family of $[n,k_j,d_j]_q$ codes, $j = 1,\dots,a$, $q = p^b$, we have:*

(a) $$\min_{x \in L_i \setminus \{0\}} |x|^2 \geq 1;$$

(b) $$\det L_i = (\det L)^n p^{-b\sum_{j=1}^{i} k_j};$$

(c) $$\nu(L_i) \geq n\nu(L) - b\sum_{j=1}^{i} k_j.$$

SKETCH OF THE PROOF. We use induction on i (for $i=0$ the claim is obvious). First check that $T^{-1}\colon L_i \to L_{i-1}$ and $pT\colon L_i \to L_i$. Let $x \in L_i$, and let

$$x = y + \sum_{j=1}^{bk_i} a_j T^i \psi_0(c_j),$$

where $a_j \in \mathbb{Z}$, $c_j \in C_i$, and $y \in L_{i-1}$. Since pT sends L_{i-1} to itself, we may assume that $0 \le a_j \le p-1$ for any j, i.e., that $x = y + T^i\psi_0(c)$ for some $c \in C_i$. If $z \in \mathbb{F}_q^*$, then $\psi_0(z) \in K \subseteq L$, and since $K/(T^{-1}L \cap K) \cong \mathbb{F}_q$, we get $\psi_0(z) \notin T^{-1}L$. Hence, $x = T^i(T^{-i}y + \psi_0(c))$, where $T^{-i}y \in T^{-1}(L_0) = (T^{-1}L)^n$. Let $c = (z_1, \dots, z_m) \neq 0$; then $\psi_0(x_j)$ has at least d_i entries that do not lie in $T^{-1}L$. Therefore,

$$\|x\| \ge t^{-2i} d_i \rho_0 \ge 1,$$

which proves (a), claim (b) follows from the isomorphism $L_i/L_{i-1} \cong C_i$, and (c) is checked by a direct computation. □

Now let $P \subset \mathbb{R}^m$ be a packing with $\rho_0 = d(P)^2 \ge 1$, and let $\nu_0 = \nu(P)$. Let P be the union of disjoint packings P_x indexed by $x \in \mathbb{F}_q^a$, $q = 2^b$. For $v = (v_1, \dots, v_i) \in \mathbb{F}_q^i$, $i \le a$, define

$$X_v = \{x = (x_1, \dots, x_a) \in \mathbb{F}_q^a \mid (x_1, \dots, x_i) = v\},$$

$$P_v^{(i)} = \bigcup_{x \in X_v} P_x, \qquad \rho_i = d(P_v^{(i)})^2.$$

Let $C_1, \dots, C_a \subseteq \mathbb{F}_q^n$ be a family of $[n, k_i, d_i]_q$ codes with $d_i\rho_i \ge 1$. Consider a new packing

$$Q = \bigcup_{c_1 \in C_1} \bigcup_{c_2 \in C_2} \cdots \bigcup_{c_a \in C_a} P_{(c_1^{(1)}, \dots, c_a^{(1)})} \times \dots \times P_{(c_1^{(n)}, \dots, c_a^{(n)})} \subseteq \mathbb{R}^{mn},$$

where $c_i = (c_i^{(1)}, \dots, c_i^{(n)})$. Then one can deduce the following bound.

PROPOSITION 10.2.6. *We have*

$$\nu(Q) \ge n\nu_0 - b\sum_{i=1}^{a} k_i.$$ □

If $P = L \subset \mathbb{R}^m$ is a (p, b)-lattice, we can set

$$P_x = L_x = L + \sum_{i=1}^{a} \psi_i(x_i), \quad x = (x_1, \dots, x_a) \in \mathbb{F}_q^a.$$

LEMMA 10.2.7. *If the codes C_i are nested, then the lattice thus obtained coincides with that of Theorem* 10.2.5. □

REMARK 10.2.8. If C_i are not nested, then Q can be a nonlattice packing.

Often $(2, m)$-lattices give very dense lattices; moreover, T-lattices are the most useful. Note that for a T-lattice the transformation T lifts as $T = (T, \dots, T)$ to the obtained lattices and there satisfies all the requirements of the definition of T-lattices; hence, all lattices thus obtained are also T-lattices.

Results. Now we consider asymptotic properties of lattices obtained by the basic construction.

Let us be given an infinite family of lattices $\{\Lambda_m\}$, and for any $\Lambda_m \subset \mathbb{R}^m$, a system of nested $[4^\ell, k_i, 4^{\ell-i+1}]_{2^m}$ codes C_i, $i = 1, \dots, \ell$. Then, considering Λ_m as

$(2,m)$-lattices, we can construct a lattice family $\{L_N\}$, $L_N \subset \mathbb{R}^N$, $N = m\cdot 4^\ell$, and we get the following.

COROLLARY 10.2.9. *For those lattices we have*

$$\lambda(\{L_N\}) \le \lambda(\{\Lambda_m\}) + \sum_{i=1}^{\infty}(1-R_i). \qquad \square$$

Similarly, for a family $\{\Lambda_m\}$ of T-lattices and nested systems of $[2^\ell, k_i, 2^{\ell-i+1}]_{2^b}$ codes, we obtain a family $\{L_N\}$ of T-lattices with $N = m\cdot 2^\ell$.

COROLLARY 10.2.10. *For these T-lattices we have*

$$\lambda(\{L_N\}) \le \lambda(\{\Lambda_m\}) + \frac{1}{2}\sum_{i=1}^{\infty}(1-R_i). \qquad \square$$

This corollary is completely analogous to the previous one; note only that the transformation $(T,\dots,T)$ in $\mathbb{R}^N$ has the same properties as T.

REMARK 10.2.11. If the codes are not nested, then a similar fact is valid for packings.

Complexity reduction. Let us use the above results to reduce the lattice construction complexity.

THEOREM 10.2.12. *We have*

$$\widetilde{\lambda}_\ell^{\mathrm{pol}} \le \widetilde{\lambda}_\ell^{\exp} + 4/3,$$
$$\widetilde{\lambda}_T^{\mathrm{pol}} \le \widetilde{\lambda}_T^{\exp} + 1.$$

SKETCH OF THE PROOF. For any $\varepsilon > 0$ there exists a lattice family $\{\Lambda_m\}$ of exponential complexity with $\lambda(\{\Lambda_m\}) \le \widetilde{\lambda}_\ell^{\exp} + \varepsilon$. Apply Corollary 10.2.9 to $\{\Lambda_m\}$ and the nested family of Reed–Solomon codes with parameters

$$\left[4^{m/2}, 4^{m/2} - 4^{m/2-i+1} + 1, 4^{m/2-i+1}\right]_{2^m},$$

and using that $\sum_{i=1}^{\infty}(1-R_i) = 4/3$, we get

$$\lambda(\{L_N\}) \le \widetilde{\lambda}_\ell^{\exp} + \varepsilon + 4/3.$$

Since Λ_m is an exponential family and therefore is polynomial in $N = m\cdot 2^m$, the Reed–Solomon codes are polynomial, and our construction is also polynomial, we conclude that the obtained family is polynomial. By letting $\varepsilon \to 0$, we get the first inequality of the theorem.

Similarly, for T-lattices Λ_m we can take nested Reed–Solomon codes with parameters

$$\left[2^{m/2}, 2^{m/2} - 2^{m/2-i+1}, 2^{m/2-i+1}\right]_{2^{m/2}},$$

and since in this case $\frac{1}{2}\sum_{i=1}^{\infty}(1-R_i) = 1$, we get the second inequality. $\square$

Asymptotic parameters. Now let us consider the asymptotic parameters. Theorems 10.1.13 and 10.1.15 imply the following.

COROLLARY 10.2.13. *We have* (1) $\widetilde{\lambda}^{\mathrm{pol}} \le 2.34$; (2) $\widetilde{\lambda}_\ell^{\mathrm{pol}} \le 2.34$. $\square$

Using the lattice $\mathbb{Z} \subset \mathbb{R}^1$ and nested code families meeting the Gilbert–Varshamov bound, we prove the following result.

THEOREM 10.2.14. *We have*

$$\widetilde{\lambda}_\ell^{\exp} \leq 1.27. \qquad \square$$

Then let us use algebraic geometry codes.

THEOREM 10.2.15. *We have*

$$\widetilde{\lambda}_T^{\exp} \leq 1.30.$$

SKETCH OF THE PROOF. Let us apply the construction to the Leech lattice Λ_{24} (recall that it is a T-lattice) and to the following nested family of 2^{12}-ary codes. For a fixed length n, let d_0 be the largest integer such that for $d_0 \geq d$ the dimension of the corresponding algebraic geometry code is greater than that given by the Gilbert–Varshamov bound. We start with embedded algebraic geometry codes $C_1 \subset C_2 \subset \ldots \subset C_a$ with $d \geq d_0$ and continue this family by the Varshamov procedure (Exercise 1.1.60), getting $C_a \subset C_{a+1} \subset \ldots \subset C_\ell$ with $d \leq d_0$. Direct computation yields the statement. $\square$

COROLLARY 10.2.16. *We have*

$$\widetilde{\lambda}_T^{\mathrm{pol}} \leq 2.30. \qquad \square$$

Applying Proposition 10.2.6 to the Leech lattice and to concatenated codes with outer algebraic geometry codes and inner trivial $[n,n,1]$ repetition codes, we obtain the following result.

THEOREM 10.2.17. *We have*

$$\widetilde{\lambda}^{\mathrm{pol}} \leq 1.31. \qquad \square$$

Presently known asymptotic results are shown in Diagram A.3 in the Appendix to this chapter.

10.2.2. Spherical Codes and Kissing Numbers

Algebraic geometry codes can also be applied for $n \to \infty$ to other classical objects, namely spherical codes and kissing numbers. First, recall some definitions.

A *spherical code* is a set X of points lying on the unit sphere

$$S^{n-1} = \{x \in \mathbb{R}^n \mid \|x\|^2 = 1\}.$$

If x and y are in S^{n-1}, we set

$$\varphi(x,y) = \arccos(x \cdot y) \quad (0 \leq \varphi \leq \pi),$$
$$\rho(x,y) = \|x-y\|^2.$$

For a spherical code $X \subset S^{n-1}$, its *minimal angle* $\varphi(X)$ is defined as

$$\varphi(X) = \min\{\varphi(x,y) \mid x,y \in X, \, x \neq y\},$$

and its minimum distance $\rho(X)$ as

$$\rho(X) = \min\{\rho(x,y) \mid x,y \in X, \, x \neq y\}.$$

Thus,

$$\rho(X) = 2 - 2\cos\varphi(X) = 4\sin^2\frac{\varphi(X)}{2}.$$

Let $M=|X|$; the numbers (n,M,ρ) are *parameters* of a spherical code X. For a spherical code $X\subset S^{n-1}$, we set

$$\dim X = n, \qquad \theta(X)=\frac{\log_2|X|}{\dim X}.$$

The *domain of spherical codes* is the set U of pairs $(\rho,R)\in[0,2]\times[0,+\infty]$ such that there exists an infinite family $\{X_i\}_{i\geq 0}$ of such codes with $\dim X_i\to\infty$ having

$$\lim_{i\to\infty}\theta(X_i)\geq R, \qquad \lim_{i\to\infty}\rho(X_i)\geq\rho.$$

We restrict our attention to families for which the above limits exist, and define

$$R(\rho)=\sup\{R\mid(\rho,R)\in U\}.$$

Here are some bounds on $R(\rho)$.

PROPOSITION 10.2.18. *We have*

(i) $R(\rho)=0$ *for* $\rho\in[2,4]$;

(ii) $R(\rho)\geq R_{CSW}(\rho):=1-\frac{1}{2}\log_2\rho(4-\rho)=\log_2\frac{1}{\sin\varphi(\rho)}$ *for* $\rho\in(0,2)$, *where* $\varphi(\rho)=\arccos\left(1-\frac{\rho}{2}\right)=2\arcsin\frac{\sqrt{\rho}}{2}$;

(iii) $R(\varphi)\leq R_{KL}(\varphi):=\frac{1+\sin\varphi}{2\sin\varphi}\log_2\frac{1+\sin\varphi}{2\sin\varphi}-\frac{1-\sin\varphi}{2\sin\varphi}\log_2\frac{1-\sin\varphi}{2\sin\varphi}$ *with* $\varphi=\varphi(\rho)$; *note that this is an* upper *bound*; *in particular*, $R(\pi/3)\leq 0.40414$. □

The bound R_{CSW} is called the Chabauty–Shannon–Wyner bound; R_{KL} is the Kabatiansky–Levenshtein bound.

For a lattice $L\subset\mathbb{R}^n$ with $d(L)=d$, the parameter

$$\tau=\tau(L):=|\{x\in L:\ \|x\|=d(L)\}|$$

is the *kissing number* of L; each sphere in the corresponding packing of spheres touches exactly τ other spheres.

Let Σ be a finite alphabet. Here we consider spherical codes $X\subset S^{m-1}$ which are images of maps $F\colon\Sigma^k\to S^{m-1}$. We say that a family $\{X_i\}_{i\geq 0}$ of spherical codes is *polynomially constructible* if for a sequence $\{F_i\}_{i\geq 0}$ of maps

$$F_i\colon\Sigma^{k_i}\to S^{m_i-1}$$

we have the following:

(i) F_i is a bijection of Σ^{k_i} onto X_i.

(ii) For every $a\in\Sigma^{k_i}$, the point $F_i(a)$ is computable from i and a in poly(m_i) time, where $m_i=\dim X_i$.

Lachaud–Stern bounds. Now let q be a prime power $q=p^a$, and consider a block code C with parameters $[n,k,d]_q$ (the code need not be linear; here $k=\log_q|C|$). Also take a spherical code X which is the image of an injective map $F\colon\mathbb{F}_q\to S^{m-1}$. The spherical code obtained by concatenation of X and C is the spherical code

$$Y\subset S^{nm-1}\subset\mathbb{R}^{nm}=\mathbb{R}^m\times\ldots\times\mathbb{R}^m,$$

whose elements are the vectors

$$y(a)=\frac{1}{\sqrt{n}}(F(a_1),\ldots,F(a_n)), \quad \text{where} \quad a=(a_1,\ldots,a_n)\in C.$$

PROPOSITION 10.2.19. *The parameters of the code Y are as follows:*

$$\dim Y = nm, \qquad |Y| = q^k, \qquad \rho(Y) \geq \frac{2}{n}\rho(X), \qquad \theta(Y) = \frac{k}{n}\theta(X). \qquad \square$$

We now define the *domain of polynomially constructible spherical codes* as the set U_{pol} of the pairs $(\rho, R) \in [0,2] \times [0,+\infty]$ such that there exists a polynomially constructible family $\{X_i\}_{i\geq 0}$ of such codes with $\dim X_i \to \infty$ having

$$\lim_{i\to\infty} \theta(X_i) \geq R, \qquad \lim_{i\to\infty} \rho(X_i) \geq \rho.$$

We set

$$R_{\mathrm{pol}}(\rho) = \sup\{R \mid (\rho, R) \in U_{\mathrm{pol}}\};$$

this is the *asymptotic bound for polynomially constructible spherical codes*. We look for lower bounds

$$R_{\mathrm{pol}}(\rho) \geq F(\rho).$$

PROPOSITION 10.2.20. *If Q is an even prime power and $|X| = Q$, define*

$$F(X,p) = \theta(X)\left(1 - \frac{\rho}{\rho(X)} - \frac{1}{\sqrt{Q}-1}\right),$$
$$R_*(\rho) = \max F(X,\rho),$$

where the maximum is over all spherical codes X of size Q. Then

$$R_{\mathrm{pol}}(\rho) \geq R_*(\rho). \qquad \square$$

Spherical codes from binary block codes. The trivial spherical code $T \subset S^0$ with two elements has $\rho(T) = 4$ as minimum distance, hence $\theta(T) = 1$. The concatenation of T with a binary $[n, 2t, d]$ block code C gives a spherical code $X(C)$ satisfying

$$\dim X(C) = n, \qquad |X(C)| = 2^{2t};$$

then Proposition 10.2.19 implies

$$\rho(X(C)) = \frac{d}{n}\rho(T) = \frac{4d}{n}, \qquad \theta(X(C)) = \frac{2t}{n}.$$

Define

$$F(X(C),\rho) = \frac{2t}{n}\left(1 - \rho\frac{n}{4d} - \frac{1}{2^t - 1}\right);$$

then

$$R_*(\rho) = \max F(X,\rho) \geq R_{KTV}(\rho).$$

Polygonal lower bound. Applying Proposition 10.2.20 to certain spherical codes X_i of small lengths, $i = 0, 1, \ldots, 12$, one can get the following result.

PROPOSITION 10.2.21. *For $0 < \rho < 2$ we have*

$$R_{\mathrm{pol}}(\rho) \geq F_*(\rho),$$

where

$$F_*(\rho) = \max_{0\leq i\leq 12} F_i(\rho) \quad \text{and} \quad F_i(\rho) := F(X_i,\rho), \quad i = 0, 1, \ldots, 12. \qquad \square$$

We do not specify these codes here; note only that the code X_7 with parameters $(16, 2^8, 3/2)$ and $F_7(\rho) = \frac{7}{15} - \frac{\rho}{3}$ is constructed from the nonlinear $[16, 8, 6]$ Nordstrom–Robinson code $\mathcal{N}_{16}$ (anyway, we do not describe its construction).

TABLE 1. The function $F_*(\rho)$

ρ	$R_{CSW}(\rho)$	$\frac{1}{2}R_{CSW}(\rho)$	$F_*(\rho)$
0.1	1.6792	0.8396	1.08966
0.3	0.9247	0.4623	0.59304
0.5	0.5963	0.2981	0.39536
0.7	0.3961	0.1980	0.24910
0.9	0.2599	0.1299	0.16666
1.0	0.2075	0.1037	0.13333
1.1	0.1632	0.0816	0.10504
1.3	0.0942	0.0471	0.05587
1.5	0.0465	0.0232	0.02380
1.7	0.0164	0.0082	0.00520
1.9	0.0018	0.0009	0.00034

Table 1 shows some values of the function $F_*(\rho)$.

Kissing number. The value $\rho = 1$ is of particular interest. If $n \geq 2$, the kissing number $\tau(n)$ is the maximum cardinality of a spherical code $X \subset S^{n-1}$ with $\varphi(X) \geq 60°$. Set

$$\theta = R(1) = \limsup_{i \to \infty} \frac{\log_2 \tau(n)}{n}.$$

Proposition 10.2.18 shows that

$$0.401414 \geq \theta \geq R_{CSW}(1) = -\frac{1}{2}\log_2 3 = 0.2075\ldots.$$

The *asymptotic polynomially constructible kissing number* θ_{pol} is defined as

$$\theta_{\text{pol}} = \sup_X \frac{\log_2 \tau_X(n)}{n}$$

when X ranges over all polynomially constructible families of codes.

THEOREM 10.2.22. *We have*

$$\theta_{\text{pol}} \geq 0.133\ldots = 2/15.$$

PROOF. Consider

$$F_7(\rho) = F(X_7, \rho) = \left(\frac{7}{15} - \frac{1}{3}\rho\right);$$

then from Proposition 10.2.20 for $\rho = 1$ we get

$$R_*(1) \geq F(X_7, \rho) = \frac{2}{15}.$$

□

We also have the following.

PROPOSITION 10.2.23. *We have*

$$R_*(\rho) \geq \frac{1}{2}R_{CSW}(\rho) \quad \textit{if} \quad \rho \leq 1.535\ldots;$$

thus,

$$R_*(\rho(\varphi)) \geq \frac{1}{2}\log_2 \frac{1}{\sin\varphi} \quad \textit{if} \quad \varphi \leq 76°\,34'.$$

□

Solé–Belfiore bounds. The above bounds can be considerably improved for very small values of ρ. This improvement uses the so-called Yaglom map. Namely, let

$$Y\colon \mathbb{R}^n \to \mathbb{R}^{n+1}, \quad Y\colon x \mapsto \big(x, \sqrt{\|x\|^2}\big),$$

be the *Yaglom map*. Then $Y(B^n) \subseteq B^{n+1}$ for the unit balls in $\mathbb{R}^n$ and $\mathbb{R}^{n+1}$, respectively.

THEOREM 10.2.24. *There are families of spherical codes constructed from lattice packings by the Yaglom map satisfying*

$$R(\rho) \geq R_L(\rho) := -(1/2)\log_2(\rho).$$

SKETCH OF THE PROOF. There are spherical codes satisfying

$$R \geq \delta + 1 + R_L(\rho),$$

where δ is the asymptotic exponent; this inequality is called the *Yaglom bound*. By the Minkowski bound, we have $\delta \geq -1$, and the result follows. $\square$

Using lattice packings constructed from codes over fields via Construction A above, we obtain the following bound.

THEOREM 10.2.25. *There are families of spherical codes obtained from Construction A lattice packings such that*

$$R(\rho) \geq R_L(\rho).$$

SKETCH OF THE PROOF. There are lattice packings constructed from codes over prime fields by Construction A having asymptotic density $\delta \geq -1$; then apply the Yaglom bound. $\square$

We can further improve the polynomial bound using algebraic geometry code lattices from Section 10.2.1.

THEOREM 10.2.26. *There are families of spherical codes of binary rate $R(\rho)$ constructed from lattice packings by the Yaglom map such that*

$$R(\rho) \geq R_L(\rho) - 1.30.$$

PROOF. By the Yaglom bound we know there are spherical codes satisfying $R \geq \delta + 1 + R_L(\rho)$, and $\delta \geq -2.30$ by Corollary 10.2.16. $\square$

COROLLARY 10.2.27. *For $\rho \leq e^{-90}$ there are spherical codes constructed from polynomial time constructible lattices with $R \geq 0.98 R_{CSW}(\rho)$.*

PROOF. Applying Theorem 10.2.26 with the approximation $R_{CSW}(\rho) \approx R_L(\rho)$, we obtain the bound for $\rho \leq 2^{-130} \approx e^{-90}$. $\square$

10.2.3. Lattices with Exponentially Large Kissing Numbers

In this section we construct lattice packings in $\mathbb{R}^n$ for growing n having exponentially large kissing numbers. Note that the random choice procedure guarantees (the Chabauty–Shannon–Wyner bound) the existence of (nonlattice) packings $P_n \subset \mathbb{R}^n$ with

$$\frac{\log_2 \tau(P_n)}{n} \geq \log_2 \frac{2}{\sqrt{3}} \simeq 0.2075\ldots,$$

the Kabatiansky–Levenshtein upper bound being 0.4041.... However, for lattice packings this procedure does not work, and the best lower bound for the maximum lattice kissing number $\tau_\ell(n)$ in n dimensions known prior to our present construction is given by the Barnes–Wall lattices BW_n with $n=2^m$ providing only the quasipolynomial bound $\tau_\ell(n)\ge n^{c\log n}$, i.e., $\log\tau_\ell(n)\ge c\log^2 n$. Our construction is based on the techniques of Sec. 4.5.2.

The main results are as follows.

THEOREM 10.2.28. *We have*

$$\frac{\log(\tau_\ell(n))}{n}\ge\frac{1}{20}\left(1-\frac{2\log 33}{31}\right)-\frac{2+2\log n}{n}$$

for $n=5\cdot 2^{10m+2}$, *where*

$$\frac{1}{20}\left(1-\frac{2\log 33}{31}\right)\simeq 0.033727\ldots. \qquad \square$$

We can improve the constant slightly, losing the effectivity.

THEOREM 10.2.29. *We have*

$$\frac{\log(\tau_\ell(n))}{n}\ge\frac{1}{20}\left(\frac{21}{31}-\log\frac{1024}{1023}\right)-o(1)\simeq 0.033800\ldots-o(1)$$

for $n=5\cdot 2^{10m+2}$. $\square$

Therefore,

$$\limsup_{n\to\infty}\frac{\log(\tau_\ell(n))}{n}\ge\frac{1}{20}\left(\frac{21}{31}-\log\frac{1024}{1023}\right).$$

For the lower limit we can prove the following.

THEOREM 10.2.30. *We have*

$$\liminf_{n\to\infty}\frac{\log(\tau_\ell(n))}{n}\ge\frac{21H(\delta_0)-4-21A}{504}\simeq 0.025476\ldots,$$

where $A=\log\frac{4096}{4095}$ *and* $\delta_0\simeq 0.75561\ldots$ *is a unique root of the equation*

$$21\delta H(\delta)=17-21A+(21A-4)\delta$$

in the interval $(0.5,1)$, $H(\delta)$ *being the binary entropy function.* $\square$

We say that a family of packings $P_n\subset\mathbb{R}^n$ is τ-*asymptotically good* whenever $\limsup_{n\to\infty}\frac{\log\tau(P_n)}{n}>0$. Since the random choice argument does not work for lattices, it is not immediately clear whether τ-asymptotically good lattice families exist.

The above results show that this is the case; however, a complete proof of the results is rather technical, and here we only give a simplest construction of lattices with exponential kissing numbers, which leads to a much worse constant. Thus, we are going to prove the following in this section.

PROPOSITION 10.2.31. *We have*

$$\limsup_{n\to\infty}\frac{\log(\tau_\ell(n))}{n}>0.0018.$$

PROOF. First recall the following result from Sec. 4.5.2 (see Theorem 4.5.18) concerning codes with many light vectors.

THEOREM 10.2.32. *Let $q = 2^{2s}$, $s = 3,4,\ldots$, be fixed. Then for any $\delta_1 < \delta < \delta_2$ there exists a sequence of binary linear codes $\{C_n\}$ of length $n = qN$, $N \to \infty$, and distance $d_n = n\delta/2$ such that*

$$\frac{\log A_{d_n}}{n} \geq \frac{E_s(\delta)}{2^{2s}} - o(1). \qquad \square$$

Here A_{d_n} is the number of minimum-weight vectors in C_n, and

$$E_s(\delta) := H(\delta) - \frac{2s}{2^s - 1} - \log_2 \frac{2^{2s}}{2^{2s} - 1}.$$

The function $E_s(\delta)$ has two zeroes, $0 < \delta_1 < 0.5 < \delta_2 < 1 - 2^{-2s}$, and is positive for $\delta_1 < \delta < \delta_2$.

The following construction is a particular elementary case of the general construction on p. 243 allowing one to construct good lattices from good codes.

Construction D. Let $C_0 = \mathbb{F}_2^n \supset C_1 \supset \ldots \supset C_a$ be a finite decreasing family of linear binary codes with parameters $[n, k_i, d_i]$ for C_i, $i = 0, \ldots, a$, where $d_i = 4^i$. We can and will consider C_0 as a subset of $\mathbb{R}^n$. Choose a basis $c_1, \ldots, c_n$ in $\mathbb{F}_2^n$ such that $c_1, \ldots, c_{k_i}$ span C_i for $i = 0, \ldots, a$. Let L be the lattice in $\mathbb{R}^n$ generated by $(2\mathbb{Z})^n$ and the vectors $\{c_j \cdot 2^{1-i}\}$ for $i = 1, \ldots, a$, $k_{i+1} + 1 \leq j \leq k_i$. Then we have the following (obviously in view of the minimum distances d_i, $i = 0, \ldots, a$).

PROPOSITION 10.2.33. *For the lattice L obtained by Construction D, its minimal square norm is $d(L)^2 = 4$.* $\square$

Here is an easy construction giving lattices with a small but strictly positive lower bound for the ratio $\log(\tau_\ell(n))/n$ using only the most elementary means.

Let us take $q = 64$, let $m \geq 3$ be odd, and let $X = X_m = X_0(T^{m+1})$ be the Drinfeld curve over $\mathbb{F}_8$ of genus $g_m = g(X_m)$ for an odd m, having $N_m \geq 2^{3m} = 8^m$ rational points over $\mathbb{F}_{64}$, while $g_m < N_m/7$. We put $K = 3m - 1 = 2\ell$, $n = 2^{K+1}$, $d = a = n/2 = 2^K = 4^\ell$, and apply Theorem 10.2.32 with $s = 3$ and $q = 64$ to the codes on the curve X. Then we get a binary $[n, k, d]$ code C_K with

$$\log A_d \geq \frac{1}{64} E_3\left(\frac{1}{2}\right) - o(n).$$

Next, construct a decreasing family $C_0 = \mathbb{F}_2^n \supset C_1 \supset \ldots \supset C_K$ inductively defining C_{K-i} for $i = 1, \ldots, K-1$ as generated by C_{K-i+1} and c_i for any binary vector $c_i \in \mathbb{F}_2^n$, $i = 1, \ldots, K-1$, with $\mathrm{wt}(c_i) = 2^{K-i}$. Then applying Construction D, we get a lattice $L_n \subset \mathbb{R}^n$ with $d(L) = 2$, and each minimum-weight vector of C_K produces a minimum-norm vector in L. Therefore, we have

$$\frac{\log \tau(L_n)}{n} \geq \frac{\log A_d}{n} \geq \frac{1}{64} E_3\left(\frac{1}{2}\right) - o(1),$$

where $\frac{1}{64} E_3(0.5) > 0.0018$, which completes the proof of Proposition 10.2.31. $\square$

Note also that one can replace the curve X_m with the curve $\widetilde{X}_m$ from the García–Stichtenoth tower $\mathcal{W}_2$, having $7 \cdot 2^{3m-3}$ rational points instead of $2^{3m} = 8 \cdot 2^{3m-3}$ points on X_m; this leads to the constant $\frac{1}{64} E_3(4/7)$, which is > 0.0015.

REMARK 10.2.34. The result in this subsection is asymptotic; however, it is interesting to compare the corresponding kissing numbers with the previously best known lattices, the Barnes–Wall lattices $BW_{2^m} \subset \mathbb{R}^{2^m}$. A simple calculation shows that for $m \leq 10$, i.e., up to dimension $2^{10} = 1024$, the BW_{2^m} lattices have more minimal vectors than (any) lattices from Theorems 10.2.28–10.2.30; for $m = 11$ they are comparable; and for $m \geq 12$, i.e., for dimensions starting from 4096, the lattices from Theorems 10.2.28–10.2.30 are much better than BW_{2^m}.

10.3. Lattices from Global Fields

Algebraic geometry codes are constructed using curves over finite fields; by using their analogues for global fields to construct codes and lattices, we can obtain seven further constructions, which will be denoted by combining the following symbols:

- we use either number (N) or function (F) fields;
- we use either additive (A) or multiplicative (M) structure;
- we obtain either lattices (L) or codes (C);
- the construction either depends on a divisor (D) or not,

resulting in the abbreviations used below. Note that the construction of algebraic geometry codes can be abbreviated in this way as FACD. For each construction we estimate parameters and produce asymptotically good families.

Section 10.3.1 is devoted to additive constructions (NAL and NAC) arising in the number field case; Sec. 10.3.2 contains multiplicative constructions of lattices (NML and FML). In Sec. 10.3.3 we discuss multiplicative congruence constructions, i.e., those depending on a divisor (FMLD, NMLD, and FMCD).

In this section, we write log for $\log_2$ and ln for $\log_e$.

10.3.1. Additive Constructions

Additive lattices (NAL). Let $\mathbb{k}$ an be algebraic number field of degree $N = s+2t$, and let $\mathcal{O}_{\mathbb{k}}$ be its ring of integers. Instead of the standard embedding

$$\sigma\colon \mathbb{k} \hookrightarrow \mathbb{R}^s \times \mathbb{C}^t$$

considered in Proposition 6.1.27, we now define a modified map

$$\widetilde{\sigma}\colon \mathbb{k} \hookrightarrow \mathbb{R}^s \times \mathbb{C}^t$$

given by

$$\widetilde{\sigma}(f) = \big(\sigma_1(f), \dots, \sigma_s(f); \sqrt{2}\sigma_{s+1}(f), \dots, \sqrt{2}\sigma_{s+t}(f)\big).$$

As usual, we take one embedding from each complex pair.

The map $\widetilde{\sigma}$ is even more natural than σ, since it is quite analogous to what we do to define the log map and the regulator; see Sec. 10.3.2 below.

By Proposition 6.1.27(1), $\sigma(\mathcal{O}_{\mathbb{k}})$ is a lattice of rank N; therefore, this is also true for $L = \widetilde{\sigma}(\mathcal{O}_{\mathbb{k}})$.

Parameters. Let us compute the density of L. Note that we have

$$\det \sigma(\mathcal{O}_{\mathbb{k}}) = 2^{-t}\sqrt{|D_{\mathbb{k}}|}$$

by Proposition 6.1.27(2); therefore,

$$\det L = \sqrt{|D_{\mathbb{k}}|},$$

and this can be taken as a definition of the discriminant of a number field.

PROPOSITION 10.3.1. *We have*

$$d(L) = \sqrt{s+2t} = \sqrt{N}.$$

PROOF. Let

$$\widetilde{\sigma}(f) = \big(x_1, \dots, x_s; \sqrt{2}(y_1 + \sqrt{-1}\,z_1), \dots, \sqrt{2}(y_t + \sqrt{-1}\,z_t)\big);$$

then

$$|\widetilde{\sigma}(f)| = \sqrt{\sum_{j=1}^{s} x_j^2 + 2\sum_{j=1}^{t}(y_j^2 + z_j^2)}.$$

On one hand, for $f = 1$, we have $|\widetilde{\sigma}(1)| = \sqrt{s+2t}$. On the other hand, the inequality of arithmetic and geometric means yields

$$\begin{aligned}\sqrt{\sum_{j=1}^{s} x_j^2 + 2\sum_{j=1}^{t}(y_j^2 + z_j^2)} &= \sqrt{\sum_{j=1}^{s} x_j^2 + \sum_{j=1}^{t}(y_j^2 + z_j^2) + \sum_{j=1}^{t}(y_j^2 + z_j^2)} \\ &\geq \sqrt{s+2t}\left(\prod_{j=1}^{s} x_j^2 \prod_{j=1}^{t}(y_j^2 + z_j^2)^2\right)^{1/(2N)} \\ &= \sqrt{s+2t}\,|N_{\mathbb{K}/\mathbb{Q}}(f)|^{1/N} \geq \sqrt{s+2t},\end{aligned}$$

since $N_{\mathbb{K}/\mathbb{Q}}(f) \in \mathbb{Z}$. □

Note that the shortest vectors of L are images of roots of unity.

Unramified towers. Now let the field $\mathbb{K}$ vary in a family with $N \to \infty$. Then

$$\lambda(L) \sim -\log\sqrt{\frac{\pi e}{2}} + \frac{1}{N}\log\sqrt{|D_{\mathbb{K}}|}.$$

If we want to construct good lattices, the last term should be bounded, and this is definitely the case if $\mathbb{K}$ runs over unramified extensions of some $\mathbb{K}_0$, and thus this term is constant. We obtain the following result.

THEOREM 10.3.2. *If a number field $\mathbb{K}_0$ of degree N_0 has an infinite unramified tower of fields $\mathbb{K} \supset \mathbb{K}_0$, then it yields an asymptotically good family of lattices $\{L_N \subset \mathbb{R}^N\}$ with*

$$\lambda(L) \sim -\log\sqrt{\frac{\pi e}{2}} + \frac{1}{N_0}\log\sqrt{|D_{\mathbb{K}_0}|}.$$ □

For $\mathbb{K}_0 = \mathbb{Q}\left(\cos\frac{2\pi}{11}, \sqrt{-46}\right)$, which has an infinite class field tower, we get $\lambda \sim 2.2218$.

On the other hand, using the Odlyzko–Serre bound one can show that for any family of fields $\mathbb{K}$ we cannot get asymptotically less than $1.193\ldots$ (or $1.694\ldots$ under the GRH).

REMARK 10.3.3. In contrast to the polynomial and exponential algorithms from Sec. 10.2.1, for lattices from Theorem 10.3.2 we only know a triple exponential algorithm.

REMARK 10.3.4. A natural generalization of this NAL construction is an NALD construction: instead of $\mathcal{O}_{\Bbbk}$ take an ideal $D \subset \mathcal{O}_{\Bbbk}$ and set $L = \widetilde{\sigma}(D)$. For a principal ideal, the density is the same. What happens for a nonprincipal D? It would be nice if we could get lattices denser than those described above.

REMARK 10.3.5. It is tempting to consider a finite set S_f of non-Archimedean places and take $\widehat{\sigma}\colon \Bbbk \to \mathbb{R}^s \times \mathbb{C}^t \times \prod_{v\in S_f} \mathbb{K}_v$. The problem is that the target is no longer a Euclidean space.

Number field codes (NAC). The above construction can be generalized using non-Archimedean places. Let $S = S_f \cup S_\infty$ with $S_f \cap S_\infty = \varnothing$ be a fixed finite set of places of a number field $\mathbb{K}$. For $v \in S_f$, let $\Bbbk(v)$ be the residue field; for $v \in S_\infty$, let $\Bbbk(v) = \mathbb{R}$ for real places and $\Bbbk(v) = \mathbb{C}$ for complex ones. Let $\mathfrak{a}$ be a fractional ideal with $S_f \cap \operatorname{Supp}(\mathfrak{a}) = \varnothing$, where

$$\operatorname{Supp}(\mathfrak{a}) = \{\text{prime } \mathfrak{p} \text{ in } \mathcal{O}_\Bbbk \mid \mathfrak{p} \supseteq \mathfrak{a}\}.$$

For any $v \in S$ there is a natural map

$$\sigma_v\colon \mathfrak{a}^{-1} = L(\mathfrak{a}) \to \Bbbk(v).$$

Together they form a map

$$\sigma_S\colon L(\mathfrak{a}) \to \bigoplus_{v \in S} \Bbbk(v).$$

However, the image of σ_S is neither a code nor a lattice, except for the case of $S = S_\infty$. Here is a way out.

Construction. Let $[\mathbb{K} : \mathbb{Q}] = N = s + 2t$. Fix two integers $q \geq r > 1$. Let S_f be the set of finite places v of $\mathbb{K}$ such that for some $c(v) \in \mathbb{Z}$

$$r \leq N(v)^{c(v)} \leq q,$$

where $N(v) = |\Bbbk(v)|$. Let $S = S_f \cup S_\infty$, where S_∞ is the set of all Archimedean places, and let

$$n = |S| = |S_f| + s + t.$$

Assume for simplicity that the field is totally real, i.e., $t = 0$.

Let

$$U = \{x \in \mathbb{R}^N \mid 0 < x_i < r^{a/N}\}.$$

Then there exists a shift U' of U such that

$$|U' \cap \sigma(\mathcal{O}_\mathbb{K})| \geq r^a / \sqrt{|D_\mathbb{K}|},$$

where $\sigma(\mathcal{O}_\mathbb{K})$ is the lattice studied above. Divide each side of the cube U' into q equal parts and identify the set of q^N small cubes with $\mathbb{F}_q^N$. Define the map

$$\varphi_\infty\colon U' \cap \sigma(\mathcal{O}_\Bbbk) \to \mathbb{F}_q^N,$$

mapping each point to the small cube in which it lies; let φ_v be the component of φ_∞ corresponding to $v \in S_\infty$. For $v \in S_f$, let φ_v be the map

$$U' \cap \sigma(\mathcal{O}_\mathbb{K}) \hookrightarrow \mathcal{O}_\mathbb{K} \to \mathcal{O}_\mathbb{K}/v^{c(v)} \hookrightarrow \mathbb{F}_q$$

for a fixed embedding of sets $\mathcal{O}_\mathbb{K}/v^{c(v)} \hookrightarrow \mathbb{F}_q$, where we identify the place $v \in S_f$ with the corresponding prime ideal.

Let φ_S be defined as

$$\varphi_S\colon U' \cap \sigma(\mathcal{O}_\mathbb{K}) \to \mathbb{F}_q^n, \quad \varphi_S = (\varphi_{v_1}, \dots, \varphi_{v_n}),$$

for all $v \in S$; then its image

$$C = \varphi_S(U' \cap \sigma(\mathcal{O}_\mathbb{K})) \subseteq \mathbb{F}_q^n$$

is a (nonlinear) code of length n.

Parameters. The parameters are estimated as follows (cf. Theorem 4.1.1).

PROPOSITION 10.3.6. *Let* $a \le n$. *Then*

(a) $d \ge n+1-a$;
(b) $k \ge a\log_q r - \log_q \sqrt{|D_{\mathbb{K}}|}$.

PROOF. (a) Let $f_1, f_2 \in U' \cap \sigma(\mathcal{O}_{\mathbb{K}})$. Set

$$A = \{v \in S_\infty \mid \varphi_v(f_1) = \varphi_v(f_2)\},$$
$$B = \{v \in S_f \mid \varphi_v(f_1) = \varphi_v(f_2)\};$$

then for any $v \in S_\infty$ we have

$$|f_1 - f_2|_v \le r^{a/N}$$

and

$$|f_1 - f_2|_v \le \frac{r^{a/N}}{q}$$

for $v \in A$. Moreover,

$$f_1 - f_2 \in v^{c(v)}$$

for any $v \in B$, and $N(v^{c(v)}) \ge r$. For $\alpha = |A|$ and $\beta = |B|$, we have

$$r^\beta \le N_{\mathbb{K}/\mathbb{Q}}(f_1 - f_2) < \frac{r^a}{q^\alpha} \le r^{a-\alpha},$$

and thus $\alpha + \beta < a$; i.e., $d > n - a$.

(b) We see that if $a \le n$, then φ_S is an embedding, and

$$|U' \cap \sigma(\mathcal{O}_{\mathbb{K}})| \ge r^a / \sqrt{|D_{\mathbb{K}}|}.$$ □

Proposition 10.3.6 is also valid for $t \ne 0$, but the proof is more involved.

Asymptotic behaviour. Fix q and r and take a family of fields $\mathbb{K}$ of growing degree n. Let

$$\gamma = \liminf_{\mathbb{K}} \frac{\log_q \sqrt{|D_{\mathbb{K}}|}}{n},$$

where $n = s + t + |S_f|$. Then we obtain a family of nonlinear codes with $n \to \infty$ and

$$R \gtrsim (1-\delta)\log_q r - \gamma;$$

for $\gamma < \log_q r$, we get asymptotically good codes.

THEOREM 10.3.7. *A family of number fields K of growing degree with*

$$\lim \frac{\log_q \sqrt{|D_{\mathbb{K}}|}}{n} = \gamma < \log_q r,$$

where $n = s + t + |S_f|$ and S_f is the set of non-Archimedean places v such that for some $c(v) \in \mathbb{Z}$ we have $r \le N_{\mathbb{K}/\mathbb{Q}}(v)^{c(v)} \le q$ (for r and q fixed), yields a family of asymptotically good nonlinear q-ary codes with any given δ and

$$R \gtrsim (1-\delta)\log_q r - \gamma.$$

Under the GRH, there exist fields with

$$\gamma = \text{const} \cdot \frac{\log q}{q^{1/4}}.$$ □

It is not difficult to see that for large q on some segment of the δ-axis these codes improve upon the Gilbert–Varshamov bound.

10.3.2. Multiplicative Constructions

Up to this moment, we have used additive groups of global fields. Now we are going to consider their multiplicative structure.

Multiplicative number field lattices (NML). We start with a number field $\mathbb{K}$ of degree $N = s+2t$ and a finite number of its places $S = S_\infty \cup S_f$, which includes all Archimedean places. Let $n = |S|$, and let $\mathcal{O}_S^*$ be the set of S-units.

There is a natural map

$$\varphi_S\colon \mathcal{O}_S^* \to \mathbb{R}^n, \quad f \mapsto \{\ln \|f\|_v\},$$

where $v \in S$ and $\|\cdot\|_v$ is the normalized absolute value, i.e., $\|f\|_v = |\sigma_v(f)|$ for real places, $\|f\|_v = |\sigma_v(f)|^2$ for complex places, and $\|f\|_v = N(v)^{-\operatorname{ord}_v(f)}$ for $v \in S$. It is clear that

$$\operatorname{Ker}\varphi_S = \mu_{\mathbb{K}}$$

is the group of roots of 1 in $\mathbb{K}$ and that

$$\operatorname{Im}\varphi_S \subset H = \{x \in \mathbb{R}^n \mid \textstyle\sum x_i = 0\}$$

due to the product formula.

Parameters. Let R be the regulator of $\mathbb{K}$, and let $h = h_{\mathbb{K}}$ be its class number. Set $\operatorname{ht}(f) = \sum\limits_v |\ln \|f\|_v|$ for $f \in \mathbb{K}^*$. This is the *height function*, and $\operatorname{ht}(f) = 0$ precisely for $f \in \mu_{\mathbb{K}}$. Set

$$\operatorname{ht}(\mathbb{K}) = \min_{f \in \mathbb{K}^* \setminus \mu_{\mathbb{K}}} \operatorname{ht}(a)$$

and call it the *height of the field* $\mathbb{K}$.

PROPOSITION 10.3.8. *Let $L_S = \varphi_S(\mathcal{O}_S^*)$. Then*

(a) $d(L_S) \geq \dfrac{1}{\sqrt{n}} \operatorname{ht}(\mathbb{K})$;

(b) $\operatorname{rk} L_S = n-1$, *and*

$$\det L_S \leq \sqrt{n} R h \prod_{v \in S_f} \ln N(v).$$

PROOF. (a) is obvious, since

$$\sqrt{\sum_{i=1}^n x_i^2} \geq \frac{1}{\sqrt{n}} \sum_{i=1}^n |x_i|.$$

(b) Let the first coordinates in $\mathbb{R}^n$ correspond to $v \in S_\infty$. Consider the orthogonal projection

$$T\colon H \to H_0 = \left\{ x \in \mathbb{R}^n \,\middle|\, \sum_{i=1}^{s+t} x_i = 0 \right\} = H_1 \oplus \mathbb{R}^{n-s-t},$$

where

$$H_1 = \{x \in \mathbb{R}^{s+t} \mid \textstyle\sum x_i = 0\};$$

note that T multiplies volumes by $\sqrt{s+t}/\sqrt{n}$. Since $H_1 \cap T(L_S)$ is the lattice of units, we get

$$\det(T(L_S) \cap H_1) = \sqrt{s+t}\,R,$$

and thus

$$\det T(L_S) = \sqrt{s+t}\,R \det(\mathrm{pr}_2 T(L_S)).$$

Using the obvious inequality

$$\left[\sum_{v \in S_f} \mathbb{Z} \cdot \ln N(v) : \mathrm{pr}_2 T(L_S)\right] \le h,$$

we get the conclusion. $\square$

Asymptotic behaviour. To obtain good families of lattices, we take unramified towers of number fields; assume for simplicity that the towers are totally real.

THEOREM 10.3.9. *If a number field $\mathbb{K}_0$ of degree n_0 has an infinite unramified tower of totally real fields, then the above construction with $S = S_\infty$ yields a family of asymptotically good multiplicative lattices $\{L_N = L_S \subset \mathbb{R}^N\}$ as $N \to \infty$ and*

$$\lambda(\{L_N\}) \le -\log\sqrt{\frac{\pi^3 e}{2}} - \log\ln\left(\frac{1+\sqrt{5}}{2}\right) + \frac{1}{n_0}\log|D_{\mathbb{K}_0}|. \qquad \square$$

If we take $K_0 = \mathbb{Q}(\sqrt{2}, \sqrt{3 \cdot 5 \cdot 7 \cdot 23 \cdot 29})$, we get $\lambda \lesssim 8.41$.

Function field lattices (FML). A direct function field analogue of the above multiplicative construction is as follows.

Construction. Let $\mathbb{K} = \mathbb{F}_q(X)$, and let

$$\mathcal{O}_{\mathcal{P}}^* = \{f \in \mathbb{K}^* \mid \mathrm{Supp}(f) \subseteq \mathcal{P} \subseteq X(\mathbb{F}_q)\}$$

for a curve X over $\mathbb{F}_q$. Let $\mathrm{Div}_{\mathcal{P}}(X)$ denote the group of divisors supported in $\mathcal{P}$, $\mathrm{Div}_{\mathcal{P}}^0(X)$ the group of those of degree 0, and $P_{\mathcal{P}}(X)$ the subgroup of principal divisors; then $J_X = \mathrm{Div}^0(X)/P(X)$ is the Jacobian of X.

We have a natural map

$$\varphi_{\mathcal{P}} \colon \mathcal{O}_{\mathcal{P}}^* \to \mathrm{Div}_{\mathcal{P}}(X) \cong \mathbb{Z}^n, \quad f \mapsto (f),$$

$\mathrm{Ker}\,\varphi_{\mathcal{P}} = \mathbb{F}_q^*$, and

$$\mathrm{Im}\,\varphi_{\mathcal{P}} \subseteq \mathrm{Div}_{\mathcal{P}}^0(X) \cong A_{n-1} = \{x \in \mathbb{Z}^n \mid \textstyle\sum x_i = 0\}.$$

Now define a sublattice $L_{\mathcal{P}}$ in $\mathbb{R}^{n-1}$ by

$$L_{\mathcal{P}} = \varphi_{\mathcal{P}}(\mathcal{O}_{\mathcal{P}}^*) \subseteq A_{n-1} \otimes \mathbb{R} \cong \mathbb{R}^{n-1}.$$

Parameters. The parameters are estimated as follows.

THEOREM 10.3.10. *We have*

(a) $d(L_{\mathcal{P}}) \ge \min\limits_{f \in \mathcal{O}_{\mathcal{P}}^* \setminus \mathbb{F}_q^*} \sqrt{2\deg f} \ge \sqrt{\dfrac{2|X(\mathbb{F}_q)|}{q+1}}$;

(b) $\mathrm{rk}\, L_{\mathcal{P}} = n-1$ *and*

$$\det L_{\mathcal{P}} \le \sqrt{n}\,|J_X(\mathbb{F}_q)| \le \sqrt{n}\left(1 + q + \frac{|X(\mathbb{F}_q)| - q - 1}{g}\right)^g.$$

PROOF. (a) Let $f \in \mathcal{O}_{\mathcal{P}}^*$, $f \notin \mathbb{F}_q^*$, and let

$$\varphi_{\mathcal{P}}(f) = (x_1, \dots, x_n) \in \mathbb{Z}^n.$$

Then

$$|\varphi_{\mathcal{P}}(f)| = \sqrt{\sum x_i^2} \geq \sqrt{\sum |x_i|} = \sqrt{2 \deg f},$$

since $x_i \in \mathbb{Z}$, $\sum x_i = 0$, and $\deg f = \sum_{x_i > 0} x_i$. Since f maps $\mathbb{F}_q$-points to $\mathbb{F}_q$-points of $\mathbb{P}^1$, we obtain the second inequality,

$$|X(\mathbb{F}_q)| \leq (q+1) \deg f.$$

(b) Since $\det A_{n-1} = \sqrt{n}$, we get

$$\det L_{\mathcal{P}} = [A_{n-1} : L_{\mathcal{P}}] \det A_{n-1};$$

moreover,

$$A_{n-1} \cong \operatorname{Div}_{\mathcal{P}}^0(X) \subset \operatorname{Div}^0(X),$$

while

$$L_{\mathcal{P}} \cong P_{\mathcal{P}}(X) = P(X) \cap \operatorname{Div}_{\mathcal{P}}^0(X).$$

Thus,

$$[A_{n-1} : L_{\mathcal{P}}] \leq [\operatorname{Div}^0(X) : P(X)] = |J_X(\mathbb{F}_q)|.$$

Then we note that

$$|J_X(\mathbb{F}_q)| \leq \left(1 + q + \frac{|X(\mathbb{F}_q)| - q - 1}{g}\right)^g. \tag{10.3.1}$$

Indeed, $|J_X(\mathbb{F}_q)| = \prod_{i=1}^{2g} (1 - \omega_i)$, where ω_i are Frobenius roots, $|\omega_i| = \sqrt{q}$, $\omega_{g+i} = \overline{\omega}_i$, and the AM–GM inequality yields

$$\prod_{i=1}^{2g} (1 - \omega_i) = \prod_{i=1}^{g} (q + 1 - \omega_i - \overline{\omega}_i) \leq \left(\frac{\sum_{i=1}^{g} (q + 1 - \omega_i - \overline{\omega}_i)}{g}\right)^g,$$

which implies the formula, since

$$-\sum_{i=1}^{g} (\omega_i + \overline{\omega}_i) = |X(\mathbb{F}_q)| - q - 1. \qquad \square$$

Asymptotic behaviour. Now take a family of curves with

$$\frac{|X(\mathbb{F}_q)|}{g} \to A \quad \text{as} \quad g = g(X) \to \infty$$

and set $\mathcal{P} = |X(\mathbb{F}_q)|$. We obtain the following result.

THEOREM 10.3.11. *A family of curves X over $\mathbb{F}_q$, $q = p^{2m}$, of growing genus g such that*

$$\frac{|X(\mathbb{F}_q)|}{g} \to A > 0$$

yields an asymptotically good family of lattices $\{L_N \subset \mathbb{R}^N\}$ with

$$\lambda(\{L_N\}) \leq -\log \sqrt{\pi e} + \log \sqrt{q+1} + A^{-1} \log(1 + q + A). \qquad \square$$

For $q = p^{2m}$ and $A = A(q) = \sqrt{q} - 1$ we can slightly improve the above bound as follows.

PROPOSITION 10.3.12. *For a family of curves X over $\mathbb{F}_q$ with*

$$\frac{|X(\mathbb{F}_q)|}{g} \to \sqrt{q} - 1$$

we have the asymptotic equality

$$\frac{1}{g} \log |J_X(\mathbb{F}_q)| \sim \log q + (\sqrt{q} - 1) \log \frac{q}{q-1}. \qquad \square$$

Therefore, we obtain the following.

THEOREM 10.3.13. *A family of curves X over $\mathbb{F}_q$ of growing genus g such that*

$$\frac{|X(\mathbb{F}_q)|}{g} \to \sqrt{q} - 1$$

yields an asymptotically good family of lattices $\{L_n \subset \mathbb{R}^N\}$ with

$$\lambda(\{L_N\}) \le -\log \sqrt{\pi e} + \log \frac{\sqrt{q}+1}{q-1} + \frac{\sqrt{q}}{\sqrt{q}-1} \log q. \qquad \square$$

For $q = 9$, we get $\lambda \lesssim 1.8687\ldots$.

REMARK 10.3.14. One can give an exponential construction algorithm for these lattices, but a polynomial one is not known.

REMARK 10.3.15. For an algebraically closed ground field $\Bbbk$ of characteristic 0, it is known that $\deg f \ge g/2$ for the "general" curve X and any $f \in \Bbbk(X)$. If this were the case for modular curves over $\mathbb{F}_q$, we would get the estimate

$$\lambda(\{L_N\}) \le -\log \sqrt{\frac{\pi e}{2}} + \log \frac{\sqrt{\sqrt{q}+1}}{q-1} + \frac{\sqrt{q}}{\sqrt{q}-1} \log q,$$

which would give $\lambda(\{L_N\}) \lesssim 1.18$ for $q = 16$.

10.3.3. Congruence Constructions

Now we shall discuss some constructions depending on a divisor.

Multiplicative congruence sublattices (FMLD). The FML construction can be slightly improved by using an auxiliary divisor.

Let D be a positive divisor on X, $D = \sum a_i P_i$, $r_i = \deg P_i$, $N(P_i) = q^{r_i}$, and

$$a = \deg D = \sum a_i r_i.$$

Assume that $\mathcal{P} \cap \operatorname{Supp} D = \varnothing$. Let

$$\mathcal{O}^*_{\mathcal{P},D} = \{f \in \mathcal{O}^*_{\mathcal{P}} \mid f \equiv 1 \pmod{D}\}$$

and consider the lattice $L_{\mathcal{P},D} = \varphi_{\mathcal{P}}(\mathcal{O}^*_{\mathcal{P},D}) \subseteq L_{\mathcal{P}}$.

Parameters. The estimates are as follows.

PROPOSITION 10.3.16. *Let $L_{\mathcal{P},D} = \varphi_{\mathcal{P}}(\mathcal{O}^*_{\mathcal{P},D})$. Then*

(a) $d(L_{\mathcal{P},D}) \ge \sqrt{2a}$;

(b) $\operatorname{rk} L_{\mathcal{P},D} = n-1$ *and*

$$\det L_{\mathcal{P},D} \leq \sqrt{n}\,|J_X(\mathbb{F}_q)|\frac{q^a}{q-1}\prod\left(1-q^{-r_i}\right).$$

SKETCH OF THE PROOF. (a) We use the first inequality of Theorem 10.3.10(a), which in our case reads

$$d(L_{\mathcal{P}}, D) \geq \min_{f\in\mathcal{O}^*_{\mathcal{P},D}\setminus\{1\}} \sqrt{2\deg f},$$

and note that $\deg f = \deg(f-1) \geq \deg D = a$.

(b) Theorem 10.3.10(b) estimates the value of $\det L_{\mathcal{P}}$, and we only have to estimate $[L_{\mathcal{P}} : L_{\mathcal{P},D}]$. A simple local calculation shows that

$$[\mathcal{O}^*_{\mathcal{P}} : \mathcal{O}^*_{\mathcal{P},D}] = (q-1)[L_{\mathcal{P}} : L_{\mathcal{P},D}]. \qquad \square$$

Asymptotic behaviour. Consider the same family of curves as in Sec. 10.3.2, let $\mathcal{P} = X(\mathbb{F}_q)$, and let D be such that

$$\lim \frac{\deg D}{|X(\mathbb{F}_q)|} = (2\ln q)^{-1}.$$

Then we get the following.

THEOREM 10.3.17. *A family of curves X over $\mathbb{F}_q$ of growing genus g such that*

$$\frac{|X(\mathbb{F}_q)|}{g} \to \sqrt{q}-1$$

with an appropriate choice of divisors yields an asymptotically good family of lattices $\{L_N \subset \mathbb{R}^N\}$ with

$$\lambda(\{L_N\}) \leq -\log\sqrt{\frac{\pi}{2}} + \frac{1}{2}\log(\ln q) + \frac{\sqrt{q}}{\sqrt{q}-1}\log q - \log(q-1). \qquad \square$$

For $q = 2209 = 47^2$ we get $\lambda \lesssim 1.3888\ldots$.

REMARK 10.3.18. Since the obtained value of $\lambda \approx 1.3888$ is not that far from 1 and since, according to Remark 10.1.11, "explicit" lattices L_N in $\mathbb{R}^N$ for $N \geq 2^{10}$ have $\lambda(L_N) > 1$, one can ask for a value of N where the FMDL lattices begin to supersede other explicit lattices. For instance, one can compare the FMLD lattices (with $q = 32^2$ or 64^2) with Elkies' Mordell–Weil lattices MW_{2^m} from Sec. 10.4.2 below. A straightforward calculation shows that the FMLD lattices supersede the MW_{2^m} lattices for $m \geq 15$, i.e., $N \geq 2^{15} = 32768$. However, for such values of N the MW_{2^m} lattices are not the best known. In fact, the FMLD lattices begin to dominate over other "explicit" constructions for N of the order of approximately $2^{30} \approx 10^9$.

Number field congruence sublattices (NMLD). Now we turn back to the number field case and discuss an analogue of FMLD lattices.

Let $\mathfrak{a} \subset \mathcal{O}_{\mathbb{K}}$ be an ideal

$$\mathfrak{a} = \prod \mathfrak{p}_i^{a_i}$$

such that $\mathfrak{p}_i \notin S_f$ (i.e., $S_f \cap \operatorname{Supp}\mathfrak{a} = \varnothing$). Let

$$\mathcal{O}^*_{S,\mathfrak{a}} = \{f \in \mathcal{O}^*_S \mid f \equiv 1 \pmod{\mathfrak{a}}\},$$

and consider the lattice $L_{S,\mathfrak{a}} = \varphi_S(\mathcal{O}^*_{S,\mathfrak{a}}) \subseteq L_S$.

Parameters. Everything is quite similar to the function field case, although the estimates are worse. Let $\mu_{\mathfrak{a}} = \mu_{\mathbb{K}} \cap \mathcal{O}^*_{S,\mathfrak{a}}$.

PROPOSITION 10.3.19. *Let* $L_{S,\mathfrak{a}} = \varphi_S(\mathcal{O}^*_{S,\mathfrak{a}})$. *Then*

(a) $d(L_{S,\mathfrak{a}}) \geq \dfrac{2}{\sqrt{n}}\big(\ln N_{\mathbb{K}/\mathbb{Q}}(\mathfrak{a}) - (s+2t)\ln 2\big)$;

(b) $\operatorname{rk} L_{S,\mathfrak{a}} = n-1$ *and*

$$\begin{aligned}\det L_{S,\mathfrak{a}} &\leq (\det L_S) N_{\mathbb{K}/\mathbb{Q}}(\mathfrak{a}) \left(\prod_i \left(1 - \frac{1}{N_{\mathbb{K}/\mathbb{Q}}(\mathfrak{p}_i)}\right)\right) [\mu_{\mathbb{K}} : \mu_{\mathfrak{a}}]^{-1} \\ &\leq \sqrt{(s+t)(n-s-t)}\, Rh \left(\prod_{v \in S_f} \ln N(v)\right) \\ &\qquad \times N_{\mathbb{K}/\mathbb{Q}}(\mathfrak{a}) \left(\prod_i \left(1 - \frac{1}{N_{\mathbb{K}/\mathbb{Q}}(\mathfrak{p}_i)}\right)\right) [\mu_{\mathbb{K}} : \mu_{\mathfrak{a}}]^{-1}.\end{aligned}$$

PROOF. (a) Let

$$\varphi_S(f) = (x_1, \dots, x_n), \quad x_i = \ln \|f\|_{v_i};$$

then

$$|\varphi_S(f)| = \sqrt{\sum x_i^2} \geq \frac{1}{\sqrt{n}} \sum |x_i| = \frac{2}{\sqrt{n}} \sum_{x_i > 0} x_i.$$

Now we have

$$\begin{aligned}\sum_{x_i>0} x_i &= \sum_{\substack{v \in S \\ \|f\|_v > 1}} \ln \|f\|_v \\ &= \sum_{\substack{v \in S_\infty \\ \|f\|_v > 1}} \ln \|f\|_v + \sum_{\substack{v \in S_f \\ \operatorname{ord}_v(f-1) < 0}} (-\operatorname{ord}_v(f-1)) \ln N(v) \\ &= \sum_{\substack{v \in S_\infty \\ \|f\|_v > 1}} \ln \|f\|_v - \sum_{\substack{v \in S_\infty \\ \|f-1\|_v > 1}} \ln \|f-1\|_v \\ &\qquad + \sum_{\substack{v \in S_\infty \\ \|f-1\|_v < 1}} (-\ln \|f-1\|_v) + \sum_{\substack{v \notin S_\infty \\ \|f-1\|_v < 1}} (\operatorname{ord}_v(f-1)) \ln N(v).\end{aligned}$$

We discard the nonnegative third term; the fourth term is at least $\ln N_{\mathbb{K}/\mathbb{Q}}(\mathfrak{a})$, since $f-1 \in \mathfrak{a}$; and the sum of the first two terms is at least $-\sum_{v \in S_\infty} \ln \|2\|_v$, since

$$\max\{0, \ln \|z\|\} - \max\{0, \ln \|1-z\|\}$$

is minimal when $z = -1$, both for $z \in \mathbb{R}$ and $z \in \mathbb{C}$.

(b) According to Proposition 10.3.8(a), we only have to estimate $[L_S : L_{S,\mathfrak{a}}]$. Note that $\operatorname{Ker} \varphi_S = \mu_{\mathbb{K}}$, $\operatorname{Ker} \varphi_S \cap \mathcal{O}^*_{S,\mathfrak{a}} = \mu_{\mathfrak{a}}$, and proceed as in the proof of Proposition 10.3.16(b). □

Asymptotic behaviour. The proof of Proposition 10.3.19(a) shows also that if $\log N_{\mathbb{K}/\mathbb{Q}}(\mathfrak{a}) > s + 2t$, then $\mu_{\mathfrak{a}} = \{1\}$, since $\varphi_S(f) = 0$ for $f \in \mu_{\mathfrak{a}}$. We choose $\mathfrak{a}$ in such a way that

$$\frac{1}{n}\log N_{\mathbb{K}/\mathbb{Q}}(\mathfrak{a}) \sim \frac{s+2t}{n} + \log e,$$

and $\mathbb{K}$ runs over an unramified tower. For $Rh/|\mu_{\mathbb{K}}|$ we use standard estimates for the residue of the zeta function of $\mathbb{K}$; cf. Sec. 8.2.1. They give

$$\frac{Rh}{|\mu_{\mathbb{K}}|} \leq \frac{2|D_{\mathbb{K}}|}{\pi^{s+2t}}.$$

Choosing S in such a way that $\dfrac{s+2t}{n}$ is constant in the tower and S and $\operatorname{Supp}\mathfrak{a}$ completely split in it, we obtain the following result.

THEOREM 10.3.20. *If a number field $\mathbb{K}_0$ of degree N_0 has an infinite unramified tower in which sets of its places S_0 and $\operatorname{Supp}\mathfrak{a}_0$ split completely, then it yields an asymptotically good family of lattices $\{L_N \subset \mathbb{R}^N\}$ with*

$$\lambda(\{L_N\}) \leq -\log\sqrt{\frac{2\pi}{e}} + \frac{1}{n_0}\log|D_{\mathbb{K}_0}| - \frac{N_0}{n_0}(\log\pi - 1) + \frac{1}{n_0}\sum_{v\in S_{0f}}\log(\ln N(v)) + \frac{1}{n_0}\sum_{\mathfrak{p}_i|\mathfrak{a}}\log\left(1 - \frac{1}{N_{\mathbb{K}_0/\mathbb{Q}}(\mathfrak{p}_i)}\right),$$

where $n_0 = |S_0|$. □

For $\mathbb{K}_0 = \mathbb{Q}\Big(\cos\frac{2\pi}{11}, \sqrt{-46}\Big)$ and $S_0 = S_\infty$ we get $\lambda \lesssim 11.1512\ldots$; for $\mathbb{K}_0 = \mathbb{Q}(\sqrt{2}, \sqrt{3\cdot 5\cdot 7\cdot 23\cdot 29})$ and $S_0 = S_\infty$ we get $\lambda \lesssim 8.80$. These are not the best choices, but what we get is always much worse than in the FMLD case.

Congruence codes (FMCD). Assume that $D = pD'$ for some positive divisor D', where $q = p^m$.

Let

$$C = L_{\mathcal{P},D}/((p\mathbb{Z})^n \cap L_{\mathcal{P},D}) \subseteq (\mathbb{Z}/p)^n = \mathbb{F}_p^n.$$

This is a p-ary code of length n. To study C, we have to use the Ω-construction. Let $C' = (X, \mathcal{P}, D)_\Omega$, and thus $C' = \operatorname{Im}(\operatorname{Res}_{\mathcal{P}})$, where

$$\operatorname{Res}_{\mathcal{P}}\colon \Omega\big(\textstyle\sum P_i - D\big) \to \mathbb{F}_q^n, \quad \omega \mapsto (\operatorname{Res}_{P_1}(\omega), \ldots \operatorname{Res}_{P_n}(\omega)).$$

Now we are ready to estimate parameters.

PROPOSITION 10.3.21. *Let $D = pD' \geq 0$, and let*

$$C = L_{\mathcal{P},D}/((p\mathbb{Z})^n \cap L_{\mathcal{P},D}) = C' \cap \mathbb{F}_p^n \subseteq \mathbb{F}_p^n.$$

Then

(a) $d \geq d' \geq a - 2g + 2$;

(b) $k \geq n - 1 - m\dfrac{p-1}{p}a$.

PROOF. (a) The first inequality is obvious; the second is proved in Sec. 4.1.1.
(b) Let $B = L_{\mathcal{P},D} \cap (p\mathbb{Z})^n$. Then

$$\begin{aligned} p^k = |C| &= [L_{\mathcal{P},D} : B] \\ &= [A_{n-1} : A_{n-1} \cap (p\mathbb{Z})^n] \cdot [A_{n-1} \cap (p\mathbb{Z})^n : B] \cdot [A_{n-1} : L_{\mathcal{P},D}]^{-1}, \end{aligned}$$

and we have $[A_{n-1} : A_{n-1} \cap (p\mathbb{Z})^n] = p_{n-1}$. Multiplication by p sends A_{n-1} onto $A_{n-1} \cap (p\mathbb{Z})^n$ and $L_{\mathcal{P},D'}$ onto B, and thus

$$\begin{aligned}[A_{n-1} \cap (p\mathbb{Z})^n : B] \cdot [A_{n-1} : L_{\mathcal{P},D}]^{-1} &= [A_{n-1} : L_{\mathcal{P},D'}] \cdot [A_{n-1} : L_{\mathcal{P},D}]^{-1} \\ &= [L_{\mathcal{P},D'} : L_{\mathcal{P},D}]^{-1}.\end{aligned}$$

Proceeding as in the proof of Proposition 10.3.16(b), we see that

$$[L_{\mathcal{P},D'} : L_{\mathcal{P},D}] \leq q^{\deg D - \deg D'} = p^{m \frac{p-1}{p} a}.$$

Summing this up, we get

$$k \geq n - 1 - m \frac{p-1}{p} a. \qquad \square$$

Asymptotic behaviour. As usual, the best results are obtained for

$$\frac{|X(\mathbb{F}_q)|}{g} \to A,$$

where $A = \sqrt{q} - 1$.

THEOREM 10.3.22. *A family of curves X over $\mathbb{F}_q$, $q = p^m$, of growing genus g such that*

$$\frac{|X(\mathbb{F}_{p^m})|}{g} \to A > \frac{2m(p-1)}{p}$$

yields a family of codes such that for any

$$\delta < \frac{p}{m(p-1)} - 2A^{-1}$$

there is an asymptotically good subfamily with parameters δ and

$$R \gtrsim 1 - m \frac{p-1}{p} (2A^{-1} + \delta). \qquad \square$$

One easily checks that this result is a particular case of Theorem 4.5.29, but we do not know whether a similar construction can be given for all codes from that theorem.

10.4. Mordell–Weil Lattices

In this section we consider another class of lattices, naturally associated with algebraic geometry objects, namely the Mordell–Weil lattices of abelian varieties over global fields. They are close analogues to the lattices from Section 10.3.2, but their study requires much more elaborate algebraic geometry and number theory means. In fact, below we only consider a special case of elliptic curves.

Thus, in this section, let $\mathbb{K}$ denote a global field, and $E/\mathbb{K}$ an elliptic curve over $\mathbb{K}$. All further constructions of Mordell–Weil lattices are based on the following famous Mordell–Weil theorem.

THEOREM 10.4.1 (Mordell–Weil). *The group $E(\mathbb{K})$ is finitely generated.* □

Thus, $E(\mathbb{K}) = L \oplus T$, where L is a free abelian group of rank N (the *rank* of E over $\mathbb{K}$) and T is a finite abelian group of torsion points. Moreover, L is equipped with a canonical positive definite quadratic form, called the *canonical height*, which describes the "magnitude" of a point $P \in E(\mathbb{K})$. The canonical height is defined as a suitable normalization of the "naive" height on $E(\mathbb{K})$. In the most simple cases $\mathbb{K} = \mathbb{Q}$ or $\mathbb{K} = \mathbb{F}_q(t)$, the latter can be defined as follows.

In the case of $\mathbb{K} = \mathbb{Q}$,

$$E: \ y^2 + a_1xy + a_3y = x^3 + a_2x^2 + a_4x + a_6$$

being an elliptic curve in its Weierstrass form, one defines the "naive" height of a point $P = (x_0, y_0) \in E(\mathbb{Q})$ simply as

$$h_n(P) = \max\{\log|p|, \log|q|, \log|r|, \log|s|\},$$

where $x_0 = p/q$ and $y_0 = r/s$ are fractions in lowest terms. If $\mathbb{K} = \mathbb{F}_q(t)$, then the "naive" height is

$$h_n(P) = \max\{\deg p, \deg q, \deg r, \deg s\}.$$

A fundamental discovery of John Tate was that the function h_n is approximately quadratic. More precisely, one has the following fundamental result.

THEOREM 10.4.2 (Néron–Tate). *There exists a unique quadratic form $\hat{h}$: $E(\mathbb{K}) \to \mathbb{R}_+$ which is positive definite on L and such that $|\hat{h}(P) - h_n(P)|$ is a bounded function on the entire $E(\mathbb{K})$. It can be constructed as*

$$\hat{h}(P) := \lim_{n\to\infty} 2^{-2n} h_n(2^n P).$$ □

The Néron–Tate form $\hat{h}(P)$ gives rise to a positive definite form $\hat{h}(P) \otimes \mathbb{R}$ on the real vector space $E(\mathbb{K}) \otimes \mathbb{R} = \mathbb{R}^N$ and thus defines the lattice $L \subset L \otimes \mathbb{R} = \mathbb{R}^N$.

In contrast to the case of characteristic 0, where these lattices are not very well understood, for $\mathbb{K} = \mathbb{F}_q(t)$ there is a quite satisfactory theory leading to new constructions of many classical lattices, as well as other very good lattices, in dimensions up to $N \leq 2^{12} = 4096$. Note, however, that these lattices become asymptotically bad as $N \to \infty$.

Parameters of these lattices in the function field case can be estimated in two alternative ways. The first possibility is to associate with a curve E an algebraic surface over a finite field and use rich theory of such surfaces. The second is to work directly with arithmetical invariants of elliptic curves over function fields, which are also well studied. The first possibility is investigated in Sec. 10.4.1 in an odd characteristic; the second is used in Sec. 10.4.2 for the most interesting case of supersingular elliptic curves over a function field in characteristic 2.

10.4.1. Shioda Lattices

Elliptic curves over a function field lead to so-called elliptic surfaces.

Elliptic surfaces. Let E be an elliptic curve defined over $\mathbb{K}=\Bbbk(C)$, the function field of a smooth projective curve C over an algebraically closed ground field $\Bbbk$. One can associate with $E/\mathbb{K}$ an algebraic surface S, called the Kodaira–Néron model of $E/\mathbb{K}$, which can be characterized by the following properties:

1. There exists a canonical morphism $f\colon S\to C$ such that all fibres of f, except for finitely many *exceptional fibres*, are elliptic curves and $\Bbbk(S)=\mathbb{K}(E)$.
2. Singular fibres are connected curves of arithmetic genus 1, and S is a minimal surface, i.e., does not contain exceptional curves with self-intersection index $(L^2)=(L,L)=-1$.

We always assume that f is *not* smooth and that f admits a section, which is taken as the zero section O. Then there is a natural symmetric bilinear pairing $\langle\cdot,\cdot\rangle$ on $E(\mathbb{K})$, whose kernel is the torsion part. The pairing is defined via the map $\varphi\colon E(\mathbb{K})\to \mathrm{NS}(S)_{\mathbb{Q}}$, the Néron–Severi group of the surface S, as the opposite of the intersection form: $\langle P,Q\rangle=-(\varphi(P)\cdot\varphi(Q))$, that is, $(\varphi(P)\cdot\varphi(Q))$ is the general number of points in the intersection $\varphi(P)\cap\varphi(Q)$ counted with multiplicities. The map φ is characterized by the two properties: $\varphi(P)\equiv(P) \bmod T_{\mathbb{Q}}$ and $\mathrm{Im}(\varphi)\perp T$, where (P) denotes the divisor of the section corresponding to $P\in E(\mathbb{K})$ and T denotes the "trivial sublattice" of $\mathrm{NS}(S)$ generated by (O) and all the irreducible components of fibres of f ($T_{\mathbb{Q}}=T\otimes\mathbb{Q}$).

Actually, $\langle P,Q\rangle$ coincides with the bilinear form associated with $\hat{h}$, but we do not need this fact and will work with φ, which is simpler in many respects.

Let $E(\mathbb{K})^0$ be the subgroup of finite index, say h, in $E(\mathbb{K})$ consisting of those P which pass the same irreducible component of every fibre as the zero section. The pairing above defines the structure of a positive-definite lattice on $E(\mathbb{K})/(\mathrm{tor})$, which is the *Mordell–Weil lattice* of $E/\mathbb{K}$, and the structure of a positive-definite even integral lattice on $E(\mathbb{K})^0$, called the *narrow Mordell–Weil lattice* of $E/\mathbb{K}$. The latter is mapped via φ isomorphically onto $T^{\perp}$, the orthogonal complement of T in $\mathrm{NS}(S)$.

Now let us consider $L=E(\mathbb{K})^0$. Below we neglect the sign of determinant, i.e., det stands for $|\det|$.

PROPOSITION 10.4.3. (1) *The rank r of $E(\mathbb{K})$ is related to $\rho=\mathrm{rk}\,\mathrm{NS}(S)$ by the formula*

$$\rho=r+\mathrm{rk}\,T=r+2+\sum_{v\in R}(m_v-1),$$

where R is the set of reducible fibres $f^{-1}(v)$ and m_v denotes the number of irreducible components in $f^{-1}(v)$.

(2) *The determinant of L is given by*

$$\det L=\det\mathrm{NS}(S)/\det T. \qquad \square$$

COROLLARY 10.4.4. *In particular, if* $\det T = 1$, *then we have*

$$\det L = \det \mathrm{NS}(S). \qquad \square$$

PROPOSITION 10.4.5. *The center density* δ *of* $L = E(\mathbb{K})^0$ *satisfies*

$$\delta \geq \frac{(\chi/2)^{r/2}}{h\sqrt{\det \mathrm{NS}(S)/\det T}},$$

where the equality holds if and only if there is at least one $P \in L$ *such that* (P) *is disjoint from* (O). *In this case, the kissing number* $\tau = \tau(L)$ *is equal to the number of such* $P \in L$.

SKETCH OF THE PROOF. For any $P \in L$, $P \neq O$, we have

$$\langle P, P \rangle = 2\chi + 2(PO),$$

where χ is the arithmetic genus of the surface S (a positive integer), $(PO) = ((P) \cdot (O))$ is the intersection number, and $(PO) \geq 0$. Thus, the minimal norm of L is at least 2χ. $\square$

Numerical invariants of the elliptic surface. The above general formulae can be efficiently applied to the elliptic surface $S = S_m$ defined by

$$y^2 = x^3 + 1 + t^m \tag{10.4.1}$$

over an algebraically closed field $\Bbbk$ of any characteristic $\neq 2, 3$.

Applying known algorithms for calculating singular fibres, one can deduce the following description of T.

PROPOSITION 10.4.6. *The trivial sublattice* T *of* $\mathrm{NS}(S)$ *is*

$$T = \langle O, F \rangle \oplus \begin{cases} 0, & m \equiv 0,\ 5 \pmod 6, \\ E_8, & m \equiv 1 \pmod 6, \\ E_6, & m \equiv 2 \pmod 6, \\ D_4, & m \equiv 3 \pmod 6, \\ A_2, & m \equiv 4 \pmod 6, \end{cases}$$

where $\langle O, F \rangle$ *is the unimodular sublattice generated by the zero section* (O) *and the fibre class* F *and where* $E_8, \ldots, A_2$ *denote the root lattices. In particular, we have*

$$\mathrm{rk}\, T = 2,\ 10,\ 8,\ 6,\ 4,\ 2 \quad \textit{for} \quad m \equiv 0,\ 1,\ \ldots,\ 5 \pmod 6$$

and

$$\det T = \begin{cases} 1, & m \equiv 0,\ \pm 1 \pmod 6, \\ 3, & m \equiv 2,\ 4 \pmod 6, \\ 4, & m \equiv 3 \pmod 6. \end{cases} \tag{10.4.2}$$

$\square$

Using the theory of elliptic surfaces, one can deduce the following invariants of the surface S.

PROPOSITION 10.4.7. *For the elliptic surface $S = S_m$, we have*

$$c_2 = \text{the Euler number} = \begin{cases} 2m, & m \equiv 0 \pmod 6, \\ 12(\lfloor m/6 \rfloor + 1) & \text{otherwise}, \end{cases},$$

$$\chi = \text{the arithmetic genus} = c_2/12 = \begin{cases} m/6, & m \equiv 0 \pmod 6, \\ \lfloor m/6 \rfloor + 1 & \text{otherwise}, \end{cases}$$

$$p_g = \text{the geometric genus} = \chi - 1,$$

$$b_2 = \text{the second Betti number} = c_2 - 2.$$

□

Next, let $\lambda = c_2 - 2 - \rho \geq 0$ be the *Lefschetz number* of S.

COROLLARY 10.4.8. *For the surface $S = S_m$, we have*

$$r + \lambda = \begin{cases} 2m - 4, & m \equiv 0 \pmod 6, \\ 2m - 2 & \text{otherwise}. \end{cases}$$

□

REMARK 10.4.9. It is known that S_m is a rational surface (i.e., S is birationally equivalent to $\mathbb{P}^2_{\mathbb{k}}$; $\mathbb{k}(S) = \mathbb{k}(x,y)$) if and only if $m \leq 6$, and in that case we have $\lambda = 0$ for any characteristic. Hence, Corollary 10.4.8 implies that $r = 0$, 2, 4, 6, 8, 8 for $m = 1, 2, \ldots, 6$.

PROPOSITION 10.4.10. *The Mordell–Weil group $E(\mathbb{K})$ of* (10.4.1) *is torsion-free for $\mathbb{K} = \mathbb{k}(t)$.*

PROOF. In general, if the order of the torsion subgroup of $E(\mathbb{K})$ is n, $\det T$ must be divisible by n^2. In our case, it follows from (10.4.2) that $n = 1$ unless $m \equiv 3 \pmod 6$, in which case $n = 1$ or 2. If $n = 2$, then $x^3 + 1 + t^m$ must be reducible as a polynomial of x over $\mathbb{k}(t)$, which is impossible. Hence, $n = 1$ in all cases. □

Supersingular case. Now let us assume that $\mathbb{k}$ is not closed but is a finite field.

The most interesting case of S_m is the supersingular one. A surface is called *supersingular* if its Lefschetz number is $\lambda = 0$. A surface is called *unirational* if there is a dominant rational map from $\mathbb{P}^2$ to it or, equivalently, if $\mathbb{k}(S)$ can be embedded in $\mathbb{k}(x,y)$. It is known that a unirational surface is supersingular.

As an example, consider $X_d(p)$, the Fermat surface of degree d in characteristic p:

$$X_d(p): \; x_0^d + x_1^d + x_2^d + x_3^d = 0.$$

Then we have the following result.

PROPOSITION 10.4.11. *If $p^e \equiv -1 \pmod d$ for some $e > 0$, then $X_d(p)$ is unirational, and hence supersingular.*

PROOF. Indeed, the map $(x_0 : x_1 : x_2 : x_3) \mapsto (x_0^{p^e} : \ldots : x_3^{p^e})$ is a regular map of plane $\mathbb{P}^2 \subset \mathbb{P}^3$ onto $X_d(p)$. □

The elliptic surface $S = S_m$ considered above is related to the Fermat surface X_d of degree $d = 6m/(6,m)$. Indeed, putting

$$x = x_1^{d/3}, \qquad y = \sqrt{-1}\, x_2^{d/2}, \qquad t = x_3^{6/(6,m)}$$

with respect to the inhomogeneous coordinates x_1, x_2, x_3 ($x_0 = 1$), we obtain a dominant rational map of X_d to S.

Let us write $S_m(p)$ for the surface S_m in characteristic p.

PROPOSITION 10.4.12. *Given m, let $d = 6m/(6,m)$ and take a prime number p such that $p^e \equiv -1 \pmod{d}$ for some $e > 0$. Then the elliptic surface $S_m(p)$ is unirational, and the rank of the Mordell–Weil lattice r is given by*

$$r = \begin{cases} 2m-4, & m \equiv 0 \pmod{6}, \\ 2m-2 & \textit{otherwise}. \end{cases}$$

PROOF. The first assertion is clear from the above, and the second follows from Corollary 10.4.8, since $\lambda = 0$. □

PROPOSITION 10.4.13. *Under the same assumption as above, the determinant of the Néron–Severi lattice $\mathrm{NS}(S)$ for $S = S_m(p)$ is given by*

$$\det \mathrm{NS}(S) = q^{2p_g}/b \le q^{2p_g},$$

where $q = p^e$ with the smallest $e > 0$ such that $p^e \equiv -1 \pmod{d}$, and p_g is the geometric genus as in Proposition 10.4.7. The integer b is equal to 1 in the case $e = 1$, but, in general, b is a p-power which is determined combinatorially by p, e, and m. In particular, if $m = q+1$, $q = p^e$ with $p \equiv 5 \pmod{6}$ and e odd, then we have

$$b \ge p^{2(e-1)n_0}, \quad n_0 = \left(\frac{p-5}{6}\right)^e,$$

with equality in the case $e = 3$. □

Density and kissing numbers. Finally, we can collect all of the above information on the surfaces S_n and their Néron–Severi lattices and obtain the following results.

THEOREM 10.4.14. *Let $\mathbb{K} = \Bbbk(t)$, let $p \equiv -1 \pmod{6}$, and let E be the elliptic curve*

$$y^2 = x^3 + 1 + t^{p+1}.$$

Then the Mordell–Weil lattice $E(\mathbb{K})$ is an even integral lattice with the following invariants:

$$\text{rank } r = 2p-2,$$
$$\det = p^{2(u-1)}, \quad u = (p+1)/6,$$
$$\text{minimal norm} = 2u = (p+1)/3.$$

Its center density is

$$\delta = \frac{\left(\frac{p+1}{12}\right)^{p-1}}{p^{(p-5)/6}},$$

and its kissing number τ satisfies

$$\tau \ge 6p(p-1).$$ □

THEOREM 10.4.15. *Let m be any positive integer, and let $d = \mathrm{lcm}\{6,m\} = 6m/(6,m)$. Let $E/\mathbb{K}$ denote the elliptic curve*

$$y^2 = x^3 + 1 + t^m,$$

and let $L = E(\mathbb{K})^0$ denote the narrow Mordell–Weil lattice of E over $\mathbb{K}$. Assume that the characteristic of $\Bbbk$ is a prime number p such that $p^e \equiv -1 \pmod{d}$ for some $e > 0$. Let $q = p^e$ with the smallest e as above, and assume that $\Bbbk$ contains $\mathbb{F}_{q^2}$. Then L is an even integral lattice with the following invariants:

$$\operatorname{rk} L = r = \begin{cases} 2m-4, & m \equiv 0 \pmod 6, \\ 2m-2 & \textit{otherwise}, \end{cases}$$

$$\det L = (cq^{u-1})^2/b,$$

$$\text{minimal norm} \geq 2u,$$

with equality for the minimal norm at least if $\gcd(m,6) > 1$, where

$$u = \begin{cases} m/6, & m \equiv 0 \pmod 6, \\ \lfloor m/6 \rfloor + 1 & \textit{otherwise}, \end{cases}$$

$$c = \begin{cases} 1, & m = 0, \ \pm 1 \pmod 6, \\ 2, & m \equiv 3 \pmod 6, \\ \sqrt{3}, & m \equiv \pm 2 \pmod 6) \end{cases}$$

for some p-power b. In particular, b is equal to 1 in case $e = 1$, and if $m = p^e + 1$, then we have

$$b \geq p^{2(e-1)n_0} \quad \textit{with} \quad n_0 = \left(\frac{p-5}{6}\right)^e, \tag{10.4.3}$$

in which equality holds if $e = 3$.

Hence, the center density is

$$\delta \geq \frac{\left(\frac{u}{2}\right)^{r/2}}{cq^{u-1}},$$

and the right-hand side should be multiplied by $(q/p)^{n_0}$ if $m = q+1$. The asymptotic behaviour of the density Δ is given by

$$\frac{1}{r}\log_2 \Delta \sim -\frac{1}{12}\log_2 r \quad \textit{as} \quad r \to \infty. \qquad \square$$

In the table below we give some examples compared with the best lattices known prior to the construction:

m	p	r	$\log_2 \delta$	known before
42	41	80	40.14	36
54	53	104	67.01	60
129	257	256	273.68	270.89
255	509	508	745.62	742.66
258	257	512	796.88	698
261	521	520	770.37	767.46

10.4.2. Elkies Lattices

Now let us consider a very specific but important case of supersingular curves in characteristic 2 where one needs to apply a rather elaborate number-theoretic technique, replacing the surface theory employed in the preceding subsection. We begin with the following construction.

A hyperelliptic curve. Fix a positive integer n, let $q = 2^n$, and let $\mathbb{k}$ now be the finite field of q^2 elements. Then the elliptic curve $E_0/\mathbb{k}$ given by

$$\begin{cases} E_0\colon Y^2+Y=X^3, & n \text{ odd}, \\ E_0\colon Y^2+Y=X^3+a_6, & n \text{ even}, \end{cases} \tag{10.4.4}$$

where $\mathrm{tr}_{\mathbb{k}/\mathbb{F}_2}(a_6) = 1$, is a supersingular curve with a scalar Frobenius endomorphism, $F = -q$, and $\mathrm{tr}(F) = -2q$. Let $C/\mathbb{k}$ be the hyperelliptic curve with affine model

$$C\colon u^2+u=t^{q+1}. \tag{10.4.5}$$

It has genus $g = q/2$, its regular differentials being $P(t)\,dt$, with $P \in \mathbb{k}[t]$ a polynomial of degree $< g$; the hyperelliptic involution of C is

$$\iota\colon (t,u) \mapsto (t,u+1).$$

PROPOSITION 10.4.16. *The curve C has $q^2+2gq+1 = 2q^2+1$ rational points over $\mathbb{k}$, and its Jacobian J_C is isogenous with E_0^g.*

PROOF. Note that $t \mapsto t^{q+1}$ is the norm map from $\mathbb{k}$ to its quadratic subfield. Thus, for all $t \in \mathbb{k}$, the element t^{q+1} of that subfield can be written as u^2+u for some two distinct $u \in \mathbb{k}$; thus, C has $2|\mathbb{k}| = 2q^2$ finite $\mathbb{k}$-rational points. Since the exponent $q+1$ in (10.4.5) is odd, C has a single point at infinity p_∞, which is $\mathbb{k}$-rational. Thus, the total number of rational points is $2q^2+1 = q^2+2gq+1$, so C is maximal and thus $F = -q$, and Tate's theorem on isogeny classes of abelian varieties implies that J_C and E_0^g are isogenous. □

A potentially constant elliptic curve. Now let L be the commutative group $\mathrm{Hom}_{\mathbb{k}}(J_C, E_0)$ of homomorphisms of abelian varieties from J_C to E_0 defined over $\mathbb{k}$. Since J_C is isogenous with E_0^g, this group has the same rank as $\mathrm{Hom}_{\mathbb{k}}(E_0^g, E_0) = (\mathrm{End}_{\mathbb{k}}(E_0))^g$. Now E_0 is supersingular, so its group of $\overline{\mathbb{k}}$-endomorphisms has rank 4, and all of these endomorphisms are defined over $\mathbb{k}$ because they commute with the Frobenius endomorphism. We conclude that the rank of $(\mathrm{End}_{\mathbb{k}}(E_0))^g$, and thus also of $L = \mathrm{Hom}_{\mathbb{k}}(J_C, E_0)$, is

$$r = 4g = 2q = 2^{n+1}.$$

We next identify L with the $\mathbb{k}$-rational maps $f\colon C \to E_0$ taking p_∞ to the origin $\mathbf{0}$ of E_0. The embedding of C in J_C taking any $p \in C$ to the class of the divisor $(p)-(p_\infty)$ lets us restrict any homomorphism from J_C to E_0 to a map $f\colon C \to E_0$ with $f(p_\infty) = 0$. Conversely, we can use the group law on E_0 to extend any map $f\colon C \to E_0$ to a function $f_+\colon \mathrm{Div}(C) \to E_0$, which must satisfy $f_+(D) = f_+(D')$ for any pair of linearly equivalent divisors D, D'; the restriction of f_+ to the degree-zero divisors then descends to a homomorphism $J_C \to E_0$ whose restriction to the embedding of C is $p \mapsto f(p) - f(p_\infty)$, and thus coincides with f if $f(p_\infty) = \mathbf{0}$. This concludes the construction, which also lets us identify pointwise addition and subtraction of maps $f\colon C \to E_0$ with the group operations on L.

In particular, it follows that any $f\colon C \to E_0$ with $f(p_\infty) = \mathbf{0}$ satisfies

$$f(\iota(p)) = -f(p), \tag{10.4.6}$$

since the divisor $(p)+(\iota(p))$ is linearly equivalent to $2(p_\infty)$. Recalling that $-(x,y) = (x,y+1)$ on E_0, this means that either f is the constant function $C \to \mathbf{0}$ or it is of the form

$$f(t,u) = (x(t), y(t) + u), \tag{10.4.7}$$

where x, y are rational functions satisfying

$$y^2 + y = x^3 + t^{q+1} + a_6 \tag{10.4.8}$$

with $a_6 = 0$ for n odd and a_6 as in (10.4.4) for n even. Conversely, any pair (x,y) of rational functions satisfying (10.4.8) yields a rational map (10.4.7) that satisfies (10.4.6) and thus (since $\mathbf{0}$ is the unique fixed point of multiplication by -1 on E_0) takes p_∞ to $\mathbf{0}$.

If an elliptic curve E over the global field $\mathbb{K} = \mathbb{k}(t)$ is given by (10.4.8), then its Mordell–Weil group $E(\mathbb{K})$ gets identified with L; it is a free abelian group of rank N. Note also that E has constant j-invariant 0, and thus is a potentially constant elliptic curve; indeed it becomes isomorphic with the constant curve E_0 under the separable quadratic extension $\mathbb{K}(u) = \mathbb{k}(C)$ of $\mathbb{K}$. Moreover, the curve has good reduction at all finite places of $\mathbb{K}$, and additive bad reduction at the infinite place $t = \infty$. The local reduction type at infinity may be determined by the change of coordinates $(x,y) = (t^{2z}x_1, t^{3z}y_1)$, where z is the smallest integer such that $6z > q$, obtaining a minimal model

$$y_1^2 + t^{-3z}y_1 = x_1^3 + t^{q+1-6z} + t^{-6z}a_6$$

with discriminant t^{-12z}.

Direct calculations with this model using Tate's algorithm show that the narrow Mordell–Weil group L_0, consisting of solutions of (10.4.8) which reduce on each fibre to points on its identity component, is of index 4 in L.

We mention that for even n every fibre of E has only one component and $L_0 = L$. For both even and odd n, points on the identity component of E are characterized by the condition that x_1 has no zero at infinity, which can be shown to be equivalent to that x has a pole at $t = \infty$ of order greater than $(q+1)/3$.

Heights and the Mordell–Weil lattice of E. In this case one can explicitly determine the Néron–Tate form $\hat{h}$.

PROPOSITION 10.4.17. *Let $P \in L$ be a nonzero point, with coordinates (x,y). Then the canonical height of P under the above normalization is* $\deg(x)$; *thus, the canonical and naive heights on $E(\mathbb{K})$ coincide. Furthermore,*

$$\hat{h}(P) \geq \frac{q+1}{3} \tag{10.4.9}$$

for all nonzero $P \in L$, and $\hat{h}(P)$ is even unless n is odd.

SKETCH OF THE PROOF. To identify the naive and canonical heights, we need only show that the naive height already satisfies the identity $\hat{h}(2P) = 4\hat{h}(P)$. Indeed, for $P = (x,y)$ we get

$$2P = (x^4, y^4 + x^6 + x^3 + 1),$$

and $\deg(x^4) = 4\deg(x)$. Every pole of $y^2 + y$ has even order, which implies that x has a pole of order at least $(q+1)/3$, from which the estimate (10.4.9) follows. The order of this pole is either even or exactly equals $(q+1)/3$, the latter being possible only for odd n. Note also that any pole of $x^3 + t^{q+1} + a_6$, and thus also of x, at a finite $t \in \overline{\mathbb{k}}$ must too have even order, and that the degree of x is its total number of poles counted with multiplicity, which gives the parity of $\deg(x) = \hat{h}(P)$. □

The sublattice L_0 has index 4 in L, and thus center density $1/4$, as large as that of L. On the other hand, its minimal norm is at least $(q+4)/(q+1)$ times larger. Thus, the normalized center density of L_0 is

$$\frac{1}{4}\left(\frac{q+4}{q+1}\right)^{r/2} = \frac{1}{4}\left(1+\frac{3}{q+1}\right)^{q}$$

times that of L. This factor is ≥ 1 for all q, with equality only when $q=2$, and tends to $e^3/4 = 5.021\ldots$ as $q \to \infty$. Thus, for odd $n>1$ the lattice L_0 always yields a denser sphere packing than L, albeit by a bounded factor.

Calculations using the L-function of E over $\mathbb{K}$ lead to the following estimate for $\Delta(L)$.

PROPOSITION 10.4.18. *We have*

$$\Delta(L) \leq q^{2\lfloor q/6 \rfloor}/c_\infty,$$

where $c_\infty = 1$ *or* 4 *according to whether* n *is even or odd.* □

Combining this with estimates on $N_{\min}$ yields the following result.

THEOREM 10.4.19. $\Delta(L)$ *is a power of* 2. *The narrow Mordell–Weil lattice* L_0 *of* $E/\mathbb{K}$ *of rank* $r = 2q = 2^{n+1}$ *has normalized center density*

$$\delta(L_0) \geq q^{-(q-4)/6}\left(\frac{q+2}{12}\right)^{q} \tag{10.4.10}$$

for n *even, and*

$$\delta(L_0) \geq \frac{1}{2}q^{-(q-2)/6}\left(\frac{q+4}{12}\right)^{q} \tag{10.4.11}$$

for n *odd. Equality holds if and only if* (10.4.8) *has solutions in polynomials* $x(t), y(t)$ *with* x *of degree* $2(\lfloor q/6 \rfloor + 1)$, *in which case these solutions give the minimal vectors of* L_0. □

The lattice L_0 gives a sphere packing in dimension 2^{n+1}, whose density equals previously known records for $n \leq 4$ and improves on those records for $5 \leq n \leq 9$. When $n=1$, the curves C and E_0 coincide, so L is just the endomorphism ring of E_0, which is the ring of Hurwitz quaternions, homothetic as a lattice with the root lattice D_4 of normalized center density $1/8$; the narrow Mordell–Weil lattice L_0 consists of the inseparable endomorphisms and is isometric with D_4. When $n=2$, the Mordell–Weil lattice is isomorphic with the root lattice E_8 (with $\delta = 1/16$). When $n=3$, one can show that L_0 is isomorphic with the Barnes–Wall lattice BW_{16} (of dimension 16) and normalized center density $1/16$. When $n=4$, L has the same density $(9/8)^8$ as Quebbemann's lattice Q_{32} but is not isomorphic to it.

For $n = 5, 6, \ldots, 11$, we find new lattices L_0; their parameters are given in the following table:

n	5	6	7	8	9	10	11
$r = 2^{n+1}$	64	128	256	512	1024	2048	4096
$\log_2 \delta \geq$	24.72	97.40	294.81	797.12	2012.25	4871.8	11439.76

Leech lattice. One can show that the Leech lattice can also be represented as a Mordell–Weil lattice in characteristic 2.

More precisely, let $\Bbbk$ be the finite field of $2^{12} = 4096$ elements, and $\mathbb{K} = \Bbbk(t)$. Fix $a_6 \in \Bbbk$ of trace 1, and let $E/\mathbb{K}$ be the potentially constant supersingular curve

$$y^2 + y = x^3 + t^{13} + a_6.$$

Let L be its Mordell–Weil lattice. Then we have the following result.

THEOREM 10.4.20. *The lattice L has rank* 24, *discriminant* 2^{24}, *minimal norm* 8, *and normalized center density* $\delta(L) = 1$, *and it is homothetic with the Leech lattice* Λ_{24}, *namely* $L = 2\Lambda_{24}$. □

Kissing number of MW_{128}. Calculating the kissing number of the MW_{2^m} lattices is a rather difficult task. However, this can be done for $m \leq 7$, and now we describe the result for the MW_{128} lattice, which has a kissing number considerably greater than any other known lattice in $\mathbb{R}^{128}$.

The elliptic curve $E_{128}/\mathbb{K}$ is given by

$$y^2 + y = x^3 + t^{65} + a_6, \quad a_6 \in \mathbb{F}_{2^{12}}, \quad \mathrm{tr}(a_6) = 1.$$

Then we have the following fact.

THEOREM 10.4.21. *Let $M = MW_{128}$ be the Mordell–Weil lattice of E_{128}. Then the minimal norm of M is* 22, *attained by (x, y) if and only if x, y are polynomials in t of degrees* 22 *and* 33 *respectively. There are*

$$218044170240 = 2^{17} \cdot 3 \cdot 5 \cdot 13 \cdot 19 \cdot 449$$

vectors of this minimal norm in M. □

The proof combines very ingenious arguments with clever computer-aided calculations.

Appendix: Parameters of Some Packings

TABLE A.1. Approximate values of parameters for the densest lattice packings in small dimensions: $A_1 = \mathbb{Z}$, A_2, $A_3 = D_3$, D_4, D_5, E_6, E_7, $E_8 = \Gamma_8$, and Λ_{24}.

N	1	2	3	4	5	6	7	8	24
$\Delta_\ell(N)$	1	0.907	0.740	0.617	0.465	0.373	0.295	0.254	$1.93 \cdot 10^{-3}$
$\delta_\ell(N)$	0.5	0.229	0.177	0.125	0.088	0.072	0.063	0.063	1
$\lambda_\ell(N)$	0	0.070	0.144	0.174	0.221	0.237	0.251	0.247	0.376
$\nu_\ell(N)$	-1	-1.792	-2.5	-3	-3.5	-3.792	-4	-4	0
$\gamma_\ell(N)$	1	1.155	1.260	1.414	1.516	1.665	1.811	2	4
$\delta_\ell^{-2}(N)$	4	12	32	64	128	192	256	256	1

TABLE A.2. Parameters of certain families

L_N	$\mathbb{Z}^N$	A_N	$D_N,\ N \geq 3$	$E_N,\ 4 \leq N \leq 8$	$\Gamma_N,\ N \geq 8,$ $N \equiv 0 \pmod 4$	Λ_{24}
$\delta(L_N)$	2^{-N}	$\dfrac{2^{-N/2}}{\sqrt{N+1}}$	$\dfrac{2^{-N/2}}{2}$	$\dfrac{2^{-N/2}}{\sqrt{9-N}}$	$2^{-N/2}$	1
$\gamma(L_N)$	1	$\dfrac{2}{\sqrt[N]{N+1}}$	$\dfrac{2}{\sqrt[N]{4}}$	$\dfrac{2}{\sqrt[N]{9-N}}$	2	4

$$
\begin{array}{ccccc}
\widetilde{\lambda}_T^{\mathrm{pol}} \leq 2.30 & \longrightarrow & \widetilde{\lambda}_T^{\mathrm{exp}} \leq 1.30 & \longrightarrow & \widetilde{\lambda}_T \leq 1.30 \\
\downarrow & & \downarrow & & \downarrow \\
\widetilde{\lambda}_\ell^{\mathrm{pol}} \leq 2.30 & \longrightarrow & \widetilde{\lambda}_\ell^{\mathrm{exp}} \leq 1 & \longrightarrow & \widetilde{\lambda}_\ell \leq 1 \\
\downarrow & & \downarrow & & \downarrow \\
\widetilde{\lambda}^{\mathrm{pol}} \leq 1.31 & \longrightarrow & \widetilde{\lambda}^{\mathrm{exp}} \leq 1 & \longrightarrow & 0.599 \leq \widetilde{\lambda} \leq 1
\end{array}
$$

DIAGRAM A.3. Asymptotic results (here $\rightarrow$ stands for $\geq$).

Historical and Bibliographic Notes

The sphere packing problem in Euclidean spaces is quite a classical one. Many generations of mathematicians have appreciated its beauty; in particular, packings in three dimensions were studied by J. Kepler (for their density) and I. Newton (for their kissing numbers). D. Hilbert included the sphere packing problem into his eighteenth problem (and this is the last unsolved question in it; see [**Mil76**]).

There are several excellent books devoted to packings, for example [**CS99**, **Rog64**, **FT72**], and we refer the interested reader to them. It is worth mentioning that [**CS99**] contains a bibliography of about 1500 titles.

The A_2 packing has been known to mankind for a long time. The fact that it is the densest in two dimensions was proved by A. Thue and L. Fejes Tóth. The fact that A_2 and A_3 are the densest lattice packings in their dimensions was proved by K.F. Gauss. The respective result for A_4 and D_5 is due to A.M. Korkine and E.I. Zolotareff, and for E_6, E_7, and E_8, to H.F. Blichfeldt. The lattice Λ_{24} was discovered by J. Leech [**Lee64**, **Lee67**] (who used codes). For more information on it, see [**CS99**, **CKM$^+$17**].

The root lattices of types A_N, D_N, and E_N appear in many different realms of mathematics (such as Lie groups and Lie algebras, singularities, rational surfaces, etc.). The Leech lattice looks no less important.

The conjecture that A_3 has the maximal packing density in $\mathbb{R}^3$ was first stated by J. Kepler (1611), who had started to study arrangements of spheres as a result of his correspondence (1606) with T. Harriot, a friend and assistant of Sir Walter Raleigh, who had set Harriot the problem of determining how best to stack cannonballs on the decks of his ships. The proof of the Kepler conjecture was given in 1998 by T. Hales [**Hal05**]; it uses heavy computer calculations, however [**HAB$^+$17**] and the algorithms therein give a formal proof, deducing the conjecture from the ZFC axioms. In 2016 Maryna Viazovska proved in a brilliant breakthrough [**Via17**] the optimality of the E_8 packing, and soon after that Viazovska, along with four other mathematicians, successfully extended her method to cover the Leech lattice, too [**CKM$^+$17**]. These proofs use modular functions and are human-controlled and quite lucid; an excellent exposition is given in [**Oes16**].

The Kabatiansky–Levenshtein bound [**KL78**] is an improvement of those by H.F. Blichfeldt ($\widetilde{\lambda} \geq 0.5$) and V.M. Sidelnikov ($\widetilde{\lambda} \geq 0.5096$). The bound of Proposition 10.1.12 was found by C.A. Rogers [**Rog64**]. Theorem 10.1.14 was proved by J.A. Rush and N.J.A. Sloane [**RS87**], Proposition 10.1.15 is taken from an elegant paper by J.A. Rush [**Rus89**]. Random packing in Sec. 10.1.3 is considered in [**ST01**]; for Theorem 10.1.17, see [**Sch58**].

The problem of explicit construction of high-dimensional packings has been widely studied (see [**CS99**]). The first explicit construction of an asymptotically good family of (nonlattice) packings is due to N.J.A. Sloane [**Slo72**]; he has proved that $\widetilde{\lambda}^{\mathrm{pol}} \leq 6$. The basic construction of Sec. 10.2.1 and Theorem 10.2.5 are due to A. Bos, J.H. Conway, and N.J.A. Sloane [**BCS82**]; they have constructed a family of lattices that are not asymptotically good but whose density exponent tends to infinity quite slowly. The nonlattice version of the construction is due to A. Bos [**Bos80**].

The complexity reduction trick (Theorem 10.2.12) was discovered by S.N. Litsyn [**LT87**]. This trick and algebraic-geometric codes led him and M.A. Tsfasman

[**LT87**] to the best known estimates $\widetilde{\lambda}^{\text{pol}} \leq 1.31$ and $\widetilde{\lambda}_\ell^{\text{pol}} \leq 2.30$, and Sec. 10.2.1 follows this paper.

Spherical codes and their basic properties are described in [**CS99**] and the references therein. In particular, for the Yaglom map and Yaglom bound, see [**CS99**, Chapter 9]. The constructions of those codes from algebraic geometry codes, as well as the bounds on kissing numbers, are given in [**LS94**]. The Solé–Belfiore bounds are proved in [**SB13**].

The results in Sec. 10.2.3 are borrowed from [**Vlă18**]; the curve X_n used in the proof of Proposition 10.2.31 has the required properties by [**Gek01**, Example 10.2], see also [**Gek80, Gek86**]. For the Barnes–Wall lattices, see [**BW59**] and [**CS99**, Sec. 4.10].

The constructions of Sec. 10.3 are quite natural, and most of them are well known. However, the quest of estimating their density was not addressed until the late 1980s. For additive number field lattices, this was first done by M.A. Tsfasman (see [**LT87**]), who proved Proposition 10.3.1 and Theorem 10.3.2. All results on additive number field codes (including Proposition 10.3.6 and Theorem 10.3.7) are due to H.W. Lenstra, Jr. [**Len86**]. Sections 10.3.2 and 10.3.3 (multiplicative lattices) are mostly based on results of M.Yu. Rosenbloom and M.A. Tsfasman [**RT90**]. The FML construction was independently discovered by N. Elkies [**Elk89a**] and by H-G. Quebbemann [**Que89**], who used his formula (10.3.1) to prove Theorem 10.3.10; this formula gives a slightly worse constant ($\lambda \sim 1.906$). The FMCD case is due to V.D. Goppa [**Gop88**]. A general reference for Sec. 10.3 is [**Tsf91**].

The Mordell–Weil lattices in Sec. 10.4 were constructed by T. Shioda and N.D. Elkies; more precisely, the results of Sec. 10.4.1 are due to Shioda [**Shi91**], and those of Sec. 10.4.2 to Elkies [**Elk94, Elk97b, Elk01b**]. For Tate's algorithm determining the type of a singular fibre of an elliptic surface, see [**Tat75**]; for Tate's theorem on isogenies of abelian varieties, see [**Tat66**].

CHAPTER 11

Codes from Multidimensional Varieties

In this chapter we describe some codes arising from multidimensional varieties, i.e., those of dimension at least 2. The general construction of such codes is to consider a projective system $\mathcal{P} \subset \mathbb{P}^{k-1}$, where $\mathcal{P} \subseteq X(\mathbb{F}_q)$ is some subset of an algebraic variety $X \subset \mathbb{P}^{k-1}$ defined over $\mathbb{F}_q$. Then $\mathcal{P}$ is an $[n,k,d]_q$ system (cf. Sec. 1.1.2), where

$$n = |\mathcal{P}|, \qquad k = \dim \mathbb{P} + 1, \qquad d = n - \max_H |\mathcal{P} \cap H|,$$

where the maximum is over all $\mathbb{F}_q$-hyperplanes $H \subset \mathbb{P}$.

The quantities n and k are often well controlled by algebraic geometry tools, so the main difficulty for $\dim X > 1$ is to control d. Recall that for curves we have a trivial bound $d \geq n - \deg X$, which is rather satisfactory in many cases, even if sometimes it is possible to (slightly) improve it. On the contrary, for $\dim X > 1$ there is no satisfactory general bound on d; moreover, usually it is rather difficult to get a reasonable estimation for d even in a particular case.

This difficulty is the main obstacle to the theory of algebraic geometry codes for $\dim X > 1$. There is a very limited number of cases where this difficulty can be surpassed. However, there are some interesting results in this direction, which we partly present in this chapter.

It is organized as follows. In Sec. 11.1 we consider the most elementary but yet highly nontrivial case of an intersection of several hypersurfaces, which is directly related to Reed–Muller codes. In Sec. 11.2 we touch upon the case of general algebraic sets, where there are some interesting results even though this case can hardly be treated in a satisfactory manner. These results are not as yet used to construct concrete classes of codes, but their existence could be important for further applications. Section 11.3 is devoted to the case of $\dim X = 2$, in which case there are some results specific for this particular dimension. However, to obtain relatively satisfactory estimates, one should use very specific surfaces, mostly rational ones, i.e., those birationally equivalent to the projective plane $\mathbb{P}^2$.

Section 11.4 is devoted to two examples of homogeneous hypersurfaces, namely Hermitian varieties and quadrics. The results here are rather satisfactory due to classical results giving minute information on the structure of these varieties.

Section 11.5 considers two closely related classes of varieties, namely the Grassmann and Schubert varieties. The structure of these classical varieties is also well investigated. However, due to their more complicated combinatorics, the results here are less complete than in Sec. 11.4. Finally, in Sec. 11.6 we consider a very large class of varieties, namely flag varieties, generalizing many of the examples given in preceding sections.

We have to note that the present chapter uses a rather wide spectrum of algebraic geometry techniques and thus cannot be made reasonably self-contained.

We are trying at least to formulate necessary results in a form comprehensible for a nonexpert.

11.1. Complete Intersections and Reed–Muller Codes

In this section we treat the case of complete intersection, i.e., consider the number of points on an intersection of hypersurfaces in a projective (or affine) space. For a single hypersurface we have an optimal upper bound. In the case of several hypersurfaces the situation is not so simple, at least for the projective space; however, one can get some very interesting bounds there. The affine case is simpler, and the results in this case are more advanced. Since Reed–Muller codes, both affine (= usual) and projective, by their very definitions are directly related to intersections of hypersurfaces, we also describe some consequences for those codes.

11.1.1. Tsfasman–Serre–Sørensen Bound

This bound concerns the case where X is isomorphic to a projective space, $X \cong \mathbb{P}^m$, and X is considered as a subvariety in $\mathbb{P}^N$, where $N = \binom{m+r}{r} - 1$ with $r \geq 1$ and the embedding $X \hookrightarrow \mathbb{P}^N$ is given by all homogeneous polynomials of degree r, i.e.,

$$x = (x_0 : x_1 : \ldots : x_m) \in X \longmapsto (P_0(x) : P_1(x) : \ldots : P_N(x)) \in \mathbb{P}^N,$$

where $\{P_0, \ldots, P_N\}$ is a basis in the space $\mathcal{P}_{m,r}$ of degree-r homogeneous polynomials in $m+1$ variables (over $\mathbb{F}_q$).

If we take $\mathcal{P} = X(\mathbb{F}_q)$, then

$$\theta_m := |\mathcal{P}| = q^m + q^{m-1} + \ldots + 1 = \frac{q^{m+1}-1}{q-1},$$

$$k = N+1 = \dim \mathcal{P}_{m,r} = \binom{m+r}{r},$$

and for d one has the following Tsfasman–Serre–Sørensen bound:

$$d \geq \theta_m - (rq^{m-1} + \theta_{m-2}) = (q-r+1)q^{m-1},$$

which is an immediately corollary of the following fact.

THEOREM 11.1.1. *If $P \in \mathcal{P}_{m,r} \setminus \{0\}$, $r \leq q+1$, then the zero locus $S_P = \{x \in \mathbb{P}^m : P(x) = 0\}$ is of cardinality*

$$s := |S_P| \leq rq^{m-1} + \theta_{m-2}. \tag{11.1.1}$$

Moreover, if $r \leq q$, the bound is attained only if S_P is a union of r hyperplanes passing through a subspace of codimension 2.

PROOF. The case $r = q+1$ is trivial, since $\theta_m = (q+1)q^{m-1} + \theta_{m-2}$. Let $r \leq q$.

For $m = 1$ the claim is also obvious. Let $m \geq 2$; we will use induction on m. Let $g_1, g_2, \ldots, g_\delta$ be different linear factors of P over $\mathbb{F}_q$, and let $G_1, \ldots, G_\delta$ be the corresponding hyperplanes in $\mathbb{P}^m$. Consider $G := \bigcup_{i=1}^{\delta} G_i \subseteq S$. We distinguish two cases.

(i) $G = S$. Then

$$|G_1 \cup \ldots \cup G_j| \leq jq^{n-1} + \theta_{n-2}, \quad \forall j = 1, \ldots, \delta, \tag{11.1.2}$$

which is clear using induction on j. Furthermore, $|G_{j+1} \cap (G_1 \cup \ldots \cup G_j)| \geq \theta_{m-2}$, and since $j \leq r$, (11.1.1) implies (11.1.2). Moreover, equality in (11.1.1) is possible only if the dimension of the span of $\{g_i\}_{i=1,\ldots,\delta}$ is 2, i.e., all G_i contain a common subspace $\mathbb{P}^{m-2}$.

(ii) $G \subset S$. Take $a \in S \setminus G$. Then the restriction of P to a hyperplane H containing a is nonzero. By induction, we have

$$|S \cap H| \leq rq^{m-2} + \theta_{m-3}. \tag{11.1.3}$$

Now let

$$T = \{(b, H) : \ b \in S \setminus \{a\} \text{ a point}, \ H \ni a, b \text{ a hyperplane}\}.$$

For a given $b \in S \setminus \{a\}$, the number of hyperplanes H such that $H \ni a, b$ equals θ_{m-2}, and thus

$$|T| = (s-1)\theta_{m-2}. \tag{11.1.4}$$

On the other hand, for a fixed H, the number of points $b \in (S \setminus \{a\}) \cap H$ is

$$|S \cap H| - 1 \leq rq^{m-2} + \theta_{m-3} - 1,$$

and therefore

$$|T| \leq \theta_{m-1}(rq^{m-2} + \theta_{m-3} - 1). \tag{11.1.5}$$

Hence, by (11.1.4) and (11.1.5) we obtain

$$s \leq 1 + \frac{\theta_{m-1}(rq^{m-2} + \theta_{m-3} - 1)}{\theta_{m-2}} = rq^{m-2} + \theta_{m-2} - \frac{(q+1-r)q^{m-2}}{\theta_{m-2}},$$

and since $q+1-r > 0$, this completes the proof:

$$s \leq rq^{m-2} + \theta_{m-2} - (q+1-r). \qquad \square$$

For hypersurfaces without linear components, the bound can be improved as follows.

THEOREM 11.1.2. *Let $X \subset \mathbb{P}^m$ be a hypersurface of degree r over $\mathbb{F}_q$ without $\mathbb{F}_q$-linear components. Then*

$$N_q(X) \leq (r-1)q^{m-1} + rq^{m-2} + \theta_{m-3}. \qquad \square$$

The results of the present section concerning a single equation (a hypersurface) are rather precise and satisfactory. We are going to discuss generalizations to several equations and more complicated varieties.

11.1.2. Generalization to Several Polynomials: Tsfasman–Boguslavsky Conjecture

A natural generalization of the situation discussed in Sec. 11.1.1 is the case of ℓ homogeneous polynomials of the same degree. Consider ℓ homogeneous polynomials $F_1, F_2, \ldots, F_\ell \in \mathbb{F}_q[x_0, \ldots, x_m]$, all of degree r, and let

$$S_\ell = S_\ell(F_1, \ldots, F_\ell) = \{a \in \mathbb{P}^m(\mathbb{F}_q) : \ F_i(a) = 0 \text{ for all } i = 1, \ldots, \ell\}.$$

The case $\ell = 2$ is rather comprehensible, and one can prove the following theorem.

THEOREM 11.1.3. *Let F_1 and F_2 be homogeneous polynomials of degree r in $m+1$ variables. Assume that they are linearly independent and $r < q-1$. Then the maximal possible number of their common zeroes in $\mathbb{P}^m(\mathbb{F}_q)$ is*

$$(r-1)q^{m-1} + q^{m-2} + \theta_{m-2}.$$

The proof uses the following general result.

THEOREM 11.1.4. *Let $X \subset \mathbb{P}^m$ be an algebraic set of degree δ and dimension s. Then*

$$|X| \le \delta\theta_s.$$

REMARK 11.1.5. The bound of Theorem 11.1.3 is tight and is attained on a pair of hypersurfaces which are unions of hyperplanes.

PROOF OF THEOREM 11.1.4. We use induction on m. For $m=1$ we have $\dim X = 0$ and $|X| \le \deg X = \delta$. Assume that we have proved the theorem for dimension $m-1$.

First, we consider case (I) below, where X is irreducible and is not contained in a hyperplane. Second, we consider case (II) where X is contained in a hyperplane. Finally, in case (III) we deduce from (I) and (II) the bound for an arbitrary X.

(I) Assume that X is irreducible and is not contained in any hyperplane. Then $X \cap H$ is an algebraic set of dimension $s-1$ and degree δ in $H \cong \mathbb{P}^{m-1}$ for any hyperplane H. By the induction hypothesis, $|X \cap H| \le \delta\theta_{s-1}$.

Now we use a construction quite similar to that used to prove Theorem 11.1.1. Consider the set in $(\mathbb{P}^m)^* \times \mathbb{P}^m$ consisting of all pairs (H, P) where H is a hyperplane and P is a point in $\mathbb{P}^m$, both defined over $\mathbb{F}_q$, such that $P \in H \cap X$. We compute the number of $\mathbb{F}_q$-points in this set in two different ways.

There are $|X|$ ways to select a point $P \in X$, and for each P there are θ_{m-1} ways to select H. On the other hand, we can first select one of the θ_m hyperplanes in $\mathbb{P}^m$ and then one of the points in the intersection $H \cap X$. Thus,

$$|X|\theta_{m-1} = \sum_H |H \cap X|.$$

Combining this with $|H \cap X| \le \delta\theta_{s-1}$, we obtain

$$|X|\theta_{m-1} \le \theta_m \delta\theta_{s-1}.$$

Thus,

$$|X| \le \delta\theta_s \frac{\theta_{s-1}\theta_m}{\theta_s\theta_{m-1}},$$

and since $\dfrac{\theta_{s-1}\theta_m}{\theta_s\theta_{m-1}} < 1$, we get

$$|X| < \delta\theta_s. \tag{11.1.6}$$

(II) Assume that X is contained in a hyperplane H. Then X is an algebraic set of dimension s and degree δ in $H \cong \mathbb{P}^{m-1}$. By the induction hypothesis,

$$|X| \le \delta\theta_s. \tag{11.1.7}$$

(III) Let X be an arbitrary algebraic set of dimension s. Then X can be decomposed into a union of (absolutely) irreducible components $X = X_1 \cup X_2 \cup \ldots \cup X_k$ of degrees $\delta_1, \delta_2, \ldots, \delta_k$, $\sum\limits_{i=1}^{k} \delta_i = \delta$. If X_i is not contained in a hyperplane, then $|X_i| \le \delta_i \theta_s$ by (11.1.6). If X_i is contained in a hyperplane, the same holds by (11.1.7). Note that it can happen that some of X_i are not defined over $\mathbb{F}_q$ and some of them have dimension less than s; in both cases we have the same bound.

Thus, we finally get

$$|X| \le \sum_{i=1}^{k} |X_i| \le \left(\sum_{i=1}^{k} \delta_i\right) \theta_s = \delta\theta_s. \qquad \square$$

Now we prove Theorem 11.1.3. Denote by

$$M_q(2,m,r) = (r-1)q^{m-1} + q^{m-2} + \theta_{m-2}$$

the number of points given in the theorem.

PROOF OF THEOREM 11.1.3. We will show that for any two effective divisors R and S of degree r in $\mathbb{P}^m$ we have

$$|R \cap S| \le M_q(2,m,r).$$

Here by $R \cap S$ we denote for simplicity the intersection ignoring multiplicities, i.e., $(\operatorname{Supp} R) \cap (\operatorname{Supp} S)$.

(I) Let X be the intersection $R \cap S$, and let Y be the maximal divisor such that $R - Y \ge 0$ and $S - Y \ge 0$. Let X' be $(R-Y) \cap (S-Y)$. Then $X = Y \cup X'$. Let $b = \deg Y$. Then $0 \le b \le r-1$. By Theorem 11.1.1, $|Y| \le bq^{m-1} + \theta_{m-2}$. Furthermore, X' is an algebraic set of codimension 2 and degree $\delta = (r-b)^2$. By Theorem 11.1.4, $|X'| \le (r-b)^2 \theta_{m-2}$.

(II) Assume that $b = 0$; then $X = X'$ and $|X| \le r^2 \theta_{m-2}$. One can easily check that

$$|X| - M_q(2,m,r) \le \frac{r-1}{q-1}\left(-q^m + (r+2)q^{m-1} - (r+1)\right) - q^{m-2}.$$

For $r < q-1$ the last expression is negative, so $|X| \le M_q(2,m,r)$.

(III) Assume that $b > 0$. Then

$$|X| \le |Y| + |X'| \le bq^{m-1} + \theta_{m-2} + (r-b)^2\theta_{m-2},$$

and after some calculations we get

$$|X| - M_q(2,m,r) \le -\frac{1}{q-1}\left(q^{m-1}(r-b-1)(q-(r-b+2)) + (r-b)^2 - q^{m-2}\right).$$

The sign of the right-hand side of this inequality is the same as the sign of

$$\gamma = -q^{m-1}(r-b-1)(q-(r-b+2)) - (r-b)^2 + q^{m-2}.$$

(IV) Assume that $0 < b < r-1$. If $(r-b-1)(q-(r-b+2)) > 0$, then

$$q^{m-1}(r-b-1)(q-(r-b+2)) \ge q^{m-1}$$

and $\gamma < 0$. We have $r-b-1 > 0$, $r-b \le r-1$. Combining this with the assumption $r < q-1$, we get $r-b < q-2$ and $(r-b-1)(q-(r-b+2)) > 0$.

Thus, $\gamma < 0$ and $|X| \le M_q(2,m,r)$.

(V) Now we consider the last case $b = r-1$. We have $\deg X' = 1$, so X' is a linear subspace of codimension 2. We cannot apply Theorem 11.1.1 directly, since we would get

$$|X| \le |Y| + |X'| \le (r-1)q^{m-1} + \theta_{m-2} + \theta_{m-2} > (r-1)q^{m-1} + q^{m-2} + \theta_{m-2}.$$

Thus, we consider two cases. If Y contains an $\mathbb{F}_q$-hyperplane H, then $H \cap X'$ contains a linear subspace of dimension $m-3$, whence

$$\begin{aligned}|X| \le |Y| + |X'| - \theta_{m-3} &\le (r-1)q^{m-1} + \theta_{m-2} + \theta_{m-2} - \theta_{m-3} \\ &= (r-1)q^{m-1} + q^{m-2} + \theta_{m-2}.\end{aligned}$$

Now assume that Y does not contain an $\mathbb{F}_q$-hyperplane; then $Y \cap H$ is a divisor on H for any H. As in the proof of Theorem 11.1.4, we will use induction on m and the same construction.

The case $m = 1$ is trivial. Now assume that we have proved the assertion for dimension $m-1$. If $Y(\mathbb{F}_q) \subset X'$, then the assertion is obvious. Otherwise, fix an $\mathbb{F}_q$-point $Q \in (Y \setminus X')$. There exists a unique hyperplane H_0 passing through Q and X', and the intersection $H_0 \cap Y$ is a divisor of degree $r-1$ on $H \cong \mathbb{P}^{m-1}$, $H_0 \cap X' = X'$. Then

$$|H_0 \cap X| \le |H_0 \cap Y| + |X'| \le (r-1)q^{m-2} + \theta_{m-3} + \theta_{m-2}.$$

For any other hyperplane $H \ne H_0$ passing through Q, the intersection $H \cap X$ is an algebraic set in $H \cong \mathbb{P}^{m-1}$ consisting of an effective divisor $H \cap Y$ of degree $r-1$ and a linear subspace $H \cap X'$ of codimension 2 in H. By the induction hypothesis, we have

$$|H \cap X| \le M_q(2, m-1, r).$$

Consider the set in $(\mathbb{P}^m)^* \times \mathbb{P}^m$ consisting of all pairs (H, P) with H a hyperplane and P a point in $\mathbb{P}^m$, both defined over $\mathbb{F}_q$, such that $Q \in H$, $P \in H \cap X$, and $P \ne Q$. We compute the number of $\mathbb{F}_q$-points in this set in two different ways.

There are $|X| - 1$ ways to select a point $P \in X$ such that $P \ne Q$, and for each P there are θ_{m-2} ways to select H passing through P and Q. On the other hand, we can first select one of the θ_{m-1} hyperplanes in $\mathbb{P}^m$ passing through Q and then one of the points in $(H \cap X) \setminus Q$.

Thus,

$$\begin{aligned}(|X| - 1)\theta_{m-2} &= \sum_{H \ne H_0} (|H \cap X| - 1) + |H_0 \cap X| - 1 \\ &\le \theta_{m-1}(M_q(2, m-1, r) - 1) + \theta_{m-2} - q^{m-3}.\end{aligned}$$

Therefore,

$$\begin{aligned}|X| &\le 1 + \frac{\theta_{m-1}}{\theta_{m-2}}(M_q(2, m-1, r) - 1) + \frac{\theta_{m-2} - q^{m-3}}{\theta_{m-2}} \\ &= M_q(2, m, r) + \frac{1}{\theta_{m-2}}(q^{m-2}(r-1-q) + \theta_{m-3}).\end{aligned}$$

We have $r \le q$, and hence $q^{m-2}(r-1-q) + \theta_{m-3} < 0$ and $|X| < M_q(2, m, r)$, which completes the proof. □

For $\ell > 2$ there is no such simple formula. However, the following conjecture seems to be natural and coincides with the above results for $\ell \le 2$.

Tsfasman–Boguslavsky Conjecture (TBC). Assume that $\ell \le \binom{m+r}{r}$ and $r < q-1$. Then the maximum number of common zeroes that a system of ℓ

linearly independent homogeneous polynomials of degree r can have in $\mathbb{P}^m(\mathbb{F}_q)$ is

$$T_\ell(r,m) := \theta_{m-2j} + \sum_{i=j}^{m} \nu_i(\theta_{m-i} - \theta_{m-i-j}),$$

where $(\nu_1,\ldots,\nu_{m+1})$ is the ℓth element in descending lexicographic order among $(m+1)$-tuples $(\alpha_1,\ldots,\alpha_{m+1})$ of nonnegative integers satisfying $\alpha_1+\ldots+\alpha_{m+1}=r$, and where $j := \min\{i : \alpha_i \neq 0\}$.

Theorems 11.1.1 and 11.1.3 confirm the TBC for $\ell=1$ and $\ell=2$. However, this conjecture, though true in many cases, is not always valid. We will explain this phenomenon.

In fact, the TBC is true for $\ell \le m+1$ and in general is false for $\ell > m+1$. More precisely, for $\ell \le m+1$ we have the following result.

THEOREM 11.1.6. *Assume that $1 \le r < q-1$ and $1 \le \ell \le m+1$. Then the maximum number of zeroes in $\mathbb{P}^m(\mathbb{F}_q)$ for a system of ℓ linearly independent homogeneous polynomials, each of degree r in $S=\mathbb{F}_q[x_0,x_1,\ldots,x_m]$, is given by the Tsfasman–Boguslavsky bound $T_\ell(r,m)$, namely,*

$$T_\ell(r,m) = \begin{cases} \theta_{m-\ell} & \text{if } r=1 \text{ and } 1\le \ell \le m+1,\\ (r-1)q^{m-1}+\theta_{m-2}+\lfloor q^{m-\ell}\rfloor & \text{if } r>1 \text{ and } 1 \le \ell \le m+1.\end{cases}$$

Moreover, if $r=1$, then the maximum value $T_\ell(r,m)$ is always attained, whereas if $r>1$ and the maximum is attained by a family $\{F_1,\ldots,F_\ell\}$ of ℓ linearly independent polynomials of degree r, then $F_1,\ldots,F_\ell$ must have a common linear factor in S. □

The case $\ell > m+1$. For $\ell > m+1$ the TBC is false even in the case of quadrics. This is implied by the following results. Denote

$$\delta_j := \binom{j+2}{2} = 1+2+\ldots+(j+1) \quad \text{for any } j \in \mathbb{Z} \text{ with } j \ge -1.$$

THEOREM 11.1.7. *Let ℓ be a positive integer $\le \delta_m$, and let $F_1,\ldots,F_\ell$ be linearly independent homogeneous polynomials of degree 2 in $\mathbb{F}_q[x_0,\ldots,x_m]$. If k is the unique integer such that $-1 \le k < m$ and $\delta_m - \delta_{k+1} < \ell \le \delta_m - \delta_k$, then*

$$|S_\ell(F_1,\ldots,F_\ell)| \le \theta_k + \lfloor q^{\varepsilon-1}\rfloor \quad \textit{for} \quad \varepsilon := \delta_m - \delta_k - \ell. \qquad \square$$

COROLLARY 11.1.8. *If $r=2$ and $m>2$, then the TBC does not hold in general. In fact, the TBC is false for at least $\binom{m-1}{2}$ values of the positive integer ℓ with $m+1 < \ell \le \delta_m$.*

PROOF. For $1 \le i \le m+1$, denote by e_i the $(m+1)$-tuple with 1 in the ith place and 0 elsewhere. Also, let k be a unique integer such that $-1 \le k < m$ and $\delta_m - \delta_{k+1} < \ell \le \delta_m - \delta_k$. We set $\ell = \delta_m - \delta_{k+1} + i$, so that $1 \le i \le k+2$. Observe that the ℓth element, in descending lexicographic order, among the exponent vectors of monomials of degree 2 in $m+1$ variables, is precisely $e_{m-k} + e_{m-k+i-1}$. In particular, its first nonzero coordinate is in the position $j := m-k$, and thus the corresponding Tsfasman–Boguslavsky bound is given by

$$\begin{aligned} T_\ell &= \theta_{m-2j} + (\theta_{m-j} - \theta_{m-2j}) + (\theta_{m-(j+i-1)} - \theta_{m-j-(j+i-1)})\\ &= \theta_{m-j} + \theta_{m-i-j+1} - \theta_{m-i-2j+1}\\ &= \theta_k + \theta_{k-i+1} - \theta_{2k-m-i+1}. \end{aligned}$$

On the other hand, the corresponding bound Z_ℓ in Theorem 11.1.7 is given by

$$Z_\ell = \theta_k + \lfloor q^{\varepsilon-1} \rfloor,$$

where

$$\varepsilon = \delta_m - \delta_k - (\delta_m - \delta_{k+1} + i) = \delta_{k+1} - \delta_k - i = k + 2 - i.$$

It follows that if $0 \le k < m-1$ and $1 \le i \le k$, then

$$T_\ell - Z_\ell = \theta_{k-i} - \theta_{2k-m-i+1} \ge q^{k-i} > 0,$$

and therefore $Z_\ell < T_\ell$. Thus, Theorem 11.1.7 implies that T_ℓ cannot be the maximum value of $|S_\ell(F_1, \dots, F_\ell)|$ for arbitrary sets of ℓ linearly independent homogeneous polynomials of degree 2 in $\mathbb{F}_q[x_0, \dots, x_m]$. The number of such values of $\ell = \delta_m - \delta_{k+1} + i$ is

$$\sum_{k=0}^{m-2} \sum_{i=1}^{k} 1 = \sum_{k=0}^{m-2} k = \binom{m-1}{2},$$

which is positive if $m > 2$. □

11.1.3. Reed–Muller Codes and the Affine Case

The results of the preceding two subsections can be applied to (projective) Reed–Muller codes and their higher weights.

Projective Reed–Muller codes. Let $\mathbb{P} = \mathbb{P}^m(\mathbb{F}_q)$; let $\mathcal{P}_{m,r}$ be the space of homogeneous polynomials of degree r in $m+1$ variables over $\mathbb{F}_q$. We define a linear map

$$\widetilde{\mathrm{Ev}}\colon L_m^0(r) \to \mathbb{F}_q^{\theta_m}$$

in the following way: first, for $x \in \mathbb{P}$ and $P \in L_m^0(r)$ we set

$$\widetilde{\mathrm{Ev}}_x(F) = \frac{P(x_0, \dots, x_m)}{x_i^r} \quad \text{if} \quad x = (x_0 : \dots : x_m), \quad x_0 = \dots = x_{i-1} = 0, \quad x_i \ne 0.$$

Clearly, the value of $\widetilde{\mathrm{Ev}}_x(P)$ does not change if we replace $(x_0 : \dots : x_m)$ by $(\lambda x_0 : \dots : \lambda x_m)$, $\lambda \ne 0$; thus, this map is well defined on $\mathbb{P}^m(\mathbb{F}_q)$. Define the map $\widetilde{\mathrm{Ev}}$ as

$$\widetilde{\mathrm{Ev}}(P) = \big(\widetilde{\mathrm{Ev}}(P)\big)_{x \in \mathbb{P}};$$

then it is injective for $r < q$ (use induction on m), and its image defines a code $\mathcal{R}_q(r, \mathbb{P}^m) \subset \mathbb{F}_q^{\theta_m}$, which we call the *projective Reed–Muller code* of order r on $\mathbb{P}^m$.

The cardinality of zero locus S_P described in Theorem 11.1.1 is equal to the number of $x \in \mathbb{P}$ such that $\widetilde{\mathrm{Ev}}_x(P) = 0$, i.e., to the number of zero coordinates of the codeword $\widetilde{\mathrm{Ev}}(P)$. Since the bound of Theorem 11.1.1 is tight and since

$$\theta_m - (rq^{m-1} + \theta_{m-2}) = (q - r + 1)q^{m-1},$$

we have the following result.

THEOREM 11.1.9. *For $r < q$, the code $\mathcal{R}_q(r, \mathbb{P}^m)$ has parameters*

$$n = \theta_m, \qquad k = \binom{r+m}{r}, \qquad d = (q - r + 1)q^{m-1}.$$

□

EXAMPLE 11.1.10. For $r=1$, the space $\mathcal{P}_{m,1}$ is the space of linear forms on $\mathbb{P}^m(\mathbb{F}_q)$, and we have

$$k=m+1, \qquad d=q^m;$$

this code attains the Plotkin bound. The code $\mathcal{R}_2(1,\mathbb{P}^{m-1})$ for $q=2$ is equal to the simplex code with parameters $[2^m-1,m,2^{m-1}]$.

EXAMPLE 11.1.11. When $m=1$, the projective Reed–Muller codes are generalized Reed–Solomon codes (cf. Example 4.1.12). More precisely, take a set $X\subset\mathbb{P}^1(\mathbb{F}_q)$ with $|X|=n>r$ and define as before

$$\widetilde{\mathrm{Ev}}\colon \mathcal{P}_{1,r}\to\mathbb{F}_q^n, \quad \widetilde{\mathrm{Ev}}\colon P\mapsto \big(\widetilde{\mathrm{Ev}}_x(P)\big)_{x\in X}.$$

Then $\widetilde{\mathrm{Ev}}$ is injective (because $|X|>r$), and we get an $[n,k,d]$ MDS code with

$$n=|X|, \qquad k=r+1, \qquad d=n-r.$$

Theorem 11.1.7 allows us to calculate certain higher weights for projective Reed–Muller codes. Indeed, if $r<q$, then for $1\le\ell\le\binom{m+r}{r}$ we have

$$d_\ell(\mathcal{R}_q(r,\mathbb{P}^m))=\theta_m-\max_{F_1,\dots,F_\ell}|V(F_1,\dots,F_\ell)|,$$

where the maximum is over linearly independent $F_1,\dots,F_\ell$ in $\mathbb{F}_q[x_0,x_1,\dots,x_m]_r$; thus, from Theorem 11.1.7 we obtain the following.

COROLLARY 11.1.12. *Assume that $r=2$ and $1\le\ell\le m+1$. Then the TBC holds. In particular,*

$$\max_{F_1,\dots,F_\ell}|V(F_1,\dots,F_\ell)|=\theta_{m-1}+\lfloor q^{m-\ell}\rfloor=\begin{cases}\theta_{m-1}+q^{m-\ell} & \text{if } \ell=1,\dots,m,\\ \theta_{m-1} & \text{if } \ell=m+1.\end{cases}$$

The corollary implies that

$$d_\ell(\mathcal{R}_q(2,\mathbb{P}^m))=q^m-\lfloor q^{m-\ell}\rfloor \quad\text{for}\quad \ell=1,\dots,m+1;$$

moreover,

$$d_{\delta_m-\ell}(\mathcal{R}_q(2,\mathbb{P}^m))=\begin{cases}\theta_m-\ell & \text{if } \ell=0,1,2,\\ \theta_m-(q+\ell-2) & \text{if } \ell=3,4.\end{cases}$$

Indeed, the last five exponent vectors of monomials of degree 2 in $m+1$ variables are $(0,\dots,0,0,2)$, $(0,\dots,0,1,1)$, $(0,\dots,0,2,0)$, $(0,\dots,1,0,1)$, and $(0,\dots,1,1,0)$. The corresponding Tsfasman–Boguslavsky bound is given, respectively, by

$$0,\quad 1,\quad 2,\quad q+1,\quad\text{and}\quad q+2;$$

and it is easy to see that the bounds are attained by taking in the first case the set, say $\mathcal{Q}$, of all monomials of degree 2 in $\mathbb{F}_q[x_0,x_1,\dots,x_m]$ and, in the remaining four cases, the set obtained from $\mathcal{Q}$ by successively dropping x_m^2, x_{m-1}^2, $x_{m-1}x_m$, and x_{m-2}^2. Therefore, the TBC is true for quadrics in $\mathbb{P}^2$, implying the desired formula.

Affine case. If we consider the affine space $\mathbb{A}^m$ instead of $\mathbb{P}^m$, the situation becomes much simpler, yet highly nontrivial, and the TBC, or rather its affine analogue, holds true. The projective Reed–Muller codes get replaced by usual Reed–Muller codes, and their study leads to the following result.

THEOREM 11.1.13. *Assume that $1 \le r < q$. Then the maximum number of zeroes in $\mathbb{A}^m(\mathbb{F}_q)$ of a system of ℓ linearly independent polynomials of degree at most r in $\mathbb{F}_q[x_1,\dots,x_m]$ is*

$$H_\ell(r,m) := q^m - \left(1 + \sum_{j=1}^{m} \alpha_j q^{m-j}\right),$$

where $(\alpha_1,\dots,\alpha_m)$ is the ℓth tuple in the set $\Lambda(r,m)$ of m-tuples $(\beta_1,\dots,\beta_m)$ with coordinates from $\{0,1,\dots,q-1\}$ satisfying $\beta_1+\dots+\beta_m \ge m(q-1)-r$, and where the m-tuples are arranged lexicographically in ascending order. In particular, if $\ell \le m+1$, then this maximum number is $(r-1)q^{m-1} + \lfloor q^{m-\ell} \rfloor$.

SKETCH OF THE PROOF. The first assertion can be translated into the question of higher weights of a (usual) affine Reed–Muller code, i.e., the image of the space $L_m(r)$ of polynomials of degree at most r in m variables under the evaluation map

$$\mathrm{Ev}\colon L_m(r) \to \mathbb{F}_q^n = \mathbb{A}^n(\mathbb{F}_q);$$

cf. Sec. 1.2.2. These higher weights can be calculated using combinatorial techniques close in spirit to that of Gröbner basis theory; however, this calculation is rather cumbersome and we do not present it here.

For the last assertion, the argument is as follows. Note that $\alpha^* := (q-1-r, q-1,\dots,q-1)$ is the least element of $\Lambda(r,m)$ and, for $1<\ell\le m$, the ℓth element is obtained from α^* by changing the first coordinate to $q-r$ and the ℓth coordinate to $q-2$, whereas the $(m+1)$st element is $(q-r,q-1,\dots,q-1)$. Consequently, $H_\ell(r,m)$ simplifies to $(r-1)q^{m-1}+q^{m-\ell}$ if $1\le\ell\le m$ and to $(r-1)q^{m-1}$ if $\ell=m+1$. □

As an application of the above result, we show how the Tsfasman–Boguslavsky bound $T_\ell(r,m)$ can be readily obtained for intersections of hypersurfaces of degree r in $\mathbb{P}^m$ having a hyperplane in common.

PROPOSITION 11.1.14. *Assume that $\ell \le m+1$ and $1<r\le q$. Let $F_1,\dots,F_\ell$ be linearly independent homogeneous polynomials in S_r having a common linear factor. Then*

$$|S_\ell(F_1,\dots,F_\ell)| \le (r-1)q^{m-1} + \theta_{m-2} + \lfloor q^{m-\ell} \rfloor.$$

PROOF. Let $H \in S$ be a common linear factor of $F_1,\dots,F_\ell$. Then H is necessarily homogeneous and we may assume without loss of generality that $H = x_0$. Thus, $x_0 \mid F_i$ for all $i=1,\dots,\ell$. Put $f_i(x_1,x_2,\dots,x_m) = F_i(1,x_1,\dots,x_m)$ for $i = 1,\dots,\ell$, and let X' denote the set of common zeroes in $\mathbb{A}^m(\mathbb{F}_q)$ of the polynomials $f_1,\dots,f_\ell \in \mathbb{F}_q[x_1,\dots,x_m]$. Note that $X' = S_\ell(F_1,\dots,F_\ell)\cap\{x_0=1\}$ and thus

$$S_\ell(F_1,\dots,F_\ell) = X' \cup X'', \quad \text{where} \quad X'' := S_\ell(F_1,\dots,F_\ell)\cap\{x_0=0\} = S_\ell(x_0).$$

Since $F_1,\dots,F_\ell$ are linearly independent, so are $f_1,\dots,f_\ell$. Also, $\deg f_i \le r-1 < q$ for each $i=1,\dots,\ell$. By Theorem 11.1.13, we have $|X'| \le (r-2)q^{m-1} + \lfloor q^{m-\ell} \rfloor$; therefore,

$$|S_\ell(F_1,\dots,F_\ell)| = |X'| + |X''| \le (r-2)q^{m-1} + \lfloor q^{m-\ell} \rfloor + \theta_{m-1},$$

which yields the desired inequality. □

11.2. General Algebraic Sets

In this section we give some bounds on the number of rational points of an arbitrary algebraic set, i.e., the set of common zeroes of several polynomials in a projective or affine space. Since the geometric structure of those sets are much more complicated than that of complete intersections treated in the preceding section, one cannot hope to get bounds similar to those in Sec. 11.1. However, applying some advanced algebraic geometry techniques, one can obtain some rather interesting results, which are considered in the present section.

In Sec. 11.2.1 we very briefly describe the techniques used along with the general bounds given by G. Lachaud. Section 11.2.2 contains a more recent bound of a more combinatorial nature, due to A. Couvreur; in particular, this bound confirms a conjecture on the equidimensional case put forth by S.R. Ghorpade and G. Lachaud.

11.2.1. Lachaud's Bounds

We begin with some rather general algebraic geometry techniques, which allows us to advance in the question under consideration.

Zeta function of an algebraic variety. We start with recalling some facts on projective varieties over finite fields, i.e., absolutely irreducible algebraic projective sets (*absolutely irreducible* means that the set remains irreducible while the base field becomes algebraically closed; see below). We also assume that our varieties are smooth, i.e., do not have singular points (over the algebraic closure of the original base field).

In this case we have very precise information on the number of rational points on them. Let X be such a variety defined over a finite field $\mathbb{F}_q$, $\operatorname{char}\mathbb{F}_q = p$, and let $d = \dim X$. Then one can define the zeta function of X by exactly the same formula as in Sec. 3.1.1 in the case of curves:

$$Z_X(t) = Z(X;t) := \exp\left(\sum_{r=1}^{\infty} \frac{N_r t^r}{r}\right),$$

where N_r is the number of $\mathbb{F}_{q^r}$-rational points of X:

$$N_r = N_r(X) = |X(\mathbb{F}_{q^r})|.$$

This defines $Z_X(t)$ as a formal series with rational coefficients, $Z_X(t) \in \mathbb{Q}[[t]]$. However, in the case of a curve we have seen in Sec. 3.1.1 that in fact $Z_X(t)$ is a (development of a) rational function of t. The same is also true for an arbitrary (smooth projective) X. Note, however, that the proof of this fact is much more complicated than in the case of curves, where there exists an elementary proof. In fact, one has the following.

THEOREM 11.2.1. *Let $X/\mathbb{F}_q$ be a smooth projective variety over $\mathbb{F}_q$ of dimension d. Then*

$$Z_X(t) = \frac{P_1(t)P_3(t)\dots P_{2d-1}(t)}{P_0(t)P_2(t)\dots P_{2d}(t)},$$

where $P_i(t) \in \mathbb{Z}[t]$, $i = 0,\dots,2d$, are polynomials,

$$P_0(t) = 1-t, \qquad P_{2d}(t) = 1-q^d t,$$

and for $i = 1, \dots, 2d-1$,

$$P_i(t) = \prod_{j=1}^{b_i} (1 - \alpha_{i,j} t),$$

where $\alpha_{i,j}$ *are algebraic integers such that*

$$|\alpha_{i,j}| = q^{i/2}, \quad \forall j = 1, \dots, b_i, \quad i = 0, \dots, 2d.$$

The numbers $b_i = b_i(X)$ *are called the* Betti numbers *of* X; $b_0 = b_{2d} = 1$; *the numbers* $b_1, b_3, \dots, b_{2d-1}$ *are even.*

Moreover, if we set

$$E := \sum_{i-0}^{2d} (-1)^i b_i(X)$$

(E *is called the* Euler characteristic *of* X), *then we have the following functional equation for* $Z(t) = Z_X(t)$:

$$Z_X(q^{-d}t^{-1}) = \pm q^{dE/2} t^E Z_X(t). \qquad \square$$

The main technical result required to prove the theorem is the so-called Lefschetz–Grothendieck trace formula. To state it, recall that if we consider the variety $\overline{X}$ over the algebraic closure $\overline{\mathbb{F}}_q$ of $\mathbb{F}_q$ corresponding to X, i.e., defined by the same equation (note that $\overline{X}$ is a variety over $\overline{\mathbb{F}}_q$, since we assumed that X is absolutely irreducible), then $X(\mathbb{F}_q)$ coincides with the set $\overline{X}(\overline{\mathbb{F}}_q)^F$ of points in $\overline{X}(\overline{\mathbb{F}}_q)$ invariant under the Frobenius morphism $F = \mathrm{Frob}_q \colon \overline{X}(\overline{\mathbb{F}}_q) \to \overline{X}(\overline{\mathbb{F}}_q)$, which acts as

$$F(x_0 : x_1 : \dots : x_N) = (x_0^q : x_1^q : \dots : x_N^q)$$

on any point $(x_0 : x_1 : \dots : x_N) \in \overline{X}(\overline{\mathbb{F}}_q)$.

The theory of étale cohomologies associates with any $\overline{X}$ and any $\ell \neq p$ a string of finite-dimensional $\mathbb{Q}_\ell$-vector spaces:

$$\begin{aligned}
&H^0(\overline{X}) = \mathbb{Q}_\ell, \\
&H^i(\overline{X}), \quad i = 1, \dots, 2d, \quad \dim_{\mathbb{Q}_\ell} H^i(\overline{X}) = b_i(X), \\
&H^j(\overline{X}) = 0, \quad j \notin \{0, \dots, 2d\},
\end{aligned}$$

in a functorial manner; i.e., for any morphism $f \colon \overline{X} \to \overline{Y}$ of projective algebraic varieties, for each i one has a $\mathbb{Q}_\ell$-linear map $f^* \colon H^i(\overline{Y}) \to H^i(\overline{X})$. Moreover, the Poincaré duality says that the $\mathbb{Q}_\ell$-vector spaces $H^j(\overline{X})$ and $H^{2d-j}(\overline{X})$ are (non-canonically) isomorphic.

One can apply the functoriality to the Frobenius morphism $F \colon \overline{X} \to \overline{X}$ to obtain the following.

THEOREM 11.2.2. *The* Lefschetz–Grothendieck trace formula *holds*:

$$|X(\mathbb{F}_q)| = \sum_{i=0}^{2d} (-1)^i \,\mathrm{Tr}(F^* \mid H^i(\overline{X})).$$

Moreover, the eigenvalues of F^* *in* $H^i(\overline{X})$ *are of absolute value* $q^{i/2}$. $\square$

Theorem 11.2.1 is a formal consequence of Theorem 11.2.2 and the definition of $Z_X(t)$.

Algebraic sets. A projective algebraic set $X \subset \mathbb{P}^N$ over a field $\mathbb{k}$ (cf. Sec. 2.1.1) is the set of common zeroes of finitely many homogeneous polynomials $F_1, \ldots, F_\ell \in \mathbb{k}[T_0, T_1, \ldots, T_N]$:

$$X = \{P = (x_0 : x_1 : \ldots : x_N) \in \mathbb{P}^N \mid F_1(x_0, \ldots, x_N) = \ldots = F_\ell(x_0, \ldots, x_N) = 0\}.$$

In Sec. 2.1.1 we defined the Zariski topology on such sets, their irreducible components, varieties and their dimensions, regular and singular points. We will apply these notions in the case of a fixed finite ground field $\mathbb{k} = \mathbb{F}_q$.

We call $X = E$ a *linear variety* if all F_i, $i = 1, \ldots, \ell$, are linear forms. A linear variety $E \subset \mathbb{P} = \mathbb{P}^N$ is *properly intersecting* an algebraic subset $X \subset \mathbb{P}$ if

$$\dim(X \cap E) = \dim X - \operatorname{codim} E \quad \text{for} \quad \operatorname{codim} E \leq \dim X + 1.$$

For any projective set X, we write $\dim X = -1$ if and only if $X = \varnothing$; thus, the preceding condition implies in particular

$$X \cap E = \varnothing \quad \text{if} \quad \operatorname{codim} E = \dim X + 1.$$

We want to give a bound for the number of $\mathbb{F}_q$-points of a projective algebraic set. For that purpose, we need the following result. Assume that

$$\operatorname{codim} X - 1 \leq k \leq \dim X - 1,$$

and let $C_k(X)$ be the open subset of the Grassmannian $\mathbb{G}_k(\mathbb{P})$ formed by linear varieties of dimension k properly intersecting X; then the set $C_k(X)(\mathbb{F}_q)$ contains the linear varieties in $C_k(X)$ that are defined over $\mathbb{F}_q$.

PROPOSITION 11.2.3. *Let $X \subset \mathbb{P}$ be an algebraic set of dimension m and of degree $r \leq q+1$ defined over $\mathbb{F}_q$. If $N - m - 1 \leq k \leq N - 1$, then $C_k(X)(\mathbb{F}_q)$ is nonempty and*

$$|C_k(X)(\mathbb{F}_q)| \geq q^k(q - r + 1).$$

PROOF. (a) We first prove that if $P_1, \ldots, P_r$ are points in $\mathbb{P}$, then there are at least $q^{N-1}(q+1-r)$ hyperplanes defined over $\mathbb{F}_q$ containing none of these points. Indeed, if $H_1, \ldots, H_r$ is a sequence of r hyperplanes in $\mathbb{P}$ and $Y = H_1 \cup \ldots \cup H_r$, then it is easy to see that

$$|Y(\mathbb{F}_q)| \leq rq^{N-1} + \theta_{N-2}.$$

Hence, the number of points of $\mathbb{P}(\mathbb{F}_q)$ not lying on any of these hyperplanes is at least

$$\theta_N - rq^{N-1} - \theta_{N-2} = q^N + q^{N-1} - rq^{N-1} = q^{N-1}(q+1-r),$$

and the desired inequality follows by duality.

(b) Now we prove that

$$|C_{N-1}(X)(\mathbb{F}_q)| \geq q^{N-1}(q - r + 1). \tag{11.2.1}$$

Let $X_1, \ldots, X_g$ be the irreducible components of X and pick $P_i \in X_i(\mathbb{F}_q)$. From (a) we deduce that there are at least $q^{N-1}(q+1-g)$ hyperplanes containing none of the points P_i; such a hyperplane H does not contain any of the varieties X_i, hence $H \cap X$ is an algebraic subset of codimension 1 in X. On the other hand we have $g \leq r$, since by Bézout's theorem we have

$$\deg X_1 + \ldots + \deg X_g = \deg X,$$

and this proves (11.2.1).

(c) The desired result can be stated as

$$|C_{N-s}(X)(\mathbb{F}_q)| \geq q^{N-s}(q-r+1)$$

for $1 \leq s \leq m+1$, and we prove it by induction on s. The case $s=1$ is (11.2.1), and if the assertion is true for s, then in particular $C_{N-s}(X)(\mathbb{F}_q)$ is nonempty; hence, there is a linear subvariety E of codimension s defined over $\mathbb{F}_q$ such that $\dim(X \cap E) = m-s$. Let $X' = X \cap E$. The number of hyperplanes L of $E \cong \mathbb{P}^{N-s}$ such that $\dim(X' \cap L) = \dim X' - 1 = m-s-1$ is at least

$$q^{\dim E - 1}(q-r+1) = q^{N-s-1}(q-r+1)$$

by (11.2.1) applied to X'. Since those hyperplanes are subvarieties of $\mathbb{P}^N$ of dimension $N-s-1$, we get

$$|C_{N-s-1}(X)(\mathbb{F}_q)| \geq q^{N-s-1}(q-r+1),$$

and the proposition is thereby proved. □

We now state a relative geometric form of Noether's normalization theorem, which generalizes Theorem 11.1.4.

PROPOSITION 11.2.4. *If $X \subset \mathbb{P}$ is an algebraic set defined over $\mathbb{F}_q$ of dimension m and of degree $r < q+1$, then there exists a nonconstant finite mapping $X \to \mathbb{P}^m$ of degree r defined over $\mathbb{F}_q$. Moreover,*

$$|X(\mathbb{F}_q)| \leq r\theta_m.$$

PROOF. The second assertion is a consequence of the first one, since such a mapping is epimorphic, and the number of inverse images of any point of $\mathbb{P}^m$ does not exceed r. Now, in view of Proposition 11.2.3, there is a linear variety E of dimension $N-m-1$ in $\mathbb{P}$ defined over $\mathbb{F}_q$ such that $X \cap L = \varnothing$. Let $\varphi_E\colon X \to \mathbb{P}^m$ be the restriction of the projection mapping $P \setminus E \to \mathbb{P}^m$ with center at E; then φ_E is finite, $\deg \varphi_E = r$, and it is defined over $\mathbb{F}_q$. □

Recall that an algebraic variety $X \subset \mathbb{P}$ is *nondegenerate* in $\mathbb{P}$ if it is not contained in a hyperplane.

The *singular locus* of a projective variety X will be denoted by $\operatorname{Sing} X$: this is the smallest subvariety $Y \subset X$ such that $X \setminus Y$ is regular. Thus, $\dim(\operatorname{Sing} X) = -1$ if X is regular, and $\dim(\operatorname{Sing} X) = 0$ if X has only isolated singularities.

The following result will be used below.

PROPOSITION 11.2.5. *Let $X \subset \mathbb{P}$ be a nondegenerate algebraic variety, and let E be a linear subvariety of $\mathbb{P}$. If*

$$\dim(\operatorname{Sing} X) \leq 2\dim X - \dim \mathbb{P} - 2\operatorname{codim} E - 1,$$

then X and E intersect properly and $X \cap E$ is nondegenerate in E. □

Projective varieties. Let $\mathbb{P} = \mathbb{P}^N = \mathbb{P}(V)$ be the projectivization of an affine space V, $\dim_{\mathbb{F}_q} V = N+1$. We denote by $P_h(V,\mathbb{F}_q)$ the vector space of forms of degree h in V with coefficients in $\mathbb{F}_q$. If $f \in P_h(V,\mathbb{F}_q)$, we define $Z(f)$ as the projective subset (in fact, to be precise here, one should speak about subschemes, but we do not want to involve details which are rather technical and so ignore this distinction in what follows) of P whose ideal is the principal ideal generated by f, and we call the subset $X \cap Z(f)$ a *degree-h section* of X. We set

$$\sigma(f) = \dim(\operatorname{Sing} X \cap Z(f))$$

and

$$B_0(f,\overline{X}) = \sum_{i=0}^{m+\sigma(f)+2} \dim H^i(\overline{X}) + \sum_{i=0}^{m+\sigma(f)} \dim H^i(\overline{X}\cap\overline{Z}(f)).$$

The last number depends only of the geometry of $\overline{X}$ and $\overline{X}\cap\overline{Z}(f)$ and does not change if we replace q by q^a, $a\geq 1$. Moreover, it can be shown that if X is the reduction modulo p of an algebraic subset defined over $\mathbb{Z}$, there is a finite set S of primes such that $B_0(f,\overline{X})$ is constant for all primes $p\notin S$.

Using the Lefschetz–Grothendieck formula and its generalization to singular varieties, one can deduce the following result.

PROPOSITION 11.2.6. *Let $X\subset\mathbb{P}$ be an algebraic variety of dimension m defined over $\mathbb{F}_q$ and choose $f\in P_h(V,\mathbb{F}_q)$. Then*

$$\begin{aligned}&|(X\cap Z(f))(\mathbb{F}_q)| - |X(\mathbb{F}_q)|\\ &\quad = q\sum_{i=0}^{m+\sigma(f)} (-1)^i\operatorname{Tr}(F\,|\,H^i(\overline{X}\cap\overline{Z}(f))) - \sum_{i=0}^{m+\sigma(f)+2}(-1)^i\operatorname{Tr}(F\,|\,H^i(\overline{X}))\end{aligned}$$

and

$$\big||(X\cap Z(f))(\mathbb{F}_q)| - |X(\mathbb{F}_q)|\big| \leq B_0(f,\overline{X})q^{(m+\sigma(f)+2)/2}.\qquad\square$$

Some more advanced techniques also allow us to deduce the following result.

THEOREM 11.2.7. *Let $X\subset\mathbb{P}$ be a smooth nondegenerate variety defined over $\mathbb{F}_q$, and let $E=Z(U)$ be a linear subvariety of $\mathbb{P}$ of codimension s. Assume that $2s\leq 2\dim X-\dim\mathbb{P}$. Then*

$$\big|q^s|(X\cap E)(\mathbb{F}_q)| - |X(\mathbb{F}_q)|\big| \leq B_0(U,\overline{X})q^{(\dim\mathbb{P}+2s-1)/2},$$

and this inequality remains true if we replace q by any of its powers. $\square$

Complete intersections. Let us look at consequences of the above general theory for complete intersections. They seem less precise than the result of Secs. 11.1.1 and 11.1.2 above; however, a direct comparison is difficult, since the results below are expressed in terms different from those of Sec. 11.1.

Let an algebraic set $X\subset\mathbb{P}$ of codimension t be a *complete intersection* of multidegree $\boldsymbol{r}=(r_1,\dots,r_t)$, i.e., its defining ideal has a system of t generators of respective degrees $r_1,\dots,r_t$. The multidegree $\boldsymbol{r}$ depends only on X but not on the chosen system of generators.

THEOREM 11.2.8. (a) *If $X\subset\mathbb{P}$ is a nondegenerate complete intersection and if $E\subset\mathbb{P}$ is a linear subvariety, then*

$$\dim(\operatorname{Sing}(X\cap E)) \leq \dim(\operatorname{Sing}X) + \operatorname{codim}E.$$

(b) *Moreover, if*

$$\dim(\operatorname{Sing}X) \leq \dim X - 2\operatorname{codim}E - 2,$$

then X and E intersect properly, and the set $X\cap E$ is a nondegenerate normal subvariety of E which is a complete intersection. $\square$

Denote by $b_0(m,\boldsymbol{r})$ the mth Betti number of all smooth complete intersections of a given dimension m and multidegree $\boldsymbol{r}$. Define $\varepsilon_m=0$ if m is odd and $\varepsilon_m=1$ if m is even, and define the *corrected* Betti number as $b(m,\boldsymbol{r})=b_0(m,\boldsymbol{r})-\varepsilon_m$.

The two principal results on the number of points on a complete intersection (Lachaud's bounds) are as follows.

PROPOSITION 11.2.9. *There exists a constant $B_1(m,\boldsymbol{r},s)$ such that for any finite field $\mathbb{F}_q$, any smooth complete intersection X of multidegree $\boldsymbol{r}$, $\dim_{\mathbb{F}_q} X = m$, and any linear variety E of codimension s such that $2s \le \dim X - 1$, we have*

$$\big||X_E(\mathbb{F}_q)| - \theta_{m-s}\big| \le B_1(m,\boldsymbol{r},s)q^{m/2},$$

with $B_1(m,\boldsymbol{r},0) = b(m,\boldsymbol{r})$ and $B_1(m,\boldsymbol{r},1) = b(m,\boldsymbol{r}) + b(m-1,\boldsymbol{r})$. □

THEOREM 11.2.10. *There exists a constant $B_2(m,\boldsymbol{r},s)$ such that for any finite field $\mathbb{F}_q$, for any smooth nondegenerate complete intersection X of multidegree $\boldsymbol{r}$, $\dim_{\mathbb{F}_q} X = m$, and for any subspace $U \subset L(V,\mathbb{F}_q)$ of dimension s such that $2s \le 2\dim X - \dim \mathbb{P}$, we have*

$$\big|q^s|X_{Z(U)}(\mathbb{F}_q)| - |X(\mathbb{F}_q)|\big| \le B_2(m,\boldsymbol{r},s)q^{(m+2s)/2}.$$

If $s = 1$, we can replace the right-hand side by

$$b(m,\boldsymbol{r})q^{m/2} + (b(m-1,\boldsymbol{r}) + b(m-2,\boldsymbol{r}))q^{m+1/2}.$$ □

Now we get the following consequence.

COROLLARY 11.2.11. *If X is a smooth nondegenerate hypersurface of dimension $m \ge 3$ and degree r defined over $\mathbb{F}_q$, $f \in L(V,\mathbb{F}_q)$, and $H = Z(f)$, then*

$$\big||X_H(\mathbb{F}_q)| - \theta_{m-1}\big| \le (r-1)^m q^{m/2}$$

and

$$\big|q|X_H(\mathbb{F}_q)| - |X(\mathbb{F}_q)|\big| \le (r-1)^m(q+r-1)q^{m/2}.$$

PROOF. We use the well-known formula

$$b(m,r) = \frac{r-1}{r} - ((r-1)^{m+1} - (-1)^{m+1});$$

thus, $b(1,r) = (r-1)(r-2)$, etc. If $r \ge 2$ and $m \ge 2$, then

$$B_1(m-1,r,1) = b(m-1,r) + b(m-2,r) = (r-1)^m,$$
$$b(m,r) \le (r-1)^{m+1},$$
$$b(m,r) + q(b(m-1,r) + b(m-2,r)) \le (r-1)^m(q+r-1);$$

hence, Proposition 11.2.9 and Theorem 11.2.10 imply the result. □

11.2.2. Couvreur's Bound

Now we describe a bound due to A. Couvreur, which depends only on the dimension of an algebraic set, its degree, and the dimension of the ambient projective space. Therefore, it can be considered as a further generalization of Theorem 11.1.4 and Proposition 11.2.4.

Let X be an affine or projective algebraic set defined over $\mathbb{F}_q$, and let

$$X = X_1 \cup X_2 \cup \ldots \cup X_\ell$$

be its *irredundant decomposition*, where $X_1,\ldots,X_\ell$ are irreducible varieties and $X_i \not\subseteq X_j$ for all $i \ne j$. For any $i \in \{1,\ldots,\ell\}$ we let $d_{X_i} = \dim X_i \ge 0$ and $r_{X_i} = \deg X_i \ge 1$. The sequences $(d_{X_i})_{i\in\{1,\ldots,\ell\}}$ and $(r_{X_i})_{i\in\{1,\ldots,\ell\}}$ are referred to as the *dimension sequence* and *degree sequence* of X. It is worth noting that the d_{X_i} need not be distinct. For instance, when X is equidimensional, all the d_{X_i} are equal. Finally, we will denote

$$D_X := \max\{d_{X_1}, \ldots, d_{X_\ell}\} = \dim X.$$

We begin with the affine case, which is rather simple.

THEOREM 11.2.12. *Let $X \subseteq \mathbb{A}^m$ be an affine variety; then*

$$|X(\mathbb{F}_q)| \leq \sum_{i=1}^{\ell} r_{X_i} q^{d_{X_i}}. \qquad \square$$

In the projective case, the result reads as follows.

THEOREM 11.2.13 (Couvreur's bound). *Let $X \subseteq \mathbb{P}^m$ be a projective variety, and let $d_{X_i} < m$ for all $i \in \{1, \ldots, \ell\}$. Then*

$$|X(\mathbb{F}_q)| \leq \left(\sum_{i=1}^{\ell} r_{X_i} (\theta_{d_{X_i}} - \theta_{2d_{X_i} - m}) \right) + \theta_{2D_X - m}.$$

The following result is a straightforward corollary of Theorem 11.2.13, since it is nothing but the equidimensional case.

COROLLARY 11.2.14 (Ghorpade–Lachaud conjecture). *Let $X \subseteq \mathbb{P}^m$ be an equidimensional projective variety of dimension $d < m$ and degree r. Then*

$$|X(\mathbb{F}_q)| \leq r(\theta_d - \theta_{2d-m}) + \theta_{2d-m}. \qquad \square$$

Actually, Ghorpade and Lachaud stated this conjecture under the additional assumption that X is a complete intersection; however, it turns out to be true without this assumption.

Before passing to the proof of Theorem 11.2.13, let us fix some notation. For $X \subseteq \mathbb{P}^m$ with given dimension and degree sequences, we denote the right-hand side of the Couvreur bound by $B_m(X)$. In what follows, we denote the set of $\mathbb{F}_q$-hyperplanes in $\mathbb{P}^n$ by $(\mathbb{P}^m)^*(\mathbb{F}_q)$. We will need the following auxiliary result.

PROPOSITION 11.2.15. *Let $X \subseteq \mathbb{P}^m$ be a projective variety with irredundant decomposition $X_1 \cup \ldots \cup X_s$. Let $H \in (\mathbb{P}^m)^*(\mathbb{F}_q)$ be a hyperplane which does not contain any of the X_i. Then*

$$B_{m-1}(X \cap H) = \left(\sum_{i=1}^{s} r_{X_i} \left(\theta_{d_{X_i}-1} - \theta_{2d_{X_i}-m-1}\right) \right) + \theta_{2D_X - m - 1};$$

in particular, it does not depend on H, and we denote it by $B_{m-1}(X \cap \cdot)$.

PROOF. For all $i \in \{1, \ldots, s\}$, define $Y_{i,1}, \ldots, Y_{i,s_i}$ as the components of the irredundant decomposition of $X \cap H$. We have

$$B_{m-1}(X \cap H) = \left(\sum_{i=1}^{s} \sum_{j=1}^{s_i} r_{Y_{i,j}} \left(\theta_{d_{Y_{i,j}}} - \theta_{2d_{Y_{i,j}} - (m-1)} \right) \right) + \theta_{2D_{X \cap H} - (m-1)}.$$

From the generalized Bézout theorem, for all i, j we have $d_{Y_{i,j}} = d_{X_i - 1}$; hence, $D_{X \cap H} = D_X - 1$ and

$$\begin{aligned} B_{m-1}(X \cap H) &= \left(\sum_{i=1}^{s} \sum_{j=1}^{s_i} r_{Y_{i,j}} \left(\theta_{d_{X_i}-1} - \theta_{2(d_{X_i}-1)-(m-1)} \right) \right) + \theta_{2(D_X-1)-(m-1)} \\ &= \left(\sum_{i=1}^{s} \left(\theta_{d_{X_i}-1} - \theta_{2d_{X_i}-m-1} \right) \sum_{j=1}^{s_i} r_{Y_{i,j}} \right) + \theta_{2D_X-m-1}. \end{aligned}$$

For all i, we have $\sum_{j=1}^{s_i} r_{Y_{i,j}} = r_{X_i \cap H}$, which equals r_{X_i} by the generalized Bézout theorem. □

Proof of Theorem 11.2.13. The proof can be given by induction on the dimension m of the ambient space. For $m = 1$ the claim is obvious, and for $m = 2$ it directly follows from the Tsfasman–Serre–Sørensen bound above.

From now on, we assume that $m \geq 3$. Consider the following two particular cases.

Case 1. No irreducible component of X is contained in a hyperplane.

Case 2. Every irreducible component of X is either linear or is not contained in a hyperplane.

We first prove the induction step in Case 1, then in Case 2, and finally treat the general case.

Proof in Case 1. First, note that the upper bound in Theorem 11.2.13 is obviously true if $X(\mathbb{F}_q) = \varnothing$. Therefore, assume from now on that $X(\mathbb{F}_q)$ is nonempty and let $P \in X(\mathbb{F}_q)$. By the condition of Case 1, no hyperplane $H \in (\mathbb{P}^m)^*(\mathbb{F}_q)$ passing through P contains an irreducible component of X. Next, let us introduce a bipartite graph Γ whose first and second vertex sets are

$$\begin{aligned} \mathcal{V}_1 &:= X(\mathbb{F}_q) \setminus \{P\}, \\ \mathcal{V}_2 &:= \{H \in (\mathbb{P}^m)^*(\mathbb{F}_q) \mid P \in H\} \end{aligned}$$

and whose edge set is

$$\mathcal{E} := \{(Q, H) \in \mathcal{V}_1 \times \mathcal{V}_2 \mid Q \in H\}.$$

Now let us count the number of edges in two distinct manners.

In what follows, the number of edges containing a given vertex is referred to as the *valency* of the vertex (but not the degree, to avoid confusion with the degree of a variety).

For the graph Γ, we have the following:

(i) $|\mathcal{V}_2| = \theta_{m-1}$.
(ii) The valency of a vertex $Q \in \mathcal{V}_1$ equals θ_{m-2}.
(iii) The valency of a vertex $H \in \mathcal{V}_2$ equals $|(X \cap H)(\mathbb{F}_q) \setminus \{P\}|$ which, by induction, is bounded above by $B_{m-1}(X \cap H) - 1$ and hence by $B_{m-1}(X \cap \cdot) - 1$ (see Proposition 11.2.15). Actually, this holds even if the intersection $X \cap H$ is not transversal.

First, assume that $B_{m-1}(X\cap\cdot)\geq\theta_{m-1}$. From Proposition 11.2.15, we have

$$\left(\sum_{i=1}^{\ell} r_{X_i}(\theta_{d_{X_i}-1}-\theta_{2d_{X_i}-m-1})\right)+\theta_{2D_X-m-1}\geq\theta_{m-1},$$

and multiplying both sides by q yields

$$\left(\sum_{i=1}^{\ell} r_{X_i}(\theta_{d_{X_i}}-\theta_{2d_{X_i}-m})\right)+\theta_{2D_X-m}\geq\theta_{m}.$$

Since $X\subseteq\mathbb{P}^m$, we have $|X(\mathbb{F}_q)|\leq\theta_m$, and the result is clear.

From now on, we assume that

$$B_{m-1}(X\cap\cdot)<\theta_{m-1}. \tag{11.2.2}$$

Then we have

$$|\mathcal{E}|=|\mathcal{V}_1|\theta_{m-1}=(|X(\mathbb{F}_q)|-1)\theta_{m-2}$$

from (ii) and

$$|\mathcal{E}|\leq\theta_{m-1}(B_{m-1}(X\cap\cdot)-1)$$

from (i) and (iii), which implies

$$|X(\mathbb{F}_q)|\leq 1+\frac{\theta_{m-1}}{\theta_{m-2}}(B_{m-1}(X\cap\cdot)-1)\leq 1+\left(q+\frac{1}{\theta_{m-2}}\right)(B_{m-1}(X\cap\cdot)-1).$$

By the definition of $B_{m-1}(X\cap\cdot)$, we get the inequality

$$|X(\mathbb{F}_q)|\leq\left(\sum_{i=1}^{\ell} r_{X_i}(\theta_{d_{X_i}}-\theta_{2d_{X_i}-m})\right)+\theta_{2D_X-m}-q+\frac{B_{m-1}(X\cap\cdot)-1}{\theta_{m-2}},$$

and (11.2.2) implies $\frac{B_{m-1}(X\cap\cdot)-1}{\theta_{m-2}}<\frac{\theta_{m-1}-1}{\theta_{m-2}}$, so we obtain $\frac{B_{m-1}(X\cap\cdot)-1}{\theta_{m-2}}<q$, which yields

$$|X(\mathbb{F}_q)|\leq\left(\sum_{i=1}^{\ell} r_{X_i}(\theta_{d_{X_i}}-\theta_{2d_{X_i}-m})\right)+\theta_{2D_X-m}.$$

Proof in Case 2. The proof in this case is done in three steps. Under the assumption that no nonlinear irreducible component of X is contained in an $\mathbb{F}_q$-rational hyperplane,

(1) We first treat the case where one of the irreducible components of X is a hyperplane.
(2) Then we treat the case where only one irreducible component L of X is linear and L is not a hyperplane.
(3) Finally, we treat the case of multiple linear irreducible components which are not hyperplanes.

For the first step, we need the following lemma.

LEMMA 11.2.16. *Let Y be an irreducible closed subvariety of $\mathbb{P}^m$ of dimension d and degree r. Let H be an $\mathbb{F}_q$-rational hyperplane in $\mathbb{P}^m$, and let Y_{aff} be the affine chart $Y\setminus H$ of Y; then Y_{aff} has degree d and dimension r.* □

<u>If X has an $\mathbb{F}_q$-rational hyperplane in its irredundant decomposition.</u> Write the irredundant decomposition of X as a union of irreducible varieties

$$X=X_1\cup X_2\cup\ldots\cup X_\ell,\quad\text{with}\quad X_1=H$$

and such that $X_i \not\subseteq X_j$ for all $i \neq j$. Note that $D_X = d_H = m-1$. Now, X is the disjoint union of H and $X_{\text{aff}} := X \setminus H$. The variety X_{aff} is affine, and its dimension and degree sequences are $d_{X_2}, \ldots, d_{X_\ell}$ and $r_{X_2}, \ldots, r_{X_\ell}$ by Lemma 11.2.16. From Theorem 11.2.12 we get

$$|X(\mathbb{F}_q)| = |H(\mathbb{F}_q)| + |X_{\text{aff}}(\mathbb{F}_q)| \leq \theta_{m-1} + \sum_{i=2}^{\ell} r_{X_i} q^{d_{X_i}}.$$

Next, note that for all i, we have

$$q^{d_{X_i}} \leq \theta_{d_{X_i}} - \theta_{2d_{X_i}-m}.$$

Indeed, since $d_{X_i} < m$, we have $2d_{X_i} - m < d_{X_i}$. On the other hand, since $r_H = 1$, we get

$$\theta_{m-1} = \theta_{d_H} = r_H(\theta_{d_H} - \theta_{2d_H-m}) + \theta_{2d_H-m}.$$

Therefore,

$$|X(\mathbb{F}_q)| \leq \left(r_H(\theta_{d_H} - \theta_{2d_H-m}) + \sum_{i=2}^{\ell} r_{X_i}(\theta_{d_{X_i}} - \theta_{2d_{X_i}-m}) \right) + \theta_{2d_H-m},$$

and the result follows, since $d_H = D_X$.

<u>If X has a single linear subvariety L with $d_L < m-1$ in its irredundant decomposition.</u> Moreover, we assume that the other irreducible components of X are nonlinear; recall that none of the other components is contained in a hyperplane. We write the irredundant decomposition of X as a union of irreducible components:

$$X = X_1 \cup X_2 \cup \ldots \cup X_\ell, \quad \text{with} \quad X_1 = L.$$

Now we consider a new incidence graph Γ_1, choose $P \in L(\mathbb{F}_q)$, and set

$$\begin{aligned}
\mathcal{V}_1 &:= (X \setminus L)(\mathbb{F}_q); \\
\mathcal{V}_2 &:= \{H \in (\mathbb{P}^m)^*(\mathbb{F}_q) \mid P \in H \text{ and } L \not\subseteq H\}; \\
\mathcal{E} &:= \{(Q,H) \in \mathcal{V}_1 \times \mathcal{V}_2 \mid Q \in H\}.
\end{aligned}$$

Here, we have

$$|\mathcal{V}_2| = \theta_{m-1} - \theta_{m-d_L-1},$$

since the set of hyperplanes in $(\mathbb{P}^m)^*(\mathbb{F}_q)$ containing L has θ_{m-d_L-1} elements. Next, note that in Γ_1 we have the following:

- The valency of a vertex of $\mathcal{V}_1$ is $\theta_{m-2} - \theta_{m-d_L-2}$.
- By induction and since no irreducible component of X but L is contained in a hyperplane, the valency of a vertex H of $\mathcal{V}_2$ equals $|(X \cap H)(\mathbb{F}_q)| - |(L \cap X)(\mathbb{F}_q)|$ which, from Proposition 11.2.15 is bounded above by

$$\sum_{i=1}^{\ell} \left(r_{X_i}(\theta_{d_{X_i}-1} - \theta_{2d_{X_i}-m-1}) \right) + \theta_{2D_X-m-1} - \theta_{d_L-1},$$

where the "$-\theta_{d_L-1}$" term corresponds to the rational points of $L \cap H$, which are not counted.

Counting as above the number of edges of Γ_1 in two different ways, we get

$$\begin{aligned}&(|X(\mathbb{F}_q)|-\theta_{d_L})(\theta_{m-2}-\theta_{m-d_L-2})\\&\le(\theta_{m-1}-\theta_{m-d_L-1})\left[\left(\sum_{i=1}^{\ell}r_{X_i}(\theta_{d_{X_i}-1}-\theta_{2d_{X_i}-m-1})\right)+\theta_{2D_X-m-1}-\theta_{d_L-1}\right],\end{aligned}$$

and therefore

$$\begin{aligned}|X(\mathbb{F}_q)|&\le\theta_{d_L}+q\left[\left(\sum_{i=1}^{\ell}r_{X_i}(\theta_{d_{X_i}-1}-\theta_{2d_{X_i}-m-1})\right)+\theta_{2D_X-m-1}-\theta_{d_L-1}\right]\\&\le\theta_{d_L}+\left(\sum_{i=1}^{\ell}r_{X_i}(\theta_{d_{X_i}}-\theta_{2d_{X_i}-m})\right)+\theta_{2D_X-m}-\theta_{d_L}\\&\le\left(\sum_{i=1}^{\ell}r_{X_i}(\theta_{d_{X_i}}-\theta_{2d_{X_i}-m})\right)+\theta_{2D_X-m}.\end{aligned}$$

If there are several linear components $L_1, L_2, \ldots, L_s$, $s > 1$. Now we have

$$X = Y \cup L_1 \cup \ldots \cup L_s,$$

where Y has no linear irreducible component. Moreover, none of the irreducible components $Y_1, \ldots, Y_\ell$ of Y is contained in a hyperplane of $\mathbb{P}^m$ (by the conditions of Case 2). Assume that $d_{L_1} \ge d_{L_2} \ge \ldots \ge d_{L_s}$. Then as above, we get

$$|(Y\cup L_1)(\mathbb{F}_q)|\le\left(\sum_{i=1}^{\ell}r_{Y_i}(\theta_{d_{Y_i}}-\theta_{2d_{Y_i}})\right)+(\theta_{d_{L_1}}-\theta_{2d_{L_1}-m})+\theta_{2D_X-m}.$$

Now note that

$$|X(\mathbb{F}_q)|\le|(Y\cup L_1)(\mathbb{F}_q)|+\sum_{i=2}^{s}(|L_i(\mathbb{F}_q)|-|(L_1\cap L_i)(\mathbb{F}_q)|).$$

Moreover, we have

$$\dim L_1\cap L_i\ge d_{L_1}+d_{L_i}-m\ge 2d_{L_i}-m$$

for all $i \in \{2, \ldots, s\}$. Putting the last two inequalities together, we get

$$\begin{aligned}|X(\mathbb{F}_q)|\le\left(\sum_{i=1}^{\ell}r_{Y_i}(\theta_{d_{Y_i}}-\theta_{2d_{Y_i}})\right)&+(\theta_{d_{L_1}}-\theta_{2d_{L_1}-m})\\&+\theta_{2D_X-m}+\sum_{i=2}^{s}(\theta_{d_{L_i}}-\theta_{2d_{L_i}-m});\end{aligned}$$

hence,

$$|X(\mathbb{F}_q)|\le\left(\sum_{i=1}^{\ell}r_{Y_i}(\theta_{d_{Y_i}}-\theta_{2d_{Y_i}})\right)+\left(\sum_{i=1}^{s}(\theta_{d_{L_i}}-\theta_{2d_{L_i}-m})\right)+\theta_{2D_X-m},$$

which completes the proof.

Proof in the general case. Now assume that

$$X = Y \cup Z = Y_1 \cup \ldots \cup Y_s \cup Z_1 \cup \ldots \cup Z_\ell$$

where the Y_i are irreducible varieties which are either linear or are not contained in any hyperplane in $(\mathbb{P}^m)^*(\mathbb{F}_q)$ and the Z_i are nonlinear and each one is contained in at least one hyperplane in $(\mathbb{P}^m)^*(\mathbb{F}_q)$. In particular, since the Z_i are nonlinear, we have $r_{Z_i} > 1$ for all $i \in \{1,\dots,\ell\}$. From the previous results we have

$$|Y(\mathbb{F}_q)| \leq \left(\sum_{i=1}^{\ell} r_{Y_i}(\theta_{d_{Y_i}} - \theta_{2d_{Y_i}-m})\right) + \theta_{2D_X-m}, \tag{11.2.3}$$

since $D_Y \leq D_X$. Next, for all $i \in \{1,\dots,\ell\}$, since Z_i is contained in a hyperplane, one can apply the induction hypothesis to this hyperplane and get

$$|Z_i(\mathbb{F}_q)| \leq B_{m-1}(Z_i) = r_{Z_i}(\theta_{d_{Z_i}} - \theta_{2d_{Z_i}-m+1}) + \theta_{2d_{Z_i}-m+1}. \tag{11.2.4}$$

LEMMA 11.2.17. *For all integers $d > 0$ and $r > 1$, we have*

$$r(\theta_d - \theta_{2d-m+1}) + \theta_{2d-m+1} \leq r(\theta_d - \theta_{2d-m}).$$

PROOF. Consider the difference

$$\begin{aligned} r(\theta_d - \theta_{2d-m}) - \left(r(\theta_d - \theta_{2d-m+1}) + \theta_{2d-m+1}\right) &= r(\theta_{2d-m+1} - \theta_{2d-m}) - \theta_{2d-m+1} \\ &= rq^{2d-m+1} - \theta_{2d-m+1} \\ &= (r-1)q^{2d-m+1} - \theta_{2d-m}. \end{aligned}$$

Since $r > 1$, it suffices to prove that $q^{2d-m+1} \geq \theta_{2d-m}$, which is obviously true if $2d - m < 0$. It also holds true if $2d - m \geq 0$, since

$$\theta_{2d-m} = \frac{q^{2d-m+1} - 1}{q - 1} \leq q^{2d-m+1} - 1.$$

□

From Lemma 11.2.17 we see that, for all $i \in \{1,\dots,\ell\}$,

$$r_{Z_i}(\theta_{d_{Z_i}} - \theta_{2d_{Z_i}-m+1}) + \theta_{2d_{Z_i}-m+1} \leq r_{Z_i}(\theta_{d_{Z_i}} - \theta_{2d_{Z_i}-m}), \tag{11.2.5}$$

since $r_{Z_i} > 1$ for all $i \in \{1,\dots,\ell\}$. Next, the obvious inequality

$$|X(\mathbb{F}_q)| \leq |Y(\mathbb{F}_q)| + \sum_{j=1}^{\ell} |Z_j(\mathbb{F}_q)|$$

together with the bounds (11.2.3), (11.2.4), and (11.2.5) imply

$$|X(\mathbb{F}_q)| \leq \left(\sum_{i=1}^{\ell} r_{Y_i}(\theta_{d_{Y_i}} - \theta_{2d_{Y_i}-m})\right) + \left(\sum_{i=1}^{\ell} r_{Z_i}(\theta_{d_{Z_i}} - \theta_{2d_{Z_i}-m})\right) + \theta_{2D_X-m},$$

which completes the proof. □

REMARK 11.2.18. One can show that the bound is optimal for equidimensional varieties; moreover, it is attained on certain unions of linear subvarieties.

REMARK 11.2.19. The Couvreur bound $C_\ell(r)$ and the TBC bound $T_\ell(r)$ are not comparable in general. In particular, $C_\ell(r)$ is sharper than $T_\ell(r)$ for $r > 2$, $\ell > 2$, whereas $T_\ell(2)$ is better than $C_\ell(2)$ for quadrics, $r = 2$.

11.3. Codes on Surfaces

In the present section we give some examples of codes on algebraic surfaces, that is, on algebraic varieties of dimension 2. This case is intermediate between the case of curves and that of general algebraic varieties, since for surfaces we have a rather comprehensive geometric classification, which is actually out of reach in higher dimensions. Note, however, that this classification is much more complicated than that of curves, where geometric properties are controlled by the genus.

This complexity of various surfaces does not allow one to give good constructions of codes except in very special cases. Actually, a relatively satisfactory analysis of codes on surfaces can only be given for surfaces close to the projective plane.

We review the surface theory in Sec. 11.3.1. The next subsection treats some cubic surfaces (which are rational) and the corresponding codes, and the last one is devoted to some other rational surfaces.

11.3.1. Some Elements of the Theory of Surfaces

First we recall some facts on algebraic surfaces which are used below for constructing codes from surfaces.

First we consider their geometry, i.e., the theory over an algebraically closed field (we need the field $\overline{\mathbb{k}} = \overline{\mathbb{F}}_q$ only, the algebraic closure of the finite field $\mathbb{F}_q$), and then some elements of their arithmetic, i.e., the theory of rational points over a finite field.

Thus, we begin with an algebraic smooth projective surface $\overline{X}$ over $\mathbb{K} = \overline{\mathbb{k}}$, $\mathbb{k}$ being a finite field. Actually, we will only use the fact that $\mathbb{K}$ is algebraically closed but not its concrete nature. Thus, $\overline{X}$ is an irreducible algebraic set of dimension 2 in some $\mathbb{P}^N$, $\overline{X} \subset \mathbb{P}^N$.

Let us discuss divisors on $\overline{X}$. First, let $f \in \mathbb{K}(\overline{X}) \setminus \{0\}$ be a nonzero rational function, i.e., $f(x) = \dfrac{P(x)}{Q(x)}$, where $P(x)$ and $Q(x)$ are homogeneous polynomials of the same degree in variables $(x_0,\dots,x_N)$ and $Q(x)|_{\overline{X}}$ is not identically zero. Note that we consider two rational functions $f(x)$ and $g(x) = \dfrac{P_1(x)}{Q_1(x)}$ to be equal as elements of $\mathbb{K}(\overline{X})$ if and only if the polynomial $P(x)Q_1(x) - P_1(x)Q(x)$ is identically zero on $\overline{X}$.

The equations $P(x) = 0$ and $Q(x) = 0$ restricted to $\overline{X}$ define curves, i.e., algebraic sets of dimension 1. Any such curve is a finite union of irreducible curves, and thus

$$P(x)|_{\overline{X}} = P_1^{e_1} \dots P_\ell^{e_\ell}|_{\overline{X}}, \qquad Q(x)|_{\overline{X}} = Q_1^{f_1} \dots Q_s^{f_s}|_{\overline{X}},$$

where $P_i = 0$ (respectively, $Q_j = 0$) define irreducible curves C_i (respectively, D_j) for any $i \in \{1,\dots,\ell\}$ and any $j \in \{1,\dots,s\}$.

Then we define the divisor on $\overline{X}$ associated to f as

$$(f) := \sum_{i=1}^{\ell} e_i C_i - \sum_{i=1}^{s} f_j D_j \in \operatorname{Div}(\overline{X}),$$

where $\operatorname{Div}(\overline{X})$ is a free abelian group generated by all irreducible curves $C \subset \overline{X}$.

PROPOSITION 11.3.1. (i) *This definition is correct, i.e., depends only on f but not on a representation $f = \dfrac{P(x)}{Q(x)}$.*

(ii) *Define the degree of a divisor by*

$$\deg D = \sum_{i=1}^{r} a_j \quad \textit{for} \quad D = \sum_{i=1}^{r} a_j C_j;$$

then $\deg(f) = 0$.

(iii) *Let* $f, g \in \mathbb{K}(\overline{X}) \setminus \{0\}$. *Then*

$$(fg) = (f) + (g); \quad \textit{in particular,} \quad (1) = 0, \quad (f^{-1}) = -(f). \qquad \square$$

If $D_1, D_2 \in \operatorname{Div}(\overline{X})$, we call D_1 *(linearly) equivalent* to D_2 if and only if there exists a rational function $f \in \mathbb{K}(\overline{X}) \setminus \{0\}$ with $D_1 - D_2 = (f)$.

PROPOSITION 11.3.2. *Let* $S = \{P_1, \dots, P_s\}$ *be a finite set of points on* $\overline{X}$, *and let* $D = \sum a_j C_j \in \operatorname{Div}(\overline{X})$. *Then there exists* $D' \in \operatorname{Div}(\overline{X})$ *and a rational function* $f \in \mathbb{K}(\overline{X}) \setminus \{0\}$ *with* $D' - D = (f)$ *and* $\operatorname{Supp} D' \cap S = \varnothing$, *where for* $D = \sum a_j C_j$ *the support is defined by* $\operatorname{Supp} D := \bigcup_{j:\, a_j \neq 0} C_j$. $\square$

Now let $\varphi\colon \overline{X} \to \overline{Y}$ be a surjective algebraic map between algebraic surfaces. Then $\varphi^{-1}(Q)$ is a finite set for any point $Q \in \overline{Y}$ outside an exceptional finite set $S := S_\varphi = \{Q_1, \dots, Q_t\}$, whereas $\varphi_{-1}(Q_i) = C_i$ is a (possibly reducible) curve on $\overline{X}$.

If $C \cap S_\varphi = \varnothing$, then we define $\varphi^*(C) = \varphi^{-1}(C)$, which is simply a (reducible) curve defined by the equation $g(\varphi(x)) = 0$, where C is defined by $g(y) = 0$. Since for $f \in \mathbb{K}(\overline{Y}) \setminus \{0\}$ we have $(\varphi^*(f)) = (f \circ \varphi)$, using Proposition 11.3.2 we can define the homomorphism

$$\varphi^*\colon \operatorname{Pic}(\overline{Y}) \to \operatorname{Pic}(\overline{X})$$

for Picard groups

$$\operatorname{Pic}(\overline{X}) = \operatorname{Div}(\overline{X})/P(\overline{X}), \qquad \operatorname{Pic}(\overline{Y}) = \operatorname{Div}(\overline{Y})/P(\overline{Y}),$$

$P(\overline{X})$ and $P(\overline{Y})$ being the subgroups of principal divisors.

Intersection index. The group $\operatorname{Div}(\overline{X})$ has also another structure, extremely important in applications. Namely, there is a map

$$(\cdot,\cdot)\colon \operatorname{Div}(\overline{X}) \times \operatorname{Div}(\overline{X}) \to \mathbb{Z}, \quad C, D \mapsto (C, D),$$

with the following properties, which characterize this map up to a multiplicative constant:

(i) If $C \in P(\overline{X})$, i.e., $C = (f)$ for $f \in \mathbb{K}(\overline{X}) \setminus \{0\}$, then $(C, D) = 0$ for any D.
(ii) $(C, D) = (D, C)$.
(iii) $(C_1 + C_2, D) = (C_1, D) + (C_2, D)$.
(iv) If $\operatorname{Supp} C \cap \operatorname{Supp} D = \varnothing$, then $(C, D) = 0$.

Informally, for distinct irreducible curves C and D, (C, D) counts the number of intersection points $P \in (C \cap D)$ with multiplicities (if C is defined by $F = 0$, then $(C, D) = \deg(F|_D)$ is the degree of the divisor cut by F on D, which uniquely determines (C, D)). Therefore, we have the property

(v) If C and D are different irreducible curves, then $(C, D) \geq 0$.

REMARK 11.3.3. On the contrary, as we will see below, there are many cases where $(C, C) = (C^2) < 0$ for an irreducible curve C.

Now define

$$\mathrm{Div}^0(\overline{X}) = \{C \in \mathrm{Div}(\overline{X}) : (C,D) = 0 \text{ for any } D \in \mathrm{Div}(\overline{X})\}$$

Clearly, $\mathrm{Div}^0(\overline{X})$ is a subgroup of $\mathrm{Div}(\overline{X})$ containing $P(\overline{X})$, which allows us to define

$$\mathrm{Pic}^0(\overline{X}) := \mathrm{Div}^0(\overline{X})/P(\overline{X})$$

and also the Néron–Severi group

$$\mathrm{NS}(\overline{X}) := \mathrm{Div}(\overline{X})/\mathrm{Div}^0(\overline{X}) = \mathrm{Pic}(\overline{X})/\mathrm{Pic}^0(\overline{X}).$$

One has the following fundamental result of the theory of surfaces.

PROPOSITION 11.3.4. *The abelian group* $\mathrm{NS}(\overline{X})$ *is finitely generated.* □

Note that the degree gives a surjection $\deg\colon \mathrm{NS}(\overline{X}) \to \mathbb{Z}$, and sometimes this is a bijection, which will be an interesting case for us.

EXAMPLE 11.3.5. Let $\overline{X} = \mathbb{P}^2$. Then any divisor $D = \sum a_i C_i$ is equivalent to aL, where $L = (x_0 = 0)$ is the line at infinity and $a = \sum a_i \deg C_i = \deg D$. Indeed, any curve $C_i \subset \mathbb{P}^2$ is defined by a homogeneous equation $F_i = 0$ of degree $\deg F_i = \deg C_i$, and $f := \prod_i (F_i)^{a_i}/(x_0)^a$ gives the linear equivalence:

$$D = (f) + aL.$$

Therefore, we conclude that $\mathrm{Pic}^0(\mathbb{P}^2) = 0$ and $\mathrm{Pic}(\mathbb{P}^2) = \mathrm{NS}(\mathbb{P}^2) \cong \mathbb{Z}$.

Canonical class. Exactly as in the case of curves, any (smooth projective) surface has an especially important class of divisors called the canonical class (cf. Sec. 2.2.1 for curves). It can be defined as the class associated to differential forms (of degree 2 on surfaces), but we do not give this definition here; instead, we use an *ad hoc* definition based on the following fact.

PROPOSITION 11.3.6. *Let* $\overline{X}$ *be an irreducible smooth projective surface (over an algebraically closed field). Then there exists a unique divisor class* $K \in \mathrm{Pic}(\overline{X})$ *such that for any smooth irreducible projective curve* $C \subset \overline{X}$ *one has the* adjunction formula

$$g(C) = \frac{(C+K,C)}{2} + 1,$$

$g(C)$ *being the genus of* C. □

Note that the uniqueness of the canonical class K is obvious. Note also that for any curve $D \subset \overline{X}$ (maybe singular and reducible) one can define

$$p_a(D) := \frac{(D+K,D)}{2} + 1 \in \mathbb{Z}_+,$$

which is called the *arithmetic genus* of D, as opposed to the *geometric genus* $g(C)$, equal to the genus of the corresponding function field $\Bbbk(C)$; for a smooth irreducible C, we have $p_a(C) = g(C)$, and in general, $p_a(C) \geq g(C)$.

EXAMPLE 11.3.7. Let us compute the canonical class of $\mathbb{P}^2$. Since an element $D \in \mathrm{Pic}\,\mathbb{P}^2$ is completely determined by its degree, for a line $D' \cong L \subset \mathbb{P}^2$ one finds

$$0 = g(L) = 1 + \frac{(L+K,L)}{2} = 1 + \frac{(L^2)}{2} + \frac{(L,K)}{2} = \frac{3}{2} + \frac{(L,K)}{2}.$$

Therefore, $(L,K) = \deg K = -3$, and $K = -3L$ as an element of $\mathrm{Pic}(\mathbb{P}^2) = \mathbb{Z}L \cong \mathbb{Z}$.

Birational geometry of surfaces. Let us consider a rational map $\varphi\colon \overline{X}' \to \overline{X}$ which is a birational isomorphism of surfaces, i.e., $\varphi^*\colon \overline{\Bbbk}(\overline{X}) \to \overline{\Bbbk}(\overline{X}')$ is an isomorphism. Then the structure of φ can be expressed via so-called blow-ups, or monoidal transformations.

PROPOSITION 11.3.8. *Let $\overline{X} \subset \mathbb{P}^N$ be a smooth surface, and let $P \in \overline{X}$. Then there exists a unique regular morphism $\sigma\colon \overline{X}' \to \overline{X}$ of smooth projective surfaces which is bijective outside P and $\sigma^{-1}(P)$, respectively, (and thus is birational) such that $L := \sigma^{-1}(P)$ is isomorphic to $\mathbb{P}^1$. Moreover, $K_{\overline{X}'} = K_{\overline{X}} + L$, $\mathrm{Pic}(\overline{X}') = \sigma^* \mathrm{Pic}(\overline{X}) \oplus L\mathbb{Z}$, and $(L^2) = -1$. The curve L is called the* exceptional curve *(or line) of σ. Conversely, for any pair $(\overline{X}', L)$ where $\overline{X}$ is a smooth surface and L is a curve on $\overline{X}'$ isomorphic to $\mathbb{P}^1$ with $(L^2) = -1$ there exists a unique $\sigma\colon \overline{X}' \to \overline{X}$ satisfying all these conditions.* □

Any *regular* birational map φ can be expressed via blow-ups.

PROPOSITION 11.3.9. *Let $\varphi\colon \overline{X}' \to \overline{X}$ be a regular map of surfaces which is birational. Then there exists a finite sequence of surfaces $\overline{X} = \overline{X}_0, \overline{X}_1, \dots, \overline{X}_m = \overline{X}'$ and of blow-ups $\sigma_i\colon \overline{X}_{i+1} \to \overline{X}_i$ such that $\varphi = \sigma_{m-1} \circ \dots \circ \sigma_0$.* □

COROLLARY 11.3.10. *In the situation of Proposition 11.3.9, one has $\mathrm{Pic}(\overline{X}') \cong \mathrm{Pic}(\overline{X}) \oplus \mathbb{Z}^m$; the number m is called the index $\mathrm{ind}(\varphi)$ of φ.* □

PROPOSITION 11.3.11. *Let $\varphi\colon \overline{X}' \to \overline{X}$ be a birational map (in general, not regular). Then there exists a regular birational map $\psi\colon \overline{X}'' \to \overline{X}'$ such that the composition $\psi \circ \varphi\colon \overline{X}'' \to \overline{X}$ is regular.* □

COROLLARY 11.3.12. *Any birational map $\varphi\colon \overline{X}' \to \overline{X}$ is a composition of blow-ups and their inverses.* □

One calls a surface $\overline{X}$ a *minimal model* if there are no exceptional lines $C \subset \overline{X}$, i.e., no nontrivial blow-ups $\overline{X} \to \overline{Y}$.

It should be noted that minimal models are not unique in their birational classes.

EXAMPLE 11.3.13. The surfaces $\mathbb{P}^1 \times \mathbb{P}^1$ and $\mathbb{P}^2$ are birationally isomorphic, minimal, but not isomorphic, since $\mathrm{Pic}(\mathbb{P}^1 \times \mathbb{P}^1) \cong \mathbb{Z} \oplus \mathbb{Z}$ and $\mathrm{Pic}(\mathbb{P}^2) \cong \mathbb{Z}$.

Surfaces over finite fields. The theory of surfaces over nonclosed (in particular, finite) fields gets more complicated. However, here we need two cases only: cubic surfaces in $\mathbb{P}^3$ and rational surfaces, i.e., those birationally isomorphic to $\mathbb{P}^2$ (the rational function field $\mathbb{F}_q(X)$ for such a surface is isomorphic to $\mathbb{F}_q(x, y)$). Moreover, cubic surfaces are in fact rational. A cubic surface $V \subset \mathbb{P}^3$ is defined by a homogeneous absolutely irreducible cubic form in four variables $(x_0 : x_1 : x_2 : x_3)$, and we assume that V is smooth, i.e., has no singular points even over the algebraic closure $\overline{\mathbb{F}}_q$. Let us give some basic facts on cubic surfaces over a finite field.

11.3.2. Cubic Surfaces over a Finite Field

The principal geometric structure of such surfaces is the family of 27 lines on each cubic surface.

THEOREM 11.3.14. *Let $V/\mathbb{F}_q$ be a (smooth irreducible) cubic surface in $\mathbb{P}^3$ defined over $\mathbb{F}_q$.*

(i) *The surface $\overline{V} = V \otimes \overline{\mathbb{F}}_q$ (i.e., V considered over the algebraic closure of $\mathbb{F}_q$) contains exactly 27 lines.*
(ii) *These lines generate the group $\mathrm{NS}(\overline{V}) = \mathrm{Pic}(\overline{V})$, and the Galois group $G = \mathrm{Gal}(\overline{\mathbb{F}}_q/\mathbb{F}_q) \cong \widehat{\mathbb{Z}}$ is generated (topologically) by the Frobenius automorphism $F \in G$ acting on $\mathrm{NS}(\overline{V})$. We denote this action by F^*.*
(iii) *Let $N = |V(\mathbb{F}_q)|$ be the number of rational points on V. Then*
$$N = q^2 + q\,\mathrm{Tr}\,F^* + 1.$$
The action of F^* on $\mathrm{NS}(\overline{V})$ completely determines the zeta function $Z_V(T)$.
(iv) *Let $t := \mathrm{Tr}\,F^*$; then $-2 \le t \le 7$ and $t \ne 6$.*
(iv′) *If $q \ne 2,3,5$, all these t are realized by some cubic surfaces.*
(iv″) *For $q = 2,3,5$, the number t satisfies $-2 \le t \le 5$, and all these values are realized.* □

It is worth noting that the theorem also allows one to give a complete list of possible zeta functions for cubic surfaces.

We are mainly interested in cubic surfaces with $\mathrm{NS}(V) := \mathrm{NS}(\overline{V})^G$ of rank 1, since many of their hypersurface sections (in particular, all hyperplane sections) are irreducible curves, which is very useful for estimating the corresponding code parameters. Cubic surfaces with q^2+2q+1 points and $\mathrm{rk}\,\mathrm{NS}(V) = 1$ are of special interest in constructing codes.

LEMMA 11.3.15. *Let $X \subset \mathbb{P}^3$ be a smooth surface of degree d defined over $\mathbb{F}_q$, $(d,q) = 1$. Assume that the rank of the Néron–Severi group of X is 1 and that its generator is an effective divisor. Let $Y \subset \mathbb{P}^3$ be an irreducible surface of degree $1 \le m < d$ defined over $\mathbb{F}_q$. Then $X \cap Y$ is irreducible.* □

The proof uses a minute analysis of divisors on V.

REMARK 11.3.16. Note that Lemma 11.3.15 is not true for $m \ge d$. Indeed, let X be given by $\{g = 0\}$, and let f_1 and f_2 be such that $\deg(f_1) + \deg(f_2) = d$. Then $X \cap \{f_1 f_2 = 0\}$ is reducible, while $Y_\lambda\colon f_1 f_2 + \lambda g = 0$ is a pencil of surfaces with an irreducible member $(X, \lambda = \infty)$, so a generic member is irreducible but $X \cap Y_\lambda = X \cap Y_0$ is reducible.

Thus, we would like to find a cubic surface $S \subset \mathbb{P}^3$ defined over $\mathbb{F}_q$ containing q^2+2q+1 points and having $\mathrm{rk}\,\mathrm{NS}(S) = 1$. Moreover, if we find such a surface with the additional condition that there exists a plane not passing through $\mathbb{F}_q$-rational points of S, we can build a code of length q^2+2q+1 by moving this plane to infinity.

PROPOSITION 11.3.17. (i) *Let $q = 7$. Then the surface V_7 defined by the cubic form*
$$f(x_0 : x_1 : x_2 : x_3) = \sum_{i,j,k,\ell} c_{ijk\ell} x_0^i x_1^j x_2^k x_3^\ell,$$
where the vector of coefficients is
$$v_f = [5,6,1,1,0,5,0,3,6,3,5,1,0,5,0,5,0,0,2,4]$$
in lexicographic order, satisfies the conditions $N = q^2+2q+1 = 64$ and $\mathrm{rk}\,\mathrm{NS}(V_7) = 1$.

(ii) *Let* $q=9$, *and let* γ *be a generator of the group* $\mathbb{F}_9^*$. *Then the cubic surface* V_9 *defined by* $f=0$ *with*

$$v_f = [\gamma,\gamma,\gamma^2,\gamma^2,\gamma^5,\gamma^7,\gamma^2,\gamma^2,1,0,\gamma,\gamma^5,2,\gamma^6,\gamma^3,\gamma,1,\gamma^6,\gamma^2,\gamma^7]$$

satisfies $N = q^2+2q+1 = 100$ *and* $\operatorname{rk}\operatorname{NS}(V_9)=1$. □

Let us consider codes over $\mathbb{F}_7$ coming from the surface V_7 and estimate their parameters. First, since $q=7$, we have $|S(\mathbb{F}_7)|=64$, which gives a code of length 64. Using sections of degree at most 2, we have that the dimension of the code is at most 10, and since S has enough points, one can see that the dimension is exactly 10. Take the linear space of sections to be generated by

$$\{x^2,xy,xz,x,y^2,yz,y,z^2,z,1\}.$$

Then, for a surface $Q\subseteq\mathbb{P}^3$ of degree 2, by Lemma 11.3.15 the intersection $S\cap Q$ is either an irreducible curve or two plane cubics. Let us estimate the maximum number of points on the intersection.

Irreducible case, smooth. Using the adjunction formula (Proposition 11.3.6), we see that such a curve has genus $g\le 4$. For $g=4$, one can use the following lemma.

LEMMA 11.3.18. (i) *For a curve* $C\in\mathbb{P}^n(\mathbb{F}_q)$ *of degree* d *and genus* g, *the number of points* N *is bounded by*

$$N\le\frac{1}{n}\big(n(n-1)(g-1)+d(q+n)\big).$$

(ii) *If* C *is a curve defined over* $\mathbb{F}_q$ *with arithmetic genus* p_a, *the number of rational points of* C *satisfies*

$$\big||C(\mathbb{F}_q)|-(q+1)\big|\le 2p_a\sqrt{q}.$$ □

In our case ($n=3$, $q=7$, $g=4$, $d=6$), Lemma 11.3.18(i) gives $N\le 26$. Observe that this is better than Serre's bound $N\le q+1+g\lfloor 2\sqrt{q}\rfloor=28$.

If $g<4$, we can use Serre's bound to obtain $N\le q+1+g\lfloor 2\sqrt{q}\rfloor\le 23$.

Irreducible case, singular. Since the arithmetic genus of a complete intersection of surfaces of degrees a and b in $\mathbb{P}^3$ is given by

$$p_a(C)=\frac{1}{2}ab(a+b-4)+1,$$

we can use Lemma 11.3.18(ii) to get in our case ($a=3$, $b=2$) that the curve has arithmetic genus 4 and at most 29 points.

Reducible case, two plane cubics. For plane curves we use Weil's bound and conclude that each curve has at most 13 points, since $g\le 1$. Thus, the pair has at most 26 rational points.

If $Q\subseteq\mathbb{P}^3$ has degree 1, then Q is a plane, and the intersection is a plane cubic, which has at most 13 points.

Thus, we obtain the following result.

PROPOSITION 11.3.19. (i) *Let* C *be a code over* $\mathbb{F}_7$ *constructed using* V_7 *with* 64 *points by evaluating functions of degree at most* 2. *Then* C *is a* $[64,\le 10,\ge 35]_7$ *code.*

(ii) *The code defined by the plane*

$$6x+4y+2z+w=0$$

is a $[64,10,38]_7$ *code.* □

For the code from Proposition 11.3.19(ii), the bound for the dimension is attained and the actual minimum distance is larger than the lower bound that we have found. Note that this is possible only if the irreducible singular curves (if any) obtained as the intersection of the cubic and quadric surfaces do not attain the upper bound of Lemma 11.3.18(ii) for the number of rational points.

For V_9 we get the following result.

PROPOSITION 11.3.20. *For the surface* V_9 *with* 100 $\mathbb{F}_9$*-rational points, the plane*

$$\gamma^6 x+2y+2z+w=0$$

has no rational points in the intersection with V_9*. Therefore, the code obtained by sending this plane to infinity and evaluating all the functions of degree* ≤ 2 *is a* $[100,10,68]_9$ *code.* □

11.3.3. Rational Surfaces for Good Codes

Let us give examples of other rational surfaces suitable for constructing codes.

Codes on elliptic quadrics. We begin with a study of codes on so-called elliptic quadric surfaces (see Sec. 11.4.2).

Let P denote a closed point of degree 2 of $\mathbb{P}^2$, i.e., $P\in\mathbb{P}^2(\mathbb{F}_{q^2})\setminus\mathbb{P}^2(\mathbf{F}_q)$. After a base field extension, P splits into two conjugate points, p and p^F, defined over $\mathbb{F}_{q^2}$, where F denotes the Frobenius map. Let L be the unique rational line of $\mathbb{P}^2$ containing P, and let $\widetilde{X}$ be the surface obtained by blowing up $\mathbb{P}^2$ at P, $\pi\colon \widetilde{X}\to\mathbb{P}^2$. Denote by $\widetilde{L}=\pi^{-1}(L)\setminus\pi^{-1}(P)$ the strict transform of L by π, and denote by $E=\pi^{-1}(P)$ the exceptional divisor. The divisor E splits over $\mathbb{F}_{q^2}$ into a union of two conjugate lines, e and e^F. We have $\widetilde{L}^2_{\widetilde{X}}=-1$, and hence this curve is the exceptional divisor of some blow-up map, and the surface X obtained by blowing down $\widetilde{X}$ at $\widetilde{L}$ is an elliptic quadric in $\mathbb{P}^3$. We denote by $\psi\colon \widetilde{X}\to X$ this blow-down map, and by Q and H the respective images of $\widetilde{L}$ and E under ψ. The divisor H is prime but splits over $\mathbb{F}_{q^2}$ into a pair of conjugated lines, denoted by h and h^F.

Now, let us summarize some properties of these surfaces.

- $\mathbb{P}^2$:
  (A) $\mathrm{Pic}_{\mathbb{F}_q}(\mathbb{P}^2)\cong\mathbb{Z}L$ and $L^2=1$;
- $\widetilde{X}$:
  (B) $\mathrm{Pic}_{\mathbb{F}_q}(\widetilde{X})\cong\mathbb{Z}E\oplus\mathbb{Z}\widetilde{L}$ and $E^2=-2$, $E\widetilde{L}=2$, $\widetilde{L}^2=-1$;
  (C) $\pi^*L=\widetilde{L}+E$;
  (D) $E=e+e^F$;
- X:
  (E) H corresponds to the cut-out of X by its tangent plane at Q;
  (F) $\mathrm{Pic}_{\mathbb{F}_q}(X)\cong ZH$, with $H^2=2$;
  (G) $\psi^*H=2\widetilde{L}+E$;
  (H) $H=h+h^F$.

Let X be a smooth projective geometrically irreducible surface over a finite field $\mathbb{F}_q$ and $\mathscr{L}$ be a line bundle over X. Let $P_1,\ldots,P_n$ be the set of rational points of X. The code $C_L(X,\mathscr{L})$ is defined as the image of the evaluation map

$$\mathrm{ev}\colon H^0(X,\mathscr{L}) \to \bigoplus \mathscr{L}_{P_i} \cong \mathbb{F}_q^n, \quad \mathrm{ev}\colon f \mapsto (f_{P_1},\ldots,f_{P_n}).$$

We will consider evaluation codes on an elliptic quadric. To estimate their minimum distance, we use the following lemma.

LEMMA 11.3.21. *Let D be an effective divisor on X which contains Q and is smooth at this point. Let s be a positive integer such that $D \sim sH$, let $\widetilde{D}$ be the strict transform of D by ψ, and let D' be the image of $\widetilde{D}$ by π. Then we have*

(i) $|D(\mathbb{F}_q)| = |D'(\mathbb{F}_q)|$;
(ii) *D' is singular at P with multiplicity $s-1$;*
(iii) $\deg D' = 2s-1$. □

Now let us study the code defined by cubics on X. For a fixed base field $\mathbb{F}_q$, the code $C_L(X,\mathscr{O}_X(3))$ defined by X and the divisor $3H$, H being the plane section, has length $n = q^2+1$, which is the number of rational points of X. To compute its dimension and minimum distance, we need Lemma 11.3.22 and Proposition 11.3.23 below, which lead to Theorem 11.3.24.

LEMMA 11.3.22. *The dimension of the space $H^0(X,\mathscr{O}_X(m))$ cut by forms of degree $m \geq 0$ on X is $(m+1)^2$.*

PROOF. Let F be a homogeneous polynomial of degree 2 with $F(x,y,z,t) = 0$ being an equation of X. The space $H^0(X,\mathscr{O}_X(m))$ corresponds to the space of homogeneous forms of degree m modulo the forms vanishing on X, i.e., multiples of F, and thus

$$\begin{aligned}\dim H^0(X,\mathscr{O}_X(m)) &= \dim H^0(\mathbb{P}^3,\mathscr{O}_{\mathbb{P}^3}(m)) - \dim H^0(\mathbb{P}^3,\mathscr{O}_{\mathbb{P}^3}(m-2)) \\ &= \binom{m+3}{3} - \binom{m+1}{3} = (m+1)^2.\end{aligned}$$

Here $H^0(\mathbb{P}^3,\mathscr{O}_{\mathbb{P}^3}(s))$ denotes the space of degree-s homogeneous polynomials in 4 variables. □

PROPOSITION 11.3.23. *Let $q \geq 5$, and let C be an effective divisor on X equivalent to $3H$. Then*

$$|C(\mathbb{F}_q)| \leq \max\bigl(3q+3, \min(4q, q+1+4\lfloor 2\sqrt{q}\rfloor)\bigr). \quad \square$$

Since $\mathrm{Pic}_{\mathbb{F}_q}(X)$ is generated by H, the proof can be done along the lines of the case of cubic surfaces above.

Finally, we obtain the following result.

THEOREM 11.3.24. *Let X be an elliptic quadric over $\mathbb{F}_q$ with $q \geq 5$. The code $C_L(X,\mathscr{O}_X(3))$ has parameters $[q^2+1, 16, \geq \delta]$, where*

$$\delta = q^2+1-\max\bigl(3q+3, \min(4q, q+1+4\lfloor 2\sqrt{q}\rfloor)\bigr).$$

That is,

$$\delta = \begin{cases} q^2+1-4q & \text{if } q \leq 7, \\ q^2-q-4\lfloor 2\sqrt{q}\rfloor & \text{if } 8 \leq q \leq 13, \\ q^2-2-3q & \text{if } q \geq 16. \end{cases}$$

SKETCH OF THE PROOF. The length has already been computed above, the estimate for the minimum distance follows from Proposition 11.3.23, and the dimension is a consequence of Lemma 11.3.22. □

Constructing rational surfaces. Let us construct further rational surfaces which yield good codes using elliptic quadrics.

First, look at the code $C_L(X, \mathcal{O}_X(n))$ determined by $D = nH$ on an elliptic quadric. By the blow-up and blow-down operation, the linear system associated to $\mathcal{O}_X(n)$ on X corresponds to a linear system in $\mathbb{P}^2$ having the closed point P as a base point, i.e., a linear system whose members all contain P. Therefore, such curves defined over $\mathbb{F}_q$ cannot contain any rational line of $\mathbb{P}^2$ except for the line L whose strict transform is contracted. Thus, the elements of the linear system cannot have *too many* $\mathbb{F}_q$-irreducible components.

This motivates the following examples. Consider the projective plane $\mathbb{P}^2$, and let P be a rational point. Denote by F the Frobenius map. Let ℓ and ℓ^F be a pair of conjugated lines defined over $\mathbb{F}_{q^2}$ and meeting at P. Denote by D the $\mathbb{F}_q$-rational conic $D := \ell \cup \ell^F$. Next, let R be a closed point of degree 4 of D, which splits over $\mathbb{F}_{q^4}$ into four points r, r^F, r^{F^2}, and r^{F^3}.

Let Y be the surface obtained from $\mathbb{P}^2$ by blowing up P and R. We denote by $\pi\colon Y \to \mathbb{P}^2$ the blow-up map, and by E and F, the exceptional divisors above P and R, respectively.

For $i \geq 4$, let Λ_i be the linear system of plane curves of degree i containing R with multiplicity at least 1 and P with multiplicity at least 2. Let $\mathcal{F}_i$ be the linear system $\pi^*\Lambda_i - 2E - F$ on Y, and $h^0(Y, \mathcal{F}_i)$ be its dimension.

THEOREM 11.3.25. *The parameters of the code* $C_L(Y, \mathcal{F}_4)$ *are*

$$[(q+1)^2, 8, q^2 - q - 2].$$

SKETCH OF THE PROOF. The code has length $n = |Y(\mathbb{F}_q)| = (q+1)^2$.

Note that the dimension of the linear system of plane quartics is 14. The interpolating condition at P imposes 3 constraints, and the vanishing condition at R imposes 4 other constraints. These 7 constraints turn to be independent, and hence the dimension of Λ_4 is 7 and $h^0(Y, \mathcal{F}_i) = 8$. This implies that the dimension of the code is also 8.

The minimum distance d can be evaluated by a minute analysis of rational curves on the surface. □

Here are some further similar results.

THEOREM 11.3.26. *Let Z be the projective plane blown up at $P \in \mathbb{P}^2(\mathbb{F}_q)$. Denote by $\pi\colon Z \to \mathbb{P}^2$ the blow-up map and by $E = \pi^{-1}(P)$ the exceptional divisor. Let $i \geq 3$ be an integer. Let Γ_i be the linear system of plane curves of degree i containing P, and let $\mathcal{L}_i$ be the linear system $\pi^*\Gamma_i - E$ on Z. Then*

- *The parameters of* $C_L(Z, \mathcal{L}_3)$ *are*
$$[q^2 + q + 1, 7, q^2 - q - 1].$$
- *The parameters of* $C_L(Z, \mathcal{L}_4)$ *are*
$$[q^2 + q + 1, 12, q^2 - 2q - 1].$$
- *The parameters of* $C_L(Z, \mathcal{L}_5)$ *are*
$$[q^2 + q + 1, 18, q^2 - 3q - 1].$$
□

REMARK 11.3.27. All the surfaces in this section are rational. In fact, one can also give relatively satisfactory bounds for surfaces of a small rank, i.e., having few divisors modulo their linear equivalence. However, for this last class of surfaces no concrete interesting families of codes are actually known.

11.4. Hermitian Varieties and Quadrics

In this section we consider two classes of hypersurfaces in projective spaces, namely Hermitian varieties and quadrics. For these varieties one can get a complete description of their intersection with linear subspaces, which leads to rather precise information on the corresponding codes.

Note that these classes are rather close in spirit, even if the degrees are quite different. Indeed, since the map $x \mapsto x^q =: \overline{x}$ is an automorphism of $\mathbb{F}_{q^2}$, one can consider the equation

$$x_0^{q+1} + x_1^{q+1} + \ldots + x_N^{q+1} = 0$$

defining a Hermitian hypersurface as a twisted quadric

$$x_0\overline{x}_0 + x_1\overline{x}_1 + \ldots + x_N\overline{x}_N = 0,$$

which explains a certain similarity between these two classes.

In what follows we denote a linear subspace of dimension s in $\mathbb{P}^N$ by Π_s. Thus, Π_0 is a point in $\mathbb{P}^N$, $\mathbb{P}^1$ is a projective line, and $\Pi_{-1} = \varnothing$.

11.4.1. Hermitian Varieties

Definitions and basic properties. Here we give the necessary results on Hermitian varieties over finite fields.

Let $q = p^{2a}$ be an even power of a prime p, and let $q_0 = p^a$, i.e., $q_0 = \sqrt{q}$. A *Hermitian variety* $\mathcal{U}_N$ is the hypersurface in $\mathbb{P}^N$ given by $F(x) = 0$, where

$$F(x) = x_0^{q_0+1} + \ldots + x_N^{q_0+1};$$

note that $\mathcal{U}_N$ is nonsingular. The map $t \mapsto \overline{t} = t^{q_0}$ gives an involution of $\mathbb{F}_q$, and thus $F(x)$ is a Hermitian form. Moreover, any Hermitian form can be written as $\sum_{i=0}^{s} x_i\overline{x}_i$ with respect to a suitable basis. Hence, a Hermitian form is uniquely defined by its rank, and the corresponding variety $\sum_{i=0}^{s} x_i\overline{x}_i = 0$ is a cone $\Pi_v\mathcal{U}_s$ joining a space Π_v, $v \geq -1$, to a nonsingular Hermitian variety $\mathcal{U}_s$ in a space Π_s, $s = N - v - 1$, skew to Π_v; i.e., $\Pi_v\mathcal{U}_s$ is the union of lines joining a point of Π_v to a point of $\mathcal{U}_s$. If $v = -1$, then $\Pi_v\mathcal{U}_s$ is a nonsingular variety $\mathcal{U}_N$.

The intersection $\mathcal{U}_N \cap \Pi_m$ is also of type $\Pi_v\mathcal{U}_s$ with $v + s = m - 1$, where $s \geq -1$; by definition, the case $s = -1$ means that $\Pi_m \subset \mathcal{U}_N$, i.e., $\Pi_v\mathcal{U}_{-1} = \Pi_m$. The section $\Pi_m \cap \mathcal{U}_N$ of type $\Pi_v\mathcal{U}_s$ exists if and only if $v + s = m - 1$, $v \leq N - 1 - m$, and $-1 \leq v \leq m$. The cone $\Pi_v\mathcal{U}_s$ has exactly

$$\omega(m, v) = \theta_{m-1} + \frac{q^m - (-q_0)^{m+v+1}}{q_0 + 1} \tag{11.4.1}$$

$\mathbb{F}_q$-points; in particular, the number of $\mathbb{F}_q$-points on $\mathcal{U}_N$ is

$$\omega(N, -1) = \theta_{N-1} + \frac{q^N - (-q_0)^N}{q_0 + 1} = \frac{\left(q_0^{N+1} - (-1)^{N+1}\right)\left(q_0^N - (-1)^N\right)}{q - 1}. \tag{11.4.2}$$

If there exists a section $\Pi_m \cap \mathcal{U}_N$ of type $\Pi_v \mathcal{U}_s$, then the number of such spaces Π_m is

$$\rho(N,m,v) = \frac{q^{(N-m-v-1)(m-v)/2} \prod_{i=m-v+1}^{N+1} (q_0^i - (-1)^i)}{\prod_{i=1}^{v+1} (q^i - 1) \prod_{i=1}^{N-m-v-1} (q_0^i - (-1)^i)}.$$

Hermitian codes. Now consider the $[n,k]$ system defined by the set of $\mathbb{F}_q$-points of the nonsingular Hermitian variety $\mathcal{U}_{k-1}$ in $\mathbb{P}^{k-1}$.

THEOREM 11.4.1. *Let* $a = \lfloor k/2 \rfloor - 1$ *and* $\varepsilon = \begin{cases} 0 & \text{for } k \text{ even,} \\ 1 & \text{for } k \text{ odd.} \end{cases}$

Then

$$n = \theta_a(q^{a+\varepsilon} q_0 + 1)$$

and

$$d_r = \begin{cases} n - \theta_{k-r-1} & \text{if } r \geq \lfloor k/2 \rfloor, \\ q^{k-r-1}\theta_{r-1} q_0 - \varepsilon q^a q_0 & \text{if } r < \lfloor k/2 \rfloor. \end{cases}$$

PROOF. In what follows, the upper sign corresponds to the case of k even and the lower sign to that of k odd. From (11.4.2) we get

$$n = \omega(k-1,-1) = \frac{(q_0^k - (-1)^k)(q_0^{k-1} - (-1)^{k-1})}{q-1}$$
$$= \frac{(q^{k/2} \mp 1)(q^{(k-1)/2} \pm 1)}{q-1} = \theta_a(q^{a+\varepsilon} q_0 + 1).$$

From the definition and (11.4.1) we obtain

$$d_r = n - \max |\mathcal{U}_{k-1} \cap \Pi_{k-1-r}| = n - \max_v |\mathcal{U}_{k-2-r-v}|$$
$$= n - \max_v \left(\theta_{k-r-2} + \frac{q^{k-r-1} - (-q_0)^{k+v-r}}{q_0 + 1} \right),$$

where $-1 \leq v \leq \min\{r-1, k-r-1\}$. If $r \geq k/2$, then $k-r-1 \leq r-1$; thus, the maximum is attained for $v = k-r-1$ and

$$d_r = n - \left(\theta_{k-r-2} + \frac{q^{k-r-1} + q_0^{2k-2r-1}}{q_0+1} \right) = n - (\theta_{k-r-2} + q^{k-r-1}) = n - \theta_{k-r-1}.$$

If $r < k/2$, then $r-1 < k-r-1$; hence, $\min\{r-1, k-r-1\} = r-1$. However, for $v = r-1$ we have $k+v-r = k-1$. Thus, the maximum occurs when $k+v-r = k-1-\varepsilon$, and after simplification we obtain

$$d_r = n - \left(\theta_{k-r-2} + \frac{q^{k-r-1} + q_0^{k-1-\varepsilon}}{q_0+1} \right)$$
$$= \begin{cases} q^{k-r-1}\theta_{r-1} q_0 & \text{for } k \text{ even,} \\ q^{k-r-1}\theta_{r-1} q_0 - q^{k/2-1} & \text{for } k \text{ odd.} \end{cases}$$

□

COROLLARY 11.4.2. *The weight hierarchy of the code $C^\perp$ dual to the code C obtained from the Hermitian $[n,k]$ system is given by*

$$\{1,\dots,\omega(k-1,-1)\}\setminus\left\{\theta_{k-r-2}+1+\frac{q^{k-r-1}+q^{(k-1-\varepsilon)/2}}{q_0+1}\;\middle|\; r=1,\dots,\lceil k/2\rceil-1\right\}$$
$$\cup\{\theta_{k-r-1}+1\mid r=\lceil k/2\rceil,\dots,k\}.$$

PROOF. This follows from Theorem 11.4.1 and the description of the $d_r^\perp$ in Theorem 1.1.38. □

A straightforward calculation yields the following result.

COROLLARY 11.4.3. *Let k be even and $r<k/2$. Then d_r of the Hermitian code meets the generalized Griesmer bound (see Corollary 1.1.69).* □

Corollary 11.4.3 shows that Hermitian varieties are in a sense the best possible for $r<k/2$, while for $r\geq k/2$ they are the worst possible, since $n-d_r$ are the maximum that can occur.

One can obtain analogous results for all singular Hermitian varieties $\Pi_{k-2-t}\mathcal{U}_t\subset\mathbb{P}^{k-1}$.

Generalized spectra. Recall the definition of the generalized spectrum in Sec. 1.1.3. For a code $C\subseteq\mathbb{F}_q^n$,

$$A_i^r=|\{D\subseteq C:\ \dim D=r,\ \|D\|=i\}|.$$

For a projective system $\mathcal{P}\subseteq\mathbb{P}$,

$$A_i^r=|\{\Pi\subseteq\mathbb{P}:\ \operatorname{codim}\Pi=r,\ |\Pi\cap\mathcal{P}|=n-i\}|.$$

The formula for $\rho(N,m,v)$ makes it possible to compute the generalized spectrum for Hermitian codes as follows.

THEOREM 11.4.4. *Let $\mathcal{P}$ be the set of $\mathbb{F}_q$-points on $\mathcal{U}_{k-1}$. Then*

$$A_i^r=\begin{cases}\rho(k-1,k-1-r,v) & \text{if } i=n-\omega(k-1-r,v) \text{ for some } v\\ & \text{with } -1\leq v\leq\min\{r-1,k-r-1\},\\ 0 & \text{otherwise.}\end{cases}$$ □

COROLLARY 11.4.5. *The number of distinct r-weights of the Hermitian system $\mathcal{P}$ is $\min\{r+1,k-r+1\}$.* □

11.4.2. Quadrics

Projective quadrics over a finite field. Here we summarize what we need on quadrics over finite fields. In the present section q is an arbitrary prime power, and it need not be a square.

A *quadric* in $\mathbb{P}^N$ is a hypersurface of degree 2, i.e., the zero set of a homogeneous quadratic form

$$F=\sum_{i=0}^{N}a_ix_i^2+\sum_{i<j}a_{ij}x_ix_j.$$

If the form is *nondegenerate*, i.e., cannot be reduced by a linear change of variables to a form in fewer variables, then the corresponding quadric $\mathcal{Q}_N$ is nonsingular. Up to projective transforms of $\mathbb{P}^N$, over a finite field $\mathbb{F}_q$ there can exist either one or two nonsingular quadrics, depending on the parity of N, namely:

- If N is even, then any nonsingular quadric $\mathcal{Q}_N$ is projectively equivalent to the quadric $\mathcal{P}_N$ defined by the equation

$$x_0^2 + x_1x_2 + \ldots + x_{N-1}x_N = 0.$$

- If N is odd, then any nonsingular quadric $\mathcal{Q}_N$ is projectively equivalent either to the quadric $\mathcal{H}_N$ defined by the equation

$$x_0x_1 + x_2x_3 + \ldots + x_{N-1}x_N = 0$$

or to the quadric $\mathcal{E}_N$ defined by the equation

$$f(x_0, x_1) + x_2x_3 + \ldots + x_{N-1}x_N = 0$$

where f is a homogeneous polynomial of degree 2 which is irreducible over $\mathbb{F}_q$.

The quadrics equivalent to $\mathcal{P}_N$ are called *parabolic*, to $\mathcal{H}_N$ *hyperbolic*, and to $\mathcal{E}_N$ *elliptic*.

Any singular quadric is a cone $\Pi_s\mathcal{Q}_t$ with vertex Π_s over a nonsingular quadric $\mathcal{Q}_t \subset \Pi_t$, where $\Pi_s \cap \Pi_t = \varnothing$, $t+s+1 = N$. This means that $\Pi_s\mathcal{Q}_t$ is the union of all lines passing through a point of Π_s and a point of $\mathcal{Q}_t$. Here $s \geq -1$; if $s = -1$, then the quadric is nonsingular.

The quadrics $\mathcal{H}_{-1}$, $\mathcal{P}_0$, and $\mathcal{E}_1$ are empty; $\mathcal{H}_1$ is a pair of points on a line; $\mathcal{P}_2$ is a conic in a plane; and $\mathcal{E}_3$ consists of q^2+1 points in $\mathbb{P}^3$, no three of which are collinear.

The (projective) *index* of a quadric $\mathcal{Q}_N$ is the projective dimension g of the largest linear subspace lying entirely in $\mathcal{Q}_N$. It equals $\dfrac{N-1}{2}$ for $\mathcal{H}_N$, $\dfrac{N-2}{2}$ for $\mathcal{P}_N$, and $\dfrac{N-3}{2}$ for $\mathcal{E}_N$. The *character* of a nonsingular quadric is the number $w = 2g - N + 3$, so quadrics of hyperbolic type have character 2, those of parabolic type have character 1, and those of elliptic type have character 0. The character of a singular quadric $\Pi_s\mathcal{Q}_t$ is assumed to be that of $\mathcal{Q}_t$.

The number of $\mathbb{F}_q$-points on a quadric $\Pi_s\mathcal{Q}_t$ of character w in $\mathbb{P}^{s+t+1} = \mathbb{P}^N$ equals

$$\tau(s,t,w) = \left|\Pi_s\mathcal{Q}_t(\mathbb{F}_q)\right| = \theta_{s+t} + (w-1)q^{s+\frac{t+1}{2}}. \tag{11.4.3}$$

In particular, the number of $\mathbb{F}_q$-points of a nonsingular quadric is

$$\left|\mathcal{Q}_N(\mathbb{F}_q)\right| = \theta_{N-1} + (w-1)q^{\frac{N-1}{2}}. \tag{11.4.4}$$

The section of a nonsingular quadric $\mathcal{Q}_N$ by a subspace Π_m is also a quadric, since the restriction of a quadratic form to a subspace is also a quadratic form (perhaps in fewer variables). For a nonsingular quadric $\mathcal{Q}_N$ of character w there exists a section $\mathcal{Q}_N \cap \Pi_m$ of type $\Pi_{m-t-1}\mathcal{Q}_t$ of character v if and only if

$$\begin{aligned} &(1) \quad t - v \text{ is odd,} \\ &(2) \quad N > m \geq t \geq \max\{2m - N + |w-v|, 1-v\}. \end{aligned} \tag{11.4.5}$$

If such sections exist, then the number of such subspaces Π_m is

$$\begin{aligned}(11.4.6)\quad \rho(m,t,v;N,w) &= q^{\frac{1}{2}(T[t+1+vw(2-v)(2-w)]-v(2-v)(w-1)^2)} \\ &\times \left[\frac{1}{2}\{T+v+w-vw+vw(2-v)(2-w)\},\frac{1}{2}(n+1-w)\right]_+ \\ &\times \left[\frac{1}{2}\{T-v-w+vw+vw(2-v)(2-w)\},\frac{1}{2}(n-1+w)\right]_- \\ &: \left(\left[v(2-v),\frac{1}{2}(t+1-v)\right]_+ \cdot \left[1,\frac{1}{2}(t-1+v)\right]_- \cdot [1,m-t]_-\right),\end{aligned}$$

where $T = N+t-2m$ and

$$[a,b]_\pm = \begin{cases}(q^a \pm 1)(q^{a+1} \pm 1)\dots(q^b \pm 1), & a \le b,\\ 1, & a > b.\end{cases}$$

Codes from quadrics. Now consider the $[n,k]$ system consisting of $\mathbb{F}_q$-points of a nonsingular quadric $\mathcal{Q}_{k-1}$ in $\mathbb{P}^{k-1}$.

THEOREM 11.4.6. *A code associated with a nonsingular quadric $\mathcal{Q}_{k-1}$ of character w has dimension k and the following parameters, where $g = \frac{k+w}{2} - 2$ is the projective index of the quadric:*

in the hyperbolic case ($w = 2$, k even),

$$(11.4.7)\qquad \begin{aligned} n &= \theta_{k-2} + q^g,\\ d_r &= \begin{cases} q^{k-r-1}\theta_{r-1}, & r \le k-1-g,\\ q^{k-r}\theta_{r-2} + q^g, & r \ge k-1-g;\end{cases}\end{aligned}$$

in the parabolic case ($w = 1$, k odd),

$$(11.4.8)\qquad \begin{aligned} n &= \theta_{k-2},\\ d_r &= \begin{cases} q^{k-r-1}\theta_{r-1} - q^g, & r \le k-1-g,\\ q^{k-r}\theta_{r-2}, & r \ge k-1-g;\end{cases}\end{aligned}$$

in the elliptic case ($w = 0$, k even),

$$(11.4.9)\qquad \begin{aligned} n &= \theta_{k-2} + q^g \cdot q,\\ d_r &= \begin{cases} q^{k-r-1}\theta_{r-1} - q^g \cdot q, & r = 1,\\ q^{k-r-1}\theta_{r-1} - q^g(q+1), & 2 \le r \le k-1-g,\\ q^{k-r}\theta_{r-2} - q^g \cdot q, & r \ge k-1-g.\end{cases}\end{aligned}$$

PROOF. The length n is determined by (11.4.4) with $N = k-1$. By the definition,

$$d_r = n - \max|\mathcal{Q}_{k-1} \cap \Pi_{k-1-r}|.$$

The intersection $\mathcal{Q}_{k-1} \cap \Pi_{k-1-r}$ is also a quadric of type $\Pi_{(k-1-r)-t-1}\mathcal{Q}_t$. Denote its character by v. Then, by (11.4.5) and (11.4.3), we obtain

$$\begin{aligned} d_r &= n - \max_{(t,v)} |\Pi_{k-r-t-2}\mathcal{Q}_t| \\ &= n - \max_{(t,v)} \left(\theta_{k-r-2} + (v-1)q^{k-r-t-2+\frac{t+1}{2}}\right) \\ &= n - \theta_{k-r-2} - q^{k-r-2} \max_{(t,v)} \frac{(v-1)}{q^{\frac{t-1}{2}}} \\ &= q^{k-r-1}\theta_{r-1} + (w-1)q^{\frac{k}{2}-1} - q^{k-r-2} \max_{(t,v)} \frac{(v-1)}{q^{\frac{t-1}{2}}}, \end{aligned} \tag{11.4.10}$$

where the maximum is over all pairs (t,v) such that

$$\begin{aligned} k-r-1 \geq t &\geq \max\{k-2r-1+|w-v|, 1-v\} \\ 2 \geq v \geq 0, &\qquad t+v \equiv 1 \pmod 2. \end{aligned} \tag{11.4.11}$$

It is obvious that $\max \frac{v-1}{q^{\frac{t-1}{2}}}$ is attained when $v=2$ if there exists at least one pair $(t,2)$ that satisfies (11.4.11). Furthermore, t should be as small as possible.

We consider the cases of $\mathcal{H}_{k-1}$, $\mathcal{P}_{k-1}$, and $\mathcal{E}_{k-1}$ separately.

I. Hyperbolic case; $w=2$, k even.

Here the case $v=2$ is possible for any r, because conditions (11.4.11) with $v=2$ take the form

$$k-r-1 \geq t \geq \max\{k-2r-1, -1\}, \qquad t \text{ odd},$$

whence

$$t = \begin{cases} k-2r-1, & r \leq \frac{k}{2}, \\ -1, & r \geq \frac{k}{2}. \end{cases}$$

Then

$$\begin{aligned} d_r &= q^{k-r-1}\theta_{r-1} + q^{\frac{k}{2}-1} - q^{k-r-2} \times \begin{cases} q^{-(\frac{k}{2}-r-1)}, & r \leq \frac{k}{2}, \\ q, & r \geq \frac{k}{2}, \end{cases} \\ &= q^{k-r-1}\theta_{r-1} + q^{\frac{k}{2}-1} - \begin{cases} q^{\frac{k}{2}-1}, & r \leq \frac{k}{2}, \\ q^{k-r-1}, & r \geq \frac{k}{2}, \end{cases} \end{aligned}$$

from which we obtain (11.4.7), since $g = \dfrac{k}{2} - 1$.

II. Parabolic case; $w=1$, k odd.

Analogously, conditions (11.4.11) under $v=2$ take the form

$$k-r-1 \geq t \geq \max\{k-2r, -1\}, \quad t \text{ odd},$$

which is also possible for any r. Hence, we have

$$t = \begin{cases} k-2r, & r \leq \frac{k+1}{2}, \\ -1, & r \geq \frac{k+1}{2}. \end{cases}$$

Thus,

$$d_r = q^{k-r-1}\theta_{r-1} - q^{k-r-2}\cdot\begin{cases} q^{-\frac{k-2r-1}{2}}, & r \le \frac{k+1}{2},\\ q, & r \ge \frac{k+1}{2},\end{cases}$$

$$= q^{k-r-1}\theta_{r-1} - \begin{cases} q^{\frac{k-1}{2}-1}, & r \le \frac{k+1}{2},\\ q^{k-r-1}, & r \ge \frac{k+1}{2},\end{cases}$$

which yields (11.4.8), since $g = \dfrac{k-1}{2} - 1$.

III. Elliptic case; $w = 0$, k even.

Conditions (11.4.11) under $v = 2$ take the form

$$k - r - 1 \ge t \ge \max\{k - 2r + 1, -1\}, \qquad t \text{ odd.}$$

For $r = 1$ there are no such t; for $r \ge 2$ we get

$$t = \begin{cases} k - 2r + 1, & 2 \le r \le \frac{k}{2} + 1,\\ -1, & r \ge \frac{k}{2} + 1.\end{cases}$$

For $r = 1$ we take $v = 1$; then

$$k - 2 \ge t \ge \max\{k - 2, -1\}, \qquad t \text{ even.}$$

Thus, $t = k - 2$ suits. Hence, the maximum in (11.4.10) is attained when

$$t = \begin{cases} k - 2, & r = 1,\\ k - 2r + 1, & 2 \le r \le \frac{k}{2} + 1,\\ -1, & r \ge \frac{k}{2} + 1.\end{cases}$$

Therefore, we get

$$d_r = q^{k-r-1}\theta_{r-1} - q^{\frac{k}{2}-1} - q^{k-r-2}\cdot\begin{cases} 0, & r = 1,\\ q^{-(\frac{k}{2}-r)}, & 2 \le r \le \frac{k}{2} + 1,\\ q, & r \ge \frac{k}{2},\end{cases}$$

$$= q^{k-r-1}\theta_{r-1} - q^{\frac{k}{2}-1} - \begin{cases} 0, & r = 1,\\ q^{\frac{k}{2}-2}, & 2 \le r \le \frac{k}{2} + 1,\\ q^{k-r-1}, & r \ge \frac{k}{2} + 1,\end{cases}$$

whence we obtain (11.4.9), since $g = \dfrac{k}{2} - 2$. □

COROLLARY 11.4.7. *For $r \le \dfrac{k}{2}$, the weights d_r for a hyperbolic quadric $\mathcal{H}_{k-1}$ meet the generalized Griesmer bound.* □

Conversely, if $r > \dfrac{k}{2}$, i.e., if $k - r - 1 < g$, then there exist subspaces Π_{k-r-1} which lie entirely on the quadric; the corresponding weights are therefore in a sense the worst.

From (11.4.7) and (11.4.8) one immediately observes the following property.

COROLLARY 11.4.8. *For a nonsingular hyperbolic or parabolic quadric, we have*

$$d_{r+1} - d_r = \begin{cases} q^{k-r-2} & \text{for } r < k - 1 - g,\\ q^{k-r-1} & \text{for } r \ge k - 1 - g.\end{cases}$$ □

One can consider as a projective system the points of a singular quadric and obtain similar results, but the parameters of the corresponding codes are worse.

Generalized spectra. Formulas (11.4.3) for τ and (11.4.6) for ρ make it possible to compute the generalized spectrum of a code associated with a nonsingular quadric.

THEOREM 11.4.9. *For a code associated to a nonsingular quadric $\mathcal{Q}_{k-1}$ of character w, we have*

$$A_i^r = \begin{cases} \rho(k-r-1,t,v;k-1,w), & \text{if } i = n-\tau(k-r-t-2,t,v) \text{ for some } t \\ & \text{and } v \text{ that satisfy (11.4.11)}, \ v \neq 1, \\ \sum\limits_t \rho(k-r-1,t,1;k-1,w), & \text{if } i = n-\theta_{k-r-2}, \\ 0, & \text{otherwise,} \end{cases}$$

where the sum is over odd t such that

$$k-r-1 \geq t \geq \max\{k-2r-1+|w-1|, 0\}.$$

PROOF. This immediately follows from the above definitions and results, taking into account that for different s and t with $s+t = \text{const}$ the function $\tau(s,t,v)$ has the same value θ_{s+t} if $v=1$ and has different values if $v=2$ or $v=0$. □

COROLLARY 11.4.10. *The number of different r-weights of a code from a nonsingular quadric $\mathcal{Q}_{k-1}$ of character w is*

$$\min\left\{\left\lceil\frac{r+w-1}{2}\right\rceil, \left\lceil\frac{k-r+1}{2}\right\rceil\right\} + \min\left\{\left\lceil\frac{r-w+1}{2}\right\rceil, \left\lceil\frac{k-r-1}{2}\right\rceil\right\} + 1. \qquad \square$$

11.5. Grassmann and Schubert Codes

Here we consider codes coming from two types of classical algebraic varieties, namely Grassmannians and Schubert varieties. In this section q is still an arbitrary prime power.

11.5.1. Grassmann Codes

In this subsection, by V_m we denote a vector space of dimension m over $\mathbb{F}_q$, and by $\boldsymbol{V}_m$ some fixed V_m.

A Grassmann variety (or a Grassmannian) $G(\ell,m)$ is the variety of all linear ℓ-subspaces of a fixed linear space $\boldsymbol{V}_m$:

$$G(\ell,m) = \{V_\ell \mid V_\ell \subseteq \boldsymbol{V}_m\}.$$

The number of $\mathbb{F}_q$-rational points of $G(\ell,m)$ is the number of ℓ-subspaces in $\boldsymbol{V}_m$, which equals the Gaussian binomial coefficient

$$\begin{bmatrix} m \\ \ell \end{bmatrix}_q = \frac{(q^m-1)(q^m-q)\dots(q^m-q^{\ell-1})}{(q^\ell-1)(q^\ell-q)\dots(q^\ell-q^{\ell-1})},$$

i.e., the number of ways to choose ℓ linearly independent vectors in $\boldsymbol{V}_m$ divided by the number of bases in $\boldsymbol{V}_\ell$.

The natural *Plücker embedding*

$$G(\ell,m) \hookrightarrow \mathbb{P}^{\binom{m}{\ell}-1}$$

is defined as follows: for a subspace L spanned by ℓ vectors $e_1,\dots,e_\ell$ consider the exterior product

$$e_1 \wedge e_2 \wedge \dots \wedge e_\ell \in \Lambda^\ell \boldsymbol{V}_m$$

and the corresponding point

$$\widetilde{L} \in \mathbb{P}\Lambda^\ell \boldsymbol{V}_m = \mathbb{P}^{\binom{m}{\ell}-1}$$

in the projectivization. Then $\widetilde{L}$ does not depend on the choice of the basis in L.

Now consider the projective system being the set of all $\mathbb{F}_q$-rational points of the image of Plücker embedding. This is an $[n,k]$ system with

$$k = \binom{m}{\ell}, \qquad n = \begin{bmatrix} m \\ \ell \end{bmatrix}_q.$$

The corresponding family of equivalent linear codes $C(\ell,m)$ with the same parameters is called the family of *Grassmann codes*.

Note that $G(1,m) = \mathbb{P}^{m-1}$, so Grassmann codes are in a sense a generalization of the first-order projective Reed–Muller codes.

Minimum distance of Grassmann codes. The minimum distance of these codes is given by the following theorem.

THEOREM 11.5.1. *The minimum distance of the Grassmann code $C(\ell,m)$ is*

$$d = q^{(m-\ell)\ell}.$$

To prove this, we need the following three facts.

PROPOSITION 11.5.2. *Any hyperplane H in $\Lambda^\ell \boldsymbol{V}_m$ can be written as*

$$H_\omega = \{\omega' \in \Lambda^\ell \boldsymbol{V}_m \mid \omega' \wedge \omega = 0\}$$

for some $\omega \in \Lambda^{m-\ell}\boldsymbol{V}_m$.

PROOF. Indeed, a hyperplane can be regarded as an element of the dual space $(\Lambda^\ell \boldsymbol{V}_m)^*$. The exterior product provides the nondegenerate pairing

$$\wedge\colon \Lambda^\ell \boldsymbol{V}_m \times \Lambda^{m-\ell}\boldsymbol{V}_m \to \mathbb{F}_q$$

and thereby the duality $(\Lambda^\ell \boldsymbol{V}_m)^* \cong \Lambda^{m-\ell}\boldsymbol{V}_m$. □

PROPOSITION 11.5.3. *Let $\omega \in \Lambda^{m-\ell}\boldsymbol{V}_m$, $e \in \boldsymbol{V}_m$. Then $e \wedge \omega = 0$ if and only if $\omega = e \wedge \omega'$ for some $\omega' \in \Lambda^{m-\ell-1}\boldsymbol{V}_m$.*

PROOF. To verify this, it suffices to complete e to a basis in $\boldsymbol{V}_m$ and consider the decomposition of ω with respect to the corresponding basis in $\Lambda^{m-\ell}\boldsymbol{V}_m$. □

PROPOSITION 11.5.4. *Let $\omega_0 \in \Lambda^{m-\ell}\boldsymbol{V}_m$, $\omega_0 \neq 0$, be* completely decomposable, *i.e., $\omega_0 = v_1 \wedge \ldots \wedge v_{m-\ell}$ for some $v_i \in \boldsymbol{V}_m$. Then*

$$\operatorname{wt}(H_\omega) \geq \operatorname{wt}(H_{\omega_0})$$

for any $\omega \in \Lambda^{m-\ell}\boldsymbol{V}_m$, $\omega \neq 0$.

PROOF. It is obvious that $\operatorname{wt}(H_{\omega_0})$ is the same for all completely decomposable forms ω_0 and depends only on ℓ; denote it by N_ℓ.

We prove the proposition by induction on ℓ. If $\ell = 1$, there is nothing to prove, since all $\omega \in \Lambda^{m-1}\boldsymbol{V}_m$ are completely decomposable. Now let $\ell > 1$. We have

$$\begin{aligned}\operatorname{wt}(H_\omega) &= \bigl|\{\langle e_1,\ldots,e_\ell\rangle \mid e_1\wedge\ldots\wedge e_\ell \wedge \omega \neq 0\}\bigr| \\ &= \frac{\bigl|\{(e_1,\ldots,e_\ell) \mid e_1\wedge\ldots\wedge e_\ell \wedge \omega \neq 0\}\bigr|}{[\ell]_q},\end{aligned} \tag{11.5.1}$$

where $\langle e_1,\ldots,e_\ell\rangle$ is the linear envelope of $(e_1,\ldots,e_\ell)$ and

$$[\ell]_q = (q^\ell - 1)(q^\ell - q)\ldots(q^\ell - q^{\ell-1})$$

is the number of bases in $\boldsymbol{V}_\ell$. Hence,

$$\begin{aligned}\operatorname{wt}(H_\omega)[\ell]_q &= \sum_{e_1\colon e_1\wedge\omega\neq 0} \bigl|\{(e_2,\ldots,e_\ell) \mid e_2\wedge\ldots\wedge e_\ell \wedge (e_1\wedge\omega) \neq 0\}\bigr| \\ &\geq \sum_{e_1\colon e_1\wedge\omega\neq 0} N_{\ell-1}\,[\ell-1]_q \\ &= N_{\ell-1}\,[\ell-1]_q\,\bigl|\{e_1 \mid e_1\wedge\omega \neq 0\}\bigr| \\ &\geq N_{\ell-1}\,[\ell-1]_q\,\bigl|\{e_1 \mid e_1\wedge\omega_0 \neq 0\}\bigr|.\end{aligned}$$

Here the first inequality is the induction hypothesis, since $e_1 \wedge \omega \in \Lambda^{m-\ell+1}\boldsymbol{V}_m$, and the second inequality follows directly from Proposition 11.5.3. Finally, if we put ω_0 for ω, then both inequalities turn into equalities, so the last expression equals

$$\operatorname{wt}(H_{\omega_0})[\ell]_q.$$ □

PROOF OF THEOREM 11.5.1. By Proposition 11.5.4, $d = \operatorname{wt}(H_{\omega_0})$ for $\omega_0 = v_1 \wedge \ldots \wedge v_{m-\ell}$; thus,

$$d = \bigl|\{V_\ell \mid V_\ell \cap \langle v_1,\ldots,v_{m-\ell}\rangle = \{0\}\}\bigr|$$

is the number of ℓ-dimensional subspaces in $\boldsymbol{V}_m$ skew (in the projective sense) to some fixed subspace $\boldsymbol{V}_{m-\ell}$. It can be computed as the number of ways to complete $(v_1,\dots,v_{m-\ell})$ to a basis in $\boldsymbol{V}_m$, divided by $[\ell]_q$. Hence,

$$d=\frac{(q^m-q^{m-\ell})(q^m-q^{m-\ell+1})\dots(q^m-q^{m-1})}{(q^\ell-1)(q^\ell-q)\dots(q^\ell-q^{\ell-1})}=q^{(m-\ell)\ell}. \qquad \square$$

Note that the exponent $(m-\ell)\ell$ equals the dimension of the Grassmannian.

COROLLARY 11.5.5. *The number of minimum-weight codewords of a Grassmann code is* $(q-1)\begin{bmatrix} m \\ \ell \end{bmatrix}_q$.

PROOF. Indeed, by the proof of the theorem, it is equal to the number of hyperplanes H_{ω_0} corresponding to completely decomposable ω_0, i.e., to the number of $V_{m-\ell}$ in $\boldsymbol{V}_m$ (points of the dual Grassmannian $G(m-\ell,m)$), which equals $\begin{bmatrix} m \\ m-\ell \end{bmatrix}_q = \begin{bmatrix} m \\ \ell \end{bmatrix}_q$. Each hyperplane corresponds to $q-1$ collinear codewords. $\square$

The spectrum of Grassmann codes for $\ell=2$. Any nonzero $\omega\in\Lambda^{m-2}\boldsymbol{V}_m$ determines a skew-symmetric bilinear form

$$\omega\wedge x\wedge y=\omega(x,y)\in\Lambda^2\boldsymbol{V}_m^*.$$

The only invariant of a skew-symmetric bilinear form under the action of $\mathrm{GL}(\boldsymbol{V}_m)$ is its rank $\operatorname{rk}\omega=2r$, $0\le r\le\lfloor m/2\rfloor$. By (11.5.1),

$$\begin{aligned}\operatorname{wt}(H_\omega)&=\frac{\left|\{(e_1,e_2)\mid e_1\wedge e_2\wedge\omega\neq 0\}\right|}{[2]_q}\\ &=\sum_{e_1\colon e_1\wedge\omega\neq 0}\frac{\left|\{e_2\mid e_2\wedge(e_1\wedge\omega)\neq 0\}\right|}{[2]_q}.\end{aligned}$$

Let

$$s=\dim\ker\omega=m-\operatorname{rk}\omega=m-2r;$$

then there are q^m-q^s vectors e_1 such that $e_1\wedge\omega\neq 0$. Here $e_1\wedge\omega\in\Lambda^{m-1}\boldsymbol{V}_m$, so $e_1\wedge\omega$ is completely decomposable and $\dim\ker(e_1\wedge\omega)=m-1$. Hence,

$$\operatorname{wt}(H_\omega)=\frac{(q^m-q^s)(q^m-q^{m-1})}{(q^2-1)(q^2-q)}=\frac{q^m-q^s}{q^2-1}q^{m-2}=q^{2(m-r-1)}\frac{q^{2r}-1}{q^2-1}.$$

Let A_i denote, as usual, the number of codewords of weight i. Then

$$A_{\operatorname{wt}(H_\omega)}=(q-1)\left|\{H_{\omega'}\mid \operatorname{rk}\omega'=2r\}\right|,$$

which is exactly equal to the number $N(m,2r)$ of all bilinear skew-symmetric forms of rank $2r$:

$$N(m,2r)=\frac{(q^m-1)(q^{m-1}-1)\dots(q^{m-2r+1}-1)}{(q^{2r}-1)(q^{2r-2}-1)\dots(q^2-1)}q^{r(r-1)}.$$

Thus, we have proved the following result.

THEOREM 11.5.6. *The spectrum of a Grassmann code $C(2,m)$ is*

$$A_i=\begin{cases}N(m,2r) & \textit{if } i=q^{2(m-r-1)}\dfrac{q^{2r}-1}{q^2-1},\quad 0\le r\le\lfloor m/2\rfloor,\\ 0 & \textit{otherwise.}\end{cases}\qquad\square$$

Generalized weights of Grassmann codes. The well-known fact that the Grassmannian $G(\ell,m)$ contains projective spaces $\mathbb{P}^s$ for $s \le \max\{\ell, m-\ell\}$ as subvarieties trivially implies that the last higher weights d_{k-a} are given by

$$d_{k-a} = n - \theta_{a-1}, \quad 0 \le a \le \max\{\ell, m-\ell\} + 1.$$

Finding the first $\max\{\ell, m-\ell\}+1$ higher weights is less simple. Let us show that the generalized weights d_r of Grassmann codes for $r \le \ell+1$ can be established by presenting some certain subspace of codimension r and then checking that its weight meets the generalized Griesmer bound (Corollary 1.1.69)

$$d_r \ge \sum_{i=0}^{r-1} \left\lceil \frac{d_1}{q^i} \right\rceil,$$

so this weight is the minimum one.

Denote by $N(\ell,a,b)$ the number of V_ℓ in $\boldsymbol{V}_a$ that are skew to a fixed $\boldsymbol{V}_b \subseteq \boldsymbol{V}_a$:

$$N(\ell,a,b) = \left|\{V_\ell \subseteq \boldsymbol{V}_a \mid V_\ell \cap \boldsymbol{V}_b = \{0\}\}\right|, \quad b+\ell \le a.$$

One can easily check that

$$N(\ell,a,b) = \frac{(q^a-q^b)(q^a-q^{b+1})\dots(q^a-q^{b+\ell-1})}{(q^\ell-1)(q^\ell-q)\dots(q^\ell-q^{\ell-1})} = q^{\ell b}\begin{bmatrix} a-b \\ \ell \end{bmatrix}_q. \tag{11.5.2}$$

Consider r linearly independent $(m-\ell)$-subspaces $\boldsymbol{V}^{(i)} = \boldsymbol{V}^{(i)}_{m-\ell}$ in $\boldsymbol{V}_m$ such that their intersection has codimension 1 in each of them, i.e.,

$$\bigcap_i \boldsymbol{V}^{(i)} = \boldsymbol{V}_{m-\ell-1},$$

so their linear span is

$$\langle \boldsymbol{V}^{(i)} \rangle = \boldsymbol{V}_{m-\ell+r-1}.$$

Choose a basis $(e_1^{(i)}, \dots, e_{m-\ell}^{(i)})$ in $\boldsymbol{V}^{(i)}$ and consider the forms $\omega_i = e_1^{(i)} \wedge \dots \wedge e_{m-\ell}^{(i)} \in \Lambda^{m-\ell}\boldsymbol{V}_m$. These forms are linearly independent, so the intersection of the corresponding hyperplanes $H_{\omega_i} \subset \mathbb{P}^{k-1}$ is a subspace $H^r = H_{\omega_1} \cap \dots \cap H_{\omega_r}$ of codimension r.

Then

$$\begin{aligned} n - \mathrm{wt}(H^r) &= \left|\{V_\ell \mid V_\ell \cap \boldsymbol{V}^{(i)}_{m-\ell} \ne \{0\}, \text{ for all } i\}\right| \\ &= \left|\{V_\ell \mid V_\ell \cap \boldsymbol{V}_{m-\ell-1} \ne \{0\}\}\right| \qquad (11.5.3) \\ &\quad + \left|\{V_\ell \mid V_\ell \cap \boldsymbol{V}_{m-\ell+r-1} = V_r,\ V_r \cap \boldsymbol{V}_{m-\ell-1} = \{0\}\}\right|, \qquad (11.5.4) \end{aligned}$$

since any V_ℓ which intersects each of $\boldsymbol{V}^{(i)}$ either intersects their common intersection or, otherwise, intersects each of them by a single 1-space (see Fig. 11.1). Here (11.5.3) equals

$$n - N(\ell, m, m-\ell-1).$$

If $r \le \ell$, then (11.5.4) equals the number of V_r in $\boldsymbol{V}_{m-\ell+r-1}$ skew to $\boldsymbol{V}_{m-\ell-1}$, times the number of complementary $V_{\ell-r}$ outside $\boldsymbol{V}_{m-\ell+r-1}$, divided by the number of $V_{\ell-r}$ that complement $\boldsymbol{V}_r$ to $\boldsymbol{V}_\ell$. Thus, (11.5.4) equals

$$N(r, m-\ell+r-1, m-\ell-1)N(\ell-r, m, m-\ell+r-1)/N(\ell-r, \ell, r).$$

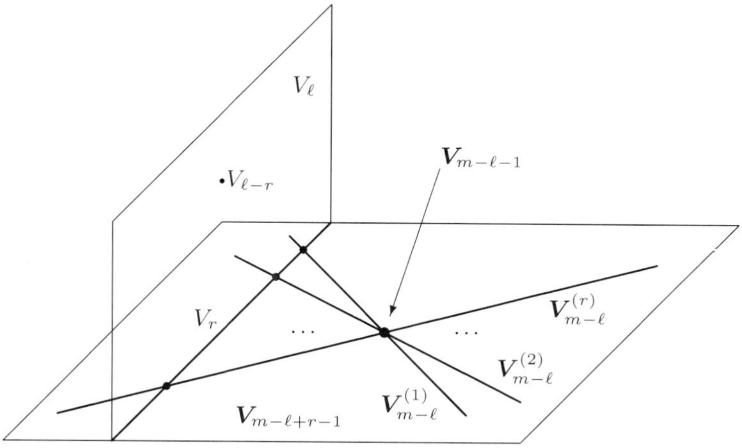

FIGURE 11.1

Therefore, using (11.5.2), we find

$$\begin{aligned}\mathrm{wt}(H^r) &= q^{\ell(m-\ell-1)}\begin{bmatrix}\ell+1\\ \ell\end{bmatrix}_q \\ &\quad - q^{r(m-\ell-1)}\begin{bmatrix}r\\ r\end{bmatrix}_q q^{(\ell-r)(m-\ell+r-1)}\begin{bmatrix}\ell-r+1\\ \ell-r\end{bmatrix}_q \Big/ \left(q^{(\ell-r)r}\begin{bmatrix}\ell-r\\ \ell-r\end{bmatrix}_q\right) \\ &= q^{\ell(m-\ell-1)}\frac{q^{\ell+1}-1}{q-1} - q^{\ell(m-\ell-1)}\frac{q^{\ell-r+1}-1}{q-1} \\ &= q^{\ell(m-\ell)}\frac{q^r-1}{q^{r-1}(q-1)}.\end{aligned}$$

If $r=\ell+1$, then (11.5.4) is zero, so

$$\mathrm{wt}(H^r) = q^{\ell(m-\ell-1)}\begin{bmatrix}\ell+1\\ \ell\end{bmatrix}_q = q^{\ell(m-\ell-1)}\frac{q^{\ell+1}-1}{(q-1)} = q^{\ell(m-\ell)}\frac{q^r-1}{q^{r-1}(q-1)}.$$

Thus, in both cases,

$$\mathrm{wt}(H^r) = d_1\frac{q^r-1}{q^{r-1}(q-1)} = d_1\left(1+\frac{1}{q}+\ldots+\frac{1}{q^{r-1}}\right),$$

which is not greater than d_r by the generalized Griesmer bound. Hence, it does equal d_r.

One can also consider the dual Grassmannian $G(m-\ell,m)$, which is isomorphic to $G(\ell,m)$, and thus obtain the same for $r \le (m-\ell)+1$. Thus, we have proved the following result.

THEOREM 11.5.7. *For $r \le \max\{\ell, m-\ell\}+1$, the generalized weights d_r of Grassmann codes meet the generalized Griesmer bound.* □

11.5.2. Schubert Codes

Now we pass to codes from Schubert varieties, which are certain natural subvarieties of Grassmannians. They can be described as follows.

Define the set

$$I(\ell,m)=\{\alpha=(\alpha_1,\dots,\alpha_\ell)\in\mathbb{Z}^\ell:\ 1\le\alpha_1<\dots<\alpha_\ell\le m\}$$

of multi-indices α. Fix any $\alpha=(\alpha_1,\dots,\alpha_\ell)\in I(\ell,m)$. Let $A_1\subset\dots\subset A_\ell$ be the canonical (partial) flag of subspaces of $\mathbb{F}_q^m$ associated with α, i.e., given by the linear span

$$A_i=\langle e_1,\dots,e_{\alpha_i}\rangle,\quad \text{for}\quad i=1,\dots,\ell,$$

where $\{e_1,\dots,e_m\}$ is a natural basis of $\mathbb{F}_q^m$. Define the *Schubert variety* associated to α (or, more precisely, to the above partial flag) in $G(\ell,m)(\mathbb{F}_q)$ by

$$\Omega_\alpha=\Omega_\alpha(\mathbb{F}_q)=\{W\in G(\ell,m)(\mathbb{F}_q):\ \dim(W\cap A_i)\ge i \text{ for } i=1,\dots,\ell\}.$$

It is well known that Ω_α is a projective algebraic variety. Indeed, if for $\beta=(\beta_1,\dots,\beta_\ell)$ and $\beta'=\beta'_1,\dots,\beta'_\ell)$ in $I(\ell,m)$ we define the relation

$$\beta\le\beta'\iff\beta_i\le\beta'_i,\quad\text{for all } i=1,\dots,\ell,$$

then we obtain a partial order on $I(\ell,m)$, and it can be shown that

$$\Omega_\alpha(\mathbb{F}_q)=\{p\in G(\ell,m)(\mathbb{F}_q):\ p_\beta=0 \text{ for all } \beta\not\le\alpha\}$$

and, moreover, that for any $\beta\le\alpha$ there exists some $p\in\Omega_\alpha(\mathbb{F}_q)$ such that $p_\beta\neq0$. In fact, the (defining) ideal $I(\Omega_\alpha)$ of Ω_α is generated by the generic Plücker coordinates p_β, $\beta\not\le\alpha$, and certain quadratic polynomials. It follows that the Plücker coordinates p_β, $\beta\le\alpha$, do not satisfy any linear relation. Thus, if we let

$$k_\alpha=|\{\beta\in I(\ell,m):\ \beta\le\alpha\}|\qquad\text{and}\qquad n_\alpha=|\Omega_\alpha(\mathbb{F}_q)|,$$

then $\Omega_\alpha(\mathbb{F}_q)$ gives rise to a nondegenerate $[n_\alpha,k_\alpha]_q$ projective system via the induced Plücker embedding and, therefore, to a linear $[n_\alpha,k_\alpha]_q$ code. This code is called a *Schubert code* and is denoted by $C_\alpha(\ell,m)$.

Note that if $\alpha=(m-\ell+1,m-\ell+2,\dots,m)$, then $\beta\le\alpha$ for all $\beta\in I(\ell,m)$, and thus $\Omega_\alpha(\mathbb{F}_q)=G(\ell,m)(\mathbb{F}_q)$. Thus, in this case $C_\alpha(\ell,m)$ equals the Grassmann code $C(\ell,m)$.

Define

$$\delta_\alpha=\sum_{i=1}^{\ell}(\alpha_i-i)=\alpha_1+\dots+\alpha_\ell-\frac{\ell(\ell+1)}{2}.$$

Let $d_1(C_\alpha(\ell,m))$ be the minimum distance of the Schubert code $C_\alpha(\ell,m)$. The following result is a simple consequence of the so-called Schubert calculus.

PROPOSITION 11.5.8. *We have* $d_1(C_\alpha(\ell,m))\le q^{\delta_\alpha}$. □

Moreover, one may hope that the following conjecture holds.

CONJECTURE 11.5.9. The minimum distance $d_1\left(C_\alpha(\ell,m)\right)$ of the Schubert code $C_\alpha(\ell,m)$ is q^{δ_α}, i.e., the inequality in Proposition 11.5.8 always holds with equality.

Actually, we have the following.

THEOREM 11.5.10. *Conjecture* 11.5.9 *is true for* $\ell=2$. □

Now let us compute the length of a Schubert code.

THEOREM 11.5.11. *The length n_α of $C_\alpha(\ell,m)$, or, in other words, the number of $\mathbb{F}_q$-rational points of Ω_α, is given by*

$$n_\alpha = \sum_{\beta\leq\alpha} q^{\delta_\beta}, \tag{11.5.5}$$

where the sum is over all $\beta\in I(\ell,m)$ with $\beta\leq\alpha$.

PROOF. Consider, as above, the subspaces A_j spanned by the first j basis vectors, where $1\leq j\leq m$. For any given $W\in G(\ell,m)$, the numbers $r_j=\dim W\cap A_j$ have the property that $0\leq r_j-r_{j-1}\leq 1$ (where $r_0=0$ by convention); this follows, for example, because the kernel of the map $W\cap A_j\to\mathbb{F}_q$ taking a vector to its jth coordinate (with respect to a fixed basis of V) is $W\cap A_{j-1}$. Also, since $r_m=\ell$, there are exactly ℓ indices for which $r_j-r_{j-1}=1$. Thus, there is a unique $\beta\in I(\ell,m)$ such that W is in

$$C_\beta := \left\{L\in G(\ell,m):\ \dim(L\cap A_{\beta_j})=j \text{ and } \dim(L\cap A_{\beta_j-1})=j-1 \text{ for } 1\leq j\leq\ell\right\}.$$

Moreover, for any $L\in C_\beta$ we have $L\in\Omega_\alpha\Leftrightarrow\beta\leq\alpha$. It follows that Ω_α is the disjoint union of the sets C_β as β varies over the elements of $I(\ell,m)$ with $\beta\leq\alpha$. Now it suffices to observe that the subspaces in C_β are in a natural one-to-one correspondence with $\ell\times m$ matrices (over $\mathbb{F}_q$) with 1 in the position (i,β_i) and with zeroes to its right and below, for $1\leq i\leq\ell$. □

Note, however, that formula (11.5.5) is simple but not very convenient in practice in view of the rather intricate summation involved. Now let us describe another formula for n_α, which is not as elegant but may also be of some interest.

We begin with an elementary definition and preliminary lemmas. Define a *consecutive block* in an ℓ-tuple $\beta=(\beta_1,\dots,\beta_\ell)\in I(\ell,m)$ as an ordered sequence of the form $\beta_i,\dots,\beta_j$ where $1\leq i\leq j\leq\ell$ and $\beta_{p+1}=\beta_p+1$ for $i\leq p<j$. Note that any $\beta\in I(\ell,m)$ always has ℓ consecutive blocks.

LEMMA 11.5.12. *Assume that $\alpha=(\alpha_1,\dots,\alpha_\ell)$ has $u+1$ consecutive blocks:*

$$\alpha=(\alpha_1,\dots,\alpha_{p_1},\ \alpha_{p_1+1},\dots,\alpha_{p_2},\ \dots,\ \alpha_{p_{u-1}+1},\dots,\alpha_{p_u},\ \alpha_{p_u+1},\dots,\alpha_\ell)$$

so that $1\leq p_1<\dots<p_u<\ell$ and $\alpha_{p_i+1},\dots,\alpha_{p_{i+1}}$ are consecutive for $0\leq i\leq u$, where, by convention, $p_0=0$ and $p_{u+1}=\ell$. Then

$$\Omega_\alpha=\{W\in G(\ell,\alpha_\ell):\ \dim(W\cap A_{\alpha_{p_i}})\geq p_i \text{ for } i=1,\dots,u\}.$$

PROOF. As in the proof of Theorem 11.5.11, for any $W\in G(\ell,\alpha_\ell)$ we have $\dim(W\cap A_{j-1})\geq\dim(W\cap A_j)-1$ for $1\leq j\leq m$. Also, $\dim(W\cap A_{\alpha_\ell})\geq\ell$ if and only if W is a subspace of A_{α_ℓ}, which implies the conclusion. □

Given any integers a,b,s,t, define

$$\lambda(a,b;s,t)=\sum_{r=s}^{t}(-1)^{r-s}q^{\binom{r-s}{2}}\begin{bmatrix}a-s\\ r-s\end{bmatrix}_q\begin{bmatrix}b-r\\ t-r\end{bmatrix}_q,$$

where $\begin{bmatrix}u\\ v\end{bmatrix}_q$ is assumed to be 0 for any $u,v\in\mathbb{Z}$ that do not satisfy $0\leq v\leq u$. In particular, $\lambda(a,b;s,t)=\begin{bmatrix}b\\ t\end{bmatrix}_q$ for $a=s=0$.

Using the Möbius inversion formula, one can prove the following lemma.

LEMMA 11.5.13. *Let B be a b-dimensional vector space over $\mathbb{F}_q$, let $G(t,b) = G_t(B)$ denote the Grassmannian of t-dimensional subspaces of B, let A be any subspace of B and S any subspace of A, and let $a = \dim A$ and $s = \dim S$. Then*

$$|\{T \in G_t(B) : T \cap A = S\}| = \lambda(a,b;s,t). \qquad \square$$

THEOREM 11.5.14. *Let u and $p_1,\dots,p_u$ be as in Lemma 11.5.12. Then the length n_α of the Schubert code $C_\alpha(\ell,m)$ is given by*

$$n_\alpha = \sum_{s_1=p_1}^{\alpha_{p_1}} \sum_{s_2=p_2}^{\alpha_{p_2}} \cdots \sum_{s_u=p_u}^{\alpha_{p_u}} \prod_{i=0}^{u} \lambda(\alpha_{p_i}, \alpha_{p_{i+1}}; s_i, s_{i+1}) \tag{11.5.6}$$

where, by convention, $s_0 = p_0 = 0$ and $s_{u+1} = p_{u+1} = \ell$.

PROOF. We use induction on u. If $u = 0$, i.e., if $\alpha_1,\dots,\alpha_\ell$ are consecutive, then $\Omega_\alpha = G(\ell,\alpha_\ell)$, and therefore $n_\alpha = \begin{bmatrix}\alpha_\ell \\ \ell\end{bmatrix}_q = \lambda(0,\alpha_\ell;0,\ell)$. Now assume that $u \geq 1$ and the result holds for all smaller values of u. Then by Lemma 11.5.12 we see that

$$\Omega_\alpha = \bigsqcup_S \{T \in G(\ell,\alpha_\ell) : \ T \cap A_{\alpha_{p_u}} = S\}$$

where the disjoint union is over the set, say Λ_u, of all subspaces S of $A_{\alpha_{p_u}}$ satisfying $\dim S \geq u$ and $\dim S \cap A_{\alpha_{p_i}} \geq p_i$ for $1 \leq i \leq u-1$. Hence, by Lemma 11.5.13 we get

$$n_\alpha = |\Omega_\alpha(\mathbb{F}_q)| = \sum_{s=p_u}^{\alpha_{p_u}} |\{S \in \Lambda_u : \dim S = s\}| \lambda(\alpha_{p_u}, \alpha_\ell; s, \ell).$$

But for any s with $p_u \leq s \leq \alpha_{p_u}$, the set of s-dimensional subspaces in Λ_u is precisely the Schubert variety in $G(s,\alpha_{p_u})$ corresponding to the tuple $(\alpha_1,\dots,\alpha_{p_u})$ with u consecutive blocks. Hence the induction hypothesis applies. $\square$

The dimension k_α of these codes can also be explicitly computed.

THEOREM 11.5.15. *The dimension k_α of the Schubert code $C_\alpha(\ell,m)$ equals the determinant of the $\ell \times \ell$ matrix whose (i,j)th entry is $\binom{\alpha_j - j + 1}{i - j + 1}$, i.e.,*

$$k_\alpha = \begin{vmatrix} \binom{\alpha_1}{1} & 1 & 0 & \dots & 0 \\ \binom{\alpha_1}{2} & \binom{\alpha_2 - 1}{1} & 1 & \dots & 0 \\ \dots & \dots & \dots & \dots & \dots \\ \binom{\alpha_1}{\ell} & \binom{\alpha_2 - 1}{\ell - 1} & \binom{\alpha_3 - 2}{\ell - 2} & \dots & \binom{\alpha_\ell - \ell + 1}{1} \end{vmatrix}. \tag{11.5.7}$$

SKETCH OF THE PROOF. Recall the above formula

$$k_\alpha = |\{\beta \in I(\ell,m) : \beta \leq \alpha\}|;$$

a vector space basis for the tth component, say R_t, of the homogeneous coordinate ring of Ω_α is indexed by the t-tuples $(\beta^{(1)},\dots,\beta^{(t)})$ of elements of $I(\ell,m)$ satisfying $\beta^{(1)} \leq \dots \leq \beta^{(t)} \leq \alpha$. Using the Hodge postulation formula, which gives the Hilbert function $h(t) = \dim R_t$, $t \in \mathbb{N}$, of this ring, we obtain

$$h(t) = \det_{1 \leq i,j \leq \ell} \left(\binom{t + \alpha_j - j}{t + i - j} \right) \quad \text{for } t \in \mathbb{N},$$

and putting $t = 1$, we get the conclusion. $\square$

COROLLARY 11.5.16. *For $\ell = 2$, we have*

$$k_\alpha = \alpha_1(\alpha_2 - 1) - \binom{\alpha_1}{2} = \frac{\alpha_1(2\alpha_2 - \alpha_1 - 1)}{2}. \qquad \square$$

THEOREM 11.5.17. *Assume that $\alpha_1, \dots, \alpha_\ell$ form an arithmetic progression: $\alpha_i = c(i-1) + d$ for $i = 1, \dots, \ell$ with $c, d \in \mathbb{Z}$. Define $\alpha_{\ell+1} = c\ell + d = \ell\alpha_2 + (1-\ell)\alpha_1$. Then*

$$k_\alpha = \frac{\alpha_1}{\ell!} \prod_{i=1}^{\ell-1} (\alpha_{\ell+1} - i) = \frac{\alpha_1}{\alpha_{\ell+1}} \binom{\alpha_{\ell+1}}{\ell}.$$

SKETCH OF THE PROOF. If $\alpha_i = c(i-1) + d$ for $i = 1, \dots, \ell$, then the (i,j)th entry of the transpose of the $\ell \times \ell$ matrix in (11.5.7) can be written as

$$\binom{c(i-1) + d - i + 1}{j - i + 1} = \binom{BL_i + A}{L_i + j}, \quad \text{where } B = 1 - c,\ L_i = 1 - i, \text{ and } A = d.$$

For determinants of this type we have the general formula

$$\frac{\prod\limits_{1 \le i < j \le \ell} (L_i - L_j)}{\prod\limits_{i=1}^{\ell} (L_i + \ell)!} \prod_{i=1}^{\ell} \frac{(BL_i + A)!}{((B-1)L_i + A - 1)!} \prod_{i=1}^{\ell} (A - Bi + 1)_{i-1},$$

where in the last product we used the shifted factorial notation, namely $(a)_0 = 1$ and $(a)_t = a(a+1)\dots(a+t-1)$ for $t \ge 1$. Substituting $B = 1 - c$, $L_i = 1 - i$, and $A = d$ and making elementary simplifications, we obtain the desired formula. $\square$

One can also obtain an alternative formula for the dimension k_α of $C_\alpha(\ell, m)$. Note first that the dimension k of the q-ary Grassmann code $C(\ell, m)$ does not depend on q and satisfies

$$\lim_{q \to 1} n(q) = k, \qquad \text{or, in other words,} \qquad \lim_{q \to 1} \begin{bmatrix} m \\ \ell \end{bmatrix}_q = \binom{m}{\ell}. \tag{11.5.8}$$

In fact, a similar relation holds in the case of Schubert codes, and one can use this relation to obtain an alternative formula for k_α.

PROPOSITION 11.5.18. *The dimension k_α of the q-ary Schubert code $C_\alpha(\ell, m)$ is independent of q and is related to the length $n_\alpha = n_\alpha(q)$ of $C_\alpha(\ell, m)$ by the formula*

$$\lim_{q \to 1} n_\alpha(q) = k_\alpha.$$

Consequently, if u and $p_1, \dots, p_u$ are as in Lemma 11.5.12, then

$$k_\alpha = \sum_{s_1 = p_1}^{\alpha_{p_1}} \sum_{s_2 = p_2}^{\alpha_{p_2}} \cdots \sum_{s_u = p_u}^{\alpha_{p_u}} \prod_{i=0}^{u} \binom{\alpha_{p_{i+1}} - \alpha_{p_i}}{s_{i+1} - s_i}$$

with $s_0 = p_0 = 0$ and $s_{u+1} = p_{u+1} = \ell$. $\square$

This result leads to the following peculiar but interesting consequence.

COROLLARY 11.5.19. *Let $R = R(q)$ and $\Delta = \Delta(q)$ denote, respectively, the rate and the relative distance of the q-ary Schubert code $C_\alpha(\ell, m)$. Then we have*

$$\lim_{q \to 1} R(q) = 1 \qquad \text{and} \qquad \lim_{q \to \infty} \Delta(q) = 1. \qquad \square$$

11.6. Codes from Flag Varieties

Flag varieties form a rather large class of varieties with many points over finite fields and also large automorphism groups, which makes them attractive to use for constructing projective systems and codes. Some codes in the preceding sections are particular cases of flag variety codes. Here we give only their general description and some most elementary examples, since their general study requires a much more advanced technique.

11.6.1. Flag Varieties

A *flag* over a finite field $\mathbb{F}_q$ is a strictly increasing sequence X of subspaces $V_{i_1} \subset V_{i_2} \subset \ldots \subset V_{i_s}$ of an m-dimensional vector space $V = (\mathbb{F}_q)^m$, $\dim V_{i_k} = i_k$ for $k = 1,2,\ldots,s$. A *flag variety* of type $(i_1,i_2,\ldots,i_s)$ is the variety of all flags $X = \{(V_{i_1},V_{i_2},\ldots,V_{i_s})\}$ with fixed $(i_1,i_2,\ldots,i_s)$.

Equivalently, we can describe it as a set G/P where $G = \mathrm{GL}_m(\mathbb{F}_q)$ and P is a parabolic subgroup (i.e., a subgroup conjugate to upper triangular block matrices). The variety of $(V_{i_1},V_{i_2},\ldots,V_{i_s})$ with $V_{i_1} \subset V_{i_2} \subset \ldots \subset V_{i_s}$ is isomorphic to the variety G/P with

$$P = \begin{pmatrix} M_1 & * & * & \ldots & * \\ 0 & M_2 & * & \ldots & * \\ 0 & 0 & M_3 & \ldots & * \\ \ldots & \ldots & \ldots & \ldots & \ldots \\ 0 & 0 & 0 & \ldots & M_{s+1} \end{pmatrix},$$

where $M_1 \in \mathrm{GL}(i_1,\mathbb{F}_q)$,

$$M_r \in \mathrm{GL}(i_r - i_{r-1},\mathbb{F}_q), \quad 2 \le r \le s,$$

and $M_{s+1} \in \mathrm{GL}(m - i_s,\mathbb{F}_q)$. The group G acts transitively on the set of flags $\{(V_{i_1},V_{i_2},\ldots,V_{i_s})\}$, and the stabilizer of the flag

$$\{(\langle e_1,e_2,\ldots,e_{i_1}\rangle,\langle e_1,e_2,\ldots,e_{i_2}\rangle,\ldots,\langle e_1,e_2,\ldots,e_{i_s}\rangle)\}$$

is the subgroup P where $(e_i)_{1\le i\le m}$ is the canonical basis of the vector space V. In that case, we shall say that P is a parabolic subgroup of the type $(i_1,i_2,\ldots,i_s)$.

For example, the flag variety of the type (ℓ) with $\ell \le m$ is simply the Grassmannian $G(\ell,m)$.

Projective systems from flag varieties. Let P be a parabolic subgroup of G, let G_1 be a subgroup of G, and let $P_1 = P \cap G_1$. We embed $X = G/P$ and the subset $X_1 = G_1/P_1$ in a projective space $\mathbb{P}^N$ by the following sequence of three embeddings.

A flag $(V_{i_1},V_{i_2},\ldots,V_{i_s})$ is a sequence of subspaces V_{i_ℓ} which are elements of the Grassmannian $G(i_\ell,m)$. This gives a morphism from the flag variety $X = G/P$ to the product of Grassmannians $G(i_1,m) \times \ldots \times G(i_s,m)$, and one can embed each Grassmannian $G(i_\ell,m)$ into $\mathbb{P}(\Lambda^\ell V_{i_\ell}) = \mathbb{P}^{r_\ell}$ with $r_\ell = \binom{m}{i_\ell} - 1$ by the Plücker embedding.

Then we can embed the product of projective spaces $\mathbb{P}^{r_1} \times \ldots \times \mathbb{P}^{r_s}$ into the projective space $\mathbb{P}^r$ with $r = (r_1+1)(r_2+1)\ldots(r_s+1)-1$ by the *Segre embedding*, which maps elements $(v_1,\ldots,v_s) \in V_{r_1+1} \times \ldots \times V_{r_s+1}$ to tensor products $v_1 \otimes \ldots \otimes v_s \in V_{r+1}$. This defines a mapping from $\mathbb{P}^{r_1} \times \ldots \times \mathbb{P}^{r_s}$ to $\mathbb{P}^r$ according to the commutative diagram

$$\begin{array}{ccccc}
(v_1,\dots,v_s) & \in & V^{\times}_{r_1+1}\times\dots\times V^{\times}_{r_s+1} & \longrightarrow & \mathbb{P}^{r_1}\times\dots\times\mathbb{P}^{r_s} \\
\downarrow & & \downarrow & & \downarrow \\
v_1\otimes\dots\otimes v_s & \in & (V_{r_1+1}\otimes\dots\otimes V_{r_s+1})^{\times} & \longrightarrow & \mathbb{P}^r,
\end{array}$$

where $V^{\times}$ denotes the vector space V minus the origin.

One can further embed $\mathbb{P}^r$ into the projective space $\mathbb{P}^N$ with $N=\binom{r+h}{h}-1$ by the *Veronese embedding* of order h, which comes from the map $V_{r+1}\to V_{N+1}$ sending an element $(u_1,u_2,\dots,u_{r+1})$ to the element $(\dots,u_1^{j_1}u_2^{j_2}\dots u_{r+1}^{j_{r+1}},\dots)$ of V_{N+1}, with $j_1+j_2+\dots+j_{r+1}=h$.

If we restrict ourselves to the subgroup G_1 of G and take $P_1=P\cap G_1$ and $X_1=G_1/P_1$, we get the following diagram:

$$\begin{array}{ccccc}
\mathrm{GL}(m)/P & \supset & G_1/P_1 & = & \{x_1,x_2,\dots,x_n\} \\
\downarrow & & \Big| & & \Big| \\
G(i_1,m)\times\dots\times G(i_s,m) & & \Big| & & \Big| \\
\downarrow & & \Big| & & \Big| \\
\mathbb{P}^{r_1}\times\dots\times\mathbb{P}^{r_s} & & \Big| & & \Big| \\
\downarrow & & \Big| & & \Big| \\
\mathbb{P}^{r} & & \Big| & & \Big| \\
\downarrow & & \downarrow & & \downarrow \\
\mathbb{P}^N & \supset & \mathbb{P}^N & \longleftarrow & V^{\times}_{N+1}\ni\{y_1,y_2,\dots,y_n\}.
\end{array} \tag{11.6.1}$$

For $f\in V_{N+1}(\mathbb{F}_q)^*$, let $\underline{\mathrm{ev}}\colon f\mapsto (f(y_1),\dots,f(y_n))$ be the evaluation map; then its image is a code with parameters

$$\big[\,|X|,\ N+1-\dim\ker\underline{\mathrm{ev}},\ |X|-\max|X\cap H|\,\big],$$

where $X=G_1/P_1$ and H runs over the set of hyperplanes in $\mathbb{P}^N$.

11.6.2. Examples

Now we give several particular examples of this general construction, relating the construction to codes described above.

Projective Reed–Muller codes. Take matrices $P=\begin{pmatrix} a & * \\ 0 & M\end{pmatrix}$, where $a\in\mathbb{F}_q^{\times}$, the symbol $*$ denotes any element of $\mathbb{F}_q^{m-1}$, and M is a square matrix of order $m-1$. Then the diagram (11.6.1) simplifies to

$$\begin{array}{ccc}
\mathrm{GL}(m)/P & = & \{x_1,x_2,\dots,x_n\} \\
\downarrow & & \Big| \\
\mathbb{P}^{m-1} & & \Big| \\
\downarrow & & \downarrow \\
\mathbb{P}^N & \longleftarrow & V_{N+1}\setminus\{0\}.
\end{array}$$

We thus obtain a code of order h whose parameters in the case where $\underline{\text{ev}}$ is injective (i.e., $h \leq q$) are

$$\left[\frac{q^m-1}{q-1}, \binom{h+m-1}{h}, (q-h+1)q^{m-2}\right],$$

where h is the order of the Veronese embedding. These codes are described in Sec. 11.1.3.

Codes on Grassmannians. Take matrices $P = \begin{pmatrix} M_1 & * \\ 0 & M_2 \end{pmatrix}$, where M_1 stands for a square matrix of order ℓ and M_2 stands for a square matrix of order $m-\ell$. We now get the diagram

$$\begin{array}{ccc} \mathrm{GL}(m)/P & = & \{x_1, x_2, \ldots, x_n\} \\ \downarrow & & \Big| \\ G(\ell, m) & & \Big| \\ \downarrow & & \downarrow \\ \mathbb{P}^N & \longleftarrow & V_{N+1} \setminus \{0\} \end{array}$$

with $N = \binom{m}{\ell} - 1$.

The parameters of the code that we obtain are

$$\left[\frac{(q^m-1)(q^m-q)\ldots(q^m-q^{\ell-1})}{(q^\ell-1)(q^\ell-q)\ldots(q^\ell-q^{\ell-1})}, \binom{m}{\ell}, q^{(m-\ell)\ell}\right];$$

cf. Sec. 11.5.1.

Codes on Hermitian hypersurfaces. Now consider the subgroup $G_1 = U(m+1, \mathbb{F}_{q^2})$ of the group $G = \mathrm{GL}(m+1, \mathbb{F}_{q^2})$ with $m \geq 2$. This subgroup is the unitary group for the Hermitian form

$$\langle x, y \rangle = x_0 y_m^q + x_1 y_{n-1}^q + \ldots + x_m y_0^q$$

on $\mathbb{F}_{q^2}^{m+1}$. Now take the parabolic subgroup

$$P = \begin{pmatrix} a & * & \ldots & * \\ 0 & & & \\ \vdots & & \begin{pmatrix} \mathrm{GL}(m) \end{pmatrix} & \\ 0 & & & \end{pmatrix}$$

where $a \in \mathbb{F}_{q^2}^{\times}$, symbols $*$ denote coefficients of a matrix in $\mathrm{GL}(m+1)$ which are allowed to take any value in $\mathbb{F}_{q^2}$, and $P_1 = U(m+1, \mathbb{F}_{q^2}) \cap P$. We get the following diagram:

$$\begin{array}{ccccc} \mathrm{GL}(m+1)/P & \supset & U(m+1)/P_1 & = & \{x_1, x_2, \ldots, x_n\} \\ \downarrow & & \Big| & & \Big| \\ \mathbb{P}^m & & \Big| & & \Big| \\ \text{Veronese} \downarrow & & \downarrow & & \downarrow \\ \mathbb{P}^N(\mathbb{F}_{q^2}) & \supset & \mathbb{P}^N(\mathbb{F}_{q^2}) & \longleftarrow & V_{N+1}(\mathbb{F}_{q^2}) \setminus \{0\}. \end{array}$$

Letting $h \leq q^2 - q$ be the order of the Veronese embedding, we get codes C_h on $\mathbb{F}_{q^2}$ with parameters

$$\left[\frac{(q^{m+1}-(-1)^{m+1})(q^m-(-1)^m)}{q^2-1}, \binom{m+h}{h}, d\right];$$

a bound for the minimum distance d can easily be computed from the general construction using Proposition 11.2.4, which gives a bound for the number of rational points over $\mathbb{F}_{q^2}$ of a hyperplane section of the set G_1/P_1. We get

$$\begin{aligned} d &\geq \frac{(q^{m+1}-(-1)^{m+1})(q^m-(-1)^m)}{q^2-1} - (q+1)h\frac{q^{2m-2}-1}{q^2-1} \\ &= \frac{(q^{m+1}-(-1)^{m+1})(q^m-(-1)^m)-(q+1)h(q^{2m-2}-1)}{q^2-1} \\ &\geq q^{2m-1} + (1-h)q^{2m-3} - hq^{2m-4} + (1-h)q^{2m-5} - \dots . \end{aligned}$$

For $h = 1$, these are precisely the Hermitian codes described in Sec. 11.4.1, and the exact minimum distance is known:

$$d = \begin{cases} q^{2m-1}, & \text{for } m \text{ odd}, \\ q^{2m-1} - q^{m-1}, & \text{for } m \text{ even}. \end{cases}$$

For $h = 2$ and $m = 3$, we get a code with parameters

$$[(q^2+1)(q^3+1), 10, \geq q^5 - q^3 - q^2 - 2q - 1],$$

and conjecturally, the parameters should be

$$[(q^2+1)(q^3+1), 10, \geq q^5 - q^3 - q^2 + q].$$

For $h = 2$ and $m = 4$, we get a code with parameters

$$[(q^5+1)(q^2+1), 15, \geq q^7 - q^5 - 2q^4 - 2q^3 - q^2 - 2q - 1],$$

which is better than a previously known bound obtained by other methods.

11.6.3. Two More Examples

Now let us give some examples which are not treated above.

Codes on the Deligne–Lusztig variety for the group $SU(5)$. As before, consider the subgroup $G_1 = U(5, \mathbb{F}_{q^2})$ of the group $G = \mathrm{GL}(5, \mathbb{F}_{q^2})$; it is the unitary group for the hermitian form

$$\langle x, y \rangle = x_0 y_4^q + x_1 y_3^q + x_2 y_2^q + x_3 y_1^q + x_4 y_0^q$$

on $\mathbb{F}_{q^2}^5$. Now take the parabolic subgroup

$$Q = \begin{pmatrix} * & * & * & * & * \\ * & * & * & * & * \\ 0 & 0 & * & * & * \\ 0 & 0 & * & * & * \\ 0 & 0 & * & * & * \end{pmatrix}$$

and $Q_1 = U(m+1, \mathbb{F}_{q^2}) \cap Q$. The variety G_1/Q_1 can be viewed as the Deligne–Lusztig variety for the group $SU(5)$.

The diagram (11.6.1) simplifies to

$$\begin{array}{ccccc} \mathrm{GL}(5)/Q & \supset & G_1/Q_1 & = & \{x_1, x_2, \dots, x_n\} \\ \downarrow & & \Big| & & \Big| \\ G(2,5) & & \Big| & & \Big| \\ {\scriptstyle \text{Plücker}}\Big\downarrow & & \Big\downarrow & & \Big\downarrow \\ \mathbb{P}^9 & = & \mathbb{P}^9(\mathbb{F}_{q^2}) & \leftarrow & V_{10}(\mathbb{F}_{q^2}) \setminus \{0\}, \end{array}$$

and we get a code with parameters

$$\left[(q^5+1)(q^3+1),\, 10,\, q^8 - q^6\right]_{q^2}$$

whose weights are $q^8 - q^6$, $q^8 - q^6 + q^5 - q^3$, $q^8 - q^6 + q^5$, $q^8 - q^6 + q^5 + q^3$, and q^8. For $q = 2$ we get a $[297, 10, 192]_4$ code with weights 256, 232, 224, 216, and 192.

The flag variety of type $(1, m-2)$. Consider the variety of flags $X = \{(V_1, V_{m-1})\}$ formed by lines V_1 and hyperplanes V_{m-1}. We identify by duality the hyperplanes V_{m-1} with the elements $V_{m-1}^{\perp}$ of the projective space associated to the dual V^*.

The commutative diagram

$$\begin{array}{ccccc} (a, \beta) & \in & V^{\times} \times (V^*)^{\times} & \longrightarrow & (V \otimes V^*)^{\times} \\ \Big\downarrow & & \Big\downarrow & & \Big\downarrow \\ (V_1, V_{m-1}^{\perp}) & \in & \mathbb{P}^{m-1} \times (\mathbb{P}^{m-1})^* & \longrightarrow & \mathbb{P}^{m^2-1}. \end{array}$$

defines the Segre embedding from $\mathbb{P}^{m-1} \times (\mathbb{P}^{m-1})^*$ to $\mathbb{P}^{m^2-1}$. A flag (V_1, V_{m-1}) belongs to X if and only if $\langle V_1, V_{m-1}^{\perp} \rangle = 0$. Therefore, a point x in $\mathbb{P}^{m^2-1}$ is in the image of X if and only if it is the image of $(a, \beta) \in V^{\times} \times (V^*)^{\times}$ with $\beta(a) = 0$. Two elements (a, β) and (a', β') of $V^{\times} \times (V^*)^{\times}$ give the same image in X if and only if $a' \in \mathbb{F}_q^{\times} a$ and $\beta' \in \mathbb{F}_q^{\times} \beta$. Let us denote by $\overline{a \otimes \beta}$ the image of (a, β) under the map $V^{\times} \times (V^*)^{\times} \to \mathbb{P}^{m^2-1}$. Define a linear form Tr on $V \otimes V^*$ by $\mathrm{Tr}(a \otimes \beta) = \beta(a)$.

Now consider the code C associated to the embedding $X \to \mathbb{P}^{m^2-1}$. Let us choose a lifting of X to $V^{\times} \times (V^*)^{\times}$: $\overline{a \otimes \beta} \mapsto (a, \beta)$. Codewords are the sequences $(f(a \otimes \beta))_{(a,\beta)}$ for (a, β) in the image of this lifting and

$$f \in \{x \in V \otimes V^* \mid \mathrm{Tr}(x) = 0\}^* = \{x \in V \otimes V^*\}^* / \mathbb{F}_q \,\mathrm{Tr}.$$

THEOREM 11.6.1. *The code C has the parameters*

$$\left[\frac{(q^m - 1)(q^{m-1} - 1)}{(q-1)^2},\, m^2 - 1,\, q^{2m-3} - q^{m-2}\right].$$

Its weights are given by

$$w = q^{m-2}\left(q^m - 1 - \sum_{\lambda \in \mathbb{F}_q} (q^{a_\lambda} - 1)\right) \Big/ (q-1),$$

where the $(a_\lambda)_{\lambda \in \mathbb{F}_q}$ are integers satisfying the following conditions:

$$0 \le a_\lambda \quad \textit{and} \quad \sum_{\lambda \in \mathbb{F}_q} a_\lambda \le m.$$

□

Historical and Bibliographic Notes

The idea of using multidimensional varieties to construct projective systems and thus codes is very natural. However, the absence of a simple bound on the minimum distance prevented their use for some time.

The first nontrivial such bound, namely the Tsfasman–Serre–Sørensen bound, was given by J-P. Serre in [**Ser91**] in response to a conjecture posed by M.A. Tsfasman; the same result was independently obtained by A.B. Sørensen in [**Sør91**]. Theorem 11.1.2 is proved in [**HK13**]. M.I. Boguslavsky was able to generalize this bound in [**Bog97**] for the case of two hypersurfaces; cf. also [**Bog98**].

Projective Reed–Muller codes were introduced by G. Lachaud [**Lac88**], and their parameters were calculated in [**Lac90, Sør91**]; see also [**Aub92**].

The paper [**Bog97**] contains also the statement of the Tsfasman–Boguslavsky conjecture (TBC). The quadratic case of the TBC (Theorem 11.1.7) was independently obtained by C. Zanella in [**Zan98**]. The general case of the TBC for $\ell \leq m+1$ was established in [**DG15**] and was refuted there for $\ell > m+1$. This led to determining some higher weights of projective Reed–Muller codes in [**DG17**] (Proposition 11.1.14). The affine case of the TBC, which corresponds to affine Reed–Muller codes, was investigated in [**HP98**], containing Theorem 11.1.13.

Lachaud's bounds in Sec. 11.2 are proved in [**Lac96**]. Proposition 11.2.5 belongs to F.L. Zak, cf. [**Zak93**]; Proposition 11.2.6 is deduced from results obtained in [**Sko92**]. Note that these results were largely developed in [**LT97, GL02b, GL02a, GL09, LR15**]; in particular, [**GL02a**] contains the formulation of the Ghorpade–Lachaud conjecture (Corollary 11.2.14). Couvreur's bound in Sec. 11.2.2 was proved in [**Cou16**]; Remark 11.2.19 is contained in [**DG17**].

The exposition in Sec. 11.3 follows [**Cou11**]; it uses elements of the general surface theory, which are exposed in many classical textbooks, e.g., [**Sha74**]. The theory of cubic surfaces was developed and exposed in [**Man74**]. For codes on surfaces of low rank, see also [**Zar07**].

Hermitian codes in Sec. 11.4.1 were first implicitly defined and investigated in a series of works [**Cha70, Cha71, Cha88, Cha91**]. Their explicit definition, description, and calculation of their higher weights is given in [**HTV94**] on the ground of the theory of Hermitian varieties developed in [**Hir79**]. The case of quadrics in Sec. 11.4.2 was developed in [**Nog93**] using [**HT91**], and independently in [**Wan94**]. In fact, some results on the minimum distance and spectra of these codes were obtained in [**Wol75**] (for nonsingular quadrics) and in [**Aub92**] (for singular ones).

Grassmann codes over $\mathbb{F}_2$ were constructed and investigated (not from a geometric point of view) in [**Rya87a, Rya87b, Rya88, RR88, RR90**]. The general case was considered by D.Yu. Nogin in [**Nog96**], which we follow in our exposition; the formula for $N(m, 2r)$ used in Theorem 11.5.6 can be found in [**HP52**, Sec. 15.2]. Spectra of two particular Grassmann codes with $\ell = 3$ are computed in [**Nog97, KP13**]. Schubert codes were introduced in [**GL00**]; their parameters were partly calculated in [**GT05**] and [**Xia08**]. Their construction uses classical Schubert calculus, treated, e.g., in [**HP52**] and [**KL72**]; for the Hodge postulation formula used in the proof of Theorem 11.5.15, see [**Hod43, Gho01**].

Codes from flag varieties were introduced and investigated by F. Rodier in [**Rod03**]. Codes on Deligne–Lusztig varieties in general were considered in [**Rod00**], and the first example in Sec. 11.6.3 is a very particular case of them. A more complete treatment of both classes—codes from flag varieties and from Deligne–Lusztig varieties—is based on a rather advanced algebraic geometry technique and is carried out in [**Rod00, Rod03**].

CHAPTER 12

Applications

Finally, in this last chapter we give some applications of algebraic geometry codes and similar objects to problems that are not directly connected with classical error-correcting codes (at least, are not connected in an obvious manner). They are broad and diversified; note, however, that we have to be very concise in our exposition below, since the applications are rather far-reaching and somewhat elaborated. Nevertheless, we consider them to be of major importance and therefore indispensable to complete the picture of the advanced theory of algebraic geometry codes. These applications are: the Chudnovsky algorithm for fast multiplication in finite fields (Sec. 12.1), two cryptographic applications to authentication codes and arithmetic secret sharing (Sec. 12.2), quantum algebraic geometry codes (Sec. 12.3), the Niederreiter–Rosenbloom–Tsfasman metric with applications to uniform nets and sequences (Sec. 12.4), and locally recoverable codes (Sec. 12.5).

We should note that a reasonably complete and self-contained exposition of these applications would need a separate volume; thus, we expose only some results which are close in spirit to our other chapters. In particular, we concentrate almost exclusively on results in an asymptotic setting, while algebraic geometry codes allow us to obtain many improvements for objects of moderate length, which we omit almost completely from our exposition.

12.1. Fast Multiplication in Finite Fields

In this section we describe algorithms for fast multiplication in finite fields based on the Chudnovsky brothers' interpolation method. We begin with some generalities on bilinear complexity and tensor rank in Sec. 12.1.1 and describe the Chudnovsky interpolation algorithm in its modern version in Sec. 12.1.2.

12.1.1. Tensor Rank and Bilinear Complexity

Tensor rank. Let K be a field, and $E_0, \dots, E_s$ be finite-dimensional K-vector spaces. A nonzero element $t \in E_0 \otimes \dots \otimes E_s$ is said to be an *elementary tensor*, or a *tensor of rank* 1, if it can be written in the form $t = e_0 \otimes \dots \otimes e_s$ for some $e_i \in E_i$. More generally, the *rank* of an arbitrary $t \in E_0 \otimes \dots \otimes E_s$ is defined as the minimal length of a decomposition of t as a sum of elementary tensors.

If

$$\alpha\colon E_1 \times \dots \times E_s \to E_0$$

is an s-linear map, the s-linear complexity of α is defined as the tensor rank of the element

$$\widetilde{\alpha} \in E_0 \otimes E_1^\vee \otimes \dots \otimes E_s^\vee$$

naturally corresponding to α by duality.

For $s = 1$, these notions are very well understood (they essentially reduce to the matrix rank). However, starting from $s = 2$, which is the case in the present section, they can be surprisingly difficult to handle.

Let $\mathcal{A}$ be a finite-dimensional K-algebra. We denote by

$$\mu(\mathcal{A}/K)$$

the bilinear complexity of the multiplication map

$$m_{\mathcal{A}}\colon \mathcal{A}\times\mathcal{A}\to\mathcal{A}$$

considered as a K-bilinear map.

Therefore, $\mu(\mathcal{A}/K)$ is the smallest integer n such that there exist linear forms $\varphi_1,\dots,\varphi_n$ and $\psi_1,\dots,\psi_n\colon \mathcal{A}\to K$ and elements $w_1,\dots,w_n\in\mathcal{A}$ such that

$$xy=\varphi_1(x)\psi_1(y)w_1+\ldots+\varphi_n(x)\psi_n(y)w_n \tag{12.1.1}$$

for all $x,y\in\mathcal{A}$, since such an expression is the same as a decomposition

$$\widetilde{m}_{\mathcal{A}}=\sum_{i=1}^{n} w_i\otimes\varphi_i\otimes\psi_i\in\mathcal{A}\otimes\mathcal{A}^\vee\otimes\mathcal{A}^\vee$$

for the multiplication tensor of $\mathcal{A}$.

In the above definitions, $\mathcal{A}$ can be arbitrary; however, below we consider algebras that are associative, commutative, and with unity.

We call a *multiplication algorithm* of length n for $\mathcal{A}/K$ a collection of φ_i,ψ_i,w_i that satisfy (12.1.1). Such an algorithm is said to be *symmetric* if $\varphi_i=\psi_i$ for all i (this can happen only if $\mathcal{A}$ is commutative). If $\mathcal{A}$ is a finite-dimensional commutative K-algebra, we define its *symmetric bilinear complexity*

$$\mu^{\mathrm{sym}}(\mathcal{A}/K)$$

as the minimal length of a symmetric multiplication algorithm for $\mathcal{A}/K$.

Equivalently, it is the minimal length of a decomposition of the multiplication tensor $\widetilde{m}_{\mathcal{A}}$ as a sum of symmetric elementary tensors, that is, of tensors of the form $w\otimes\varphi\otimes\varphi\in\mathcal{A}\otimes\mathcal{A}^\vee\otimes\mathcal{A}^\vee$.

The following properties are simple consequences of the definitions.

LEMMA 12.1.1. *Let $\mathcal{A}$ be a K-algebra of a finite dimension $d=\dim_K\mathcal{A}$. Then*

(i) *We have*
$$\mu^{\mathrm{sym}}(\mathcal{A}/K)\le\frac{d(d+1)}{2}.$$

(ii) *If* $\operatorname{char}K\neq 2$, *then also*
$$\mu^{\mathrm{sym}}(\mathcal{A}/K)\le 2\mu(\mathcal{A}/K).$$

(iii) *We have*
$$\mu(\mathcal{A}/K)\le\mu^{\mathrm{sym}}(\mathcal{A}/K).$$

(iv) *If $\mathcal{A}$ admits a unit element, then*
$$\mu(\mathcal{A}/K)\ge d.$$

(v) *If $\mathcal{A}$ has no zero-divisor, then*
$$\mu(\mathcal{A}/K)\ge 2d-1.$$ □

We also have the following properties of μ under natural operations.

LEMMA 12.1.2. (i) *Let $\mathcal{A}$ be a finite-dimensional K-algebra and L an extension field of K. Then*

$$\mu(\mathcal{A}\otimes_K L/L) \leq \mu(\mathcal{A}/K).$$

(ii) *If $[L:K] < +\infty$, then*

$$\mu(\mathcal{A}/K) \leq \mu(\mathcal{A}/L)\mu(L/K).$$

(iii) *Let $\mathcal{A}$ and $\mathcal{B}$ be two finite-dimensional K-algebras. Then*

$$\mu(\mathcal{A}\oplus\mathcal{B}/K) \leq \mu(\mathcal{A}/K)+\mu(\mathcal{B}/K),$$
$$\mu(\mathcal{A}\otimes_K \mathcal{B}/K) \leq \mu(\mathcal{A}/K)\mu(\mathcal{B}/K).$$

Moreover, for commutative algebras, all the claims of the lemma remain valid with μ replaced by μ^{sym}. □

Note that one can connect multiplication algorithms with codes; namely, given a multiplication algorithm as in (12.1.1), one can associate to it linear codes C_φ and $C_\psi \subset K^n$ as follows: let $C_\varphi = \operatorname{Im}(\varphi)$ for

$$\varphi\colon \mathcal{A}\to K^n, \quad x\mapsto (\varphi_1(x),\ldots,\varphi_n(x)),$$

and $C_\psi = \operatorname{Im}(\psi)$ for

$$\psi\colon \mathcal{A}\to K^n, \quad y\mapsto (\psi_1(y),\ldots,\psi_n(y)).$$

Interpolation systems. The Chudnovsky method is based on interpolation on algebraic curves over finite fields. Their interpolation scheme can be described as follows.

If $\mathcal{B}$ is a K-algebra and if $E_1, E_2 \subset \mathcal{B}$ are two linear subspaces, by E_1E_2 we denote the *linear span* of the products e_1e_2 in $\mathcal{B}$, for $e_1\in E_1$ and $e_2\in E_2$.

Let $\mathcal{A}$ and $\mathcal{A}'$ be two finite-dimensional K-algebras. Then an *interpolation system* for $\mathcal{A}'$ by $\mathcal{A}$ is given by the following data:

- a K-algebra $\mathcal{B}$ equipped with two K-algebra morphisms $f\colon \mathcal{B}\to\mathcal{A}$ and $f'\colon \mathcal{B}\to\mathcal{A}'$;
- two linear subspaces $E_1, E_2\subset\mathcal{B}$

satisfying the following conditions:

(i) the restriction $f|_{E_1E_2}\colon E_1E_2\to\mathcal{A}$ is injective;
(ii) the resrictions $f'|_{E_1}\colon E_1\to\mathcal{A}'$ and $f'|_{E_2}\colon E_2\to\mathcal{A}'$ are surjective;

note that the case $\dim\mathcal{B}=\infty$ is not excluded.

This can be summarized with the following diagram:

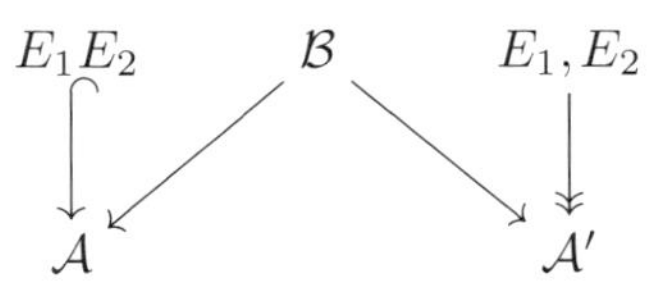

Such an interpolation system is said to be *symmetric* if $E_1 = E_2$.

PROPOSITION 12.1.3. *Let $\mathcal{A}$ and $\mathcal{A}'$ be two finite-dimensional K-algebras. Assume there exists an interpolation system for $\mathcal{A}'$ by $\mathcal{A}$. Then*

$$\mu(\mathcal{A}'/K) \leq \mu(\mathcal{A}/K).$$

Moreover, if $\mathcal{A}$ and $\mathcal{A}'$ are commutative and the interpolation system is symmetric, then also $\mu^{\text{sym}}(\mathcal{A}'/K)\leq\mu^{\text{sym}}(\mathcal{A}/K)$.

PROOF. Let $\varphi_1, \dots, \varphi_n$, $\psi_1, \dots, \psi_n : \mathcal{A} \to K$, and $w_1, \dots, w_n \in \mathcal{A}$ define a multiplication algorithm for $\mathcal{A}/K$, where $n = \mu(\mathcal{A}/K)$.

Assume that we are given an interpolation system for $\mathcal{A}'$ by $\mathcal{A}$. Thanks to properties (i) and (ii) above, we can choose:

- a retraction $\rho \colon \mathcal{A} \to E_1E_2$ of $f|_{E_1E_2}$;
- sections $\sigma_1 : \mathcal{A}' \to E_1$ of $f'|_{E_1}$ and $\sigma_2 : \mathcal{A}' \to E_2$ of $f'|_{E_2}$.

Now, for $1 \le i \le n$, let

- $\varphi'_i = \varphi_i \circ f|_{E_1} \circ \sigma_1 : \mathcal{A}' \to K$;
- $\psi'_i = \psi_i \circ f|_{E_2} \circ \sigma_2 : \mathcal{A}' \to K$;
- $w'_i = f'(\rho(w_i)) \in K$.

Then $\varphi'_1, \dots, \varphi'_n$, $\psi'_1, \dots, \psi'_n$, and $w'_1, \dots, w'_n \in \mathcal{A}$ define a multiplication algorithm for $\mathcal{A}'/K$. Indeed, for any $x', y' \in \mathcal{A}'$, if we let $x = f(\sigma_1(x'))$ and $y = f(\sigma_2(y'))$, then

$$\begin{aligned}
\sum \varphi'_i(x')\psi'_i(y')w'_i &= \sum \varphi_i(x)\psi_i(y) f'(\rho(w_i)) \\
&= f'(\rho(\sum \varphi_i(x)\psi_i(y)w_i)) \\
&= f'(\rho(xy)) \\
&= f'(\rho(f(\sigma_1(x'))f(\sigma_2(y')))) \\
&= f'(\rho(f(\sigma_1(x')\sigma_2(y')))) \\
&= f'(\sigma_1(x')\sigma_2(y')) \\
&= f'(\sigma_1(x'))f'(\sigma_2(y')) = x'y'.
\end{aligned}$$

Thus, $\mu(\mathcal{A}'/K) \le n$, as claimed.

For the last assertion, assuming $E_1 = E_2$, note that if we start with a symmetric algorithm for $\mathcal{A}/K$ and choose $\sigma_1 = \sigma_2$, then the construction gives a symmetric algorithm for $\mathcal{A}'/K$. □

COROLLARY 12.1.4. *If $\mathcal{A}$ is a finite-dimensional K-algebra and if $\mathcal{A}'$ is a subalgebra of $\mathcal{A}$ or a quotient algebra of $\mathcal{A}$, then*

$$\mu(\mathcal{A}'/K) \le \mu(\mathcal{A}/K).$$

If $\mathcal{A}$ is commutative, then also $\mu^{\mathrm{sym}}(\mathcal{A}'/K) \le \mu^{\mathrm{sym}}(\mathcal{A}/K)$.

PROOF. If $\mathcal{A}'$ is a subalgebra of $\mathcal{A}$, define an interpolation system by taking $E_1 = E_2 = \mathcal{B} = \mathcal{A}'$, f the natural inclusion, and $f' = \mathrm{id}_{\mathcal{A}'}$.

If $\mathcal{A}'$ is a quotient algebra of $\mathcal{A}$, take $E_1 = E_2 = \mathcal{B} = \mathcal{A}$, $f = \mathrm{id}_{\mathcal{A}}$, and f' the natural projection. □

12.1.2. The Extended Chudnovsky Algorithm

Now we assume that $K = \mathbb{F}_q$; we will only consider algebras that are associative, commutative, and with unity.

In fact, we will be particularly interested in the following $\mathbb{F}_q$-algebras. For any integers $m, \ell \ge 1$, define

$$\mathcal{A}_q(m,\ell) = \mathbb{F}_{q^m}[t]/(t^\ell), \quad \dim_{\mathbb{F}_q} \mathcal{A}_q(m,\ell) = m\ell,$$

and let

$$\mu_q(m,\ell) = \mu(\mathcal{A}_q(m,\ell)/\mathbb{F}_q)$$

be its bilinear complexity over $\mathbb{F}_q$.

Of special significance are the following two cases: when $\ell = 1$, we denote the bilinear complexity of multiplication in $\mathbb{F}_{q^m}$ over $\mathbb{F}_q$ by

$$\mu_q(m) := \mu_q(m,1),$$

and when $m = 1$, we set

$$\widehat{M}_q(\ell) := \mu_q(1,\ell).$$

LEMMA 12.1.5. *With the above notation, we have*

$$\mu_q(m,\ell) \leq \mu_q(m)\widehat{M}_{q^m}(\ell).$$

PROOF. This is a direct consequence of Lemma 12.1.2(ii). □

The Chudnovsky algorithm uses interpolation on an algebraic curve X over a finite field $\mathbb{F}_q$ to construct an interpolation algebra $\mathcal{B}$.

For a divisor D on X, we use the standard notation $L(D)$, $\ell(D) = \dim L(D)$, and $i(D) = \ell(K_X - D)$ (the *index of specialty* of D), and thus the Riemann–Roch theorem reads

$$\ell(D) - i(D) = \deg D + 1 - g.$$

Now consider the evaluation maps

$$E\colon L(D_1 + D_2) \to \prod_{i=1}^{n} \mathcal{A}_q(d_i, u_i),$$

$$F_1\colon L(D_1) \to \mathcal{A}_q(m,\ell), \qquad F_2\colon L(D_2) \to \mathcal{A}_q(m,\ell)$$

constructed as follows: we consider the image of $f \in L(D_1 + D_2)$ in the local algebra $\mathcal{O}_{P_i}/P_i^{u_i}$ (cf. Sec. 2.5.1), which is isomorphic to $\mathcal{A}_q(d_i, u_i)$ for any point $P_i \in X$ of degree $d_i \geq 1$ and any $u_i \geq 1$. The map E is simply the product of these evaluations. Similarly, $F_i(h)$ for $h \in L(D_i)$ is the image of h in $\mathcal{O}_Q/Q^\ell$ for a point $Q \in X$ of degree $m \geq 1$ and any $\ell \geq 1$.

THEOREM 12.1.6. *Let X be a curve of genus g over $\mathbb{F}_q$, and let $m, \ell \geq 1$ be two integers. Assume that X admits a closed point Q of degree $\deg Q = m$. Let $G = u_1 P_1 + \ldots + u_n P_n$ be an effective divisor, where the P_i are pairwise distinct closed points of degree $\deg P_i = d_i$. Assume also that for two divisors D_1, D_2 on X disjoint from G we have the following:*

(i) *The natural evaluation map E is injective.*
(ii) *The natural evaluation maps F_1 and F_2 are surjective.*

Then

$$\mu_q(m,\ell) \leq \sum_{i=1}^{n} \mu_q(d_i, u_i). \tag{12.1.2}$$

If $D_1 = D_2$, this inequality remains valid for μ^{sym}.

SKETCH OF THE PROOF. Our conditions yield

$$\mu_q(m,\ell) \leq \mu\left(\prod_{i=1}^{n} \mathcal{A}_q(d_i, u_i)/\mathbb{F}_q\right),$$

and then we can apply Lemma 12.1.2(iii) to get (12.1.2). □

We can give the following sufficient numerical criteria.

PROPOSITION 12.1.7. *A sufficient condition for the existence of Q of degree m on X is that $2g+1 \leq q^{(m-1)/2}(q^{1/2}-1)$, while sufficient conditions for* (i) *and* (ii) *of Theorem* 12.1.6 *are:*

(i$'$) *The divisor D_1+D_2-G is zero-dimensional:*

$$\ell(D_1+D_2-G)=0.$$

(ii$'$) *The divisors D_1-lQ and D_2-lQ are nonspecial:*

$$i(D_1-lQ)=i(D_2-lQ)=0.$$

More precisely, (i) *and* (i$'$) *are equivalent, while* (ii$'$) *only implies* (ii) *a priori.* □

The numerical sufficient condition stated here for the existence of Q is a simple counting argument.

Asymptotic behaviour. Now we will be interested in the asymptotic behaviour of the algorithm. Note, however, that there are many interesting applications of the algorithm in the nonasymptotic setting.

If q is a prime power, we let

$$m_q=\liminf_{n\to\infty}\frac{1}{n}\mu_q(n),$$
$$M_q=\limsup_{n\to\infty}\frac{1}{n}\mu_q(n),$$

and their symmetric counterparts m_q^{sym} and M_q^{sym} are defined likewise.

Before establishing our main asymptotic bounds, we need the following fact.

PROPOSITION 12.1.8. *Let X be a curve of genus g over $\mathbb{F}_q$, where q is any prime power, and let $m,\ell\geq 1$ be two integers. Assume that X admits a closed point Q of degree $\deg Q=m$. Define*

$$e_q=\begin{cases}2 & \textit{if } q=2,\\ 1 & \textit{if } q=3,4,5,\\ 0 & \textit{if } q>5.\end{cases}$$

Now consider a collection of integers $n_{d,u}\geq 0$ (for $d,u\geq 1$) almost all being zero and such that, for any d,

$$n_d=\sum_u n_{d,u}\leq B_d(X/\mathbb{F}_q).$$

Then, provided that

$$\sum_{d,u} n_{d,u}du\geq 2m\ell+2e_q+2g-1,$$

we have

$$\mu_q(m,\ell)\leq\sum_{d,u} n_{d,u}\mu_q(d,u).$$ □

The proof of this result is rather elaborated and is based on a minute study of the number of special divisors on X of a given degree.

Using this result, we can establish the main asymptotic bounds.

THEOREM 12.1.9. *If* $A(q) > 1$, *then*

$$m_q \leq 2\left(1 + \frac{1}{A(q)-1}\right).$$

Moreover, if $A(q) > 5$, *then also* $m_q^{\mathrm{sym}} \leq 2\left(1 + \frac{1}{A(q)-1}\right)$.

SKETCH OF THE PROOF. Consider a family of curves X_s over $\mathbb{F}_q$, of genus g_s going to infinity, with

$$\lim_{s\to\infty} \frac{|X_s(\mathbb{F}_q)|}{g_s} = A(q).$$

Given an integer s, let

$$n(s) = \left\lfloor \frac{1}{2}(|X_s(\mathbb{F}_q)| - g_s - 5) \right\rfloor;$$

hence

$$\lim_{s\to\infty} \frac{n(s)}{g_s} = \frac{A(q)-1}{2}.$$

Then for s large enough we have $2g_s + 1 \leq q^{(n(s)-1)/2}(q^{1/2} - 1)$ and we can apply Proposition 12.1.8 with $\ell = 1$ and $m = n(s)$, and with all $n_{d,u}$ zero except $n_{1,1} = 2n(s) + g_s + 5$, to get

$$\mu_q(n(s)) \leq 2n(s) + g_s + 5,$$

which allows us to conclude.

If $A(q) > 5$, then $|X_s(\mathbb{F}_q)| > 5g_s$ for s large enough, and we can conclude likewise. □

To get the next result, one needs an optimal family X_i of curves, $g(X_i) = g_i$, such that their genera satisfy the condition

$$\lim_{i\to\infty} \frac{g_i}{g_{i+1}} = 1,$$

i.e., the following proposition.

PROPOSITION 12.1.10. *Let* $q = p^{2m} \geq 9$. *Then there exists a family* $\{X_s\}$ *of (absolutely irreducible smooth) curves of genus* $g(X_s) = g_s$ *with*

$$\lim_{s\to\infty} \frac{N_1(X_s)}{g_s} = \sqrt{q} - 1$$

and

$$\lim_{s\to\infty} \frac{g_{s+1}}{g_s} = 1.$$ □

For $q = p^2$ the proposition is clear: one sets $X_s = X_0(11\ell_s)$ where ℓ_s is the sth prime number. Then $g_s = \ell_s$ and $N_1(X_s) \geq (p-1)(\ell_s + 1)$. For $q = p^{2m}$ with $m \geq 2$ one can use reductions of certain Shimura curves defined by quaternion algebras over totally real abelian extensions $\mathbb{E}_m$ of $\mathbb{Q}$ of degree m with the condition that p is inert in $\mathbb{E}_m$. Then direct computations close to (but more complicated than) those of Sec. 5.1.1 give the result.

Another possibility is to take the sequence of Drinfeld modular curves (over $\mathbb{F}_q$) $X_i = X_0(F_i) \otimes \mathbb{F}_q$ for a suitable sequence of monic polynomials $F_i \in \mathbb{F}_q[T]$.

This additional condition $g_{s+1}/g_s \to 1$ allows us to replace m_q with M_q in our main result.

THEOREM 12.1.11. *If $q = p^{2r} \geq 9$ is a square, then*

$$M_q \leq 2\left(1 + \frac{1}{\sqrt{q}-2}\right).$$

Moreover, if $q = p^{2r} \geq 49$, then also $M_q^{\mathrm{sym}} \leq 2\Big(1 + \dfrac{1}{\sqrt{q}-2}\Big)$.

SKETCH OF THE PROOF. Consider the curves X_i from Proposition 12.1.10. Thus, $\lim_{s\to\infty} N_1(X_s)/g_s = \sqrt{q}-1$ and $\lim_{s\to\infty} g_{s+1}/g_s = 1$. Given an integer n, let $s(n)$ be the smallest integer such that

$$|X_{s(n)}(\mathbb{F}_q)| \geq 2n + g_{s(n)} - 1;$$

hence,

$$g_{s(n)} = \frac{2n}{\sqrt{q}-2} + o(n). \tag{12.1.3}$$

Then this gives $2g_{s(n)} + 1 \leq q^{(n-1)/2}(q^{1/2}-1)$ for n large enough, and we can conclude similarly to the argument in Theorem 12.1.9. □

REMARK 12.1.12. Using similar arguments, one can also get the bounds $M_p \leq 3\Big(1 + \dfrac{2}{p-2}\Big)$ for p prime, and $M_q \leq 3\Big(1 + \dfrac{p}{q-2}\Big)$ for $q = p^r$, $r \geq 3$ odd.

12.2. Cryptographic Applications

In the present section we give two cryptographic application of algebraic geometry codes, namely authentication and arithmetic secret sharing schemes.

Authentication codes (A-codes for short) allow two trusting parties to communicate in the presence of an opponent who may construct a fraudulent message, and/or substitute a transmitted message with a fraudulent one.

On can apply these codes in the study of distributed authentication systems in which the functionality of authentication is to be distributed among a number of participants. The extra algebraic property allows a more efficient construction of such distributed systems. These codes are considered in Sec. 12.2.1.

Section 12.2.2 is devoted to secret sharing schemes, which are important to secure multi-party communications. There are many versions of such schemes, but here we are only concerned with a special but important scheme called an $(n,t,d,n-t)$ arithmetic secret sharing scheme. Note that this scheme is very close in spirit to the fast multiplication algorithms of the preceding section, though the relation between them is rather indirect.

12.2.1. Authentication Codes from Algebraic Curves

Recall that in the conventional model for the unconditionally secure authentication system there are three participants: a *transmitter*, a *receiver*, and an *opponent*. The transmitter wants to communicate a message to a receiver using a public channel which is subject to active attacks. That is, the opponent may impersonate the transmitter and insert a message into the channel, or replace a transmitted message with a fraudulent one. To protect against these attacks, the transmitter and the receiver share a secret key which is used to choose an authentication rule from an A-code.

Now we give some generalities on (systematic) A-codes and their relation to error-correcting (EC) codes.

A-codes and EC codes. A *systematic* A-code (or A-code *without secrecy*) is a code in which a *message* sent over the channel consists of a *source state* (i.e., plaintext) concatenated with an *authenticator* (or a *tag*). Such a code is a triple $(\mathcal{S},\mathcal{E},\mathcal{Z})$ of finite sets together with an (authentication) function $f\colon \mathcal{S}\times\mathcal{E}\to\mathcal{Z}$. We sometimes denote the A-code by $(\mathcal{S},\mathcal{E},\mathcal{Z},f)$, and thus, for any $e\in\mathcal{E}$ and $s\in\mathcal{S}$, the message $m=(s,z)\in\mathcal{S}\times\mathcal{Z}$ is defined by letting $z=f(s,e)$. The component z is called the *tag* of the message m, and $\log|\mathcal{Z}|$ is referred to as the *tag size*. The set

$$\mathcal{M}=\{m=(s,f(s,e)):\ s\in\mathcal{S},\ e\in\mathcal{E}\}$$

is the set of all messages. We see that the source state s appears in the message in an uncoded form; we write the encoding as $z=e(s)$, i.e., every e describes a mapping from $\mathcal{S}$ to $\mathcal{Z}$. Therefore, $\mathcal{S}$ is the set of source states, $\mathcal{E}$ is the set of keys, and $\mathcal{Z}$ is the set of authenticators. When the transmitter wants to send the source state $s\in\mathcal{S}$ using a key $e\in\mathcal{E}$, which is secretly shared with the receiver, he transmits the message (s,z), where $z=f(s,e)\in\mathcal{Z}$. When the receiver receives (s,z), he checks the authenticity of the message by verifying if $z=f(s,e)$ holds, using the secret key $e\in\mathcal{E}$. If the equality holds, he accepts s as authentic.

Assume that the opponent has the ability to insert messages into the channel and/or to modify the existing messages. An *impersonation attack* is when the opponent inserts a new message (s',z') into the channel. A *substitution attack* is when the opponent sees a message (s,z) and changes it to (s',z') where $s \neq s'$. A message (s,z) is said to be *valid* if there exists a key e such that $z = f(s,e)$. We assume that there is a probability distribution on the source states, which is known to all the participants. Given the probability distribution on the source states, the receiver and the transmitter will choose a probability distribution for $\mathcal{E}$. We will denote the probability of success of the opponent in the impersonation and substitution attack by P_I and P_S, respectively.

In what follows, we assume that $\Pr(s) \neq 0$ for all $s \in \mathcal{S}$ and adopt the following notation:

$$\forall (s,z) \in (\mathcal{S},\mathcal{Z}): \quad \mathcal{E}(s,z) = \{e \in \mathcal{E} : e(s) = z\}.$$

The following properties show that for any given s the sets $\mathcal{E}(s,z)$ form a partition of $\mathcal{E}$:

$$\mathcal{E}(s,z) \cap \mathcal{E}(s,z') = \varnothing, \quad z \neq z', \qquad \text{(uniqueness of encoding)}$$

$$\forall s \in \mathcal{S}: \quad \mathcal{E} = \bigcup_z \mathcal{E}(s,z).$$

The probability of a successful impersonation attack P_I is given by the formula

$$P_I = \max_{\substack{s \in \mathcal{S} \\ z \in \mathcal{Z}}} \Pr\big((s,z) \text{ is valid}\big) = \max_{\substack{s \in \mathcal{S} \\ z \in \mathcal{Z}}} \frac{|\mathcal{E}(s,z)|}{|\mathcal{E}|}.$$

For the substitution attack, the opponent observes the message $m = (s,z)$ and replaces this with another message $m' = (s',z')$ such that $s' \neq s$. We assume that the opponent will use an optimal substitution strategy, i.e., he uses a pair (s',z') that will optimize his chances of success. Thus we introduce

$$\begin{aligned} P_S(s,z) &= \max_{\substack{(s',z') \\ s' \neq s}} \Pr\big((s',z') \text{ is valid} \mid (s,z) \text{ is observed}\big) \\ &= \max_{\substack{(s',z') \\ s' \neq s}} \frac{\Pr(\mathcal{E}(s',z') \cap \mathcal{E}(s,z))}{\Pr(\mathcal{E}(s,z))}. \end{aligned}$$

The probability of a successful substitution attack P_S equals

$$P_S = \max_{(s,z)} P_S(s,z)$$

$$= \max_{\substack{s,s' \neq s \in \mathcal{S} \\ z,z' \in \mathcal{Z}}} \frac{\Pr(\mathcal{E}(s',z') \cap \mathcal{E}(s,z))}{\Pr(\mathcal{E}(s,z))} \tag{12.2.1}$$

$$= \max_{\substack{s,s' \neq s \in \mathcal{S} \\ z,z' \in \mathcal{Z}}} \frac{|\mathcal{E}(s',z') \cap \mathcal{E}(s,z)|}{|\mathcal{E}(s,z)|} \tag{12.2.2}$$

$$= \max_{\substack{s,s' \neq s \in \mathcal{S} \\ z,z' \in \mathcal{Z}}} \frac{|\{e \in \mathcal{E} : e(s) = z,\, e(s') = z'\}|}{|\{e \in \mathcal{E} : e(s) = z)\}|}. \tag{12.2.3}$$

First of all, we observe the following fundamental property.

PROPOSITION 12.2.1. *For any systematic A-code one has* $P_S \geq P_I$.

PROOF. From the uniqueness of encoding it immediately follows that

$$\sum_z \Pr(\mathcal{E}(s,z)) = 1.$$

Let (s',z') be a message maximizing $P_I = \Pr(\mathcal{E}(s',z'))$.

Let us replace (s,z) with (s',z'), $s' \neq s$. From (12.2.1) we have for this case

$$P_S \geq \frac{\Pr(\mathcal{E}(s,z) \cap \mathcal{E}(s',z'))}{\Pr(\mathcal{E}(s,z))} \tag{12.2.4}$$

for any $s \neq s'$ and any z and z'. Thus,

$$\begin{aligned} P_S = 1 \cdot P_S = \sum_z \Pr(\mathcal{E}(s,z))P_S &\geq \sum_z \Pr(\mathcal{E}(s,z)) \frac{\Pr(\mathcal{E}(s,z) \cap \mathcal{E}(s',z'))}{\Pr(\mathcal{E}(s,z))} \\ &= \sum_z \Pr(\mathcal{E}(s,z) \cap \mathcal{E}(s',z')), \end{aligned}$$

where the inequality follows from (12.2.4). Now

$$\sum_z \Pr(\mathcal{E}(s,z) \cap \mathcal{E}(s',z')) = \Pr(\mathcal{E} \cap \mathcal{E}(s',z')) = \Pr(\mathcal{E}(s',z')) = P_I,$$

and the proposition is proved. □

In the sequel we only consider A-codes with the additional property that

$$\forall s \in \mathcal{S},\ \forall z \in \mathcal{Z}: \quad P_I = \frac{|\mathcal{E}(s,z)|}{|\mathcal{E}|},$$

and we call such codes *I-equitable A-codes*. Such codes are optimal against the impersonation attack in the sense that only these codes meet the trivial bound $P_I \geq |\mathcal{S}|/|\mathcal{M}|$ with equality.

For I-equitable codes we have a rather tight relation between (systematic I-equitable) A-codes and EC codes. Indeed, let us put $n = |\mathcal{E}|$, $q = |\mathcal{Z}|$, and let us assume that q is a prime power. In particular, we have $P_I = 1/n$. Under this assumption there exists a two-way connection between A-codes and EC codes: one can construct an EC code from a given A-code, and under an additional (rather mild) condition it is possible to construct A-codes using EC codes.

This is done as follows. Let $(\mathcal{S},\mathcal{E},\mathcal{Z})$ be a systematic I-equitable code; we write $z = e(s)$ hereafter. Consider the following map:

$$\mathcal{S} \to \mathcal{Z}^n, \quad s \mapsto \boldsymbol{v}^{(s)} = (v_1(s),\dots,v_n(s)),$$

where $v_j(s) = e_j(s)$. The image $V \subseteq \mathcal{Z}^n$ is referred to as the *(EC) code related to the A-code*. Under our assumptions (i.e., for systematic I-equitable A-codes) one has

$$P_I = 1/q, \qquad P_S = 1 - D(V)/n,$$

where the *minimum A-distance* $D(V)$ of V is defined as follows:

$$D(V) := \min\{D(\boldsymbol{v},\boldsymbol{v}') : \boldsymbol{v} \neq \boldsymbol{v}' \in V\}$$

with

$$D(\boldsymbol{v}^{(s)},\boldsymbol{v}^{(s')}) := n - q\gamma(\boldsymbol{v}^{(s)},\boldsymbol{v}^{(s')})$$

and

$$\gamma(\boldsymbol{v}^{(s)},\boldsymbol{v}^{(s')}) := \max\{|\{j : v_j(s) = \alpha,\ v_j(s') = \beta\}| : \alpha,\beta \in \mathcal{Z}\}.$$

We get the following fundamental relation:

$$D(V) = n(1-P_S),$$

and our systematic A-codes can therefore be interpreted as q-ary codes with the minimum A-distance $n(1-P_S)$. However, note that the A-distance is not a metric, since it does not satisfy the triangle inequality; nevertheless, we have

$$D(V) \leq d(V),$$

where $d(V)$ is the usual minimum Hamming distance of V. Therefore, one can obtain upper bounds on the cardinality of (systematic I-equitable) A-codes from known upper bounds for EC codes.

Now let C be an (n, M, d) EC code over $\mathbb{F}_q$ (C does not need to be linear) satisfying the following additional condition:

$$\forall c \in C,\ \forall \lambda \in \mathbb{F}_q : \quad c + \lambda \mathbf{1} \in C, \tag{12.2.5}$$

where $\mathbf{1} = (1, \ldots, 1)$ is the all-one vector. For a linear C, this condition is equivalent to

$$\mathbf{1} \in C. \tag{12.2.6}$$

Consider the following equivalence relation on C:

$$c \equiv c' \iff \exists \lambda \in \mathbb{F}_q : \quad c - c' = \lambda \mathbf{1},$$

and let $U \subset C$ be a set of representatives for this equivalence relation. Then U is an $(n, M/q, d)$ code. Now consider the code V over $\mathbb{F}_q$ of length nq defined as follows:

$$V := \{(u, u+\alpha_2 \mathbf{1}, \ldots, u+\alpha_q \mathbf{1}) : u \in U\},$$

where $\mathbb{F}_q = \{\alpha_1 = 0, \alpha_2, \ldots, \alpha_q\}$. The code V is called the *q-twisted code* obtained from C. One can see that V corresponds to an A-code by the above construction. Moreover, one can prove that in this case $D(V) = d(C)$, thus giving the following result.

THEOREM 12.2.2. *Let C be an (n, M, d) code over $\mathbb{F}_q$ satisfying condition (12.2.5). Then the q-twisted construction applied to C produces an A-code with the following parameters:*

$$|\mathcal{S}| = M/q, \qquad |\mathcal{E}| = nq, \qquad P_I = 1/q, \qquad P_S = 1 - d/n. \qquad \square$$

Similarly to the theory of EC codes, we want to study the maximal possible number of source states for given n, P_S, and q, which leads to the following definitions:

$$m(n, \varepsilon, q) := \max\{|\mathcal{S}| : |\mathcal{E}| = q,\ P_S \leq \varepsilon\},$$

where the maximum is over all systematic I-equitable A-codes with the given conditions; the quantity

$$R_A(n, \varepsilon, q) := \frac{1}{n} \log_q m(n, \varepsilon, q)$$

is called the *rate* of systematic I-equitable A-codes; and the *asymptotic rate* is defined by

$$R_A(\varepsilon, q) := \limsup_{n \to \infty} R_A(n, \varepsilon, q).$$

Asymptotic bounds and algebraic geometry constructions. We are mainly interested in the asymptotic behaviour of A-codes. Using their connections with EC codes, one gets both lower and upper bounds on their asymptotic rate. Namely, it is not difficult to prove that there exist codes satisfying condition (12.2.5) which asymptotically meet the Gilbert–Varshamov bound. This gives the following result.

THEOREM 12.2.3 (semiconstructive lower bound). *We have*

$$R_A(\varepsilon,q) \geq (1-H_q(1-\varepsilon))/q,$$

where

$$H_q(x) = x\log_q(q-1) - x\log_q x - (1-x)\log_q(1-x)$$

is the usual q-ary entropy function. □

The application of asymptotic upper bounds for EC codes gives the following asymptotic upper bound for A-codes.

THEOREM 12.2.4. *We have*

$$R_A(\varepsilon,q) \leq F(\varepsilon,q),$$

where

$$F(\varepsilon,q) := H_2\left(\frac{1-\sqrt{1-u^2}}{2}\right) \Big/ \log_2 q$$

with

$$u = \frac{2(\varepsilon-1)}{q} + \frac{2}{q}\sqrt{\varepsilon(q-2+\varepsilon)}.$$ □

The function $F(\varepsilon,q)$ is increasing on the segment $[1/q,1]$, and its values at the endpoints of the segment are 0 and

$$\frac{1}{q} + \frac{q-1}{q}\log_q\frac{q}{q-1},$$

respectively.

Note, however, that the semiconstructive bound, which is also an increasing function on $[1/q,1]$, has the value $1/q$ at 1. Applying arguments of the Gilbert–Varshamov bound type to A-codes directly (not passing through EC codes), one gets the following result.

THEOREM 12.2.5 (Gilbert–Varshamov bound). *We have*

$$R_A(\varepsilon,q) \geq \frac{1}{q} + \frac{q-1}{q}\log_q\frac{q-1}{q} - \frac{1}{q\log_2 q}H_2(\varepsilon) - \frac{q-1}{q\log_2 q}H_2\left(\frac{1-\varepsilon}{q-1}\right).$$ □

Note that $R_A(1/q,q) = F(1/q,q)$, i.e., the value at 1 coincides with the upper bound.

Now we can apply the preceding construction using algebraic geometry codes, namely we have the following trivial statement, which is nevertheless very useful when applied to A-codes.

PROPOSITION 12.2.6. *For any curve X with positive number N of $\mathbb{F}_q$-rational points and any nonnegative integer a, there exists an $\mathbb{F}_q$-rational divisor D on X such that condition (12.2.6) is satisfied for the code $C = (X,\mathcal{P},D)_L$, where the set $\mathcal{P}$ has at least $N-1$ elements.*

PROOF. Indeed, it is sufficient to set $D = aP$ and $\mathcal{P} = X(\mathbb{F}_q) \setminus \{P\}$ for an $\mathbb{F}_q$-rational point P on X. Then the space $L(D)$ contains constants, which gives the proposition. □

Applying Theorem 12.2.2 and Proposition 12.2.6, we immediately get the following asymptotic bound.

THEOREM 12.2.7 (algebraic geometry bound for A-codes). *Let q be an even prime power. Then*

$$R_A(\varepsilon, q) \geq \frac{1}{q}\left(\varepsilon - \frac{1}{\sqrt{q}-1}\right). \qquad \square$$

Moreover, we have the following.

THEOREM 12.2.8. *Let $q \geq 81$ be an even prime power. Then the equation*

$$\frac{1}{q}\left(\varepsilon - \frac{1}{\sqrt{q}-1}\right) = \frac{1}{q} + \frac{q-1}{q}\log_q \frac{q}{q-1} - \frac{1}{q\log_2 q} H_2(\varepsilon) - \frac{q-1}{q\log_2 q} H_2\left(\frac{1-\varepsilon}{q-1}\right)$$

has two roots, ε_1 and ε_2, with $\gamma_q < \varepsilon_1 < \varepsilon_2 < 1$, and for $\varepsilon \in (\varepsilon_1, \varepsilon_2)$ the algebraic geometry bound of Theorem 12.2.7 lies above the Gilbert–Varshamov bound of Theorem 12.2.5.

SKETCH OF THE PROOF. Since the Gilbert–Varshamov bound is convex, it suffices to show that the conclusion of the theorem is valid at the point ε_0 where the derivative of the Gilbert–Varshamov bound equals $1/q$. For $\varepsilon_0 = \varepsilon_0(q)$ one has the equation

$$(q-1)\varepsilon_0^2 - (3q+2)\varepsilon_0 + q = 0,$$

and thus

$$\varepsilon_0(q) = \frac{3-\sqrt{5}}{2} + \left(\frac{25-13\sqrt{5}}{10}\right)\frac{1}{q} + O\left(\frac{1}{q^2}\right).$$

For the first few values of $q \geq 81$ one can check the conclusion by brute-force computations, and for $q \geq 256$ it suffices to apply computations to $\varepsilon_0' = (3-\sqrt{5})/2$ using known asymptotics of $H_2(x)$ for $x \to 0$ with $x = (1-\varepsilon)/(q-1)$. □

Therefore, applying algebraic geometry codes to A-codes, one gets an improvement of asymptotic lower bounds for their rate, similarly to other classes of codes. Moreover, since algebraic geometry codes can be efficiently constructed (cf. Secs. 5.1.3, 5.2.3, and 7.4.3), we get efficiently constructible codes from algebraic geometry codes.

In fact, condition (12.2.6) is not too restrictive, which allows one to apply the expurgation method of Sec. 4.5.2.

PROPOSITION 12.2.9. *Let $q \geq 49$ be a square; let R_0 be a unique root of the equation*

$$\log_2(q-1) - (1-R)\log_2 q = H_2(R+\gamma_q) + \gamma_q \log_2 q$$

satisfying $2\gamma_q \leq R_0 \leq 3\gamma_q$ for $\gamma_q := (\sqrt{q}-1)^{-1}$; and let $R > R_0$. Then for any sufficiently long algebraic geometry code $C = (X, \mathcal{P}, D)_L$ with $R + \delta \geq 1 - \gamma_q$ there exists a code C' satisfying condition (12.2.6) and equivalent to C (in particular, having the same parameters).

Recall that two codes of length n are equivalent if they can be obtained from each other by the action of the group $(\mathbb{F}_q^*)^n$. Informally speaking, this proposition shows that condition (12.2.6) is automatic for asymptotically optimal algebraic geometry codes with not too small rates.

PROOF. To begin with, note that for $R = 2\gamma_q$ the right-hand side of the equation given in the proposition is greater that the left-hand side, and for $R = 3\gamma_q$ this is not the case, which is easily checked by brute-force computations. Thus, $2\gamma_q < R_0 < 3\gamma_q$.

To prove the proposition, it suffices to prove that under its conditions there exists a word in $(X,\mathcal{P},D)_L$ of weight n, i.e., with no zero entries. Indeed, then there exists a code C' equivalent to C and satisfying (12.2.6). To prove this, one can apply the following result on the number $A_n = A_n(C)$ of maximum-weight codewords in C (in what follows, g is the genus of X).

LEMMA 12.2.10. *We have*

$$|A_n - \widetilde{A}_n| \le n2^{nH_2(\frac{d^*+g}{n})} + gq^{g-1}2^{nH_2(\frac{d^*+2g-2}{n})}$$

under the condition $n \ge (\sqrt{q}+1)(d^*+2g)/\sqrt{q}$, *where*

$$\widetilde{A}_n = (q-1)^n q^{-r^n}$$

is the number of words of weight n *in a random code of length* n *and dimension* k^*, *and* $r^* = n - k^* = d^* + g - 1$ *is its redundancy.* □

The proof of the lemma is based on the expurgation method; note also that the necessary positivity of $A_n(C)$ can be deduced directly by applying expurgation.

Then it is sufficient to check that for $R^* > R_0$ we have

$$\widetilde{A}_n - n2^{nH_2(\frac{d^*+g}{n})} - gq^{g-1}2^{nH_2(\frac{d^*+2g-2}{n})} > 0.$$

To this end, note that under our condition $R^* > R_0$ we have

$$n\log_2(q-1) - r^*\log_2 q \ge \max\left\{nH_2\Big(\frac{d^*+g}{n}\Big), nH_2\Big(\frac{d^*+2g-2}{n}\Big) + (g-1)\log_2 q\right\} + cn,$$

which is equivalent to

$$\log_2(q-1) - (1-R^*)\log_2 q \ge \max\{H_2(R^*), H_2(R^*+\gamma_q) + \gamma_q\log_2 q\} + c$$

for some $c > 0$, which is the case by the very definition of R_0. Indeed,

$$\frac{d^*+2g-2}{n} = R^* + \gamma_q - \frac{2}{n}, \qquad \frac{d^*+g}{n} = R^*.$$

For $R^* < 1/2$ the maximum is attained on the second term, and for $R^* \ge 1/2$ the inequality is obvious (under our condition $q \ge 49$). □

Applying Theorem 4.5.13, we get the following result.

THEOREM 12.2.11. *Let* $q \ge 49$ *be a square, let* ε_1 *and* ε_4 *be the roots of*

$$H_q(1-\varepsilon) + \frac{q}{q-1}\varepsilon = 1 + \gamma_q,$$

and let ε_2 *and* ε_3 *be the roots of*

$$H_q(1-\varepsilon) + \varepsilon\log_q(q-1) = 1 + \gamma_q,$$

$1/q < \varepsilon_1 < \varepsilon_2 < \varepsilon_3 < \varepsilon_4 < 1$. *Then*

$$R_A(\varepsilon,q) \geq \frac{1}{q} R_V(\varepsilon,q)$$

for $\varepsilon \geq \varepsilon_0$, *where* ε_0 *is a unique root of the equation* $R_V(\varepsilon,q) = R_0$ *and where*

$$R_V(\varepsilon,q) = 1 - H_q(1-\varepsilon)$$

for $1/q < \varepsilon < \varepsilon_1$ *or* $\varepsilon_4 < \varepsilon < 1$,

$$R_V(\varepsilon,q) = \varepsilon - \gamma_q$$

for $\varepsilon_2 < \varepsilon < \varepsilon_3$, *and for* $\varepsilon_1 < \varepsilon < \varepsilon_2$ *or* $\varepsilon_3 < \varepsilon < \varepsilon_4$ *the function* $R_V(\varepsilon,q)$ *is given by the implicit equation*

$$(R_V(\varepsilon,q)+\gamma_q)H_q\Big(\frac{1-\varepsilon}{R_V(\varepsilon,q)+\gamma_q}\Big) + H_q(1-\varepsilon) = 1+\gamma_q. \qquad \square$$

12.2.2. Arithmetic Secret Sharing Schemes

These schemes, first based on Shamir's scheme and later abstracted and generalized, are fundamental to secure multi-party computation. An $(n,t,d,n-t)$ arithmetic secret sharing scheme (with uniformity) for $\mathbb{F}_q^k$ over $\mathbb{F}_q$ is an $\mathbb{F}_q$-linear secret sharing scheme where $k,n,t \geq 1$, $d \geq 2$, the secret is selected from $\mathbb{F}_q^k$, and each of the n shares is an element of $\mathbb{F}_q$. Moreover, there is t-privacy (in addition, any t shares are uniformly random in $\mathbb{F}_q^t$) and, if one considers the d-fold "componentwise" product of any d sharings, then the d-fold componentwise product of the d respective secrets is $(n-t)$-wise uniquely determined by it.

Below we describe constructions of asymptotically good $(n,t,d,n-t)$ arithmetic secret sharing schemes for $\mathbb{F}_q^k$ over $\mathbb{F}_q$ for (almost) all finite fields $\mathbb{F}_q$. As in many other applications, they are based on asymptotically good towers (or simply sequences) of curves over $\mathbb{F}_q$. Another ingredient which is crucial in these constructions is the "asymptotic rate of d-torsion", i.e., the asymptotic rank of the d-torsion subgroup of the Jacobian in the corresponding tower.

Arithmetic secret sharing. Let us begin with some basic definitions.

Let k,n be integers with $k,n \geq 1$. Consider the $\mathbb{F}_q$-vector space $\mathbb{F}_q^k \times \mathbb{F}_q^n$, where $\mathbb{F}_q$ is an arbitrary finite field.

Let $\pi_0 \colon \mathbb{F}_q^k \times \mathbb{F}_q^n \to \mathbb{F}_q^k$ be the $\mathbb{F}_q$-linear projection

$$(s_1,\dots,s_k,c_1,\dots,c_n) \mapsto (s_1,\dots,s_k).$$

Similarly, for $i \in \{1,\dots,n\}$, $\pi_i \colon \mathbb{F}_q^k \times \mathbb{F}_q^n \to \mathbb{F}_q$ is defined by

$$(s_1,\dots,s_k,c_1,\dots,c_n) \mapsto c_i,$$

and for a nonempty $A \subset \{1,\dots,n\}$, $\pi_A \colon \mathbb{F}_q^k \times \mathbb{F}_q^n \to \mathbb{F}_q^{|A|}$ is given by

$$(s_1,\dots,s_k,c_1,\dots,c_n) \mapsto (c_i)_{i\in A}.$$

For $\boldsymbol{v} \in \mathbb{F}_q^k \times \mathbb{F}_q^n$, it is sometimes convenient to denote $\pi_0(\boldsymbol{v}) \in \mathbb{F}_q^k$ by $\boldsymbol{v}_0$ and $\pi_A(\boldsymbol{v}) \in \mathbb{F}_q^{|A|}$ by $\boldsymbol{v}_A$. We write $\mathcal{I} = \{1,\dots,n\}$, and we sometimes refer to $\boldsymbol{v}_0$ as the *secret-component* of $\boldsymbol{v}$ and to $\boldsymbol{v}_{\mathcal{I}}$ as its *shares-component*. An *n-code for* $\mathbb{F}_q^k$ *(over* $\mathbb{F}_q$*)* is an $\mathbb{F}_q$-vector space $C \subset \mathbb{F}_q^k \times \mathbb{F}_q^n$ such that

(i) $\pi_0(C) = \mathbb{F}_q^k$;
(ii) $(\operatorname{Ker} \pi_{\mathcal{I}}) \cap C \subset (\operatorname{Ker} \pi_0) \cap C$.

For $\boldsymbol{c} \in C$, $\boldsymbol{c}_0 \in \mathbb{F}_q^k$ is the *secret* and $\boldsymbol{c}_{\mathcal{I}} \in \mathbb{F}_q^n$ are the *shares*.

The first condition means that, in C, the secret can take any value in $\mathbb{F}_q^k$. Moreover, for a uniformly random vector $\boldsymbol{c} \in C$, the secret $\boldsymbol{c}_0$ is uniformly random in $\mathbb{F}_q^k$. This follows from the fact that the projection $(\pi_0)|_C$ is a surjective $\mathbb{F}_q$-linear map. The second condition means that the shares uniquely determine the secret. Indeed, if there are $\boldsymbol{c}, \boldsymbol{c}' \in C$ such that their shares coincide but not their secrets, then condition (ii) fails by linearity. Note that these two conditions imply that $k \leq n$ and that an n-code with the stronger condition $(\operatorname{Ker} \pi_{\mathcal{I}}) \cap C = (\operatorname{Ker} \pi_0) \cap C$ is a k-dimensional error-correcting code of length n.

An n-code C for $\mathbb{F}_q^k$ is *r-reconstructing* $(1 \leq r \leq n)$ if

$$(\operatorname{Ker} \pi_A) \cap C \subset (\operatorname{Ker} \pi_0) \cap C$$

for each $A \subset \mathcal{I}$ with $|A| = r$; i.e., any r shares uniquely determine the secret.

An n-code C for $\mathbb{F}_q^k$ is *t-disconnected* if $t = 0$ or else if $1 \leq t < n$, and the projection

$$\pi_{0,A} \colon C \to \mathbb{F}_q^k \times \pi_A(C), \quad \boldsymbol{c} \mapsto (\pi_0(\boldsymbol{c}), \pi_A(\boldsymbol{c})),$$

is surjective for each $A \subset \mathcal{I}$ with $|A| = t$; if $\pi_A(C) = \mathbb{F}_q^t$, we say that C is *t-uniform*.

For $t > 0$, t-disconnectedness means the following. Let $A \subset \mathcal{I}$ with $|A| = t$. Then, for uniformly randomly $\boldsymbol{c} \in C$, the secret $\boldsymbol{c}_0$ is independently distributed from the t shares $\boldsymbol{c}_A$.

A *generator* (k_0, σ) *of an n-code* C consists of a positive integer k_0 and a surjective $\mathbb{F}_q$-linear map $\sigma \colon \mathbb{F}_q^k \times \mathbb{F}_q^{k_0} \to C$ such that $\pi_0(\sigma(\boldsymbol{s}, \boldsymbol{z})) = \boldsymbol{s}$ for all $(\boldsymbol{s}, \boldsymbol{z}) \in \mathbb{F}_q^k \times \mathbb{F}_q^{k_0}$.

Let $m \in \mathbb{N}$. For $\boldsymbol{x}, \boldsymbol{x}' \in \mathbb{F}_q^m$, define their *product* $\boldsymbol{x} * \boldsymbol{x}' \in \mathbb{F}_q^m$ as $(x_1 x_1', \ldots, x_m x_m')$. Let d be a positive integer. If C is an n-code for $\mathbb{F}_q^k$, then $C^{*d} \subset \mathbb{F}_q^k \times \mathbb{F}_q^n$ is the $\mathbb{F}_q$-linear span of all terms of the form $\boldsymbol{c}^{(1)} * \ldots * \boldsymbol{c}^{(d)}$ with $\boldsymbol{c}^{(1)}, \ldots, \boldsymbol{c}^{(d)} \in C$.

An (n,t,d,r) *arithmetic secret sharing scheme for* $\mathbb{F}_q^k$ is an n-code C for $\mathbb{F}_q^k$ such that

(i) $t \geq 1$, $d \geq 2$;
(ii) C is t-disconnected;
(iii) C^{*d} is in fact an n-code for $\mathbb{F}_q^k$; and
(iv) C^{*d} is r-reconstructing.

An n-code C has *uniformity* if, in addition, it is t-uniform.

PROPOSITION 12.2.12. *Let C be an (n,t,d,r) arithmetic secret sharing scheme for $\mathbb{F}_q^k$ over $\mathbb{F}_q$. As a linear secret sharing scheme for $\mathbb{F}_q^k$ over $\mathbb{F}_q$, C has t-privacy and $(r-(d-1)t)$-reconstruction. Hence, $dt + k \leq r$, which in particular gives $3t + 1 \leq n$ for $k = 1$, $d = 2$, and $r = n - t$.* □

We note that arithmetic secret sharing schemes enjoy efficient recovery of the secret in the presence of faulty shares. More precisely, we have the following.

THEOREM 12.2.13. *Let i, j be integers with $1 \leq i < j$. Assume that C and C^{*j} are n-codes for $\mathbb{F}_q^k$. If the n-code C^{*i} is t-disconnected and if C^{*j} has $(n-t)$-reconstruction, then, given a generator for C, there is an efficient algorithm for the*

n-code $C^{*(j-i)}$ that, on input $\widetilde{\boldsymbol{a}} := \boldsymbol{c}_{\mathcal{I}} + \boldsymbol{e} \in \mathbb{F}_q^n$ (faulty shares) with $\boldsymbol{c} \in C^{*(j-i)}$ and $\boldsymbol{e} \in \mathbb{F}_q^n$ of Hamming weight at most t, outputs $\boldsymbol{c}_0 \in \mathbb{F}_q^k$ (correct secret). □

Two applications. We briefly sketch two applications. First, consider an $(n,t,2,n)$ arithmetic secret sharing scheme C for $\mathbb{F}_q^k$ over $\mathbb{F}_q$. Such a scheme can be used to reduce n-party secure multiplication to secure addition in the case of an *honest-but-curious adversary*, at the cost of one round of interaction. From the definition, it follows that there is an $\mathbb{F}_q$-linear map $\psi\colon \mathbb{F}_q^n \to \mathbb{F}_q^k$ such that, for all $\boldsymbol{c}, \boldsymbol{c}' \in C$, $\psi(c_1c_1', \dots, c_nc_n') = \boldsymbol{c}_0 * \boldsymbol{c}_0'$. The reduction works as follows. Let $\boldsymbol{c}_{\mathcal{I}}, \boldsymbol{c}_{\mathcal{I}}' \in \mathbb{F}_q^n$ be secret-sharings, with respective secrets $\boldsymbol{c}_0, \boldsymbol{c}_0' \in \mathbb{F}_q^k$ ($\boldsymbol{c}, \boldsymbol{c}' \in C$). Using a generator for C, player P_i secret-shares $(\lambda_{i1}c_ic_i', \dots, \lambda_{ik}c_ic_i') \in \mathbb{F}_q^k$, where the coefficient vector is the "ith row of the matrix representing ψ in the standard basis" ($i = 1, \dots, n$). Next, player P_j sums the n received shares ($j = 1, \dots, n$). This gives a secret-sharing of $\boldsymbol{c}_0 * \boldsymbol{c}_0'$ according to C.

Second, consider an $(n,t,2,n-t)$ arithmetic secret sharing scheme C for $\mathbb{F}_q^k$ over $\mathbb{F}_q$. Such a scheme can be used for "zero-knowledge verification of secret multiplications" with the following main idea. Assume that a prover puts forward commitments to secrets $\boldsymbol{x}_0, \boldsymbol{y}_0, \boldsymbol{z}_0 \in \mathbb{F}_q^k$ and claims that $\boldsymbol{x}_0 * \boldsymbol{y}_0 = \boldsymbol{z}_0$. To prove his claim, he gives ("coordinatewise") commitments to random $\boldsymbol{x}, \boldsymbol{y} \in C$ where the respective secrets are the $\boldsymbol{x}_0, \boldsymbol{y}_0$ from the input, and a ("coordinatewise") commitment to a random $\boldsymbol{z} \in C^{*2}$ where the secret is the $\boldsymbol{z}_0 \in \mathbb{F}_q^k$ from the input. If the commitment scheme is $\mathbb{F}_q$-linear, then it is easy to enforce that indeed $\boldsymbol{x}, \boldsymbol{y} \in C$ and $\boldsymbol{z} \in C^{*2}$, and that the respective secrets are indeed the ones from the input. Now, if $\boldsymbol{z} = \boldsymbol{x} * \boldsymbol{y}$, then indeed $\boldsymbol{x}_0 * \boldsymbol{y}_0 = \boldsymbol{z}_0$, but inspection of any t "share-triples" (x_i, y_i, z_i) gives no information on the "secret-triple" $(\boldsymbol{x}_0, \boldsymbol{y}_0, \boldsymbol{x}_0\boldsymbol{y}_0)$. Yet, $z_i = x_iy_i$ for each of those t share-triples. On the other hand, assume that $\boldsymbol{z}_0 \neq \boldsymbol{x}_0 * \boldsymbol{y}_0$. Then there are at most $n-t-1$ share-triples (x_i, y_i, z_i) such that $z_i = x_iy_i$, and hence there are at least $t+1$ share-triples for which an inconsistency could show up. These facts together give a handle to checking that $\boldsymbol{z}_0 = \boldsymbol{x}_0\boldsymbol{y}_0$ in several different application scenarios.

Now we are going to give some results on curves (function fields) over $\mathbb{F}_q$, $q = p^m$, and their Jacobians, which will be applied to construct our secret sharing schemes.

Torsion point limits. For a sequence $\mathcal{X}$ of curves X we as usual define the limit

$$A(\mathcal{X}) = \limsup_{X \in \mathcal{X}} \frac{|X(\mathbb{F}_q)|}{g(X)}.$$

Let $X/\mathbb{F}_q$ be a (smooth, projective, absolutely irreducible) curve, and let $\mathcal{J}_X$ be its Jacobian. For a positive integer r greater than 1, define the *r-torsion point group* in $\mathcal{J}_X$,

$$\mathcal{J}_X[r] := \{[D] \in \mathcal{J}_X : \ r[D] = 0\};$$

the cardinality of $\mathcal{J}_X[r]$ is denoted by $J_X[r]$.

For each family $\mathcal{X} = \{X/\mathbb{F}_q\}$ of curves with $g(X) \to \infty$, we define the asymptotic limit

$$J_r(\mathcal{X}) := \liminf_{X \in \mathcal{X}} \frac{\log_q |\mathcal{J}_X[r]|}{g(X)}.$$

We need to define an asymptotic notion involving both $J_r(\mathcal{X})$ and $A(\mathcal{X})$. For a prime power q, an integer $r>1$, and a real $a \leq A(q)$, let $\mathfrak{X}_a$ be the set of families $\{\mathcal{X}\}$ of curves over $\mathbb{F}_q$ such that the genus in each family tends to ∞ and $A(\mathcal{X}) \geq a$ for every $\mathcal{X} \in \mathfrak{X}_a$. Then the asymptotic quantity $J_r(q,a)$ is defined by

$$J_r(q,a) = \liminf_{\mathcal{X} \in \mathfrak{X}_a} J_r(\mathcal{X}).$$

THEOREM 12.2.14. *Let $\mathbb{F}_q$ be a finite field, and let $r>1$ be a prime.*

(i) *If $r \mid (q-1)$, then $J_r(q,A(q)) \leq \dfrac{2}{\log_r q}$.*

(ii) *If $r \nmid (q-1)$, then $J_r(q,A(q)) \leq \dfrac{1}{\log_r q}$.*

(iii) *If q is square and $r \mid q$, then $J_r(q,\sqrt{q}-1) \leq \dfrac{1}{(\sqrt{q}+1)\log_r q}$.*

Claims (i) and (ii) are simple consequences of the general theory of abelian varieties. The proof of (iii) is outlined below.

COROLLARY 12.2.15. *Let $\mathbb{F}_q$ be a finite field of characteristic p.*

(i) *If $r \geq 2$ is an integer, then $J_r(q,A(q)) \leq \log_q(dr)$, where $d = \gcd(r,q-1)$.*

(ii) *Write r as $p^\ell m$ for some $\ell \geq 0$ and a positive integer m coprime to p. If q is a square, then*

$$J_r(q,\sqrt{q}-1) \leq \frac{\ell}{\sqrt{q}+1}\log_q(p) + \log_q(cm),$$

where $c = \gcd(m,q-1)$. □

THEOREM 12.2.16. *For every prime power $q \geq 8$ except perhaps for $q = 11$ or 13, there exists a family $\mathcal{X}$ of curves over $\mathbb{F}_q$ such that $A(\mathcal{X})$ exists and satisfies $A(\mathcal{X}) > 1 + J_2(\mathcal{X})$.*

PROOF. We prove this in two steps. The first one is to prove that the result is true for all $q \geq 17$ by using class field theory. The second step is to show that the result holds for $q = 8, 9, 16$ by looking at each individual q. For $q \geq 17$, we prove the result only for odd q. For even q, we can similarly get it by considering Artin–Schreier extensions. Choose 7 nonzero square elements $t_1, \ldots, t_7$ in $\mathbb{F}_q$ (this is possible since $(q-1)/2 \geq 7$). For each i, consider the extension $\mathbb{K}_i = \mathbb{F}_q(x,y_i)$, where $y_i^2 = x + t_i$. Then the place x splits completely in $\mathbb{K}_i$. Let $\mathbb{K}$ be the field $\mathbb{F}_q(x,y)$, where $y^2 = \prod_{i=1}^{7}(x+ti)$; then $\mathbb{K}$ is a subfield of $\mathbb{K}_1 \ldots \mathbb{K}_7/\mathbb{F}_q(x)$ such that $[\mathbb{K} : \mathbb{F}_q(x)] = 2$ and $\mathbb{K}_1 \ldots \mathbb{K}_7/\mathbb{K}$ is an unramified abelian extension. The three places, ∞ and those lying above x, split completely in the compositum $\mathbb{K}_1 \ldots \mathbb{K}_7/\mathbb{K}$. Since the 2-rank of the Galois group of $\mathbb{K}_1 \ldots \mathbb{K}_7/\mathbb{K}$ is 6, which is equal to $2+2\sqrt{3+1}$, $\mathbb{K}$ has an infinite $(2,S)$-Hilbert class field tower $\mathcal{F}$, where S consists of the three places, ∞ and those lying above x. This yields

$$A(\mathcal{F}) \geq \frac{3}{g(\mathbb{K})-1} = \frac{3}{2}.$$

Now we have

$$A(\mathcal{F}) \geq \frac{3}{2} > 1 + \frac{2}{\log_2 17} \geq 1 + \frac{2}{\log_2 q} \geq 1 + J_2(\mathcal{F}),$$

which completes the proof for $q \geq 17$.

For $q=8$ we know that there exists a family $\mathcal{F}$ over $\mathbb{F}_8$ such that $A(\mathcal{F}) \geq 3/2$, since $A(q^3) \geq \dfrac{2(q^2-1)}{q+2}$ for a prime power q. Thus,

$$A(\mathcal{F}) \geq \frac{3}{2} > 1+\frac{1}{3} \geq 1+J_2(\mathcal{F}).$$

For $q=9$, $A(q)=2$, and thus there exists a family $\mathcal{F}$ over $\mathbb{F}_9$ such that $A(\mathcal{F})=2$. Therefore,

$$A(\mathcal{F}) = 2 > 1+\frac{2}{\log_2 9} \geq 1+J_2(\mathcal{F}).$$

For $q=16$, there exists a family $\mathcal{F}$ over $\mathbb{F}_{16}$ such that $A(\mathcal{F})=3$, since $A(16)=3$. Therefore,

$$A(\mathcal{F}) = 3 > 1+\frac{1}{4} \geq 1+J_2(\mathcal{F}). \qquad \square$$

SKECTH OF THE PROOF OF THEOREM 12.2.14(iii). Let q be a square, more precisely, $q=p^{2m}$, and let $\{F^{(n)}=F_n\}$ be the García–Stichtenoth $\mathcal{W}_2$ tower (see Sec. 7.4.3). Then we have the following.

THEOREM 12.2.17. *Let $\gamma(F^{(n)}) = \dim_{\mathbb{F}_p}(J_{X_n}[p])$ be the p-rank of the function field $\mathbb{F}_q(X_n)=F^{(n)}$. Then*

$$\gamma(F^{(n)}) = \begin{cases} (q^{\frac{n}{4}}-1)^2 & \text{if } n \equiv 0 \pmod 2, \\ (q^{\frac{n-1}{4}}-1)(q^{\frac{n+1}{4}}-1) & \text{if } n \equiv 1 \pmod 2. \end{cases}$$

It follows from Proposition 7.4.17 together with Theorem 12.2.7 that

$$\lim_{n\to\infty} \frac{g(F^{(n)})}{\gamma(F^{(n)})} = \sqrt{q}+1,$$

and the proof is completed. $\square$

The formula for $\gamma(F^{(n)})$ uses the following theorem.

THEOREM 12.2.18 (Deuring–Shafarevich). *Let $f\colon X\to Y$ be a Galois covering of curves over $\overline{\mathbb{F}}_q$. Assume that its Galois group is a p-group. Then*

$$\gamma(X)-1 = [X:Y](\gamma(Y)-1) + \sum_{P\in\mathbb{P}(X)} \sum_{\substack{Q\in\mathbb{P}(Y)\\ f(Q)=P}} (e(Q\,|\,P)-1). \qquad \square$$

SKECTH OF THE PROOF OF THEOREM 12.2.17. Without loss of generality we can assume that the constant fields of the function fields are $\overline{\mathbb{F}}_q$. Consider the extension $F^{(n)}/F^{(n-1)}$. Since this is an Artin–Schreier extension, it is Galois and its Galois group is a p-group. By the Hurwitz formula and the Deuring–Shafarevich theorem above, respectively, we get

$$2g(F^{(n)})-2 = \sqrt{q}\big(2g(F^{(n-1)})-2\big) + (\sqrt{q}+2) \sum_{P\in\mathbb{P}(F^{(n-1)})} \sum_{\substack{Q\in\mathbb{P}(F^{(n)})\\ Q\,|\,P}} (e(Q\,|\,P)-1),$$

$$\gamma(F^{(n)})-1 = \sqrt{q}\big(\gamma(F^{(n-1)})-1\big) + \sum_{P\in\mathbb{P}(F^{(n-1)})} \sum_{\substack{Q\in\mathbb{P}(F^{(n)})\\ Q\,|\,P}} (e(Q\,|\,P)-1);$$

combining these two equations, we find

$$\gamma(F^{(n)}) = \sqrt{q}\gamma(F^{(n-1)}) + \frac{2g(F^{(n)}) - 2\sqrt{q}g(F^{(n-1)}) - q + \sqrt{q}}{\sqrt{q}+2}.$$

The result is obtained by using the fact that $\gamma(F^{(0)}) = 0$ and applying induction. □

Application of algebraic curves. Now let us apply the above results on curves to construct infinite families of arithmetic secret sharing schemes. First we state—without proof, which is rather long and cumbersome—the main result connecting algebraic curves and arithmetic secret sharing schemes. Recall that M_r denotes the number of ($\mathbb{F}_q$-rational) effective divisors of degree $r \geq 0$.

THEOREM 12.2.19. *Let $X/\mathbb{F}_q$ be a curve. Let $d,k,t,n \in \mathbb{Z}$ with $d \geq 2$, $n > 1$, and $1 \leq t < n$. Assume that $Q_1,\dots,Q_k,P_1,\dots,P_n \in X(\mathbb{F}_q)$ are pairwise distinct. If there is $s \in \mathbb{Z}$ such that*

$$h > \binom{n}{t}(M_{r_1} + M_{r_2}|\mathcal{J}_X[d]|)$$

where $r_1 := 2g - s + t + k - 2$ and $r_2 := ds - n + t$, then there exists an $(n,t,d,n-t)$ arithmetic secret sharing scheme for $\mathbb{F}_q^k$ over $\mathbb{F}_q$ with uniformity. □

From this one can derive the following result.

THEOREM 12.2.20. *Let $\mathbb{F}_q$ be a finite field. Assume that $\varkappa \in [0,1/3)$, $\tau \in (0,1]$, and $0 < A \leq A(q)$ are real numbers such that*

$$A > \frac{1+\varkappa}{1-3\varkappa}(1+J_2(q,A))$$

and

$$\tau + \frac{H_2(\tau)}{\log q} < \frac{1}{3}\left(1 - 3\varkappa - \frac{(1+J_2(q,A))(1+\varkappa)}{A}\right).$$

Then there exists an infinite family of $(n,t,2,n-t)$ arithmetic secret sharing schemes for $\mathbb{F}_q^k$ over $\mathbb{F}_q$ with uniformity where n is unbounded, $k = \lfloor\varkappa n\rfloor + 1$, and $t = \lfloor\tau n\rfloor$. □

This implies the following main theorem.

THEOREM 12.2.21. *Let $\mathbb{F}_q$ be a finite field and $d \geq 2$ an integer. If*

$$A > d - 1 + J_d(q,A)$$

for some $A \in (0, A(q)]$, then there exists an infinite family of $(n,t,d,n-t)$ arithmetic secret sharing schemes for $\mathbb{F}_q^k$ over $\mathbb{F}_q$ with t-uniformity as $n \to \infty$, $k = \Omega(n)$, and $t = \Omega(n)$. □

For $d = 2$, this yields the following result.

THEOREM 12.2.22. *For $q = 8,9$ and for all prime powers $q \geq 16$ there exists an infinite family of $(n,t,2,n-t)$ arithmetic secret sharing schemes for $\mathbb{F}_q^k$ over $\mathbb{F}_q$ with t-uniformity for $n \to \infty$, $k = \Omega(n)$, and $t = \Omega(n)$.* □

Let us briefly outline how to derive Theorem 12.2.20 from Theorem 12.2.19. The numerical condition in Theorem 12.2.19 includes the following ingredients: $J_X[2]$, which is controlled by the torsion limit $J_2(q,A)$; binomial coefficients, given by Stirling's formula; and the quotients of the form M_r/h, h being the class number of X. The latter can be bounded as follows, using the equality $P(1) = h$, the functional equation for $P(t)$, and passing to the limit as $t \to q^{-1}$, where $P(t) = \prod_{i=1}^{2g}(1 - \omega_i t)$ is the numerator of the zeta function of X.

PROPOSITION 12.2.23. *Let $X/\mathbb{F}_q$ be a curve with $g \geq 1$. Then, for any integer r, $0 \leq r \leq g-1$, we have*

$$\frac{M_r}{h} \leq \frac{g}{q^{g-r-1}(\sqrt{q}-1)^2}.$$

□

Combining the upper bounds for $J_2(q, A)$, Stirling's formula, and the inequalities of Proposition 12.2.23, we can use Theorem 12.2.19 to obtain Theorem 12.2.20. Note, however, that computations, being straightforward, are rather long.

12.3. Quantum Codes

In this section we give some applications of algebraic geometry codes to quantum codes. The latter are a necessary part of purported quantum computers, since all known schemes for their operation involve error correction as an indispensable fundamental principle. First we give some general definitions, properties, and constructions in Sec. 12.3.1; then in Sec. 12.3.2 we discuss applications of algebraic geometry codes to constructing asymptotically good quantum codes. The codes in Secs. 12.3.1 and 12.3.2 are binary in a natural sense; Sec. 12.3.3 is devoted to the nonbinary case.

12.3.1. Quantum Error-Correcting Codes

We start with the notion of a qubit, basic in quantum error correction. Information is transmitted as vectors of the Hilbert space $\mathbb{C}^k$. It is natural to ask about the minimum dimension t that is necessary for transmission of one bit of (classical) information. According to quantum mechanics, two vectors from the space $\mathbb{C}^t$ are completely distinguishable by a measurement if and only if they are orthogonal. In $\mathbb{C}^2$ there exists an orthogonal basis, say v_0, v_1, and we can assign logical 0 to v_0 and logical 1 to v_1. Thus, the space $\mathbb{C}^2$ allows us to encode one bit of information; any of its element is called a qubit (quantum bit).

Note that in accordance with quantum-mechanical principles, if we want to transmit n bits of information, we need the space $\mathbb{C}^{2^n} = \mathbb{C}^2 \otimes \ldots \otimes \mathbb{C}^2$, where $\otimes$ is the tensor product. In $\mathbb{C}^{2^n}$ we can choose 2^n orthogonal vectors from the space that allows the transmission of n bits of information. In fact, it is not necessary to restrict ourselves to using only mutually orthogonal vectors; any vectors from $\mathbb{C}^{2^n}$ could be sent over a channel (by a vector we always mean a column vector).

In the process of transmission over a channel the information can be altered by an error, and here we consider the most popular channel model, namely, the completely depolarized channel. In this model a vector $v \in \mathbb{C}^2$ can be altered by one of the following error operators:

$$\sigma_x = \begin{bmatrix} 0 & 1 \\ 1 & 0 \end{bmatrix}, \qquad \sigma_z = \begin{bmatrix} 1 & 0 \\ 0 & -1 \end{bmatrix}, \qquad \sigma_y = \begin{bmatrix} 0 & -i \\ i & 0 \end{bmatrix}$$

(where $i = \sqrt{-1}$), called the Pauli matrices. An error operator acting in $\mathbb{C}^{2^n}$, say E, can therefore be written in the following form:

$$E = \sigma_1 \otimes \ldots \otimes \sigma_n,$$

where $\sigma_i \in \{I_2, \sigma_x, \sigma_z, \sigma_y\}$, $i = 1, \ldots, n$, I_2 being the 2×2 identity matrix. The number of nonidentity matrices in this tensor product is called the *weight* of E and is denoted by $\mathrm{wt}(E)$. If a vector $\boldsymbol{v}$ is affected by an error E, then the result is $\boldsymbol{w} = E\boldsymbol{v}$.

A *quantum code* Q of length n and size K, referred to as an $((n, K))$ *code* (or an $[[n, k]]$ code for $k = \log_2 K$), is a K-dimensional linear subspace of the space $\mathbb{C}^{2^n}$.

We say that an error E is detectable if any two orthogonal vectors from the code remain orthogonal (i.e., distinguishable) after one of them has been altered by E. This yields the following definition.

An error E is *detectable* if and only if

$$\boldsymbol{v}^\dagger E \boldsymbol{w} = 0 \tag{12.3.1}$$

for all orthogonal vectors $\boldsymbol{v}$ and $\boldsymbol{w}$ from the code Q, $\boldsymbol{v}^\dagger$ being the Hermitian conjugation of $\boldsymbol{v}$ (i.e., $\boldsymbol{v}^\dagger$ is the transpose of v conjugated in $\mathbb{C}$).

Let us give yet another definition of error detectability. We say that E is *detectable* if and only if

$$PEP = \lambda_E P \tag{12.3.2}$$

for the orthogonal projection operator P on the code subspace, where λ_E is a constant that depends only on E.

PROPOSITION 12.3.1. *The definitions* (12.3.1) *and* (12.3.2) *are equivalent.* □

The *minimum distance* of the code Q is the maximum integer w such that any error E with $\mathrm{wt}(E) \leq w-1$ is detectable.

Note that if we use similar definitions for classical codes, then we get the usual definition of the minimum distance. Moreover, as in the case of classical codes, one can introduce enumerators of a quantum $((n,K))$ code. For an $((n,K))$ quantum code Q we can define two enumerators, $(B_i)_{i=0}^n$ and $(B_i^\perp)_{i=0}^n$, given by

$$B_i = \frac{1}{K^2} \sum_{\mathrm{wt}(E)=i} (\mathrm{Tr}(EP))^2, \qquad B_i^\perp = \frac{1}{K} \sum_{\mathrm{wt}(E)=i} \mathrm{Tr}(EPEP).$$

Note that according to the definition of the error set $\{\sigma_x, \sigma_z, \sigma_y\}$, an error operator E is Hermitian. Hence, $(\mathrm{Tr}(EP))^2$ and $\mathrm{Tr}(EPEP)$ are real numbers, and so are the numbers B_i and $B_i^\perp$.

In the case of a quantum stabilizer code (to be defined later), say Q, the numbers B_i and $B_i^\perp$ are the weight spectra of codes C and $C^\perp$ over $\mathbb{F}_4$ associated with Q. Thus, in this case B_i and $B_i^\perp$ are connected by the MacWilliams identities, which are valid for the enumerators of general quantum codes, and define the minimum distance of a quantum code.

Let

$$B(x,y) = \sum_{i=0}^n B_i x^{n-t} y^t, \qquad B^\perp(x,y) = \sum_{i=0}^n B_i^\perp x^{n-t} y^t.$$

THEOREM 12.3.2. *The numbers B_i and $B_i^\perp$ are related by the quaternary MacWilliams identities. That is,*

$$B(x,y) = \frac{1}{2^n K} B^\perp(x+3y, x-y). \qquad \square$$

THEOREM 12.3.3. *Let Q be a quantum code with enumerators B_i and $B_i^\perp$. Then*

(i) $B_i^\perp \geq B_i \geq 0$, $i = 1, \dots, n$, *and* $B_0 = B_0^\perp = 1$.
(ii) *The minimum distance of Q equals $t+1$, where $t = \max\{i : B_i = B_i^\perp\}$.* □

For quantum codes one can obtain upper bounds parallel to those for the usual error-correcting codes, but they are much more cumbersome than the usual ones, and here we only give some of their corollaries which are more comprehensible.

THEOREM 12.3.4 (Singleton-type bound). *For any $((n,K))$ quantum code of minimum distance w, we have*

$$K \leq 2^{n-2w+2}. \qquad \square$$

THEOREM 12.3.5 (Hamming-type bound). *The usual Hamming bound is valid for quantum codes, i.e., if Q is an $((n,K))$ quantum code, then*

$$\frac{\log_2 K}{n} \leq 1 - \frac{\delta}{2}\log_2 3 - H\Big(\frac{\delta}{2}\Big) + o(1). \qquad \square$$

THEOREM 12.3.6 (linear programming bound). *If $0 \leq \delta \leq 0.3152$, then the usual first linear programming bound is valid for quantum codes:*

$$\frac{\log_2 K}{n} \leq H\Big(\frac{3}{4} - \frac{1}{2}\delta - \frac{1}{2}\sqrt{\delta(1-\delta)}\Big) + \Big(\frac{3}{4} - \frac{1}{2}\delta - \frac{1}{2}\sqrt{\delta(1-\delta)}\Big)\log_2 3 - 1 + o(1). \qquad \square$$

Note that when $\delta = 0.3152$, $\frac{\log_2 K}{n} \approx 0.0028$. Thus, for all $0.0028 \leq \frac{\log_2 K}{n} \leq 1$, the usual first linear programming bound is valid.

Stabilizer codes. Probably the most interesting and important class of quantum codes are quantum stabilizer codes. These codes can be viewed as natural analogues of classical linear codes. To define a quantum stabilizer code, we first introduce a class of usual (nonquantum) codes.

Let $T = \mathbb{F}_4$, and let $x \mapsto \overline{x}$ be the nontrivial automorphism of $\mathbb{F}_4$ over $\mathbb{F}_2$. We fix a (symplectic) form on T^n given by $\omega(x,y) = \mathrm{Tr}(x\overline{y})$ and fix a usual $\mathbb{F}_4$ Hamming norm on T^n. A *small symplectic code* $F \subset T^n$ is an ω-isotropic $\mathbb{F}_2$-subspace in T^n; i.e., $\omega(x,y) = 0$ for any $x,y \in F$. The *minimum distance* $d = d_F$ is defined as the minimum $\mathbb{F}_4$ Hamming norm of a nontrivial vector in F, and the dimension $k = k_F$ is its $\mathbb{F}_2$-dimension; in particular, $k \leq n$. The ω-dual F^ω of a small symplectic code F is called a *large symplectic code*; for a large symplectic code we have $n \leq k_{F^\omega} \leq 2n$. Note that $F \subset F^\omega$, since ω is symplectic.

Let $F \subset T^n$ be a small symplectic code with parameters $[n,k,d]$. We are going to define a standard stabilizer code $Q_F \subset \mathbb{C}^{2^n}$ corresponding to F. Let $\mathbb{F}_4 = \{0, 1, \varepsilon, \overline{\varepsilon}\}$. Set

$$\sigma(0) = I_2, \qquad \sigma(\varepsilon) = \sigma_x, \qquad \sigma(\overline{\varepsilon}) = \sigma_z, \qquad \sigma(1) = \sigma_y$$

and put

$$\sigma(t) = \sigma(t_1) \otimes \ldots \otimes \sigma(t_n)$$

for $t = (t_1, \ldots, t_n) \in T^n$. We get a map (of sets) $\sigma\colon T^n \to L(\mathbb{C}^{2^n})$, where $L(\mathbb{C}^{2^n})$ is the space of linear operators on $\mathbb{C}^{2^n}$. Being restricted to a small symplectic code $F \in T^n$, the map σ becomes almost a group homomorphism, namely for $f_1, f_2 \in F$ we have

$$\sigma(f_1)\sigma(f_2) = \sigma(f_2)\sigma(f_1) = \pm\sigma(f_1 + f_2),$$

in particular $\sigma(f_1)$ and $\sigma(f_2)$ commute. This makes it possible to consider the subspace of $\mathbb{C}^{2^n}$ fixed by $\sigma(F)$ in the following way. Let $\mathcal{F} = \{f_1, \ldots, f_k\}$ be an $\mathbb{F}_2$-basis of F, and let $\mu = \{\mu_1, \ldots, \mu_k\}$ with $\mu_i \in \{\pm 1\}$ for $i = 1, \ldots, k$.

Define $Q_{\mathcal{F},\mu}$ as follows:

$$Q_{\mathcal{F},\mu} = \{x \in \mathbb{C}^{2^n} \mid \sigma(f_i)(x) = \mu_i x \text{ for any } i = 1, \ldots, k\}.$$

The quantum code $Q_{\mathcal{F},\mu}$ is called a *stabilizer code*. For any $f \in F$, the operator $\sigma(f)$ acts on $Q_{\mathcal{F},\mu}$ as ± 1.

The small symplectic code F being fixed, we get 2^k different codes $Q_{\mathcal{F},\mu}$. We need only their properties that do not depend on $\mathcal{F}$ or μ, so we write $Q_{\mathcal{F},\mu} = Q_F$.

PROPOSITION 12.3.7. *The parameters of the thus obtained quantum codes are*

$$K_{Q_F} = 2^{n-k_F}, \qquad d_{Q_F} = \min_{f \in F^\omega \setminus F} \|f\| \geq d_{F^\omega}. \qquad \square$$

In the asymptotic setting most interesting to us, we let $k_Q = \log_2 K_Q$ and denote

$$R_Q = \frac{k_Q}{n} \qquad \text{and} \qquad \delta_Q = \frac{d_Q}{n}.$$

We are interested in

$$R(\delta) = \limsup_{n \to \infty} R_Q,$$

where the limit is over all codes with $\delta_Q \geq \delta$.

For stabilizer codes there is a Gilbert–Varshamov-type lower bound.

THEOREM 12.3.8. *There exist quantum codes of length n, dimension K, and relative distance δ such that*

$$\frac{\log_2 K}{n} \geq 1 - \delta \log_2 3 - H(\delta),$$

where $H(x) = -s\log_2 x - (1-x)\log_2(1-x)$ is the binary entropy function. $\square$

12.3.2. Quantum Codes from Algebraic Geometry Codes

Quantum codes can be obtained from algebraic geometry in several steps, and the main result is as follows.

THEOREM 12.3.9. *For any $\delta \in \left(0, \frac{1}{18}\right)$ and R lying on the broken line given by the piecewise linear function*

$$R(\delta) = 1 - \frac{2}{2^m - 2} - \frac{10}{3} m\delta \quad \textit{for} \quad \delta \in [\delta_m, \delta_{m-1}],$$

where $m = 3, 4, 5, \ldots$, $\delta_2 = \frac{1}{18}$, and

$$\delta_m = \frac{3}{5} \frac{2^m}{(2^m - 2)(2^{m+1} - 2)} \quad \textit{for} \quad m = 3, 4, 5, \ldots,$$

there exist polynomially constructible families of quantum codes with $n \to \infty$ and asymptotic parameters greater than or equal to (δ, R). $\square$

Now we describe principal steps of the construction leading to this result.

From binary codes to symplectic codes. We start with a triple $D' \supset D \supset D^\perp$ of binary codes, where D is an $[n, k, d]$ code containing its dual $D^\perp$, and D' is a larger $[n, k']$ code with $k' \geq k + 2$. Let G be a generator matrix of D, and let G' be a matrix such that

$$\begin{pmatrix} G \\ G' \end{pmatrix}$$

is a generator matrix of D'. Denote by d'_2 the second generalized weight of D'. Form the code $C \subset \mathbb{F}_2^{2n}$ with the generator matrix

$$\begin{pmatrix} G & 0 \\ 0 & G \\ G' & G'' \end{pmatrix},$$

where G'' is obtained from G' by a permutation of rows with no fixed positions.

Fix the following $\mathbb{F}_2$-linear isomorphism between $\mathbb{F}_2^{2n}$ and $\mathbb{F}_4^n$: first map $(x_1,\dots,x_n,y_1,\dots,y_n) \in \mathbb{F}_2^{2n}$ to $((x_1,y_1),\dots,(x_n,y_n)) \in (\mathbb{F}_2^2)^n$ and then identify $\mathbb{F}_2^2$ and $\mathbb{F}_4$ by $(0,0)=0$, $(0,1)=\varepsilon$, $(1,0)=\overline{\varepsilon}$, $(1,1)=1$ (this is the well-known Gray map). The image of C under this map is $F \subset \mathbb{F}_4^n$. Let us estimate its parameters.

PROPOSITION 12.3.10. *The code $F \subset \mathbb{F}_4^n$ is a large symplectic code, i.e., $F \supset F^{\omega}$. Its parameters are $k_F = k + k'$ and $d_F \geq \min(d, d_2')$.* □

COROLLARY 12.3.11. *The parameters of the corresponding quantum stabilizer code Q_F satisfy*

$$k_{Q_F} = k + k' - n, \qquad d_{Q_F} \geq \min(d, d_2') \geq \min(d, \frac{3}{2}d'). \qquad \square$$

From nonbinary to binary codes. The following result is a basis for the construction.

PROPOSITION 12.3.12. *Let C be a code over $\mathbb{F}_{2^m}$ and $C^{\perp} \subset C$. Let α_i, $i = 1,\dots,m$, be a self-dual basis of $\mathbb{F}_{2^m}$ over $\mathbb{F}_2$, i.e.,*

$$\mathrm{Tr}(\alpha_i\alpha_j) = \delta_{ij}.$$

Let D and $D^{\perp}$ be codes obtained by the symbolwise binary expansion of codes C and $C^{\perp}$ in the basis α_i. Then $D^{\perp} \subset D$, and $D^{\perp}$ is the binary dual of D. □

From curves to codes. Now we need to construct a triple $C' \supset C \supset C^{\perp}$ of codes over $\mathbb{F}_{2^m}$ with good parameters.

In fact, we have the following statement.

THEOREM 12.3.13. *If there exists a curve X over $\mathbb{F}_q$ of genus g with $|X(\mathbb{F}_q)| \geq n' \geq 4g$, then for any $n \leq n' - g$ and any $a = 2g-1,\dots,\frac{n}{2}+g-1$ there exists an $[n,k,d]_q$ code C with parameters*

$$k = n - a + g - 1,$$
$$d \geq a - 2g + 2$$

such that $C \supseteq C_w^{\perp}$ for some $w \in (\mathbb{F}_q^*)^n$.

Moreover, if q is a power of 2, there is such a code with $C \supseteq C_w^{\perp}$.

PROOF. Let $a \in \mathbb{N}$. We take an effective divisor D with $\deg D = a$, $\mathcal{P}' = \{P_1,\dots,P_{n'}\}$, $P_i \in X(\mathbb{F}_q)$, and $C := (X,\mathcal{P},D)_L$.

If $a \leq \frac{n'}{2} + \frac{g}{2} - 1$, then Riemann–Roch gives $0 \neq \omega \in \Omega(\boldsymbol{P}' - 2D - E)$ for any $E \geq 0$, $\deg E = n' + g - 2 - 2a$. Let $\mathcal{P} = \{P_1,\dots,P_n\} \subseteq \mathcal{P}'$ be the set of poles of ω; then $n \geq n' - g$. Let $\boldsymbol{P} = P_1 + \dots + P_n$; then $\omega \in \Omega(\boldsymbol{P} - 2D - E)$, and we put $w = (\mathrm{Res}(\omega_1),\dots,\mathrm{Res}(\omega_n))$. For any $f,g \in L(D)$ we have

$$\sum_i w_i f(P_i) g(P_i) = \sum_i \mathrm{Res}_{P_i}(\omega) = 0$$

by the residue formula, and thus $C \supseteq C_w^{\perp}$.

Now let $q = 2^m$; in this case $w_i = v_i^2$ for some $v_i \in \mathbb{F}_q$. Then we put $C' = g_v(C)$, g_v being the coordinatewise multiplication by $v = (v_1,\dots,v_n)$. Clearly, $C' \supset C'^{\perp}$, and the parameters of C' give the theorem. □

Applying the result to any family of asymptotically optimal curves over $\mathbb{F}_q$, we obtain the following result.

COROLLARY 12.3.14. *Let q be an even power of a prime. Then for any*

$$\alpha \in \left(\frac{2}{\sqrt{q}-2}, \frac{1}{2}+\frac{1}{\sqrt{q}-2}\right)$$

there exist families of codes with asymptotic parameters

$$R = 1-\alpha+\frac{1}{\sqrt{q}-2},$$
$$\delta \geq \alpha - \frac{2}{\sqrt{q}-2},$$

with the auxiliary property $C \supseteq C_w^\perp$ for some $w \in (\mathbb{F}_q^*)^n$.

If q is an even power of 2, a stronger property $C \supseteq C^\perp$ holds. □

Now we construct a triple $C' \supset C \supset C^\perp$ which will be used to construct quantum codes.

If we take two divisors $D' \leq D$, then $(X,\mathcal{P},D')_L \subseteq (X,\mathcal{P},D)_L$, and we have the opposite inclusion for duals. The differential form ω with the above properties good for D is also good for D'. Taking $D = aP_0$ and $D' = a'P_0$ with $a' < a$, we prove the following result.

COROLLARY 12.3.15. *Let $q = 2^{2m}$. Then for any pair of real numbers (α',α) such that $\frac{2}{2^m-2} \leq \alpha' \leq \alpha \leq \frac{1}{2}+\frac{1}{2^m-2}$ there exist families of triples of 2^{2m}-ary codes $C' \supset C \supseteq C^\perp$ with asymptotic parameters*

$$R' = 1-\alpha'+\frac{1}{2^m-2}, \qquad \delta' \geq \alpha' - \frac{2}{2^m-2},$$
$$R = 1-\alpha+\frac{1}{2^m-2}, \qquad \delta \geq \alpha - \frac{2}{2^m-2},$$

R' being the asymptotic rate of the codes C', and R and δ being the asymptotic parameters of C. □

By the results of Sec. 7.4.3, these codes are polynomially constructible if we choose the $\mathcal{W}_1$ tower to use as our optimal family.

To finish the construction, and thus prove Theorem 12.3.9, we have to sum up the above constructions.

To construct an asymptotically good quantum code Q, we start with a family of curves X over $\mathbb{F}_{2^{2m}}$ with $\frac{|X(\mathbb{F}_q)|}{g(X)} \to 2^m - 1$. Each curve gives a triple $C' \supset C \supset C^\perp$ of algebraic geometry codes C over $\mathbb{F}_{2^{2m}}$ as described above. Let C be an $[n,k,d]$ code and let C' be an $[n,k',d']$ code. Binary expansions of C and C' with respect to a self-orthogonal basis give a triple $D' \supset D \supset D^\perp$ of binary codes with $n_{D'} = n_D = 2mn$, $k_{D'} = 2mk'$, $k_D = 2mk$, $d_{D'} \geq d'$, and $d_D \geq d$. These codes lead to symplectic codes F, their parameters being $[2mn, 2m(k+k'), \geq \min\{d, \frac{3}{2}d'\}]$. In turn, those give quantum stabilizer $[[2mn, 2m(k+k'-n), \geq \min\{d, \frac{3}{2}d'\}]]$ codes Q, and the corresponding asymptotic parameters are

$$R_Q = R + R' - 1, \qquad \delta_Q \geq \min\{\delta, \frac{3}{2}\delta'\},$$

where R, R', δ, and δ' are the parameters of algebraic geometry $\mathbb{F}_{2^{2m}}$-ary codes.

Now use Corollary 12.3.15, putting $\alpha' = \frac{2}{3}(\alpha+\gamma)$, where $\gamma = \frac{1}{2^m-2}$; this choice of α' is optimal. The restrictions $2\gamma \le \alpha' < \alpha \le \frac{1}{2}+\gamma$ are equivalent to $2\gamma \le \alpha \le \frac{1}{2}+\gamma$, and the asymptotic parameters of the algebraic geometry codes are as follows:

$$R = 1-\alpha+\gamma, \qquad \delta \ge \alpha - 2\gamma,$$
$$R' = 1-\frac{2}{3}\alpha+\frac{1}{3}\gamma, \qquad \delta' \ge \frac{2}{3}\alpha - \frac{4}{3}\gamma.$$

Their binary expansions have the same R and R', and the estimates for their δ and δ' get divided by $2m$; thus, by Corollary 12.3.11 we obtain

$$R_Q = R+R'-1 = 1+\frac{4}{3}\gamma-\frac{5}{3}\alpha, \qquad \delta_Q \ge \frac{1}{2m}(\alpha-2\gamma).$$

Therefore, for any $m \ge 3$ we get a polynomial bound

$$R_Q = 1-\frac{2}{2^m-2}-\frac{10}{3}m\delta_Q$$

with the restriction

$$\delta_Q \le \frac{1}{2m}\left(\frac{1}{2}-\frac{1}{2^m-2}\right),$$

i.e.,

$$1-\frac{2}{2^m-2} \ge R_Q \ge \frac{1}{6}-\frac{1}{3}\frac{1}{2^m-2}.$$

Improvement. Now we show that by applying a concatenated construction one can further improve the bound of Theorem 12.3.9. More precisely, we have the following.

THEOREM 12.3.16. *Let $t \ge 3$, and let $\delta \in (0,\delta_t)$, where*

$$\delta_t = \frac{2}{3}\frac{2^t-3}{(2t+1)(2^t-1)}.$$

Then there exist polynomially constructible families of quantum codes with $n \to \infty$ and asymptotic parameters $(\delta, R_1(\delta))$, where $R_1(\delta) = 3t(\delta_t-\delta)$.

COROLLARY 12.3.17. *Using $t=3$ and the equation $R+9\delta = \frac{30}{49}$, the bound in Theorem 12.3.16 is defined in $\left(0,\frac{10}{147}\right)$ (which is greater than $\left(0,\frac{1}{18}\right)$ in Theorem 12.3.9) and exceeds the bound of Theorem 12.3.9 in the interval $\left(\frac{8}{147},\frac{1}{18}\right)$.*

SKETCH OF THE PROOF. The first part of the statement is clear from Theorem 12.3.16. For the second part, it is clear that, in the interval $(8/147, 1/18)$, the bound of Theorem 12.3.9 is defined by $R(\delta) = 2/3 - 10\delta$ and the bound in Theorem 12.3.16 is given by $R_1(\delta) = 30/49 - 9\delta$. These equations define two straight lines, and it is easy to check that $R_1(1/18) > R(1/18)$ and $R_1(8/147) = R(8/147)$. Thus, the second part of the statement is proved. □

An outline of the proof of Theorem 12.3.16 is as follows. We begin with an asymptotically good family of curves over $\mathbb{F}_{2^{2t}}$ satisfying the Drinfeld–Vlăduţ bound, admitting a polynomially constructible code, e.g., the $\mathcal{W}_1$ tower. From each of these curves, we construct a pair of algebraic geometry codes over $\mathbb{F}_{2^{2t}}$, which are then used to yield, via concatenation, a pair of binary codes C_1 and C_2 such that $C_1 \subset C_2$. The Calderbank–Shor–Steane construction (see Proposition 12.3.18 below) is then applied to these pairs of codes to yield the desired family of quantum codes with good asymptotic parameters.

Now recall the Calderbank–Shor–Steane construction. For a classical (maybe nonlinear) code C, by $\mathrm{wt}(C)$ we denote the minimum Hamming weight of all codewords of C.

PROPOSITION 12.3.18. *Let C_1 and C_2 be two binary codes with parameters, respectively, $[n,k_1]$ and $[n,k_2]$. Assume that $C_1 \subset C_2$. Then a quantum $[[n,k_2-k_1,d]]$ code can be constructed, where*

$$d = \min\{\mathrm{wt}(C_2 \setminus C_1), \mathrm{wt}(C_1^\perp \setminus C_2^\perp)\}. \qquad \square$$

Now let us study the dual of a concatenated code.

Let C be an $[s,t,d]$ code over $\mathbb{F}_{q^k}$ and, for $i=1,2,\ldots,s$, let $\pi_i\colon \mathbb{F}_{q^k} \to \mathbb{F}_q^{n_i}$ be an $\mathbb{F}_q$-linear injective map whose image $C_i = \mathrm{Im}(\pi_i)$ is an $[n_i,k,d_i]$ code over $\mathbb{F}_q$. The image $\pi(C)$ of the $\mathbb{F}_q$-linear injective map

$$\pi\colon C \to \mathbb{F}_q^{n_1+\ldots+n_s}, \quad (c_1,\ldots,c_s) \mapsto (\pi_1(c_1),\ldots,\pi_s(c_s)),$$

is a linear (concatenated) $[n_1+\ldots+n_s, tk]_q$ code. One can easily verify the following.

LEMMA 12.3.19. *Assume that the images $\mathrm{Im}(\pi_i)$, $1 \le i \le s$, are identical and have parameters $[n,k,w]$. Then $\pi(C)$ is a linear $[ns,tk]_q$ code with minimum distance at least wd.* $\square$

Next we describe the dual code of $\pi(C)$. Proposition 12.3.20 below states that it is a direct sum $D \oplus \pi'(C^\perp)$, where D and $\pi'(C^\perp)$ are two codes to be described.

Let $C^\perp \subset \mathbb{F}_q^{n_i}$ be the dual code of C_i, and let D be the direct sum $C_1^\perp \oplus \ldots \oplus C_s^\perp$. It is clear that $D \subset \mathbb{F}_q^{n_1} + \ldots + n_s$ is a linear

$$[n_1+\ldots+n_s, n_1+\ldots+n_s-sk]_q$$

code.

To describe $\pi'(C^\perp)$, we first define $\mathbb{F}_q$-linear injective maps $\pi_i'\colon \mathbb{F}_{q^k} \to \mathbb{F}_q^{n_i}$, for $1 \le i \le s$. Let $\{e_1,\ldots,e_k\}$ be an $\mathbb{F}_q$-basis of $\mathbb{F}_{q^k}$. A set $\{e_1',\ldots,e_k'\}$ in $\mathbb{F}_{q^k}$ is called the dual basis of $\{e_1,\ldots,e_k\}$ if we have $\mathrm{Tr}_{\mathbb{F}_{q^k}/\mathbb{F}_q}(e_ie_j') = \delta_{ij}$. We say that a basis is self-dual if it is its own dual.

Now we choose an $\mathbb{F}_q$-basis $\{e_1,\ldots,e_k\}$ in $\mathbb{F}_{q^k}$, and let $\{e_1',\ldots,e_k'\}$ be its dual basis. For each $1 \le i \le s$, we define the $\mathbb{F}_q$-linear injective map $\pi_i'\colon \mathbb{F}_{q^k} \to \mathbb{F}_q^{n_i}$ by first defining the images $\pi_i'(e_j')$ for $1 \le j \le k$ and then extending by $\mathbb{F}_q$-linearity. For each $1 \le j \le k$ and $1 \le \ell \le k$, we want $\pi_i'(e_j')$ to satisfy

$$\pi_i(e_\ell) \cdot \pi_i'(e_j') = \delta_{\ell j},$$

and since ℓ runs through all values from 1 to k, this gives a linearly independent system of k equations in n_i variables. The system admits a solution since $k \le n_i$, and we call this solution $\pi_i'(e_j')$. In general, this choice of $\pi_i'(e_j')$ is not unique, but is unique up to addition by a vector in $C_i^\perp$.

It is clear that π_i' is an $\mathbb{F}_q$-linear injective map whose image is a linear $[n_i,k]_q$ code.

Let $C^\perp \subset \mathbb{F}_{q^k}^s$ be the dual of C; this is a linear $[s,s-t]_q$ code. We define $\pi'(C^\perp)$ to be the concatenated code defined through $C^\perp$ and $\pi_1',\ldots,\pi_s'$, similar to the way that $\pi(C)$ was defined through C and $\pi_1, \ldots, \pi_s$; this is a linear $[n_1 + \ldots + n_s, (s-t)k]_q$ code.

PROPOSITION 12.3.20. *The dual code $\pi(C)^\perp$ of $\pi(C)$ is the direct sum $D \oplus \pi'(C^\perp)$.* □

This result is a refinement of Proposition 12.3.12.

COROLLARY 12.3.21. *Assume that $n_1 = \ldots = n_s = k$ and, for each $1 \leq i \leq s$, π_i is the expansion with respect to some $\mathbb{F}_q$-basis T of $\mathbb{F}_q$, so that $\pi(C)$ is the expansion of C with respect to the basis T. Then the dual of $\pi(C)$ is the expansion of $C^\perp$ with respect to the dual basis $T^\perp$.* □

PROOF OF THEOREM 12.3.16. Let $q = 2^{2t}$, and let $X_n/\mathbb{F}_q$ be an optimal tower of curves over $\mathbb{F}_q$; denote $g(n) = g(X_n)$.

For any n, choose a rational point $P_n \in X_n(\mathbb{F}_q)$ and two divisors,

$$D_1(n) = m_1(n)P_n, \qquad D_2(n) = m_2(n)P_n,$$

such that

$$2g(n) - 2 < m_2(n) < m_1(n) < N(n),$$

where $N(n) = |X_n(\mathbb{F}_q)|$. Let $\mathcal{P}_n := X_n(\mathbb{F}_q) \setminus \{P_n\}$, and let

$$T_i = (X_n, \mathcal{P}_n, D_i(n))_L, \quad i = 1, 2,$$

be the corresponding two algebraic geometry codes; we have $T_2 \subset T_1$. Clearly, T_i is a linear

$$[N(n), m_i(n) - g(n) + 1, N(n) - m_i(n)]_q$$

code. The dual of T_i is the algebraic geometry code $T_i^\perp = (X_n, \mathcal{P}_n, D_i(n))_\Omega$, which is a linear

$$[N(n), N(n) - m_i(n) + g(n) - 1, m_i(n) - 2g(n) + 2]_q$$

code.

For any $t \in \{3, 4, 5, \ldots\}$, let π_* be an $\mathbb{F}_2$-linear injective map from $\mathbb{F}_{2^{2t}}$ into $\mathbb{F}_2^{2t+1}$ whose image C_* is the trivial MDS $[2t+1, 2t, 2]_2$ code. Then, with $\pi_1 = \ldots = \pi_{N(n)} = \pi_*$, we have the concatenated codes $C_2 = \pi(T_2) \subset C_1 = \pi(T_1)$. From Lemma 12.3.19, C_i is a

$$[(2t+1)N(n), 2t(m_i(n) - g(n) + 1), 2(N(n) - m_i(n))]_2$$

code for $i = 1, 2$. The dual of C_i is $C_i^\perp = D \oplus (\pi'((X_n, \mathcal{P}_n, D_i(n))_\Omega))$ from Proposition 12.3.20.

Now we claim that for any vector $\boldsymbol{x} \in (C_2^\perp \setminus C_1^\perp)$ we have $\mathrm{wt}(\boldsymbol{x}) \geq m_2(n) - 2g(n) + 2$. Indeed, let

$$\boldsymbol{x} = \pi'(\boldsymbol{x}') + \boldsymbol{b} = \big(\pi'_*(x_1) + b_1, \ldots, \pi'_*(x_{N(n)}) + b_{N(n)}\big)$$

with

$$\boldsymbol{x}' = (x_1, \ldots, x_{N(n)}) \in T_2^\perp, \qquad \boldsymbol{b} = (b_1, \ldots, b_{N(n)}) \in D.$$

It is clear that $\boldsymbol{x}'$ is a nonzero vector since $\boldsymbol{x} \notin C_1^\perp$; thus, $\mathrm{wt}(\boldsymbol{x}') \geq m_2(n) - 2g(n) + 2$. At each position j with $x_j \neq 0$, we have $\pi'_*(x_j) + b_j \neq 0$, since $b_j \in C_*^\perp$ and $C_*^\perp \cap \mathrm{Im}(\pi'_*) = 0$. Thus, the claim is proved.

Using the Calderbank–Shor–Steane construction (see Proposition 12.3.18), we get, for each n, a

$$[[(2t+1)N(n), 2t(m_1(n)-m_2(n)), \geq \min\{2(N(n)-m_1(n)), m_2(n)-2g(n)+2\}]]$$

quantum code $\mathcal{A}_n$. Let $k(n) = m_1(n) - m_2(n)$; it is clear that $k(n)$ can take any integer value in $(0, N(n) - 2g(n)]$. For any such $k(n)$, taking

$$m_1(n) = \lfloor (2N(n) + 2g(n) + k(n) - 2)/3 \rfloor,$$

we get

$$2(N(n) - m_1(n)) \geq m_2(n) - 2g(n) + 2 \geq \frac{2}{3}(N(n) - 2g(n) - k(n) + 1).$$

Thus, $\mathcal{A}_n$ has parameters

$$[[(2t+1)N(n), 2tk(n), \geq 2/3(N(n) - 2g(n) - k(n) + 1)]].$$

For a fixed $\lambda \in (0, 1 - 2/(\sqrt{q} - 1))$, we let $k(n)/N(n) \to \lambda$ as $n \to \infty$. Put

$$R := \lim_{n\to\infty} \frac{2tk(n)}{(2t+1)N(n)} = \frac{2t}{2t+1}\lambda$$

and

$$\delta := \limsup_{n\to\infty} \frac{2(N(n) - 2g(n) - k(n) + 1)}{3(2t+1)N(n)} = \frac{2}{3(2t+1)}\left(1 - \frac{2}{2^t - 1} - \lambda\right).$$

Then (δ, R) satisfies

$$R_1(\delta) = \frac{2t}{2t+1}\left(1 - \frac{2}{2^t - 1}\right) - 3t\delta. \qquad \square$$

12.3.3. Nonbinary Case

As well as for usual error-correcting codes, algebraic geometry constructions are especially good for nonbinary quantum codes, which we are going to define.

Let q be a prime power, let $\mathbb{C}^q$ be a complex vector space of dimension q, and let $\{v_a : a \in \mathbb{F}_q\}$ be an orthonormal basis of $\mathbb{C}^q$ with respect to the Hermitian inner product $\langle \cdot | \cdot \rangle$. A q-ary quantum state $v \in \mathbb{C}^q \setminus \{0\}$ can be written as

$$v = \sum_{c \in \mathbb{F}_q} \alpha_c v_c, \quad \alpha_c \in \mathbb{C}.$$

For $n \geq 1$, the tensor product $(\mathbb{C}^q)^{\otimes n} = \mathbb{C}^{q^n}$ has a basis

$$\{v_{\boldsymbol{c}} = v_{c_1} \otimes \ldots \otimes v_{c_n} : \boldsymbol{c} = (c_1, \ldots, c_n) \in \mathbb{F}_q^n\};$$

thus, any q-ary n-system state $\boldsymbol{v} \in (\mathbb{C}^q)^{\otimes n} \setminus \{0\}$ can be written as

$$\boldsymbol{v} = \sum_{\boldsymbol{c} \in \mathbb{F}_q^n} \alpha_{\boldsymbol{c}} \boldsymbol{v}_{\boldsymbol{c}}, \quad \alpha_{\boldsymbol{c}} \in \mathbb{C}.$$

Let ζ be the pth primitive root of unity: $\zeta = e^{\frac{2\pi i}{p}} \in \mathbb{C}$; define two $p \times p$ matrices $T = (t_{ij})$ and $R = (r_{ij})$ with $t_{ij} = \delta_{i,j-1 \bmod p}$ and $r_{ij} = \zeta^i \delta_{ij}$.

Fix a basis $\{\alpha_1, \ldots, \alpha_m\}$ of $\mathbb{F}_q$ as a vector space over $\mathbb{F}_p$. For two elements $a = \sum_{i=1}^{m} a_i \alpha_i$ and $b = \sum_{i=1}^{m} b_i \alpha_i$ in $\mathbb{F}_q$, consider the error operator

$$T_a R_b = T^{a_1} R^{b_1} \otimes \ldots \otimes T^{a_m} R^{b_m}.$$

Then we have

$$(T_aR_b)(T_cR_d) = \zeta^{-\langle b,c\rangle} T_{a+c}R_{b+d},$$

where $\langle \cdot,\cdot \rangle$ stands for the usual inner product of $\mathbb{F}_q \cong \mathbb{F}_p^m$ over $\mathbb{F}_p$ with respect to the basis $\{\alpha_1, \dots, \alpha_m\}$, and

$$(T_aR_b)(T_cR_d) = \zeta^{\langle a,d\rangle - \langle b,c\rangle}(T_cR_d)(T_aR_b).$$

The action of T_aR_b on a basis element v_c of $\mathbb{C}^q$ with $c \in \mathbb{F}_q$ is as follows:

$$T_aR_bv_c = \zeta^{\langle b,c\rangle}(v_c - v_a).$$

Now for two vectors

$$\boldsymbol{a} = (a_1, \dots, a_n) \qquad \text{and} \qquad \boldsymbol{b} = (b_1, \dots, b_n)$$

in $\mathbb{F}_q^n$, we define the error operator on $(\mathbb{C}^q)^{\otimes n}$ by

$$E_{\boldsymbol{ab}} = T_{a_1}R_{b_1} \otimes \dots \otimes T_{a_n}R_{b_n}.$$

The set of all quantum errors

$$\mathcal{E} := \left\{\zeta^\lambda E_{\boldsymbol{ab}} : \ \boldsymbol{a}, \boldsymbol{b} \in \mathbb{F}_q^n, \ \lambda \in \mathbb{F}_p\right\}$$

forms a nonabelian group of order p^{2mn+1} under the group law

$$E_{\boldsymbol{ab}}E_{\boldsymbol{uv}} = \zeta^{-\langle \boldsymbol{b},\boldsymbol{u}\rangle} E_{\boldsymbol{a}+\boldsymbol{u}}E_{\boldsymbol{b}+\boldsymbol{v}},$$

where $\langle \boldsymbol{b}, \boldsymbol{u}\rangle = \sum_{i=1}^{n} \langle b_i, u_i\rangle$.

It is easy to verify that the center $C(\mathcal{E})$ of $\mathcal{E}$ is $\{\zeta^\lambda I : \lambda \in \mathbb{F}_p\}$.

The *quantum weight* of $E_{\boldsymbol{ab}}$ is defined by

$$w_Q(E_{\boldsymbol{ab}}) = \left|\left\{i : \ 1 \le i \le n, \ (a_i, b_i) \ne (0,0) \in \mathbb{F}_q^2\right\}\right|.$$

A complex subspace $Q \ne \{\boldsymbol{0}\}$ of $(\mathbb{C}^q)^{\otimes n} = \mathbb{C}^{q^n}$ is called a q-ary *quantum code of length* n.

We denote by $K = \dim_{\mathbb{C}} Q$ the dimension of Q over $\mathbb{C}$ and put $k = \log_q K$, $0 \le k \le n$.

The *minimum distance* of a quantum code is defined to be the largest positive integer d satisfying the following condition:

> For any $\boldsymbol{u}, \boldsymbol{v} \in Q \setminus \{0\}$ with $\langle \boldsymbol{u}|\boldsymbol{v}\rangle = 0$ and any $\boldsymbol{e} \in \mathcal{E}$ with quantum weight $\le d-1$, we have $\boldsymbol{uev} = 0$.

A q-ary quantum code Q with length n, dimension K, and minimum distance d is denoted by $((n,K,d))_q$ or by $[[n,k,d]]_q$ for $k = \log_q K$. A quantum $((n,K,d))_q$ code Q is said to be *pure* if $\boldsymbol{uev} = 0$ for all $\boldsymbol{e} \in \mathcal{E}$ with $1 \le w_Q(\boldsymbol{e}) \le d-1$ and $\boldsymbol{u}, \boldsymbol{v} \in Q$ (note that we do not require $\langle \boldsymbol{u}|\boldsymbol{v}\rangle = 0$ here).

One can generalize the result of Sec. 12.3.1 to the nonbinary case, obtaining the following results.

PROPOSITION 12.3.22. *Let C be a classical linear $[n,k]_q$ code. If the minimum distance $d^\perp$ of the dual code $C^\perp$ is at least d, then for each partition $\{1,2,\dots,n\} = A \cup B$ with $|A| = d-1$ and $|B| = n-d+1$, and any $\boldsymbol{c}_A \in \mathbb{F}_q^{d-1}$ and $\boldsymbol{v} \in \mathbb{F}_q^n$, we have*

$$\left|\left\{\boldsymbol{c}_B \in \mathbb{F}_q^{n-d+1} : \ (\boldsymbol{c}_A, \boldsymbol{c}_B) \in \boldsymbol{v} + C\right\}\right| = q^{k-d+1}. \qquad \square$$

PROPOSITION 12.3.23. *Let C be a q-ary classical linear code of length n, and let $V=\{\boldsymbol{v}_i\}_{i=0}^K$ be a set of K distinct vectors in $\mathbb{F}_q^n$. Put*

$$d_v := \min\{\mathrm{wt}(\boldsymbol{v}_i-\boldsymbol{v}_j+\boldsymbol{c}):\ 1\le i\ne j\le K \text{ and } \boldsymbol{c}\in C\}$$

and $d=\min\{d_v,d^\perp\}$, where wt *denotes the Hamming weight. If $d>0$, then there exists a q-ary $((n,K,d))$ quantum code.* □

Now, let $\alpha_q^Q(\delta)$ be the exact asymptotic bound for q-ary quantum codes. Its existence can be proved analogously to the case of usual error-correcting codes. Then the quantum Gilbert–Varshamov bound reads as

$$\alpha_q^Q(\delta)\ge 1-\delta\log_q(q+1)-H_q(\delta).$$

Applying the algebraic geometry construction, we obtain the following.

PROPOSITION 12.3.24. *If there is an algebraic curve $X/\mathbb{F}_q$ with at least $n+1$ ($\mathbb{F}_q$-rational) points and genus g, then for any $2g-2<m<n<2m-2g+2$ there exists a q-ary $((n,q^{n-2m+2g-2},m-2g+2))$ quantum code.* □

Thus, we have the following result.

THEOREM 12.3.25. *For a prime power q, one has*

$$\alpha_q^Q(\delta)\ge 1-2\delta-\frac{2}{A(q)}.$$ □

REMARK 12.3.26. The bound of Theorem 12.3.25 is better than the quantum Gilbert–Varshamov bound in an interval for some large q. In particular, when q is a square prime power greater than or equal to 19^2, the bound of Theorem 12.3.25 improves on the Gilbert–Varshamov bound in an interval around $(q^2-1)/(2q^2-1)$. For instance, for $q=19^2$, it improves on the Gilbert–Varshamov bound in the interval $(0.3515,0.4445)$.

Applying constructions from Sec. 4.6.2, one can also obtain the following refinement.

THEOREM 12.3.27. *For a prime power q, one has*

$$\alpha_q^Q(\delta)\ge 1-2\delta-\frac{2}{A(q)}+\log_q\left(1+\frac{1}{q^3}\right).$$ □

12.4. Niederreiter–Rosenbloom–Tsfasman Metric

In the present section we consider a generalization of the Hamming metric which was independently discovered by H. Niederreiter in the context of uniform distributions of points in Euclidean spaces and later by M. Rosenbloom and M. Tsfasman in the context of coding theory, especially that of algebraic geometry codes.

After defining the Niederreiter–Rosenbloom–Tsfasman (NRT) metric, we give some upper and lower bounds for it in Sec. 12.4.1 and then some examples of algebraic geometry origin and corresponding asymptotic bounds in Sec. 12.4.2. In Sec. 12.4.3 we explain very briefly the relation of the NRT metric with uniform sequences and nets used in quasi-Monte Carlo methods. Note, however, that while our discussion of algebraic geometry codes for the NRT metric is reasonably self-contained, the discussion of uniform sets is very rudimentary and cannot give an adequate picture of this rich domain.

In what follows, we fix a finite field $\mathbb{F}_q$.

12.4.1. NRT Metric Spaces: Definitions and Bounds

NRT metric. Consider the n-dimensional coordinate vector space $V \simeq \mathbb{F}_q^n$ over a finite field of q elements; thus, its nonempty subsets C are codes, and its $\mathbb{F}_q$-linear subspaces are linear codes. Assume that $n = ms$ for positive integers m and s. We will consider elements of V as matrices $\|v_{ij}\|$, $1 \le i \le m$, $1 \le j \le s$, with entries in $\mathbb{F}_q$. The (NRT) weight of such a matrix is defined as

$$\mathrm{wt}(\|v_{ij}\|) = \sum_{1\le j\le s} (m - \max(i \mid v_{kj} = 0 \text{ for any } k \le i)),$$

$0 \le \mathrm{wt}(v) \le n$; it determines a metric on V if we set $d(v,v') = \mathrm{wt}(v - v')$. Note that for $m = 1$ it is just the usual Hamming metric.

There exists a very convenient interpretation of the metric thus defined in terms of so-called ordered Hamming spaces, which we are going to explain briefly. This interpretation is useful for obtaining various bounds on the metric, as well as for clarifying its connections with uniform distribution problems. Below, we often denote the space V with the metric d by $Q^{s,m}$ and call d the NRT metric. One can reformulate its definition as follows.

Let $\boldsymbol{v} = \{v_{11}, \dots, v_{1s}; \dots; v_{m1}, \dots, v_{ms}\}$ be fixed, and let e_i for $i = 1, \dots, s$ be the number of s-blocks of $\boldsymbol{v}$ whose rightmost nonzero entry is in the ith position from the beginning of the block. The s-vector $\mathrm{sh}(\boldsymbol{v}) := (e_1, \dots, e_s) = e$ is called the *shape* of $\boldsymbol{v}$. For $\boldsymbol{v}, \boldsymbol{w} \in Q^{s,m}$, one writes $\boldsymbol{v} \sim_e \boldsymbol{w}$ if $\mathrm{sh}(\boldsymbol{v} - \boldsymbol{w}) = e$. A shape vector $e = (e_1, \dots, e_s)$ defines a partition of a number $N \le m$ into a sum of s parts. Let

$$\Delta_{s,m} = \left\{ e \in (\mathbb{Z}_+ \cup \{0\})^s : \sum_i e_i \le n \right\}$$

be the set of all such partitions. We define

$$|e| = \sum_i e_i, \qquad |e|' = \sum_i ie_i, \qquad e_0 = n - |e|,$$

and then for any $\boldsymbol{v} \in Q^{s,m}$ with $\mathrm{sh}(\boldsymbol{v}) = e$ we get

$$\mathrm{wt}(\boldsymbol{v}) = |e|'.$$

One can also call this weight the *ordered Hamming weight* and $Q^{s,m}$ the *ordered Hamming space*; we use the expressions "ordered Hamming space" and "NRT space" as synonyms.

An (ms, M, d) *ordered code* (NRT code) $C \subset Q^{s,m}$ is an arbitrary subset of M vectors in $Q^{s,m}$ such that the NRT distance between any two distinct vectors in C is no less than d; C is *linear* if it is an $\mathbb{F}_q$-subspace of $Q^{s,m}$. Note also that $Q^{s,m}$ and C can be defined over any q-ary alphabet; however, we assume here that q is a prime power and all our codes are linear.

One can give a model of an information transmission channel for which the NRT metric is a natural quality characteristic of a code. Namely, assume that a sender transmits messages, each being an s-tuple of m-tuples of q-ary symbols, through m parallel channels. There is an interfering noise of the following nature: sometimes several consecutive channels, starting from the last (mth) one, are occupied by a priority user, who sets them free as he needs them no more, in the order reverse to the order of occupation. The degree of interference is measured by the total number of q-ary symbols sent by the priority user "over" the main message; then the minimum distance of a code in the sense of the NRT metric characterizes its stability under such interference.

Bounds. Throughout this section, distance will always be understood in the sense of the NRT metric.

THEOREM 12.4.1 (Singleton bound). *Let $C \subset Q^{s,m}$ be a code of cardinality K with minimum distance $d = \min\limits_{\substack{v,v' \in C \\ v \neq v'}} \mathrm{wt}(v - v')$. Then*

$$K \leq q^{n-d+1}.$$

PROOF. In the matrix $\|v_{ij}\|$, mark the first $d-1$ positions in lexicographic order. Two elements of C never coincide in all other positions because of the minimum distance condition. Hence, there are at most q^{n-d+1} of them. □

THEOREM 12.4.2 (Plotkin bound). *Let $C \subset Q^{s,m}$ be a code of cardinality K with minimum distance d; let $\delta = \dfrac{d}{n}$ be the relative minimum distance. Then*

$$\delta(K-1) \leq \delta_{\mathrm{crit}} K,$$

where $\delta_{\mathrm{crit}} = 1 - \dfrac{1}{m} \sum\limits_{1 \leq i \leq m} q^{-i}$.

PROOF. On one hand,

$$\sum_{v,v' \in C} d(v, v') = \sum_{\substack{v,v' \in C \\ v \neq v'}} d(v, v') \geq dK(K-1). \tag{12.4.1}$$

On the other hand, let

$$d(v_j, v_j') = m - \max(i \mid v_{kj} = v_{kj}' \text{ for any } k \leq i);$$

then

$$\sum_{v,v' \in C} d(v, v') = \sum_{v,v' \in C} \sum_{1 \leq j \leq s} d(v_j, v_j') = \sum_{1 \leq j \leq s} \sum_{X,Y} d(X,Y) t_j(X) t_j(Y).$$

Here X and Y run over all columns of length m in the $\mathbb{F}_q$-alphabet, $t_j(X)$ being the number of code words whose jth column equals X. The last sum on the right is at most $s \max \sum_{X,Y} d(X,Y)t_X t_Y$, the maximum being taken over the strings $(t_X, X \in \mathbb{F}_q^m)$ with $t_X \geq 0$, $\sum_X t_X = K$. To complete the proof, we need the following estimate.

LEMMA 12.4.3. *The quadratic form $F = \sum_{X,Y} d(X,Y)t_X t_Y$ considered on the hyperplane $\sum_X t_X = K$ has its global maximum at the point T_0: $\{t_X = Kq^{-m}$ for every $X\}$.*

Assuming this to be true, the last sum is at most

$$sK^2q^{-2m}\sum_{X,Y} d(X,Y) = sK^2q^{-2m}q^m\sum_{Y} d(0,Y) = sK^2q^{-m}\left(mq^m - \sum_{1\leq i\leq m} q^{i-1}\right),$$

which, together with (12.4.1), proves the theorem after being divided by $n = ms$. □

PROOF OF LEMMA 12.4.3. The point T_0 is clearly a critical point of the form F on the hyperplane $\sum_X t_X = K$; therefore it suffices to check that the matrix $\|d(X,Y)\|$ is negative definite on the hyperplane $\sum_X t_X = 0$. Let $q = p^r$. Choose an $\mathbb{F}_p$-basis in $\mathbb{F}_q = \mathbb{F}_{p^r}$. Then the Abelian group G of columns of length m in the $\mathbb{F}_q$-alphabet is written as

$$G = \mathbb{F}_q^m = \bigoplus_{1\leq i\leq m}\bigoplus_{1\leq \ell\leq r} \mathbb{Z}/p\mathbb{Z},$$

and its group algebra over $\mathbb{C}$ is

$$\mathbb{C}[G] = \bigotimes_{1\leq i\leq m}\bigotimes_{1\leq \ell\leq r} \mathbb{C}[t]/(t^p - 1).$$

Since the distance $d(X,Y)$ is invariant under shifts by elements of G, the matrix $\|d(X,Y)\|$ is the matrix of multiplication by $h = \sum_Y \mathrm{wt}(Y)t_Y$ in the group algebra $\mathbb{C}[G]$ with respect to the basis $\{t^a\}$, where $a \in \mathbb{F}_q^m$, and $t^a = \prod_{\substack{1\leq i\leq m\\ 1\leq \ell\leq r}} t_{il}^{a_{il}}$, $0 \leq a_{il} \leq p-1$. Then we have

$$\begin{aligned} h &= \sum_Y \mathrm{wt}(Y)t_Y \\ &= \sum_{1\leq i\leq m} i\left(\prod_{1\leq \ell\leq r}\sum_{0\leq k\leq p-1} t^k_{m-i+1,\ell} - 1\right)\prod_{m-i+2\leq i'\leq m}\prod_{1\leq \ell\leq r}\sum_{0\leq k\leq p-1} t^k_{i'\ell} \\ &= -1 - \sum_{2\leq i\leq m}\prod_{i\leq i'\leq m}\prod_{1\leq \ell\leq r}\sum_{0\leq k\leq p-1} t^k_{i'\ell} + \prod_{1\leq i\leq m}\prod_{1\leq \ell\leq r}\sum_{0\leq k\leq p-1} t^k_{il}. \end{aligned}$$

On the other hand, the algebra $\mathbb{C}[G]$ is the sum of copies of $\mathbb{C}$, the projection on each copy being given by putting t_{il} equal to some pth roots of 1 (including 1 itself). Note that $\sum_k t_{il}^k$ equals zero for $t_{il} \neq 1$, and p for $t_{il} = 1$. The condition

$$0 = \sum_X t_X = \prod_{\substack{1\leq i\leq m\\ 1\leq \ell\leq r}}\sum_{0\leq k\leq p-1} t^k_{il}$$

defines a linear span of all copies of $\mathbb{C}$, except for the one given by $t_{il} = 1$ for every i and ℓ. The last term in the formula for h projects to zero on all such copies; hence, for $\lambda \geq 0$, the operator $h-\lambda$ is invertible on this subspace, whence the lemma follows. □

THEOREM 12.4.4 (Hamming bound). *Let $C \subset Q^{s,m}$ be a code of cardinality K with minimum distance d. Then*

$$K \operatorname{vol}\left(S_{\left\lfloor \frac{d-1}{2} \right\rfloor}\right) \leq q^n.$$ □

The volume $\operatorname{vol}(S_d)$ of a ball of radius d in $Q^{s,m} \simeq \mathbb{F}_q^{ms}$ is given by the following straightforward formula.

PROPOSITION 12.4.5. *We have*

$$\operatorname{vol}(S_d) = \sum_{\substack{k_1,\dots,k_m \geq 0 \\ \sum_{1\leq i\leq m} k_i \leq s \\ \sum_{1\leq i\leq m} ik_i \leq d}} \frac{s!}{\prod_{1\leq i\leq m} k_i! \left(s - \sum_{1\leq i\leq m} k_i\right)!} \left(\frac{q-1}{q}\right)^{\sum_{1\leq i\leq m} k_i} q^{\sum_{1\leq i\leq m} ik_i}.$$ □

There is also a kind of the Bassalygo–Elias type bound, which we state without a proof.

THEOREM 12.4.6. *Let C be an (ms, M, d) code and let $n = ms$. Then*

$$M \leq q^n \frac{d}{S_w\left(d - 2w + \frac{w^2}{n\delta_{\text{crit}}}\right)}$$

for any $w \leq n\delta_{\text{crit}}\left(1 - \sqrt{1 - d/(n\delta_{\text{crit}})}\right)$. □

As usual, we also have a random choice (Gilbert–Varshamov) existence bound.

THEOREM 12.4.7 (Gilbert–Varshamov bound). *For a space $Q^{s,m}$, $\dim Q^{s,m} = n = ms$, and for any d, $0 < d < n$, there exists a linear code $C \subset Q^{s,m}$ with minimum distance d whose dimension k satisfies*

$$k + \log_q \operatorname{vol}(S_{d-1}) \geq n.$$

PROOF. Since the distance $d(v, v')$ is invariant under simultaneous multiplication of v and v' by a nonzero element of $\mathbb{F}_q$, the required code can be constructed by the usual procedure, namely, choosing the elements of its basis one by one. □

12.4.2. Examples and Asymptotic Bounds

Reed–Solomon NRT codes. Let C be the space of polynomials over $\mathbb{F}_q$ of degree at most a. Let us define a linear map

$$C \xrightarrow{\text{Ev}} Q^{s,m} = \mathbb{F}_q^{mq}$$

mapping a polynomial to the matrix of values of its derivatives of orders from 0 to $(m-1)$ at all points of the affine line over $\mathbb{F}_q$. Since the total number of zeroes of a nonzero polynomial, counted with multiplicities, is at most a, we see that for $0 \leq a \leq mq$ the code $\text{Ev}(C)$ has NRT metric parameters

$$[mq, a+1, mq-a]_q$$

lying on the Singleton bound, while the length of the code $\text{Ev}(C)$ is m times larger than that of the usual Reed–Solomon code.

Adding to $\mathrm{Ev}(P)$ a column of m leading coefficients of P, we obtain a code with parameters

$$[m(q+1), a+1, m(q+1)-a]_q,$$

nondegenerate whenever $a \geq m-1$.

Note that using concatenation, i.e., mapping elements of $\mathbb{F}_{q^m}$ to $\mathbb{F}_q$-columns of length m, we can consider the usual Reed–Solomon code over $\mathbb{F}_{q^m}$ as an $\mathbb{F}_q$-code in the NRT metric, but the code that we get has parameters $[mq^m, a+1, q^m-a]_q$, lying well below the Singleton bound.

First-order Reed–Muller NRT codes. Let us choose a nonzero vector x_i in each subspace of dimension 1 in $\mathbb{F}_q^k$.

Let C be the space of linear forms in k variables over $\mathbb{F}_q$. Choose m linear operators

$$A_1 = \mathrm{Id}, \dots, A_m, \quad A_i \in \mathrm{Aut}_{\mathbb{F}_q} C,$$

such that all their linear combinations over $\mathbb{F}_q$ are invertible, let $s = \dfrac{q^k-1}{q-1}$, and let the map $C \overset{\mathrm{Ev}}{\to} Q^{s,m}$ be given by $\ell \mapsto (A_i\ell(x_j))$. This map is injective and provides a code that is a natural analogue of the usual first-order Reed–Muller code.

THEOREM 12.4.8. *The parameters of the code* $\mathrm{Ev}(C)$ *in the NRT metric are*

$$\left[m\frac{q^k-1}{q-1}, k, \frac{mq^k}{q-1}\delta_{\mathrm{crit}}\right]_q;$$

in particular, they meet the Plotkin bound.

PROOF. We have to compute the minimum distance; by definition, the weight of a codeword is

$$\mathrm{wt}(\ell) = \sum_{1\leq i\leq m} (m-i+1)\,\left|\{x \mid A_{i'}x = 0 \text{ for } i < i', \text{ and } A_i x \neq 0\}\right|.$$

The operators A_i are chosen so that the forms $(A_i\ell, 1 \leq i \leq m)$ are linearly independent for any form ℓ; hence, the last sum equals

$$\sum_{1\leq i\leq m} (m-i+1)q^{k-i} = q^{k-m} \sum_{1\leq i\leq m} iq^{i-1} = \frac{mq^k}{q-1}\left(1 - \frac{1}{m}\sum_{1\leq i\leq m} q^{-i}\right). \qquad \square$$

The required operators $A_1, \dots, A_m$ do exist for any $m \leq k$, since we can identify C with the extension $\mathbb{F}_{q^k}$ of the ground field $\mathbb{F}_q$ and take the operators of multiplication by the elements of a given basis of this extension containing 1.

Algebraic geometry NRT codes. Let $X/\mathbb{F}_q$ be a smooth, projective, absolutely irreducible algebraic curve of genus g and let $D \in \mathrm{Div}_{\mathbb{F}_q}(X)$ be a divisor of degree a. Put $C = L(D)$, and let $P_1 \dots P_s$ be different $\mathbb{F}_q$-points of X such that $P_j \notin \mathrm{Supp}\, D$ for every j. Fix some local parameters t_j at P_j defined over $\mathbb{F}_q$. For m and s with $ms > a$, set

$$Q^{s,m} = \bigoplus_{1\leq j\leq s} \mathscr{O}_{P_j}/(t_j)^m \simeq \mathbb{F}_q^{ms},$$

where $\mathscr{O}_{P_j}$ is the local ring of P_j. Then the map $C \overset{\mathrm{Ev}}{\to} Q^{s,m}$ sending a function from $L(D)$ to the strings of the first m coefficients of its Taylor series at the points P_j with respect to the chosen local parameters t_j is nondegenerate (since $a < ms$).

THEOREM 12.4.9. *The NRT metric parameters of the code* $\mathrm{Ev}(C)$ *satisfy the inequalities*

$$[ms, \geq a-g+1, \geq ms-a]_q;$$

in particular,

$$k+d \geq n-g+1. \qquad \square$$

The proof easily follows from the Riemann–Roch theorem.

Note that, as above for usual algebraic geometry codes, the conditions $P_j \notin \operatorname{Supp} D$ are in fact not necessary and can be treated using local uniformizers.

Asymptotic bounds. Let us look at possible limit values of the transmission rate $R = \frac{1}{n}\log_q |C|$ and the relative minimum distance $\delta = \frac{1}{n} d_{\min}(C)$ for families of codes of length n tending to infinity. First we note that if the divisor m of $n = ms$ tends to infinity, the situation becomes trivial.

PROPOSITION 12.4.10. *Fix* δ, $0 \leq \delta \leq 1$. *Consider spheres of radii* d_ℓ *in the space* $Q_\ell^{s,m} = \mathbb{F}_q^{m_\ell s_\ell}$. *Let* $\lim_{\ell\to\infty} m_\ell = \infty$, $\lim_{\ell\to\infty} \frac{d_\ell}{m_\ell s_\ell} = \delta$. *Then*

$$\lim_{\ell\to\infty} \frac{1}{m_\ell s_\ell} \log_q \operatorname{vol}(S_{d_\ell}) = \delta.$$

PROOF. There are $m+1$ possibilities for the number of the last zero position in a column of an $m \times s$ matrix, and the number of possible values of a matrix entry is at most q. Thus, for the number of matrices of weight at most d_ℓ we have

$$\operatorname{vol}(S_{d_\ell}) \leq (m+1)^{s_\ell} q^{d_\ell}.$$

On the other hand, in the formula

$$\operatorname{vol}(S_d) = \sum_{\substack{k_1,\dots,k_m \geq 0 \\ \sum_{1\leq i\leq m} k_i \leq s \\ \sum_{1\leq i\leq m} ik_i \leq d}} \frac{s!}{\prod_{1\leq i\leq m} k_i! \left(s - \sum_{1\leq i\leq m} k_i\right)!} \left(\frac{q-1}{q}\right)^{\sum_{1\leq i\leq m} k_i} q^{\sum_{1\leq i\leq m} ik_i},$$

the term corresponding to $\left\{k_{\min\left(m_\ell, \left\lfloor \frac{d_\ell}{s_\ell} \right\rfloor\right)} = s_\ell, \text{ the rest of } k_i = 0\right\}$ equals

$$\left(\frac{q-1}{q}\right)^{s_\ell} q^{s_\ell \min\left(m_\ell, \left\lfloor \frac{d_\ell}{s_\ell} \right\rfloor\right)}.$$

Therefore, we have

$$\begin{aligned} \frac{1}{m_\ell}\log_q \frac{q-1}{q} + \min\left(1, \frac{1}{m_\ell}\left\lfloor \frac{d_\ell}{s_\ell} \right\rfloor\right) &\leq \frac{1}{m_\ell s_\ell} \log_q \operatorname{vol}(S_{d_\ell}) \\ &\leq \frac{1}{m_\ell}\log_q(m_\ell+1) + \frac{d_\ell}{m_\ell s_\ell}, \end{aligned}$$

and for $\ell \to \infty$ the result follows. $\square$

COROLLARY 12.4.11. *As* $m \to \infty$, *the Gilbert–Varshamov lower bound tends to the Singleton upper bound.* $\square$

Now assume that m is fixed.

THEOREM 12.4.12 (asymptotic Plotkin bound). *Let C_ℓ be a sequence of codes of length ms_ℓ, $s_\ell \to \infty$, such that a limit $\lim\limits_{\ell\to\infty} \frac{1}{ms_\ell}(k(C_\ell), d_{\min}(C_\ell)) = (R,\delta)$ exists. Then*

$$R + \frac{\delta}{\delta_{\text{crit}}} \le 1.$$

PROOF. Theorem 12.4.2 gives $\delta \le \delta_{\text{crit}}$. The rest follows from the "spoiling lemma" (Lemma 1.1.39). □

THEOREM 12.4.13. (i) *For $0<\delta<1$, the equation*

$$\delta = \frac{\frac{q-1}{q}\sum\limits_{1\le i\le m} iz^i}{m\left(1+\frac{q-1}{q}\sum\limits_{1\le i\le m} z^i\right)} \tag{12.4.2}$$

has a unique positive root z_0.

(ii) *Fix δ, $0<\delta<1$. Consider spheres of radii d_s in $Q^{s,m} = \mathbb{F}_q^{ms}$. Let $\lim\limits_{s\to\infty} \frac{d_s}{ms} = \delta$. Then*

$$\lim_{s\to\infty}\frac{1}{ms}\log_q \operatorname{vol}(S_{d_s}) = \begin{cases} 1 & \text{for } \delta \ge \delta_{\text{crit}},\\ H_{q,m}(\delta) & \text{for } 0\le\delta\le\delta_{\text{crit}},\\ 0 & \text{for } \delta = 0,\end{cases}$$

where $H_{q,m}(\delta) = \delta - \delta\log_q z_0 + \frac{1}{m}\log_q\left(1+\frac{q-1}{q}\sum\limits_{1\le i\le m} z_0^i\right)$, z_0 being the unique positive root of equation (12.4.2).

REMARK 12.4.14. The function $H_{q,m}(\delta)$ is called the (q-ary) m-entropy. For $m=1$, we have $z_0 = \frac{q}{q-1}\frac{\delta}{1-\delta}$, $\delta_{\text{crit}} = 1-\frac{1}{q}$, and

$$H_{q,1}(\delta) = \delta\log_q(q-1) - \delta\log_q\delta - (1-\delta)\log_q(1-\delta) = H_q(\delta)$$

is the usual q-ary entropy.

PROOF OF THEOREM 12.4.13. (i) The right-hand side of (12.4.2) tends to 0 for $z\to 0$, and to 1 for $z\to\infty$, its derivative being

$$\frac{q-1}{qm\left(1+\frac{q-1}{q}\sum\limits_{1\le i\le m} z^i\right)^2}\left(\sum_{1\le j\le m} j^2 z^{j-1} + \frac{q-1}{q}\sum_{\substack{1\le i\le m\\1\le j\le m}} (j^2-ij)z^{i+j-1}\right),$$

and has no zeroes for $z>0$, since the numerator is in fact a polynomial with positive coefficients.

(ii) For $\delta>0$, to find the summand in the equality of Proposition 12.4.5 determining the asymptotic behaviour as $s\to\infty$, we have to maximize

$$\Phi(\chi) = \sum_{1\le i\le m}\chi_i \ln\frac{q-1}{q} + \sum_{1\le i\le m} i\chi_i\ln q - \sum_{1\le i\le m}\chi_i\ln\chi_i - \left(1-\sum_{1\le i\le m}\chi_i\right)\ln\left(1-\sum_{1\le i\le m}\chi_i\right)$$

under the conditions

$$\left\{ \sum_{1\le i\le m} \chi_i \le 1;\ \chi_i \ge 0 \text{ for any } i;\ \sum_{1\le i\le m} i\chi_i \le m\delta \right\};$$

in asymptotics, this function equals $\frac{1}{s}\ln\operatorname{vol}(S_{d_s})$, as we see by setting $\chi_i = \frac{1}{s}k_i$ and using the Stirling formula. Denote the domain in $\mathbb{R}^m$ given by the latter conditions by $\overline{\Delta}$; its interior by Δ; the domain given by the first two conditions and its interior, respectively, by $\overline{\Gamma}$ and Γ; and the hyperplane $\sum\limits_{1\le i\le m} i\chi_i = m\delta$ by H. The function $\Phi(\chi)$ bounded in Γ and continuous on its boundary has in Γ a unique critical point X with coordinates $\chi_i(X) = (q-1)q^{i-m-1}$. One checks that the matrix of second derivatives of $\Phi(\chi)$ is

$$\|B_{ij}\| = \frac{1}{1-\sum\limits_{1\le k\le m}\chi_k}\|A_{ij}\|,$$

where

$$A_{ij} = \begin{cases} -1 & \text{for } i\ne j,\\ -1-\dfrac{1-\sum\limits_{1\le k\le m}\chi_k}{\chi_i} & \text{for } i=j, \end{cases}$$

and is negative definite in the whole of Γ. Hence, the critical point X is nondegenerate and is the point of global maximum for $\Phi(\chi)$ in Γ. For $\delta \ge \delta_{\text{crit}}$ it belongs to $\overline{\Delta}$, and the volume of the sphere S_{d_s} for $s\to\infty$ asymptotically equals the volume of the whole space $Q^{s,m}$. For $\delta<\delta_{\text{crit}}$, the point X is out of $\overline{\Delta}$. The critical point of $\Phi(\chi)$ on the hyperplane H must have the coordinates

$$\chi_i = \frac{\frac{q-1}{q}z_0^i}{1+\frac{q-1}{q}\sum\limits_{1\le k\le m} z_0^k},$$

where z_0 is the positive root of (12.4.2).

From the first part of the theorem, we see that this point Y exists and is unique. Since the matrix of second derivatives of $\Phi(\chi)$ is globally negative definite, Y is the global maximum point of the restriction of $\Phi(\chi)$ to $H\cap\overline{\Gamma}$. Since the level surfaces of $\Phi(\chi)$ are connected in $\overline{\Gamma}$, and the point X lies in its interior, while for $\delta<\delta_{\text{crit}}$ the point X and the domain $\overline{\Delta}$ lie on different sides of the hyperplane H, the point of global maximum of $\Phi(\chi)$ in $\overline{\Delta}$ for $\delta<\delta_{\text{crit}}$ lies on H and hence coincides with the point Y. Since $S_{d_s}\subset S_{d'_s}$ for $d_s\le d'_s$ and $\lim\limits_{\delta\to 0} H_{q,m}(\delta)=0$, we see that the statement of the theorem is also valid for $\delta=0$. □

Using Theorems 12.4.4 and 12.4.13, one can prove the following result.

THEOREM 12.4.15 (asymptotic Hamming bound). *Let C_ℓ be a sequence of codes as in Theorem 12.4.12. Then*

$$R + H_{q,m}\left(\frac{\delta}{2}\right) \le 1.$$ □

The asymptotic version of Theorem 12.4.6 is as follows.

THEOREM 12.4.16 (asymptotic Bassalygo–Elias bound). *For $0 \le \delta \le \delta_{\text{crit}}$, the asymptotic rate and distance of any sequence of codes satisfy*

$$R \le 1 - H_{q,m}\big(\delta_{\text{crit}}(1 - \sqrt{1-\delta/\delta_{\text{crit}}})\big). \qquad \square$$

This bound is better than the Hamming bound for all $\delta \in (0, \delta_{\text{crit}}]$. It is also often better than the Plotkin bound. For instance, for $q = 2$ and $m = 2$ this bound is better than the Plotkin bound for all $\delta \in (0, \delta_{\text{crit}})$. For larger q and m the improvement is attained only for low values of δ, since the right-hand side becomes $\cap$-convex close to δ_{crit}. For instance, for $q = 3$ and $m = 4$ this range is $(0, 0.54)$, etc.

Applying Theorems 12.4.7 and 12.4.13, we obtain the following result.

THEOREM 12.4.17 (asymptotic Gilbert–Varshamov bound). *For any $\delta \in (0,1)$ there exists a sequence of linear codes C_ℓ of length ms_ℓ, $s_\ell \to \infty$, such that*

$$\lim_{\ell\to\infty} \frac{1}{ms_\ell}(k(C_\ell), d_{\min}(C_\ell)) = (R, \delta)$$

and

$$R \ge \begin{cases} 0 & \text{for } \delta \ge \delta_{\text{crit}}, \\ 1 - H_{q,m}(\delta) & \text{for } \delta < \delta_{\text{crit}}. \end{cases} \qquad \square$$

Theorem 12.4.9 easily implies the following.

THEOREM 12.4.18 (asymptotic algebraic geometry bound). *For any $\delta \in (0,1)$ there exists a sequence of linear codes C_ℓ of length ms_ℓ, $s_\ell \to \infty$, such that*

$$\lim_{\ell\to\infty} \frac{1}{ms_\ell}(k(C_\ell), d_{\min}(C_\ell)) = (R, \delta)$$

and

$$R \ge 1 - \delta - \frac{1}{mA(q)},$$

where $A(q) = \limsup(|X(\mathbb{F}_q)|/g(X))$, the maximum being over all smooth absolutely irreducible curves $X/\mathbb{F}_q$. $\square$

THEOREM 12.4.19. *The asymptotic Gilbert–Varshamov bound intersects the asymptotic algebraic geometry bound if and only if the condition*

$$\log_q\left(1 + m\frac{q-1}{q}\right) \ge \frac{1}{A(q)}$$

is satisfied.

PROOF. The derivative of $H_{q,m}(\delta) - \delta$ equals $-\log_q z_0$, where z_0 is given by (12.4.2). Since $z_0(\delta)$ is increasing monotone, this function (just as the m-entropy itself) is convex and its maximum is attained if $H'_{q,m}(\delta) = 1$. In this case,

$$z_0 = 1, \qquad \delta = \frac{(m+1)(q-1)}{2(mq+q-m)}, \qquad H_{q,m}(\delta) - \delta = \frac{1}{m}\log_q\left(1 + m\frac{q-1}{q}\right),$$

and the result follows from Theorems 12.4.17 and 12.4.18. $\square$

COROLLARY 12.4.20. (i) *The algebraic geometry bound intersects the Gilbert–Varshamov bound*

- *for $m = 1$ and $q = p^{2\ell} \ge 49$ (the Hamming metric case),*
- *for $m = 2$ and $q = p^{2\ell} \ge 16$,*

- *for* $m=3$ *and* $q=p^{2\ell}\geq 9$,
- *for* $m=4$ *and* $q=p^{2\ell}\geq 4$

(*for* $m=q=4$ *the bounds are tangent at the point* $\delta=15/32$).

(ii) *There exists* m_0 *such that, for* $m\geq m_0$, *the algebraic geometry bound intersects the Gilbert–Varshamov bound for any* q (*not necessarily of the form* $q=p^{2\ell}$).

PROOF. (i) For $q=p^{2\ell}$, we have $A(q)=\sqrt{q}-1$ (Theorem 3.2.3).

(ii) There exists an absolute constant c such that $A(q)\geq c\log_2 q$ for any q (see Theorem 3.2.4). Then it suffices to put $m_0=2^{c^{-1}+1}-2$. □

12.4.3. Uniform Nets and Sequences

In fact, the NRT metric was initially introduced by H. Niederreiter to construct point sets suitable for numerical (quasi-Monte Carlo) integration.

Let us briefly discuss some appropriate notions. Let $I^s=[0,1]^s$ be a unit cube, let f be a continuous function of bounded variation, and let $\mathcal{N}=\{x_1,\dots,x_N\}$ be a set of N points in I^s, called a *net*. The quasi-Monte Carlo method uses the following approximation in the context of numerical integration over I^s:

$$\int_{I^s} f(x)\,dx \approx \frac{1}{N}\sum_{i=1}^{N} f(x_i).$$

Its error can be bounded as follows (the *Koksma–Hlawka inequality*):

$$\left|\int_{I^s} f(x)\,dx - \frac{1}{N}\sum_{x\in\mathcal{N}} f(x)\right| \leq V(f)D(\mathcal{N}),$$

where $V(f)$ is the total variation of f on I^s and $D(\mathcal{N})$ is the (star) *discrepancy factor* of the net, which we will simply call the *discrepancy* in what follows. The parameter $D(\mathcal{N})$ measures the deviation of the net from a uniformly distributed set of points. We do not give the general definition of this quantity, but we will give some bounds for it in the situations considered below.

One can find nets $\mathcal{N}$ satisfying

$$D(\mathcal{N})\leq \frac{C(\log N)^{s-1}}{N},\qquad N\geq 2,$$

for an absolute constant C; it is believed (and known for $s=1$ and $s=2$) that this bound is the best possible up to a constant. A point set $\mathcal{N}$ achieving this bound is called a *low-discrepancy point set*.

Another closely related notion is that of a *low-discrepancy sequence*, which is an infinite sequence $\mathcal{S}$ of points in I^s such that for all $N\geq 2$ the discrepancy $D(\mathcal{S}_N)$ of the first N elements of $\mathcal{S}$ satisfies

$$D(\mathcal{S}_N)\leq C\frac{(\log N)^s}{N}$$

for an absolute constant C.

Powerful constructions of low-discrepancy point sets and sequences are based on the theory of (t,m,s)-nets and (t,s)-sequences, and we briefly discuss these notions.

(t,m,s)-nets. The first basic concept in the theory is that of a (t,m,s)-net, which is a special case of the notion of a *uniform point set*; note that a point set here is actually a multiset. Informally speaking, a (t,m,s)-net is a set of a special form with a small value of $D(\mathcal{N})$.

More formally, let $(X,\mathcal{B},\mu)$ be an arbitrary probability space and let $\mathcal{M}$ be a nonempty subset of $\mathcal{B}$. A point set $P=\{\boldsymbol{x}_1,\dots,\boldsymbol{x}_N\}$ of $N\geq 1$ elements of X is said to be $(\mathcal{M},\mu)$-*uniform* if

$$\frac{1}{N}\sum_{n=1}^{N}\chi_M(\boldsymbol{x}_n)=\mu(M)\quad\text{for all}\quad M\in\mathcal{M},$$

where χ_M denotes the characteristic function of M.

Let $s\geq 1$, $b\geq 2$, and $0\leq t\leq m$ be integers, and let λ_s be the restriction of the s-dimensional Lebesgue measure to I^s. Fix $b\geq 2$, and let $\mathcal{J}_{b,m,t}^{(s)}$ be the collection of all subintervals J of I^s of the form

$$J=\prod_{i=1}^{s}\big[a_ib^{-d_i},(a_i+1)b^{-d_i}\big)$$

with integers $d_i\geq 0$ and $0\leq a_i<b^{d_i}$ for $1\leq i\leq s$ and with $\lambda_s(J)=b^{t-m}$. Then a $(\mathcal{J}_{b,m,t}^{(s)},\lambda_s)$-uniform point set consisting of b^m points in I^s is called a (t,m,s)-*net in base* b.

It is important to note that the smaller the value of t for given b, m, and s, the larger the family $\mathcal{J}_{b,m,t}^{(s)}$ of intervals, and so the stronger the uniform point set property. In particular, a (t,m,s)-net in base b is automatically a (u,m,s)-net in base b for any integer u with $t\leq u\leq m$. Thus, the primary interest is in (t,m,s)-nets in base b with the value of t as small as possible. The number t is often called the *quality parameter* of a (t,m,s)-net in base b.

The main result relating (t,m,s)-nets and low-discrepancy point sets is as follows.

THEOREM 12.4.21. *The discrepancy $D(\mathcal{N})$ of a (t,m,s)-net $\mathcal{N}$ in base b satisfies*

$$D(\mathcal{N})\leq\frac{Cb^t(\log N)^{s-1}}{N},\quad\textit{for all } N\geq 2.\qquad\square$$

(t,s)-sequences. The second basic notion is that of a sequence analogue of a uniform set. A (t,s)-sequence in base b is an infinite sequence in which certain b-adic segments form (t,m,s)-nets with t independent of m. More precisely, given a real number $x\in[0,1]$, let

$$x=\sum_{j=1}^{\infty}y_jb^{-j}\quad\text{with all } y_j\in\mathbb{Z}/b\mathbb{Z}:=\{0,1,\dots,b-1\}$$

be a b-adic expansion of x, where the case $y_j=b-1$ for all but finitely many j is allowed. For an integer $m\geq 1$, we define the truncation

$$[x]_{b,m}=\sum_{j=1}^{m}y_jb^{-j}.$$

Note that this truncation operates on the *expansion* of x and not on x itself, since it may yield different results depending on which b-adic expansion of x is used.

If $\boldsymbol{x} = (x^{(1)}, \dots, x^{(s)}) \in I^s$ and the $x^{(i)}$, $1 \le i \le s$, are given by prescribed b-adic expansions, then we define

$$[\boldsymbol{x}]_{b,m} = \big([x^{(1)}]_{b,m}, \dots, [x^{(s)}]_{b,m}\big).$$

Let $s \ge 1$, $b \ge 2$, and $t \ge 0$ be integers. A sequence $\boldsymbol{x}_0, \boldsymbol{x}_1, \dots$ of points in I^s is a *(t,s)-sequence in base b* if for all integers $k \ge 0$ and $m > t$ the points $[\boldsymbol{x}_n]_{b,m}$ with

$$kb^m \le n < (k+1)b^m$$

form a (t,m,s)-net in base b. Here the coordinates of all points $\boldsymbol{x}_n$, $n = 0, 1, \dots$, are given by prescribed b-adic expansions.

As in the case of (t,m,s)-nets, smaller values of t mean stronger uniformity properties of a (t,s)-sequence in base b. We call t the *quality parameter* of a (t,s)-sequence in base b.

In fact, we have the following result relating (t,s)-sequences and low-discrepancy sequences.

THEOREM 12.4.22. *For a (t,s)-sequence $\mathcal{S}$ in base b we have*

$$D(\mathcal{S}_N) \le \frac{Cb^t (\log N)^s}{N}, \quad \textit{for all } N \ge 2. \qquad \square$$

Digital method and generating matrices. One of the most powerful techniques for constructing (t,m,s)-nets and (t,s)-sequences is the *digital method*, which we are going to describe using the notion of generating matrices.

Let us fix a prime power q and integers $m \ge 1$ and $s \ge 1$, and let a map $\varphi_m \colon \mathbb{F}_q^m \to [0,1]$ be constructed by selecting a bijection $\eta \colon \mathbb{F}_q \to \mathbb{Z}/q\mathbb{Z}$ and putting

$$\varphi_m(\boldsymbol{h}) = \sum_{j=1}^{m} \eta(h_j) q^{-j}$$

for any column vector $\boldsymbol{h} = (h_1, \dots, h_m)^T$ in $\mathbb{F}_q^m$. Furthermore, we choose $m \times m$ matrices $C^{(1)}, \dots, C^{(s)}$ over $\mathbb{F}_q$, the so-called *generating matrices*. Finally, we define the point set consisting of the q^m points

$$\big(\varphi_m(C^{(1)}\boldsymbol{h}), \dots, \varphi_m(C^{(s)}\boldsymbol{h})\big) \in I^s$$

with the column vector $\boldsymbol{h}$ running through $\mathbb{F}_q^m$. If such a point set forms a (t,m,s)-net in base q, then it is called a *digital (t,m,s)-net over $\mathbb{F}_q$*.

Such a point set always forms a digital (t,m,s)-net over $\mathbb{F}_q$ for some value of t, since for instance $t = m$ satisfies the definition of a digital (t,m,s)-net. The question is how to choose the parameters in the above construction so as to obtain nontrivial values $t < m$ for the quality parameter. Note that the quality parameter of a digital (t,m,s)-net over $\mathbb{F}_q$ depends only on the generating matrices $C^{(1)}, \dots, C^{(s)}$.

Indeed, let k, m, and s be positive integers and let d be an integer with $0 \le d \le \min(k, ms)$. The system $\{\boldsymbol{c}_j^{(i)} \in \mathbb{F}_q^k : 1 \le j \le m, 1 \le i \le s\}$ of vectors is called a *(d,k,m,s)-system over $\mathbb{F}_q$* if for any integers $d_1, \dots, d_s$ with $0 \le d_i \le m$ for $1 \le i \le s$ and $\sum_{i=1}^{s} d_i = d$ the system $\{\boldsymbol{c}_j^{(i)} \in \mathbb{F}_q^k : 1 \le j \le d_i, 1 \le i \le s\}$ is linearly independent over $\mathbb{F}_q$ (the empty system is considered linearly independent). A (d,m,m,s)-system over $\mathbb{F}_q$ is also called a *(d,m,s)-system over $\mathbb{F}_q$*.

Let $C^{(1)},\dots,C^{(s)}$ be the generating matrices as above. For $1 \le i \le s$, let $\boldsymbol{c}_j^{(i)} \in \mathbb{F}_q^m$, $1 \le j \le m$, be the row vectors of $C^{(i)}$. Then we have the following result.

THEOREM 12.4.23. *The point set* $\left(\varphi_m(C^{(1)}\boldsymbol{h}),\dots,\varphi_m(C^{(s)}\boldsymbol{h})\right)$ *is a digital* (t,m,s)*-net over* $\mathbb{F}_q$ *if and only if the system* $\{c_j^{(i)} \in \mathbb{F}_q^m : 1 \le j \le m, 1 \le i \le s\}$ *of row vectors of the generating matrices is a* (d,m,s)*-system over* $\mathbb{F}_q$ *with* $d = m-t$. □

The digital method for the construction of sequences is as follows. Fix a prime power q and a dimension $s \ge 1$ and define a map $\varphi_\infty\colon \mathbb{F}_q^\infty \to [0,1]$ by selecting a bijection $\eta\colon \mathbb{F}_q \to \mathbb{Z}/q\mathbb{Z}$ and putting

$$\varphi_\infty(\boldsymbol{h}) = \sum_{j=1}^{\infty} \eta(h_j) q^{-j}$$

for any infinite-dimensional column vector $\boldsymbol{h} = (h_1, h_2, \dots)^T$ in $\mathbb{F}_q^\infty$. If one chooses *generating matrices* $C^{(1)},\dots,C^{(s)}$ over $\mathbb{F}_q$ with infinitely many rows and columns, then, for $n = 0,1,\dots,$ let

$$n = \sum_{\ell=0}^{\infty} a_\ell(n) q^\ell$$

be the digit expansion of n in base q, where $a_\ell(n) \in \mathbb{Z}/q\mathbb{Z}$ for $\ell \ge 0$ and $a_\ell(n) = 0$ for all sufficiently large ℓ. Choose a bijection $\psi\colon \mathbb{Z}/q\mathbb{Z} \to \mathbb{F}_q$ with $\psi(0) = 0$ and associate with n the infinite-dimensional column vector

$$\boldsymbol{n} = (\psi(a_0(n)), \psi(a_1(n)), \dots)^T \in \mathbb{F}_q^\infty.$$

Now we define the sequence $\boldsymbol{x}_0, \boldsymbol{x}_1, \dots$ of points in I^s by

$$\boldsymbol{x}_n = \left(\varphi_\infty(C^{(1)}\boldsymbol{n}),\dots,\varphi_\infty(C^{(s)}\boldsymbol{n})\right) \quad \text{for} \quad n = 0,1,\dots.$$

Note that the matrix-vector products $C^{(i)}\boldsymbol{n}$ are well defined, since n has only finitely many nonzero entries.

If the sequence $\boldsymbol{x}_0, \boldsymbol{x}_1, \dots$ forms a (t,s)-sequence in base q, then it is called a *digital* (t,s)*-sequence over* $\mathbb{F}_q$.

Connection with codes.

THEOREM 12.4.24. *Given a linear* $[n,k,d]$ *code over* $\mathbb{F}_q$ *with* $d \ge 3$*, we can construct a digital* $(n-k-d+1, n-k, s)$*-net over* $\mathbb{F}_q$ *with* $s = \lfloor (n-1)/h \rfloor$ *if* $d = 2h+2$ *is even and* $s = \lfloor n/h \rfloor$ *if* $d = 2h+1$ *is odd.* □

Another type of construction principle uses a linear code and a digital net to get a new digital net. Let us give two constructions in this vein.

THEOREM 12.4.25. *Assume that there exists a linear* $[n, \ge n-h, \ge d+1]$ *code over* $\mathbb{F}_q$ *with integers* $1 \le h \le n$ *and* $d \ge 0$*. Then from a digital* (t,m,s)*-net over* $\mathbb{F}_{q^h}$ *one can construct a digital* (u, hm, ns)*-net over* $\mathbb{F}_q$ *with*

$$u = hm - \min(d, m-t). \qquad \square$$

THEOREM 12.4.26. *Let* m *be a positive integer and assume that there exists a linear* $[n, \ge n-h, \ge d+1]$ *code over* $\mathbb{F}_{q^m}$ *with integers* $1 \le h \le n$ *and* $d \ge 0$*.*

Then from a digital (t,m,s)-net over $\mathbb{F}_q$ one can construct a digital (u,hm,ns)-net over $\mathbb{F}_q$ with

$$u = hm - \min(d, m-t).$$ □

Ordered orthogonal arrays and NRT codes. One can also give another notion essentially equivalent to that of a (t,m,s)-net, namely that of an ordered orthogonal array (OOA); these are objects dual to NRT codes.

To define ordered orthogonal arrays, let us call a subset of coordinates $\mathcal{I} \subset \{1,\dots,ms\}$ left-adjusted if with any coordinate $im+j$, $0 \le i \le s-1$, $1 \le j \le m$, in the ith block it also contains all the coordinates $(im+1,\dots,im+j-1)$ of the same block. A subset $A \subset Q^{m,s}$ of cardinality $|A| = M$ is called a (t,s,m,q) *ordered orthogonal array* (OOA) of strength t if its projection on any left-adjusted set of t coordinates contains all the q^t rows the same number, say θ, of times. The parameter θ is called the *index* of A, and thus $M = \theta q^t$; if C is a linear $[ms,k,d]$ code, then the code orthogonal to it with respect to the usual inner product is a $(d-1,s,m,q)$ linear OOA of index $\theta = q^{ms-k-d+1}$. If Q is equipped with the structure of an additive group, then one can construct additive OOAs. If Q is a finite field, we can consider linear OOAs.

The following theorem shows that OOAs and (t,m,s)-nets are essentially equivalent.

THEOREM 12.4.27. *There exists a (t,m,s)-net if and only if there exists an $(m-t,s,m-t,q)$ OOA of index q^t and size $M = q^m$.* □

One can use (error-correcting) codes to construct OOAs, but in general, bounds on the former do not imply lower bounds on the latter. However, there are some special cases where these two types of objects are equivalent. For example, if C is an $[ms,k,d]$ linear code over $\mathbb{F}_q$, then the code

$$C' := \left\{ y \in \mathbb{F}_q^{ms} : \sum_{i=1}^{ms} x_i y_i = 0 \text{ for all } x \in C \right\}$$

is a $(d-1,s,m,q)$ linear OOA.

One can use codes in the NRT metric to construct OOAs and nets; this leads to the following result, which can also be deduced using the code $\mathcal{N}^\perp$ dual to $\mathcal{N}$.

THEOREM 12.4.28. *Let $m \ge 1$ and $s \ge 2$ be integers. Then from any $\mathbb{F}_q$-linear subspace $\mathcal{N}$ of $\mathbb{F}_q^{ms}$ with $\dim(\mathcal{N}) \ge ms-m$ we can construct a digital (t,m,s)-net over $\mathbb{F}_q$ with $t = m+1-\delta_m(\mathcal{N})$, where $\delta_m(\mathcal{N})$ is the NRT minimum distance of $\mathcal{N}$.* □

Connection with algebraic geometry codes. Theorems 12.4.24–12.4.26 and 12.4.28 have many applications, but here we only show how they work for algebraic geometry codes in the NRT metric.

For any prime power q and any given integer $s \ge 1$, define the quantity

$$V_q(s) = \min\{g : N_q(g) \ge s+1\}.$$

Then, applying to digital (t,m,s)-nets an algebraic geometry construction essentially very close to that of Theorem 12.4.9 and using Theorem 12.4.23, we can obtain the following result.

THEOREM 12.4.29. *For any prime power q, any $s \ge 1$, and any $m \ge V_q(s)$ there exists a (digital) $(V_q(s),m,s \ge 1)$-net.* □

Applying Theorem 12.4.29 to algebraic curves with many points gives many nets with very good parameters, both in the asymptotic and finite-length settings.

This result can be transposed for (t,s)-sequences; namely, we have the following.

THEOREM 12.4.30. *Let X be a curve over $\mathbb{F}_q$ with $N_q(X)\geq s+1$ and assume that there exists a positive divisor D on X with $\deg(D)=g(X)$ and $\ell(D)=1$. Then the construction below yields a digital (t,s)-sequence constructed over $\mathbb{F}_q$ with $t=g(X)$.* □

Construction. Note that a divisor $D\in \mathrm{Div}_{\mathbb{F}_q}(X)$ satisfying the conditions in Theorem 12.4.30 exists if $g(X)=0$, or $q=2$ and $g(X)\geq 4$, or $q\geq 4$ and $g(X)\geq 2$. We construct infinite matrices $C^{(1)},C^{(2)},\ldots,C^{(s)}$ as follows.

Let $P_\infty,P_1,P_2,\ldots,P_s\in X(\mathbb{F}_q)$, and let $D\in \mathrm{Div}_{\mathbb{F}_q}(X)$ satisfy the condition of the theorem. Then by the Riemann–Roch theorem we see that for any pair (i,j), $1\leq i\leq s$, $j\geq 1$, we can find $k_j^{(i)}\in L(D+jP_i)$ such that

$$v_{P_i}\left(k_j^{(i)}\right)=-v_{P_i}(D)-j.$$

Since $v_{P_\infty}(k_j^{(i)})\geq -v_{P_\infty}(D)=:-v$, at P_∞ we have expansions of the form

$$k_j^{(i)}=z^{-v}\sum_{r=0}^{\infty}b_{j,r}^{(i)}z^r,\quad \text{for } 1\leq i\leq s \text{ and } j\geq 1,$$

where $z=t_\infty$ is a local parameter at P_∞ and $b_{j,r}^{(i)}\in\mathbb{F}_q$. We can now construct $C^{(i)}=\left(c_{j,r}^{(i)}\right)$ by setting

$$c_{j,r}^{(i)}=\begin{cases} b_{j,r}^{(i)} & \text{for } 0\leq r\leq v-1,\\ b_{j,r+1}^{(i)} & \text{for } r\geq v,\end{cases}$$

for any $1\leq i\leq s$, $j\geq 1$.

Note that this construction is very close in spirit and essentially equivalent to that of Theorem 12.4.9.

12.5. Locally Recoverable Codes

In the present section we discuss codes motivated by recent applications of coding to increasing the reliability of distributed storage systems. These are the so-called locally recoverable codes, or simply LRC codes. First we describe some bounds and an optimal construction for these codes and an optimal construction for them, which is parallel to Reed–Solomon codes, and then its generalization to some algebraic curves, giving rather interesting results. In particular, Sec. 12.5.3 contains asymptotic results comparing LRC codes from optimal towers and codes obtained by an analogue of the Gilbert–Varshamov procedure.

12.5.1. Optimal LRC Codes

We begin with some definitions. We say that a code $\mathcal{C} \subset \mathbb{F}_q^n$ is *LRC* with *locality* r if for every $i \in [n] := \{1, 2, \ldots, n\}$ there exists a subset $A_i \subset [n] \setminus \{i\}$ of cardinality $|A_i| \le r$ and a function $\varphi_i \colon \mathbb{F}_q^r \to \mathbb{F}_q$ such that for every codeword $x \in \mathcal{C}$ we have

$$x_i = \varphi_i(\{x_j, j \in A_i\}).$$

In other words, given $a \in \mathbb{F}_q$, consider the sets of codewords

$$\mathcal{C}(i,a) = \{x \in \mathcal{C} : x_i = a\}, \quad i \in [n];$$

then the code $\mathcal{C}$ is said to have locality r if for every $i \in [n]$ there exists a subset $I_i \subset [n] \setminus i$ with $|I_i| \le r$ such that the restrictions of the sets $\mathcal{C}(i,a)$ to the coordinates in I_i for different a are disjoint:

$$\mathcal{C}_{I_i}(i,a) \cap \mathcal{C}_{I_i}(i,a') = \varnothing, \quad a \ne a'.$$

We use the notation (n,k,r) to refer to the parameters of an LRC code of length n, cardinality q^k, and locality r. The code $\mathcal{C}_{A_i \cup \{i\}}$ is called a *local code* of the code $\mathcal{C}$.

The next proposition is the first result showing how the locality condition influences the rate.

PROPOSITION 12.5.1. *Let $\mathcal{C}$ be an (n,k,r) LRC code of cardinality q^k. Then the rate of $\mathcal{C}$ satisfies*

$$\frac{k}{n} \le \frac{r}{r+1}. \qquad \square$$

LRC codes satisfy also the following Singleton-type bound.

PROPOSITION 12.5.2. *Let $\mathcal{C}$ be an (n,k,r) LRC code of cardinality q^k, and let d be its minimum distance. Then*

$$d \le n - k - \left\lceil \frac{k}{r} \right\rceil + 2. \qquad \square$$

Below we call codes whose parameters attain this bound with equality *optimal LRC codes*.

Optimal LRC codes. A construction of optimal LRC codes can be given in a spirit very close to that of Reed–Solomon codes. Indeed, we are going to construct an LRC code over $\mathbb{F}_q$ with the parameters (n,k,r), where $n \le q$. For simplicity we additionally assume that $(r+1) \mid n$ and $r \mid k$, but in fact this is not essential. Let $A = \{P_1, \ldots, P_n\} \subset \mathbb{F}_q$ be a subset of points of $\mathbb{F}_q$, and let $g(x) \in \mathbb{F}_q[x]$ be a polynomial of degree $r+1$ such that there exists a partition $\mathcal{A} = \{A_1, \ldots, A_{\frac{n}{r+1}}\}$ of the set A into subsets of size $r+1$ with g being constant on each of the sets $A_i \in \mathcal{A}$.

Consider a k-dimensional linear subspace $V \subset \mathbb{F}_q[x]$ generated by the set of polynomials

$$\Big(g(x)^j x^i,\ i=0,\dots,r-1,\ j=0,\dots,\frac{k}{r}-1\Big). \tag{12.5.1}$$

Given

$$a=\Big(a_{ij},\ i=0,\dots,r-1,\ j=0,\dots,\frac{k}{r}-1\Big)\in\mathbb{F}_q^k,$$

define

$$f_a(x)=\sum_{i=0}^{r-1} f_i(x)x^i, \tag{12.5.2}$$

where

$$f_i(x)=\sum_{j=0}^{\frac{k}{r}-1} a_{ij}g(x)^j,\quad i=0,\dots,r-1; \tag{12.5.3}$$

the polynomials f_i are called coefficient polynomials. Now define the code $\mathcal{C}$ as the image of the linear evaluation map

$$e\colon V\to\mathbb{F}_q^n,\quad f_a\mapsto\big(f_a(P_i),\ i=1,\dots,n\big). \tag{12.5.4}$$

We call the elements of the set A *locations* and the elements of the vector $(f_a(\alpha))$ *symbols* of the codeword.

The local recovery of an erased symbol is then accomplished as follows. Assume that the erased symbol corresponds to the location $\alpha\in A_j$, where A_j is one of the sets in the partition $\mathcal{A}$. Let $(c_\beta,\beta\in A_j\setminus\alpha)$ denote the remaining r symbols in the locations of the set A_j. To find the value $c_\alpha=f_a(\alpha)$, find a unique polynomial $\delta(x)$ of degree less than r such that $\delta(\beta)=c_\beta$ for all $\beta\in A_j\setminus\alpha$, namely

$$\delta(x)=\sum_{\beta\in A_j\setminus\alpha} c_\beta \prod_{\beta'\in A_j\setminus\{\alpha,\beta\}}\frac{x-\beta'}{\beta-\beta'}, \tag{12.5.5}$$

and set $c_\alpha=\delta(\alpha)$. We call $\delta(x)$ the *decoding polynomial* for the symbol c_α. Thus, to find one erased symbol, we need to perform polynomial interpolation from r known symbols in its recovery set.

Moreover, the LRC codes thus constructed are optimal.

THEOREM 12.5.3. *The linear code $\mathcal{C}$ defined in* (12.5.4) *has dimension k and is an optimal (n,k,r) LRC code.*

PROOF. Note that for $i=0,\dots,r-1$ and $j=0,\dots,\frac{k}{r}-1$ the k polynomials $g(x)^jx^i$ all are of distinct degrees, and therefore are linearly independent over $\mathbb{F}$. In other words, the mapping $a\mapsto f_a(x)$ is injective. By (12.5.2) and (12.5.3) the degree of the polynomial $f_a(x)$ is at most

$$\Big(\frac{k}{r}-1\Big)(r+1)+r-1=k+\frac{k}{r}-2\le n-2,$$

where the last inequality follows from Proposition 12.5.1. This means that two distinct encoding polynomials f_a and f_b give rise to two distinct code vectors, so the dimension of the code is k. Since the encoding is linear, for the distance we get

$$d(\mathcal{C}) \geq n - \max_{f_a\colon a\in\mathbb{F}_q^k} \deg(f_a) = n - k - \frac{k}{r} + 2,$$

which together with Proposition 12.5.2 completes the proof of distance optimality.

Let us prove the locality property. Let A_j be a member of the partition $\mathcal{A}$ and assume that the lost symbol of the codeword equals $c_\alpha = f_a(\alpha)$, where $\alpha \in A_j$ is a field element. Define the *decoding polynomial*

$$\partial(x) = \sum_{i=0}^{r-1} f_i(\alpha)x^i$$

for the coefficient polynomials $f_i(x)$ in (12.5.3). Let us show that $\partial(x)$ coincides with $\delta(x)$ defined in (12.5.5). Since $f_i(x)$ is a function of g, it is constant on the set A_j, and thus we have

$$f_i(\beta) = f_i(\alpha)$$

for any $\beta \in A_j$ and any $i = 1, \ldots, r-1$. Therefore, for any β in A_j we have

$$\partial(\beta) = \sum_{i=0}^{r-1} f_i(\alpha)\beta^i = \sum_{i=0}^{r-1} f_i(\beta)\beta^i = f_a(\beta).$$

Since $\partial(x)$ is of degree at most $r-1$, it can be interpolated from the r symbols c_β, $\beta \in A_j \setminus \alpha$, and then we find the lost symbol as $\partial(\alpha)$. □

There is a natural algebraic method for constructing a polynomial g, namely the following simple proposition gives a good polynomial using any subgroup of either the multiplicative group $\mathbb{F}_q^*$ or the additive group $\mathbb{F}_q^+$.

PROPOSITION 12.5.4. *Let H be a subgroup of $\mathbb{F}_q^*$ or $\mathbb{F}_q^+$. Then the polynomial*

$$g(x) = \prod_{h\in H} (x-h)$$

is constant on each coset of H.

PROOF. For a multiplicative subgroup H, consider two elements a and $a\overline{h}$ in the coset aH, where $\overline{h} \in H$. Then

$$g(a\overline{h}) = \prod_{h\in H} (a\overline{h} - h) = \overline{h}^{|H|} \prod_{h\in H} (a - h\overline{h}^{-1}) = \prod_{h\in H} (a-h) = g(a).$$

The proof for an additive subgroup is completely analogous. □

12.5.2. Locally Recoverable Codes on Algebraic Curves

Basic construction. Let us start with a straightforward algebraic geometry interpretation of the above optimal construction of LRC codes. The polynomial g is just a regular map $g\colon \mathbb{P}^1 \to \mathbb{P}^1$ of degree $\deg g = r+1$ such that all fibres $g^{-1}(\alpha)$ consist of $\mathbb{F}_q$-rational points for any $\alpha \in g(A)$, where $A \subseteq \mathbb{F}_q \subset \mathbb{P}^1(\mathbb{F}_q)$ is a subset of cardinality n. This naturally leads to the following *basic construction* if we replace two copies of $\mathbb{P}^1$ with two algebraic curves.

Let X and Y be smooth projective absolutely irreducible curves over $\Bbbk = \mathbb{F}_q$. Let $g\colon X \to Y$ be a regular separable map of degree $\deg g = r+1$. As usual, denote by $\Bbbk(X)$ (respectively, $\Bbbk(Y)$) the field of rational functions on X (respectively, Y). Let $g^*\colon \Bbbk(Y) \to \Bbbk(X)$ be the corresponding field embedding, i.e., $g^*(f)(P) = f(g(P))$, for $f \in \Bbbk(Y)$ and $P \in X$ (see Secs. 2.1.1 and 2.5.1). We identify $\Bbbk(Y)$ with its image $g^*(\Bbbk(Y)) \subset \Bbbk(X)$.

Since g is separable, the primitive element theorem implies that there exists a function $x \in \Bbbk(X)$ such that $\Bbbk(X) = \Bbbk(Y)(x)$and we can choose for x an equation of the form

$$x^{r+1} + b_r x^r + \ldots + b_0 = 0$$

with $b_i \in \Bbbk(Y)$. The function x can be considered as a regular map $x\colon X \to \mathbb{P}^1_{\Bbbk}$; denote its degree by h. The quantity h is crucial for what follows.

The codes that we construct are given by an evaluation construction, as in many cases above. Let $S = \{P_1, \ldots, P_s\} \subset Y(\Bbbk)$ be a subset of $\mathbb{F}_q$-rational points of Y, and let Q_∞ be a positive divisor of degree $\ell \geq 1$ whose support is disjoint from S. For instance, we can assume that $Q_\infty \subset \pi^{-1}(\infty)$ for a projection $\pi\colon Y \to \mathbb{P}^1_{\Bbbk}$. Let us assume that S and g satisfy the following conditions:

$$\begin{aligned} A := g^{-1}(S) = \{P_{ij},\ i = 0, \ldots, r,\ j = 1, \ldots, s\} &\subseteq X(\Bbbk), \\ g(P_{ij}) = P_j \text{ for all } i, j, \\ b_i \in L(n_i Q_\infty), \quad i = 0, 1, \ldots, r, \end{aligned} \tag{12.5.6}$$

for some natural numbers n_i.

Now let $D = tQ_\infty$ be a positive divisor on Y, and let $\{f_1, \ldots, f_m\} \subset \Bbbk(Y)$ be a basis of the linear space $L(D)$. Then the functions f_i are constant on the fibres of the map g while we consider them as elements of $\Bbbk(X)$. By the Riemann–Roch theorem we have $m \geq t\ell - g_Y + 1$, where g_Y is the genus of Y. Below we assume that $m = t\ell - g_Y + 1$. Let $V \subset \Bbbk(X)$, $\dim V = rm$, be the linear span of the functions

$$\{f_j x^i,\ i = 0, \ldots, r-1,\ j = 1, \ldots, m\}$$

(cf. (12.5.1)). Since Q_∞ is disjoint from S, the evaluation map

$$e = \mathrm{ev}_A\colon V \to \Bbbk^{(r+1)s}, \quad F \mapsto \big(F(P_{ij}),\ i = 0, \ldots, r,\ j = 1, \ldots, s\big)$$

is well defined, and its image is a linear code in $\mathbb{F}_q^{(r+1)s}$, which we denote by $\mathcal{C}(D, g)$. The code coordinates are naturally partitioned into s subsets $A_j = \{P_{ij}, i = 0, \ldots, r\}$, $j = 1, \ldots, s$, of size $r+1$ each; see (12.5.6). Assume throughout what follows that, for any fixed j, the function x takes different values at the points in the set $(P_{ij}, i = 0, \ldots, r)$.

THEOREM 12.5.5. *The subspace $\mathcal{C}(D, g) \subset \mathbb{F}_q$ forms an (n, k, r) linear LRC code with the parameters*

$$\begin{aligned} n &= (r+1)s, \\ k &= rm \geq r(t\ell - g_Y + 1), \\ d &\geq n - t\ell(r+1) - (r-1)h, \end{aligned}$$

provided that the right-hand side of the inequality for d is a positive integer. Local recovery of an erased symbol $c_{ij} = F(P_{ij})$ can be performed by polynomial interpolation through the points of the recovery set A_j.

PROOF. The length and the inequality for the distance are immediate, since the function $f_j x^i$, $f_j \in L(D)$, evaluated on A can have at most $t\ell(r+1)+(r-1)h$ zeroes. Since $d \geq 1$, the mapping ev_T is injective, which gives the dimension. Finally, the functions f_i are constant on the fibres $(P_{ij}, i=0,\dots,r-1)$; therefore, on each subset A_j the codeword is obtained as an evaluation of the polynomial of the variable x of degree $\rho \leq r-1$. This representation accounts for the fact that coordinate c_P, $P \in A_j$, of the codeword can be found by interpolating a polynomial of degree at most $r-1$ through the remaining points of A_j. □

Now let us give some examples, first for finite lengths and then in asymptotics.

LRC codes from Hermitian curves. Let $q=q_0^2$, where q_0 is a prime power, let $\Bbbk=\mathbb{F}_q$, and let $X := H$ be the Hermitian curve

$$X: x^{q_0}+x=y^{q_0+1},$$

having $q_0^3+1=q\sqrt{q}+1$ rational points of which one is the infinite point and the remaining q_0^3 are located in the affine plane. There are two slightly different ways of constructing Hermitian LRC codes.

Projection on y. Here we construct q-ary $(n,k,r=q_0-1)$ LRC codes. Take $Y=\mathbb{P}^1(\Bbbk)$ and take g to be the natural projection defined by $g(x,y):=y$; then the degree of g is $q_0=r+1$ and the degree of x is $h=q_0+1$. We can write $X(\mathbb{F}_q)=g^{-1}(\mathbb{F}_q)\cup Q'_\infty$, where $Q'_\infty \in X$ is the unique point over $\infty \in Y$.

Now take $S=\Bbbk \subset \mathbb{P}^1$, $Q_\infty=\infty$, $\ell=1$, and $D=tQ'_\infty$ for some $t \geq 1$. We have

$$L(D)=\left\{\sum_{i=0}^{t} a_i y^i\right\} \subset \Bbbk[y],$$

and the above general construction yields the following result.

PROPOSITION 12.5.6. *The construction of Theorem* 12.5.5 *gives a family of q-ary Hermitian LRC codes with the parameters*

$$n=q_0^3, \qquad k=(t+1)(q_0-1), \qquad r=q_0-1,$$
$$d \geq n-tq_0-(q_0-2)(q_0+1).$$ □

Computing the gap to the Singleton bound of Proposition 12.5.2, we obtain

$$\begin{aligned} d+\frac{k}{r}(r+1) &\geq q_0^3-tq_0-(q_0-2)(q_0+1)+q_0(t+1)\\ &= q_0^3-q_0^2+2q_0+2\\ &= n-q+2\sqrt{q}+2. \end{aligned} \tag{12.5.7}$$

For codes that meet the Singleton bound, we would have $d+k(r+1)/r=n+2$, so the Singleton gap of the Hermitian LRC codes is at most $q-2\sqrt{q}=q_0(q_0-2)$. Of course, these codes cannot be Singleton-optimal, because their length is much greater than the alphabet size, but the gap in this case is still reasonably small.

Projection on x. Again take $Y=\mathcal{P}^1$, and let $g'(x,y):=x$ be the second natural projection on $\mathcal{P}^1$. There are q_0 points on $\mathcal{P}^1$ that are fully ramified, namely the points in the set

$$M=\{a \in \mathbb{F}_q : a^{q_0}+a=0\}.$$

Therefore, every fibre of g' over $\mathbb{F}_q \setminus M$ consists of $\mathbb{F}_q$-rational points, since there are in total

$$|\mathbb{F}_q \setminus M|(q_0+1) = q_0^3 - q_0$$

rational points in those fibres. Obviously, $|g^{-1}(a) \cap g'^{-1}(b)| \le 1$ for all $a, b \in \mathbb{F}_q$.

Take $S = \mathbb{F}_q \setminus M$; then $r = q_0$, and clearly $h = \deg(y) = q_0$. We obtain the following result.

PROPOSITION 12.5.7. *The construction of Theorem* 12.5.5 *gives a family of q-ary Hermitian LRC codes with the parameters*

$$n = q_0^3 - q_0, \qquad k = (t+1)q_0, \qquad r = q_0,$$
$$d \ge n - t(q_0+1) - q_0(q_0-1). \qquad \square$$

Calculating similarly to (12.5.7), we get a quantity one less than in that case:

$$d + \frac{k}{r}(r+1) = n - q + 2\sqrt{q} + 1.$$

REMARK 12.5.8. Hermitian LRC codes are in a certain sense almost optimal for our construction. Note that most known close-to-maximal curves have the property that for a natural projection $g\colon X \to \mathcal{P}^1$ the point $\infty \in \mathcal{P}^1$ is totally ramified (cf., for example, the towers $\mathcal{W}_1$ and and $\mathcal{W}_2$ in Sec. 7.4.2). In this case the quantity h satisfies $h \ge n/q$ (a gonality estimate; cf. the proof of Theorem 10.3.10), whereas for Hermitian curves we have $h = n/q$ or $n/q+1$. Recall also that Hermitian curves are absolutely maximal, i.e., attain the Weil bound.

The parameters of LRC codes from Hermitian curves can be considered as satisfactory; however, their locality is rather large, and usually one prefers to get smaller locality r, preferably 2 or 3. This is often possible if we modify the curve $Y = \mathbb{P}^1$ in the above construction. Indeed, one can replace $\mathbb{P}^1$ by another curve Y with a projection $X \to Y$ of degree $r+1$ for some small r. This is possible, e.g., in the following two cases:

1. either $(r+1) \mid q_0$, and we consider a projection $X \to Y$ such that the natural composition $X \to Y \to \mathbb{P}^1$ gives our y-projection $X \to \mathbb{P}^1$
2. or $(r+1) \mid (q_0+1)$, and $X \to Y \to \mathbb{P}^1$ gives the above x-projection of degree q_0+1.

In the first case, $\Bbbk(Y) = \Bbbk(z,y)$ with $z^r + z = y^{q_0+1}$; in the second case, $\Bbbk(Y) = \Bbbk\big(x, y^{\frac{q_0+1}{r+1}}\big)$. One can easily calculate the parameters of the thus obtained LRC codes of a low locality r, which remain satisfactory with respect to the Singleton bound. Note that the value $r = 2$ can be obtained, e.g., for $q_0 = 3^k$ and for $q_0 \equiv 2 \bmod 3$.

12.5.3. Asymptotic Behaviour

Now let us pass to asymptotic constructions. Here García–Stichtenoth towers predictably give very impressive results.

García–Stichtenoth towers. Thus, let us use the García–Stichtenoth towers $\mathcal{W}_1$ and $\mathcal{W}_2$ from Sec. 7.4.2. We start with $\mathcal{W}_2$.

Let $q = q_0^2$ be a square, and let $l \geq 2$ be an integer. Define the curve X_l and the functions x_l, z_l inductively as follows:

$$x_0 := 1 \quad X_1 := \mathbb{P}^1, \quad \Bbbk(X_1) = \Bbbk(x_1),$$
$$X_l : z_l^{q_0} + z_l = x_{l-1}^{q_0+1}, \quad \text{where} \quad x_{l-1} := \frac{z_{l-1}}{x_{l-2}} \in \Bbbk(X_{l-1}) \quad (\text{if } l \geq 3),$$

where $\Bbbk = \mathbb{F}_q$ (this is the $\mathcal{W}_2$ tower of Section 7.4.2 in a slightly modified notation). In particular, $X_2 = H$ is the Hermitian curve, and since this family generalizes Hermitian curves, we can expect that it gives rise to families of codes which extend the above constructions of LRC codes from Hermitian curves. This is indeed the case.

To use the general construction of Theorem 12.5.5, we take the map $g_l \colon X_l \to X_{l-1}$ to be the natural projection of degree $q_0 = r+1$ and note that

$$g_l^* : \Bbbk(X_{l-1}) \to \Bbbk(X_l) = \Bbbk(X_{l-1})(z_l)$$

is identical on $\Bbbk(X_{l-1})$. To describe rational points of the curve X_l, let $\psi_l \colon X_l \to \mathbb{P}^1$ be the natural projection of degree q_0^{l-1}, i.e., the map $\psi_l = g_l \circ g_{l-1} \circ \ldots \circ g_2$. Then all the points in the preimage $P_l := \psi_l^{-1}(\mathbb{F}_q^*)$ are $\mathbb{F}_q$-rational, and there are $n_l = q_0^{l-1}(q_0^2 - 1)$ such points. The genus of the curve X_l, given explicitly in Proposition 7.4.17(a), can be bounded above as

$$G_l \leq q_0^l + q_0^{l-1} = q_0^{l-1}(q_0+1) = \frac{n_l}{q_0 - 1},$$

and we obtain the following result.

PROPOSITION 12.5.9. *There exists a family of q-ary $(n, k, r = q_0 - 1)$ LRC codes on the curve X_l, $l \geq 2$, with the parameters*

$$n = n_l = q_0^{l-1}(q_0^2 - 1),$$
$$k \geq r\Big(t - \frac{n_{l-1}}{q_0 - 1} + 1\Big),$$
$$d \geq n_l - tq_0 - \frac{2n_l(q_0 - 2)}{q_0^2 - 1},$$

where t is any integer such that $G_{l-1} \leq t \leq n_{l-1}$.

PROOF. We apply the construction of Theorem 12.5.5 to the curves $X := X_l$ and $Y := X_{l-1}$, taking the map $g := g_l$, $Q_\infty := P_{\infty,l}$, $D = tQ_\infty$, and $\ell = 1$.

The function x in the general construction in this case is $x = z_l$. To estimate the distance of the code $\mathcal{C}(D, g)$ using Theorem 12.5.5, we need to find the degree $h = \deg(z_l)$. Since $z_l = x_l x_{l-1}$, we get

$$\deg(z_l) = \deg(x_l) + \deg(x_{l-1}).$$

Let $(x_l)_0^{(l)}$ be the divisor of zeroes of x_l on X_l; then by Lemma 7.4.18 we have $(x_l)_0^{(l)} = q_0^{l-1} Q_l$, where Q_l is the unique common zero of $x_1, z_2, \ldots, z_l$. Therefore, $\deg(x_l)_0^{(l)} = q_0^{l-1}$ and $\deg(x_{l-1})_0^{(l-1)} = q_0^{l-2}$. Since the map $X_l \to X_{l-1}$ is of degree q_0, we obtain $\deg(x_{l-1})_0^{(l)} = q_0 \deg(x_{l-1})_0^{(l-1)} = q_0^{l-1}$, and finally,

$$h = 2q_0^{l-1} = \frac{2n_l}{q_0^2 - 1}.$$

Now the result is obtained from Theorem 12.5.5 by direct computation. □

Now consider the second natural projection of curves in the tower. Namely, let Y_l correspond to the function field $\Bbbk(z_2,\dots,z_l)$ and consider the field embedding

$$(g_l')^*: \Bbbk(Y_l) \to \Bbbk(X_l) = \Bbbk(x_1,z_2,\dots,z_l).$$

Note that g_2' is the projection $g'\colon X_2 \to \mathbb{P}^1$ described in the case of Hermitian curves. The curves $\{Y_l, l=2,3,\dots\}$ form another optimal tower of curves given by the recursive equations

$$Y_l\colon z_l^q + z_l = \frac{z_{l-1}^q}{z_{l-1}^{q-1}+1}, \quad l \geq 3, \qquad Y_2 := \mathbb{P}^1,$$

i.e., the $\mathcal{W}_1$ tower from Sec. 7.4.2 in slightly modified notation. In geometric terms, the embedding $(g_l')^*$ implies that the curve X_l is the fibre product of X_2 and Y_l over $Y_2 = \mathbb{P}^1$, i.e., $X_l = X_2 \times_{Y_2} Y_l$, which in turn implies that the projection $g_l'\colon X_l \to Y_l$ shares the main properties of $g' = g_2'$. Indeed, we have the following:

1. The genus of Y_l satisfies $G_l' < q_0^{l-1}$ (the exact value of G_l' is given in Proposition 7.4.22(b)).
2. Let $\pi_l\colon Y_l \to Y_2$ be the natural projection of degree $\deg(\pi_l) = q_0^{l-2}$. All the points in $S_l := \pi_l^{-1}(\mathbb{F}_q \setminus M)$ are $\mathbb{F}_q$-rational and
$$|S_l| = q_0^{l-2}(q_0^2 - q_0) = q_0^{l-1}(q_0-1) = n_l/(q_0+1).$$
3. The point $\infty \in Y_2 = \mathbb{P}^1$ is totally ramified, i.e., $\pi_l^{-1}(\infty) = P_{\infty,l}'$ for a rational point $P_{\infty,l}' \in Y_l$.
4. We have $(g_l')^{-1}(S_l) = (\psi_l)^{-1}(\mathbb{F}_q)$ and $|(g_l')^{-1}(S_l)| = n_l$, and all points in $(g_l')^{-1}(S_l)$ are $\mathbb{F}_q$-rational. The degree of the projection g_l' is $\deg(g_l') = q_0+1$. The fibres of g_l' are transversal with those of g_l, i.e., they intersect at a unique point.
5. The degree of $x_1\colon X_l \to \mathbb{P}^1$ equals $h := \deg(x_1) = \deg(\pi_l)\deg\big((x_1)_0^{(2)}\big) = q_0^{l-1}$.

Therefore, we obtain the following.

PROPOSITION 12.5.10. *There exists a family of q-ary $(n,k,r=q_0)$ LRC codes on the curve X_l, $l \geq 2$, with the parameters*

$$\begin{aligned} n &= n_l = q_0^{l-1}(q_0^2-1), \\ k &\geq r(t - q_0^{l-1}+1), \\ d &\geq n_l - t(q_0+1) - (q_0-1)q_0^{l-1}, \end{aligned}$$

where t is any integer such that $G_{l-1} \leq t \leq n_{l-1}$.

PROOF. Put $r = q_0$, $\ell = 1$, and apply the construction of Theorem 12.5.5 to

$$X := X_l, \qquad Y := Y_l, \qquad g := g_l', \qquad Q_\infty := P_{\infty,l}', \qquad D_t = tQ_\infty. \qquad \square$$

Modifications of the main construction for towers. The above constructions yield infinite families of q-ary LRC codes with good parameters, but the locality parameter r is fixed and is close to $\sqrt{q}$. As above for Hermitian curves, we can modify the construction to attain small values of r, e.g., $r=2$.

For example, for the García–Stichtenoth tower $\mathcal{W}_2$ we construct codes with locality r, where $(r+1) \mid (q_0+1)$. Let $X := X_l$, and let $Y := Y_{l,r}$ be the curve with the function field

$$\Bbbk(Y_{l,r}) := \Bbbk(x_1^{r+1}, z_2, \dots, z_l).$$

Considering the natural projection $g\colon X \to Y$ given by $x_1 \mapsto x_1^{r+1}$ and using the pair (X,Y) in Theorem 12.5.5, we obtain the following result.

PROPOSITION 12.5.11. *Let $(r+1) \mid (q_0+1)$. There exists a family of q-ary (n,k,r) LRC codes on the curve X_l, $l \geq 2$, with the parameters*

$$\begin{aligned} n &= n_l = q_0^{l-1}(q_0^2-1), \\ k &\geq r\Big(t - q_0^{l-1}\frac{q_0+1}{r+1} + 1\Big), \\ d &\geq n_l - t(r+1) - (r-1)q_0^{l-1}, \end{aligned}$$

where t is any integer such that $g_Y \leq t \leq n_{l-1}$.

PROOF. Recall that $g_X \geq 1 + N(g_Y - 1)$ by the Hurwitz formula for any map $f\colon X \to Y$ of degree N. Applying this inequality to our pair of curves we get an estimate of the genus of Y, and the parameters of the code are obtained directly from Theorem 12.5.5. □

The same method can be applied to $\mathcal{W}_1$.

The LRC codes from optimal towers are especially good in the asymptotics of growing length. Now we give an asymptotic bound for LRC codes which is obtained by an appropriate Gilbert–Varshamov procedure.

Asymptotic lower (Gilbert–Varshamov) bound. To introduce the asymptotic parameters, consider the sequence of LRC codes C_i, $i = 1,2,\dots$, of length n_i, dimension k_i, and distance d_i. We assume the locality parameter r to be fixed. Let $n_i \to \infty$, $i = 1,2\dots$, and assume that there exist limits $R = \lim\limits_{i\to\infty} k_i/n_i$ and $\delta = \lim\limits_{i\to\infty} d_i/n_i$. In this case we say that the code sequence (C_i) has asymptotic parameters (R,δ).

A bound analogous to the GV bound in the LRC case is as follows.

THEOREM 12.5.12. *There exists a sequence of q-ary linear r-LRC codes with the asymptotic parameters (R,δ) as long as*

$$R < \frac{r}{r+1} - \min_{0<s\leq 1}\Big\{\frac{1}{r+1}\log_q b(s) - \delta \log_q s\Big\},$$

where

$$b(s) = \frac{1}{q}\big((1+(q-1)s)^{r+1} + (q-1)(1-s)^{r+1}\big).$$ □

Using our construction of LRC codes on asymptotically maximal curves, we can construct sequences of codes whose parameters surpass the bound of Theorem 12.5.12 for sufficiently large q.

Now let us compute the asymptotic relation between the parameters of LRC codes on the García–Stichtenoth curves constructed above. Propositions 12.5.9 and 12.5.10 lead to the following asymptotic results.

PROPOSITION 12.5.13. *Let $q = q_0^2$, where q_0 is a prime power. There exist families of LRC codes with locality r whose rate and relative distance satisfy*

$$R \geq \frac{r}{r+1}\Big(1-\delta-\frac{3}{\sqrt{q}+1}\Big), \quad r=\sqrt{q}-1, \tag{12.5.8}$$

$$R \geq \frac{r}{r+1}\Big(1-\delta-\frac{2\sqrt{q}}{q-1}\Big), \quad r=\sqrt{q}. \tag{12.5.9}$$

PROOF. For instance, let us check the first inequality of the propositions. From Proposition 12.5.9 we obtain

$$\begin{aligned} d+\frac{k(r+1)}{r} &\geq n_l - \frac{q_0 n_{l-1}}{q_0-1} - \frac{2n_l(q_0-2)}{q_0^2-1} + q_0 \\ &\geq n_l\Big(1-\frac{1}{q_0-1}-\frac{2q_0-4}{q_0^2-1}\Big) = n_l\Big(1-\frac{3}{q_0+1}\Big). \end{aligned}$$

Letting $\delta = d/n_l$ and $R = k/n_l$, $l \to \infty$, we obtain (12.5.8). □

REMARK 12.5.14. Recall that without the locality constraint the relation between R and δ for codes on asymptotically optimal curves (for instance, on the curves X_l, $l = 2,3,\ldots$) takes the form $R \geq 1-\delta-\frac{1}{\sqrt{q}-1}$; see Corollary 4.5.3 in *Basic Notions*.

By Theorem 4.5.4, the classic AG codes improve the Gilbert–Varshamov (GV) bound starting with $q = 49$. Similarly, the bound given in (12.5.9) (i.e., the code family constructed in Proposition 12.5.10) improves upon the GV-type bound of Theorem 12.5.12 for large alphabets. For instance, for $q_0 = 23$ the code rate (12.5.9) is better than the GV-type bound for $\delta \in [0.413, 0.711]$, and the length of this interval increases as q_0 becomes greater. Similar conclusions can be made for the codes in the family of Proposition 12.5.9.

Turning to codes with small locality, let us derive an asymptotic estimate of the parameters of the code family in Proposition 12.5.11. The curve Y is a quotient of an asymptotically optimal curve, so it is asymptotically optimal itself. Therefore, the genus g_Y is asymptotic to $G_{X_l}/(r+1)$, and the genus G_{X_l} is asymptotic to $q_0^{l-1}(q_0+1) = \frac{n_l}{q_0-1}$.

PROPOSITION 12.5.15. *Let $(r+1) \mid (q_0+1)$, and let $q = q_0^2$, where q_0 is a prime power. There exists a family of q-ary LRC codes with locality r whose asymptotic rate and relative distance satisfy the bound*

$$R \geq \frac{r}{r+1}\Big(1-\delta-\frac{q_0+r}{q_0^2-1}\Big).$$ □

For instance, take $r = 2$ and let $3 \mid (q_0+1)$. We obtain the bound

$$R \geq \frac{2}{3}\Big(1-\delta-\frac{1}{\sqrt{q}-1}-\frac{1}{q-1}\Big). \tag{12.5.10}$$

Asymptotic bounds obtained above are shown in Fig. 12.1 both for locality $r = q_0$ and for $r = 2$.

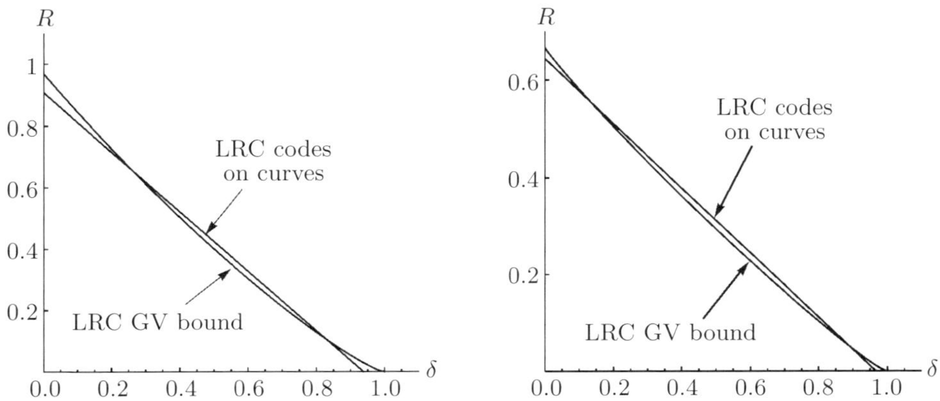

FIGURE 12.1. *Left plot*: The bound (12.5.8) shown together with the Gilbert–Varshamov type bound of Theorem 12.5.12 ($q_0 = 32$). *Right plot*: The bound (12.5.10) together with the GV-type bound, $r = 2$, $q_0 = 32$.

Historical and Bibliographic Notes

The general scheme of the Chudnovsky algorithm (Sec. 12.1) is given by the Chudnovsky brothers in [**CC88**]. In [**STV92**] the Chudnovsky algorithm was generalized, presented in a concise form, and applied to modular and Shimura curves, which led to some upper bounds on multiplicative complexity. Unfortunately, this paper contained a serious gap, and therefore the bounds stated there were not rigorously established. The situation was corrected by H. Randriambololona in [**Ran12**] (see also [**Ran11, Ran15**]), who greatly improved the technique used in the Chudnovsky algorithm and, in particular, definitely proved the bounds stated in [**STV92**] (essentially, Theorem 12.1.11). We follow the paper [**Ran12**] rather closely in Secs. 12.1.1 and 12.1.2. Many related results are contained in [**BR04, Bal08a, Bal08b, BLBR10, BRR10, BP11**]. The result of Proposition 12.1.10 in its Drinfeld curves version is given in [**Vlă18**, Proposition 6.1].

The subject of A-codes (Sec. 12.2) was introduced in a landmark paper [**GMS74**] by E.N. Gilbert, F.J. MacWilliams, and N.J.A. Sloane. They were inspired by the work of G.J. Simmons, who later introduced a mathematical model for authentication [**Sim82, Sim85**]. The connection between A-codes and error-correcting codes was discovered by G.A. Kabatiansky, B. Smeets, and T. Johansson, who were inspired by an idea of Simmons [**Sim92**]. They applied this connection to obtain many constructions and bounds [**KSJ96**], in particular, Proposition 12.2.1 and Theorems 12.2.2–12.2.5.

All other results in Sec. 12.2.1 are from [**Vlă98a**]. A very close algebraic geometry construction was independently discovered by C. Xing [**XWL00**] and applied to a number of curves with many rational points; see also [**Xin01, NWX07**].

Shamir's scheme of secret sharing was introduced in [**Sha79**]. It was widely generalized and applied to secure multi-party communications in [**BOGW88, CCD88, FY92, CDM00**]. The idea of an application of algebraic curves to arithmetic secret sharing schemes was given in [**CCdHC08, CCCX09**]. It was widely perfected in [**CCX11**], which we follow in Sec. 12.2.2. For further progress, see [**CCXY12, CCX13**]. Theorem 12.2.18 was first stated in [**Deu37**]; it was proved in a particular case in [**Sha47**] and then proved completely in [**Sub75**].

The idea of quantum computations was put forward by Yu.I. Manin [**Man77**] and R.P. Feynman [**Fey82**]. Interest to them exploded after P.W. Shor's discovery of an efficient quantum algorithm for (presumably difficult for ordinary computers) problems of prime factorization and discrete logarithms [**Sho94, Sho97**]. However, it became clear that quantum computations should use some form of error correction to be realizable on a physical device. This leads to numerous papers on quantum error correction; see, e.g., [**RP00, KLM07**] and the references therein.

The first example of asymptotically good quantum error-correcting codes from algebraic geometry codes was given by A. Ashikhmin, S. Litsyn, and M. Tsfasman in [**ALT01**], which we follow in Secs. 12.3.1 and 12.3.2. The fundamental Proposition 12.3.12 is due to T. Kasami and S. Lin [**KL88**]. The improvement at the end of Sec. 12.3.2 was obtained in [**CLX01**]. Proposition 12.3.18 is given in [**CRSS98, CS96, Ste99**]. Nonbinary quantum codes were defined in [**Rai99**]. Section 12.3.3 is based on [**FLX06**].

The study of uniformly distributed point sets was initiated by H. Weyl and E. Hlawka. I.M. Sobol [**Sob67**] developed the notion of such sets and gave the first constructions of nets with bounded discrepancy factor D. The main motivation for this work was to provide excellent deterministic sample points for quasi-Monte Carlo methods; a review of this aspect is given, e.g., in [**Nie92**, **Nie04b**].

The study of nearly uniform point sets was taken up by H. Niederreiter [**Nie86**], who put forward the notion of (t, m, s)-nets and derived a bound on $D(\mathcal{N})$ via its parameters. In this paper Niederreiter essentially introduced the NRT metric for the first time. In [**Nie87**] he introduced the digital method for the construction of (t, m, s)-nets and (t, s)-sequences. The notion of a (d, k, m, s)-system was introduced in [**Nie91**]. The idea of algebraic geometry constructions for (t, m, s)-nets and (t, s)-sequences was introduced in [**NX95**, **XN95**] and then widely developed in [**NX96**, **NX98**, **NX01**, **NX02**, **NÖ02**, **NÖ07**]. Connections of the theory of (t, m, s)-sets with linear codes was first observed in [**LMMS96**]; see also [**Nie04a**].

M.Yu. Rosenbloom and M.A. Tsfasman [**RT97**] defined the NRT metric, which they called the m-metric, independently of works by Niederreiter. Their motivation was to generalize the Goppa construction of algebraic geometry codes using not only values of rational functions at given points but also their higher-order derivatives.

M.M. Skriganov further investigated applications of the m-metric to uniform distributions [**Skr01**, **Skr07**]; in [**DS02**], a version of the MacWilliams duality for the NRT metric was given.

The notion of ordered orthogonal arrays (also called hypercubic designs) was introduced in [**Law96**, **MS96**] as a combinatorial equivalent of (t, m, s)-nets in base b. The paper [**BP09**] contains general bounds on OAAs and thus on NRT metric spaces.

More complete information on the subject can be found in [**Nie05**, **Nie92**, **NX01**, **Mat99**] and the references therein. In our exposition in Sec. 12.4 we touched upon the subject in a very superficial manner, which is but natural, since the number of papers on the subject that appeared over the last 20 years amounts to several hundreds. Let us only point out that Theorem 12.4.24 was obtained in [**LMMS96**], and the construction from Theorems 12.4.25 and 12.4.26 was given in [**NX98**].

The notion of LRC codes appeared recently [**HLM07**, **HCL07**], motivated mainly by the fact that in today's storage systems by far the most common scenario is a failure of a single node. However, for several years all the constructions of LRC codes used code alphabets of the size exponential in the code length. This drawback was overcome by I. Tamo and A. Barg in the paper [**TB14**], giving the above-described optimal construction (Sec. 12.5.1), which had won the Best Paper Award of the IEEE Information Theory Society. The constructions and results of Sec. 12.5.2 were given in [**BTV15**, **BTV17**]; note that the latter contains some other constructions and examples of algebraic geometry LRC codes. Theorem 12.5.12 was proved in [**TBF16**].

APPENDIX

Some Basic Facts from Volume 1

In this appendix we formulate some basic definitions and results from our book *Algebraic Geometry Codes: Basic Notions*, which constitutes the first volume of the present book. We keep the notation and numbering of results, which would facilitate using this appendix together with the first volume. The presentation is highly concise, but we hope that it, however, will be useful for the reader of the present volume.

A.1. Codes and Projective Systems

This section contains some basic definitions concerning codes and their geometric counterpart, namely projective systems.

A.1.1. Codes and Their Parameters

Let A be a finite set, called an *alphabet*, of cardinality $q=|A|$. The set $A^n = A\times\ldots\times A$ is equipped with the *Hamming metric*: the distance $d(a,b)$ is the number of coordinates in which a and b differ, i.e.,

$$d((a_1,\ldots,a_n),(b_1,\ldots,b_n)) = |\{i \mid a_i \neq b_i\}|.$$

A nonempty subset $C\subseteq A^n$ is called a *q-ary code of length n*. The *cardinality* of a code is $M=|C|\in\mathbb{N}$, and the log-*cardinality* is $k=\log_q|C|\in\mathbb{R}$. The *minimum distance* of C is the nonnegative integer

$$d=\min\{d(a,b)\mid a,b\in C,\ a\neq b\}.$$

A code with these parameters is called an *$[n,k,d]_q$ code*; note, however, that k may be not an integer. Elements of C are called *code vectors* or *codewords*; their components are called *coordinates* or *positions*.

Sometimes, the following relative parameters are convenient to use: the *code rate* (*transmission rate* or simply *rate*) $R=k/n$ and the *relative minimum distance* $\delta=d/n$. Of course, $0\leq R\leq 1$ and $0\leq\delta\leq 1$.

If $A=\mathbb{F}_q$ is a finite field, then $q=p^m$ for a prime $p=\mathrm{char}(\mathbb{F}_q)$. A *linear code C* of length n is a linear subspace in $\mathbb{F}_q^n$. For a linear code we have

$$k=\dim C,\qquad d=\min\{\|a\|\mid a\in C,\ a\neq 0\},$$

where $\|a\|=|\{i\mid a_i\neq 0\}|$ is the *weight* (*Hamming weight*, or *norm*) of a vector from $\mathbb{F}_q^n$. For the weight, we often use the notation $\mathrm{wt}(a)$. Thus, k is the *dimension* of a linear code.

A choice of a basis in C gives an embedding $C\colon \mathbb{F}_q^k\to\mathbb{F}_q^n$, which we usually denote by the same symbol as the code itself. The matrix of this map in the standard bases, usually denoted by G, is called a *generator matrix* of the code;

linear combinations of its rows are precisely the code vectors. The map C sits in a short exact sequence

$$0 \longrightarrow \mathbb{F}_q^k \xrightarrow{C} \mathbb{F}_q^n \xrightarrow{H} \mathbb{F}_q^{n-k} \longrightarrow 0,$$

i.e., C is an injection, H is a surjection, and the kernel of H coincides with the image of C. It would be even more correct to say that we have $H\colon \mathbb{F}_q^n \to V$, where $\dim_{\mathbb{F}_q} V = n-k$; to define H as a matrix, that is, to write $H\colon \mathbb{F}_q^n \to \mathbb{F}_q^{n-k}$, we should fix a basis in V.

The matrix of the map H, also denoted by H, is called a *parity-check matrix* of C, since the condition $x \in C$ is equivalent to the equality $Hx^T = 0$.

Equivalence. Denote by $\mathcal{A}$ the subgroup in the group of linear automorphisms of $\mathbb{F}_q^n$ generated by permutations of coordinates and multiplications of coordinates x_i by elements $a_i \in \mathbb{F}_q^*$. The group $\mathcal{A}$ acts on subsets of $\mathbb{F}_q^n$, i.e., on codes. Two codes, C and C', are called *equivalent* if $C' = A(C)$ for some $A \in \mathcal{A}$; note, however, that this notion is often understood in the literature in a more restrictive manner, for $\mathcal{A}$ generated only by multiplications (permutations are not allowed).

A.1.2. $[n,k,d]_q$ Systems

Let V be a linear space over a finite field $\mathbb{F}_q$. By an $[n,k,d]_q$ *system* we refer to an ordered finite family $\mathcal{P}$ of points in V (in general, points of $\mathcal{P}$ need not be distinct) such that $\mathcal{P}$ does not lie in a hyperplane; hence, $|\mathcal{P}| \geq \dim V$. The parameters of the system are defined as a triple of integers $[n,k,d]$ with

$$n = |\mathcal{P}|, \qquad k = \dim V, \qquad d = n - \max_H |\mathcal{P} \cap H| \geq 1,$$

where the maximum is taken over all hyperplanes $H \subset V$, and points are counted with their multiplicities.

Two $[n,k,d]_q$ systems, $\mathcal{P}$ and $\mathcal{P}'$, in V and V', respectively, are called *equivalent* if there is an isomorphism $V \xrightarrow{\sim} V'$ taking $\mathcal{P}$ to $\mathcal{P}'$, and we mainly consider classes of this equivalence relation.

PROPOSITION 1.1.4. *There is a one-to-one correspondence between the set of classes of $[n,k,d]_q$ systems and the set of linear $[n,k,d]_q$ codes.* □

Projective systems. Now let $\mathbb{P} = \mathbb{P}(V)$ be a projective space over $\mathbb{F}_q$. A *projective $[n,k,d]_q$ system* is a finite unordered family $\mathcal{P}$ of points in $\mathbb{P}$ that does not lie in any (projective) hyperplane H (note that $|\mathcal{P}| \geq \dim \mathbb{P} + 1 = \dim V$). By abuse of notation, we write $\mathcal{P} \subset \mathbb{P}$, though, in general, there are multiplicities. The parameters $[n,k,d]$ are defined as follows:

$$n = |\mathcal{P}|, \qquad k = \dim \mathbb{P} + 1, \qquad d = n - \max_H |\mathcal{P} \cap H| \geq 1.$$

Just as for $[n,k,d]_q$ systems, two projective systems, $\mathcal{P} \subset \mathbb{P}$ and $\mathcal{P}' \subset \mathbb{P}'$, are called *equivalent* if there is a linear projective isomorphism $\mathbb{P} \cong \mathbb{P}'$ that takes $\mathcal{P}$ to $\mathcal{P}'$.

We call a linear code $C \subset \mathbb{F}_q^n$ *degenerate* if $C \subseteq \mathbb{F}_q^{n-1} \subset \mathbb{F}_q^n$, where $\mathbb{F}_q^{n-1}$ is the subspace of vectors with 0 at some fixed position. Otherwise, we say that the code is *nondegenerate*.

THEOREM 1.1.6. *Let $k \geq 1$ and $d \geq 1$. There is a one-to-one correspondence between the set of equivalence classes of nondegenerate linear $[n,k,d]_q$ codes and the set of equivalence classes of projective $[n,k,d]_q$ systems.* □

Generalized weights. Generalized Hamming weights are another family of parameters of a linear code.

Let $D \subset \mathbb{F}_q^n$ be a linear subspace of dimension r. The Hamming weight, defined on vectors in $\mathbb{F}_q^n$, is naturally generalized to subspaces, namely

$$\|D\| = |\mathrm{Supp}(D)|, \quad \mathrm{Supp}(D) = \{i : \exists\, v \in D,\ v_i \neq 0\}.$$

For a linear code C, define its *rth generalized weight* d_r, $r = 1, \ldots, k$, as

$$d_r = d_r(C) = \min\{\|D\| : D \subset C,\ \dim D = r\}.$$

Obviously, $d_1 = d$ is the minimum distance of the code.

In geometric terms, the definition of higher weights $d_r = d_r(\mathcal{P})$ for projective systems reads as

$$n - d_r = \max\{|\mathcal{P} \cap \Pi^r| : \Pi^r \text{ is a projective subspace of codimension } r \text{ in } \mathbb{P}\}.$$

A.1.3. Bounds

The code parameters are subject to numerous bounds; we list some of them below.

Singleton bound: For any linear $[n,k,d]_q$ code, we have

$$n \geq k + d - 1.$$

Griesmer bound: For a linear $[n,k,d]_q$ code, we have

$$n \geq \sum_{i=0}^{k-1} \left\lceil \frac{d}{q^i} \right\rceil.$$

Plotkin bound: For any $[n,k,d]_q$ code, we have

$$d \leq \frac{nq^k(q-1)}{(q^k-1)q}.$$

Bassalygo–Elias bound: For any $[n,k,d]_q$ code and any integer w with $1 \leq w \leq n$ and

$$A = d - 2w + \frac{qw^2}{(q-1)n} > 0,$$

we have the following inequality:

$$n - k \geq \log_q \binom{n}{w} + w \log_q(q-1) - \log_q d + \log_q A.$$

The above bounds are possibility (upper) bounds; there is also the following existence (lower) bound.

Gilbert–Varshamov bound: Whenever

$$q^{n-k} > \sum_{i=0}^{d-2} \binom{n-1}{i} (q-1)^i,$$

there always exists a linear $[n,k,d]_q$ code.

Note that this is a purely existential result, which does not give a construction of the corresponding codes.

A.1.4. Asymptotic Problems

Main asymptotic problem. Fix the alphabet cardinality q. To a q-ary code C, assign its relative parameters $(\delta(C), R(C))$, i.e., a point in the unit square $[0,1]^2$ on the plane with coordinates (δ, R). All such points form a family of code points (code domain) V_q. Denote by U_q the set of limit points of this family. In other words, $(\delta, R) \in U_q$ if and only if there exists an infinite sequence of different codes C_i such that

$$\lim_{i\to\infty} (\delta(C_i), R(C_i)) = (\delta, R);$$

since there are only finitely many codes of any given length, for any sequence we have $n(C_i) \to \infty$. If for a sequence (family) of codes C_i we have $\delta > 0$ and $R > 0$, it is called *asymptotically good*.

THEOREM 1.3.1. *There exists a decreasing continuous function $\alpha_q(\delta)$, $\delta \in [0,1]$, such that*

$$U_q = \{(\delta, R) \mid 0 \leq R \leq \alpha_q(\delta)\}.$$

Moreover, $\alpha_q(0) = 1$, $\alpha_q(\delta) = 0$ for $(q-1)/q \leq \delta \leq 1$, and $\alpha_q(\delta)$ strictly decreases on the segment $[0, (q-1)/q]$. □

The same is true for the set U_q^{lin} defined using linear codes only. The corresponding function is denoted by $\alpha_q^{\text{lin}}(\delta)$; thus,

$$U_q^{\text{lin}} = \{(\delta, R) \mid 0 \leq R \leq \alpha_q^{\text{lin}}(\delta)\}.$$

Clearly, $\alpha_q^{\text{lin}}(\delta) \leq \alpha_q(\delta)$ for any $\delta \in [0,1]$.

One can also define the functions $\alpha_q^{\text{pol}}(\delta) \leq \alpha_q(\delta)$ and $\alpha_q^{\text{lin,pol}}(\delta) \leq \alpha_q^{\text{lin}}(\delta)$ describing asymptotic parameters of polynomially constructible codes (unrestricted and linear, respectively).

Asymptotic upper bounds. One can translate the bounds presented in Sec. A.1.3 above into the following estimates for $\alpha_q(\delta)$.

Singleton bound:

$$\alpha_q(\delta) \leq 1 - \delta.$$

Griesmer bound:

$$\alpha_q^{\text{lin}}(\delta) \leq 1 - \frac{q}{q-1}\delta.$$

Plotkin bound:

$$\alpha_q(\delta) \leq R_P(\delta) = 1 - \frac{q}{q-1}\delta.$$

Denote by

$$H_q(x) = x \log_q(q-1) - x \log_q x - (1-x)\log_q(1-x)$$

the *q-ary entropy function.*

Hamming bound:

$$\alpha_q(\delta) \leq R_H(\delta) = 1 - H_q(\delta/2).$$

Bassalygo–Elias bound:

$$\alpha_q(\delta) \le R_{BE}(\delta) = 1 - H_q\left(\frac{q-1}{q} - \frac{q-1}{q}\sqrt{1 - \frac{q\delta}{q-1}}\right).$$

McEliece–Rodemich–Rumsey–Welch bound:

$$\alpha_q(\delta) \le R_{MRRW}(\delta) = H_q\left(\frac{(q-1) - \delta(q-2) - 2\sqrt{(q-1)\delta(1-\delta)}}{q}\right).$$

Asymptotic lower bound. Here we get the following.

Gilbert–Varshamov bound:

$$\alpha_q^{\mathrm{lin}}(\delta) \ge R_{GV}(\delta) = 1 - H_q(\delta).$$

A.1.5. Some Code Constructions and Their Parameters

Here by ℓ and m we denote arbitrary integers such that for the resulting codes we have $0 \le k \le n$ and $1 \le d \le n$.

Construction	Parameters of original codes	Parameters of new codes
Lengthening by zeroes	$[n,k,d]_q$	$[n+\ell,k,d]_q$
Omitting parity checks	$[n,k,d]_q$	$[n-\ell,k-\ell,d]_q$
Projection	$[n,k,d]_q$	$[n-\ell,k,d-\ell]_q$
Direct sum	$[n_1,k_1,d_1]_q$ $[n_2,k_2,d_2]_q$	$[n_1+n_2,k_1-k_2,d]_q$ $d=\min\{d_1,d_2\}$
Tensor product	$[n_1,k_1,d_1]_q$ $[n_2,k_2,d_2]_q$	$[n_1n_2,k_1k_2,d_1d_2]_q$
Repetition	$[n,k,d]_q$	$[\ell n,k,\ell d]_q$
Parity check	$[n,k,2t+1]_2$	$[n+1,k,2t+2]_2$
Subfield restriction	$[n,k,d]_{q^m}$	$[n,\ge n-m(n-k),\ge d]_q$
Concatenation	$[n,k,d]_{q^m}$ $[N,m,D]_q$	$[nN,km,\ge dD]_q$
Field descent	$[n,k,d]_{q^m}$	$[mn,mk,d]_q$
Alphabet extension	$[n,k,d]_r, \quad r \le q$	$[n,k\log_q r,d]_q$
Alphabet restriction	$[n,k,d]_r, \quad r \ge q$	$[n,k_r,\ge d]_q$ $k_r \ge n-(n-k)\log_q r$

A.2. Curves over Finite Fields

In this section we give a brief description of algebraic curve theory, especially in the case of a finite ground field.

A.2.1. Algebraic Curves

First let us give some basic definitions and results concerning algebraic curves over a fixed algebraically closed field $\mathbb{k}$.

Quasiprojective varieties. Let $\mathbb{A}^n$ be the n-dimensional affine space over $\mathbb{k}$ and $\mathbb{P}^n$ be the projective space containing $\mathbb{A}^n$; thus, $\mathbb{A}^n \hookrightarrow \mathbb{P}^n$ is given by

$$(x_1,\ldots,x_n) \mapsto (1 : x_1 : \ldots : x_n).$$

A *closed set* X in $\mathbb{A}^n$ (an *affine closed set*) is the set of common zeroes of a finite number of polynomials $F_1,\ldots,F_s \in \mathbb{k}[T_1,\ldots,T_n]$:

$$X = \{P = (x_1,\ldots,x_n) \in \mathbb{A}^n \mid F_1(x_1,\ldots,x_n) = \ldots = F_s(x_1,\ldots,x_n) = 0\}$$

(note that if $s = 0$, then $X = \mathbb{A}^n$). A subset U in $\mathbb{A}^n$ is *open* if its complement $\mathbb{A}^n \setminus U$ is closed in $\mathbb{A}^n$. A *closed* subset Y in $\mathbb{P}^n$ (a *projective closed set*) is the set of common zeroes of a finite number of homogeneous polynomials (*forms*) $G_1,\ldots,G_r \in \mathbb{k}[T_0 : \ldots : T_n]$:

$$Y = \{Q = (y_0 : \ldots : y_n) \in \mathbb{P}^n \mid G_1(y_0 : \ldots : y_n) = \ldots = G_r(y_0 : \ldots : y_n) = 0\}.$$

A complement $\mathbb{P}^n \setminus Y$ of a closed set Y is called *open* in $\mathbb{P}^n$. To any closed affine set $X \subseteq \mathbb{A}^n$ we assign a projective closed set $\widetilde{X} \subseteq \mathbb{P}^n$ containing X, called the *projective closure* of X, which is obtained as follows. Let $F_1 = \ldots = F_s = 0$ be equations defining X, where $F_1,\ldots,F_s \in \mathbb{k}[T_1,\ldots,T_n]$; define the forms $\widetilde{F}_i \in \mathbb{k}[T_0,\ldots,T_n]$ by

$$\widetilde{F}_i(T_0,\ldots,T_n) = \sum_{j=0}^{\deg F_i} T_0^{\deg F_i - j} F_{ij}(T_1,\ldots,T_s), \quad i = 1,\ldots,s,$$

where F_{ij} are forms of degree j involved in the decomposition

$$F_i(T_1,\ldots,T_n) = \sum_{j=0}^{\deg F_i} F_{ij}(T_1,\ldots,T_n)$$

of the polynomial F_i into a sum of homogeneous polynomials. Then $\widetilde{X}$ is defined by

$$\widetilde{F}_1(T_0 : \ldots : T_n) = \ldots = \widetilde{F}_s(T_0 : \ldots : T_n) = 0.$$

Now let $S \subseteq \mathbb{P}^n$ be an arbitrary set. A subset $T \subseteq S$ is called *closed in* S if there exists a projective closed set $X \subseteq \mathbb{P}^n$ such that $T = S \cap X$. A subset $U \subseteq S$ is called *open in* S if its complement $S \setminus U$ is closed in S. This defines a topology on S, known as the *Zariski topology*.

An open subset U of a closed projective set X is called a *quasiprojective set*. In particular, any affine closed set is quasiprojective. A quasiprojective set $X \subseteq \mathbb{P}^n$ is called *reducible* if there exists a pair of nonempty proper closed subsets $X_1, X_2 \subset X$ such that $X = X_1 \cup X_2$. Otherwise, X is called *irreducible*. An irreducible quasiprojective set is referred to as a *quasiprojective variety*. Most often, we omit the term *quasiprojective* and just speak about a *variety*. If a variety is closed in $\mathbb{P}^n$, it is called a *projective* variety.

THEOREM 2.1.3. *Any quasiprojective set X can be uniquely represented as a union of finitely many nonempty varieties X_i such that X_i does not contain X_j for $i \neq j$.* □

The varieties X_i are called *irreducible components* of X.

Dimension. For a variety X, its *dimension* $n = \dim X$ is the largest integer n such that there exists a strictly descending chain of algebraic varieties

$$X = X_0 \supset \ldots \supset X_n \neq \varnothing$$

with X_i closed in X_{i-1} for $i = 1, \ldots, n$.

Rational functions. Let $X \subseteq \mathbb{P}^n$ be a quasiprojective variety. Let $F, G \in \Bbbk[T_0, \ldots, T_n]$ be forms of equal degree in homogeneous coordinates in $\mathbb{P}^n$, and let $G(P) \neq 0$ for at least one point $P \in X$. Then the fraction $F/G \in \Bbbk(T_0 : \ldots : T_n)$ is called a *rational function* on X. Fractions F/G and F'/G' define the same rational function if the form $F'G - FG'$ vanishes at all points $P \in X$. The set of rational functions on X is denoted by $\Bbbk(X)$; it is a field with respect to operations of pointwise addition and multiplication.

A function $f \in \Bbbk(X)$ is called *regular* at $P \in X$ if it admits a representation $f = F/G$ with $G(P) \neq 0$. In this case, $f(P) = F(P)/G(P)$ is called the *value* of f at P.

Rational maps. An n-tuple of rational functions $f = (f_1, \ldots, f_n)$, where $f_i \in \Bbbk(X)$, defines a *rational map* $f\colon X \to \mathbb{A}^n$ from a variety $X \subseteq \mathbb{P}^m$ to $\mathbb{A}^n$. A map f is called *regular at* $P \in X$ if each f_i is regular at P. If a rational map is regular at each $P \in X$, it is called a *regular map* from X to $\mathbb{A}^n$.

A rational map regular at all points $P \in X$ is called a *regular map* (or *morphism*) of varieties. Thus, a rational map of X is a regular map of an open subset of X. If there exists a variety $Y \subseteq \mathbb{P}^n$ such that $F(P) \in Y$ for any $P \in U_f$, we write $f\colon X \to Y$ and refer to f as a *rational map from X to Y*.

Let $f\colon X \to Y$ be a rational map. The set $\operatorname{Im} f = f(U_f)$ is called the *image* of f. If the image of f is dense in Y (i.e., the Zariski closure of $\operatorname{Im} f$ in Y coincides with Y), then f is called a *dominant* map.

Let $f\colon X \to Y$ be a dominant rational map, and let $g \in \Bbbk(Y)$. Define a function $f^*(g)$ by

$$f^*(g)(P) = g(f(P)),$$

where $P \in X$ is a point such that $P \in U_f$ and $f(P) \in U_g$. Thus, a rational dominant map $f\colon X \to Y$ defines a field embedding $f^*\colon \Bbbk(Y) \hookrightarrow \Bbbk(X)$, which is identical on $\Bbbk$.

If the fields $\Bbbk(X)$ and $\Bbbk(Y)$ of two varieties X and Y are isomorphic as $\Bbbk$-algebras, then X and Y are called *birationally isomorphic*.

Local ring of a point. Let P be a point on a variety X. The set of rational functions that are regular at P is denoted by $\mathscr{O}_P$; obviously, $\mathscr{O}_P$ is a ring. When we want to emphasize the dependence of this set on X, we denote it by $\mathscr{O}_{X,P}$. The ring $\mathscr{O}_P$ has a unique maximal ideal $\mathfrak{m}_P$, which consists of functions $f \in \mathscr{O}_P$ such that $f(P) = 0$. The ring $\mathscr{O}_P$ is called the *local ring of* P, and $\mathscr{O}_P/\mathfrak{m}_P = \Bbbk$. A point $P \in X$ is called *smooth* or *regular* if $\dim_{\Bbbk}(\mathfrak{m}_P/\mathfrak{m}_P^2) = \dim X$; otherwise, P is said to be *singular*.

Quasiprojective curves. A *quasiprojective curve* (*algebraic curve* or simply a *curve*) is a quasiprojective variety of dimension 1. A curve can be defined as a variety whose proper closed subvarieties are only points. If a curve is closed in some $\mathbb{P}^n$, it is called *projective* or *complete*.

Until the end of the Appendix X will be a curve.

A point $P \in X$ is nonsingular if and only if the ideal $\mathfrak{m}_P$ in the local ring $\mathscr{O}_P$ is *principal*, i.e., $\mathfrak{m}_P = t_P\mathscr{O}_P$ for some $t_P \in \mathscr{O}_P$. A curve X has only finitely many singular points. If P is a nonsingular point, then any function $t_P \in \mathscr{O}_P$ such that $\mathfrak{m}_P = t_P\mathscr{O}_P$ is called a *local parameter* at P.

Now let $P \in X \cap \mathbb{A}^n$, and let $F_1(x) = \ldots = F_N(x) = 0$ be a complete system of equations defining the affine curve $X \cap \mathbb{A}^n$. Consider the Jacobi matrix

$$\frac{\partial F}{\partial x}(P) = \left(\frac{\partial F_i}{\partial x_j}(P)\right)_{\substack{i=1,\ldots,N\\ j=1,\ldots,n}}.$$

PROPOSITION 2.1.42. *A point $P \in X$ is nonsingular if and only if the rank of $\dfrac{\partial F}{\partial x}(P)$ equals $n-1$.* □

In particular, for a plane curve with affine equation $F(x_1,x_2) = 0$, $P \in X$ is nonsingular if and only if at least one of the partial derivatives $\partial F/\partial x_1$ and $\partial F/\partial x_2$ does not vanish at P.

Smooth complete curves. Smooth complete curves are the most interesting; they possess the following useful property.

THEOREM 2.1.46. *Smooth complete curves are isomorphic if and only if they are birationally isomorphic.* □

Fields $\mathbb{K}$ that occur as $\Bbbk(X)$ for smooth complete curves X have the following description.

PROPOSITION 2.1.49. *A field $\mathbb{K}$ containing $\Bbbk$ is isomorphic to $\Bbbk(X)$ for some smooth complete curve X if and only if $\mathbb{K}$ is of transcendence degree 1 over $\Bbbk$ and is finitely generated over $\Bbbk$.* □

One can characterize all points of a smooth projective curve X as follows. Recall that a ring $\mathcal{O}$, $\Bbbk \subsetneqq \mathcal{O} \subsetneqq \Bbbk(X)$, is called a *valuation ring* if $f \in \Bbbk(X) \setminus \mathcal{O}$ implies $f^{-1} \in \mathcal{O}$.

THEOREM 2.1.50. *Let X be a smooth projective curve. Then the correspondence $P \mapsto \mathscr{O}_P$ is a bijection between the set of points $P \in X$ and the set of valuation rings in $\Bbbk(X)$.* □

The set of valuation rings is called the set of *places* of $\Bbbk(X)$. Therefore, Theorem 2.1.50 establishes an equivalence between places of $\Bbbk(X)$ and points of X.

Divisors. Let X be a smooth complete curve over $\Bbbk$. A *divisor* D on X is a formal finite sum of the form $D = \sum a_P P$, where P are points of X and a_P are integers. The *support* $\operatorname{Supp} D$ of a divisor D is the set $\{P \in X \mid a_P \neq 0\}$. Sometimes we write $D = \sum a_i P_i$ instead of $D = \sum a_P P$, where $a_i = a_{P_i}$. The set of divisors on a curve X is denoted by $\operatorname{Div}(X)$. It is an abelian group with the operation of addition of divisors: if $D = \sum a_P P$ and $E = \sum b_P P$, then $D \pm E = \sum (a_P \pm b_P)P$. The *degree* $\deg P$ of a divisor $D = \sum a_P P$ is the integer $\sum a_P$. The degree map $\operatorname{Div}(X) \to \mathbb{Z}$ is surjective; its kernel, i.e., the subgroup of divisors of degree zero on X, is denoted by $\mathrm{Div}^0(X)$. If all a_P for a divisor $D = \sum a_P P$ are nonnegative, D is called an *effective* divisor; we write $D \geq 0$. If D is effective and $D \neq 0$, it is called a *positive* divisor. The set of effective divisors is denoted by $\mathrm{Div}^+(X)$. This definition induces a partial order on $\mathrm{Div}(X)$; namely, $D \geq F$ if $D - F \in \mathrm{Div}^+(X)$. Note that any divisor is a difference of two effective ones.

Now to any $f \in \Bbbk(X)^*$ we assign the divisor

$$(f) = \sum \mathrm{ord}_P(f)P,$$

called the *divisor of the function* f, where $\mathrm{ord}_P(f)$ is the unique k such that $f \in \mathfrak{m}_P^k \setminus \mathfrak{m}_P^{k+1}$.

Divisors of the form (f) are called *principal*. We have $(fg^{\pm 1}) = (f) \pm (g)$, and therefore principal divisors form a subgroup $P(X)$ in $\mathrm{Div}(X)$. Divisors D_1 and D_2 such that $D_1 - D_2 \in P(X)$ are called *linearly equivalent* (or simply *equivalent*); we write $D_1 \sim D_2$.

THEOREM 2.1.52. *The degree of a principal divisor equals zero.* □

Thus, $P(X) \subseteq \mathrm{Div}^0(X)$. It follows from the theorem that all divisors in a given linear equivalence class are of the same degree.

Let D be a divisor on a curve X. Denote

$$L(D) = \{f \in \Bbbk(X)^* \mid (f) + D \geq 0\} \cup \{0\};$$

then $L(D)$ is a vector space over $\Bbbk$. It is called the *space associated to the divisor* D, and its dimension is denoted by $\ell(D)$.

THEOREM 2.1.53. *The dimension of* $L(D)$ *is finite for any* $D \in \mathrm{Div}(X)$. □

Riemann–Roch theorem. Here is the principal result on algebraic curves.

THEOREM 2.2.17 (Reimann–Roch theorem). *Let* X *be a smooth projective curve. There exists a unique class* K *of divisors on* X *such that for any divisor* $D \in \mathrm{Div}(X)$ *we have*

$$\ell(D) - \ell(K - D) = \deg D - g + 1,$$

where the number $g = \ell(K)$ *is the* genus *of* X. □

The class $K = K_X$ is called the *canonical class* of X. Note that $D \in K_X$ if and only if $D = (\omega)$ is a divisor associated to a rational differential form ω, i.e., an expression of the form $f\,dg$ for $f, g \in \Bbbk(X)^*$, which satisfies the usual rules of differential algebra, namely the Leibniz formula $d(g_1 g_2) = g_1\,dg_2 + g_2\,dg_1$, etc. The divisor $(\omega) = \sum a_P P$ for $\omega = f\,dg$ is defined as follows: $a_P = \mathrm{ord}_P(f)$ if $g \in \mathcal{O}_P$ and it is possible to choose such a representation for any $P \in X$. One can also define the residue $\mathrm{res}_P(\omega) \in \Bbbk$ in the usual way, as the coefficient a_{-1} in the expansion $\omega = \sum_{i=-N}^{\infty} a_i t_P^i$; it does not depend on t_P.

A.2.2. Algebraic Function Fields

Sometimes it is more convenient to work with an algebraic avatar of curves, namely with algebraic function fields. Note that the theory of algebraic function fields is equivalent to that of algebraic curves. Now let our ground field $\Bbbk$ be arbitrary, but perhaps not algebraically closed.

Function fields. An *algebraic function field* $\mathbb{K}/\Bbbk$ *of one variable over* $\Bbbk$ (or simply a *function field* $\mathbb{K}/\Bbbk$) is any field $\mathbb{K} \supseteq \Bbbk$ which is a finite algebraic extension of $\Bbbk(x)$, where x is an element of $\mathbb{K}$ transcendental over $\Bbbk$ or, equivalently, a finite-generated extension of $\Bbbk$ of transcendence degree 1.

The set of elements of $\mathbb{K}$ that are algebraic over $\Bbbk$ is a subfield of $\mathbb{K}$, which is called the *field of constants* of $\mathbb{K}/\Bbbk$ and is denoted by $\widetilde{\Bbbk}$. We have $\Bbbk \subseteq \widetilde{\Bbbk} \subset \mathbb{K}$, and it is easily verified that $\mathbb{K}/\widetilde{\Bbbk}$ is also a function field over $\widetilde{\Bbbk}$. We say that $\Bbbk$ is *algebraically closed in* $\mathbb{K}$ (or is the *full constant field* of $\mathbb{K}$) if $\widetilde{\Bbbk} = \Bbbk$, and below we always assume that this is the case.

Recall that a *valuation ring* of the function field $\mathbb{K}/\Bbbk$ is a proper subring $\mathscr{O} \subset \mathbb{K}$ such that $\Bbbk \subset \mathscr{O}$, and for any $f \in \mathbb{K}$ at least one of the elements f and f^{-1} belongs to $\mathscr{O}$. The set $\mathscr{O}^* = \{f \in \mathbb{K} \mid f \in \mathscr{O}, f^{-1} \in \mathscr{O}\}$ is called the *group of units* of $\mathscr{O}$.

PROPOSITION 2.5.1. *Let $\mathscr{O}$ be a valuation ring of a function field $\mathbb{K}/\Bbbk$. Then*

(a) *$\mathscr{O}$ is a local ring; i.e., $\mathscr{O}$ has a unique maximal ideal P, and $P = \mathscr{O} \setminus \mathscr{O}^*$.*
(b) *If $f \in \mathbb{K}$, $f \neq 0$, then $f \in P$ if and only if $f^{-1} \notin \mathscr{O}$.*
(c) *$\widetilde{\Bbbk} \subseteq \mathscr{O}$ and $\widetilde{\Bbbk} \cap P = \{0\}$.*
(d) *P is a principal ideal.*
(e) *If $P = t\mathscr{O}$, then any nonzero element $f \in \mathbb{K}$ has a unique representation of the form $f = t^n u$, where $n \in \mathbb{Z}$ and $u \in \mathscr{O}^*$.*
(f) *$\mathscr{O}$ is a principal ideal domain; namely, if I is a nontrivial proper ideal in $\mathscr{O}$ and $P = t\mathscr{O}$, then $I = t^n\mathscr{O}$ for some $n \in \mathbb{N}$.* □

A ring obeying these properties is called a *discrete valuation ring*.

Places of a function field. A *place* P of a function field $\mathbb{K}/\Bbbk$ is the maximal ideal of some valuation ring $\mathscr{O}$ of $\mathbb{K}/\Bbbk$. Any element $t \in P$ such that $P = t\mathscr{O}$ is called a *local parameter at* P (also called a *prime element* or *uniformizing variable*). The set of all places of a function field $\mathbb{K}$ is denoted by $\mathbb{P}_{\mathbb{K}}$; note that $\mathbb{P}_{\mathbb{K}}$ is always nonempty.

Since a valuation ring $\mathscr{O}$ is uniquely determined by its maximal ideal P, namely $\mathscr{O} = \{f \in \mathbb{K} \mid f^{-1} \notin P\}$, $\mathscr{O} = \mathscr{O}_P$ is called the *valuation ring of the place* P. Note that Theorem 2.1.50 establishes a bijection between places of $\Bbbk(X)$ and points of X in the case of algebraically closed $\Bbbk$.

A *discrete valuation* of $\mathbb{K}/\Bbbk$ is a function $v\colon \mathbb{K}^* \to \mathbb{Z}$ with the following properties:

(1) $v(fg) = v(f) + v(g)$ for any $f, g \in \mathbb{K}$;
(2) $v(f+g) \geq \min\{v(f), v(g)\}$ for any $f, g \in \mathbb{K}$;
(3) there exists an element $f \in \mathbb{K}$ with $v(f) = 1$;
(4) $v(c) = 0$ for any $c \in \Bbbk \setminus \{0\}$.

We extend v to the whole $\mathbb{K}$, putting by definition $v(0) = \infty$.

For each place $P \in \mathbb{P}_{\mathbb{K}}$, the choice of a local parameter t yields a unique representation of every nonzero element $f \in \mathbb{K}$ in the form $f = t^n u$, $u \in \mathscr{O}^*$. The function $v_P\colon \mathbb{K} \setminus \{0\} \to \mathbb{Z}$ equal to the exponent n in this representation is well defined.

THEOREM 2.5.5. *Let $\mathbb{K}/\mathbb{k}$ be a function field.*

(a) *For any place $P \in \mathbb{P}_{\mathbb{K}}$, the function υ_P is a discrete valuation. Moreover, we have*

$$\mathscr{O}_P = \{f \in \mathbb{K} \mid \upsilon_P(f) \geq 0\},$$
$$\mathscr{O}_P^* = \{f \in \mathbb{K} \mid \upsilon_P(f) = 0\},$$
$$P = \{f \in \mathbb{K} \mid \upsilon_P(f) > 0\},$$

and an element $t \in \mathbb{K}$ is a local parameter for P if and only if $\upsilon_P(t) = 1$.

(b) *Conversely, let υ be a discrete valuation of the function field $\mathbb{K}/\mathbb{k}$. Then the set $P = \{f \in \mathbb{K} \mid \upsilon(f) > 0\}$ is a place of $\mathbb{K}$, and $\mathscr{O}_P = \{f \in \mathbb{K} \mid \upsilon(f) \geq 0\}$ is the corresponding valuation ring.*

(c) *Any valuation ring $\mathscr{O}$ of $\mathbb{K}/\mathbb{k}$ is a maximal proper subring of $\mathbb{K}$.* □

Degree of a place. We have $\mathbb{k} \subseteq \mathscr{O}_P$ and $\mathbb{k} \cap P = \{0\}$, so the canonical residue class map $\mathscr{O}_P \to \mathscr{O}_P/P$ induces a canonical embedding $\mathbb{k} \hookrightarrow \mathbb{k}_P$, and we call $\deg P = [\mathbb{k}_P : \mathbb{k}]$ the *degree* of a place P. The degree of a place is always finite; moreover, we have the following.

PROPOSITION 2.5.7. *Let P be a place of a function field $\mathbb{K}/\mathbb{k}$, and let x be a nonzero element of P. Then*

$$\deg P \leq [\mathbb{K} : \mathbb{k}(x)].$$ □

A.2.3. Finite Field Case

Now let us assume that the base field $\mathbb{k}$ is finite, $\mathbb{k} = \mathbb{F}_q$. In this case the number of points of any curve is finite, and its study leads to many results expressed mostly in terms of zeta functions of curves.

Zeta function. Now let $\mathbb{k} = \mathbb{F}_q$, $q = p^a$, be a finite field. Define $N_r = N_r(X) = |X(\mathbb{F}_{q^r})|$; in particular, $N = N_1$ is the number of $\mathbb{F}_q$-points of X or of degree-one places of $\mathbb{F}_q(X)$. It is convenient to consider the generating function of the numbers N_r, i.e., a formal power series

$$Z(t) = Z(X;t) = \exp\left(\sum_{r=1}^{\infty} \frac{N_r t^r}{r}\right) \in \mathbb{Q}[[t]],$$

called the *zeta function of the curve* X over $\mathbb{F}_q$.

In addition to N_r, it is sometimes important to consider B_r, the number of points of degree r of X, which satisfies the equation

$$N_r = \sum_{d|r} dB_d.$$

Let M_n be the number of effective divisors of degree n on X; note that $M_0 = 1$.

PROPOSITION 3.1.3. *For the zeta function of a curve X, we have the identities*

$$Z(t) = \prod_{d=1}^{\infty} \left(1 - t^d\right)^{-B_d} = \sum_{n=0}^{\infty} M_n t^n.$$ □

An immediate consequence is as follows.

PROPOSITION 3.1.6. *We have*

$$N_r = \frac{1}{(r-1)!} \frac{d^r}{dt^r} (\log Z(t)) \bigg|_{t=0}.$$ □

The *class number* h of the function field $\mathbb{F}_q(X)$ is the order of the group $\mathrm{Pic}^0_{\mathbb{F}_q}(X)$ of linear equivalence classes of degree-zero $\mathbb{F}_q$-divisors.

PROPOSITION 3.1.7. *The class number is finite.* □

Rationality.

PROPOSITION 3.1.9. *The power series* $Z(t)=\sum_{n=0}^{\infty} M_n t^n$ *converges for* $|t|<q^{-1}$. *Moreover, for* $|t|<q^{-1}$ *we have the following equalities*:

(a) *For a curve of genus* 0 (*i.e., for the projective line*), *we have*

$$Z(t)=\frac{1}{(1-t)(1-qt)}.$$

(b) *For a curve of genus* $g\geq 1$, *the zeta function is of the form* $Z(t)=F(t)+G(t)$ *with*

$$F(t)=\frac{1}{q-1}\sum_{\substack{D\in \mathrm{Pic}(X)\\ 0\leq \deg D\leq 2g-2}} q^{\ell(D)}t^{\deg D},$$

$$G(t)=\frac{h}{q-1}\left(q^g t^{2g-1}\frac{1}{1-qt}-\frac{1}{1-t}\right).$$ □

COROLLARY 3.1.10. *The function* $Z(t)$ *can be extended to a rational function on* $\mathbb{C}$. *It has a simple pole at* $t=1$. □

Functional equation. One of the key properties of the zeta function of a curve is its functional equation.

THEOREM 3.1.15. *The zeta function satisfies the functional equation*

$$Z(t)=q^{g-1}t^{2g-2}Z\Big(\frac{1}{qt}\Big).$$ □

In particular, the zeta function is of the form

$$Z(t)=\frac{P(t)}{(1-t)(1-qt)},$$

where $P(t)$ is a polynomial of degree $2g$ with integer coefficients.

PROPOSITION 3.1.16. *The polynomial* $P(t)$ *has the following properties*:

(a) $P(0)=1$, *and* $P(1)=h$ *is the class number.*

(b) (*functional equation*)

$$P(t)=q^g t^{2g}P\Big(\frac{1}{qt}\Big).$$

(c) *Let* $P(t)=a_0+a_1t+\ldots+a_{2g}t^{2g}$. *Then*

$$a_{2g-i}=q^g a_i,\quad 0\leq i\leq g;$$

in particular, $a_{2g}=q^g$, *whence* $\deg P(t)=2g$.

(d) $a_1=N-(q+1)$, *where* $N=N_1$ *is the number of* $\mathbb{F}_q$*-points of the curve.* □

PROPOSITION 3.1.19. (a) *In the expansion*

$$P(t) = \prod_{i=1}^{2g} (1 - \omega_i t),$$

the inverse roots ω_i of $P(t)$ are algebraic integers, and they can be arranged in such a way that $\omega_i \omega_{g+i} = q$ holds for $i = 1, \dots, g$.

(b) *The polynomial $P_r(t)$ corresponding to the zeta function $Z_r(t)$ of a curve X over $\mathbb{F}_{q^r}$ can be written as*

$$P_r(t) = \prod_{i=1}^{2g} (1 - \omega_i^r t)$$

with the same ω_i for all $i = 1, \dots, 2g$. □

COROLLARY 3.1.20. *For any $r \geq 1$,*

$$N_r = q^r + 1 - \sum_{i=1}^{2g} \omega_i^r.$$

In particular,

$$|X(\mathbb{F}_q)| = q + 1 - \sum_{i=1}^{2g} \omega_i = q + 1 - a_1.$$ □

Weil theorem and its corollaries. There is the following fundamental result.

THEOREM 3.1.21 (Weil theorem). *We have*

$$|\omega_i| = \sqrt{q}, \quad i = 1, \dots, 2g.$$ □

This theorem is often referred to as the *Riemann hypothesis for algebraic function fields*.

PROPOSITION 3.1.23. *For the class number h (the number of $\mathbb{F}_q$-points of the Jacobian), we have the bound*

$$(\sqrt{q} - 1)^{2g} \leq h \leq (\sqrt{q} + 1)^{2g}.$$ □

The formula for N_r in Corollary 3.1.20 yields the following bound:

$$|N_r - (q^r + 1)| \leq 2q^{r/2} g.$$

An important particular case is as follows.

THEOREM 3.1.24 (Weil bound). *For the number of points $N = N_1 = |X(\mathbb{F}_q)|$, we have*

$$|N - (q + 1)| \leq 2g\sqrt{q}.$$ □

In fact, this bound can be slightly improved, the improvement being notable for large g.

THEOREM 3.1.25 (Serre bound). *For the number of points N, we have*

$$|N - (q + 1)| \leq g \lfloor 2\sqrt{q} \rfloor.$$ □

Explicit formula. It is sometimes convenient to represent ω_j as $\omega_j = \sqrt{q}\alpha_j$ with $\alpha_j = e^{i\theta_j}$, $\theta_j \in (-\pi, \pi]$; in these terms, we can write the following formula for the number of points on a curve.

THEOREM 3.1.27 (explicit formula). *Let $c_1, c_2, \ldots$ be arbitrary real numbers. Consider the functions*

$$f_m(\varphi) = 1 + 2\sum_{r=1}^{m} c_r \cos(r\varphi)$$

and

$$\psi_m(t) = \sum_{r=1}^{m} c_r t^r.$$

Then

$$N\psi_m(q^{-1/2}) = \psi_m(q^{1/2}) + \psi_m(q^{-1/2}) + g - \sum_{i=1}^{g} f_m(\theta_i) - \sum_{r=1}^{m}(N_r - N)c_r q^{-r/2},$$

where $\omega_1, \ldots, \omega_{2g}$ are indexed in such a way that $\theta_{g+i} = -\theta_i$. □

Asymptotic behaviour and the Drinfeld–Vlăduţ theorem. For a fixed finite field $\mathbb{F}_q$, consider the behaviour (as $g \to \infty$) of $N_q(g)$, the maximal number of points on a curve of genus g, introducing the parameter

$$A(q) = \limsup_{g\to\infty} \frac{N_q(g)}{g}.$$

The Serre bound (Theorem 3.1.25) yields

$$A(q) \leq \lfloor 2\sqrt{q} \rfloor,$$

but in fact the following much tighter bound holds.

THEOREM 3.1.29 (Drinfeld–Vlăduţ bound). *We have*

$$A(q) \leq \sqrt{q} - 1.$$ □

Lower asymptotic bounds. Remarkably, the upper bound for $A(q)$ is attained if q is a square.

THEOREM 3.2.3. *Let $q = p^{2m}$; then*

$$A(q) = \sqrt{q} - 1 = p^m - 1.$$ □

The answer for odd powers is not known, and lower bounds are rather far from $\sqrt{q} - 1$.

THEOREM 3.2.4. *For any $\mathbb{F}_q$, $q = p^e$, we have*

(1) $A(q) \geq \frac{1}{96}\log_2 q$.

(2) *For $e > 1$,* $A(q) \geq \frac{2}{q-2}$.

(3) *For $e = 3$,* $A(q) \geq \frac{2(q^{2/3} - 1)}{q^{1/3} + 2}$.

(4) *If ℓ is a prime, $\ell \mid (q-1)$, and $q > 4\ell + 1$, then*

$$A(q^\ell) \geq \frac{\sqrt{\ell(q-1)} - 2\ell}{\ell - 1}.$$

(5) *If q is odd and is not a square, then*

$$A(q^m) \geq \frac{2q+2}{\lceil 2\sqrt{2q+3}\rceil + 1}.$$

(6) *If $q \geq 8$ is even and $m \geq 3$ is odd, then*

$$A(q^m) \geq \frac{q+1}{\lceil 2\sqrt{2q+2}\rceil + 2}.$$

(7) $A(2) \geq 81/317 = 0.2555\ldots.$

(8) $A(3) \geq 8/17 = 0.4705\ldots.$

(9) $A(5) \geq 8/11 = 0.7272\ldots.$ □

A.3. Algebraic Geometry Codes

Here we give some basic facts on algebraic geometry codes.

A.3.1. Constructions and Their Parameters

***L*-construction.** Let X be a curve such that $X(\mathbb{F}_q) \neq \varnothing$. Let $\mathcal{P} \subseteq X(\mathbb{F}_q)$, $|\mathcal{P}| = n$, and let $D \in \mathrm{Div}(X)$. Assume that $\mathrm{Supp}\, D \cap \mathcal{P} = \varnothing$.

Consider the *evaluation map*

$$\mathrm{Ev}_{\mathcal{P}}\colon L(D) \to \mathbb{F}_q^n, \quad f \mapsto (f(P_1), \dots, f(P_n)),$$

where $\mathcal{P} = \{P_1, \dots, P_n\}$. We get a code $C = \mathrm{Ev}_{\mathcal{P}}(L(D)) \subseteq \mathbb{F}_q^n$. For this code, we use the notation $C = (X, \mathcal{P}, D)_L$.

Assume that D is chosen in such a way that any function $f \in L(D)$ has at most b zeroes at $\mathbb{F}_q$-points of X. If $n > b$, then $\mathrm{Ev}_{\mathcal{P}}$ is an embedding, and we have

$$k = \ell(D), \qquad d \geq n - b.$$

The Riemann–Roch theorem allows us to estimate the parameters of C.

THEOREM 4.1.1. *Let X be a curve of genus g, and let $0 \leq \deg D = a < n = |\mathcal{P}|$. Then $C = (X, \mathcal{P}, D)_L$ is an $[n,k,d]_q$ code with*

$$k \geq a - g + 1, \qquad d \geq n - a. \qquad \square$$

EXAMPLE 4.1.5. Let $X = \mathbb{P}^1$ and $D = a \cdot \infty$; then $L(D)$ is the space of polynomials of degree at most a. If we take for $\mathcal{P}$ all $\mathbb{F}_q$-points of $\mathbb{P}^1$ except for ∞, i.e., elements of the field $\mathbb{F}_q$, then the obtained $[q, a+1, q-a]_q$ code is a Reed–Solomon code.

***P*-construction.** Since the equivalence class of a nondegenerate linear $[n,k,d]_q$ code C is uniquely determined by a projective $[n,k,d]_q$ system $\mathcal{P}$, we can give the following construction.

If a variety X is given with a fixed embedding $X \subset \mathbb{P}^m$, then, choosing any $\mathcal{P} \subseteq X(\mathbb{F}_q)$, we obtain a projective $[n,k,d]_q$ system with $n = |\mathcal{P}|$, $k = m+1$, and $d = n - \max_H\{|H \cap \mathcal{P}|\}$, where the maximum is taken over all hyperplanes $H \subset \mathbb{P}^m$ defined over $\mathbb{F}_q$.

If X is a curve (with a fixed embedding), then by definition its degree $\deg X$ is the number of $\overline{\mathbb{F}}_q$-points (counted with appropriate multiplicities) in the intersection $X \cap H$ for an arbitrary hyperplane H. Thus, $\max_H\{|H \cap \mathcal{P}|\} \leq \deg X$.

PROPOSITION 4.1.18. *Let a curve $X \subset \mathbb{P}^m$ be given, and let $N = |X(\mathbb{F}_q)| > 0$. Then for any n in the interval $N \geq n > \max\{m, \deg X\}$ there exists a nondegenerate linear $[n,k,d]_q$ code with $k = m+1$ and $d \geq n - \deg X$.* $\square$

In fact, for divisors D with $\deg D \geq 2g+1$, the above L-construction can be obtained from P-construction, since D leads to an embedding $X \hookrightarrow \mathbb{P}L(D)$ of X into the projectivization of the space $L(D)$. In this situation we can define the code $(X, \mathcal{P}, D)_P$ via its P-construction.

A.3.2. Example: Hermitian Curves and Codes

A remarkable example of algebraic geometry codes is given by codes on Hermitian curves. Let $q = r^2$. The Hermitian curve X_r over $\mathbb{F}_q$ is given by the projective equation

$$X^{r+1}+Y^{r+1}=Z^{r+1}$$

and is isomorphic to the curve X'_r with the affine equation

$$v^r+v=u^{r+1}.$$

It has r^3+1 points over $\mathbb{F}_q$, namely, the unique point at infinity (the common pole of the functions u and v) and points with coordinates (α,β), where α runs over all elements of $\mathbb{F}_q$ and β runs over all solutions of $\beta^r+\beta=\alpha^{r+1}$.

We denote the common pole of u and v by Q, and a point with coordinates (α,β) by $P_{\alpha,\beta}$. The following properties (Exercises 4.4.35–4.4.40) hold.

(a) For the principal divisors of the functions $u-\alpha$ and $v-\beta$ we have the following expressions:

$$(u-\alpha)=\sum_{\beta^r+\beta=\alpha^{r+1}} P_{\alpha,\beta}-rQ;$$

$$(v-\beta)=\begin{cases}(r+1)P_{0,\beta}-(r+1)Q & \text{if } \beta^r+\beta=0,\\ \sum\limits_{\alpha^{r+1}=\beta^r+\beta} P_{\alpha,\beta}-(r+1)Q & \text{if } \beta^r+\beta\neq 0.\end{cases}$$

(b) The principal divisor of $z=u^q-u$ is $(z)=\boldsymbol{P}-r^3Q$, where $\boldsymbol{P}=\sum\limits_{\alpha,\beta} P_{\alpha,\beta}$; the residue of the differential $\omega=dz/z$ at any point $P_{\alpha,\beta}$ is 1, and the divisor (ω) is

$$(\omega)=(r^3+r^2-r-2)Q-\boldsymbol{P}.$$

(c) The $\mathbb{F}_q$-automorphisms of X'_r of the form $\sigma(\gamma,\delta)$, where $\sigma(\gamma,\delta)$ is given by

$$\sigma u=u+\gamma, \qquad \sigma v=v+\gamma^r u+\delta,$$

for any $\gamma,\delta\in\mathbb{F}_q$ such that $\delta^r+\delta=\gamma^{r+1}$, constitute a subgroup G_r of order r^3 in $\mathrm{Aut}_{\mathbb{F}_q}(X'_r)$ which acts transitively on the set $\{P_{\alpha,\beta}\}$ and stabilizes the point Q.

Let $n=r^3$, and let m be an arbitrary positive integer. Denote by C_m the following q-ary code:

$$C_m=(X'_r,\mathcal{P},mQ)_L,$$

where $\mathcal{P}=\operatorname{Supp}\boldsymbol{P}=\{P_{\alpha,\beta}\}=X'_r(\mathbb{F}_q)\setminus\{Q\}$ is the set of affine $\mathbb{F}_q$-points of X'_r.

For any $m\le r^3+r^2-r-2$, the codes C_m and $C_{r^3+r^2-r-2-m}$ are dual. In particular, if q is even and $m=(r^3+r^2-r-2)/2$, the code C_m is self-dual.

Now let us describe the dimension k_m of the code C_m.

The space $L(mQ)$ has a basis of the following form:

$$\{u^iv^j \mid 0\le i,\ 0\le j\le r-1,\ ir+j(r+1)\le m\}.$$

For $m\in\mathbb{N}$, define an integer $\nu(m)$ as follows: let $\widetilde{m}$ be the largest integer of the form $ir+j(r+1)$, $i\ge 0$, $0\le j\le r-1$, not greater than m, and let

$$\widetilde{m}=tr+s, \quad 0\le s\le r-1.$$

Define

$$\nu(m)=1+\frac{t(t+1)}{2}+\min\{t,s\}.$$

Then $k_m = \dim C_m$ is given by

$$k_m = \begin{cases} \nu(m) & \text{if } m \le r^2 - r - 2, \\ m+1-\dfrac{r^2-r}{2} & \text{if } r^2-r-2 < m < r^3, \\ r^3 - \nu(r^3+r^2-r-2-m) & \text{if } m \ge r^3. \end{cases}$$

A generator matrix of C_m can be calculated as follows. Denote by M_m the $\nu(m) \times n$ matrix whose rows are the following vectors $u(\alpha,\beta)$ of length $\nu(m)$:

$$u(\alpha,\beta) = (\alpha^i \beta^j), \quad ir + j(r+1) \le m, \quad 0 \le i, \quad 0 \le j \le r-1,$$

for all $\alpha, \beta \in \mathbb{F}_q$ such that $\alpha^{r+1} = \beta^r + \beta$.

PROPOSITION 4.4.41. (a) *For $0 \le m < n = r^3$, M_m is a generator matrix of C_m.*

(b) *For $r^2 - r - 2 < m \le r^3 + r^2 - r - 2$, $M_{r^3+r^2-r-2-m}$ is a parity-check matrix of C_m.* □

For the minimum distance d_m of C_m, we have the usual estimate $d_m \ge r^3 - m$. In some cases, it is tight:

PROPOSITION 4.4.42. *Let $m = ir + j(r+1) \le r^3 - 1$ for some $i \ge 0$ and $0 \le j \le r-1$. If either $j = 0$ or $i \le r^2 - r - 1$ (in particular, if $m \le r^3 - r^2$), then $d_m = r^3 - m$.* □

A.3.3. Asymptotic Results

The basic algebraic geometry bound and its variants. The basic algebraic geometry construction applied to families of curves of growing genus yields the following results.

THEOREM 4.5.1. *If over $\mathbb{F}_q$ there is a family of curves X_i of genus $g_i \to \infty$ such that*

$$\gamma = \liminf \frac{g_i}{N_i} < 1,$$

where $N_i = |X_i(\mathbb{F}_q)|$, then

$$\alpha_q^{\text{lin}}(\delta) \ge 1 - \gamma - \delta.$$ □

Recall that

$$A(q) = \limsup \frac{N}{g},$$

where the limit is taken over all curves defined over $\mathbb{F}_q$; introduce the quantity

$$\gamma_q = \frac{1}{A(q)}.$$

COROLLARY 4.5.2. *We have*

$$\alpha_q^{\text{lin}} \ge 1 - \gamma_q - \delta.$$ □

COROLLARY 4.5.3. *Let q be an even power of a prime. Then*

$$\alpha_q^{\text{lin}}(\delta) \ge R_{TVZ}(\delta) = 1 - (\sqrt{q}-1)^{-1} - \delta.$$ □

The bound $R_{TVZ}(\delta)$ is known as the *basic algebraic geometry bound* or *the Tsfasman–Vlăduţ–Zink bound*.

THEOREM 4.5.4. *The basic algebraic geometry bound R_{TVZ} totally lies below the Gilbert–Varshamov bound R_{GV} for $q=p^{2m}<49$. For $q=p^{2m}\geq 49$ these bounds intersect, and the basic algebraic geometry bound improves the Gilbert–Varshamov bound on the segment (δ_1,δ_2), where δ_1 and δ_2 are the roots of the equation*

$$H_q(\delta)-\delta=(\sqrt{q}-1)^{-1}. \qquad \square$$

Expurgation bound. Applying the expurgation method to algebraic geometry codes, namely expurgating over the set of divisors of a fixed degree, we can prove the following.

THEOREM 4.5.13. *Let q be an even power of a prime. Fix a family $\{X_i\}$ of curves of genus $g_i\to\infty$ over $\mathbb{F}_q$ such that $\lim\frac{|X_i(\mathbb{F}_q)|}{g_i}=\sqrt{q}-1$, and let $\beta_q(\delta)$ be the lower asymptotic bound for linear codes of the form $C=(X_i,X_i(\mathbb{F}_q),D)_P$.*

(a) *If $q<49$, then*

$$\alpha_q^{\mathrm{lin}}(\delta)\geq\beta_q(\delta)\geq R_{GV}(\delta).$$

(b) *If $q\geq 49$, put $\gamma=(\sqrt{q}-1)^{-1}$, and let δ_1' and δ_4' be the two roots of the equation*

$$H_q(\delta)+\frac{q}{q-1}(1-\delta)=1+\gamma$$

and δ_2' and δ_3' be the two roots of

$$H_q(\delta)+(1-\delta)\log_q(q-1)=1+\gamma,$$

$$0<\delta_1'<\delta_2'<\delta_3'<\delta_4'<\frac{q-1}{q}.$$

Then

$$\alpha_q^{\mathrm{lin}}(\delta)\geq\beta_q(\delta)\geq R_V(\delta),$$

where

$$R_V(\delta)=\begin{cases}R_{GV}(\delta) & \text{for } \delta\in[0,\delta_1']\cup\left[\delta_4',\frac{q-1}{q}\right],\\ R_{TVZ}(\delta)=1-\gamma-\delta & \text{for } \delta\in[\delta_2',\delta_3'],\\ R_V^0(\delta) & \text{for } \delta\in[\delta_1',\delta_2']\cup[\delta_3',\delta_4']\end{cases}$$

with the function $R_V^0(\delta)$ given by the implicit equation

$$(R_V^0(\delta)+\gamma)H_q\Big(\frac{1-\delta}{R_V^0(\delta)+\gamma}\Big)+H_q(\delta)=1+\gamma. \qquad \square$$

The bound R_V is always not worse than both R_{GV} and R_{TVZ}.

Concatenation bound. The following result is valid for any prime power q.

THEOREM 4.5.22. *Define*

$$R_{KTV(\mathrm{lin})}(\delta)=\max\left\{\left(1-(q^{k/2}-1)^{-1}\right)\frac{k}{n}-\frac{k}{d}\delta\right\},$$

where the maximum is taken over all linear q-ary $[n,k,d]_q$ codes such that if q is an odd power of a prime, then k is even. Then

$$\alpha_q^{\mathrm{pol,lin}}(\delta)\geq R_{KTV(\mathrm{lin})}(\delta). \qquad \square$$

Bibliography

[Ahl73] Rudolf Ahlswede, *Channel capacities for list codes*, J. Appl. Probability **10** (1973), no. 4, 824–836, DOI 10.2307/3212384. MR0449876

[ALT01] A. Ashikhmin, S. Litsyn, and M.A. Tsfasman, *Asymptotically good quantum codes*, Phys. Rev. A **63**, (2001), no. 3, p. 032311.

[Atk67] A. O. L. Atkin, *Weierstrass points at cusps* $\Gamma_o(n)$ (German), Ann. of Math. (2) **85** (1967), 42–45, DOI 10.2307/1970524. MR0218561

[Aub92] Yves Aubry, *Reed-Muller codes associated to projective algebraic varieties*, Coding theory and algebraic geometry (Luminy, 1991), Lecture Notes in Math., vol. 1518, Springer, Berlin, 1992, pp. 4–17, DOI 10.1007/BFb0087988. MR1186411

[Bal08a] Stéphane Ballet, *A note on the tensor rank of the multiplication in certain finite fields*, Algebraic geometry and its applications, Ser. Number Theory Appl., vol. 5, World Sci. Publ., Hackensack, NJ, 2008, pp. 332–342, DOI 10.1142/9789812793430_0017. MR2484062

[Bal08b] Stéphane Ballet, *On the tensor rank of the multiplication in the finite fields*, J. Number Theory **128** (2008), no. 6, 1795–1806, DOI 10.1016/j.jnt.2007.06.010. MR2419195

[BBGS14] Alp Bassa, Peter Beelen, Arnaldo Garcia, and Henning Stichtenoth, *Galois towers over non-prime finite fields*, Acta Arith. **164** (2014), no. 2, 163–179, DOI 10.4064/aa164-2-6. MR3224833

[BBGS15] Alp Bassa, Peter Beelen, Arnaldo Garcia, and Henning Stichtenoth, *Towers of function fields over non-prime finite fields* (English, with English and Russian summaries), Mosc. Math. J. **15** (2015), no. 1, 1–29, 181. MR3427409

[BCS82] A. Bos, J. H. Conway, and N. J. A. Sloane, *Further lattice packings in high dimensions*, Mathematika **29** (1982), no. 2, 171–180 (1983), DOI 10.1112/S0025579300012262. MR696873

[Ber67] E. R. Berlekamp, *Factoring polynomials over finite fields*, Bell System Tech. J. **46** (1967), 1853–1859, DOI 10.1002/j.1538-7305.1967.tb03174.x. MR0219231

[Ber68] Elwyn R. Berlekamp, *Algebraic coding theory*, McGraw-Hill Book Co., New York-Toronto, Ont.-London, 1968. MR0238597

[BH08] Peter Beelen and Tom Høholdt, *The decoding of algebraic geometry codes*, Advances in algebraic geometry codes, Ser. Coding Theory Cryptol., vol. 5, World Sci. Publ., Hackensack, NJ, 2008, pp. 49–98, DOI 10.1142/9789812794017_0002. MR2509121

[BK75] B.J. Birch and W. Kuyk, eds., *Modular functions of one variable. IV*, vol. 476 of Lecture Notes in Math., Berlin: Springer, 1975.

[BLBR10] Stéphane Ballet, Dominique Le Brigand, and Robert Rolland, *On an application of the definition field descent of a tower of function fields* (English, with English and French summaries), Arithmetics, geometry, and coding theory (AGCT 2005), Sémin. Congr., vol. 21, Soc. Math. France, Paris, 2010, pp. 187–203. MR2856567

[Bli86] V. M. Blinovskiĭ, *Bounds for codes in decoding by a list of finite length* (Russian), Problemy Peredachi Informatsii **22** (1986), no. 1, 11–25. MR838684

[Bli97] Volodia Blinovsky, *Asymptotic combinatorial coding theory*, The Kluwer International Series in Engineering and Computer Science, vol. 415, Kluwer Academic Publishers, Boston, MA, 1997. MR1600244

[Bog97] M. Boguslavsky, *On the number of solutions of polynomial systems*, Finite Fields Appl. **3** (1997), no. 4, 287–299, DOI 10.1006/ffta.1997.0186. MR1478830

[Bog98] M. I. Boguslavskiĭ, *Sections of del Pezzo surfaces, and generalized weights* (Russian, with Russian summary), Problemy Peredachi Informatsii **34** (1998), no. 1, 18–29; English transl., Problems Inform. Transmission **34** (1998), no. 1, 14–24. MR1654814

[BOGW88] M. Ben-Or, S. Goldwasser, and A. Wigderson, *Completeness theorems for non-cryptographic fault-tolerant distributed computation*, in Proceedings of * the 20th Annual ACM Symposium on Theory of Computing (STOC'88, Chicago, Illinois, * USA), pp. 1–10, New York: ACM, 1988.

[Bos80] A. Bos, *Sphere packings in high dimensional space*, preprint, 1980.

[Bos98] N. Boston, *The unramified Fontaine–Mazur conjecture*, in Arithmetical Applications of Modular Forms (Proc. Research Conf. on Number Theory and Arithmetical Geometry, San Feliu de Guixols, Spain, 1997), G. Frey, ed., Inst. für Experimentelle Mathematik, Essen, Germany, 1998, pp. 26–28.

[BP09] Alexander Barg and Punarbasu Purkayastha, *Bounds on ordered codes and orthogonal arrays* (English, with English and Russian summaries), Mosc. Math. J. **9** (2009), no. 2, 211–243. MR2568437

[BP11] Stéphane Ballet and Julia Pieltant, *On the tensor rank of multiplication in any extension of* $\mathbb{F}_2$, J. Complexity **27** (2011), no. 2, 230–245, DOI 10.1016/j.jco.2011.01.008. MR2776494

[BR04] S. Ballet and R. Rolland, *Multiplication algorithm in a finite field and tensor rank of the multiplication*, J. Algebra **272** (2004), no. 1, 173–185, DOI 10.1016/j.jalgebra.2003.09.031. MR2029030

[BR18] A. Bassa and C. Ritzenthaler, *Good recursive towers over prime fields exist*, arXiv:1807.05714 [math.NT], (2018).

[Bro77] Michel Broué, *Codes correcteurs d'erreurs auto-orthogonaux sur le corps à deux éléments et formes quadratiques entières définies positives à discriminant* +1 (French), Discrete Math. **17** (1977), no. 3, 247–269, DOI 10.1016/0012-365X(77)90159-5. MR0449863

[BRR10] S. Ballet, C. Ritzenthaler, and R. Rolland, *On the existence of dimension zero divisors in algebraic function fields defined over* $\mathbb{F}_q$, Acta Arith. **143** (2010), no. 4, 377–392, DOI 10.4064/aa143-4-4. MR2652586

[BS66] A. I. Borevich and I. R. Shafarevich, *Number theory*, Translated from the Russian by Newcomb Greenleaf. Pure and Applied Mathematics, Vol. 20, Academic Press, New York-London, 1966. MR0195803

[BS85] Z.I. Borevich and I.R. Shafarevich, *Teoriya chisel*, Moscow: Nauka, 3rd ed., 1985. English translation of the 1st ed.: see [**BS66**].

[BTV15] A. Barg, I. Tamo, and S. Vlăduţ, *Locally recoverable codes on algebraic curves*, arXiv:1501.04904v2 [cs.IT], 2015.

[BTV17] Alexander Barg, Itzhak Tamo, and Serge Vlădaţ, *Locally recoverable codes on algebraic curves*, IEEE Trans. Inform. Theory **63** (2017), no. 8, 4928–4939, DOI 10.1109/TIT.2017.2700859. MR3683544

[BW59] E. S. Barnes and G. E. Wall, *Some extreme forms defined in terms of Abelian groups*. part 1, J. Austral. Math. Soc. **1** (1959/1961), no. part 1, 47–63. MR0106893

[Car38] Leonard Carlitz, *A class of polynomials*, Trans. Amer. Math. Soc. **43** (1938), no. 2, 167–182, DOI 10.2307/1990037. MR1501937

[Cas59] J. W. S. Cassels, *An introduction to the geometry of numbers*, Die Grundlehren der mathematischen Wissenschaften in Einzeldarstellungen mit besonderer Berücksichtigung der Anwendungsgebiete, Bd. 99 Springer-Verlag, Berlin-Göttingen-Heidelberg, 1959. MR0157947

[CC88] D. V. Chudnovsky and G. V. Chudnovsky, *Algebraic complexities and algebraic curves over finite fields*, J. Complexity **4** (1988), no. 4, 285–316, DOI 10.1016/0885-064X(88)90012-X. MR974928

[CCCX09] Ignacio Cascudo, Hao Chen, Ronald Cramer, and Chaoping Xing, *Asymptotically good ideal linear secret sharing with strong multiplication over any fixed finite field*, Advances in cryptology—CRYPTO 2009, Lecture Notes in Comput. Sci., vol. 5677, Springer, Berlin, 2009, pp. 466–486, DOI 10.1007/978-3-642-03356-8_28. MR2556974

[CCD88] D. Chaum, C. Crépeau, and I. Damgård, *Multiparty unconditionally secure protocols*, in Proceedings of the Twentieth Annual ACM Symposium on Theory of Computing (STOC'88, Chicago, Illinois, USA), ACM, New York, 1988, pp. 11–19.

[CCdHC08] Hao Chen, Ronald Cramer, Robbert de Haan, and Ignacio Cascudo Pueyo, *Strongly multiplicative ramp schemes from high degree rational points on curves*, Advances in cryptology—EUROCRYPT 2008, Lecture Notes in Comput. Sci., vol. 4965, Springer, Berlin, 2008, pp. 451–470, DOI 10.1007/978-3-540-78967-3_26. MR2606348

[CCX11] Ignacio Cascudo, Ronald Cramer, and Chaoping Xing, *The torsion-limit for algebraic function fields and its application to arithmetic secret sharing*, Advances in cryptology—CRYPTO 2011, Lecture Notes in Comput. Sci., vol. 6841, Springer, Heidelberg, 2011, pp. 685–705, DOI 10.1007/978-3-642-22792-9_39. MR2874882

[CCX13] Ignacio Cascudo, Ronald Cramer, and Chaoping Xing, *Bounds on the threshold gap in secret sharing and its applications*, IEEE Trans. Inform. Theory **59** (2013), no. 9, 5600–5612, DOI 10.1109/TIT.2013.2264504. MR3096944

[CCXY12] Ignacio Cascudo, Ronald Cramer, Chaoping Xing, and An Yang, *Asymptotic bound for multiplication complexity in the extensions of small finite fields*, IEEE Trans. Inform. Theory **58** (2012), no. 7, 4930–4935, DOI 10.1109/TIT.2011.2180696. MR2949863

[CDM00] Ronald Cramer, Ivan Damgård, and Ueli Maurer, *General secure multi-party computation from any linear secret-sharing scheme*, Advances in cryptology—EUROCRYPT 2000 (Bruges), Lecture Notes in Comput. Sci., vol. 1807, Springer, Berlin, 2000, pp. 316–334, DOI 10.1007/3-540-45539-6_22. MR1772026

[CF67] J. W. S. Cassels and A. Fröhlich (eds.), *Algebraic number theory*, 2nd ed., London Mathematical Society, London, 2010. Papers from the conference held at the University of Sussex, Brighton, September 1–17, 1965; Including a list of errata. MR3618860

[Cha70] I. M. Chakravarti, *Strongly regular graphs (two-class association schemes) and block designs from Hermitian varieties in a finite projective space* $\mathrm{PG}(3, q^2)$, Proc. Second Chapel Hill Conf. on Combinatorial Mathematics and its Applications (Univ. North Carolina, Chapel Hill, N.C., 1970), Univ. North Carolina, Chapel Hill, N.C., 1970, pp. 58–66. MR0276112

[Cha71] I. M. Chakravarti, *Some properties and applications of Hermitian varieties in a finite projective space $PG(N, q^2)$ in the construction of strongly regular graphs (two-class association schemes) and block designs*, J. Combinatorial Theory Ser. B **11** (1971), 268–283. MR0285414

[Cha88] I. M. Chakravarti, *The generalized Goppa codes and related discrete designs from Hermitian surfaces in* $\mathrm{PG}(3, s^2)$, Coding theory and applications (Cachan, 1986), Lecture Notes in Comput. Sci., vol. 311, Springer, Berlin, 1988, pp. 116–124, DOI 10.1007/3-540-19368-5_12. MR960713

[Cha91] I. M. Chakravarti, *Association schemes, orthogonal arrays and codes from nondegenerate quadrics and Hermitian varieties in finite projective geometries*, Calcutta Statist. Assoc. Bull. **40** (1990/91), no. 157-160, 89–96, DOI 10.1177/0008068319900507. MR1172636

[CKM+17] Henry Cohn, Abhinav Kumar, Stephen D. Miller, Danylo Radchenko, and Maryna Viazovska, *The sphere packing problem in dimension 24*, Ann. of Math. (2) **185** (2017), no. 3, 1017–1033, DOI 10.4007/annals.2017.185.3.8. MR3664817

[CLX01] Hao Chen, San Ling, and Chaoping Xing, *Asymptotically good quantum codes exceeding the Ashikhmin-Litsyn-Tsfasman bound*, IEEE Trans. Inform. Theory **47** (2001), no. 5, 2055–2058, DOI 10.1109/18.930941. MR1842540

[CM15] Viveck R. Cadambe and Arya Mazumdar, *Bounds on the size of locally recoverable codes*, IEEE Trans. Inform. Theory **61** (2015), no. 11, 5787–5794, DOI 10.1109/TIT.2015.2477406. MR3418933

[Cou11] Alain Couvreur, *Construction of rational surfaces yielding good codes*, Finite Fields Appl. **17** (2011), no. 5, 424–441, DOI 10.1016/j.ffa.2011.02.007. MR2831703

[Cou16] Alain Couvreur, *An upper bound on the number of rational points of arbitrary projective varieties over finite fields*, Proc. Amer. Math. Soc. **144** (2016), no. 9, 3671–3685, DOI 10.1090/proc/13015. MR3513530

[CRSS98] A. Robert Calderbank, Eric M. Rains, P. W. Shor, and Neil J. A. Sloane, *Quantum error correction via codes over* GF(4), IEEE Trans. Inform. Theory **44** (1998), no. 4, 1369–1387, DOI 10.1109/18.681315. MR1665774

[CS96] A.R. Calderbank and P.W. Shor, *Good quantum error-correcting codes exist*, Phys. Rev. A, vol. 54, no. 2, pp. 1098–1105, 1996.

[CS99] J. H. Conway and N. J. A. Sloane, *Sphere packings, lattices and groups*, 3rd ed., Grundlehren der Mathematischen Wissenschaften [Fundamental Principles of Mathematical Sciences], vol. 290, Springer-Verlag, New York, 1999. With additional contributions by E. Bannai, R. E. Borcherds, J. Leech, S. P. Norton, A. M. Odlyzko, R. A. Parker, L. Queen and B. B. Venkov. MR1662447

[Deu37] Max Deuring, *Automorphismen und Divisorenklassen der Ordnung ι in algebraischen Funktionenkörpern* (German), Math. Ann. **113** (1937), no. 1, 208–215, DOI 10.1007/BF01571630. MR1513087

[DG15] Mrinmoy Datta and Sudhir R. Ghorpade, *On a conjecture of Tsfasman and an inequality of Serre for the number of points of hypersurfaces over finite fields*, Mosc. Math. J. **15** (2015), no. 4, 715–725. MR3438829

[DG17] Mrinmoy Datta and Sudhir R. Ghorpade, *Number of solutions of systems of homogeneous polynomial equations over finite fields*, Proc. Amer. Math. Soc. **145** (2017), no. 2, 525–541, DOI 10.1090/proc/13239. MR3577858

[Die78] J. Dieudonné, ed., *Abrégé d'histoire des mathématiques 1700–1900*, Tomes I, II: Algèbre, analyse classique, théorie des nombres; Fonctions elliptiques, analyse fonctionnelle, topologie, géométrie différentielle, probabilités, logique mathématique, Paris: Hermann, 1978.

[DK73] Pierre Deligne and Willem Kuyk (eds.), *Modular functions of one variable. II*, Lecture Notes in Mathematics, Vol. 349, Springer-Verlag, Berlin-New York, 1973. MR0330050

[DL76] P. Deligne and G. Lusztig, *Representations of reductive groups over finite fields*, Ann. of Math. (2) **103** (1976), no. 1, 103–161, DOI 10.2307/1971021. MR0393266

[DM13] Iwan Duursma and Kit-Ho Mak, *On lower bounds for the Ihara constants $A(2)$ and $A(3)$*, Compos. Math. **149** (2013), no. 7, 1108–1128, DOI 10.1112/S0010437X12000796. MR3078640

[DR73] P. Deligne and M. Rapoport, *Les schémas de modules de courbes elliptiques* (French), Modular functions of one variable, II (Proc. Internat. Summer School, Univ. Antwerp, Antwerp, 1972), Lecture Notes in Math., Vol. 349, Springer, Berlin, 1973, pp. 143–316. MR0337993

[Dri74] V. G. Drinfel′d, *Elliptic modules* (Russian), Mat. Sb. (N.S.) **94(136)** (1974), 594–627, 656. MR0384707

[Dri85] V. G. Drinfel'd, Unpublished manuscript on the Langlands conjecture for $GL(2)$, 1985.

[DRTR01] I. M. Duursma, C. Rentería, and H. Tapia-Recillas, *Reed-Muller codes on complete intersections*, Appl. Algebra Engrg. Comm. Comput. **11** (2001), no. 6, 455–462, DOI 10.1007/s002000000047. MR1831940

[DS02] Steven T. Dougherty and Maxim M. Skriganov, *MacWilliams duality and the Rosenbloom-Tsfasman metric* (English, with English and Russian summaries), Mosc. Math. J. **2** (2002), no. 1, 81–97, 199. MR1900585

[Duu93a] Iwan Maynard Duursma, *Decoding codes from curves and cyclic codes* (English, with Dutch summary), Technische Universiteit Eindhoven, Eindhoven, 1993. Dissertation, Technische Universiteit Eindhoven, Eindhoven, 1993. MR1226447

[Duu93b] Iwan M. Duursma, *Majority coset decoding*, IEEE Trans. Inform. Theory **39** (1993), no. 3, 1067–1070, DOI 10.1109/18.256518. MR1237731

[Edo07] Frédéric A. B. Edoukou, *Codes defined by forms of degree 2 on Hermitian surfaces and Sørensen's conjecture*, Finite Fields Appl. **13** (2007), no. 3, 616–627, DOI 10.1016/j.ffa.2006.07.001. MR2332489

[Edo09] Frédéric A. B. Edoukou, *Codes defined by forms of degree 2 on non-degenerate Hermitian varieties in $\mathbb{P}^4(\mathbb{F}_q)$*, Des. Codes Cryptogr. **50** (2009), no. 1, 135–146, DOI 10.1007/s10623-008-9219-1. MR2480674

[EHK+04] Noam D. Elkies, Everett W. Howe, Andrew Kresch, Bjorn Poonen, Joseph L. Wetherell, and Michael E. Zieve, *Curves of every genus with many points. II. Asymptotically good families*, Duke Math. J. **122** (2004), no. 2, 399–422, DOI 10.1215/S0012-7094-04-12224-9. MR2053756

[Ehr93] Dirk Ehrhard, *Achieving the designed error capacity in decoding algebraic-geometric codes*, IEEE Trans. Inform. Theory **39** (1993), no. 3, 743–751, DOI 10.1109/18.256485. MR1237713

[Eli57] Peter Elias, *List decoding for noisy channels*, Research Laboratory of Electronics, Massachusetts Institute of Technology, Cambridge, Mass., Rep. No. 335, 1957. MR0099261

[Eli91] Peter Elias, *Error-correcting codes for list decoding*, IEEE Trans. Inform. Theory **37** (1991), no. 1, 5–12, DOI 10.1109/18.61123. MR1087881

[Elk89a] N. D. Elkies, 1989, private communication.

[Elk89b] N. D. Elkies, Letters to N.J.A. Sloane, Aug. 15, Sep. 15, 1989.

[Elk90] N. D. Elkies, *On Mordell–Weil lattices*, Math. Arbeitstagung, Max-Planck-Institut für Mathematik Bonn, 1990.

[Elk94] Noam D. Elkies, *Mordell-Weil lattices in characteristic* 2. *I. Construction and first properties*, Internat. Math. Res. Notices **8** (1994), 343 ff., approx. 18 pp., DOI 10.1155/S1073792894000395. MR1289579

[Elk97a] N. D. Elkies, *Explicit modular towers*, in Proceedings of the 35th Annual Allerton Conference on Communication, Control and Computing, 1997, pp. 23–32.

[Elk97b] Noam D. Elkies, *Mordell-Weil lattices in characteristic 2. II. The Leech lattice as a Mordell-Weil lattice*, Invent. Math. **128** (1997), no. 1, 1–8, DOI 10.1007/s002220050133. MR1437492

[Elk98] Noam D. Elkies, *Shimura curve computations*, Algorithmic number theory (Portland, OR, 1998), Lecture Notes in Comput. Sci., vol. 1423, Springer, Berlin, 1998, pp. 1–47, DOI 10.1007/BFb0054850. MR1726059

[Elk01a] Noam D. Elkies, *Explicit towers of Drinfeld modular curves*, European Congress of Mathematics, Vol. II (Barcelona, 2000), Progr. Math., vol. 202, Birkhäuser, Basel, 2001, pp. 189–198. MR1905359

[Elk01b] Noam D. Elkies, *Mordell-Weil lattices in characteristic 2. III. A Mordell-Weil lattice of rank 128*, Experiment. Math. **10** (2001), no. 3, 467–473. MR1917431

[ELX11] Frédéric A. B. Edoukou, San Ling, and Chaoping Xing, *Structure of functional codes defined on non-degenerate Hermitian varieties*, J. Combin. Theory Ser. A **118** (2011), no. 8, 2436–2444, DOI 10.1016/j.jcta.2011.05.006. MR2834185

[EOR91] N. D. Elkies, A. M. Odlyzko, and J. A. Rush, *On the packing densities of superballs and other bodies*, Invent. Math. **105** (1991), no. 3, 613–639, DOI 10.1007/BF01232282. MR1117154

[Fey82] Richard P. Feynman, *Simulating physics with computers*, Internat. J. Theoret. Phys. **21** (1981/82), no. 6-7, 467–488, DOI 10.1007/BF02650179. Physics of computation, Part II (Dedham, Mass., 1981). MR658311

[FLX06] Keqin Feng, San Ling, and Chaoping Xing, *Asymptotic bounds on quantum codes from algebraic geometry codes*, IEEE Trans. Inform. Theory **52** (2006), no. 3, 986–991, DOI 10.1109/TIT.2005.862086. MR2238065

[FM95] Jean-Marc Fontaine and Barry Mazur, *Geometric Galois representations*, Elliptic curves, modular forms, & Fermat's last theorem (Hong Kong, 1993), Ser. Number Theory, I, Int. Press, Cambridge, MA, 1995, pp. 41–78. MR1363495

[FR93] Gui Liang Feng and T. R. N. Rao, *Decoding algebraic-geometric codes up to the designed minimum distance*, IEEE Trans. Inform. Theory **39** (1993), no. 1, 37–45, DOI 10.1109/18.179340. MR1211489

[FR94] Gui Liang Feng and T. R. N. Rao, *A simple approach for construction of algebraic-geometric codes from affine plane curves*, IEEE Trans. Inform. Theory **40** (1994), no. 4, 1003–1012, DOI 10.1109/18.335972. MR1301416

[FR95] Gui-Liang Feng and T. R. N. Rao, *Improved geometric Goppa codes. I. Basic theory*, IEEE Trans. Inform. Theory **41** (1995), no. 6, 1678–1693, DOI 10.1109/18.476241. Special issue on algebraic geometry codes. MR1391027

[FT72] László Fejes Tóth, *Lagerungen in der Ebene auf der Kugel und im Raum* (German), Springer-Verlag, Berlin-New York, 1972. Zweite verbesserte und erweiterte Auflage; Die Grundlehren der mathematischen Wissenschaften, Band 65. MR0353117

[FT91] Gui Liang Feng and Kenneth K. Tzeng, *A generalization of the Berlekamp-Massey algorithm for multisequence shift-register synthesis with applications to decoding cyclic codes*, IEEE Trans. Inform. Theory **37** (1991), no. 5, 1274–1287, DOI 10.1109/18.133246. MR1136665

[FT94] G.L. Feng and K.K. Tzeng, *A new procedure for decoding cyclic and BCH codes up to actual minimum distance*, IEEE Trans. Inform. Theory, vol. 40, no. 5, pp. 1364–1374, 1994.

[FWRT94] Gui Liang Feng, Victor K. Wei, T. R. N. Rao, and Kenneth K. Tzeng, *Simplified understanding and efficient decoding of a class of algebraic-geometric codes*, IEEE Trans. Inform. Theory **40** (1994), no. 4, 981–1002, DOI 10.1109/18.335973. MR1301415

[FY92] M. Franklin and M. Yung, *Communication complexity of secure computation (extended abstract)*, in Proceedings of the 24th Annual ACM Symposium on Theory of Computing (STOC'92, Victoria, British Columbia, Canada), ACM, New York, 1992, pp. 699–710.

[Gek80] Ernst-Ulrich Gekeler, *Drinfeld-Moduln und modulare Formen über rationalen Funktionenkörpern* (German), Bonner Mathematische Schriften [Bonn Mathematical Publications], 119, Universität Bonn, Mathematisches Institut, Bonn, 1980. Dissertation, Rheinische Friedrich-Wilhelms-Universität, Bonn, 1979. MR594434

[Gek86] Ernst-Ulrich Gekeler, *Drinfel′d modular curves*, Lecture Notes in Mathematics, vol. 1231, Springer-Verlag, Berlin, 1986. MR874338

[Gek01] Ernst-Ulrich Gekeler, *Invariants of some algebraic curves related to Drinfeld modular curves*, J. Number Theory **90** (2001), no. 1, 166–183, DOI 10.1006/jnth.2000.2642. MR1850880

[Gho01] Sudhir R. Ghorpade, *A note on Hodge's postulation formula for Schubert varieties*, Geometric and combinatorial aspects of commutative algebra (Messina, 1999), Lecture Notes in Pure and Appl. Math., vol. 217, Dekker, New York, 2001, pp. 211–219. MR1824230

[GKT06] Massimo Giulietti, Gábor Korchmáros, and Fernando Torres, *Quotient curves of the Suzuki curve*, Acta Arith. **122** (2006), no. 3, 245–274, DOI 10.4064/aa122-3-3. MR2239917

[GL00] Sudhir R. Ghorpade and Gilles Lachaud, *Higher weights of Grassmann codes*, Coding theory, cryptography and related areas (Guanajuato, 1998), Springer, Berlin, 2000, pp. 122–131. MR1749453

[GL02a] Sudhir R. Ghorpade and Gilles Lachaud, *Étale cohomology, Lefschetz theorems and number of points of singular varieties over finite fields*, Mosc. Math. J. **2** (2002), no. 3, 589–631. Dedicated to Yuri I. Manin on the occasion of his 65th birthday. MR1988974

[GL02b] Sudhir R. Ghorpade and Gilles Lachaud, *Number of solutions of equations over finite fields and a conjecture of Lang and Weil*, Number theory and discrete mathematics (Chandigarh, 2000), Trends Math., Birkhäuser, Basel, 2002, pp. 269–291. MR1962145

[GL09] Sudhir R. Ghorpade and Gilles Lachaud, *Corrigenda and addenda: Étale cohomology, Lefschetz theorems and number of points of singular varieties over finite fields [MR1988974]*, Mosc. Math. J. **9** (2009), no. 2, 431–438. MR2568444

[GMS74] E. N. Gilbert, F. J. MacWilliams, and N. J. A. Sloane, *Codes which detect deception*, Bell System Tech. J. **53** (1974), 405–424, DOI 10.1002/j.1538-7305.1974.tb02751.x. MR0432317

[Gop88] V. D. Goppa, *Geometry and codes*, Mathematics and its Applications (Soviet Series), vol. 24, Kluwer Academic Publishers Group, Dordrecht, 1988. Translated from the Russian by N. G. Shartse. MR1029027

[Gos80] David Goss, *The algebraist's upper half-plane*, Bull. Amer. Math. Soc. (N.S.) **2** (1980), no. 3, 391–415, DOI 10.1090/S0273-0979-1980-14751-5. MR561525

[Gos96] David Goss, *Basic structures of function field arithmetic*, Ergebnisse der Mathematik und ihrer Grenzgebiete (3) [Results in Mathematics and Related Areas (3)], vol. 35, Springer-Verlag, Berlin, 1996. MR1423131

[Gro90] Benedict H. Gross, *Group representations and lattices*, J. Amer. Math. Soc. **3** (1990), no. 4, 929–960, DOI 10.2307/1990907. MR1071117

[GRS00] Oded Goldreich, Ronitt Rubinfeld, and Madhu Sudan, *Learning polynomials with queries: the highly noisy case*, SIAM J. Discrete Math. **13** (2000), no. 4, 535–570, DOI 10.1137/S0895480198344540. MR1797839

[GS64] E. S. Golod and I. R. Šafarevič, *On the class field tower* (Russian), Izv. Akad. Nauk SSSR Ser. Mat. **28** (1964), 261–272. MR0161852

[GS95] Arnaldo García and Henning Stichtenoth, *A tower of Artin-Schreier extensions of function fields attaining the Drinfel′d-Vlăduţ bound*, Invent. Math. **121** (1995), no. 1, 211–222, DOI 10.1007/BF01884295. MR1345289

[GS96a] Arnaldo Garcia and Henning Stichtenoth, *Asymptotically good towers of function fields over finite fields* (English, with English and French summaries), C. R. Acad. Sci. Paris Sér. I Math. **322** (1996), no. 11, 1067–1070. MR1396642

[GS96b] Arnaldo Garcia and Henning Stichtenoth, *On the asymptotic behaviour of some towers of function fields over finite fields*, J. Number Theory **61** (1996), no. 2, 248–273, DOI 10.1006/jnth.1996.0147. MR1423052

[GS99] Venkatesan Guruswami and Madhu Sudan, *Improved decoding of Reed-Solomon and algebraic-geometry codes*, IEEE Trans. Inform. Theory **45** (1999), no. 6, 1757–1767, DOI 10.1109/18.782097. MR1720630

[GS00] Venkatesan Guruswami and Madhu Sudan, *List decoding algorithms for certain concatenated codes*, Proceedings of the Thirty-Second Annual ACM Symposium on Theory of Computing, ACM, New York, 2000, pp. 181–190, DOI 10.1145/335305.335327. MR2114531

[GS01] Venkatesan Guruswami and Madhu Sudan, *On representations of algebraic-geometry codes*, IEEE Trans. Inform. Theory **47** (2001), no. 4, 1610–1613, DOI 10.1109/18.923745. MR1830110

[GS05] Arnaldo Garcia and Henning Stichtenoth, *Asymptotics for the genus and the number of rational places in towers of function fields over a finite field*, Finite Fields Appl. **11** (2005), no. 3, 434–450, DOI 10.1016/j.ffa.2005.04.004. MR2158772

[GS07] Arnaldo Garcia and Henning Stichtenoth, *Explicit towers of function fields over finite fields*, Topics in geometry, coding theory and cryptography, Algebr. Appl., vol. 6, Springer, Dordrecht, 2007, pp. 1–58, DOI 10.1007/1-4020-5334-4_1. MR2278034

[GT05] Sudhir R. Ghorpade and Michael A. Tsfasman, *Schubert varieties, linear codes and enumerative combinatorics*, Finite Fields Appl. **11** (2005), no. 4, 684–699, DOI 10.1016/j.ffa.2004.09.002. MR2181414

[Gur02] Venkatesan Guruswami, *Limits to list decodability of linear codes*, Proceedings of the Thirty-Fourth Annual ACM Symposium on Theory of Computing, ACM, New York, 2002, pp. 802–811, DOI 10.1145/509907.510022. MR2121529

[Gur04] V. Guruswami, *List decoding of error-correcting codes. Winning thesis of the 2002 ACM Doctoral Dissertation Competition*, vol. 3282 of Lecture Notes in Comput. Sci., Berlin: Springer, 2004.

[Gur06] Venkatesan Guruswami, *Algorithmic results in list decoding*, Found. Trends Theor. Comput. Sci. **2** (2006), no. 2, 107–195, DOI 10.1561/0400000007. MR2453147

[GV04] Lucio Guerra and Rita Vincenti, *On the linear codes arising from Schubert varieties*, Des. Codes Cryptogr. **33** (2004), no. 2, 173–180, DOI 10.1023/B:DESI.0000035470.05639.2b. MR2080363

[GX12] Venkatesan Guruswami and Chaoping Xing, *Folded codes from function field towers and improved optimal rate list decoding [extended abstract]*, STOC'12—Proceedings of the 2012 ACM Symposium on Theory of Computing, ACM, New York, 2012, pp. 339–350, DOI 10.1145/2213977.2214009. MR2961515

[GX13] Venkatesan Guruswami and Chaoping Xing, *List decoding Reed-Solomon, algebraic-geometric, and Gabidulin subcodes up to the Singleton bound*, STOC'13—Proceedings of the 2013 ACM Symposium on Theory of Computing, ACM, New York, 2013, pp. 843–852, DOI 10.1145/2488608.2488715. MR3210846

[GX17] V. Guruswami and C. Xing, *Optimal rate list decoding over bounded alphabets using Hayalgebraic-geometric codes*, arXiv:1708.01070 [cs.CC], (2017).

[HAB+17] Thomas Hales, Mark Adams, Gertrud Bauer, Tat Dat Dang, John Harrison, Le Truong Hoang, Cezary Kaliszyk, Victor Magron, Sean McLaughlin, Tat Thang Nguyen, Quang Truong Nguyen, Tobias Nipkow, Steven Obua, Joseph Pleso, Jason Rute, Alexey Solovyev, Thi Hoai An Ta, Nam Trung Tran, Thi Diep Trieu, Josef Urban, Ky Vu, and Roland Zumkeller, *A formal proof of the Kepler conjecture*, Forum Math. Pi **5** (2017), e2, 29, DOI 10.1017/fmp.2017.1. MR3659768

[Hal05] Thomas C. Hales, *A proof of the Kepler conjecture*, Ann. of Math. (2) **162** (2005), no. 3, 1065–1185, DOI 10.4007/annals.2005.162.1065. MR2179728

[Han92] Johan P. Hansen, *Deligne-Lusztig varieties and group codes*, Coding theory and algebraic geometry (Luminy, 1991), Lecture Notes in Math., vol. 1518, Springer, Berlin, 1992, pp. 63–81, DOI 10.1007/BFb0087993. MR1186416

[Han95] Søren Have Hansen, *Rational points on curves over finite fields*, Lecture Notes Series, vol. 64, Aarhus Universitet, Matematisk Institut, Aarhus, 1995. MR1467481

[Han02] Johan P. Hansen, *Toric varieties Hirzebruch surfaces and error-correcting codes*, Appl. Algebra Engrg. Comm. Comput. **13** (2002), no. 4, 289–300, DOI 10.1007/s00200-002-0106-0. MR1953195

[Has26] Helmut Hasse, *Bericht über neuere Untersuchungen und Probleme aus der Theorie der algebraischen Zahlkörper. Teil I: Klassenkörpertheorie. Teil Ia: Beweise zu Teil I* (German), Zweite, durchgesehene Auflage, Physica-Verlag, Würzburg-Vienna, 1965. MR0195847

[Hay74] D. R. Hayes, *Explicit class field theory for rational function fields*, Trans. Amer. Math. Soc. **189** (1974), 77–91, DOI 10.2307/1996848. MR0330106

[Hay79] David R. Hayes, *Explicit class field theory in global function fields*, Studies in algebra and number theory, Adv. in Math. Suppl. Stud., vol. 6, Academic Press, New York-London, 1979, pp. 173–217. MR535766

[Hay92] David R. Hayes, *A brief introduction to Drinfel′d modules*, The arithmetic of function fields (Columbus, OH, 1991), Ohio State Univ. Math. Res. Inst. Publ., vol. 2, de Gruyter, Berlin, 1992, pp. 1–32. MR1196509

[HCL07] C. Huang, M. Chen, and J. Li, *Pyramid codes: Flexible schemes to trade space for access efficiency in reliable data storage systems*, in Proc. 6th IEEE Int. Sympos. on Network Computing and Applications (NCA'2007), Cambridge, MA, July 12–14, 2007, IEEE, 2007, pp. 79–86.

[Hei73] H. Heilbronn, *On real zeros of Dedekind ζ-functions*, Canad. J. Math. **25** (1973), 870–873, DOI 10.4153/CJM-1973-090-3. MR0327719

[Hir79] J. W. P. Hirschfeld, *Projective geometries over finite fields*, The Clarendon Press, Oxford University Press, New York, 1979. Oxford Mathematical Monographs. MR554919

[HJR07] Johan P. Hansen, Trygve Johnsen, and Kristian Ranestad, *Schubert unions in Grassmann varieties*, Finite Fields Appl. **13** (2007), no. 4, 738–750, DOI 10.1016/j.ffa.2007.06.003. MR2359314

[HK13] Masaaki Homma and Seon Jeong Kim, *An elementary bound for the number of points of a hypersurface over a finite field*, Finite Fields Appl. **20** (2013), 76–83, DOI 10.1016/j.ffa.2012.11.002. MR3015353

[HLM07] J. Han and L. Lastras-Montao, *Reliable memories with subline accesses*, in Proc. 2007 IEEE Int. Sympos. on Information Theory (ISIT'2007), Nice, France, June 24–29, 2007, IEEE, 2007, pp. 2531–2535.

[HM01] Farshid Hajir and Christian Maire, *Tamely ramified towers and discriminant bounds for number fields*, Compositio Math. **128** (2001), no. 1, 35–53, DOI 10.1023/A:1017537415688. MR1847664

[HM02] Farshid Hajir and Christian Maire, *Tamely ramified towers and discriminant bounds for number fields. II*, J. Symbolic Comput. **33** (2002), no. 4, 415–423, DOI 10.1006/jsco.2001.0514. MR1890578

[HN99] T. Høholdt and R. Refslund Nielsen, *Decoding Hermitian codes with Sudan's algorithm*, Applied algebra, algebraic algorithms and error-correcting codes (Honolulu,

HI, 1999), Lecture Notes in Comput. Sci., vol. 1719, Springer, Berlin, 1999, pp. 260–270, DOI 10.1007/3-540-46796-3_26. MR1846502

[Hod43] W. V. D. Hodge, *Some enumerative results in the theory of forms*, Proc. Cambridge Philos. Soc. **39** (1943), 22–30. MR0007739

[HP52] W. V. D. Hodge and D. Pedoe, *Methods of algebraic geometry. Vol. II. Book III: General theory of algebraic varieties in projective space. Book IV: Quadrics and Grassmann varieties*, Cambridge, at the University Press, 1952. MR0048065

[HP93] Johan P. Hansen and Jens Peter Pedersen, *Automorphism groups of Ree type, Deligne-Lusztig curves and function fields*, J. Reine Angew. Math. **440** (1993), 99–109. MR1225959

[HP95] Tom Høholdt and Ruud Pellikaan, *On the decoding of algebraic-geometric codes*, IEEE Trans. Inform. Theory **41** (1995), no. 6, 1589–1614, DOI 10.1109/18.476214. Special issue on algebraic geometry codes. MR1391018

[HP98] Petra Heijnen and Ruud Pellikaan, *Generalized Hamming weights of q-ary Reed-Muller codes*, IEEE Trans. Inform. Theory **44** (1998), no. 1, 181–196, DOI 10.1109/18.651015. MR1486657

[HS89] J. Hansen and H. Stichtenoth, *Group codes on algebraic curves associated to the Sylow-2-subgroups of the Suzuki groups*, Preprint no. 7, Aarhus Univ. Mat. Inst., 1989.

[HS90] Johan P. Hansen and Henning Stichtenoth, *Group codes on certain algebraic curves with many rational points*, Appl. Algebra Engrg. Comm. Comput. **1** (1990), no. 1, 67–77, DOI 10.1007/BF01810849. MR1325513

[HT91] J. W. P. Hirschfeld and J. A. Thas, *General Galois geometries*, Oxford Mathematical Monographs, The Clarendon Press, Oxford University Press, New York, 1991. Oxford Science Publications. MR1363259

[HTV94] J. W. P. Hirschfeld, M. A. Tsfasman, and S. G. Vlăduţ, *The weight hierarchy of higher dimensional Hermitian codes*, IEEE Trans. Inform. Theory, **40** (1994), no. 1, pp. 275–278.

[Iha81] Yasutaka Ihara, *Some remarks on the number of rational points of algebraic curves over finite fields*, J. Fac. Sci. Univ. Tokyo Sect. IA Math. **28** (1981), no. 3, 721–724 (1982). MR656048

[Iha83] Yasutaka Ihara, *How many primes decompose completely in an infinite unramified Galois extension of a global field?*, J. Math. Soc. Japan **35** (1983), no. 4, 693–709, DOI 10.2969/jmsj/03540693. MR714470

[IR90] Kenneth Ireland and Michael Rosen, *A classical introduction to modern number theory*, 2nd ed., Graduate Texts in Mathematics, vol. 84, Springer-Verlag, New York, 1990. MR1070716

[JH01] Jørn Justesen and Tom Høholdt, *Bounds on list decoding of MDS codes*, IEEE Trans. Inform. Theory **47** (2001), no. 4, 1604–1609, DOI 10.1109/18.923744. MR1830109

[JLJ+89] Jørn Justesen, Knud J. Larsen, H. Elbrønd Jensen, Allan Havemose, and Tom Høholdt, *Construction and decoding of a class of algebraic geometry codes*, IEEE Trans. Inform. Theory **35** (1989), no. 4, 811–821, DOI 10.1109/18.32157. MR1013689

[JLJH92] J. Justesen, K. J. Larsen, H. Elbrønd Jensen, and T. Høholdt, *Fast decoding of codes from algebraic plane curves*, IEEE Trans. Inform. Theory **38** (1992), no. 1, 111–119, DOI 10.1109/18.108255. MR1146072

[KL72] S. L. Kleiman and Dan Laksov, *Schubert calculus*, Amer. Math. Monthly **79** (1972), 1061–1082, DOI 10.2307/2317421. MR0323796

[KL78] G. A. Kabatjanskiĭ and V. I. Levenšteĭn, *Bounds for packings on the sphere and in space* (Russian), Problemy Peredači Informacii **14** (1978), no. 1, 3–25. MR0514023

[KL88] Tadao Kasami and Shu Lin, *The binary weight distribution of the extended* $(2^m, 2^m - 4)$ *code of the Reed-Solomon code over* $\mathrm{GF}(2^m)$ *with generator polynomial* $(x - \alpha)(x - \alpha^2)(x - \alpha^3)$, Linear Algebra Appl. **98** (1988), 291–307, DOI 10.1016/0024-3795(88)90169-3. MR919120

[KLM07] Phillip Kaye, Raymond Laflamme, and Michele Mosca, *An introduction to quantum computing*, Oxford University Press, Oxford, 2007. MR2311153

[KM85] Nicholas M. Katz and Barry Mazur, *Arithmetic moduli of elliptic curves*, Annals of Mathematics Studies, vol. 108, Princeton University Press, Princeton, NJ, 1985. MR772569

[Köt96] R.Kötter, *On algebraic decoding of algebraic-geometric and cyclic codes*, Linköping Studies in Science and Technology. Dissertations. No. 419, Department of Electrical Engineering, Linköping University, Linköping, Sweden, 1996.

[KP13] Krishna V. Kaipa and Harish K. Pillai, *Weight spectrum of codes associated with the Grassmannian* $G(3,7)$, IEEE Trans. Inform. Theory **59** (2013), no. 2, 986–993, DOI 10.1109/TIT.2012.2219497. MR3015710

[Kra88] V.Y. Krachkovsky, *Decoding of codes on algebraic curves*, preprint, 1988 (in Russian).

[KS73] Willem Kuyk and J.-P. Serre (eds.), *Modular functions of one variable. III*, Lecture Notes in Mathematics, Vol. 350, Springer-Verlag, Berlin-New York, 1973. MR0323724

[KSJ96] Gregory A. Kabatianskii, Ben Smeets, and Thomas Johansson, *On the cardinality of systematic authentication codes via error-correcting codes*, IEEE Trans. Inform. Theory **42** (1996), no. 2, 566–578, DOI 10.1109/18.485725. MR1381365

[Kuy73] Willem Kuyk (ed.), *Modular functions of one variable. I*, Lecture Notes in Mathematics, Vol. 320, Springer-Verlag, Berlin-New York, 1973. MR0323723

[Lac88] Gilles Lachaud, *Projective Reed-Muller codes* (English, with French summary), Coding theory and applications (Cachan, 1986), Lecture Notes in Comput. Sci., vol. 311, Springer, Berlin, 1988, pp. 125–129, DOI 10.1007/3-540-19368-5_13. MR960714

[Lac89] Gilles Lachaud, *Distribution of the weights of the dual of the Melas code*, Discrete Math. **79** (1989/90), no. 1, 103–106, DOI 10.1016/0012-365X(90)90059-Q. MR1032638

[Lac90] Gilles Lachaud, *The parameters of projective Reed-Muller codes* (English, with French summary), Discrete Math. **81** (1990), no. 2, 217–221, DOI 10.1016/0012-365X(90)90155-B. MR1054981

[Lac96] Gilles Lachaud, *Number of points of plane sections and linear codes defined on algebraic varieties*, Arithmetic, geometry and coding theory (Luminy, 1993), de Gruyter, Berlin, 1996, pp. 77–104. MR1394928

[Lan56] Serge Lang, *Unramified class field theory over function fields in several variables*, Ann. of Math. (2) **64** (1956), 285–325, DOI 10.2307/1969975. MR0083174

[Lan70] Serge Lang, *Algebraic number theory*, Addison-Wesley Publishing Co., Inc., Reading, Mass.-London-Don Mills, Ont., 1970. MR0282947

[Lan71] Serge Lang, *On the zeta function of number fields*, Invent. Math. **12** (1971), 337–345, DOI 10.1007/BF01403312. MR0294268

[Lan76] Serge Lang, *Introduction to modular forms*, Springer-Verlag, Berlin-New York, 1976. Grundlehren der mathematischen Wissenschaften, No. 222. MR0429740

[Lau96] Kristin Lauter, *Ray class field constructions of curves over finite fields with many rational points*, Algorithmic number theory (Talence, 1996), Lecture Notes in Comput. Sci., vol. 1122, Springer, Berlin, 1996, pp. 187–195, DOI 10.1007/3-540-61581-4_54. MR1446511

[Lau99] Kristin Lauter, *Deligne-Lusztig curves as ray class fields*, Manuscripta Math. **98** (1999), no. 1, 87–96, DOI 10.1007/s002290050127. MR1669591

[Law96] K. Mark Lawrence, *A combinatorial characterization of* (t,m,s)*-nets in base* b, J. Combin. Des. **4** (1996), no. 4, 275–293, DOI 10.1002/(SICI)1520-6610(1996)4:4⟨275::AID-JCD5⟩3.0.CO;2-C. MR1391810

[Lee64] John Leech, *Some sphere packings in higher space*, Canad. J. Math. **16** (1964), 657–682, DOI 10.4153/CJM-1964-065-1. MR0167901

[Lee67] John Leech, *Notes on sphere packings*, Canad. J. Math. **19** (1967), 251–267, DOI 10.4153/CJM-1967-017-0. MR0209983

[Len86] H. W. Lenstra Jr., *Codes from algebraic number fields*, Mathematics and computer science, II (Amsterdam, 1986), CWI Monogr., vol. 4, North-Holland, Amsterdam, 1986, pp. 95–104. MR878574

[LMMS96] K. Mark Lawrence, Arijit Mahalanabis, Gary L. Mullen, and Wolfgang Ch. Schmid, *Construction of digital* (t,m,s)*-nets from linear codes*, Finite fields and applications (Glasgow, 1995), London Math. Soc. Lecture Note Ser., vol. 233, Cambridge

Univ. Press, Cambridge, 1996, pp. 189–208, DOI 10.1017/CBO9780511525988.017. MR1433149

[LMSE02] Wen-Ching W. Li, Hiren Maharaj, Henning Stichtenoth, and Noam D. Elkies, *New optimal tame towers of function fields over small finite fields*, Algorithmic number theory (Sydney, 2002), Lecture Notes in Comput. Sci., vol. 2369, Springer, Berlin, 2002, pp. 372–389, DOI 10.1007/3-540-45455-1_30. MR2041098

[LN97] Rudolf Lidl and Harald Niederreiter, *Finite fields*, 2nd ed., Encyclopedia of Mathematics and its Applications, vol. 20, Cambridge University Press, Cambridge, 1997. With a foreword by P. M. Cohn. MR1429394

[LO77] J. C. Lagarias and A. M. Odlyzko, *Effective versions of the Chebotarev density theorem*, Algebraic number fields: L-functions and Galois properties (Proc. Sympos., Univ. Durham, Durham, 1975), Academic Press, London, 1977, pp. 409–464. MR0447191

[LR15] Gilles Lachaud and Robert Rolland, *On the number of points of algebraic sets over finite fields* (English, with English and French summaries), J. Pure Appl. Algebra **219** (2015), no. 11, 5117–5136, DOI 10.1016/j.jpaa.2015.05.008. MR3351576

[LS94] Gilles Lachaud and Jacques Stern, *Polynomial-time construction of codes. II. Spherical codes and the kissing number of spheres*, IEEE Trans. Inform. Theory **40** (1994), no. 4, 1140–1146, DOI 10.1109/18.335961. MR1301424

[LT87] S. N. Litsyn and M. A. Tsfasman, *Constructive high-dimensional sphere packings*, Duke Math. J. **54** (1987), no. 1, 147–161, DOI 10.1215/S0012-7094-87-05411-1. MR885781

[LT97] Gilles Lachaud and Michael A. Tsfasman, *Formules explicites pour le nombre de points des variétés sur un corps fini* (French), J. Reine Angew. Math. **493** (1997), 1–60. MR1491808

[Mad93] Y. Madelung, *Implementation of a decoding algorithm for AG-codes from the Hermitian curve*, Report IT-93-137, Institute for Circuit Theory and Telecommunication, Technical University of Denmark, Lyngby, 1993.

[Mai96] Christian Maire, *Finitude de tours et p-tours T-ramifiées modérées, S-décomposées* (French, with English and French summaries), J. Théor. Nombres Bordeaux **8** (1996), no. 1, 47–73. MR1399946

[Man74] Yu. I. Manin, *Cubic forms*, 2nd ed., North-Holland Mathematical Library, vol. 4, North-Holland Publishing Co., Amsterdam, 1986. Algebra, geometry, arithmetic; Translated from the Russian by M. Hazewinkel. MR833513

[Man77] Yu. I. Manin, *A course in mathematical logic*, Springer-Verlag, New York-Berlin, 1977. Translated from the Russian by Neal Koblitz; Graduate Texts in Mathematics, Vol. 53. MR0457126

[Mar78] Jacques Martinet, *Tours de corps de classes et estimations de discriminants* (French), Invent. Math. **44** (1978), no. 1, 65–73, DOI 10.1007/BF01389902. MR0460281

[Mas69] James L. Massey, *Shift-register synthesis and* BCH *decoding*, IEEE Trans. Information Theory **IT-15** (1969), 122–127. MR0242556

[Mat99] Jiří Matoušek, *Geometric discrepancy*, Algorithms and Combinatorics, vol. 18, Springer-Verlag, Berlin, 1999. An illustrated guide. MR1697825

[MH73] John Milnor and Dale Husemoller, *Symmetric bilinear forms*, Springer-Verlag, New York-Heidelberg, 1973. Ergebnisse der Mathematik und ihrer Grenzgebiete, Band 73. MR0506372

[Mil75] J. S. Milne, *On a conjecture of Artin and Tate*, Ann. of Math. (2) **102** (1975), no. 3, 517–533, DOI 10.2307/1971042. MR0414558

[Mil76] J. Milnor, *Hilbert's problem 18: on crystallographic groups, fundamental domains, and on sphere packing*, Mathematical developments arising from Hilbert problems (Proc. Sympos. Pure Math., Northern Illinois Univ., De Kalb, Ill., 1974), Amer. Math. Soc., Providence, R. I., 1976, pp. 491–506. Proc. Sympos. Pure Math., Vol. XXVIII. MR0430101

[Mil13] J. S. Milne, *Class field theory*, Course Notes, Univ. of Michigan, v4.02, 2013. Available at `http://www.jmilne.org/math/CourseNotes/cft.html`.

[Mil14] J. S. Milne, *Algebraic number theory*, Course Notes, Univ. of Michigan, v3.06, 2014. Available at `http://www.jmilne.org/math/CourseNotes/ant.html`.

[MS96] Gary L. Mullen and Wolfgang Ch. Schmid, *An equivalence between (t, m, s)-nets and strongly orthogonal hypercubes*, J. Combin. Theory Ser. A **76** (1996), no. 1, 164–174, DOI 10.1006/jcta.1996.0098. MR1406000

[Neu99] Jürgen Neukirch, *Algebraic number theory*, Grundlehren der Mathematischen Wissenschaften [Fundamental Principles of Mathematical Sciences], vol. 322, Springer-Verlag, Berlin, 1999. Translated from the 1992 German original and with a note by Norbert Schappacher; With a foreword by G. Harder. MR1697859

[Nie86] Harald Niederreiter, *Low-discrepancy point sets*, Monatsh. Math. **102** (1986), no. 2, 155–167, DOI 10.1007/BF01490206. MR861937

[Nie87] Harald Niederreiter, *Point sets and sequences with small discrepancy*, Monatsh. Math. **104** (1987), no. 4, 273–337, DOI 10.1007/BF01294651. MR918037

[Nie91] Harald Niederreiter, *A combinatorial problem for vector spaces over finite fields*, Discrete Math. **96** (1991), no. 3, 221–228, DOI 10.1016/0012-365X(91)90315-S. MR1139449

[Nie92] Harald Niederreiter, *Random number generation and quasi-Monte Carlo methods*, CBMS-NSF Regional Conference Series in Applied Mathematics, vol. 63, Society for Industrial and Applied Mathematics (SIAM), Philadelphia, PA, 1992. MR1172997

[Nie00] Harald Niederreiter, *Constructions of (t, m, s)-nets*, Monte Carlo and quasi-Monte Carlo methods 1998 (Claremont, CA), Springer, Berlin, 2000, pp. 70–85. MR1849843

[Nie03] Harald Niederreiter, *Error bounds for quasi-Monte Carlo integration with uniform point sets*, J. Comput. Appl. Math. **150** (2003), no. 2, 283–292, DOI 10.1016/S0377-0427(02)00665-9. MR1947748

[Nie04a] Harald Niederreiter, *Digital nets and coding theory*, Coding, cryptography and combinatorics, Progr. Comput. Sci. Appl. Logic, vol. 23, Birkhäuser, Basel, 2004, pp. 247–257. MR2090654

[Nie04b] Harald Niederreiter, *High-dimensional numerical integration*, Applied mathematics entering the 21st century, SIAM, Philadelphia, PA, 2004, pp. 337–351. MR2296274

[Nie05] Harald Niederreiter, *Constructions of (t, m, s)-nets and (t, s)-sequences*, Finite Fields Appl. **11** (2005), no. 3, 578–600, DOI 10.1016/j.ffa.2005.01.001. MR2158777

[NÖ02] Harald Niederreiter and Ferruh Özbudak, *Constructions of digital nets using global function fields*, Acta Arith. **105** (2002), no. 3, 279–302, DOI 10.4064/aa105-3-5. MR1931794

[NÖ07] Harald Niederreiter and Ferruh Özbudak, *Low-discrepancy sequences using duality and global function fields*, Acta Arith. **130** (2007), no. 1, 79–97, DOI 10.4064/aa130-1-5. MR2354150

[Nog93] D. Yu. Nogin, *Generalized Hamming weights for codes on multidimensional quadrics* (Russian, with Russian summary), Problemy Peredachi Informatsii **29** (1993), no. 3, 21–30; English transl., Problems Inform. Transmission **29** (1993), no. 3, 218–227 (1994). MR1239374

[Nog96] D. Yu. Nogin, *Codes associated to Grassmannians*, Arithmetic, geometry and coding theory (Luminy, 1993), de Gruyter, Berlin, 1996, pp. 145–154. MR1394931

[Nog97] D. Yu. Nogin, *The spectrum of codes associated with the Grassmannian variety $G(3,6)$* (Russian, with Russian summary), Problemy Peredachi Informatsii **33** (1997), no. 2, 26–36; English transl., Problems Inform. Transmission **33** (1997), no. 2, 114–123. MR1663924

[NP01] Harald Niederreiter and Gottlieb Pirsic, *Duality for digital nets and its applications*, Acta Arith. **97** (2001), no. 2, 173–182, DOI 10.4064/aa97-2-5. MR1824983

[NWX07] Harald Niederreiter, Huaxiong Wang, and Chaoping Xing, *Function fields over finite fields and their applications to cryptography*, Topics in geometry, coding theory and cryptography, Algebr. Appl., vol. 6, Springer, Dordrecht, 2007, pp. 59–104, DOI 10.1007/1-4020-5334-4_2. MR2278035

[NX95] Harald Niederreiter and Chao Ping Xing, *Low-discrepancy sequences obtained from algebraic function fields over finite fields*, Acta Arith. **72** (1995), no. 3, 281–298, DOI 10.4064/aa-72-3-281-298. MR1347491

[NX96] Harald Niederreiter and Chaoping Xing, *Quasirandom points and global function fields*, Finite fields and applications (Glasgow, 1995), London Math. Soc. Lecture Note Ser., vol. 233, Cambridge Univ. Press, Cambridge, 1996, pp. 269–296, DOI 10.1017/CBO9780511525988.022. MR1433154

[NX97] Harald Niederreiter and Chaoping Xing, *Cyclotomic function fields, Hilbert class fields, and global function fields with many rational places*, Acta Arith. **79** (1997), no. 1, 59–76, DOI 10.4064/aa-79-1-59-76. MR1438117

[NX98] Harald Niederreiter and Chaoping Xing, *Nets, (t, s)-sequences, and algebraic geometry*, Random and quasi-random point sets, Lect. Notes Stat., vol. 138, Springer, New York, 1998, pp. 267–302, DOI 10.1007/978-1-4612-1702-2_6. MR1662844

[NX01] Harald Niederreiter and Chaoping Xing, *Rational points on curves over finite fields: theory and applications*, London Mathematical Society Lecture Note Series, vol. 285, Cambridge University Press, Cambridge, 2001. MR1837382

[NX02] Harald Niederreiter and Chaoping Xing, *Constructions of digital nets*, Acta Arith. **102** (2002), no. 2, 189–197, DOI 10.4064/aa102-2-7. MR1889629

[Odl75] A. M. Odlyzko, *Some analytic estimates of class numbers and discriminants*, Invent. Math. **29** (1975), no. 3, 275–286, DOI 10.1007/BF01389854. MR0376613

[Odl76] A. M. Odlyzko, *Lower bounds for discriminants of number fields*, Acta Arith. **29** (1976), no. 3, 275–297, DOI 10.4064/aa-29-3-275-297. MR0401704

[Odl77a] A. M. Odlyzko, *Lower bounds for discriminants of number fields. II*, Tôhoku Math. J. **29** (1977), no. 2, 209–216, DOI 10.2748/tmj/1178240652. MR0441918

[Odl77b] A. M. Odlyzko, *On conductors and discriminants*, Algebraic number fields: L-functions and Galois properties (Proc. Sympos., Univ. Durham, Durham, 1975), Academic Press, London, 1977, pp. 377–407. MR0453701

[Odl90] A. M. Odlyzko, *Bounds for discriminants and related estimates for class numbers, regulators and zeros of zeta functions: a survey of recent results* (English, with French summary), Sém. Théor. Nombres Bordeaux (2) **2** (1990), no. 1, 119–141. MR1061762

[Oes85] Joseph Oesterlé, *Le théorème des classes jumelles de Zimmert et les formules explicites de Weil* (French), Séminaire de théorie des nombres, Paris 1983–84, Progr. Math., vol. 59, Birkhäuser Boston, Boston, MA, 1985, pp. 181–197. MR902832

[Oes16] J. Oesterlé, Densité maximale des empilements de sphères en dimensions 8 et 24, (d'après Maryna S. Viazovska et al.), Séminaire Bourbaki, Vol. 2016/2017 69ème année, Exp. No. 1133, 2016, pp. 1–22.

[Ogg74] Andrew P. Ogg, *Hyperelliptic modular curves*, Bull. Soc. Math. France **102** (1974), 449–462. MR0364259

[ÖT07] Ferruh Özbudak and Burcu Gülmez Temur, *Fibre products of Kummer covers and curves with many points*, Appl. Algebra Engrg. Comm. Comput. **18** (2007), no. 5, 433–443, DOI 10.1007/s00200-007-0047-8. MR2342563

[Ped92] Jens Peter Pedersen, *A function field related to the Ree group*, Coding theory and algebraic geometry (Luminy, 1991), Lecture Notes in Math., vol. 1518, Springer, Berlin, 1992, pp. 122–131, DOI 10.1007/BFb0087997. MR1186420

[Poi77a] Georges Poitou, *Minorations de discriminants (d'après A. M. Odlyzko)*, Séminaire Bourbaki, Vol. 1975/76 28ème année, Exp. No. 479, Springer, Berlin, 1977, pp. 136–153. Lecture Notes in Math., Vol. 567. MR0435033

[Poi77b] Georges Poitou, *Sur les petits discriminants* (French), Séminaire Delange-Pisot-Poitou, 18e année: (1976/77), Théorie des nombres, Fasc. 1 (French), Secrétariat Math., Paris, 1977, pp. Exp. No. 6, 18. MR551335

[Que89] H-G. Quebbemann, *Lattices from curves over finite fields*, preprint, 1989.

[Rai99] Eric M. Rains, *Nonbinary quantum codes*, IEEE Trans. Inform. Theory **45** (1999), no. 6, 1827–1832, DOI 10.1109/18.782103. MR1720636

[Ran11] H. Randriam, *Diviseurs de la forme $2D - G$ sans sections et rang de la multiplication dans les corps finis (Divisors of the form $2D - G$ without sections and bilinear complexity of multiplication in finite fields)*, arXiv:1103.4335v2 [math.AG], (2011).

[Ran12] Hugues Randriambololona, *Bilinear complexity of algebras and the Chudnovsky-Chudnovsky interpolation method*, J. Complexity **28** (2012), no. 4, 489–517, DOI 10.1016/j.jco.2012.02.005. MR2925903

[Ran15] Hugues Randriambololona, *On products and powers of linear codes under componentwise multiplication*, Algorithmic arithmetic, geometry, and coding theory, Contemp. Math., vol. 637, Amer. Math. Soc., Providence, RI, 2015, pp. 3–78, DOI 10.1090/conm/637/12749. MR3364442

[RI10] C. Rovi and M. Izquierdo, *Maximal curves covered by the Hermitian curve*, Preprint submitted to Disertaciones del Seminario de Matemáticas Fundamentales, 2010. Available at `http://www.manypoints.org/upload/802784935.pdf`.

[Rod00] F. Rodier, *Nombre de points des surfaces de Deligne et Lusztig* (French, with English and French summaries), J. Algebra **227** (2000), no. 2, 706–766, DOI 10.1006/jabr.1999.8246. MR1759842

[Rod03] F. Rodier, *Codes from flag varieties over a finite field*, J. Pure Appl. Algebra **178** (2003), no. 2, 203–214, DOI 10.1016/S0022-4049(02)00188-3. MR1952426

[Rog64] C. A. Rogers, *Packing and covering*, Cambridge Tracts in Mathematics and Mathematical Physics, No. 54, Cambridge University Press, New York, 1964. MR0172183

[Roq67] Peter Roquette, *On class field towers*, Algebraic Number Theory (Proc. Instructional Conf., Brighton, 1965), Thompson, Washington, D.C., 1967, pp. 231–249. MR0218331

[Ros54] Maxwell Rosenlicht, *Generalized Jacobian varieties*, Ann. of Math. (2) **59** (1954), 505–530, DOI 10.2307/1969715. MR0061422

[Ros02] Michael Rosen, *Number theory in function fields*, Graduate Texts in Mathematics, vol. 210, Springer-Verlag, New York, 2002. MR1876657

[RP00] E. Rieffel and W. Polak, *An introduction to quantum computing for non-physicists*, ACM Comput. Surv., vol. 32, no. 3, pp. 300–335, 2000.

[RR88] Charles T. Ryan and Kevin M. Ryan, *An application of geometry to the calculation of weight enumerators*, Congr. Numer. **67** (1988), 77–89. Nineteenth Southeastern Conference on Combinatorics, Graph Theory, and Computing (Baton Rouge, LA, 1988). MR992921

[RR90] Charles T. Ryan and Kevin M. Ryan, *The minimum weight of the Grassmann codes $C(k,n)$*, Discrete Appl. Math. **28** (1990), no. 2, 149–156, DOI 10.1016/0166-218X(90)90112-P. MR1067604

[RS87] Jason A. Rush and N. J. A. Sloane, *An improvement to the Minkowski-Hlawka bound for packing superballs*, Mathematika **34** (1987), no. 1, 8–18, DOI 10.1112/S0025579300013231. MR908835

[RS08] François Rodier and Adnen Sboui, *Highest numbers of points of hypersurfaces over finite fields and generalized Reed-Muller codes*, Finite Fields Appl. **14** (2008), no. 3, 816–822, DOI 10.1016/j.ffa.2008.02.001. MR2435064

[RT90] Michael Yu. Rosenbloom and Michael A. Tsfasman, *Multiplicative lattices in global fields*, Invent. Math. **101** (1990), no. 3, 687–696, DOI 10.1007/BF01231520. MR1062800

[RT97] M. Yu. Rozenblyum and M. A. Tsfasman, *Codes for the m-metric* (Russian, with Russian summary), Problemy Peredachi Informatsii **33** (1997), no. 1, 55–63; English transl., Problems Inform. Transmission **33** (1997), no. 1, 45–52. MR1441528

[Rus89] J. A. Rush, *A lower bound on packing density*, Invent. Math. **98** (1989), no. 3, 499–509, DOI 10.1007/BF01393834. MR1022304

[Rya87a] Charles Ryan, *An application of Grassmannian varieties to coding theory*, Congr. Numer. **57** (1987), 257–271. Sixteenth Manitoba conference on numerical mathematics and computing (Winnipeg, Man., 1986). MR889714

[Rya87b] Charles T. Ryan, *Projective codes based on Grassmann varieties*, Congr. Numer. **57** (1987), 273–279. Sixteenth Manitoba conference on numerical mathematics and computing (Winnipeg, Man., 1986). MR889715

[Rya88] Charles T. Ryan, *The weight distribution of a code associated to intersection properties of the Grassmann variety $G(3,6)$*, Congr. Numer. **61** (1988), 183–198. Seventeenth Manitoba Conference on Numerical Mathematics and Computing (Winnipeg, MB, 1987). MR961652

[Sak88] Shojiro Sakata, *Finding a minimal set of linear recurring relations capable of generating a given finite two-dimensional array*, J. Symbolic Comput. **5** (1988), no. 3, 321–337, DOI 10.1016/S0747-7171(88)80033-6. MR946587

[Sak90] Shojiro Sakata, *Extension of the Berlekamp-Massey algorithm to N dimensions*, Inform. and Comput. **84** (1990), no. 2, 207–239, DOI 10.1016/0890-5401(90)90039-K. MR1035866

[SAK+01] Kenneth W. Shum, Ilia Aleshnikov, P. Vijay Kumar, Henning Stichtenoth, and Vinay Deolaikar, *A low-complexity algorithm for the construction of algebraic-geometric codes better than the Gilbert-Varshamov bound*, IEEE Trans. Inform. Theory **47** (2001), no. 6, 2225–2241, DOI 10.1109/18.945244. MR1873198

[SB13] Patrick Solé and Jean-Claude Belfiore, *Constructive spherical codes near the Shannon bound*, Des. Codes Cryptogr. **66** (2013), no. 1-3, 17–26, DOI 10.1007/s10623-012-9633-2. MR3016553

[Sch58] Wolfgang M. Schmidt, *The measure of the set of admissible lattices*, Proc. Amer. Math. Soc. **9** (1958), 390–403, DOI 10.2307/2032994. MR0096638

[Sch66] Laurent Schwartz, *Théorie des distributions* (French), Publications de l'Institut de Mathématique de l'Université de Strasbourg, No. IX-X. Nouvelle édition, entiérement corrigée, refondue et augmentée, Hermann, Paris, 1966. MR0209834

[Sch90] R. Schoof, *Algebraic curves and coding theory*, Universita di Trento. Preprint 336, 1990. Available at `http://swc.math.arizona.edu/aws/2000/00SchoofN.pdf`.

[Sch92] René Schoof, *Algebraic curves over* $\mathbf{F}_2$ *with many rational points*, J. Number Theory **41** (1992), no. 1, 6–14, DOI 10.1016/0022-314X(92)90079-5. MR1161141

[Ser59] Jean-Pierre Serre, *Groupes algébriques et corps de classes* (French), Publications de l'institut de mathématique de l'université de Nancago, VII. Hermann, Paris, 1959. MR0103191

[Ser75] J-P. Serre, *Minorations de discriminants*, note of Oct. 1975, published in [**Ser86**], 1975.

[Ser83] Jean-Pierre Serre, *Sur le nombre des points rationnels d'une courbe algébrique sur un corps fini* (French, with English summary), C. R. Acad. Sci. Paris Sér. I Math. **296** (1983), no. 9, 397–402. MR703906

[Ser84] J-P. Serre, *Resume des cours de 1983–1984*, Annuaire du College de France, 1984. pp. 79–83. Reprinted in [**Ser86**]; see also [**Ser85**].

[Ser85] J-P. Serre, *Rational points on curves over finite fields*, Notes by Fernando Q. Gouvêa of lectures at Harvard University, 1985.

[Ser86] Jean-Pierre Serre, *Œuvres. Vol. III*, 1972–1984, (French), Springer-Verlag, Berlin, 1986. MR926691

[Ser88] Jean-Pierre Serre, *Algebraic groups and class fields*, Graduate Texts in Mathematics, vol. 117, Springer-Verlag, New York, 1988. Translated from the French. MR918564

[Ser91] Jean-Pierre Serre, *Lettre à M. Tsfasman* (French, with English summary), Astérisque **198-200** (1991), 11, 351–353 (1992). Journées Arithmétiques, 1989 (Luminy, 1989). MR1144337

[Ser18] J-P. Serre, *Distribution asymptotique des valeurs propres des endomorphismes de Frobenius (d'après Abel, Chebyshev, Robinson, ...)*, Séminaire Bourbaki, Vol. 2017/2018 70ème année, Exp. No. 1146, 2018, pp. 1–32.

[SGB67] C. E. Shannon, R. G. Gallager, and E. R. Berlekamp, *Lower bounds to error probability for coding on discrete memoryless channels. II*, Information and Control **10** (1967), 522–552. MR0216899

[Sha47] I. Shafarevitch, *On p-extensions* (Russian, with English summary), Rec. Math. [Mat. Sbornik] N.S. **20(62)** (1947), 351–363. MR0020546

[Sha63] I. R. Šafarevič, *Extensions with prescribed ramification points* (Russian, with French summary), Inst. Hautes Études Sci. Publ. Math. **18** (1963), 71–95. MR0176979

[Sha74] I. R. Shafarevich, *Basic algebraic geometry*, Springer-Verlag, New York-Heidelberg, 1974. Translated from the Russian by K. A. Hirsch; Die Grundlehren der mathematischen Wissenschaften, Band 213. MR0366917

[Sha79] Adi Shamir, *How to share a secret*, Comm. ACM **22** (1979), no. 11, 612–613, DOI 10.1145/359168.359176. MR549252

[She92] Ba Zhong Shen, *Algebraic-geometric codes and their decoding algorithm* (English, with Dutch summary), Technische Universiteit Eindhoven, Eindhoven, 1992. Dissertation, Technische Universiteit Eindhoven, Eindhoven, 1992. MR1181677

[She93] B.-Z. Shen, *On encoding and decoding of the codes from Hermitian curves*, Cryptography and coding, III (Cirencester, 1991), Inst. Math. Appl. Conf. Ser. New Ser., vol. 45, Oxford Univ. Press, New York, 1993, pp. 337–357. MR1275849

[Shi71] Goro Shimura, *Introduction to the arithmetic theory of automorphic functions*, Publications of the Mathematical Society of Japan, No. 11. Iwanami Shoten, Publishers, Tokyo; Princeton University Press, Princeton, N.J., 1971. Kanô Memorial Lectures, No. 1. MR0314766

[Shi89a] Tetsuji Shioda, *Mordell-Weil lattices and Galois representation. I*, Proc. Japan Acad. Ser. A Math. Sci. **65** (1989), no. 7, 268–271. MR1030197

[Shi89b] Tetsuji Shioda, *Mordell-Weil lattices and Galois representation. II, III*, Proc. Japan Acad. Ser. A Math. Sci. **65** (1989), no. 8, 296–303. MR1030204

[Shi89c] Tetsuji Shioda, *Mordell-Weil lattices and sphere packings*, Amer. J. Math. **113** (1991), no. 5, 931–948, DOI 10.2307/2374791. MR1129298

[Shi90] Tetsuji Shioda, *On the Mordell-Weil lattices*, Comment. Math. Univ. St. Paul. **39** (1990), no. 2, 211–240. MR1081832

[Shi91] Tetsuji Shioda, *Mordell-Weil lattices and sphere packings*, Amer. J. Math. **113** (1991), no. 5, 931–948, DOI 10.2307/2374791. MR1129298

[Sho94] Peter W. Shor, *Algorithms for quantum computation: discrete logarithms and factoring*, 35th Annual Symposium on Foundations of Computer Science (Santa Fe, NM, 1994), IEEE Comput. Soc. Press, Los Alamitos, CA, 1994, pp. 124–134, DOI 10.1109/SFCS.1994.365700. MR1489242

[Sho97] Peter W. Shor, *Polynomial-time algorithms for prime factorization and discrete logarithms on a quantum computer*, SIAM J. Comput. **26** (1997), no. 5, 1484–1509, DOI 10.1137/S0097539795293172. MR1471990

[Sim82] G. J. Simmons, *A game theory model of digital message authentication*, in Numerical mathematics and computing (Proc. 11th Manitoba Conf., Winnipeg, MB, 1981), Congr. Numer., vol. 34, 1982, pp. 413–424.

[Sim85] G. J. Simmons, *Authentication theory – coding theory*, in Advances in cryptology (Proc. Workshop CRYPTO'84, Santa Barbara, CA, 1984), G. Blakely and D. Chaum, eds., vol. 196 of Lect. Notes Comput. Sci., pp. 411–431, New York: Springer, 1985.

[Sim92] G. J. Simmons, *A survey of information authentication*, Contemporary cryptology, IEEE, New York, 1992, pp. 379–419. MR1205136

[SJH95] Shojiro Sakata, Helge Elbrønd Jensen, and Tom Høholdt, *Generalized Berlekamp-Massey decoding of algebraic-geometric codes up to half the Feng-Rao bound*, IEEE Trans. Inform. Theory **41** (1995), no. 6, 1762–1768, DOI 10.1109/18.476248. Special issue on algebraic geometry codes. MR1391034

[SJM+95a] Shajiro Sakata, Jørn Justesen, Y. Madelung, Helhe Elbrønd Jensen, and Tom Høholdt, *Fast decoding of algebraic-geometric codes up to the designed minimum distance*, IEEE Trans. Inform. Theory **41** (1995), no. 6, 1672–1677, DOI 10.1109/18.476240. Special issue on algebraic geometry codes. MR1391026

[SJM+95b] S. Sakata, J. Justesen, Y. Madelung, H. E. Jensen, and T. Høholdt, *A fast decoding method of AG codes from Miura-Kamiya curves C_{ab} up to half the Feng-Rao bound*, Finite Fields Appl. **1** (1995), no. 1, 83–101, DOI 10.1006/ffta.1995.1006. MR1334627

[Sko92] Alexei N. Skorobogatov, *Exponential sums, the geometry of hyperplane sections, and some Diophantine problems*, Israel J. Math. **80** (1992), no. 3, 359–379, DOI 10.1007/BF02808077. MR1202578

[Skr01] M. M. Skriganov, *Coding theory and uniform distributions* (Russian, with Russian summary), Algebra i Analiz **13** (2001), no. 2, 191–239; English transl., St. Petersburg Math. J. **13** (2002), no. 2, 301–337. MR1834866

[Skr07] M. M. Skriganov, *On linear codes with large weights simultaneously for the Rosenbloom-Tsfasman and Hamming metrics*, J. Complexity **23** (2007), no. 4-6, 926–936, DOI 10.1016/j.jco.2007.02.004. MR2372000

[Slo72] N. J. A. Sloane, *Sphere packings constructed from BCH and Justesen codes*, Mathematika **19** (1972), 183–190, DOI 10.1112/S0025579300005635. MR0314769

[SNWX01] Rei Safavi-Naini, Huaxiong Wang, and Chaoping Xing, *Linear authentication codes: bounds and constructions*, Progress in cryptology—INDOCRYPT 2001 (Chennai),

Lecture Notes in Comput. Sci., vol. 2247, Springer, Berlin, 2001, pp. 127–135, DOI 10.1007/3-540-45311-3_13. MR1934491

[Sob67] I. M. Sobol′, *Distribution of points in a cube and approximate evaluation of integrals* (Russian), Ž. Vyčisl. Mat. i Mat. Fiz. **7** (1967), 784–802. MR0219238

[Sør91] Anders Bjært Sørensen, *Projective Reed-Muller codes*, IEEE Trans. Inform. Theory **37** (1991), no. 6, 1567–1576, DOI 10.1109/18.104317. MR1134296

[ST01] Senya Shlosman and Michael A. Tsfasman, *Random lattices and random sphere packings: typical properties*, Mosc. Math. J. **1** (2001), no. 1, 73–89. MR1852935

[Sta74] H. M. Stark, *Some effective cases of the Brauer-Siegel theorem*, Invent. Math. **23** (1974), 135–152, DOI 10.1007/BF01405166. MR0342472

[Sta75] H. M. Stark, *The analytic theory of algebraic numbers*, Bull. Amer. Math. Soc. **81** (1975), no. 6, 961–972, DOI 10.1090/S0002-9904-1975-13873-0. MR0444611

[Ste99] Andrew M. Steane, *Enlargement of Calderbank-Shor-Steane quantum codes*, IEEE Trans. Inform. Theory **45** (1999), no. 7, 2492–2495, DOI 10.1109/18.796388. MR1725135

[Sti06] Henning Stichtenoth, *Transitive and self-dual codes attaining the Tsfasman-Vlăduţ-Zink bound*, IEEE Trans. Inform. Theory **52** (2006), no. 5, 2218–2224, DOI 10.1109/TIT.2006.872986. MR2234479

[Sti09] Henning Stichtenoth, *Algebraic function fields and codes*, 2nd ed., Graduate Texts in Mathematics, vol. 254, Springer-Verlag, Berlin, 2009. MR2464941

[Sti11] Henning Stichtenoth, *Curves with a prescribed number of rational points*, Finite Fields Appl. **17** (2011), no. 6, 552–559, DOI 10.1016/j.ffa.2011.03.004. MR2846291

[STV92] Igor E. Shparlinski, Michael A. Tsfasman, and Serge G. Vladut, *Curves with many points and multiplication in finite fields*, Coding theory and algebraic geometry (Luminy, 1991), Lecture Notes in Math., vol. 1518, Springer, Berlin, 1992, pp. 145–169, DOI 10.1007/BFb0087999. MR1186422

[Sub75] Doré Subrao, *The p-rank of Artin-Schreier curves*, Manuscripta Math. **16** (1975), no. 2, 169–193, DOI 10.1007/BF01181639. MR0376693

[Sud97] Madhu Sudan, *Decoding of Reed Solomon codes beyond the error-correction bound*, J. Complexity **13** (1997), no. 1, 180–193, DOI 10.1006/jcom.1997.0439. MR1449766

[Sud00] M. Sudan, *List decoding: Algorithms and applications*, SIGACT News, 2000, vol. 31, pp. 16–27.

[Sud01] M. Sudan, *Algorithmic tntroduction to coding theory*, A course taught at MIT, Fall 2001. Available at `http://people.csail.mit.edu/madhu/FT01/`.

[SV90] Alexei N. Skorobogatov and Sergei G. Vlăduţ, *On the decoding of algebraic-geometric codes*, IEEE Trans. Inform. Theory **36** (1990), no. 5, 1051–1060, DOI 10.1109/18.57204. MR1066592

[SW99] M. Amin Shokrollahi and Hal Wasserman, *List decoding of algebraic-geometric codes*, IEEE Trans. Inform. Theory **45** (1999), no. 2, 432–437, DOI 10.1109/18.748993. MR1677008

[SZ77a] J-P. Serre and D.B. Zagier, eds., *Modular functions of one variable. V*, vol. 601 of Lecture Notes in Math., Berlin: Springer, 1977.

[SZ77b] J-P. Serre and D.B. Zagier, eds., *Modular functions of one variable. VI*, vol. 627 of Lecture Notes in Math., Berlin: Springer, 1977.

[Tak77] Kisao Takeuchi, *Commensurability classes of arithmetic triangle groups*, J. Fac. Sci. Univ. Tokyo Sect. IA Math. **24** (1977), no. 1, 201–212. MR0463116

[Tat66] John Tate, *Endomorphisms of abelian varieties over finite fields*, Invent. Math. **2** (1966), 134–144, DOI 10.1007/BF01404549. MR0206004

[Tat75] J. Tate, *Algorithm for determining the type of a singular fiber in an elliptic pencil*, Modular functions of one variable, IV (Proc. Internat. Summer School, Univ. Antwerp, Antwerp, 1972), Springer, Berlin, 1975, pp. 33–52. Lecture Notes in Math., Vol. 476. MR0393039

[Tat95] John Tate, *On the conjectures of Birch and Swinnerton-Dyer and a geometric analog*, Séminaire Bourbaki, Vol. 9, Soc. Math. France, Paris, 1995, pp. Exp. No. 306, 415–440. MR1610977

[TB14] Itzhak Tamo and Alexander Barg, *A family of optimal locally recoverable codes*, IEEE Trans. Inform. Theory **60** (2014), no. 8, 4661–4676, DOI 10.1109/TIT.2014.2321280. MR3245347

[TBF16] Itzhak Tamo, Alexander Barg, and Alexey Frolov, *Bounds on the parameters of locally recoverable codes*, IEEE Trans. Inform. Theory **62** (2016), no. 6, 3070–3083, DOI 10.1109/TIT.2016.2518663. MR3506725

[Tem01] Alexandre Temkine, *Hilbert class field towers of function fields over finite fields and lower bounds for $A(q)$*, J. Number Theory **87** (2001), no. 2, 189–210, DOI 10.1006/jnth.2000.2596. MR1824142

[Tsf91] Michael A. Tsfasman, *Global fields, codes and sphere packings*, Astérisque **198-200** (1991), 373–396 (1992). Journées Arithmétiques, 1989 (Luminy, 1989). MR1144339

[Tsf92] Michael A. Tsfasman, *Some remarks on the asymptotic number of points*, Coding theory and algebraic geometry (Luminy, 1991), Lecture Notes in Math., vol. 1518, Springer, Berlin, 1992, pp. 178–192, DOI 10.1007/BFb0088001. MR1186424

[Tsf18] M. Tsfasman, *Serre's theorem and measures corresponding to abelian varieties over finite fields*, Moscow Math. J., to appear.

[TV91] M. A. Tsfasman and S. G. Vlăduţ, *Algebraic-geometric codes*, Mathematics and its Applications (Soviet Series), vol. 58, Kluwer Academic Publishers Group, Dordrecht, 1991. Translated from the Russian by the authors. MR1186841

[TV97] M. A. Tsfasman and S. G. Vlăduţ, *Asymptotic properties of zeta-functions*, J. Math. Sci. (New York) **84** (1997), no. 5, 1445–1467, DOI 10.1007/BF02399198. Algebraic geometry, 7. MR1465522

[TV02] M. A. Tsfasman and S. G. Vlăduţ, *Infinite global fields and the generalized Brauer-Siegel theorem*, Mosc. Math. J. **2** (2002), no. 2, 329–402. Dedicated to Yuri I. Manin on the occasion of his 65th birthday. MR1944510

[TVZ82] M. A. Tsfasman, S. G. Vlăduţ, and Th. Zink, *Modular curves, Shimura curves, and Goppa codes, better than Varshamov-Gilbert bound*, Math. Nachr. **109** (1982), 21–28, DOI 10.1002/mana.19821090103. MR705893

[VD83] S. G. Vlèduts and V. G. Drinfel′d, *The number of points of an algebraic curve* (Russian), Funktsional. Anal. i Prilozhen. **17** (1983), no. 1, 68–69. MR695100

[vdG09] Gerard van der Geer, *Hunting for curves with many points*, Coding and cryptology, Lecture Notes in Comput. Sci., vol. 5557, Springer, Berlin, 2009, pp. 82–96, DOI 10.1007/978-3-642-01877-0_9. MR2836234

[vdGvdV92] Gerard van der Geer and Marcel van der Vlugt, *Reed-Muller codes and supersingular curves. I*, Compositio Math. **84** (1992), no. 3, 333–367. MR1189892

[vdGvdV00] Gerard van der Geer and Marcel van der Vlugt, *Tables of curves with many points*, Math. Comp. **69** (2000), no. 230, 797–810, DOI 10.1090/S0025-5718-99-01143-6. MR1654002

[Via17] Maryna S. Viazovska, *The sphere packing problem in dimension 8*, Ann. of Math. (2) **185** (2017), no. 3, 991–1015, DOI 10.4007/annals.2017.185.3.7. MR3664816

[Vlă83] S.G. Vlăduţ, "On the polynomiality of codes on classical modular curves." Preprint, 1983.

[Vlă91] S. G. Vlăduţ, *Kronecker's Jugendtraum and modular functions*, Studies in the Development of Modern Mathematics, vol. 2, Gordon and Breach Science Publishers, New York, 1991. Translated from the Russian by M. Tsfasman. MR1121266

[Vlă98a] Serge G. Vlăduţ, *A note on authentication codes from algebraic geometry*, IEEE Trans. Inform. Theory **44** (1998), no. 3, 1342–1345, DOI 10.1109/18.669432. MR1616673

[Vla98b] S. G. Vladuts, *Secant spaces and Clifford's theorem over finite fields*, Finite Fields Appl. **4** (1998), no. 1, 101–111, DOI 10.1006/ffta.1997.0201. MR1612013

[Vlă18] S. Vlăduţ, *Lattices with exponentially large kissing numbers*, arXiv:1802.00886v2 [math.NT], (2018).

[VM84] S. G. Vlèduts and Yu. I. Manin, *Linear codes and modular curves* (Russian), Current problems in mathematics, Vol. 25, Itogi Nauki i Tekhniki, Akad. Nauk SSSR, Vsesoyuz. Inst. Nauchn. i Tekhn. Inform., Moscow, 1984, pp. 209–257. MR770943

[VZ10] José Felipe Voloch and Marcos Zarzar, *Algebraic geometric codes on surfaces* (English, with English and French summaries), Arithmetics, geometry, and coding theory (AGCT 2005), Sémin. Congr., vol. 21, Soc. Math. France, Paris, 2010, pp. 211–216. MR2856569

[Wan94] Zhe Xian Wan, *The weight hierarchies of the projective codes from nondegenerate quadrics*, Des. Codes Cryptogr. **4** (1994), no. 3, 283–300, DOI 10.1007/BF01388457. MR1277945

[Wei67] André Weil, *Basic number theory*, Die Grundlehren der mathematischen Wissenschaften, Band 144, Springer-Verlag New York, Inc., New York, 1967. MR0234930

[Wei84] André Weil, *Number theory: An approach through history; From Hammurapi to Legendre*, Birkhäuser Boston, Inc., Boston, MA, 1984. MR734177

[Wei98] Edwin Weiss, *Algebraic number theory*, Dover Publications, Inc., Mineola, NY, 1998. Reprint of the 1963 original. MR1635455

[Wol75] J. Wolfmann, *Codes projectifs à deux ou trois poids associés aux hyperquadriques d'une géométrie finie* (French), Discrete Math. **13** (1975), no. 2, 185–211, DOI 10.1016/0012-365X(75)90017-5. MR0381840

[Woz58] J. M. Wozencraft, *List decoding*, in Quarterly progress report, MIT Research Laboratory of Electronics, vol. 48, 1958, pp. 90–95.

[WXSN03] Huaxiong Wang, Chaoping Xing, and Rei Safavi-Naini, *Linear authentication codes: bounds and constructions*, IEEE Trans. Inform. Theory **49** (2003), no. 4, 866–872, DOI 10.1109/TIT.2003.809567. MR1984479

[Xia08] Xu Xiang, *On the minimum distance conjecture for Schubert codes*, IEEE Trans. Inform. Theory **54** (2008), no. 1, 486–488, DOI 10.1109/TIT.2007.911283. MR2446778

[Xin01] Chaoping Xing, *Authentication codes and algebraic curves*, European Congress of Mathematics, Vol. II (Barcelona, 2000), Progr. Math., vol. 202, Birkhäuser, Basel, 2001, pp. 239–244. MR1905364

[XN95] Chao Ping Xing and Harald Niederreiter, *A construction of low-discrepancy sequences using global function fields*, Acta Arith. **73** (1995), no. 1, 87–102, DOI 10.4064/aa-73-1-87-102. MR1358190

[XWL00] Chaoping Xing, Huaxiong Wang, and Kwok Yan Lam, *Constructions of authentication codes from algebraic curves over finite fields*, IEEE Trans. Inform. Theory **46** (2000), no. 3, 886–892, DOI 10.1109/18.841168. MR1763467

[XY07] Chaoping Xing and Sze Ling Yeo, *Algebraic curves with many points over the binary field*, J. Algebra **311** (2007), no. 2, 775–780, DOI 10.1016/j.jalgebra.2006.12.029. MR2314733

[Zak93] F. L. Zak, *Tangents and secants of algebraic varieties*, Translations of Mathematical Monographs, vol. 127, American Mathematical Society, Providence, RI, 1993. Translated from the Russian manuscript by the author. MR1234494

[Zan98] Corrado Zanella, *Linear sections of the finite Veronese varieties and authentication systems defined using geometry*, Des. Codes Cryptogr. **13** (1998), no. 2, 199–212, DOI 10.1023/A:1008286614783. MR1600208

[Zar07] Marcos Zarzar, *Error-correcting codes on low rank surfaces*, Finite Fields Appl. **13** (2007), no. 4, 727–737, DOI 10.1016/j.ffa.2007.05.001. MR2359313

[Zim81] Rainer Zimmert, *Ideale kleiner Norm in Idealklassen und eine Regulatorabschätzung* (German), Invent. Math. **62** (1981), no. 3, 367–380, DOI 10.1007/BF01394249. MR604833

[ZP81] V. V. Zyablov and M. S. Pinsker, *List cascade decoding*; Russian transl., Problems Inform. Transmission **17** (1981), no. 4, 29–33. MR673744

[Zyw13] David Zywina, *Explicit class field theory for global function fields*, J. Number Theory **133** (2013), no. 3, 1062–1078, DOI 10.1016/j.jnt.2012.02.009. MR2997787

List of Names

Index